MW00844005

Combustion Engines Development

Günter P. Merker • Christian Schwarz
Rüdiger Teichmann
Editors

Combustion Engines Development

Mixture Formation, Combustion, Emissions and Simulation

Editors
Univ.-Prof. Dr.-Ing. habil.
Günter P. Merker
Schöneckstr. 30
88060 Tettnang
Germany
g.p.merker@t-online.de

Apl. Prof. Dr.-Ing. habil.
Christian Schwarz
BMW-Group
EA 31
Hufelandstr. 8a
80788 München
Germany
christian.schwarz@bmw.de

Dr.-Ing. Rüdiger Teichmann
AVL List GmbH
Hans-List-Platz 1
8020 Graz
Austria
ruediger.teichmann@avl.com

Original german edition published by Vieweg & Teubner, Wiesbaden, 2009.

ISBN 978-3-642-02951-6 e-ISBN 978-3-642-14094-5
DOI 10.1007/978-3-642-14094-5
Springer Heidelberg Dordrecht London New York

Library of Congress Control Number: 2011936001

Printed on acid-free paper

Springer is part of Springer Science+Business Media (www.springer.com)

Preface

Nowadays commercial computer programs are normally used for engine and vehicle development for e.g. the simulation of the transient performance of vehicles or of the complete power train as well as for the simulation of high-grade transient processes in the engine combustion chamber.

As in general no source code for these computer programs is available and due to time-consuming studies of the corresponding descriptions users often have insufficient knowledge concerning the physical and chemical contents of the models used in these programs. Therefore it is our special concern to reveal different physical and chemical approaches and to show options and limits of the models used. So the focus of this book is definitely on the thermodynamical, fluid mechanical and chemical base of modeling of motor-driven processes.

The book is composed of three parts. Part I describes the operation and the thermodynamically base of combustion engines, the combustion and reaction kinetics as well as emissions building in six chapters. This chapter is completed by a proceeding concerning combustion measurement. Part II with Chaps. 7–9 is dedicated to the operating process calculation, the charging process and the modeling of exhaust after-treatment systems. Part III contains Chaps. 10–15 and starts with the overall process analysis. Part III is preferentially about the calculation possibilities in the combustion engine for e.g. phenomenological combustion models, simulation of injection and combustion and 3D simulations for fluid flow and supercharging.

We hope that we succeeded in showing a comprehensive and current description of the simulation of motor-driven processes and we would appreciate if all academical and technical users will profit from this book.

We thank all authors for their constructive and diligent co-operation. All authors and their employers or institutions as well as their responsibility for the proceedings are mentioned in the prefix of the book. We say special thanks to AVL List GmbH for their professional and substantial support during the compilation of this book. We are much obliged to Mr. Gary Patterson who made the final correction of the English translation and we are also very thankful for his critical remarks concerning the precise terminology. Many thanks also to the Springer-Verlag especially to Mrs. Ulrike Butz for the kind and good co-operation.

This book is a translation of the 4th German edition which was published by the Vieweg-Teubner Verlag in 2009. For the present English edition the text was updated and extended by an overview description regarding the 3D calculation program FIRE which is developed and sold by AVL.

Tettnang, Germany Günter P. Merker
Graz, Austria Rüdiger Teichmann

Contents

Part I Foundations of Thermodynamics and Chemistry

Part III Simulation of Combustion and Charging

Chapters, Contributions and Authors

Univ.-Prof. Dr.-Ing. habil. Günter P. Merker Chapter 1, 2
Leibniz Universität, Hannover, g.p.merker@t-online.de www.uni-hannover.de

Dr.-Ing. Rüdiger Teichmann Section 3.1, 3.3
AVL LIST GmbH, Graz, Austria, ruediger.teichmann@avl.com www.avl.com

Ao.Univ.-Prof. Dr. techn. Andreas Wimmer Section3.1
Technical University Graz, andreas.wimmer@lec.tugraz.at, www.tugraz.at

apl. Prof. Dr.-Ing. habil. Christian Schwarz Chapter 7,10; Section 3.2
BMW AG, München, christian.schwarz@bmw.de, www.bmw.de

Dr. Ernst Winklhofer Section 3.3
AVL LIST GmbH, Graz, Austria, ernst.winklhofer@avl.com, www.avl.com

Dr.-Ing. Sebastian Rakowski Chapter 6; Section 4.1,4.2,11.3
WABCO Development GmbH, sebastian.rakowski.sr@googlemail.com, www.wabco.de

Dr.-Ing. Peter Eckert Chapter 6; Section 4.2, 5.2, 11.3
IAV GmbH, Berlin, peter.eckert@iav.de, www.iav.de

Prof. Dr. techn. Andreas Witt Section 4.3
BMW AG, München, andreas.witt@bmw.de, www.bmw.de

Prof. Dr.-Ing. habil. Gunnar Stiesch Section 5.1, 5.2, 11.2
MAN Diesel SE, Augsburg, gunnar.stiesch@man.eu, www.mandiesel.com

Dr.-Ing. Claus Reulein Chapter 8
BMW AG, München, claus.reulein@bmw.de, www.bmw.de

Dr. Reinhard Tatschl Chapter 9; Appendix
AVL LIST GmbH, Graz, Austria, reinhard.tatschl@avl.com, www.avl.com

Dr. Johann Wurzenberger Chapter 9
AVL LIST GmbH, Graz, Austria, johann.wurzenberger@avl.com, www.avl.com

Dr. rer. nat. Frank Otto Chapter 12, 13, 14
Daimler AG, Stuttgart, frank.otto@daimler.com, www.daimler.de

Dr.-Ing. Christian Krüger Chapter 12, 13, 14
Daimler AG, Stuttgart, christian.c.krueger@daimler.com, www.daimler.de

Dr.-Ing. Bodo Durst Chapter 15
BMW AG, München, bodo.durst@bmw.de, www.bmw.de

List of Companies and Universities

Companies

AVL LIST GmbH, Graz, Austria	Dr. Reinhard Tatschl Dr.-Ing. Rüdiger Teichmann Dr. Ernst Winklhofer Dr. Johann Wurzenberger
BMW AG, München	Dr.-Ing. Bodo Durst Dr.-Ing. Claus Reulein apl. Prof. Dr.-Ing. habil. Christian Schwarz Dr. techn. Andreas Witt
Daimler AG, Stuttgart	Dr.-Ing. Christian Krüger Dr. rer. nat. Frank Otto
IAV GmbH, Berlin	Dr.-Ing. Peter Eckert Dr.-Ing. Sebastian Rakowski
MAN Diesel SE, Augsburg	apl. Prof. Dr.-Ing. habil. Gunnar Stiesch
WABCO Development GmbH	Dr.-Ing. Sebastian Rakowski

Universities

Leibniz University Hannover	Univ.-Prof. Dr.-Ing. habil. Günter P. Merker
Technical University Graz, Austria	Ao. Univ.-Prof. Dr. techn. Andreas Wimmer

The Editors

Univ.-Prof. Dr.-Ing. habil. Günter P. Merker was born in Augsburg/Germany in 1942. He studied mechanical engineering at the Technical University in Munich and received his diploma in 1969. He worked then as an assistant lecturer at this Institute and got his PhD in 1974 with a thesis on free convection problems. From 1975 to 1976, he has been as a research fellow at the Californian Institute of Technology in Pasadena/CA. Having returned to the Munich Technical University, he qualified as a university lecturer in 1978. From 1978 to 1980, he was with MTU GmbH in Munich. In 1980 he accepted an associate professorship for refrigeration technology at the University of Karlsruhe. In 1986 he started work at MTU Friedrichshafen, where he was head of the department analytics and engine calculation. In 1994 he accepted a full professorship for combustion engines at the University of Hannover. There he was head of the Institute of Technical Combustion until he retired in 2005. He was adviser to 43 PhD students. He is editor and co-editor of more than 140 technical scientific publications as well as six technical books on thermal transmission, fluid mechanics and combustion engines. He is a member of the Scientific Association in Braunschweig as well as of different technical committees. In addition to that, he works as an independent adviser for engine manufactories and service industries.

Apl. Prof. Dr.-Ing. habil. Christian Schwarz was born in Regensburg in 1964. From 1983 to 1988, he studied mechanical engineering at the Munich Technical University. From 1989 to 1993, he worked as an assistant lecturer at the chair of internal combustion engines and vehicles at the Munich Technical University (Prof. Woschni). During that time he worked on unsteady, 0-dimensional process calculation for charged and quickly working diesel engines and got his PhD in 1993 on his thesis "Simulation of the transient reaction of charged diesel engines during operation". Still at the same university, he was head of a team that worked on a part of the EU research project "Hybrid Zero Emission Mobility" (1993–1997). In 1998 he qualified as a university lecturer, his treatise was on "Theory and Simulation of Charged Combustion Engines". He was appointed private lecturer by the University of Hannover in 1999, and he got the venia legendi in combustion engines. Since then he has been teaching the subject "simulation of processes in combustion engines". In 2004 the Hannover University conferred him the title "apl. Prof.".

In 1997 he started work in the department for advance development at BMW AG. In 2001 he was appointed head of this department and was responsible for the development of new combustion techniques for spark ignition (SI) engines, such as direct hydrogen ignition, as well as of new concepts for charged SI engines. Since 2004 he has been working in series development, being responsible for the development of combustion techniques of the SI engines produced in series. He has supervised several dissertations on combustion engines. Besides, he is the author and co-author of several publications and lectures, as well as of a technical book about combustion techniques and the charging of combustion engines.

In the research association about internal combustion engines, he is head of the planning team for total processes, and he is also part of its scientific advisory board.

Dr.-Ing. Rüdiger Teichmann was born in Nordhausen in 1960. He studied mechanical engineering at the Technical University Dresden from 1982 till 1987. After his study he became a research student and a scientific staff member till 1990. In 1991 he got his PhD with a thesis on combustion development for truck diesel engines.

In the same year he started his professional carrier in the department for advance development at BMW AG in Munich. His special fields of work were thermodynamic, combustion development, load exchange for gasoline engines and calibration in series development as well. In 1999 he became the head of product management for the complete indicating measurement chain at AVL List GmbH in Graz. After 3 years he was nominated as segment leader "Indicating Measurement" and global business segment leader for "Combustion Measurement" in 2005. In addition to this responsibility he is in charge of the vehicle measurement activities from AVL since 2007. Mr. Teichmann is author and co-author of several publications and tutor of diploma thesis.

Abbreviations

BDC Bottom Dead Center
BMEP Brake mean effective pressure
BV Burning function
CNG Compressed Natural Gas
CR Common Rail
DI Direct injection
DME Dimethylether
EGR Exhaust gas recirculation
EV Injection valve
FAME Fatty Acid Methyl Ester
GDI Gasoline direct injection
HE Hydro-erosive (cavitation)
ID Ignition delay
IMEP Indicated mean effective pressure
IR Injection rate
IV Injection valve
LDA Laser Doppler Anemomentry
LIF Laser induced Fluorescence
LPG Liquified Petroleum Gas
MOZ Engine octan number
PAK Polycyclic aromatic hydrocarbons
PCV Pressure Control Valve
PDA Phase Doppler Anemometry
PIV Particle image velocimetry
RG Residual gas
RME Rape methylester
RoHR Rate of heat release
ROZ Research octan number
TC Turbo charged; Turbocharger

TDC	Compression Top Dead Center
UV	Ultraviolet
VI	Unit injector
V-Soot	Soot number based on flame analysis
ZZP	Start of ignition

Symbols

A	Flame front surface [m^2]
B_m, B_λ	Coefficient to laminar flame velocity
C	Damkoehler-Coefficient
c	Constant
c_p	Specific heat at constant pressure
c_v	Specific heat at constant volume
C_d	Flow coefficient
c_L, c_T	Integral length scale coefficient
c_m	Medium piston velocity [m/s]
E	Energy
Ex	Expansion factor describing heat release (premixed)
h	Specific enthalpy [J/kg]
H_s	Specific heating value [J/kg]
H_U, H_u	Lower heating value [J/kg]
K	Cavitation number
k	Turbulent kinetic energy [m^2/s^2]
Ki	Cavitation index
l_I	Integrale length scale [m]
l_T	Taylor length scale
m	Mass [kg]
n	General parameter
p	Pressure [bar]
P_{me}	Mean effective pressure [bar]
Q	Heat release [W]
s	Flame velocity [m/s]
T	Temperature [K]
t	Time
u	Turbulent velocity fluctuation[m/s]
V	Volume [m^3]; volume flow [m^3/s]
X_1, X_2, X_3	Parameters for Livengood–Wu Integral
η_v	Pipe flow coefficient

Indices

0	Quiescent condition; standard reference
1	inlet
2	exit
ad	Adiabatic
crit	critical
dif	Diffusion
e or *i*	Inlet
f	Fuel
F	Fuel
ID	Ignition delay
iv	Inlet valve
l	Laminar
m, M	Mean
nock	If knocking occurs
p	constant pressure
pre	Pre-mixed
R or *RG*	Residual gas
s	isentropic
SOI	Start of ignition
t	Turbulent
u	Environment; Unburned
v	constant volume
vap	Vapour pressure (in combination with pressure)

(Greek) Symbols

Δh_i	Reaction enthalpy of i-th reaction
Δp_{ak}	Partial pressure difference across ash thickness
Δp_{rk}	Partial pressure difference across soot filtration layer
Δp_{rk}	Partial pressure difference across soot thickness
Δp_w	Partial pressure loss across the wall
$A_{F,n}$	Cross-section of pipe
a_{geo}	Geometric surface of monolith
c_g	Gas composition
$c_{j,g}$	Concentration of species j in gas
c^L	Surface concentration of species
$c_{p,s}$	Heat capacity of solid phase
D_{eff}	Effective diffusion coefficient

d_{hyd}	Hydraulic diameter of the pipe
d_i	Diameter of the pipe
F_n	Friction coefficient
Gz_h	Graetz number for heat transfer
Gz_m	Graetz number to mass transfer
h_j	Enthalpy of species j under gaseous condition
k_w,k_{rt},k_{ak},k_{rk}	Permeability of considered layers
MG_j	Molar mass of species j
$m_{russ,in}$	Intake mass flow of soot
p_g	Gas pressure
$p_{g,n}$	Pressure in pipe
$P_{S,n}$	Circumference of channel
r_i	Reaction rate of species j
Sc	Schmidt-number
S_{rk}	Soot filtration coefficient
t	Characteristic time [s]
t	Time
T_g	Gas temperature
T_s	Wall temperature
v_g	Gas phase velocity
$v_{g,n}$	Free flow velocity in pipe
v_w	Wall velocity
$v_{w,n}$	Normalized wall velocity
$v_{w,n}$	Velocity in porous wall
wg,1	Wall velocity
$w_{j,g}$	Mass fraction of species j
x	Length coordinate
z	Length coordinate
α	Heat transfer coefficient between gaseous and solid condition
α	General parameter
β	General parameter
β_j	Mass transfer coefficient of species (j)
Γ	Free surface
δ_{ak}	Ash thickness
δ_{rk}	Soot thickness
κ	Isentropic exponent
λ	Air/fuel ratio
λ_g	Heat conductivity of the gas
λ_s	Heat transfer coefficient of wall
μ	Dynamic viscosity; First viscosity coefficient [Ns/m^2]
$\nu_{i,j}$	Stoichiometric coefficient of species (j) of i-th reaction
ξ	Contraction number; pipe friction number
ρ	Density [kg/m^3]

ρ_g	Density of the gas
$\rho_{g,n}$	Density in pipe
ρ_s	Density of solid phase
τ_{fluid}, τ_{chem}	Physical and chemical mixing time [s]
φ	Crankshaft angle [°CA]
Φ	Ratio of equivalence

Part I
Foundations of Thermodynamics and Chemistry

Chapter 1
Introduction

Günter P. Merker

1.1 Preface

One of the central tasks of engineering sciences is the most possibly exact description of technical processes with the goal of understanding the dynamic behavior of complex systems, of recognizing regularities, and thereby of making possible reliable statements about the future behavior of these systems. With regard to combustion engines as the power source for propulsion systems for land, water, and air vehicles, for permanent and emergency generating sets, as well as for air conditioning and refrigeration, the analysis of the entire process thus acquires particular importance.

In the case of model-based parameter-optimization, engine behavior is described with a mathematical model. The optimization does not occur in the real engine, but rather in a model, which attempts to take into account all effects relevant for the concrete task of optimization. The advantages of this plan are a drastic reduction of the experimental cost and thus a clear saving of time in developmental tasks; see Kuder and Kruse (2000).

The prerequisite for simulation are mechanical, thermodynamic, and chemical models for the description of technical processes, whereby the understanding of thermodynamics and of chemical reaction kinetics are an essential requirement for the modeling of Internal Combustion Engine processes.

1.2 Model-Building

The first step in numeric simulation consists in the construction of the model describing the technical process. Model-building is understood as a goal-oriented simplification of reality through abstraction. The prerequisite for this is that the real process can be divided into single processual sections and thereby broken down into partial problems. These partial problems must then be physically describable and mathematically formulatable.

G.P. Merker et al. (eds.), *Combustion Engines Development*,
DOI 10.1007/978-3-642-14094-5_1, © Springer-Verlag Berlin Heidelberg 2012

A number of demands must be placed upon the resulting model:

- The model must be formally correct, i.e. free of inconsistencies. As regards the question of "true or false", it should be noted that models can indeed be formally correct but still not describe the process to be investigated or not be applicable to it. There are also cases in which the model is physically incorrect but nevertheless describes the process with sufficient exactness, e.g. the Ptolemaic model for the simulation of the dynamics of the solar system, i.e. the calculation of planetary and lunar movement.
- The model must describe reality as exactly as possible, and, furthermore, it must also be mathematically solvable. One should always be aware that every model is an approximation to reality and can therefore never perfectly conform to it.
- The cost necessary for the solution of the model with respect to the calculation time must be justifiable in the context of the setting of the task.
- With regard to model-depth, this demand is applicable: as simple as possible and as complex as necessary. So-called universal models are to be regarded with care.

It is only by means of the concept of model that we are in the position truly to comprehend physical processes.

In the following, we will take a somewhat closer look into the types of models with regard to the combustion engine. It must in the first place be noted that both the actual thermodynamic cycle process (particularly combustion) and the change of load of the engine are unsteady processes. Even if the engine is operated in a particular operating condition (i.e. load and rotational speed are constant), the thermodynamic cycle process runs unsteadily. With this, it becomes obvious that there are two categories of engine models, namely, such that describe the operating condition of the engine (total-process models) and such that describe the actual working process (combustion models).

With respect to types of models, one distinguishes between:

- *Empirical models*, i.e. a rule-based method built upon empirically grounded rules, which cannot be grasped by mathematical equations
- *Mathematical models*, i.e. a method resting on mathematical formalism

Empirical models have become known in recent times under the concepts "expert systems" and "fuzzy-logic models". Yet it should thereby be noted that rule-based methods can only interpolate and not extrapolate. We will not further go into this type of model.

Mathematical models can be subdivided into:

- Parametric
- Non-parametric

Parametric models are compact mathematical formalisms for the description of system behavior, which rests upon physical and chemical laws and show only relatively few parameters that are to be experimentally determined. These models are typically described by means of a set of partial or normal differential equations.

Non-parametric models are represented by tables that record the system behavior at specific test input signals. Typical representatives of this type of model are step responses or frequency responses. With the help of suitable mathematical methods, e.g. the Fourier transformation, the behavior of the system can be calculated at any input signal.

Like empirical models, non-parametric models can only interpolate. Only mathematical models are utilized for the simulation of combustion engine processes. But because the model parameters must be adjusted to experimental values in the case of these models as well, they are fundamentally error-prone. These errors are to be critically evaluated in the analysis of simulation results. Here too, it becomes again clear that every model represents but an approximation of the real system under observation.

1.3 Simulation

For the construction of parametric mathematical models for the simulation of temporally and spatially variable fluid, temperature, and concentration fields with chemical reactions, the knowledge of thermodynamics, fluid dynamics, and of combustion technology is an essential prerequisite (Fig. 1.1).

With respect to the simulation of fluid fields with chemical reactions, it should be noted that physical and chemical processes can progress at very different temporal and linear scales. The description of these process progressions is usually simpler when the time scales are much different, because then simplifying assumptions can be made for the chemical or physical process, and it is principally very complex when the time scales are of the same order of magnitude. This is made clear by means of the examples in Fig. 1.2.

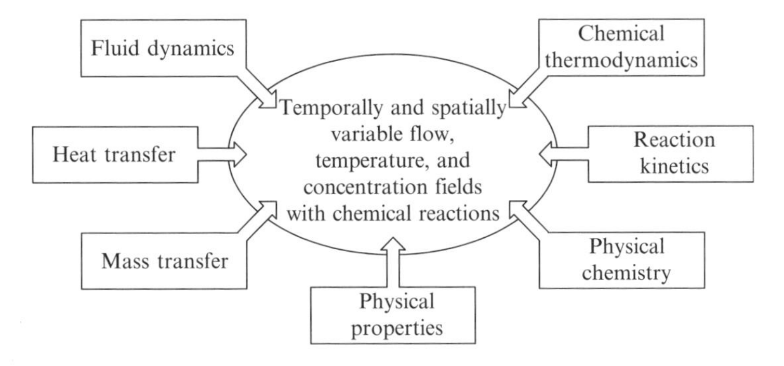

Fig. 1.1 Area of knowledge important for process simulation

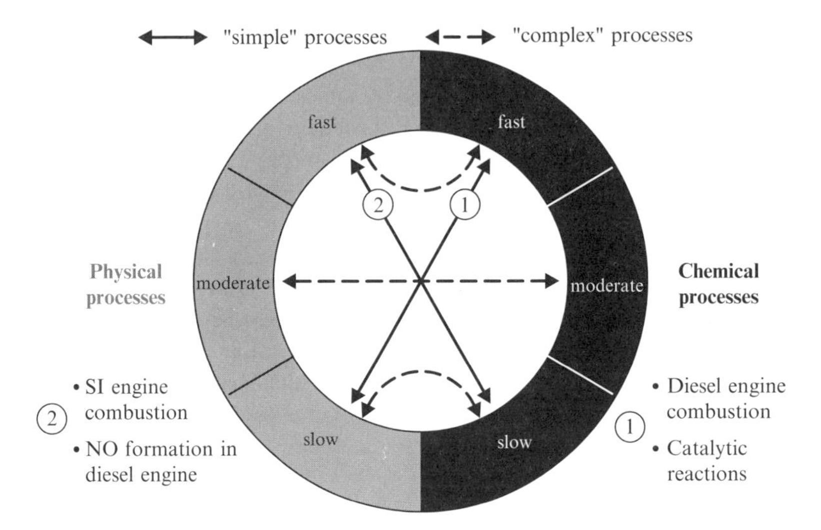

Fig. 1.2 Time scales of physical and chemical processes influencing process simulation

Yet in addition, knowledge of modeling methods is also necessary. Although some universally valid rules can be given for this, this step allows a lot of free room for the creativity and imagination of the modeler.

Essentially, the modeling procedure can be subdivided into the following steps:

First step: define the system and boundaries from the environment; determine the relevant reservoirs as well as the mass and energy flow between them.

Second step: draw up balance sheets according to the unified scheme: temporal change of the reservoir is equal to the inflow minus the outflow.

Third step: describe the mass and energy flows with the help of physical laws.

Fourth step: simplify the resulting model, if necessary by neglecting secondary influences.

Fifth step: integrate the model numerically, i.e. execute the simulation.

Sixth step: validate the model, compare the calculated data with experimentally obtained data.

In the utilization of an existing simulation program for the solution of new tasks, the prerequisites which were met in the creation of the model must always be examined. It should thereby be clarified whether and to what extent the existing program is actually suitable for the solution of the new problem. One should in such cases always be aware of the fact that "pretty, colorful pictures" exert an enormous power of suggestion upon the "uncritical" observer.

Various process simulation models are used in the development of heat engines and motor vehicles depending on the task at hand, whereby the physical and chemical processes that take place in the combustion chamber are always described in a more or less detailed fashion.

In order to simulate, for example, the acceleration behavior of an automobile, overall process analysis involves the use of *zero-dimensional models*, while the processes that take place in the combustion chamber itself are described with the help of single or multi-zone models. These purely thermodynamic models are based on the "fill and empty" method, the individual zones are assumed to be ideally mixed and the flow field in the combustion chamber is thus completely disregarded. Heat release via combustion and heat transfer to the surfaces delimiting the combustion chamber are described with simple semi-empirical approaches, partial systems such as controls or charging systems either with those as well or using characteristic maps. In order to calculate processes in the intake and exhaust pipes, the one-dimensional gas dynamics are often consulted.

In order to investigate detailed processes in the combustion chamber on the other hand – such as NO or soot production – *phenomenological multi-zone models* are employed. In these models, the combustion chamber is subdivided into hundreds of cells that are likewise assumed to be ideally mixed. Heat release due to combustion within and transport processes between these individual cells are described with detailed physical and chemical methods. With such multi-zone models, mixture formation, heat release by combustion and pollutant formation can in many cases be described with sufficient accuracy. They may indeed disregard with flow field in the combustion chamber, but they are much more elaborate with respect to modeling depth than simpler multi-zone models. They thus inhabit a position between the multi-zone models used for overall process analysis and the subsequently described 3D-CRFD *(Computational Reaction Fluid Dynamics)* models.

CRFD models are used for even more detailed investigations of processes involved during combustion, whereby in this case the flow field in the combustion chamber must be calculated. A number of CFD codes are commercially available for this purpose that are based on the Navier-Stokes equations but must be expanded with models for describing mixture formation, ignition, combustion and pollutant formation. However, these models are much more complex than those used in phenomenological models. The continuous gas phase in the Euler formulation has to be combined with the disperse phase in the Lagrange formulation. This combination, the correct selection of the grid to be used and the choice of a suitable turbulence model make enormous demands with regard to memory capacity and computing time.

Part C of this book will take a closer look at significantly more detailed models such as the *large eddy simulation (LES)* or *direct numerical simulation (DNS)*. Yet these models still play a minor role in engine development.

Numeric simulation opens up myriad possibilities. We can get an idea of what is to be expected in this field if we bear in mind the rapid development in the information sector and compare the present condition of "email" and the "internet" with that of the 1980s and 1990s.

With respect to technological progress and the ecological perspectives related to it, the reader is referred to Jischa (1993). Also, Kaufmann and Smarr (1994) have provided interesting insight into the topic of simulation.

References

Jischa M (1993) Herausforderung Zukunft: Technischer Fortschritt und ökologische Perspektiven. Spektrum Akad, Heidelberg

Kaufmann WJ, Smarr LL (1994) Simulierte Welten. Spektrum-Verlag, Heidelberg

Kuder J, Kruse Th (2000) Parameteroptimierung an Ottomotoren mit Direkteinspritzung. Motortechnische Zeitschrift MTZ 61. Franckh-Kosmos Verlags-GmbH, Stuttgart, pp 378–384

Chapter 2
Reciprocating Engines

Günter P. Merker

2.1 Energy Conversion

In energy conversion, we can distinguish hierarchically between general, thermal, and overall engine energy conversion.

Under *general energy* conversion is understood the transformation of primary into secondary energy through a technical process in an energy conversion plant (Fig. 2.1).

Thermal energy conversion is subject to the laws of thermodynamics and can be described formally, as is shown in Fig. 2.2.

The *internal combustion engine* and the *gas turbine* are specialized energy conversion plants, in which the chemical energy bound in the fuel is at first transformed into thermal energy in the combustion space or chamber, this being then transformed into mechanical energy by the motor. In the case of the stationary gas turbine plant, the mechanical energy is then converted into electrical energy by the secondary generator (Fig. 2.3).

2.2 Geometry of the Crankshaft Drive

Internal combustion engines as considered in this text are piston machines, whereby one distinguishes, according to the design of the combustion space or the pistons, between reciprocating engines and rotary engines with a rotating piston movement. Figure 2.4 shows principle sketches of possible structural shapes of reciprocating engines, whereby today only variants 1, 2, and 4 are, practically speaking, still being built.

For an extensive description of other models of the combustion engine, see Basshuysen and Schäfer (2003), Maas (1979), and Heywood (1988).

The motor transforms the oscillating movement of the piston into the rotating movement of the crankshaft, see Fig. 2.5. The piston reverses its movement at the

G.P. Merker et al. (eds.), *Combustion Engines Development*,
DOI 10.1007/978-3-642-14094-5_2, © Springer-Verlag Berlin Heidelberg 2012

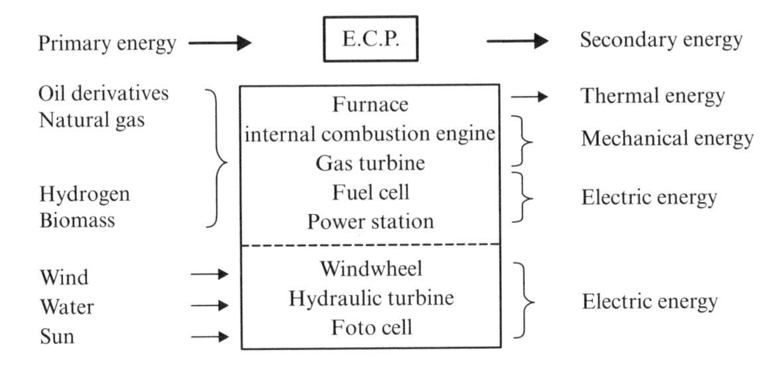

Fig. 2.1 Diagram of general energy conversion

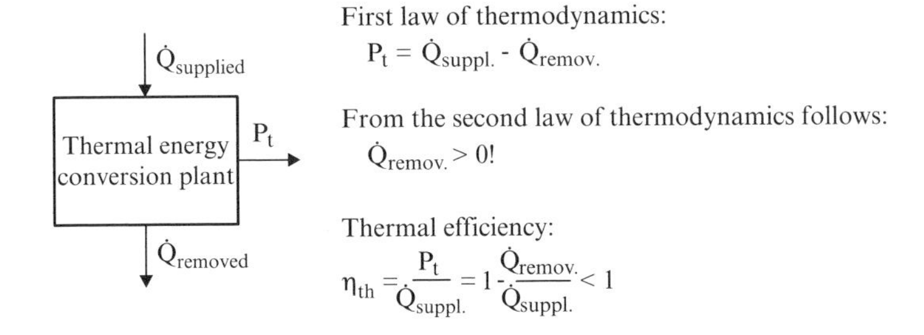

Fig. 2.2 Diagram of thermal energy conversion

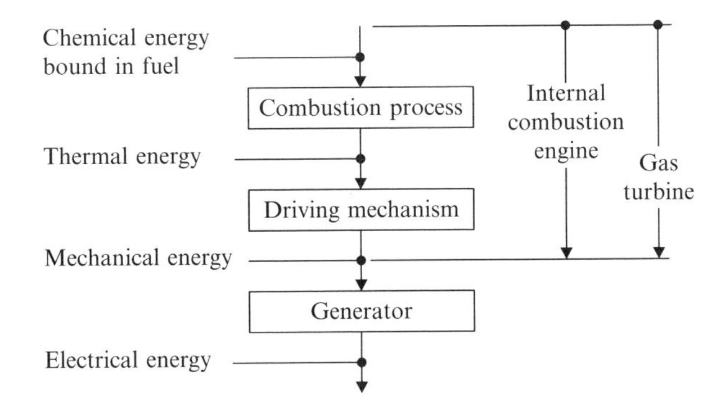

Fig. 2.3 Diagram of energy conversion in an internal combustion engine or gas turbine

top dead center (TDC) and at the bottom dead center (BDC). At both of these dead point positions, the speed of the piston is equal to zero, whilst the acceleration is at the maximum. Between the top dead center and the underside of the cylinder head, the compression volume Vc remains (also the so-called dead space in the case of reciprocating compressors).

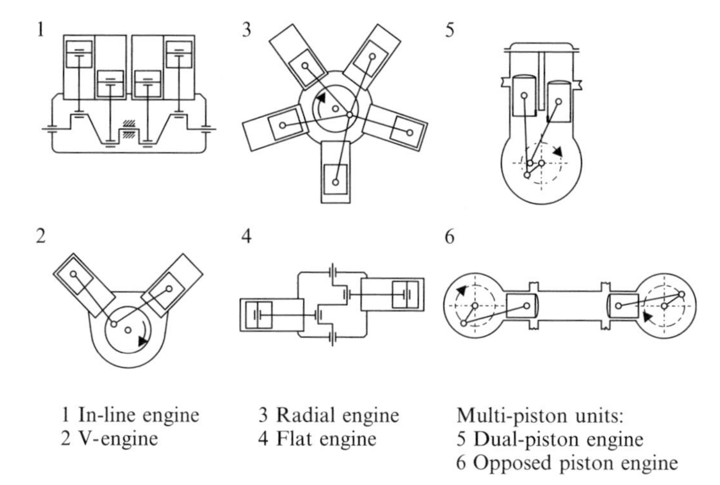

Fig. 2.4 Types of reciprocating engines

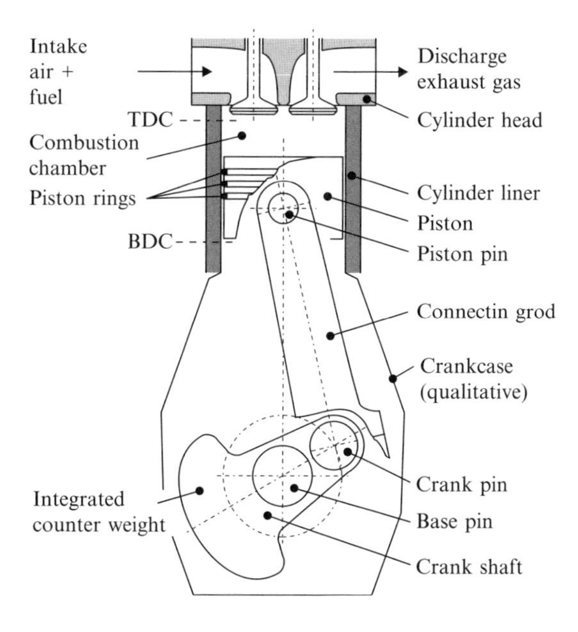

Fig. 2.5 Assembly of the reciprocating engine

Figure 2.6 shows the kinematics of a crankshaft drive with crossing, in which the longitudinal crankshaft axis does not intersect with the piston wrist pin axis, but rather is displaced by the length e.

For the piston path Π s, it follows from Fig. 2.6:

$$s(\varphi) = c_3 - c_2 - r\cos(\varphi - \beta) \tag{2.1}$$

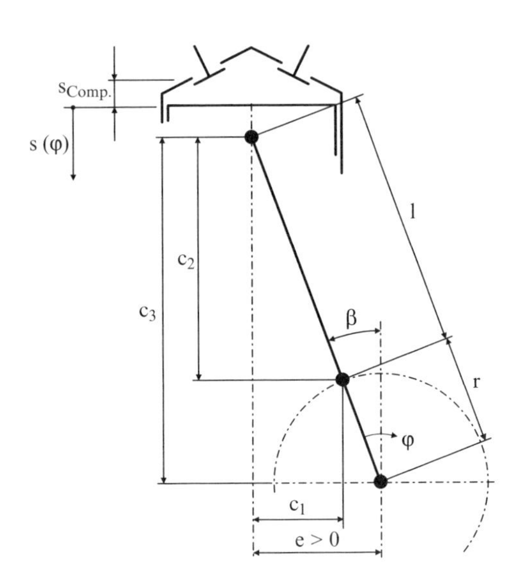

Fig. 2.6 Kinematics of the crankshaft drive

from which with

$$\sin\beta = \frac{e}{r+l} \quad \text{bzw.} \quad \beta = arc\sin\left(\frac{e}{r+l}\right),$$

$$c_1 = \mathrm{e} - r\sin\left(\beta - \varphi\right),$$

$$c_2 = \sqrt{l^2 - {c_1}^2} \qquad \text{und}$$

$$c_3 = \sqrt{(r+l)^2 - e^2}$$

finally

$$s\left(\varphi\right) = \sqrt{(r+l)^2 - e^2} - \sqrt{l^2 - \left[e + r\sin\left(\varphi - \beta\right)\right]^2} - r\cos\left(\varphi - \beta\right) \tag{2.2}$$

results. The derivative provides for the piston speed the relation

$$\frac{\mathrm{d}s}{\mathrm{d}\phi} = r\,\sin\left(\phi - \beta\right) + \frac{r\left[e + r\,\sin\left(\phi - \beta\right)\right]\,\cos\left(\phi - \beta\right)}{\sqrt{l^2 - \left[e + r\,\sin\left(\phi - \beta\right)\right]^2}}. \tag{2.3}$$

With the definition of the cylinder volume

$$V\left(\varphi\right) = V_{comp} + D^2\,\frac{\pi}{4}\,s\left(\varphi\right) \tag{2.4}$$

follows for the alteration of cylinder volume

$$\frac{dV}{d\varphi} = D^2 \frac{\pi}{4} \frac{ds}{d\varphi}. \tag{2.5}$$

With the eccentric rod relation $\lfloor e = r\, l$, it follows finally for the limiting case $e = 0$

$$s(\varphi) = r\left\{[1 - \cos(\varphi)] + \frac{1}{\lambda_s}\left[1 - \sqrt{1 - \lambda_s^2 \sin^2(\varphi)}\right]\right\} \tag{2.6}$$

and

$$\frac{ds}{d\varphi} = r\left[\sin(\varphi) + \frac{\lambda_s}{2} \frac{\sin(2\varphi)}{\sqrt{1 - \lambda_s^2 \sin^2(\varphi)}}\right] \tag{2.7}$$

For small λ_s, the expression under the root in (2.6) corresponding to

$$\sqrt{1 - \lambda_s^2 \sin^2(\varphi)} = 1 - \frac{\lambda_s^2}{2} \sin^2(\varphi) - \frac{\lambda_s^4}{8} \sin^4(\varphi) - \ldots$$

$\lambda_s = 0.25$ already becomes smaller than 0.00048 and can thus be neglected as a rule. With the help of trigonometric transformations, one finally obtains for the *piston path*

$$\frac{s}{r} = 1 - \cos(\varphi) + \frac{\lambda_s}{4}(1 - \cos(2\varphi)). \tag{2.8}$$

With the angular velocity ω

$$\frac{d\varphi}{dt} = \omega$$

one obtains for the *piston speed*

$$\frac{ds}{dt} = \frac{ds}{d\varphi}\frac{d\varphi}{dt} = \omega \frac{ds}{d\varphi}$$

the expression

$$\frac{ds}{dt} = r\omega\left[\sin(\varphi) + \frac{\lambda_s}{2}\sin(2\varphi)\right] \tag{2.9}$$

and for the *piston acceleration*

$$\frac{d^2 s}{dt^2} = \frac{d^2 s}{d\varphi^2}\left(\frac{d\varphi}{dt}\right)^2 = \omega^2 \frac{d^2 s}{d\varphi^2}$$

finally

$$\frac{d^2 s}{dt^2} = r\omega^2 \left[\cos(\varphi) + \lambda_s \cos(2\varphi)\right]. \tag{2.10}$$

With regard to charge changing in the reciprocating engine, one distinguishes between the four-stroke and the two-stroke methods and in reference to the combustion process between diesel and spark-ignition (SI) engines. In the case of the *four-stroke-procedure*, see also Fig. 2.7 (left), the charge changing occurs in both strokes, expulsion and intake, which is governed by the displacement effect of the piston and by the valves. The intake and exhaust valves open before and close after the dead point positions, whereby an early opening of the exhaust valve indeed leads to losses during expansion, but also reduces exhaust work. With increasing valve intersection, the scavenging losses increase, and the operative efficiency decreases. Most four-stroke engines are equipped, with two intakes and two exhaust valves.

In the case of the *two-stroke engine*, the charge changing occurs while the piston is near the BDC. With such so-called piston ported engines, the exhaust gas is expelled out of the cylinder by the in-flowing fresh air, if the piston sweeps over the intake and exhaust sections arranged in the lower area of the cylinder. In the case of larger engines, exhaust valves are mostly used instead of exhaust ports, which are then housed in the cylinder head. Instead of so-called loop scavenging, one then has the fundamentally more effective uniflow scavenging. For more details, see Merker and Gerstle (1997).

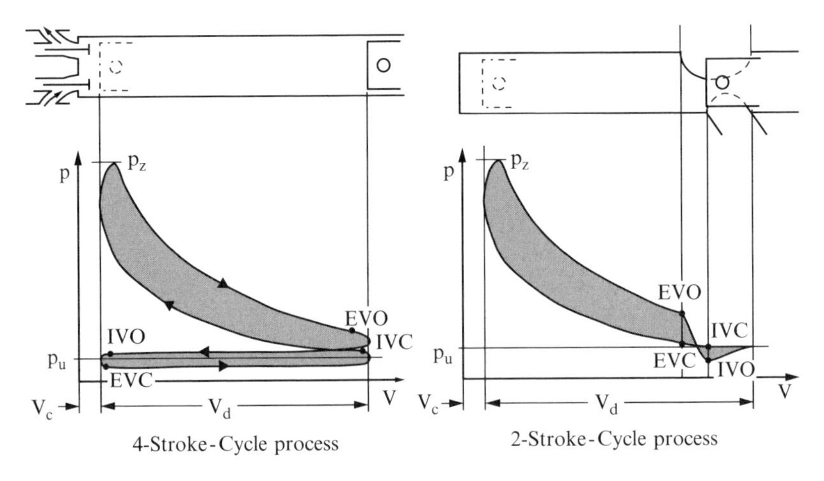

Fig. 2.7 *p*, *V* diagram for the four-stroke and two-stroke processes

2.3 Thermodynamics of the Internal Combustion Engine

2.3.1 *Foundations*

Our goal in this chapter will be to explain the basic foundations of thermodynamics without going into excessive detail. Extensive presentations can be found in Baehr (2000), Hahne (2000), Lucas (2001), Stephan and Mayinger (1998, 1999) and Heywood (1988).

For the simulation of combustion-engine processes, the internal combustion engine is separated into single components or partial systems, which one can principally view either as closed or open thermodynamic systems. For the balancing of these systems, one uses the *mass balance* (equation of continuity)

$$\frac{dm}{dt} = \dot{m}_1 - \dot{m}_2 \tag{2.11}$$

and the *energy balance* (first law of thermodynamics)

$$\frac{dU}{dt} = \dot{Q} + \dot{W} + \dot{E}_1 + \dot{E}_2 \tag{2.12}$$

With

$$\dot{E} = \dot{m}\left(h + \frac{c^2}{2}\right)$$

for the *open, stationary flooded system* shown in Fig. 2.8 (flow system), or

$$\frac{dU}{dt} = \dot{Q} + \dot{W} \tag{2.13}$$

for the closed system shown in Fig. 2.9 (combustion chamber).

In closed systems, no mass, and with that no enthalpy flows over the system limits. Neglecting the blow-by losses, the *combustion chamber* (cylinder) can be

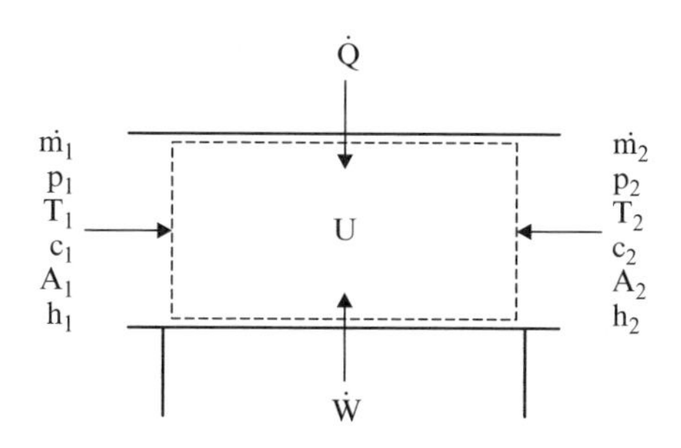

Fig. 2.8 Open thermodynamic system (*dashed lines* system boundaries)

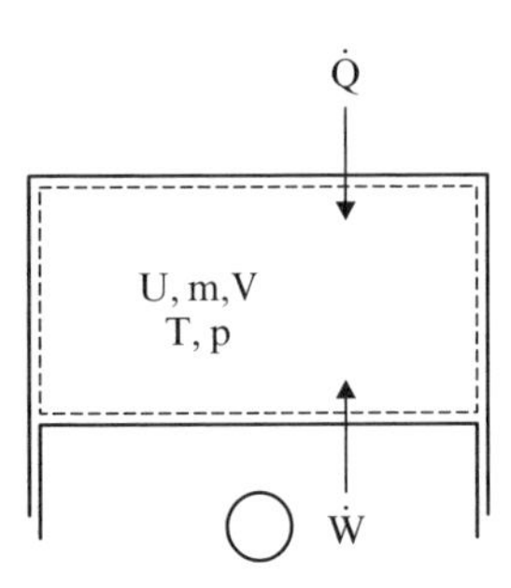

Fig. 2.9 Closed thermodynamic system (*dashed lines* system boundaries)

viewed as a closed system during the so-called high pressure process (compression and expansion act). In contrast, in the case of an open system, e.g. a *reservoir* or a *line section*, masses can flow over the system boundaries.

Neglecting the friction or dissipation of mechanical work into heat, one obtains for the *volume work*

$$\dot{W} = -p\,\frac{\mathrm{d}V}{\mathrm{d}t}. \tag{2.14}$$

In the open system, one summarizes the thermal energy transferred to the system boundaries and the intake and expulsion work practically as enthalpy

$$h \equiv u + pv \tag{2.15}$$

The *thermal state equation*

$$f\,(p, T, v) = 0 \tag{2.16}$$

ties together the three thermal condition magnitudes of pressure, temperature, and volume and the *caloric state equation*

$$\begin{aligned} u &= u\,(T, v) \text{ bzw.} \\ h &= h\,(p, T) \end{aligned} \tag{2.17}$$

describes the internal energy as a function of temperature and volume, or the enthalpy as a function of pressure and temperature. We will in the following view the materials under consideration first as *ideal gases*, for which the thermal state equation

$$p\,v = R\,T \tag{2.18}$$

is applicable. Because the inner energy of ideal gas is only dependent on temperature, follows from (2.15) with (2.18), that this is also valid for enthalpy. Thus for differential alteration of caloric magnitudes of the ideal gas we have:

$$\begin{aligned} du &= c_v\,(T)\,dT \text{ bzw.} \\ dh &= c_p\,(T)\,dT. \end{aligned} \tag{2.19}$$

For ideal gas

$$R = c_p\,(T) - c_v\,(T) \tag{2.20}$$

and

$$\kappa = \frac{c_p}{c_v}. \tag{2.21}$$

are applicable. For reversible condition alterations, the second law of thermodynamics holds in the form

$$T\,ds = dq. \tag{2.22}$$

With that, it follows from (2.13) with (2.14)

$$du = -p dv + T\,ds. \tag{2.23}$$

With (2.19), it follows for the rise of the isochores of a ideal gas

$$\left(\frac{dT}{ds}\right)_s = \frac{T}{c_v}. \tag{2.24}$$

Analogous to this, it follows for the rise of isobars

$$\left(\frac{dT}{ds}\right)_s = \frac{T}{c_p},$$

and for the isotherms and isentropes follows

$$\frac{dp}{dv} = -\frac{p}{v} \quad \text{bzw.} \quad \frac{dp}{dv} = -\kappa\,\frac{p}{v}.$$

Figure 2.10 shows the progression of simple state changes in the p,v and T, s diagram.

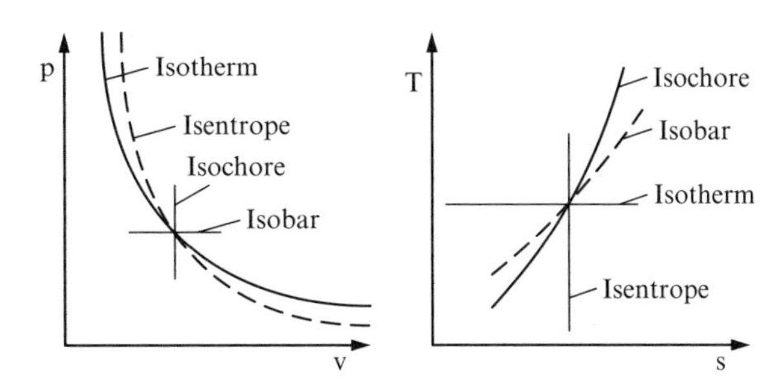

Fig. 2.10 Course of a simple change of state in the p, v- and in the T, s diagram

With the relations above, one finally obtains for *the energy balance of the closed system*

$$m\,c_v\frac{\mathrm{d}T}{\mathrm{d}t}=\frac{\mathrm{d}Q}{\mathrm{d}t}-p\,\frac{\mathrm{d}v}{\mathrm{d}t}. \tag{2.25}$$

Under consideration of the enthalpy flows and the transferred kinetic energy to the system boundaries, one obtains for the *energy balance of the open system*

$$m\,c_v\frac{\mathrm{d}T}{\mathrm{d}t}+c_v\,T\,\frac{\mathrm{d}m}{\mathrm{d}t}=\frac{\mathrm{d}Q}{\mathrm{d}t}+\frac{\mathrm{d}W}{\mathrm{d}t}+\dot{m}_1\left(h_1+\frac{c_1^2}{2}\right)-\dot{m}_2\left(h_2+\frac{c_2^2}{2}\right). \tag{2.26}$$

For *stationary* flooded open systems, it follows for the case that no work is transferred

$$\dot{m}\left[(h_2-h_1)+\left(\frac{c_2^2}{2}-\frac{c_1^2}{2}\right)\right]=\frac{\mathrm{d}Q}{\mathrm{d}t}. \tag{2.27}$$

With this relation, the flow or outflow equation for the calculation of the mass flows through throttle locations or valves can be derived. We consider an outflow process from an infinitely large reservoir and presume that the flow proceeds adiabatically. With the indices "0" for the interior of the reservoir and "1" for the outflow cross section, it follows with $c_0=0$ from (2.27)

$$\frac{c_1^2}{2}=h_0-h_1. \tag{2.28}$$

With the adiabatic relation

$$\frac{T_1}{T_0}=\left(\frac{p_1}{p_0}\right)^{\frac{\kappa-1}{\kappa}} \tag{2.29}$$

it first follows

$$\frac{c_1^2}{2}=c_p\,T_0\left(1-\frac{T_1}{T_0}\right)=c_p\,T_0\left[1-\left(\frac{p_1}{p_0}\right)^{\frac{\kappa-1}{\kappa}}\right] \tag{2.30}$$

and furthermore for the velocity c_1 in the outflow cross section

$$c_1=\sqrt{\frac{2\kappa}{\kappa-1}R\,T_0\left[1-\left(\frac{p_1}{p_0}\right)^{\frac{\kappa-1}{\kappa}}\right]}. \tag{2.31}$$

With the equation for ideal gas, it follows for the density ratio from (2.29)

$$\frac{\rho_1}{\rho_0} = \left(\frac{p_1}{p_0}\right)^{\frac{1}{\kappa}}. \tag{2.32}$$

With this results for the mass flow

$$\dot{m} = A_1\,\rho_1\,c_1$$

in the outflow cross section the relation

$$\dot{m} = A_1\,\sqrt{\rho_0\,p_0}\;\Psi\left(\frac{p_1}{p_0},\,\kappa\right) \tag{2.33}$$

Where

$$\Psi\left(\frac{p_1}{p_0},\,\kappa\right) = \sqrt{\frac{2\kappa}{\kappa-1}\left[\left(\frac{p_1}{p_0}\right)^{\frac{2}{\kappa}} - \left(\frac{p_1}{p_0}\right)^{\frac{\kappa+1}{\kappa}}\right]} \tag{2.34}$$

is the so-called outflow function, which is solely contingent upon the pressure ratio p_1/p_0 and from the isentropic exponent κ. Figure 2.11 shows the progression of the outflow function for various isentropic exponents.

The maximums of the outflow function result from the relation

$$\frac{\partial\Psi}{\partial\left(\frac{p_1}{p_0}\right)} = 0 \quad \text{für} \quad \Psi = \Psi_{max}. \tag{2.35}$$

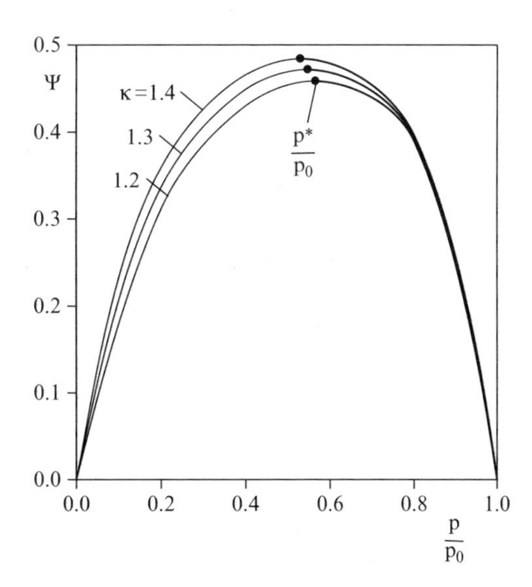

Fig. 2.11 Outflow function $\Psi\left(\frac{p_1}{p_0},\,\kappa\right)$

With this, one obtains for the so-called critical pressure ratio the relation

$$\left(\frac{p_1}{p_0}\right)_{krit} = \left(\frac{2}{\kappa+1}\right)^{\frac{\kappa}{\kappa-1}} \quad \text{bzw.} \quad \left(\frac{T_1}{T_0}\right)_{krit} = \frac{2}{\kappa+1}. \tag{2.36}$$

If we put this relation into (2.31) for the isentropic outflow velocity, then

$$c_{1,krit} = \sqrt{\kappa R T_1}. \tag{2.37}$$

finally follows. From (2.32) follows

$$\frac{\mathrm{d}p}{\mathrm{d}\rho} = \kappa\frac{p}{\rho} = \kappa R T. \tag{2.38}$$

for isentropic flows. With the definition of sound speed

$$a \equiv \sqrt{\frac{\mathrm{d}p}{\mathrm{d}\rho}} \tag{2.39}$$

it follows

$$a_1 = \sqrt{\kappa R T_1}. \tag{2.40}$$

for the velocity in the outflow cross section. The flow velocity in the narrowest cross section of a throttle location or in the valve can thereby reach maximum sonic speed.

2.3.2 *Closed Cycles*

The simplest models for the actual engine process are closed, internally reversible cycles with heat supply and removal, which are characterized by the following properties:

- The chemical transformation of fuel as a result of combustion are replaced by a corresponding heat supply
- The charge changing process is replaced by a corresponding heat removal
- Air, seen as a ideal gas, is chosen as a working medium

2.3.2.1 The Carnot Cycle

The Carnot cycle, represented in Fig. 2.12, is the cycle with the highest thermal efficiency and thus the ideal process. Heat supply results from a heat bath of

Fig. 2.12 Carnot cycle

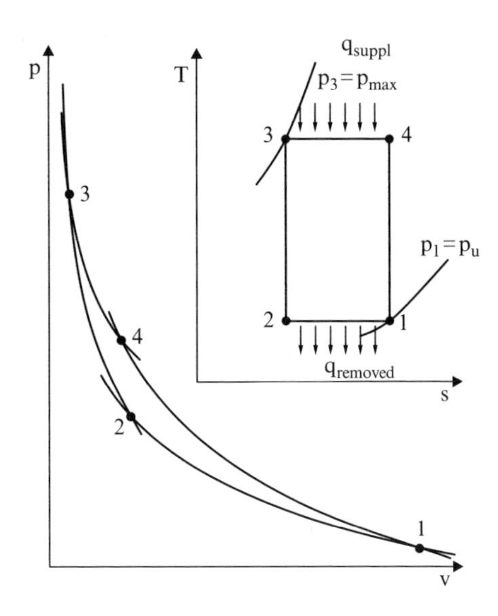

temperature 3 T, heat removal to a heat bath with temperature T1. The compression of 2 $\rightarrow$ 3 and 4 $\rightarrow$ 1 always takes place isentropically.

With the definition of the thermal efficiency

$$\eta_{th} = 1 - \frac{q_{out}}{q_{in}}$$

we obtain the well-known relation

$$\eta_{th,c} = 1 - \frac{T_1}{T_3} = f\left(\frac{T_1}{T_3}\right) \tag{2.41}$$

for it for the Carnot cycle.

The Carnot cycle cannot however be realized in internal combustion engines, because

- The isothermal expansion with $q_{supplied}$ at T_3 = const. and the isothermal compression with $q_{removed}$ at T_1 = const. are not practically feasible
- The surface in the p, v diagram and thus the internal work is extremely small even at high pressure ratios

In accordance with the definition, for the medium pressure of the process is applicable

$$p_m = \frac{w}{v_1 - v_3}. \tag{2.42}$$

For the supplied and removed heat fluxes in isothermal compression and expansion it follows

$$q_{zu} = q_{34} = R\,T_3\,\ln\frac{p_3}{p_4}\,,$$
$$q_{ab} = q_{12} = R\,T_1\,\ln\frac{p_2}{p_1}\,.$$

With the thermally and calorically ideal gas, we obtain

$$\frac{p_3}{p_2} = \left(\frac{T_3}{T_2}\right)^{\frac{\kappa}{\kappa-1}} \quad \text{und} \quad \frac{p_4}{p_1} = \left(\frac{T_4}{T_1}\right)^{\frac{\kappa}{\kappa-1}},$$

for the isentropy, from which follows

$$\frac{p_3}{p_2} = \frac{p_4}{p_1} \quad \text{bzw.} \quad \frac{p_3}{p_4} = \frac{p_2}{p_1}$$

because $T_1 = T_2$ and $T_3 = T_4$. At first we obtain for the medium pressure

$$p_m = \frac{R\,(T_3 - T_1)\,\ln\dfrac{p_3}{p_4}}{v_1 - v_3}$$

and finally by means of simple conversion

$$\frac{p_m}{p_1} = \frac{\dfrac{p_3}{p_1}\left(\dfrac{T_3}{T_1} - 1\right)\left(\ln\dfrac{p_3}{p_1} - \dfrac{\kappa}{\kappa-1}\ln\dfrac{T_3}{T_1}\right)}{\dfrac{p_3}{p_1} - \dfrac{T_3}{T_1}}. \tag{2.43}$$

The relation

$$\frac{p_m}{p_1} = f\left(\frac{T_3}{T_1}, \frac{p_3}{p_1}, \kappa\right)$$

is graphically presented in Fig. 2.13 for $\kappa = 1.4$.

While the thermal efficiency in an optimally run process at a pressure ratio of 200 with 0.6 achieves relatively high values, the reachable medium pressure still amounts only to $pm = 3.18\ p_1$. The work to be gained is thus so small that an engine realizing the Carnot cycle could in the best scenario overcome internal friction and can therefore deliver practically no performance.

The Carnot cycle is thus only of interest as a theoretical comparative process. In this context, we can only point out to its fundamental importance in connection to energy considerations.

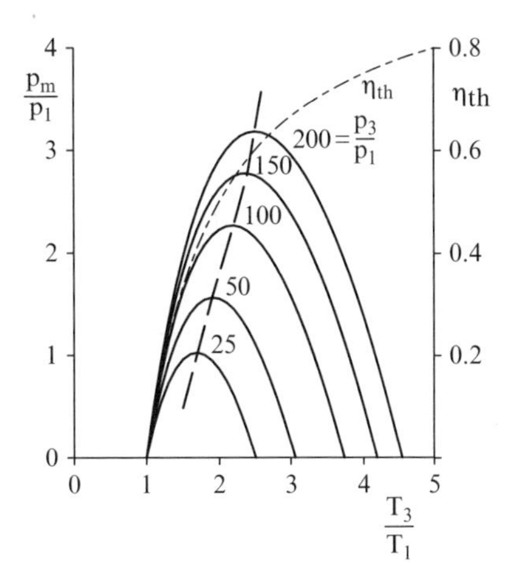

Fig. 2.13 Medium pressure and efficiency of the Carnot cycle

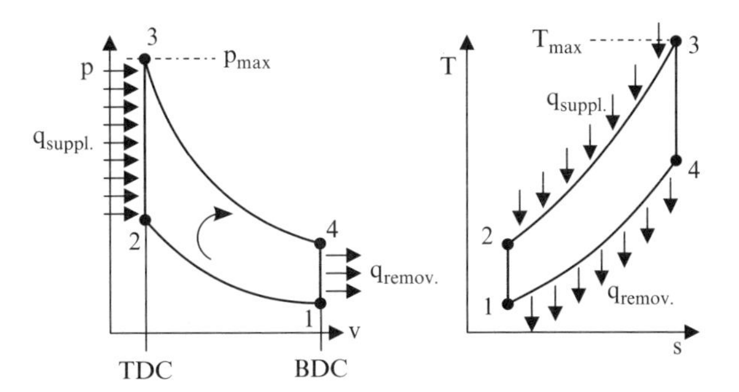

Fig. 2.14 Representation of the constant-volume cycle in the p, v and T, s diagram

2.3.2.2 The Constant-Volume Process

The constant-volume process is thermodynamically efficient and, in principle, feasible cycle (see Fig. 2.14). In contrast to the Carnot process, it avoids isothermal expansion and compression and the unrealistically high pressure ratio. It consists of two isentropes and two isochores.

It is called the constant-volume process because the heat supply (instead of combustion) ensues in constant space, i.e. under constant volume. Because the piston moves continuously, the heat supply would have to occur infinitely fast, i.e. abruptly. However, that is not realistically feasible. For the thermal efficiency of this process follows

$$\eta_{th,v} = 1 - \frac{q_{out}}{q_{in}} = 1 - \frac{c_v\,(T_4 - T_1)}{c_v\,(T_3 - T_2)} = 1 - \frac{T_1}{T_2}\,\frac{\dfrac{T_4}{T_1} - 1}{\dfrac{T_3}{T_2} - 1}.$$

With the relations for the adiabatic

$$\left.\begin{aligned}\frac{T_1}{T_2} &= \left(\frac{v_1}{v_2}\right)^{\kappa-1}\\ \frac{T_3}{T_4} &= \left(\frac{v_4}{v_3}\right)^{\kappa-1} = \left(\frac{v_1}{v_2}\right)^{\kappa-1}\end{aligned}\right\} = \frac{T_4}{T_1} = \frac{T_3}{T_2}$$

and the compression $\varepsilon = v_1/v_2$ follows finally for the thermal efficiency of the constant-volume process

$$\eta_{th,\,v} = 1 - \left(\frac{1}{\varepsilon}\right)^{\kappa-1}. \tag{2.44}$$

This relation, represented in Fig. 2.15, makes it clear that, after a certain compression ratio, no significant increase in the thermal efficiency is achievable.

2.3.2.3 The Constant-Pressure Process

In the case of high-compressing engines, the compression pressure p_2 is already very high. In order not to let the pressure climb any higher, heat supply (instead of combustion) is carried out at constant pressure instead of constant volume. The process is thus composed of two isentropes, an isobar, and an isochore (Fig. 2.16).

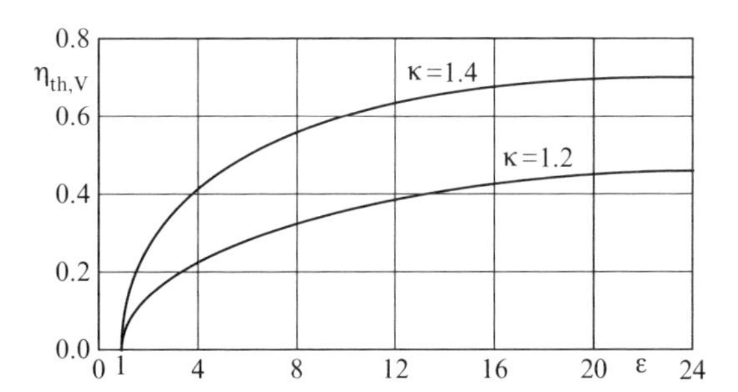

Fig. 2.15 Efficiency of the constant-volume cycle

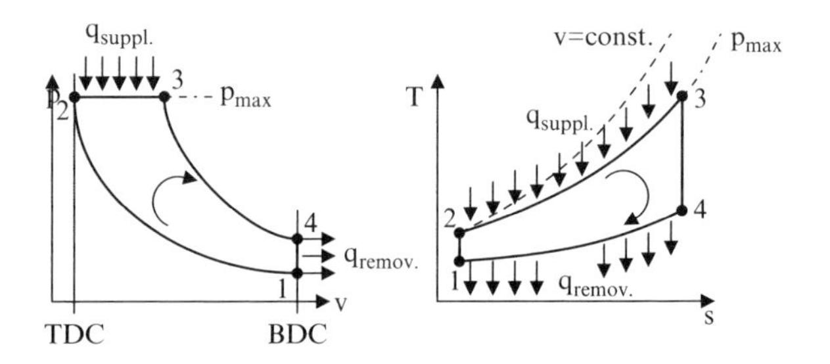

Fig. 2.16 The constant-pressure cycle in the p, v and T, s diagram

Again, for the thermal efficiency applies

$$\eta_{th,p} = 1 - \frac{q_{out}}{q_{in}} = 1 - \frac{c_v\,(T_4 - T_1)}{q_{in}}.$$

As opposed to the constant-volume process, there now appear however three prominent volumes. Therefore, a further parameter for the determination of $\eta_{th,p}$ is necessary. Pragmatically, we select

$$q* = \frac{q_{in}}{c_p\,T_1}.$$

With this, we first obtain

$$\eta_{th,p} = 1 - \frac{c_v\,(T_4 - T_1)}{c_p\,T_1\,q*} = 1 - \frac{1}{\kappa\,q*}\left(\frac{T_4}{T_1} - 1\right).$$

And finally after a few conversions

$$\eta_{th,p} = 1 - \frac{1}{\kappa\,q*}\left[\left(\frac{q*}{\varepsilon^{\kappa-1}} + 1\right)^{\kappa} - 1\right]. \tag{2.45}$$

The thermal efficiency profile of the constant-pressure process in contingency on ε and $q*$ is represented in Fig. 2.17.

2.3.2.4 The Seiliger Cycle

The Seiliger cycle, demonstrated in Fig. 2.18, represents a combination of the constant-volume and the constant-pressure processes.

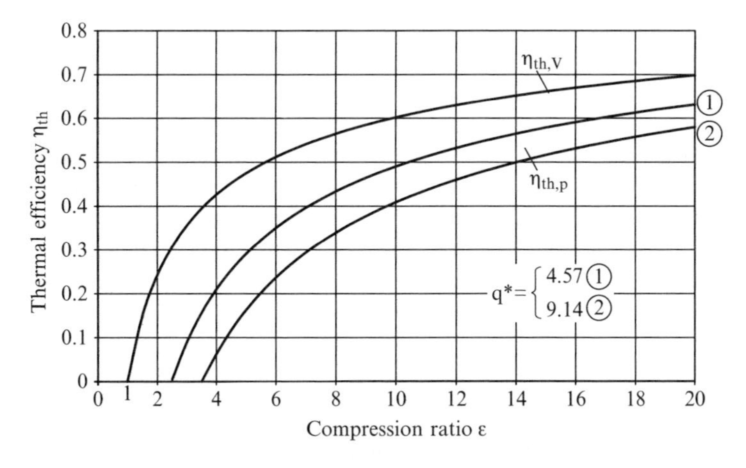

Fig. 2.17 Thermal efficiency of the constant-pressure cycle

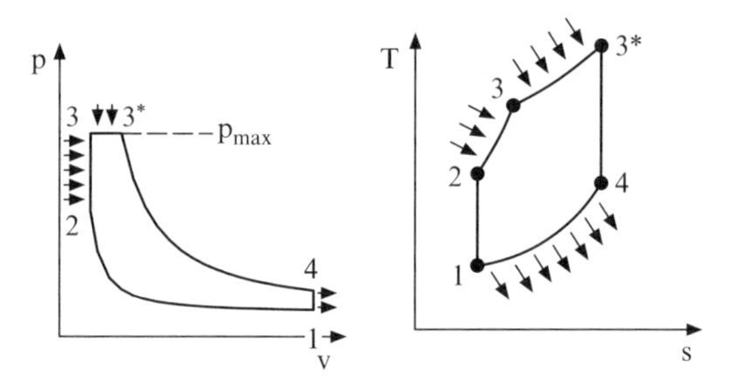

Fig. 2.18 The Seiliger cycle in the p, v and T, s diagram

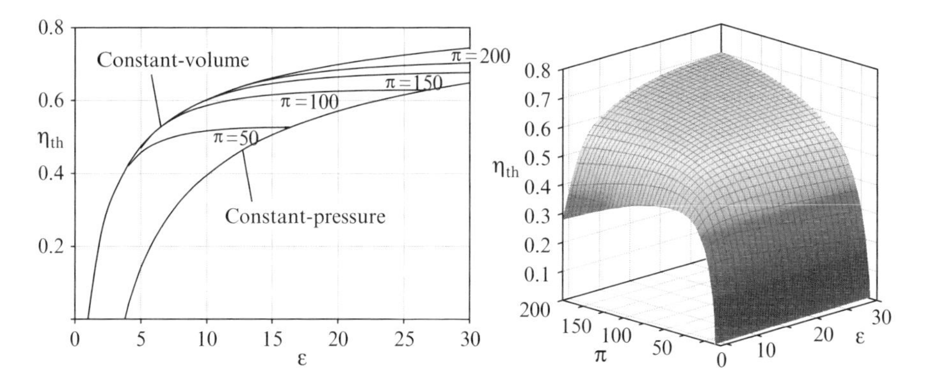

Fig. 2.19 Thermal efficiency of the Seiliger cycle

One utilizes this comparative process when, at a given compression ratio, the highest pressure must additionally be limited. Heat is supplied (instead of combustion) isochorically and isobarically. With the pressure ratio $\pi = p_3/p_1$, we finally obtain the relation

$$\eta_{th,\,vp} = 1 - \frac{1}{\kappa\, q*}\left\{\left[q * - \frac{1}{\kappa\,\varepsilon}\left(\pi - \varepsilon^{\kappa}\right) + \frac{\pi}{\varepsilon}\right]^{\kappa}\left(\frac{1}{\pi}\right)^{\kappa-1} - 1\right\}, \tag{2.46}$$

for the thermal efficiency, which is graphically represented in Fig. 2.19. From this it becomes clear that, at a constant given compression ratio ε, it is the constant-volume process, and at a constant given pressure ratio π, it is the constant-pressure process which has the highest efficiency. The adiabatic exponent κ is 1.4 for air at 25°C and 1.31 at 1,400°C while for exhaust it is about 1.33, so its influence cannot be neglected.

2.3.2.5 Comparison of the Cycles

For the efficiency of the particular comparative processes, the following contingencies result

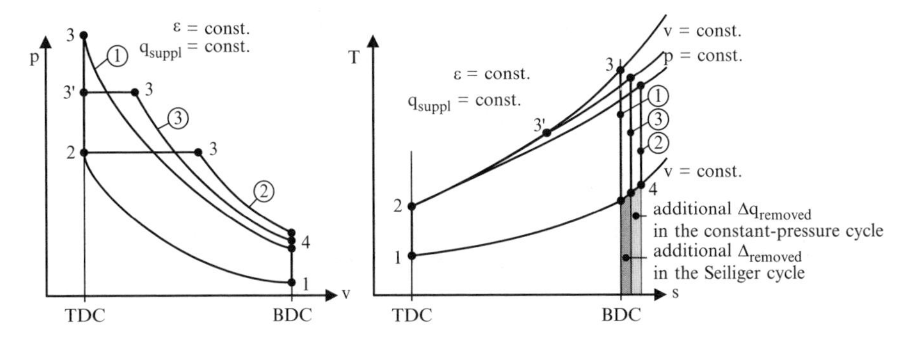

Fig. 2.20 Comparison of the closed cycles, ① = constant-volume, ② = constant pressure, ③ = Seiliger cycle

$$\eta_{th,c} = f\left(\frac{T_1}{T_3}, \kappa\right) \quad \text{Carnot,}$$

$$\eta_{th,v} = f(\varepsilon, \kappa) \quad \text{constant volume,}$$

$$\eta_{th,p} = f(\varepsilon, q^*, \kappa) \quad \text{constant pressure,}$$

$$\eta_{th,vp} = f\left(\varepsilon, q^*, \frac{p_3}{p_1}, \kappa\right) \quad \text{Seiliger.}$$

In Fig. 2.20, the constant-volume process, the constant-pressure process, and the Seiliger process are represented in a p, v- and T, s- diagram. The constant-volume process has the highest, while the constant-pressure process has the lowest efficiency. The efficiency of the Seiliger cycle lies between them. In this comparison, the compression ratio and the supplied heat amount are equal for all three cycles. This shows clearly that, in the case of the Seiliger cycle some, in the case of the constant-pressure process unmistakably more heat must be removed as in the constant-volume process and therefore that the thermal efficiency of these processes is lower.

2.3.3 *Open Comparative Processes*

2.3.3.1 The Process of the Perfect Engine

The simple cycles depart partially, yet significantly from the real engine process, such that no detailed statements are possible about the actual engine process, see Urlaub (1994). Therefore, for further investigations we also consider open comparative processes, which consider the chemical transformation involved in combustion instead of the heat supply and removal of the closed cycles.

As opposed to closed cycles, the open comparative cycles allow a charge changing, and they calculate the high pressure process in a stepwise fashion and are thus more or less realistically. However, the charge changing is, as a rule, also not considered more closely. The essential differences to the closed cycles are:

- Compression and expansion are either considered isentropically, as previously, or they are described via polytropic state changes
- Energy release via combustion is calculated gradually, if also with certain idealizations with reference to the combustion itself
- An approach to the consideration of energy losses resulting from heat transfer is made

Figure 2.21 shows an open comparative process in the T, s diagram. Specifically, the fresh load "appears" at point 1 and the exhaust gas, which is viewed as mixture burned to an arbitrary extent, "disappears" at point 4.

Charge changing is not considered here any further. For additional details, refer to Pischinger et al. (2002).

2.3.3.2 Heat Release Through Combustion

A distinction is made between incomplete/complete and perfect/imperfect combustion.

For air ratios $\lambda \geq 1$, the fuel could in principle burn completely, i.e. the added energy $m_{fuel}\ lhv$ is completely converted into thermal energy

$$Q_{\max} = Q_{th} = m_f H_u .$$

For complete combustion of hydrocarbons, the total reaction equations

$$\begin{aligned} H_2 + \tfrac{1}{2}O_2 &= H_2O \\ C + O_2 &= CO_2 \ . \end{aligned}$$

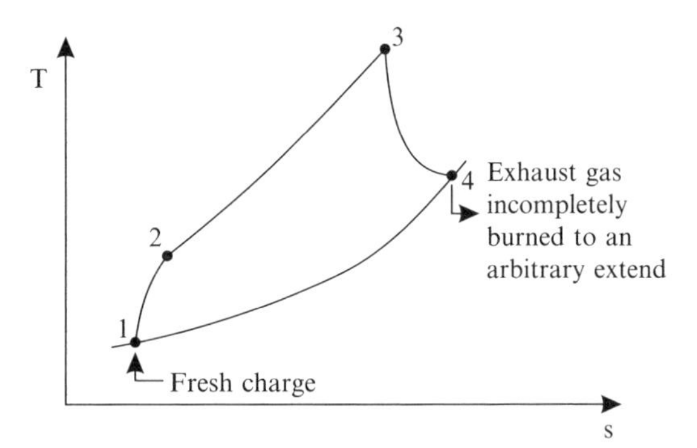

Fig. 2.21 Ideal process for internal combustion engines

Thus, only the products water and carbon dioxide are formed.

In actuality, however, combustion advances maximally until chemical equilibrium for air/fuel ratios $\lambda \geq 1$ as well, thus always incompletely.

For air ratios $\lambda < 1$, the fuel cannot completely burn as a result of a lack of O_2. In the case of such incomplete combustion, the combustion proceeds, in the best case, until chemical equilibrium.

Under all air ratios, combustion can further progress imperfectly, in the case that the oxygen present is not sufficiently optimally distributed (mixture formation), or in the case that single reactions progress slowly and thus that chemical equilibrium is never reached. In the exhaust, we therefore find not only CO_2 and H_2O, but also carbon monoxide, unburned hydrocarbons, soot particles, H_2 and nitrogen compounds.

The degree of conversion is defined as:

$$\eta_u = 1 - \frac{Q_{uv}}{m_f H_u}.$$

According to Pischinger et al. (2002), the total degree of conversion can be written thus:

$$\eta_{u,ges} = \eta_{u,ch} \cdot \eta_u.$$

On the basis of reaction-kinetic estimations, provide the term

$$\eta_{u,ch} = \begin{cases} 1 & \text{for} \quad \lambda \geq 1 \\ 1,3773\,\lambda - 0,3773 & \text{for} \quad \lambda \leq 1 \end{cases}$$

for the degree of conversion $\eta_{conv,ch}$. These conditions are clarified visually in Fig. 2.22.

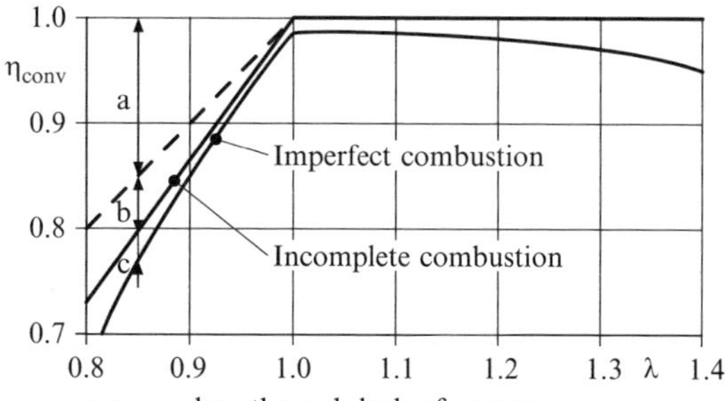

Fig. 2.22 Release of energy and degree of conversion

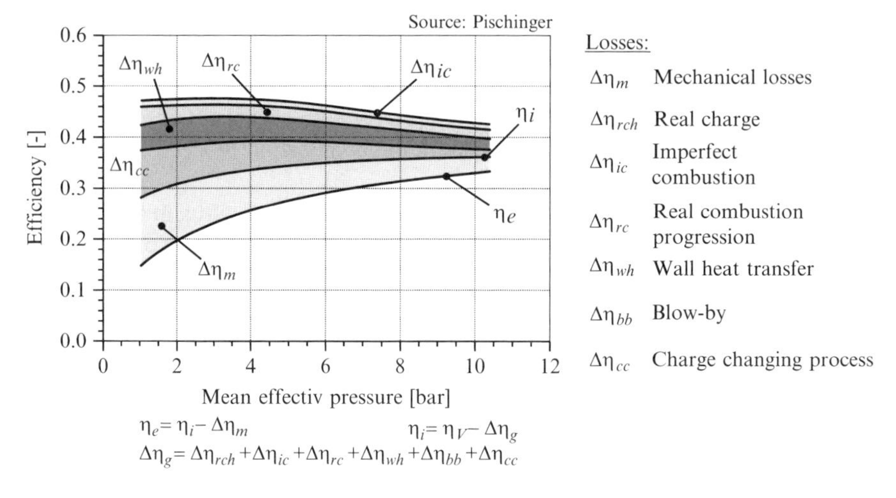

Fig. 2.23 Division of losses in the real internal combustion engine

2.3.3.3 The Real Engine Process

Proceeding from the perfect engine process, the actual efficiency of the real engine process can be established through systemactic quantification of particular idealizations. Practically, particular losses are thereby considered via corresponding reductions in efficiency, e.g.

Volumetric efficiency $\Delta\eta_{rL}$ Loss with respect to the entire engine because the actual cylinder filling at "intake closes" is smaller than that of the ideal engine

Combustion $\Delta\eta_{uv}$ Loss as a result of incomplete or imperfect combustion

Heat transfer $p_{m,W}$ Heat losses through heat transfer to the combustion chamber walls

Charge changing $\Delta\eta_{LW}$ Charge changing losses

Blow-by $\Delta\eta_{Bb}$ Leakage

Friction $\Delta\eta_m$ Mechanical losses due to engine friction (piston-piston rings cylinder liner, bearing) and to accessory drives (valve train, oil and water pump, injection pump).

Particular losses are visually clarified in Fig. 2.23. We will not go into any further detail, one reason being that these simple considerations are of increasingly less importance.

2.4 Parameters and Characteristic Values

Parameters of internal combustion engines are important with reference to the interpretation and determination of engine measurements, the examination and establishment of the actual performance and the evaluation and comparison of various combustion machines.

The *medium pressure* is an important characteristic quantity for the evaluation of performance and of the technological standing of an internal combustion engine. From the definition of piston work

$$dW = p\,A_k\,dx = p\,dV$$

we obtain through integration over a working cycle for the indicated work per cycle

$$W_i = \oint p\,dV$$

and with the definition

$$W_i = p_{m,i}\,V_h$$

for the indicated mean effective pressure

$$p_{m,i} = \frac{1}{V_h}\oint p\,dV. \tag{2.47}$$

For the indicated or internal performance of a multi-cylinder engine follows

$$P_i = P_{i,z}\,z = z\;n_A\,p_{m,i}\,V_h.$$

With the number of working cycles per time

$$n_A = i \cdot n \quad \text{and} \quad i = \begin{cases} 0,5 & \text{for } 4 - \text{Takt} \\ 1 & \text{for } 2 - \text{Takt} \end{cases}$$

we finally obtain for the indicated power for

$$P_i = i\,z\,n\,p_{m,i}\,V_h. \tag{2.48}$$

Analogous to this, we obtain with the actual mean effective pressure *mep*

$$P_e = i\,z\,n\,p_{m,e}\,V_h. \tag{2.49}$$

for the effective power. The effective power is the difference from indicated power and friction power

$$P_e = P_i - P_r, \tag{2.50}$$

from which follows the relation

$$p_{m,r} = p_{m,i} - p_{m,e} \tag{2.51}$$

for the friction mean effective pressure.

The effective power of an engine follows from the so-called indicator diagram. The effective power follows from

$$P_e = M\,2\,\pi\,n, \tag{2.52}$$

where the torque M and the speed n are determined on an engine test stand.

The *efficiency* of a thermal energy conversion machine is quite generally the relation of benefit to expenditure. In the case of the internal combustion engine, the benefit the indicated or actual engine performance and the expenditure the energy added with the fuel mass flow $\dot{m}$ *fuellhv*. With this, it follows

$$\eta_{i,e} = \frac{P_{i,e}}{\dot{m}_f H_u}. \tag{2.53}$$

The ratio of the effective to the indicated power is the mechanical efficiency

$$\eta_m = \frac{\eta_e}{\eta_i} = \frac{P_e}{P_i} = \frac{p_{m,e}}{p_{m,i}}. \tag{2.54}$$

The *specific fuel consumption* is the fuel consumption related to the engine performance

$$b_e = \frac{\dot{m}_f}{P_e} = \frac{1}{\eta_e H_u}. \tag{2.55}$$

With a medium value for the lower heating value of gasoline and diesel oil of about

$$H_u \approx 42.000 \left[\frac{\text{kJ}}{\text{kg}}\right]$$

one obtains the simple rule-of-thumb between the specific fuel consumption and the actual efficiency

$$b_e \approx \frac{86}{\eta_e} \left[\frac{\text{g}}{\text{kWh}}\right].$$

An actual efficiency of $\eta_e = 40\%$, for example, leads to a specific fuel consumption of $b_e = 215\,\text{g/kWh}$.

We distinguish between upper and lower heating values. In the determination of the upper heating value, the combustion products and cooled back to intake temperature, the water contained in it is condensed out and is thus liquid. As opposed to this, in the determination of the lower heating value, the water is not condensed out and is thus in a vapor state. For internal combustion engines, the lower heating value must be used because of the relatively high exhaust gas temperature.

A so-called mixture heating value is occasionally utilized, under which is meant the added energy flow in reference to the fresh charge. For SI and diesel engines, we retain different expressions for this, because one takes in a gasoline-air mixture and the other pure air.

$$H_G = \frac{m_f H_u}{V_G} \quad \text{with} \quad V_G = \frac{m_{air} + m_f}{\rho_G} \quad \text{ST engine}$$
$$H_G = \frac{m_f H_u}{V_{air}} \quad \text{with} \quad V_{air} = \frac{m_{air}}{\rho_{air}} \quad \text{Diesel engine}\,.$$

The ratio of added fresh charge to the theoretically possible charge mass m_{th} is designated as *air expenditure*,

$$\lambda_a = \frac{m_G}{m_{th}} = \frac{m_G}{V_h\,\rho_{th}}\,. \tag{2.56}$$

The theoretical charge density is the density of the intake valve.

As opposed to the air expenditure, the *volumetric efficiency* designates the ratio of charge mass actually found in the cylinder after charge changing in comparison to the theoretically possible charge mass.

$$\lambda_l = \frac{m_z}{m_{th}} = \frac{m_z}{V_h\,\rho_{th}}\,. \tag{2.57}$$

Volumetric efficiency λ_l is contingent above all on valve overlap in the charge changing-TDC. An optimization of the volumetric efficiency can, with constant control times, only ensue for one speed. With variable valve timing (e.g. by retarding the camshaft), volumetric efficiency can be optimized across the entire speed range. For four-stroke engines with little valve overlap, $\lambda_l \approx \lambda_a$ is valid.

In addition to the above-cited quantities, a few other characteristic quantities are used. The medium piston speed is a characteristic speed for internal combustion engines,

$$c_m = 2\,s\,n. \tag{2.58}$$

The maximum piston speed is dependant on stroke/rod ratio and lies in the range $c_{\max} = (1.6\text{–}1.7)\,c_m$.

The compression ratio is the total cylinder volume in reference to the compression volume,

$$\varepsilon = 1 + \frac{V_h}{V_c}\,. \tag{2.59}$$

The displacement (cubic capacity) is the difference between total volume and compression volume. With the piston path s we obtain

$$V_h = \frac{\pi}{4} D^2 s. \tag{2.60}$$

As a further characteristic quantity, the bore/stroke ratio

$$\frac{s}{D} \tag{2.61}$$

is used.

2.5 Engine Maps

2.5.1 Spark Ignition Engines

In the case of the conventional *spark ignition (SI) engine*, gasoline is sprayed in the intake port directly in front of the intake valve (multi point injection). Through this, a mixture of air and fuel is taken in and compressed after the closure of the intake valve. Before arrival at the top dead center, the compressed mixture is ignited by means of a sparkplug (spark ignition). Because two phases, the intake and compression strokes, are available for mixture formation, the mixture is nearly homogenous at the end of compression. In conventional engines, the amount of inducted air is regulated by means of a throttle valve positioned in the suction line. At light throttle conditions, this throttle valve is almost closed, while at full loads is it completely open. The amount of inducted air is measured and the fuel injected in proportion to the amount of air, normally such that the medium air ratio $\lambda = 1$ is maintained. Because the amount of mixture is regulated in SI engines, we speak of a quantity regulation. So that no autoignition takes place in the compressed mixture, the compression ratio ε has to be restricted.

Figure 2.24 shows the p, v diagram for a four-stroke SI engine at partial load (left) and full load (right). The throttle valve, nearly closed at partial load, results in

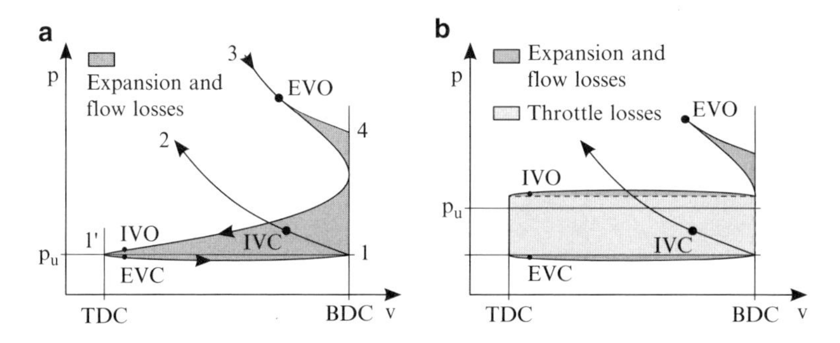

Fig. 2.24 p, v diagram for a four-stroke SI engine under (**a**) full load and (**b**) partial load

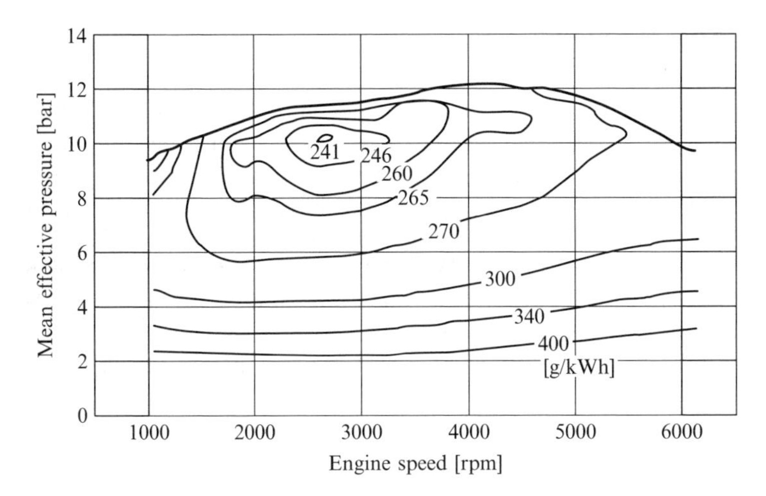

Fig. 2.25 Engine map for a four-stroke SI engine with lines of constant consumption b_e [g/kWh]

high pressure losses in the suction hose. It thus leads to a "large charge changing slide" and finally to inferior efficiency.

Figure 2.25 shows the fuel consumption (b_e) map for a four-stroke SI engine. The characteristic map is restricted by the idle and limit speeds as well as by the maximum torque line. Because $P \sim T{\cdot}n$, the lines of constant performance are hyperbolas in the engine map. The so-called conidial curves are lines of constant specific consumption.

The behavior and characteristics of the internal combustion engine can be read off from the engine map. From the relations for performance and for torque (load)

$$P_e = i\, n\, z\, p_{m,\,e}\, V_h = M_e\, 2\,\pi\, n$$

we obtain the contingencies

$$\begin{aligned} \text{Power:}\quad & P_e \sim n\, p_{m,\,e} \\ \text{Torque:}\quad & M_e \sim p_{m,\,e}\,. \end{aligned}$$

The load thereby corresponds to the torque and not to performance!

For the ratio of actual to indicated fuel consumption follows

$$\frac{b_e}{b_i} = \frac{P_i}{P_e} = \frac{P_e + P_r}{P_e} = 1 + \frac{P_r}{P_e} = 1 + \frac{p_{m,\,r}}{p_{m,\,e}}.$$

The friction mean effective pressure *fmep increases with engine* speed. Therefore at a constant speed, b_e must increase with as $P_{m,e}$ is reduced.

2.5.2 Diesel Engines

In the case of older conventional diesel engines, only air is taken in and compressed. In newer diesel engines with exhaust gas recycling up to 40% exhaust gas is taken in and compressed. The fuel (diesel oil) is injected just before the top dead center into the hot air. Because of the high compression ratio, the temperature of the compressed air is higher than the autoignition temperature of the fuel, and after the so-called ignition lag time autoignition begins. In contrast to the SI engine, only a small amount of homogeneous mixture can form in the short time between injection start and autoignition: injection, mixture formation, and combustion therefore proceed almost simultaneous. The torque output of the diesel engine is regulated by with the amount of injected fuel, one thus refers to a quality regulation. While in the conventional SI engine the air ratio is always near $\lambda = 1$ at part load, it varies in the case of the diesel engine with the load and moves within the region $1.1 \leq \lambda \leq 100$.

Figure 2.26 shows the fuel consumption (b_e) map of a four-stroke diesel engine. We recognize that the speed spread is clearly narrower and that the actual P_e is evidently higher than in the case of normal aspirated SI engine.

With reference to their speeds, internal combustion engines are divided into high-speed engines, medium-speed engines, and slow-speed engines, with reference to their construction sizes, however, into vehicle, industrial, and large engines as well. On the other hand, engines built for racing have a special position, particularly

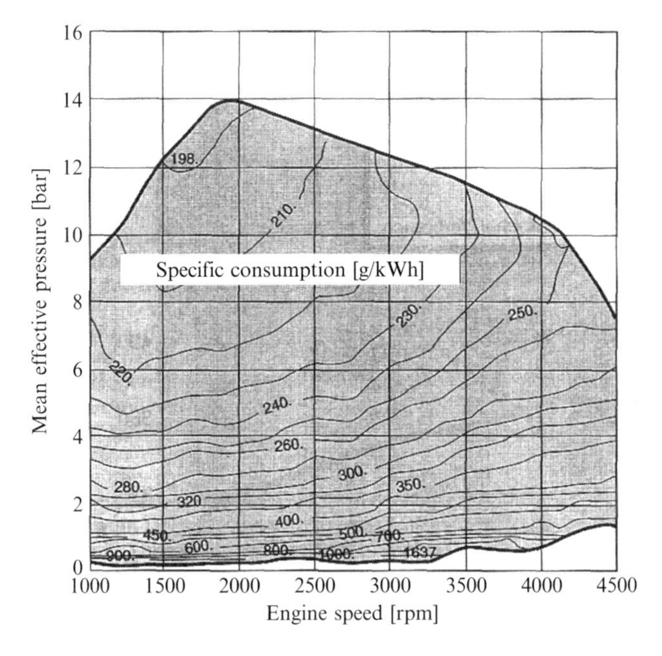

Fig. 2.26 Engine map of a four stroke diesel engine

Type	$\frac{n}{rpm}$	$\frac{mep}{bar}$	η_e	ε	$\frac{c_m}{m/s}$
Passenger cars - SI	< 7000	8 - 13	0.25 - 0.35	6 - 12	9 - 20
Passenger cars - diesel	< 5000	7 - 22	0.30 - 0.40	16 - 22	9 - 16
Trucks - diesel	< 3000	15 - 25	0.30 - 0.45	10 - 22	9 - 14
High-speed engines	1000 - 2500	10 - 30	0.30 - 0.45	11 - 20	7 - 12
Medium-speed engines	150 - 1000	15 - 25	< 0.5	11 - 15	5 - 10
Slow-speed engines	50 - 150	9 - 15	< 0.55	11 - 15	5 - 7
Racing engines		12 - 35	- 0.3	7 - 11	< 25

Fig. 2.27 Classification of internal combustion engines

because extremely light construction and high performance are principally stressed. With that we have the classification provided in Fig. 2.27.

This classification is certainly not compulsory and is not without a certain amount of arbitrariness, yet it is still practical and comprehensible. We could, in principle, expand it further in consideration of the categories mini-engines (model air planes), small engines (chain saws), and motorcycle engines.

References

Baehr HD (2000) Thermodynamik, 10. überarb. Aufl. Springer, Berlin

Hahne EWP (2000) Technische Thermodynamik, 3rd edn. Oldenbourg Wissenschaftsverlag GmbH, München, Wien

Heywood JB (1988) Internal combustion engine fundamentals. McGraw-Hill, New York

Lucas K (2001) Thermodynamik. Die Grundgesetze der Energie- und Stoffumwandlungen. 3., korr. u. aktualisierte Aufl., Springer, Berlin

Maas H (1979) Gestaltung und Hauptabmessungen der Verbrennungskraftmaschine. Die Verbrennungskraftmaschine, Neue Folge, Band 1, Springer, Heidelberg

Merker GP, Gerstle M (1997) Evaluation on two stroke engines scavenging models, SAE-Paper 970358

Pischinger R, Klell M, Sams Th (2002) Thermodynamik der Verbrennungskraftmaschine. 2., überarbeitete Auflage, Springer, Wien

Stephan K, Mayinger F (1998) Thermodynamik, 15. Aufl., Band 1: Einstoffsysteme. Grundlagen und technische Anwendungen. Springer, Berlin

Stephan K, Mayinger F (1999) Thermodynamik. Aufl. 14, Band 2, Mehrstoffsysteme und chemische Reaktionen, Springer, Berlin

Urlaub A (1994) Verbrennungsmotoren, 2nd edn. Springer, Berlin

van Basshuysen R, Schäfer F Hrsg (2003) Handbuch Verbrennungsmotor – Grundlagen, Komponenten Systeme, Perspektiven, 2. Aufl., Verlag Friedrich Vieweg & Sohn, Braunschweig

Chapter 3
Combustion Diagnostics

Rüdiger Teichmann, Andreas Wimmer, Christian Schwarz, and Ernst Winklhofer

3.1 Basics of Pressure Measurements

3.1.1 Preface

Pressure measurement is as old as the development of internal combustion engines themselves. A confirmation of this can be seen in Fig. 3.1, which shows a cylinder pressure profile recorded with a mechanical indicator by August Nikolaus SI on 18th of May, 1876. This moment in time is often considered to be the starting point of the historical development of pressure indication, see Hohenberg (1994). After over 135 years of development, this measurement technique has become universally established and is used as the standard measurement method on most engine test benches, Frommelt et al. (2008).

High-pressure indicating (measuring cylinder pressure) and low-pressure indicating (measuring the pressure in the inlet and outlet system) have since been developed into valuable, highly sophisticated analytic methods in combustion optimization. Both the sensors and computer-supported data capture have reached a stage of development which not only allows for the use of indication as an industrial measurement technology, but also satisfies demands for accuracy which make it possible to reach far-reaching conclusions from measured pressure profiles (see Fig. 3.2):

- Indication is the developmental tool for optimizing combustion engines quickly and with high quality.
- No other measurement technique delivers such a wealth of information about internal engine processes.
- When used competently, indication is a safe measurement technique and is thus applicable as a standard measurement method on the developmental test bench.

High-pressure indication in the combustion chamber is usually executed with piezoelectric pressure sensors that are either directly built into the combustion chamber or into the spark or glow plug by means of an adapter. Analysis of the measured pressure profiles in the combustion chamber permits a diverse and comprehensive assessment of internal engine processes. Based on the measured

G.P. Merker et al. (eds.), *Combustion Engines Development*,
DOI 10.1007/978-3-642-14094-5_3, © Springer-Verlag Berlin Heidelberg 2012

Fig. 3.1 Indicator diagram by August Nikolaus SI from May 18th, 1876

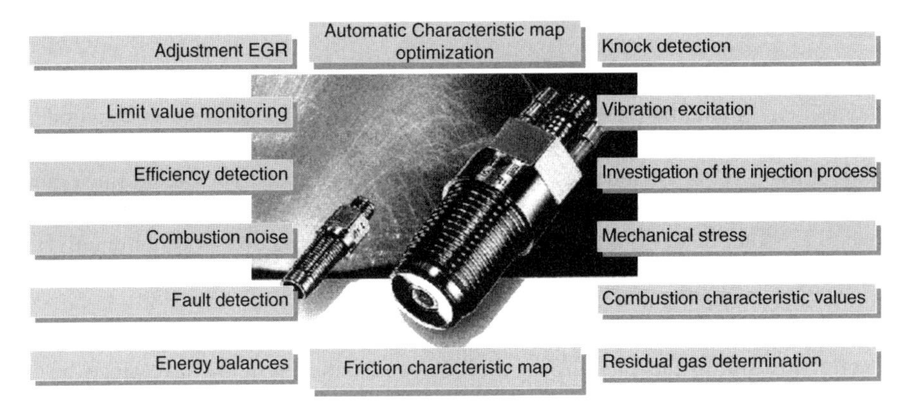

Fig. 3.2 Areas of application of indicator technology

pressure profile, a plethora of important data can be calculated. These resultant values are generally called combustion diagnostic metrics and can basically be subdivided into two categories:

- Direct
- Indirect indicator characteristic values

Direct indicator characteristic values are ascertained directly from the profile of cylinder pressure p across the working cycle and already allow for a variety of conclusions (Table 3.1). Indirect indicator characteristic values are quantities which can only be determined by means of intermediate calculations based on the pressure profile. These include above all those parameters that are determined by the combustion profile calculated in the context of thermodynamic analysis. Example of this include ignition delay, initiation of combustion and the mass fraction burned points (10%, 50% and 90% as a rule) that permit an assessment of the combustion process.

Table 3.1 Direct indicator characteristic values

Direct indicator characteristic value	Conclusion/Assessment
Indicated mean pressure p_{mi} ($p_{mi\text{-}HD}$, $p_{mi\text{-}LW}$)	– Internal work – Internal efficiency – Friction mean pressure p_{mr}
Standard deviation of the indicated mean pressure σ_{pmi}	– Combustion stability – Engine operation smoothness – Number of misfires
Point pressure p_{max}	– Component strain – Acoustic assessment
Location of the point pressure α (p_{max})	– Position of combustion relative to the optimal position – Amount of knock limitation – Rate of turnover (α (p_{max}) – ZZP)
Maximum pressure increase $(dp/d\alpha)_{max}$	– Rate of turnover – Acoustic assessment
Rate pressure increase $dp/d\alpha^2$	– Acoustic assessment

Modern indicating systems are capable of calculating many hundreds of combustion metrics in real time. They are thus available along with other measurement variables directly at the test bench. This makes it possible to make examinations such as

- Fault detection (misfire and pre ignition)
- Automatic characteristic map optimization
- Knock detection
- Leaning adjustment
- Adjustment of exhaust gas recirculation
- Optimization of combustion noise directly at the test bench

For low-pressure indicating, piezoresistive and piezoelectric pressure sensors are used. In conjunction with the cylinder pressure profile, the measured pressure profiles in the inlet and outlet system form the foundation for the analysis of charge changing, which involves:

- The design of the intake and exhaust ducting
- The design of the controls (control times, cam profiles)
- Assessment of the work of gas exchange
- The analysis of the inlet and outlet mass flows (charge, residual gas, reverse flows)

To calculate the gas exchange, it is also necessary to know the exact valve lift profiles and flow coefficients. For charge changing analysis, it is important that low-pressure indication is carried out in conjunction with high-pressure indication. That means that three pressures have to be measured simultaneously per cylinder: the intake port pressure, the combustion chamber pressure and the exhaust port pressure. Only in this configuration can a complete combustion profile and charge changing analysis be performed.

3.1.2 The Piezoelectric Measurement Chain

Figure 3.3 shows the basic structure of the piezoelectric pressure measurement chain. Essentially, it consists of the following components (Wimmer and Glaser 2002):

- The piezoelectric pressure sensor
  The piezoelectric pressure sensor is based on the functional principle that certain crystals generate or absorb a charge under mechanical strain. It thus represents an active measuring element, whereby charge is proportional to the strain, i.e. to the applied pressure.
- Charge amplifier
  With the help of a charge amplifier, the low level charge created by the piezoelectric pressure sensor is amplified and integrated into a voltage signal that is then digitized by the combustion analyzer for further processing.
- Measurement wiring
  Measurement wiring serves to transfer the charge and voltage signals. Due to the low electric charge released by the piezoelectric pressure sensor (picocoulombs), the connection between the sensor and the charge amplifier has an especially decisive role. Very high insulating values (10^{14} Ω), low noise levels, robustness and simple handling are required both for the measurement wiring and for the charge amplifier.

The following additional components are also necessary to take measurements at the combustion engine.

- Angle Sensor
  For a detection of the pressure profile that is supported by the crank angle, an angle sensor mounted on the crankshaft of the combustion engine provides the

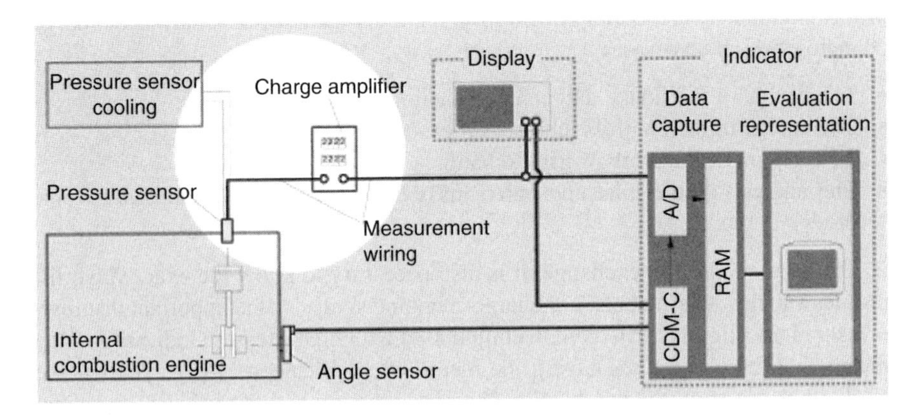

Fig. 3.3 Schematic design of the piezoelectric pressure measurement chain with additional components

temporal and angular basis. This is required to calculate the cylinder volume at each sample, which is needed for most calculations.

- Indicating system (Combustion Analyzer)
  With the help of an indicating system, the analog output signal of the charge amplifier is digitalized and recorded. An indicating system contains at least one A/D converter, a trigger unit for data capture on an angular basis and fast memory for temporary storage of the raw data. Processing of the data from memory as well as the calculation, storage and representation of the parameters is accomplished with a computer, which in the case of some advanced devices is built into the device in the form of RTP's (real time processors).

These further items can also be required:

- Pressure sensor cooling
  Water-cooled pressure sensors are still used today, with quartz sensors, which need to be constantly cooled during operation and thus require the use of a corresponding cooling system. Guaranteeing that water-cooled pressure sensors are sufficiently cooled in every engine operation condition is decisive for the quality of the measurement and serviceability of these sensors. Newer piezo-electric materials have enabled very accurate cylinder pressure measurements without water cooling.
- Additional displays
  For certain measurement tasks (e.g. operation of the engine at the knock limit), it is sometimes necessary to utilize an additional system for displaying indicated data in order to guarantee complete monitoring as the measurement data is recorded.
- Principle of measurement
  Piezoelectricity is generally defined as a linear interaction between the mechanical and electrical state in crystals that have no centers of symmetry. A distinction is drawn between the reciprocal piezoelectric effect (an external electric field leads to mechanical strain proportional to the field which deforms the piezocrystal) and the direct piezoelectric effect (a mechanical deformation of the piezoelectric body causes a proportional change to the electric polarization). It is the latter of these which is relevant for pressure measurement.

For the electrically free state of a piezocrystal (experimentally, we reach this state in the simplest manner by means of short-circuited acceptance electrodes), the direct piezoelectric effect can be described phenomenologically with (3.1).

$$D_i = d_{i\mu} \cdots T_\mu \tag{3.1}$$

D_i *($i = 1$ to 3)* vector of electric flow density

$d_{i\mu}$ tensor of the piezoelectric coefficient from (3.2)

T_μ *($\mu = 1$ to 6)* tensor of mechanical stress (with T_1 to T_3 for the normal stresses σ_x, σ_y, σ_z and T_4 to T_6 for the shear stresses τ_{yz}, τ_{xz} and τ_{xy})

$$d_{i\mu} = \begin{pmatrix} d_{11} & d_{12} & d_{13} & d_{14} & d_{15} & d_{16} \\ d_{21} & d_{22} & d_{23} & d_{24} & d_{25} & d_{26} \\ d_{31} & d_{32} & d_{33} & d_{34} & d_{35} & d_{36} \end{pmatrix} \tag{3.2}$$

Every single piezoelectric coefficient $d_{i\mu}$ determines the relation of a certain stress tensor coordinate T_μ to a certain vector coordinate of the electric flow density. Finally, the size of the charge Q of the surface of the crystal element covered by the electrodes is calculated with (3.3):

$$Q = A \cdots D_i \cdots n_i \tag{3.3}$$

A surface area

n_i ($i = 1$ to 3) components of the normal vector of the surface

Depending on the direction of piezoelectric polarization with respect to the direction of the introduced strain, we differentiate between several types of piezoelectric effect. Among these, especially the transversal and longitudinal effects are of importance for use in pressure sensors.

3.1.2.1 Longitudinal Effect

In this case, the measuring elements are usually designed as discs and the charge is delivered on the force application surfaces themselves (Fig. 3.4, left). If the crystallographic x-axis and the direction of force application are identical, the delivered charge can be calculated by (3.4) assuming a single-axis stress state.

$$Q = A \cdots d_{11} \cdots \sigma_x = A \cdots d_{11} \cdots \frac{F}{A} = d_{11} \cdots F \tag{3.4}$$

The charge delivered in the longitudinal effect is therefore not dependent on the geometry of the measuring element but only on the force F applied. In order to increase the size of the charge (sensitivity), several discs can be connected in series and electrically parallel in a force-locking manner (Fig. 3.5). One advantage of this design is the compact and resistant measuring element. Moreover, the charge is delivered directly to the compressed surfaces, practically eliminating the possibility of contact errors during charge dissipation.

3.1.2.2 Transversal Effect

The measuring elements are designed as rods and the charge is delivered perpendicular to the force application surfaces (Fig. 3.4, right). Assuming a single-axis stress state, the size of the charge can be determined with (3.5) in the case of a pure transversal cross-section (the length l of the rod has the exact direction of the crystallographic y-axis, the width w the direction of the z-axis and the charge is delivered to the surfaces normal to the x-axis).

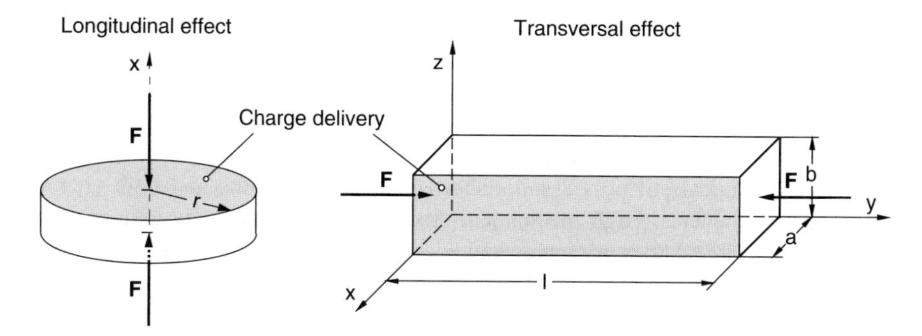

Fig. 3.4 Measuring elements for the longitudinal and transversal effect

Fig. 3.5 Increasing the size of the charge in the longitudinal effect

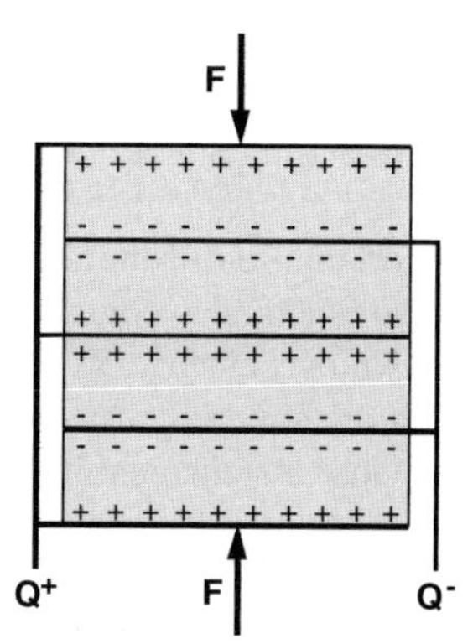

$$Q = A \cdots d_{12} \cdots \sigma_y = l \cdots b \cdots d_{12} \cdots \frac{F}{a \cdot b} = d_{12} \cdots F \cdots \frac{l}{a} \tag{3.5}$$

Thus, besides the electric properties of the piezomaterial used, the charge yield is determined above all by the slenderness of the measuring element l/a. A favorable corner ratio l/a can therefore lead to larger polarization charges in the case of the transversal effect, yet the mechanical strength of the piezomaterial used places a practical limit on the configuration of the measurement.

Piezomaterials

Materials for piezoelectric measuring elements in pressure sensors are expected to have the following properties above all (Wimmer and Glaser 2002; Wimmer 2000):

- High temperature resistance and metrological properties that are independent of temperature

- High piezoelectric sensitivity
  Piezoelectric sensitivity is determined by the piezoelectric coefficient $d_{i\mu}$, the matrix of which is contingent on the crystal symmetry. The coefficients relevant to the production of the measurement signal should be as large as possible
- High mechanical stiffness and strength
  High mechanical stiffness assures short measurement paths and high characteristic frequencies. High mechanical strength is an essential prerequisite for measuring large forces, pressures and accelerations as well as for the resistance of the sensor against mechanical impact
- High electric insulation resistance
  High electric insulation resistance make is possible to perform quasistatic measurements with piezoelectric sensors (electric drift)
- Linear relation between the measurement and the charge delivered
- Long-term stability of the important material properties and a low level of dependence of these properties on external influences
  Reaching a level of piezoelectric sensitivity that is independent of temperature and mechanical stress is a very difficult task. Besides the choice of material, the orientation of the piezoelectric elements with reference to the crystallographic axes can also contribute to the solution of this problem
- No pyroelectric effect
- Low material costs
- Good mechanical machinability

3.1.2.3 Quartz (SiO_2)

The conventional piezoelectric material for pressure sensors is quartz (SiO_2), see Fig. 3.6. Quartz appears in several crystalline structures made up of silicon/oxygen tetrahedra. For piezoelectric use, one uses the low-temperature lattice appearing below 573°C, designated as α-quartz (low quartz). Increasing temperature over 573°C leads to a phase change into β-quartz (high quartz).

Due to unavoidable inclusions and impurities in naturally occurring quartz, cultivated quartzes are used exclusively (hydrothermal synthesis) to obtain a consistently high level of quality.

Temperature has a large effect on the piezoelectric properties of quartz. Figure 3.8 shows the piezoelectric constant d_{11} as a function of temperature. Above about 300°C, d_{11} begins to drop significantly and finally disappears altogether at the conversion temperature of 573°C. Moreover, the load limit reduced with increasing temperature due to twinning. In the case of α-quartz, high stresses can lead to "Dauphiné twinning" (secondary twinning), which exhibit partially altered piezoelectric coefficient signs and thus lead to a reduction in sensitivity. While the formation of these twins at room temperature only initiates at stresses of about 5×10^8 to 9×10^8 Pa, twinning begins at lower stresses when the temperature is increased, and shortly below the conversion temperature of 573°C one can observe twinning even when the quartz is unstressed. It has been determined that

Fig. 3.6 A quartz crystal

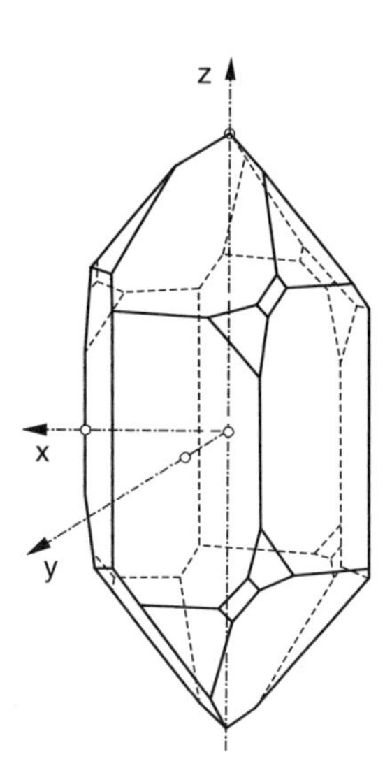

twins that have developed under strain can vanish again completely when the strain is removed. If the mechanical strain is of a longer duration however, stable twins can also develop that result in a permanent reduction of the piezoelectric sensitivity.

Common measuring elements made of quartz can therefore only be used up to temperatures of about 200–250°C, which necessitates the use of a corresponding measuring element cooling system for internal combustion engine applications, where temperatures of more than 400°C at the measurement points are completely likely.

3.1.2.4 Gallium Orthophosphate ($GaPO_4$)

The engineered piezomaterial gallium orthophosphate ($GaPO_4$) was specially developed for high-temperature applications (Krempl et al. 1997). It is characterized by high piezoelectric sensitivity that is largely independent of temperature.

The crystal structure of gallium orthophosphate can be derived from quartz by replacing silicon alternately with gallium and phosphorus (Fig. 3.7). α-gallium orthophosphate is stable up to 933°C and converts above that temperature to the high cristobalite type.

To improve temperature properties, it is possible in the case of the transversal effect (as opposed to the longitudinal effect) to configure the crystal section such that the effective piezoelectric coefficient remains relatively independent of temperature within a certain temperature range. In this way, measuring elements can be manufactured that can be employed up to a temperature of about 350°C. Such crystal sections generally also have a smaller propensity to form twins.

The following attributes of gallium orthophosphate are especially notable:

- Temperature-resistant up to over 900°C
- Sensitivity about twice as large as that of quartz, which remains unchanged until much higher than 500°C (Fig. 3.8)

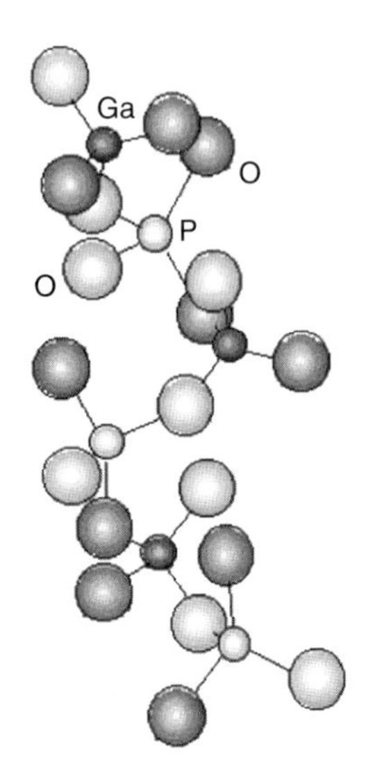

Fig. 3.7 Crystal structure of gallium orthophosphate

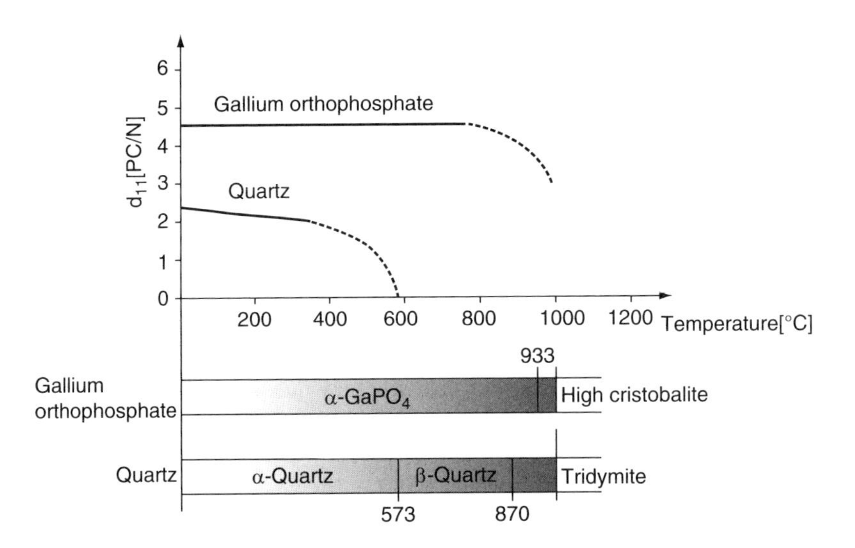

Fig. 3.8 Temperature dependence of the piezoelectric constant d_{11} for quartz and gallium orthophosphate

- High electric ionization resistance up to high temperatures
- Stable with respect to strain-induced twinning
- No pyroelectric effect

Due to its excellent temperature properties and high sensitivity, gallium orthophosphate is especially suitable for the construction of uncooled miniature pressure sensors.

3.1.2.5 Other Piezomaterials

There are several other piezomaterials besides quartz and gallium orthophosphate, but because of various disadvantages (especially due to the temperature-dependence of certain properties) they are of limited use in engine pressure measurement. The most important of these additional materials include:

- Tourmaline
- Langasite
- Lithium niobate ($LiNbO_3$) and lithium tantalate ($LiTaO_3$)
- Piezoceramics (barium titanate etc.)

Structure of Piezoelectric Pressure Sensors

The excellent properties of certain piezoelectric materials such as quartz and gallium orthophosphate are a very good basis for making precise pressure measurements in the combustion chamber of an internal combustion engine. But only an optimal combination of piezomaterials with an expertly designed and developed housing makes a precise pressure sensor possible.

Due to variety of demands involved in the use of pressure sensors in the combustion chamber (installation space, temperature stress, mechanical deformation of the installation location, accelerations etc.), different designs have been devised. But the basic structure of the sensor is always similar and will be shown using the example of the water-cooled quartz pressure sensor for the longitudinal effect (Fig. 3.9). The pressure to be measured p acts on the piezoelectric measuring element via a flexible membrane and a short, stiff pressure plate. The pressure plate has the task of producing a state of mechanical stress in the measuring element that is as consistent as possible. The compensation disc, inserted between the measuring element and the pressure plate, balances differences in thermal expansion. The measuring element and the membrane are surrounded by a water jacket and are cooled directly during operation. In this way, the measuring element is only slightly warmer than the coolant during use in the engine (typically up to about 10–20°C hotter than the coolant).

To increase the charge, several disc-shaped quartz measuring elements are employed. The individual quartz discs are electroplated with precious metal at each end so that a parallel electrical connection is ensured – by means of connecting bridges with contact lugs one the one hand and insulation zones on the other. The pressure sensor housing is connected electrically with the plus electrode of the measuring element; it thus represents the electric ground. The negative electrode is connected with a plug connection highly insulated against the housing on the back of the sensor, by means of which the electric charge is delivered.

Because of their compact construction, modern internal combustion engines using multi-valve technology and direct injection utilize the available installation space optimally with the help of miniature pressure sensors.

Figure 3.10 shows the structure of such a miniature pressure sensor (transversal effect). Miniature pressure sensors generally have to do without water-cooling,

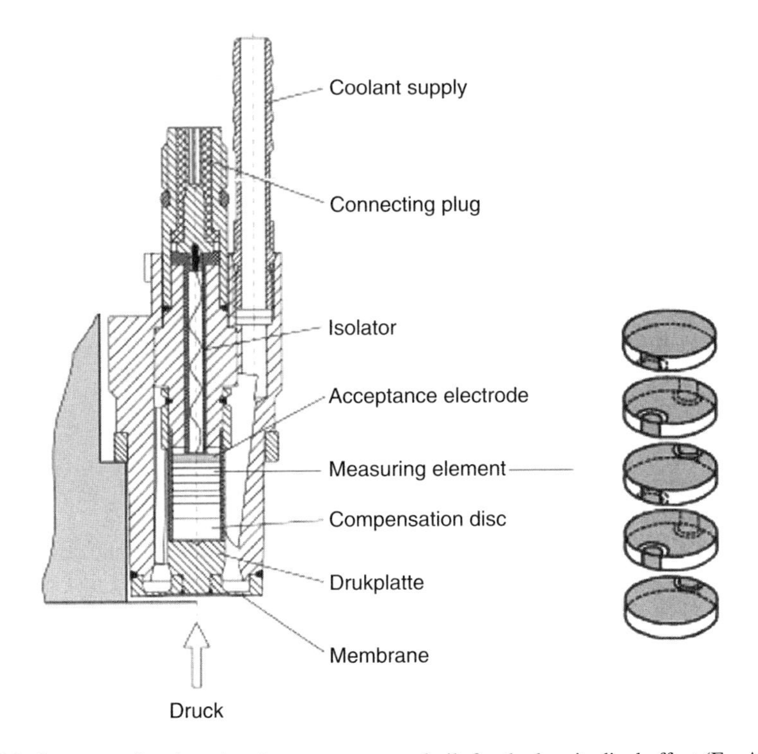

Fig. 3.9 Structure of a piezoelectric pressure sensor built for the longitudinal effect (Fa. AVL)

resulting in very high demand above all on the piezomaterials (high temperature resistance and independence from metrological properties in a broad temperature range) and on the sensor's design. Miniature pressure sensors available today have reached a quality comparable to that of highly precise water-cooled pressure sensors, making it possible to use them for precise thermodynamical analyses.

Pressure Sensor Cooling

Quartz pressure sensors are designed as a rule for operation with water-cooling. Direct water-cooling of the membrane and measuring element results in the following main advantages:

- Prevention of over-heating of the measuring element
- Reduction of the effect of temperature on sensitivity (quartz) and thermal drift
- No reduction of insulation resistance due to high temperatures
- Direct water-cooling permits a flush installation of the pressure sensor, even in the case of thermally highly stressed measuring locations

It is important that the pressure sensor is cooled constantly and without pulsation, i.e. that no vibrations are transferred to the coolant either from the coolant

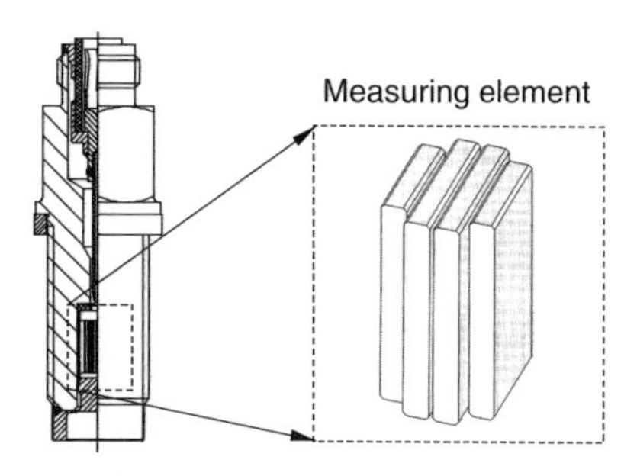

Fig. 3.10 Uncooled miniature pressure sensor using the transversal effect (AVL)

pump itself-or from external sources, as these would manifest themselves as errors in the measurement signal.

Charge amplifier and Electrical Drift

In the charge amplifier, the charge delivered by the pressure sensor is converted into a voltage signal. As shown in the circuit diagram (Fig. 3.11 left), a charge converter consists fundamentally of an amplifier V with very high internal charge amplification and a degenerative feedback capacitor C_G to integrate the charge

If a charge is delivered by the piezoelectric pressure sensor DA, there is a small increase of charge at the input of the amplifier V. This charge increase appears at the output highly amplified and negative – the output charge thus exhibits a negative polarity compared with the input charge. Feedback capacitor, which has been negatively pre-loaded in this way, accepts a corresponding charge from the input and thus keeps the charge increase at the input of the amplifier small.

At the output of the amplifier V, exactly that voltage U_A will appear that, via the feedback capacitor, subtracts the charge from the input until the remaining input voltage amplified by V results exactly in U_A.

Since the amplification factor of V is very large (up to about 100,000), the input voltage U_E is almost zero. The charge delivered by the pressure sensor is not used for charging (i.e. to increase the voltage at the input capacitances) but is drawn off by the feedback capacitor. For this reason, changes to the input capacitance – e.g. because of different cables with varying cable capacitances C_K – have practically no effect on the measurement results.

The output voltage U_A of the amplifier is directly proportional to the delivered charge Q of the sensor as well as inversely proportional to the capacitance of the feedback capacitor C_G, so arbitrarily staged measurement ranges (range setting) can be made available by adjusting the capacitance values accordingly [see relation (3.6)].

$$\mathrm{U_A} \sim -\mathrm{Q}/\mathrm{C_G} \tag{3.6}$$

If the measurement variable is constant, i.e. at constant pressure on the sensor, a constant voltage would be expected at the output of the charge amplifier.

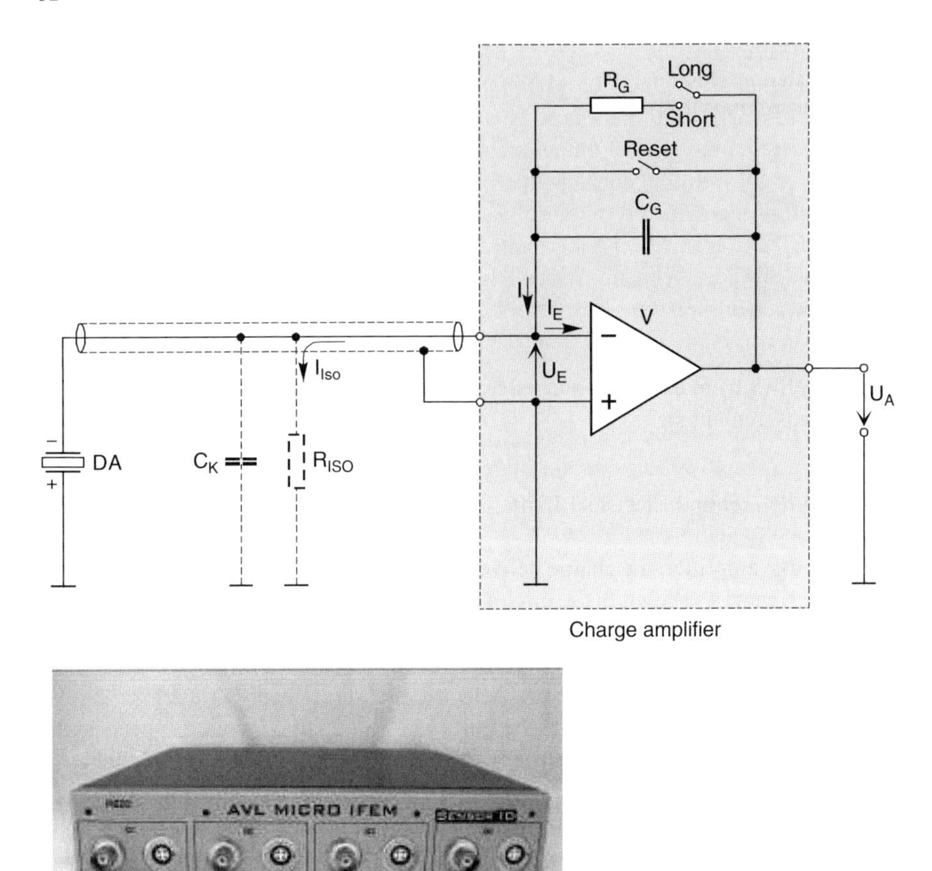

Fig. 3.11 Circuit diagram and design of a charge amplifier (AVL)

However, the functional principle of the piezoelectric measurement chain results in output signal drift. Since this drift is of electrical origin, we refer to is as electrical drift. The causes of electrical drift are essentially the charge flow arising according to the polarity of the input voltage and the discharge of the feedback capacitor.

The most crucial measure against electrical drift is the highly insulating design of the amplifier input, the pressure sensor and the measurement cables including the plug connections (insulation values 10^{13} Ω). Moreover, by means of parallel connection of an additional feedback resistance R_G, the drift can be limited to a certain value and a migration into saturation prevented. In this operation mode, measurements can be made even in the case of poorer insulation, but it is suitable only for, basic monitoring and not thermodynamic analysis since the pressure signal is phase-shifted and diminished in amplitude as a function of speed and the range setting. Finally, drift can also be compensated by supplying a compensation current that is equally large as the sum of the currents discharging over the insulation resistances that are realized in the input of the operation amplifier.

Modern charge amplifiers now have the capacity to recognize sensors and to supply the required data for automatic parameterization of the measuring channel. Because of the sensors must in this case be utilized in a hot environment, we use either semiconductor memory connected with the sensor via a measuring cable (TEDS – Transducer Electronic Data Sheet) or temperature-resistant SAW elements (Surface Acoustic Waves) built into the sensor and equipped with a hard-coded identification. The latter have the advantage that the same electrical connection can be used as the cable required for the transmission of the measurement signal. In addition to the increase in measurement accuracy due to the clear recognition of the attached sensor and the resultant possibility of an automatic parameterization of the measuring channel, the error rate is also reduced. In the meantime, specially equipped amplifiers can directly calculate the evaluation variables on the basis of the measurement values (e.g. cylinder peak pressure and location, noise characteristic values etc.).

Further Influences and Important Parameters

Besides the basic parameters of piezoelectric pressure sensors such as measurement range, sensitivity, linearity deviation etc., especially those changes to the metrological properties are significant that can lead to a significant influence of the measurement result because of external effects active during operation. Figure 3.12 shows the most important of these influences.

Besides pressure itself, these include above all:

- *Temperature and heat flow*
  Due to the very high working gas temperatures during combustion (up to >2,700 K), pressure sensors are exposed to very high heat flow stresses during operation. In extreme cases, cyclical heat fluxes arise with an amplitude of more than 1,000 W/cm^2 (e.g. in the case of knocking combustion) and average heat flows up to 50 W/cm^2. The resultantly high temperatures can reach 500°C at the diaphragm area, up to 400°C at the measuring element and up to 200°C at the plug connection with uncooled transducers. In the case of cooled sensors, the temperature level is naturally much lower. With intensive cooling, the temperature at the measuring element is about 40°C and about 20°C at the plug connection.
- *Accelerations*
  Not only structure-borne sound but also general vibrations of the engine cause accelerations at the pressure sensor. While the acceleration values caused by engine vibration is limited to about 200 g, impact stresses can cause up to 1,000 g to arise at the pressure sensor. Especially in the case of high-speed racing engines and when determining pressure profiles in the inlet and exhaust system, the effects of acceleration can have a considerable influence on the quality of the measurement.

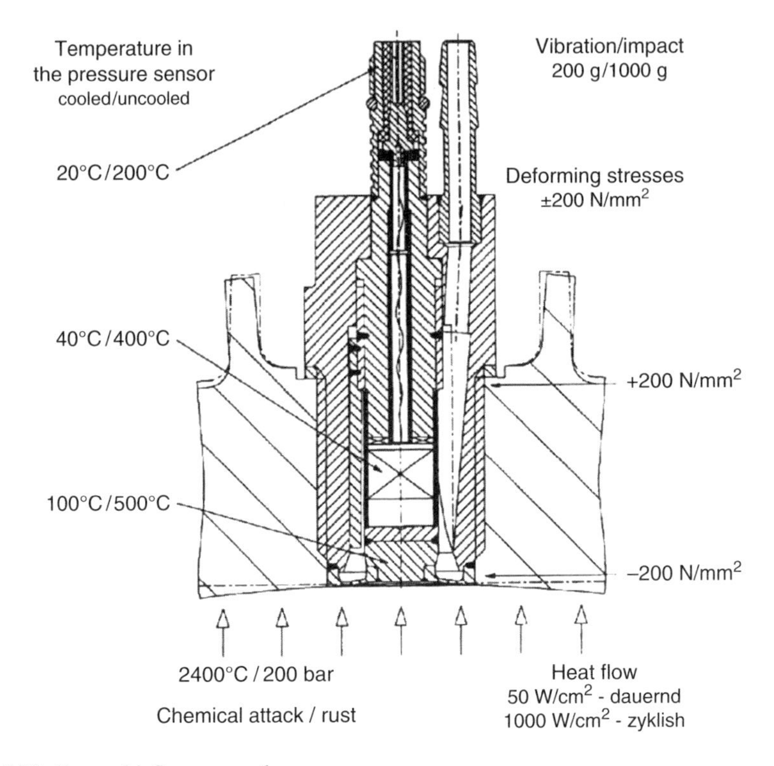

Fig. 3.12 External influences on the pressure sensor

- *Deforming stresses*
  Gas and mass forces as well as thermal stress lead to deforming stresses of up to 200 N/mm^2 at the installation location of the pressure sensor. The resulting deformations in the pressure sensor housing can lead to errors in the pressure signal. Not only the design, but especially the installation type is of decisive importance for the deformation sensitivity of pressure sensors. Plug-in pressure sensors are generally less susceptible to deformation than screw-types, but they react more strongly to temperature influences. Newer screw types use double shell construction to mitigate mounting bore deformation.
- *Chemical effects and precipitations*
  Combustion products generated in the engine can lead to corrosion damage to the pressure sensor. Moreover both spark ignition and diesel engines are subject to combustion deposits on the surface of the combustion chamber and thus on the pressure sensor as well. Corrosion damage and deposits can affect the measurement result by changing the diaphragm stiffness.

In order to describe the reaction to these individual influences and to judge their effects on piezoelectric pressure measurements in combustion engines, a series of additional parameters are indicated in addition to the basic specifications. Special test methods have been developed for many of these parameters.

Temperature and heat flow stress has an especially important effect on the measurement fidelity of pressure sensors. Not only temperature but also

- The change of sensitivity with temperature
- Thermal drift (cyclical temperature drift and load change drift) are of particular importance

Temperature-Related Sensitivity Change

The sensitivity of every piezoelectric pressure sensor changes to some extent with its average temperature. Depending on the piezomaterial used, influences of varying size exist:

- Quartz with a cut normal to the axis (x-cut)
  Quartz with a cut that runs normal to the axis (x-cut) exhibits a relatively strong dependence of sensitivity on temperature, which is why it is only used in water-cooled pressure sensors. By means of suitable cooling, the measurement element is constantly held within a temperature range of 10–20°C above the temperature of the coolant, even under extreme conditions.
- Quartz cut in a temperature-stable manner
  Due to a special load angle (appropriate cut selection with reference to the crystallographic axes), the dependence of sensitivity on temperature is much smaller than in the case of quartz with a cut normal to the axis. Temperature-stable quartz is thus suitable for use in uncooled sensors as well. The working temperature of these sensors is limited however to about 350°C.
- Alternative piezomaterials
  For example, the sensitivity of gallium orthophosphate has a small dependence on temperature and is suitable for high-temperature application of up to well over 500°C (Fig. 3.8).

Changes in the sensitivity of piezoelectric pressure sensors are described as a rule with the temperature coefficients of sensitivity, which indicates the arising sensitivity change as a percentage of the nominal sensitivity per°C within a certain temperature range.

In the case of small temperature changes – e.g. when water-cooler pressure sensors are utilized – the sensitivity change can be neglected or taken into consideration by means of the average temperature coefficient. Otherwise, it is possible to consider the effect of temperature if the pressure sensor is calibrated at working temperature, i.e. at the average measurement element temperature arising during operation, or if corresponding manufacturer information is utilized for correction.

Thermal Drift

With reference to the use of pressure sensors for measurements of internal combustion engines, thermal drift is understood as the "pressure change" that is caused by temperature changes at the pressure sensor and installation location alone. Thermal drifts represent the decisive measurement error in many measurements.

Besides the design of the pressure sensor, thermal drift is determined by the range and rate of the pressure sensor temperature change, which is highly contingent on the installation details of the pressure sensor mounting.

Two thermal drift phenomena are observable during pressure indications of internal combustion engines that must be clearly differentiated:

- Cyclical temperature drift
- Load change drift

Cyclical Temperature Drift

This concerns the faulty pressure indication within one cycle arising as a result of the cyclical heating of the pressure sensor. Cyclical temperature drift is also called short-term temperature drift or thermoshock.

Since cyclical temperature drift is active in many cases across a broad range of crank angles, its influence is considerable in the case of parameters that are integrated during one working cycle (e.g. the indicated mean effective pressure p_{mi}). Figure 3.13 shows the influence of a characteristic profile of the cyclical temperature drift on the determination of the indicated mean effective pressure and the energy balance (comparison of the fuel energy introduced per cylinder and working cycle with the energy converted in the combustion chamber calculated by the engine process computation under consideration of conversion losses). In comparison, a pressure-proportional error of the same scale – such as caused by a change in sensitivity for example – has a much smaller effect.

The characteristic values for describing cyclical temperature drift are determined on the one hand by means of a zero-pressure, cyclical heating of the sensor in a special test apparatus or on the other by means of comparative measurements in actual engine operation.

- Determination by zero-pressure, cyclical heating
  To this end, the test apparatus shown in Fig. 3.14 above (Glaser 1983) is employed, which makes it possible to subject the unstressed pressure sensor to a cyclical heating of similar size and frequency as arise in thc engine. Opposite the sensor to be tested is a radiant heat surface which is covered and again uncovered by a rotating diaphragm wheel so that the heat flow profile also shown in Fig. 3.14 is generated. The radiant heat surface is heated electrically to temperatures of over 2,000°C, with which heat flows of more than 100 W/cm^2 can be produced. The pressure sensor is installed flushly in a solid cooled steel plate.
  The characteristic value used for the cyclical temperature drift is the maximum pressure deviation arising within one cycle at a certain frequency of the diaphragm wheel and at a certain radiant heating.

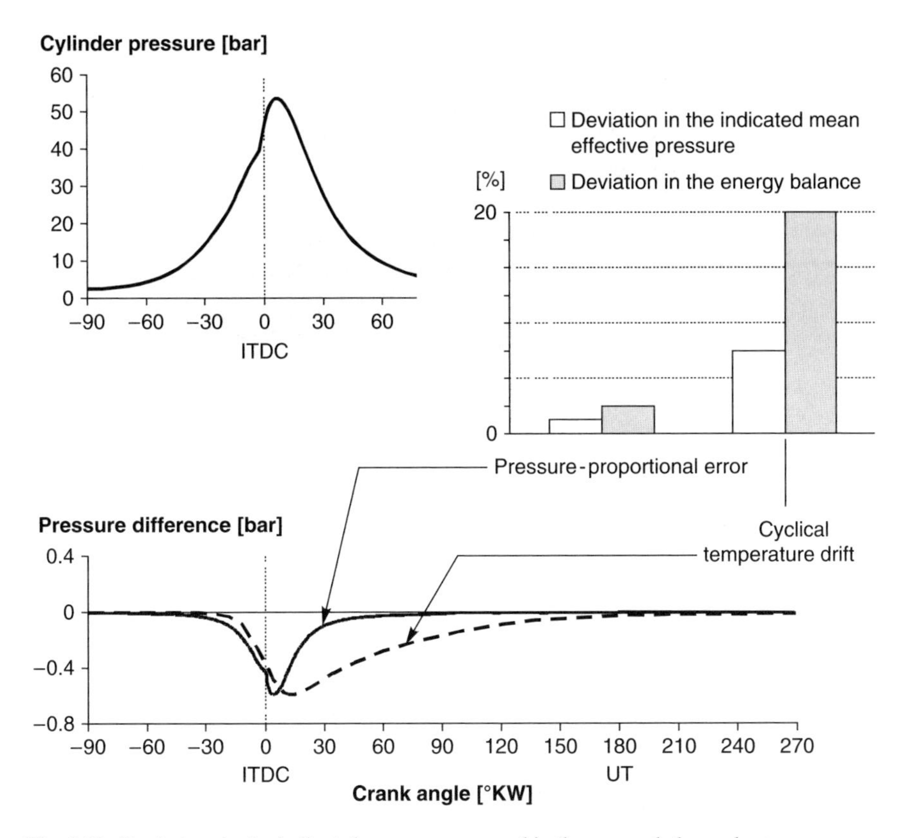

Fig. 3.13 Deviations in the indicated mean pressure and in the energy balance due to a pressure-proportionate error and cyclical temperature drift

This method is suitable for a basic evaluation of sensor attributes and for the preselection of sensors with especially low cyclical temperature drift. The results thereby obtained are not directly applicable to engine operation (i.e. radiant vs. convective heat flux), which is why the assessment of sensors should also be carried out during actual engine operation.

- Determination during engine operation with a reference pressure sensor
  In this case, a characteristic value for cyclical temperature drift is determined from the difference between the pressure profile of the sensor under investigation and that of a reference sensor. The reference sensor must satisfy very high demands for accuracy. As a rule, water-cooled sensors with special treatments to produce very low thermoshock-sensitivity are utilized. Also, the membranes of the reference sensor can be coated with silicon rubber, which significantly reduces the heat flow into the membranes caused by combustion. The maximum pressure deviation within a working cycle in a certain load state of the engine is given as the characteristic value for cyclical drift. Wimmer and Glaser (1996) provide an example of the use of this assessment method on uncooled miniature pressure sensors.

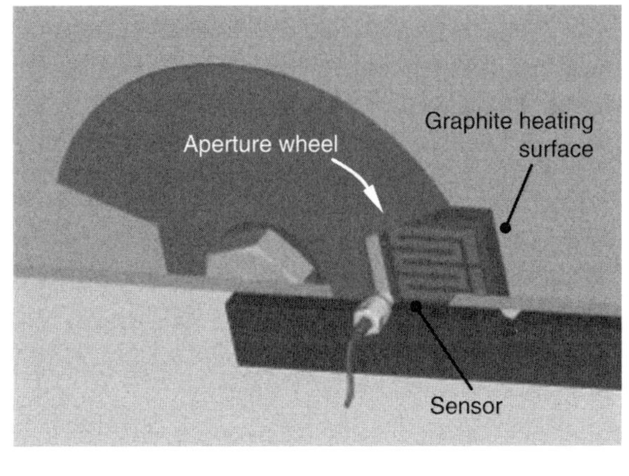

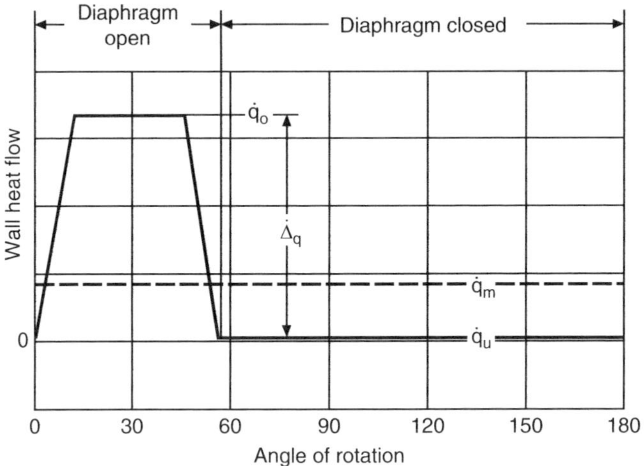

Fig. 3.14 Dynamic test apparatus to determine cyclical temperature drift

3.1.2.6 Measures to Reduce Cyclical Drift

Although the effect of cyclical temperature drift has been reduced considerably in the case of modern pressure sensors by means of special membrane and housing designs, it is still one of the most common sensor-related sources of error.

Manufacturers are aiming to optimize sensor design, in particular that of the membranes, such that the cyclical temperature drift is as low as possible. Presently, the use of the finite element method is particularly helpful in sensor design (Karst 2000). The prerequisite for this is, besides a realistic modeling, the exact knowledge of the marginal conditions – especially the knowledge of the heat flows arising on the side of the combustion chamber.

Besides selecting pressure sensors with as little cyclical temperature drift as possible, there are also some possibilities on the side of the operator. Since the cyclical temperature drift is caused by the heat transfer to the sensor, it can be reduced considerably by

- Choosing a favorable installation position
- Choosing a set-back installation
- Using thermal protection

A detailed description of these measures can be found in Wimmer and Glaser (2002) and Wimmer (2000).

Load Change Drift

Load change drift is the slow drifting of the pressure signal after a load change, i.e. a change in temperature over a series of cycles. Load pressure drift is also referred to as long-term drift or long and medium-term drift.

Load change drift manifests itself-as a relatively slow shift in pressure level caused by a temperature change of the entire pressure sensor. Only when the pressure sensor does not experience any more temperature change on average does this change in level cease. The characteristic values used to describe load change drift are determined during actual engine operation; after operation, at a certain load point, one shifts to drag operation by switching off the fuel supply, thereby causing a rapid change of the heat flux to the pressure sensor (load jump, see Fig. 3.15). As a rule, the maximum zero-point gradient and the remaining pressure level change after a certain time (20 s) are taken from the captured measurement signal as parameters for specifying the load change drift. This deviation can generally be corrected effectively in contemporary charge amplifiers with cyclic drift compensation.

3.1.3 Assembly Variants

Modern combustion chamber indication is carried out – if possible – by means of mounting ports in the head. In some cases, adapters are used that contain uncooled miniature pressure sensors or probes and are exchanged with original components such as glow or spark plugs. In this way, it is possible to minimize the cost of adaptation considerably.

3.1.3.1 Spark Plug Adaptation

In the case of spark ignition engines, we can make indications without making contact with the cylinder head by using a measuring spark plug. This measuring spark plug has a dual function: it ignites the mixture found in the cylinder like a ocommon spark plug on the one hand and houses the pressure sensor on the other.

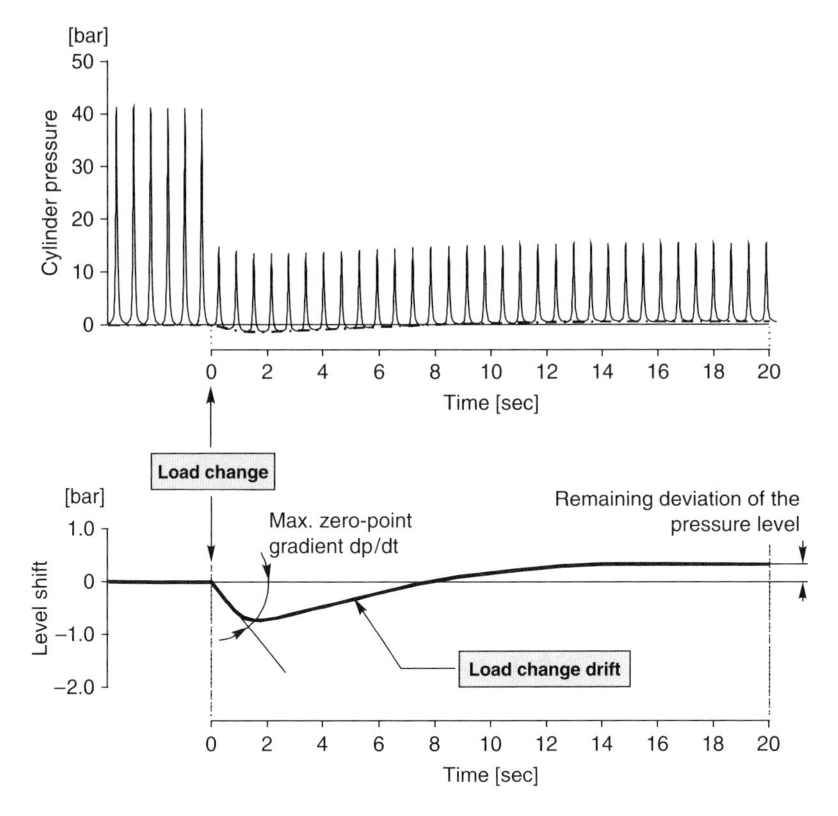

Fig. 3.15 Load change drift

Due to the cramped space available, the assembly of the sensor is usually only possible with an eccentrically arranged centre electrode.

Since measuring spark plugs are primarily used as substitutes for standard spark plugs, one must take care that one selects characteristic values such as heat rating, spark position, electrode shape and electrode distance in accordance with the engine.

As seen in Fig. 3.16, there are basically three design variants of measuring spark plugs:

- Measuring spark plugs with miniature pressure sensors
  The set-back assembly of an uncooled miniature pressure sensor in the plug head at the level of the hexagonal is almost obsolete today (Fig. 3.16 left). Here, the connection between the combustion chamber and the sensor is a relatively long channel. The advantage of this variant, stemming from the lower cyclical temperature drift due to the lower heat flow stress, cannot nearly compensate for the disadvantage of the occurrence of whistling vibrations.
- Measuring spark plugs with pressure measuring probes
  Installation of a pressure measurement probe that is flush with the combustion chamber has only been made possible by the use of new piezomaterials

(e.g. $GaPO_4$) (Fig. 3.16 center). Due to the high temperature stress with the narrowest available space, piezoelectric quartz pressure sensors were not suitable for this adaptation. Even at high speeds, only minimal pipe oscillations occur, while the cyclical temperature drift is minimized by means of a good heat transition between the pressure measuring probe, adapter and cylinder head.

- Direct measuring spark plugs
  Smaller spark plug bores can only be accommodated if the adapter, the insulator and also the pressure sensor are further miniaturized. The pressure sensor's attributes typically suffer from miniaturization. For this reason, the best compromise is to integrate the pressure sensor directly into the body of the spark plug (Fig. 3.16 right). Inclination the membranes leads to a measurement location flush with the combustion chamber that is not influenced by passage acoustic oscillations. At the same time, eccentricity of the ignition location is largely avoided in comparison to the adaptation with the pressure measuring probe described above.

As the two designs shown in the middle and right of Fig. 3.16 show, the external connection sizes of measuring spark plugs today are ideally identical to those of original spark plugs as well.

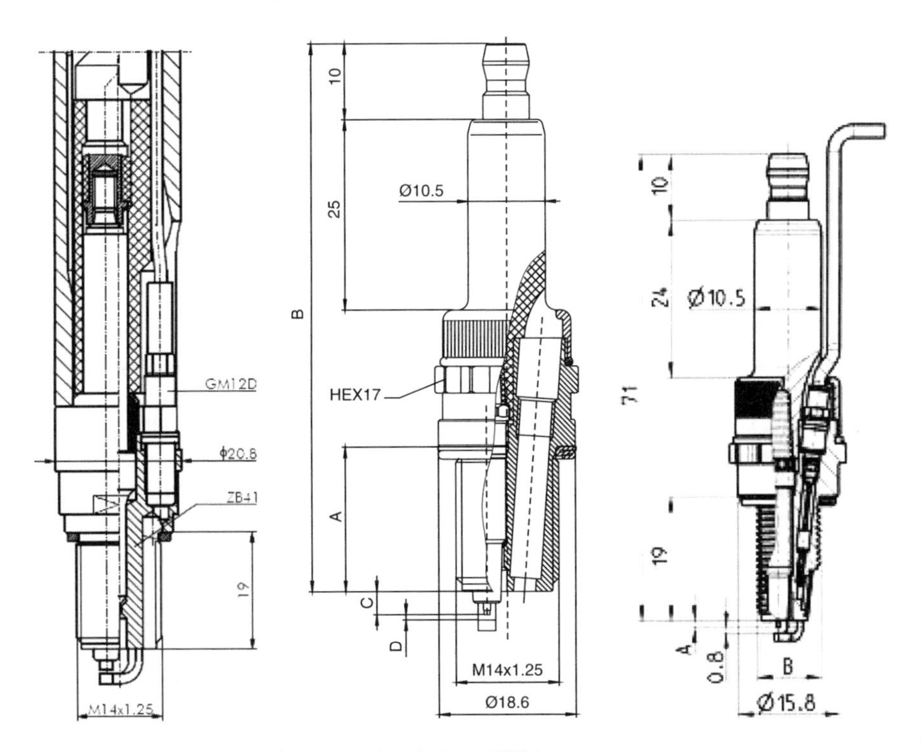

Fig. 3.16 Different measuring spark plug designs (AVL)

3.1.3.2 Glow Plug Adaptation

The glow plug bore is the most convenient measurement location in the case of diesel engines with glow plugs. This adaptation variant is determined above all by the geometrical dimensions of the bore. If possible, the pressure sensor membrane should be close to the combustion chamber in order to minimize passage. In the case of long, slender glow plugs, it is sensible to use specially developed pressure measuring probes in order to obtain a position close to the combustion chamber and thus to gain an unadulterated measurement signal (Fig. 3.17 left).

This requirement can currently be met up to the smallest common diameters, so that the pressure measuring probe is also used without an adapter and replicates the exact shape of the original glow plug by means of a special formation of the point (Fig. 3.17 right).

If the pressure measuring probe must be set back due to an excessively small bore diameter or in combination with a glow function behind the sealing face, specially shaped damping volumes in the adapter see to it that potential signal adulterations due to whistling vibrations in the gas channel are significantly reduced.

The highest measurement quality can only be attained with optimally adjusted adapters. In this context, the gap between the glow plug bore and the adapter has an especially decisive role with respect to the thermal load on the sensor. The larger the gap is, the stronger the temperature increase in the area of the sensor and the stronger the effect on the measurement signal. For this reason, glow plug adapters are manufactured according to the exact dimensions of the glow plug bore and the specific application.

3.1.3.3 Pressure Indication Direct Mount

In this type of adaptation, there are a number of potential ways to combine the sensor and the installation variant at different measurement locations. Due to mutual dependences, the correct choice of sensor and installation variant as well as of the measurement location is however a very complex one. In order to obtain useful measurement results, many potential sources of error must be taken into consideration. Suggestions concerning the selection of a suitable measurement location can be found in the subsequent chapter.

Figure 3.18 shows the assembly of two uncooled pressure sensors in a cylinder head by means of adapter sleeves, whereby in one case only the coolant jacket and in the other both the coolant and oil chambers are crossed. In such a case, one must also consider that coolant circulation is upset by the installation of sensors, potentially leading to negative local effects on cooling efficiency because of the altered flow conditions, particularly at operation close to rated power engine.

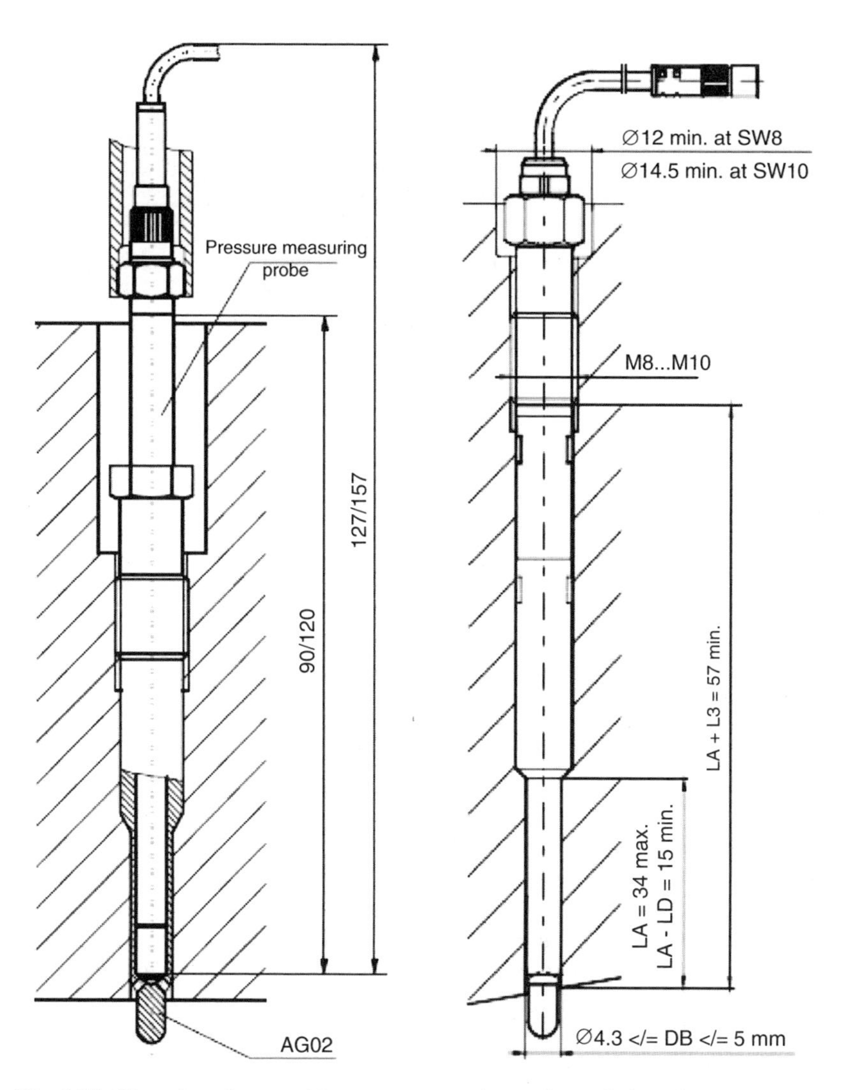

Fig. 3.17 Glow plug adapters with pressure measuring probes and glow plug sensors (AVL)

3.1.4 Selection of the Measurement Location

When choosing a measurement location, it is crucial to take heed that the pressure in the combustion chamber is not the same throughout (subdivided combustion chambers, quench areas, etc.). For the sake of a precise pressure measurement in thermodynamic investigations therefore, the location in which the pressure sensor is

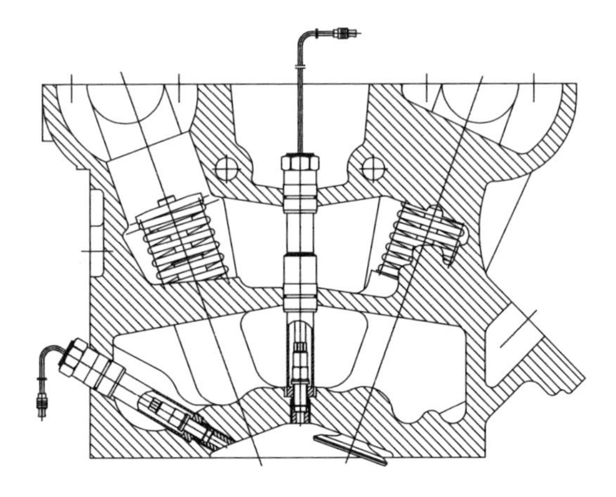

Fig. 3.18 Cylinder head with two integrated pressure sensors

installed in the cylinder head is important. The choice of an assembly location and the design of the installation site must guarantee that

- The permissible operating temperature of the pressure sensor is not exceeded,
- Heat flow stress does not become to high,
- Temperature deviations in the pressure sensor are as small as possible,
- Errors caused by the installation are kept to a minimum (passage oscillations etc.) and
- The effects of fouling the electrical connection between the sensor and the cabel (oil, water etc.) are avoided.

To make matters worse, thermally favorable installation sites are often discarded due to lacking space (cooling jacket, 4-valve technology, dual ignition, direct injectors, etc.).

Engine manufacturers rarely take the installation situation of pressure sensors into account when designing cylinder heads, since such sensors are mostly used only during the development phase and should not affect the optimal design of the series product. For this reason, the compact design of modern engines often makes it necessary to make compromises with reference to the installation location. It is thus not always possible to eliminate all potential disruptive factors.

The idcal measurement site is located at a position, at which the local pressure is representative for the respective measurement task. Arrangements with a (long) indicator channel, over a squeeze gap, over the point of impact of the injection spray and in thermally highly stressed areas should be avoided if possible. Cool installation areas (e.g. near the inlet valve) are preferable.

3.1.4.1 Indicator Channel

The addition of an indicator passage invariably causes an increase of the combustion chamber volume. This leads to a reduction in the compression ratio, which can affect engine behavior. Moreover, the indicator channel represents an acoustic resonator, which is excited by pressure changes to produce "passage oscillations". To clarify this effect, Fig. 3.19 shows the pressure profiles measured in a spark ignition engine based on a spark plug adaptation at different induction channel lengths (2.7, 25 and 37 mm). Five individual cycles are shown for each, whereby the level of the individual pressure curves was fixed such that a more clearly arranged depiction was obtained. As shown, significant whistling vibrations appear even at an indicator channel length of <4 mm is a good design target.

This interference frequency depends on the state of the gas and can thus not be exactly determined in advance. For this reason, even the subsequent use of frequency filters to eliminate whistling vibrations is discarded because of the considerable calculative and experimental cost, and distortion of the pressure signal (phase shift and amplitude attenuation).

3.1.4.2 Arrangement over the Squeeze Gap

The gasses in the squeeze gap in the squish area at top dead center (TDC) are strongly accelerated by the piston motion. These gas flows are accompanied by considerable pressure differences in the squeeze gap, which can lead to correspondingly erroneous data when assessing a pressure measurement.

The squeeze gap flow is varies in time and position, particularly in the case of an eccentric position of the piston bowl and when the cross-section is increased near the valve recess. The gas in the combustion chamber bowl and that in the squeeze gap can execute a linked gas vibration when excited (ignition). Depending on the position of the measuring site therefore, the pressure sensor can be exposed to gas

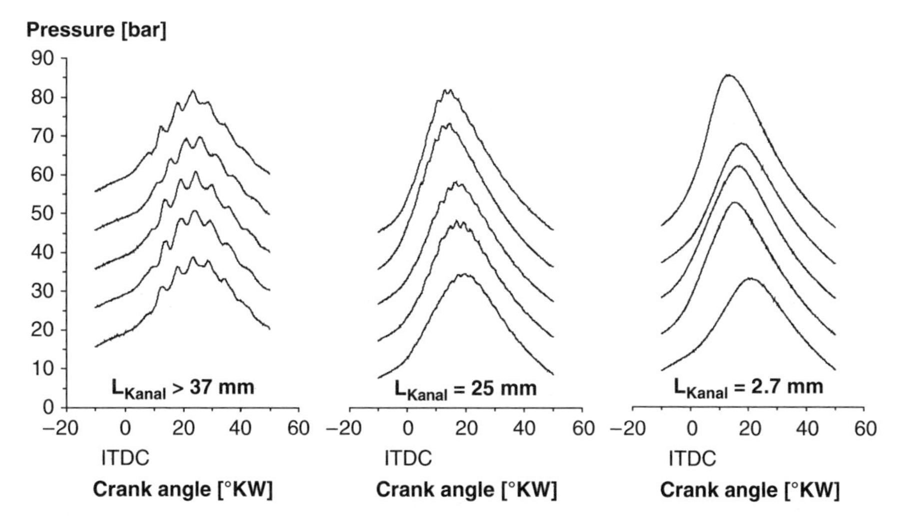

Fig. 3.19 Effect of the length of the indicator channel on the measured pressure profile

vibrations of varying frequency and amplitude, which can develop significantly if the combustion has a high rate of pressure increase. In the case of eccentric bowls, it is always advantageous to arrange the pressure sensor on the side with a short squeeze gap length so as to prevent superimposed gas vibration in the measurement signal.

3.1.4.3 Non-perpendicular Installation

In general, the pressure sensor membrane should be as closely adjusted to the contour of the cylinder head wall as possible, i.e. the longitudinal axis of the pressure sensor should if possible be perpendicular to the cylinder head wall. However, usually a slightly inclined assembly of the pressure sensor is achieved because of cylinder head design constraints. Because of the non-perpendicular diaphragm a highly inclined installation could be disadvantageous due to the minor enlargement of the combustion chamber volume and disruption of the flow conditions.

3.1.4.4 Arrangement Near the Valve

An arrangement of the measuring location near the valve results in a small local pressure change during gas exchange due to the Bernoulli effect, which above all can cause some distortion in the low-pressure loop. High flow velocities result on the surface of the pressure sensor membrane as well as an associated increased heat flow stress (cyclical temperature drift), especially in the immediate vicinity of the outlet valve during blowdown. Near the outlet valve, the wall of the cylinder head is generally the hottest. The pressure sensor thus also reaches a high operating temperature, which can result in a change in sensitivity as well as a reduction of the service life of the pressure sensor.

3.1.4.5 Arrangement over the Point of Impact of the Injection Spray: DI Engines

If the measurement location is arranged directly over the point of impact of the injection spray, errors associated with cyclical temperature drift can occur due to the impact of the cold fuel (alteration of the heat flow stress). For this reason, such measurement site arrangements should be avoided as much as possible.

3.1.5 Adjusting the Pressure Level: Zero Level Correction (Pegging)

In accordance with their functional principle, piezoelectric pressure sensors can only be used to measure the share of pressure that is changing, but not the physically correct absolute pressure p. The measured pressure profile p_{Mess} must therefore be corrected according to (3.7) by the amount Δp_n (zero-line shift).

$$p(\varphi) = p_{Mess}(\varphi) + \Delta p_n \tag{3.7}$$

Since a correct pressure profile is the basis for an accurate calculation of the combustion profile, finding the zero-line is of central importance. A positive zero-point error ($+\Delta p$) results in excessive cylinder pressure, which leads to smaller conversion rates before the TDC and to larger ones after the TDC. In the case of negative zero-point errors ($-\Delta p$), the situation is reversed. Since most of the energy conversion takes place after the TDC, most of the changes occur in this part of the combustion sequence. Figure 3.20 shows the typical effects of a zero-point error on the energy balance, on the left for a diesel engine and on the right for a spark-ignition engine (Pischinger et al. 2002).

A number of different methods can be used to determine the pressure level. One possible method is to use switch adapters, in which case a reference pressure (usually the ambient pressure) is applied for a short time to the pressure sensor. Switch adapters are rarely used for cylinder pressure measurement because they require a lot of space and can result in high pressures during the switching process, but they are very generally suitable for measuring in the inlet and outlet areas (Sect. 3.1.9).

When assigning the level of the cylinder pressure profile, we basically distinguish between reference behavior and the thermodynamic determination of the zero-line.

3.1.5.1 The Reference Method

In the case of the fixed-reference value method, the measured cylinder pressure p_{Mess} to a predetermined pressure p_{Fix} (reference point) for a certain crank angle φ_{Ref} (reference angle). In order to minimize the influence of interference

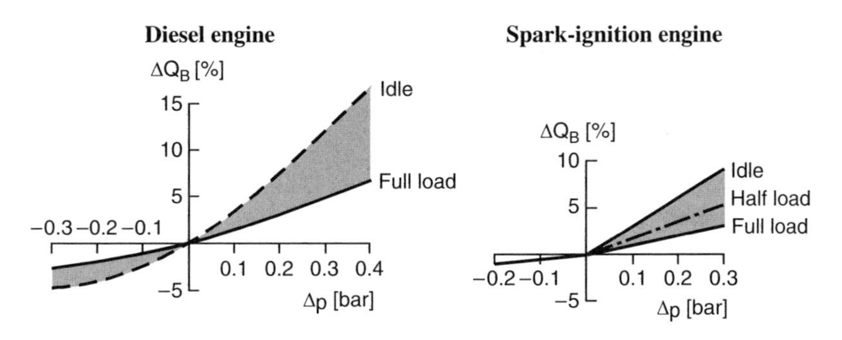

Fig. 3.20 Effect of a zero-point error on the energy balance

signals/outliers on the process when measuring the cylinder pressure profile, an averaging of the signal of the measured pressure profile p_{Mess} in the area of the reference angle φ_{Ref} is carried out [see (3.8)]. See Davis and Patterson (2006) and Patterson and Davis (2009) for best practices for zero level correction.

$$\Delta p_n = p_{Fix} - \frac{1}{N+1} \sum_{\varphi=\varphi_{\mathrm{Ref}}}^{\varphi_{\mathrm{Ref}}+N} p_{Mess}(\varphi) \tag{3.8}$$

Frequently, the ambient pressure, equated with the cylinder pressure in the charge change TDC, serves as the reference value. This assumption only leads to a sufficiently exact level assignment in the case of unthrottled, naturally aspirated engines (only at full load) with low inlet tuning.

Another possibility is based on the assumption that cylinder pressure during the suction phase corresponds approximately to the pressure of the inlet port. Thus, the cylinder pressure near the inlet stroke bottom dead centre can be made equal in accordance with (3.9) to the averaged pressure in the inlet port.

$$\Delta p_n = \bar{p}_{Saug} - \frac{1}{N+1} \sum_{\phi=LUT}^{LUT+N} p_{Mess}(\varphi) \tag{3.9}$$

The accuracy of the determination of the zero-line with the help of the inlet port pressure can be increased significantly if we use the individual cylinder inlet port pressure profile instead of the average engine pressure to determine individual zero-lines. In the case of modern indicating systems, the range of the averaging can be freely selected so that the determination of the zero-line can be optimally adjusted to the circumstances, but near BDC after the inlet stroke has consistently proven to be effective.

3.1.5.2 Thermodynamic Zero-Line Correction

Thermodynamic determination of the zero-line is based on a comparison between of the measured and calculated pressure profiles. A relative simple way to accomplish this was introduced by Hohenberg (1983), proceeding from a constant polytrope exponent n within a certain crank angle range.

Equation (3.10) is true for polytrope compression:

$$\frac{p_{2Mess} + \Delta p_n}{p_{1Mess} + \Delta p_n} = \left(\frac{V_1}{V_2}\right)^n = C \tag{3.10}$$

This results subsequently for Δp_n in (3.11):

$$\Delta p_n = \frac{C \cdot p_{1Mess} - p_{2Mess}}{1 - C} \tag{3.11}$$

The following values are suitable as the polytrope exponent n:

- For diesel engines and spark-ignition engine in motoring operation

$$n = 1.37 \cdots 1.40$$

- For spark-ignition engines with stoichiometric A/F ratio

$$n = 1.32 \cdots 1.33$$

The following ranges are recommended for the crank angle values of φ_1 and φ_2:

φ_1 = *100°CA...80°CA before TDC for Diesel Engines*
φ_1 = *80°CA...60°CA before TDC for Spark Ignition Engines*
φ_2 = *40°CA...30°CA before TDC for Diesel Engines*
φ_2 = *40°CA...20°CA before TDC for Spark Ignition Engines*

This type of zero-level correction is often used in practice, as it provides, despite its simplicity, a high level of accuracy and speed in the calculation and comprehends cylinder to cylinder tuning differences. The main uncertianity of this method stems from the use of a constant polytropic exponent. To minimize this influence, the a crank angle interval must be set as large as possible. A further weak point arises when signal disturbances (i.e. due to structure-borne sound, i.e. from inlet valve closing) is superimposed over the pressure profile in this crank angle range. Therefore φ_1 must be delayed until after the inlet valve closing noise attenuates (at the latest valve timing possible on VVT engines), while φ_2 should be at the last possible point before measurable heat release begins.

On the other hand, the thermodynamic adjustment method introduced in Feßler (1988) is based on a much more detailed calculation of the motored pressure profile, taking into consideration both leakage and the masses found in the cylinder. By minimizing the surface between the calculated and measured pressure profile in a certain crank angle range, we can assign the pressure level. The crank angle range should be set as high as possible, i.e. in the case of motored pressure profiles the entire high-pressure range and in the case of sparked operation the range between the end of the intake process and the initiation of combustion.[1] This method is also particularly suitable for assigning the measured pressure profile to top dead center (Sect. 3.1.8).

Table 3.2 shows an assessment of individual methods with respect to their applicability.

[1]In the case of engines with spark-ignition direct fuel injection, a vaporization model must be considered that takes into consideration the different material values and the enthalpy removed from the working gas by the vaporization of the fuel must be introduced and/or the adjustment range until injection start shortened, which has a negative effect on the accuracy of the method. The like is also true for DI diesel engines when a pilot injection is used.

Table 3.2 Evaluation of methods of determining the zero-line

	Fixed-point correction	Average inlet port pressure	Individual cylinder inlet port pressure	Fixed polytrope exponent	Calculated motored pressure profile
Additional metrological cost	No	(Yes)	Yes	No	No
Accuracy	Poor	Moderate	Very good	Good	Very good
Real-time capable	Yes	Yes	Yes	Yes	No

3.1.6 Methods That Measure Absolute Pressure

In the field of internal combustion engines, piezoresistive or fiber-optical pressure sensors are used for absolute pressure measurement.

3.1.6.1 Piezoresistive Pressure Sensors

Piezoresistive pressure sensors make use of the fact that electric resistances are altered by the influence of pressure. Such pressure sensors can be used both for static and dynamic applications. They are passive and thus require a excitation. Usually metallic or semiconductor strain gauges are used.

Metallic strain gauges are generally etched from a metal foil (e.g. constantan) superimposed on a support film and are used to measure deformations. On the one hand, expansion of the measuring body causes a reduction of cross-section, resulting in an increase in resistance (geometrical influence). On the other, the specific resistance also changes with expansion (material influence).

While in the case of metallic strain gauges, the geometrical influence has a larger effect on resistance change, the material influence predominates in the case of the semiconductor strain gauge. In comparison to metallic strain gauges, the piezoresistive effect is much larger (by a factor of about 50). It depends on the orientation of the semiconductor monocrystal and on the doping (type, density and distribution of the foreign atoms that determine conductivity). Semiconductor strain gauges are either applied to such a structure as well or are directly sputtered, guaranteeing an intimate connection, which is the prerequisite for an absence of hysteresis, for temperature resistance and for ageing resistance. Although the piezoresistive effect is not limited to this group, the term "piezoresistive pressure sensor" has become commonly applied to those sensors, in which both the elastic structure deforming under pressure and the resistances are integrated into one chip.

The behavior of a piezoresistive pressure sensor changes with temperature. While temperature-dependent zero-line shifts are obvious and easy to detect and monitor, temperature-related linearity and sensitivity changes are less obvious and thus often overlooked.

The cause of zero-line shifting can be the sum of various effects, e.g.:

- Different resistance values or different temperature coefficients of the individual resistances in the measuring bridge
- Mechanical stresses due to the assembly of the measurement cell on its support, which change with temperature
- In case oil is utilized as the pressure transmitting medium, the expansion of the oil in connection with the stiffness of the steel membrane can lead to a pressure build-up in the sensor

Sensitivity changes are caused by the reduction of the piezoresistive factor with increasing temperature. The bridge circuits are designed in practice in such a way that this reduction is automatically compensated. Within a temperature range of about 10–80°C, sensitivity change can be limited to less than 1%. Linearity also changes somewhat with temperature, but this can usually be neglected in practice.

Due to the limited temperature range and their size, piezoresistive pressure sensors are not very suitable for cylinder pressure measurement. Piezoresistive pressure sensors are often used however for unsteady pressure measurement in the inlet/exhaust system. As a result of the high heat transfers, especially in the outlet system, corresponding measures to reduce the stress of temperature on the sensor are unavoidable (switch adapters or cooling adapters, set-back installation).

To minimize these disadvantages, there has recently been a series of developments aimed at measuring pressure directly, i.e. without a transfer element, by means of a Wheatstone bridge and simultaneously expanding the temperature range of application. This was made possible, for example, by packing the semiconductor resistances in silicon oxide, the latter serving simultaneously as an insulator. This technology (silicon on silicon) permits the construction of smaller sensors (Kulite 2009). The effect of temperature on the measurement that exists nevertheless could be improved significantly by using digital error compensation.

3.1.6.2 Fiber-Optic Pressure Sensors

Fiber-optic pressure sensors basically consist of a metallic membrane, fused with the sensor housing, and two fiber optic cables. An optic signal is send via one of these cables, which is reflected on the back of the membrane and sent to a receiver via the second cable. Due to the stress of pressure, the membrane is shifted, which leads to a change in the received light intensity. Only a very narrow range is used for pressure measurement, typically about 20 μm, see Wlodarczyk (1999). Fiber-optic pressure sensors are used for engine monitoring, but they are not generally suitable for highly accurate cylinder pressure indication.

3.1.7 Angle and Trigger Marking

Due to irregularities in crankshaft speed angular time based data cannot be used as the basis for measuring combustion data. Instead, the non time-proportional crank angle can be and is almost exclusively used as the clock. Due to the type of data capture and processing, digital angle measurement is practically the only kind that comes into question. As a rule, trigger marks are used for synchronization (triggering) after every rotation and a series of angle marks (usually 360 or 720 marks per rotation) for the angle information. Electronic circuits multiply these coarse marks, thus permitting a resolution of up to 0.025°KW.

To create the angle signal, different principles of signal creation are utilized depending on the requirements:

3.1.7.1 Hall Sensors

The Hall effect is based on the fact that electric charge carriers are diverted in a magnetic field. The "Hall voltage" is tapped on a conductive disc that has a current flowing through it perpendicular to the current and the magnetic field. Hall sensors are also suitable for low rotation speeds, but the precision of these sensors is relatively low.

3.1.7.2 Inductive Sensors

Inductive sensors consist of a permanent magnet arranged behind a soft-iron core with a coil (Fig. 3.21 left). If we move a magnetically conductive marked support (e.g. a gearwheel) in front of the sensor, the magnetic flow in the soft-iron core changes, thereby inducing a voltage in the coil.

Inductive sensors are used mostly in conjunction with gearwheel already present in the engine. The obtainable accuracy and angle resolution strongly depends on the mechanical design of the detector wheels. For this reason, this type of sensor is mainly used for monitoring and/or rotation speed measurements, not the least because relative motions of the sensor head and marked support affect the signal. Furthermore, in the case of inductive sensors, the output voltage is highly dependent on the rotation speed, so that lower speeds can no longer be detected. The phase also shifts somewhat with engine speed, and must be corrected.

3.1.7.3 Optical Sensors

Here, the measuring principle is that of the light barrier (Fig. 3.21 right). Depending on the arrangement, either the transmitted light or reflection light method is used. Optical angle sensors met high accuracy requirements (even with small dimensions) because the marked plate can be manufactured very precisely. A further advantage is the low susceptibility to electrical interference, especially if one is working with fiber optics near strong interference fields. The angle resolution is usually less than 1°CA.

The function of angle sensors is to measure the crank shaft rotation movement relative to the crankcase as precisely as possible. They can be basically installed either on the free end of the crankshaft or on the end of the crankshaft on the output

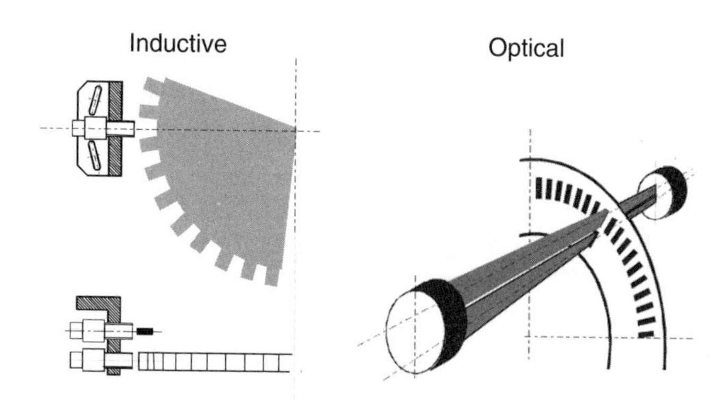

Fig. 3.21 Measuring principles of angle sensors

side towards the consumer. From this we derive the following possibilities for assembling the sensor near and marked disc:

3.1.7.4 Marked Disc Fixed on the Crankshaft: Sensor Head Fixed on the Crankcase

This type of assembly is customarily used for the position on the output side toward the consumer, since it permits a feed-through of the shaft toward the brake assembly. Relative motions between the marked disc and the sensor head cause angle errors (Fig. 3.22 left). It is advantageous to use a marked disc with a diameter that is as large as possible in order to minimize angle errors.

3.1.7.5 Marked Disc Fixed on the Crankshaft: Sensor Head Fixed Axially and Radially on the Crankshaft

Direct mounting of the angle sensor housing (with anti-twist safeguard toward the crankcase) on the free end of the crankshaft minimizes all relative movements between the marked disc and the sensor head (Fig. 3.23 right).

Both basic assembly possibilities lead – as shown in Fig. 3.23 – to two generally differing angle sensor designs. In the case of the angle sensor with self-mounting (shown left), a piece of the shaft is flange-mounted on the free end of the crankshaft, on which the marked disc is also fixed. The sensor head is revolvable and mounted on the flanged shaft piece, supporting itself-against the crankcase. The marked disc can be scanned by the sensor head either using the transmitted light or reflection light method. If the angle sensor is installed on the output side of the engine, then the sensor head is fastened on the engine block separately from the marked disc or with another support (Fig. 3.23 right).

One can obtain the best results with a angle sensor mounted on the free end of the crankshaft. It is important that one takes the measurement on the cylinder closest to

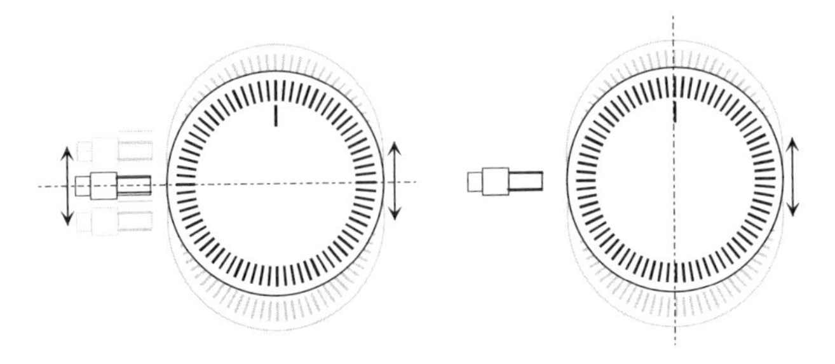

Fig. 3.22 The effect of relative motion between the marked disc and the sensor head

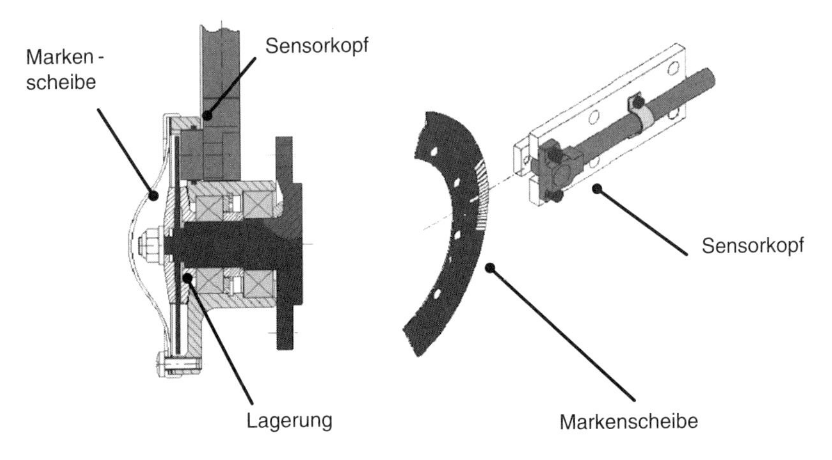

Fig. 3.23 Examples of angle sensor designs (AVL)

the angle sensor. In this case, one can expect minimal crankshaft torsion and thus minimal angle error. Assembly on the camshaft or on an intermediate drive should be avoided due to deformations and play in the connection to the crankshaft. To avoid measurement errors, one must also take care that the support of the angle sensor to the engine block is located on an area with little vibration and is designed as stiff as possible.

3.1.7.6 Using Engine Control Unit Crank Sensor

It is possible to directly use the angle marks generated for the electronic engine control unit (ECU). Typical tooth configurations are 60 minus 2 or 30 minus 1 tooth. Intelligent electronic circuits (i.e.: phase lock loop pulse multiplication) make these coarse resolutions sufficient for indication work (1.0° or 0.5°). Corresponding algorithms are necessary, especially in transient operation, so that the necessary number of crank angle marks for one working cycle is guaranteed. To this end, a good interaction between angle preparation and the indicator is required.

This kind of crank angle generation represents a considerable simplification, but is well suited for indicating work in vehicles.

3.1.8 TDC Assignment

In the thermodynamic analysis of cylinder pressure profiles, the exact determination of the top dead center is of decisive importance. The left of Fig. 3.24 shows the influence of an angle error on the energy balance and the right shows the effect on the friction mean pressure in the case of a diesel engine (acc. to Pischinger et al. (2002)).

A TDC position that is too early, which signifies a right-shifting of the cylinder pressure profile, causes an excessively low pressure in the case of an upwardly moving piston and excessively high pressure in the case of a downwardly moving piston. A seemingly longer afterburning phase and increased energy conversion are the result. In the case of a set-back TDC position, the conditions are reversed. The effects of angle shift also manifest themselves in the indicated work (indicated mean pressure p_{mi}), which becomes larger if the TDC is to early and vice versa. One then obtains dramatically (0.1°~ 10% error in FMEP) larger or smaller friction mean pressures.

The trigger marking can be assigned to the top dead center using different methods:

- *Static dead center determination*
  In this method, a marking flag (1) is first fixed on the engine block. This flag extends over a rotating section of the engine with the largest possible diameter (e.g. flywheel) (Fig. 3.25). Then the crankshaft is turned until the shaft crank and the connecting rod of the cylinder, on which the measurement is made, roughly form a right angle (position a). In this position, the height of the piston is measured. To this end, a dial indicator pin (2) is fixed on the piston

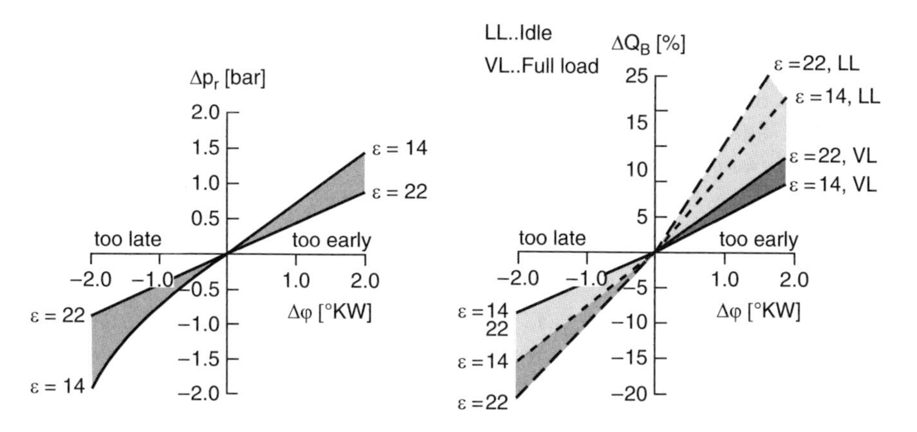

Fig. 3.24 Influence of angle error on the energy balance (*left*) and the friction mean pressure (*right*) (diesel engine)

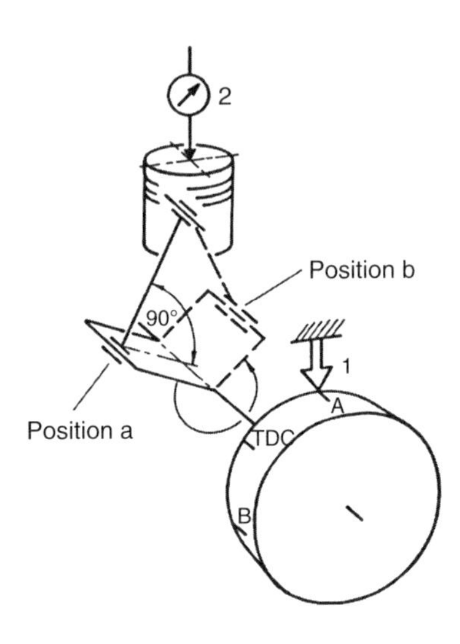

Fig. 3.25 Static TDC determination

surface – through the spark plug bore for example. This position opposite the marking flag is marked on the flywheel (A). Now the piston is lowered and lifted again by further turning of the crankshaft until the same display as before appears on the dial indicator (position b). This position is also marked on the flywheel (B). After that, the distance between both markings on the flywheel is halved. If one now turns the crankshaft until the half-way point is at rest exactly at the marking flag, the piston is now in the top dead centre.

- *Dead center determination using the thermodynamic loss angle*
  Due to losses of heat and leakage, the pressure maximum is reached before top dead center in drag engine operation. This difference angle between pressure maximum and TDC is called the thermodynamic loss angle (Pischinger et al. 2002; Lancaster et al. 1975). The thermodynamic loss angle depends on the construction of the engine and on the rotation speed. At higher speeds, the thermodynamic loss angle gets smaller, since heat loss is reduced because of lack of time available. For TDC assignment, the engine is motored, the pressure maximum determined from the measured pressure profile, and the pressure profile shifted in accordance with the loss angle.
- *Thermodynamic Adjustment*, Feßler (1988)
  As seen in Sect. 3.1.5, concerning the assignment of the pressure level, it is possible to assign the TDC by comparing the measured pressure profile with the calculated profile.
- *Dead center determination with capacitive sensors*
  Capacitive TDC sensors provide an exact determination of the top dead center (Fig. 3.26). This is possible because the piston movement is measured directly. In this way, degrees of precision of $\pm 0.1°$KW can be readily obtained. The determination process takes place in motored operation. The TDC sensor is mounted in the cylinder head for the measurement. For this, pre-existing bores intended for the spark plug, injection nozzle or pressure sensor are used.

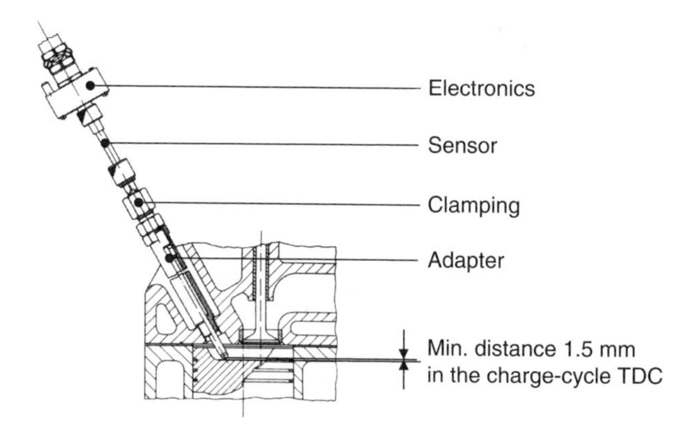

Fig. 3.26 Structure and assembly of a capacitive TDC sensor (AVL)

The functional principle is based on a capacitive measuring method in which the sensor measures the capacitance changes between the piston and the sensor head. The capacitance changes monotonically with the distance of the piston to the sensor head.

In the case of a four stroke engine, the output signal of the TDC sensor exhibits a localmaximum both in the charge-cycle TDC and the ignition TDC. Due to clearance in the bearings and the low cylinder pressure during gas exchange, the signal of the gas exchange TDC will be higher. Despite the higher signal in the charge-cycle TDC, the signal at compression TDC should be used for maximum precision. The reason for this is, on the one hand, so that the deformations arising in real operation can be taken into consideration and, on the other hand, that the valve movements influence the capacitance of the respective piston positions during gas exchange and can thus distort the result.

Because of the flat signal profile near the TDC and the resolution of the signal sampling, one cannot determine the actual TDC position from the output signal of the sensor beyond the signal maximum. Therefore, the following algorithm is used: in the rising branch, the amplitude is measured at a certain angle (about 15–5°KW before TDC), and the corresponding angle at the same signal amplitude in the falling branch is determined. Halving the angle range gives the position of the TDC. The process is repeated several times at different thresholds and an average value is determined.

The methods described for assigning to the TDC differ fundamentally with respect to cost and obtainable precision – for details see Wimmer and Glaser (2002) and Wimmer et al. (2000 and 2002) and Lancaster et al. (1975). Figure 3.27 shows a comparison. The static determination of the TDC leads to inaccuracies, especially due to the not ideal rigidity mechanical structure of internal combustion machines, and it is also relatively costly. The thermodynamic loss angle method is simple, but one must rely on experiential data for the magnitude of the loss angle, which are not always available and involve subjective interpretation.

A more exact assignment to the top dead center is possible with capacitive TDC sensors and thermodynamic adjustment, but the cost is high. Thermodynamic

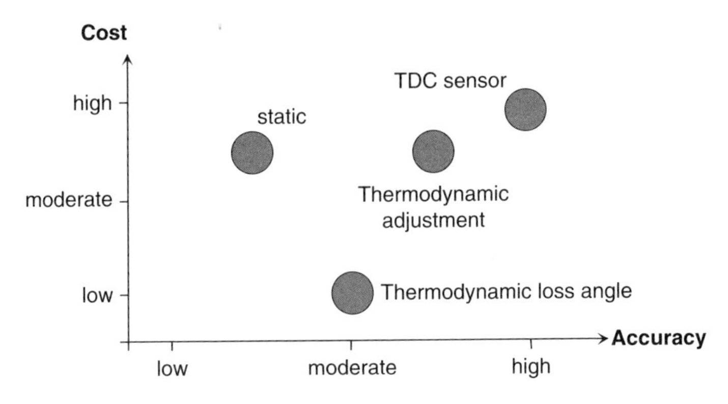

Fig. 3.27 Comparison of cost and accuracy of different TDC determination methods

adjustment is based on a calculation of the drag pressure profile and requires exact knowledge of the charge weight as well as of heat transfer and leakage. The consideration of heat transfer in particular leads to uncertainty.

3.1.9 Pressure Indication in the Inlet and Outlet System

Analyzing the gas exchange process requires not only combustion chamber pressure data during the valve opening times that is as exact as possible, but also an accurate measurement of pressure profiles in the inlet and exhaust system, see Wimmer et al. (2000). Both piezoelectric and piezoresistive pressure sensors are used for measuring charge-cycle pressure profiles:

3.1.9.1 Piezoelectric Sensors for Low Pressure Indicating

The piezoelectric pressure sensor is installed with a damping adapter in order to prevent for the most part the transfer of vibrations to the sensor. This is especially important when acceleration-sensitive pressure sensors are used. In order to reduce the temperature load, damping adapters are generally designed with cooling. Figure 3.28 shows examples of damping adapters for different pressure sensor designs.

This arrangement makes it possible to make the shortest possible connection from the measurement location to the pressure sensor membrane, so pipe oscillations and shifts in the phase position of the measured pressure profile can thus be largely prevented. When it is used in the outlet system however, it is subject to a high heat flow stress. For this reason, sensors with low cyclical temperature drift should be used. To determine the absolute pressure level, a corresponding attachment is provided on the adapter for pressure referencing. The reference pressure is

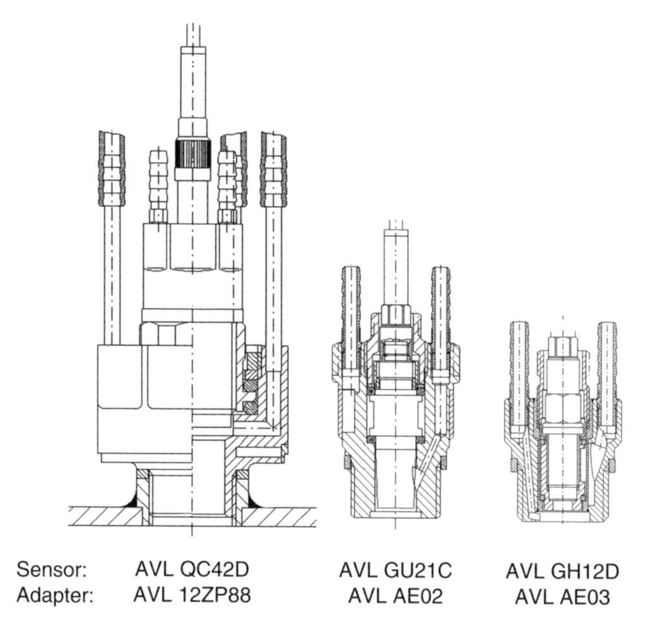

Fig. 3.28 Damping adapters for low-pressure indication with piezoelectric pressure sensors (AVL)

measured remotely at a large distance from the measurement location, with an accurate absolve pressure sensors.

Switching adapters are another way to protect the pressure sensor. These permit a short-time exposure of the piezoelectric pressure sensor alternately to ambient pressure then port pressure. This kind of adapter is also used with piezoresistive sensors in order to make the correction of the zero-level – necessary despite the absolute measurement principle – at highly stressed measurement sites where zero drift is a problem.

3.1.9.2 Piezoresistive Pressure Sensors

Piezoresistive pressure sensors have the advantage that the absolute pressure can be determined, but their metrological properties are highly dependent on temperature (zero-point drift, cyclical temperature drift). In the exhaust system, they can thus only be used with long passage set-back, or in combination with special switching adapters (see Bertola et al. (2008) and Fig. 3.29). In this case, the pressure sensor is exposed to the hot combustion gases only for a certain number of cycles (about 100) with the help of a mechanical valve. The gas channel in the switch valve is then closed again, giving the pressure sensor the chance for a thermal regeneration between the two measurement phases, as well as a zero level check.

The following influences must be considered above all when indicating in the inlet and outlet system:

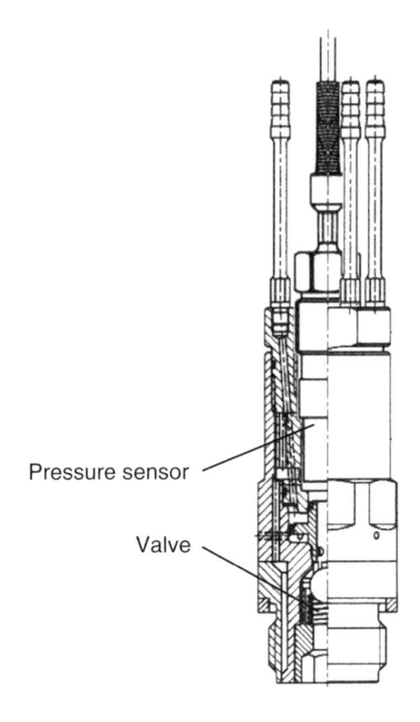

Fig. 3.29 Cooled switch adapter for low-pressure indication with piezoresistive pressure sensors (Kistler)

- Acceleration effects
- The effects of temperature and heat flow stress (especially in the outlet system): thermal zero-point drift (piezoresistive sensors) and cyclic temperature drift
- The effects of a passage mounted sensor installation (organ pipe oscillations and changes in the phase position and amplitude)

In principle, accurate measurements can be carried out with either piezoresistive or piezoelectric pressure sensors. If piezoresistive absolute pressure sensors are utilized, one must take care that the temperature and heat flow stress on the sensor remains low. This can be accomplished by means of a passage mount installation, which results in amplitude and phase errors of the pressure signal. The arrangement with the switch adapter described above, with which the pressure sensor is released for only a small number of cycles, corresponds to this type. The resultant indication channel has a length in which the phase position and amplitude are minimally influenced. Depending on the operation state of the engine however, organ pipe oscillations can affect the quality of the measurement.

Piezoelectric pressure sensors are less restricted with respect to their application limits (temperature range up to 400°C). Due to their lower sensitivity to acceleration, the use of uncooled sensors is recommended. Modern uncooled pressure sensors already exhibit a very low cyclical temperature drift and are therefore also suitable for flush or only slightly set-back installations in the outlet system. Zero level correction of the pressure trace can be accomplished simply by measuring with an absolute pressure sensor that is remotely mounted from the measuring point and thus is exposed to a low amount of thermal stress, or by using a switching adapter.

3.1.10 Data Capture

Indicating systems convert the analog output signals from transducers (or where applicable amplifiers) into discrete digital values. The number of quantization levels is determined by the resolution n of the A/D (analog to digital) converter, given in bits, as 2^n. For a resolution typical of indicator units of 14 bits we thus obtain $2^{14} = 16{,}384$ quantization levels, which, if the input voltage range of the data capture unit is -10 V to $+10$ V, lead to a resolution of about 1.2 mV. In order to obtain precise results, the range of the A/D converter should be used as completely as possible by adjusting the amplification.

The sampling rate is also critical when selecting a combustion analysis system. One approach to the question of how high the sampling rate should be in order to capture a predetermined signal is provided by Nyquist's sampling theorem. Whereby, the sampling rate must be at least twice as high as the highest occurring signal frequency. Typical sampling rates of modern indication devices are in the area of 800 kHz per channel. At a $0.1°$ crank angle resolution, engine speeds up to 13,000 1/min can be measured without a problem with this sampling rate.

An indicating system contains a trigger unit, which uses crank angle marks (with pulse multifaction if desired) to trigger the A/D conversions. The resolution of the crank angle determines the number of measurement points. To reduce the amount of data in the case of high-resolution measuring tasks (e.g. in knock investigations), only the relevant range in the cycle (in this case the range, in which high-frequency pressure deviations from the knocking process are to be expected) is captured with correspondingly high resolution.

The duration of a measurement is determined either by the memory depth of the indicating system or the speed of the interface data capture unit-PC (IEEE 1394, Ethernet). If the communication link is fast enough, the measurements can be further processed on the PC during the measurement.

Parameterization of the entire measurement chain, measurement control and visualization of the measurement data and calculated values are all carried out with PC-based software. Besides the management of the actual measurement function, indicating software generally handles the interface and data traffic to higher systems and/or subsystems (e.g. test bench automation, engine or vehicle calibration systems with or without automated support).

3.2 Pressure Trace Analysis and Loss Distribution

Despite progress in optical measurement techniques, the analysis of the cylinder pressure trace cannot be ignored in the current development of internal combustion engines. Firstly, the cylinder pressure course is the most important quantity in recognizing knocking combustion online on the test stand. Secondly, important insight with respect to combustion (inflammation duration, ignition delay, heat losses and heat release rate) as well as to so-called loss distribution can be determined from thermodynamic analysis. Furthermore, the pressure trace provides information about peak pressure, the indicated work, the charge change behavior of the engine, and the residual gas in the combustion chamber.

3.2.1 Determination of the Heat Release Rate

3.2.1.1 Pressure Signal Acquisition

In determining the cylinder pressure, pressure transducers are used, which function according to the piezoelectric measurement principle. These possess a high mechanical robustness despite small structural size and infinitesimal resolution. The pressure transducer is connected to the combustion chamber with a membrane, which is acted upon by the combustion chamber pressure and leads to a force, proportional to cylinder pressure, in the piezoelements. Proceeding from the cylinder pressure, the pressure transducer creates a charge that is transformed into an electric voltage proportional to this charge by the charge amplifier. The cylinder pressure is thus available for thermodynamic evaluation.

The allocation of the pressure signal to the engine process (triggering) results by means of a so-called encoder, which is mounted to the engine crankshaft and supported by the crank housing. A resolution of approx. 1°CA is sufficient for thermodynamic evaluations. Since the position of the encoder at TDC cannot be exactly determined, the distance between the index pulse and the piston is usually measured with a capacitive sensor while the engine is motored. The precision requirement for this process amounts to $\pm$ 0.1°CA.

In spark ignited engines (SI engines), considerable differences in combustion phasing occur due to mixture inhomogeneities at the spark plug, since combustion starts some degrees of crank angle earlier or later depending of the mixture condition at the spark plug. For simplicity these cyclic fluctuations are sometimes smoothed for thermodynamic evaluation by averaging over a large number of cycles. In the SI engine, an averaging of up to 500 cycles is desirable. Because of autoignition, these fluctuations are less marked in diesel engines, which is why an averaging over fewer then 50 cycles is usually sufficient. Note that the shape of the heat release curve for this averaged curve is systematically different than for individual cycles with similar combustion phasing (MFB50%), so care must be taken to fully understand this effect on the results.

3.2.1.2 Evaluation of the Pressure Signal

If we define the combustion chamber of the internal combustion engine as a control volume, then the conditions of the enclosed gas – i.e. pressure, temperature, and internal energy – can be clearly described via the thermal equation of state and the mass and energy balance (see Sect. 7.1).

The gas mass enclosed within the combustion chamber can be determined in the simplest case via the measurement of the fresh gas mass. The problem arises however that the degree of capture is usually insufficiently known. In addition, the allocation of fresh gas mass to the single cylinders in multi-cylinder engines proves problematic, since this quantity can usually only be integrally measured across all cylinders but cylinder to cylinder differences are inevitable. A gas exchange calculation can help as a back-up, for which a measurement of the pressure traces in the intake and exhaust ports is necessary. The pressure signals are imposed as boundary conditions onto a simplified 1D model, which, via the gas-dynamic relations described in Sect. 7.4, describes the pipe system between the measurement location on the intake side and on the exhaust side.

In engines with low amounts of internal residual gas and high external exhaust gas recirculation, a measurement of CO_2 concentration in the exhaust gas and in the intake mixture has proven practical for the determination of the residual gas mass within the cylinder. The internal exhaust gas recycling rate can, practically speaking, only be determined via the above-described gas exchange calculation.

Since combustion normally only occurs during the high pressure phase (exception: later burning through, e.g., post-injection), we can consider the combustion chamber for the high-pressure phase as a closed system. With this, the enthalpy flows over the system boundaries are zero, and the blow-by losses and evaporation enthalpy in gasoline direct injection can also at first be roughly set to zero.

$$\frac{\mathrm{d}Q_f}{\mathrm{d}t} = \frac{\mathrm{d}U}{\mathrm{d}t} - \frac{\mathrm{d}Q_w}{\mathrm{d}t} + p\,\frac{\mathrm{d}V}{\mathrm{d}t}\left[-\frac{\mathrm{d}m_{bb}}{\mathrm{d}t}\,h_{bb}\left(-\frac{\mathrm{d}m_{f.,evap.}}{\mathrm{d}t}\,\Delta h_{evap.}\right)\right] \tag{3.12}$$

The internal energy in (3.12) can be described in relation to pressure, temperature, and gas composition. The wall heat losses of the piston, cylinder head, and liners can also be represented as functions of pressure and temperature. In describing heat flow in the combustion chamber walls, the wall temperatures are necessary, which can be determined through measurement or calculation. The cylinder volume is only contingent upon slider crank geometry and angular rotation. The affiliated physical regularities are extensively described in Sect. 7.1.

The average gas temperature can be easily determined with knowledge of the instantaneous combustion chamber volume, pressure, and the total gas mass in the chamber via the thermodynamic first law equation of state. The pressure in the cylinder remains the only unknown for the determination of the heat release rate. The question in establishing the heat release rate is thus aimed at the determination of the pressure in the cylinder, which has been described above.

Pressure sensors that function according to the piezoelectric principle cannot measure absolute pressures. Therefore the pressure level of the measured pressure signal must be corrected using a method described earlier.

The highest precision is possible with the above described charge changing calculation under the provision of the measured intake and exhaust pressures. Upon completion of the charge changing calculation, the measured cylinder pressure is adjusted to the pressure of the charge changing calculation at “intake valve closing”. Moreover, with this method, an exact determination of the residual gas portion and thus an exact determination of the cylinder mass become possible, which can be used as an additional advantage for the improvement of evaluation precision.

However, the energy balance is an essential quantity in judging the quality of the heat release rate evaluation. It is produced from the quotient of the energy quantity determined by the heat release rate evaluation and the maximum energy quantity released by the fuel.

The heat release rate is thereby integrated over the entire combustion duration and represents the numerator of the quotient of the energy balance.

The maximum released energy quantity – the denominator – can be calculated from the product of the fuel injected per cycle and the lower heating value, whereby – especially in the case of the SI engine – the energy of unburned exhaust gas components must be subtracted.

$$EB = \frac{\int_{\varphi SOC}^{\varphi EOC} \frac{dQ_f}{d\varphi}}{m_f I_{hv} - Q_{ub}} = \frac{\int_{\varphi SOC}^{\varphi EOC} \frac{dQ_f}{d\varphi}}{m_f I_{hv} \eta_{conv.}}, \tag{3.13}$$

$$\eta_{conv.} = 1 - \frac{Q_{ub}}{m_f I_{hv}}. \tag{3.14}$$

The following relation is thereby valid for unburned components in the exhaust gas like CO, H2, HC, and soot, to correct Hu for combustion efficiency:

$$Q_{ub} = m_{CO} I_{hv,\,CO} + m_{H_2} I_{hv,\,H_2} + m_{C_3H_8} I_{hv,\,C_3H_8} + m_C I_{hv,\,C} \tag{3.15}$$

A fluctuation in the energy balance in the range of 95–105%, thus in the range of ±5%, can be viewed as very good in the context of achievable precision in measurement and in the description of thermodynamic relations.

Besides the heat release rate, other quantities important for the characterization of the combustion path can also be determined. These are portrayed in Fig. 3.30.

The time between ignition or injection in the diesel engine and start of combustion, which can be vary from. 3 to 5% of the total heat release rates, is called ignition delay. The time between start of combustion and end of combustion is designated as combustion duration. The center of the heat release rate (50% mass fraction burned, complies 50 mfb) is defined as the degree of crankshaft, at which 50% of the total heat released has been converted. Almost independent of the type of engine and the combustion process, consumption-optimal operating points result at 50 mfb position of approx. 8°CA after the ignition TDC.

Figure 3.31 shows the evaluation of the heat release rate for a conventional SI engine at a speed of 1,000 rpm and a load of $p_i = 1$ bar. Represented are the heat release rate and the single components there of in accordance with the first law of

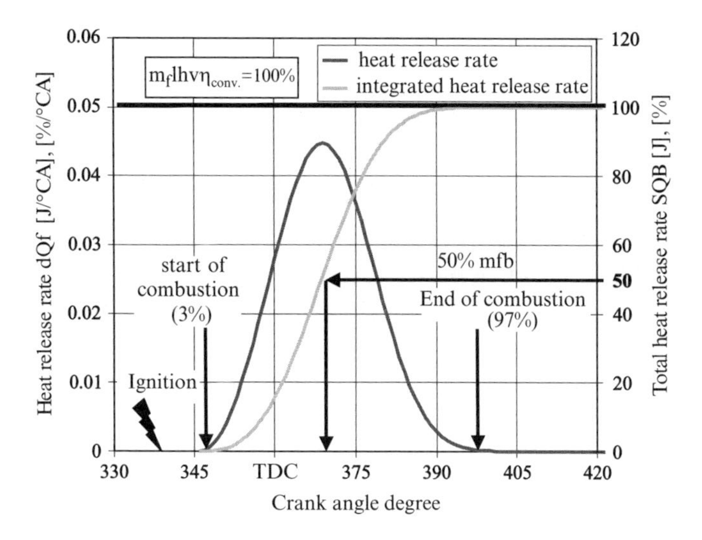

Fig. 3.30 Characteristic values of the combustion profile

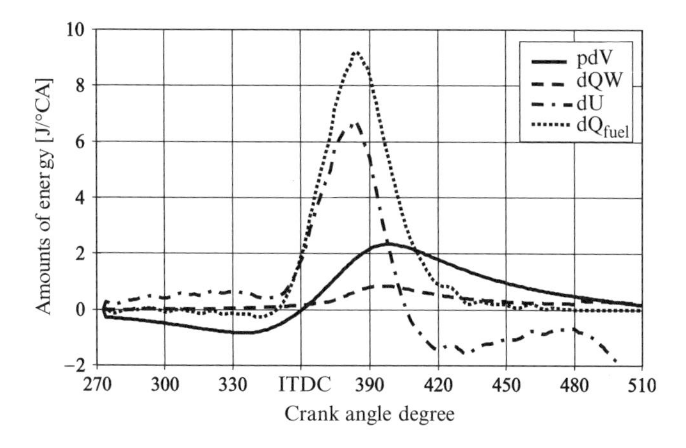

Fig. 3.31 Pressure profile and heat release rate for a SI engine at 1,000 rpm and i_{mep} = 1 bar

thermodynamics, see (3.12). The exact procedure in determining the heat release rate is described by Witt (1999), Rassweiler and Withrow (1938), Krieger and Borman (1966), Gatowski et al. (1984) and others.

In summary, one can say that for a thermodynamically correct evaluation, a high precision in pressure indication and determination of all measured quantities is necessary. If all these prerequisites are fulfilled, then it is possible to determine not only the indicated mean effective pressure, but also the temporal release of heat release rate as a decisive requirement for an efficient simulation.

In the case of a combination of charge changing calculation and pressure profile analysis, the amount of residual gas can also be determined very accurately by means of low pressure indications.

3.2.2 Loss Distribution

In order to be able to assess various combustion processes with respect to their potentials, we make use of so-called loss distribution. Individual loss fractions are systematically calculated proceeding from the perfect engine process and reconstructed for the real engine process.

The perfect engine process resembles the constant-volume process, as in the latter the total energy is added and removed at the top dead center as well. However, for the perfect engine, several deviating assumptions are made, which Witt (1999) summarizes as follows:

- Calculation with ideal gas and real physical characteristics (c_v, c_p, $\kappa = f(T)$)
- An equal equivalence air ratio as in the real process
- Combustion progresses to the point of chemical equilibrium with consideration of dissociation
- Idealized heat release rate (heat supply at the TDC in the SI engine)
- No wall heat losses
- No friction
- No flow losses
- The valve control times lie at the top and bottom dead centers (EVO in the BDC, EVC and IVO in the TDC and IVC in the BDC)
- Pressure and temperature at start of compression are fixed such that the same comparison line between the perfect and the real process results
- The charge-mass is the same as in the real process
- Equal amount of residual gas as in the real process

Figure 3.32 shows the efficiency of the perfect engine contingent upon the compression ratio and on the global air-fuel equivalence ratio according to Pischinger et al. (2002).

The real process distinguishes itself-from the perfect process through losses from incomplete/imperfect combustion, through combustion losses, wall heat losses, charge changing losses, and friction losses. In order to quantify these losses, the cycle is calculated again in consideration of the respective sources of loss, and the difference to the previous cycle is evaluated.

3.2.2.1 Losses from Incomplete/Imperfect Combustion

We understand under losses from incomplete combustion those losses, which arise due to substoichiometric combustion – i.e. due to lack of oxygen. These losses are already considered in the perfect engine, since in this case only the conversion of fuel until chemical equilibrium is taken into consideration anyhow. Losses from imperfect combustion arise when the fuel does not burn until chemical equilibrium. From this imperfect combustion, additional exhaust gas components result like CO, H_2, HC, and soot, which go beyond the level of combustion with a lack of oxygen. These components from incomplete and imperfect combustion are disclosed by

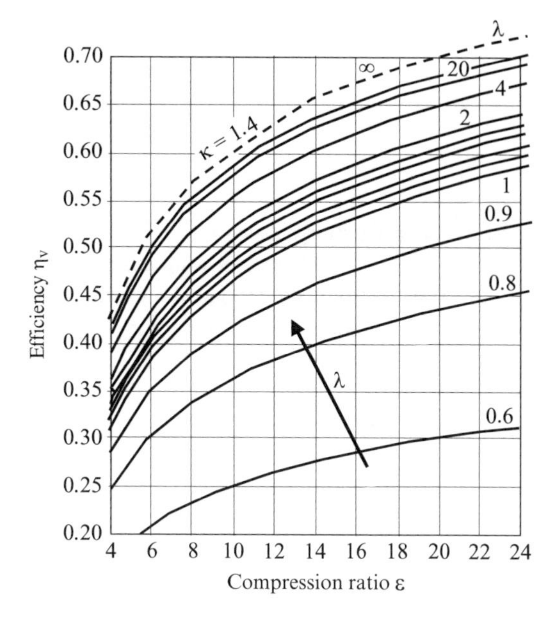

Fig. 3.32 Efficiency of the perfect engine contingent on the compression ratio and on the global air-fuel equivalence ratio, acc. to Pischinger et al. (2002)

exhaust gas analysis in its entirety. Losses from imperfect combustion reduce the heat quantity supplied to the process (related to 1 kg mixture mass) and are to be quantified as follows:

$$q_{ub,\,incomplete} = q_{ub,\,total} - q_{ub,\,chem}. \tag{3.16}$$

Thereby, for the total losses from incomplete and imperfect combustion, the following is applicable from exhaust analysis

$$q_{ub,\,total} = \left(\nu_{\mathrm{CO}}\, I_{hv,\,\mathrm{CO}} + \nu_{\mathrm{H_2}}\, I_{hv,\,\mathrm{H_2}} + \nu_{\mathrm{C_3H_8}}\, I_{hv,\,\mathrm{C_3H_8}} + \nu_{\mathrm{C}}\, I_{hv,\,\mathrm{C}}\right)\frac{1}{M_V} \tag{3.17}$$

With

$$\begin{aligned} I_{hv,\,\mathrm{CO}} &= 282.900\ \mathrm{kJ\,/\,kmol}, \\ I_{hv,\,\mathrm{H_2}} &= 241.700\ \mathrm{kJ\,/\,kmol}, \\ I_{hv,\,\mathrm{C_3H_8}} &= 406.900\ \mathrm{kJ\,/\,kmol}, \\ I_{hv,\,\mathrm{C}} &= 2.041.367\,\mathrm{kJ\,/\,kmol}, \\ M_V &= 28,905\ \mathrm{kg\ /\,kmol}\ . \end{aligned}$$

For incomplete combustion until chemical equilibrium, the following relation is valid according to Vogt (1975)

$$q_{ub,chem} = \left[1 - (1,3733\,\lambda - 0,3733)\right] h_u^* \tag{3.18}$$

with

$$h_u^* = I_{hv} \frac{1}{\lambda L_{\min} + 1}.$$

3.2.2.2 Combustion Losses

Combustion losses arise in that the combustion heat is not supplied at constant volume – i.e. in an infinitely short time span – in the real process, but rather in the form of the heat release rate (see Sect. 3.2.1) which covers several degrees of crank angle. The heat quantity supplied before the TDC is thereby acting against the compression while the heat quantity supplied after the TDC will not act during the entire expansion. This loss can be determined through twice-repeated cyclical calculation – once with instantaneous heat supply and once with the provision of the real combustion duration. One should hereby recognize that a reduction of the combustion losses compared to constant volume of the combustion, which always goes along with an increase of wall heat losses in real engine operation. This is why the total optimum from combustion and wall heat losses does not lie in isochoric combustion.

3.2.2.3 Wall Heat Losses

Two process calculations are necessary in determining wall heat losses as well. The wall heat flow is thereby calculated via known correlations, e.g. according to Woschni or Bargende (see Sect. 7.1).

3.2.2.4 Gas Exchange Losses

The perfect engine possesses, according to its definition, no gas exchange losses, since the cycle from BDC to BDC takes place with heat removal. According to Witt (1999), in order to take the gas exchange losses into exact consideration, a definition of the gas exchange losses according to the BDC-BDC method under additional consideration of the expansion and compression losses must be chosen. The reduction of the working area in the p, V diagram through the sudden fall in pressure due to the opening of the exhaust valve before the BDC is thereby taken into consideration. This is the case in the closing of the intake valve occurring after TDC as well. Corresponding compression losses must hereby be considered. These losses are, according to their respective cause added to the gas exchange losses. The consideration of gas exchange losses leads to lower net indicated mean effective pressure and thus to the indicated efficiency. Losses such as piston ring and valve leakages only marginally affect the result of the loss distribution.

3.2.3 *Case Study: Comparison of Various Combustion Processes*

In this section, the heat release rate and loss distributions for various example combustion processes will be presented. A throttled SI engine with multi point injection (MPI), a SI engine with fully variable mechanical valve lift control and multi point injection (VVH), a direct injecting SI engine with a spray-guided combustion process ($DISI_{spray}$), a SI engine with controlled autoignition (CAI), and a hydrogen engine with intake-port injection (H_2) are compared. A speed of 2,000 rpm and an indicated mean effective pressure of approx. 2 bar is common to all combustion processes.

3.2.3.1 Comparison of the Heat Release Rate of Various Combustion Processes

Figure 3.33 shows the heat release rates for the combustion processes described above. The difference in heat release rates between the throttled and unthrottled operation with fully variable valve lift control is only marginal in comparison with the heat release rates for controlled autoignition.

One clearly recognizes that the combustion duration in controlled autoignition is only approx. 10–16°CA, which is approximately one third the combustion duration of the gasoline MPI example. This is caused by the many ignition points throughout the combustion space, whose surrounding mixture burns extremely quickly.

The $DISI_{spray}$ combustion process possesses a relatively early center position. One recognizes here the trade-off between mixture formation for securing a

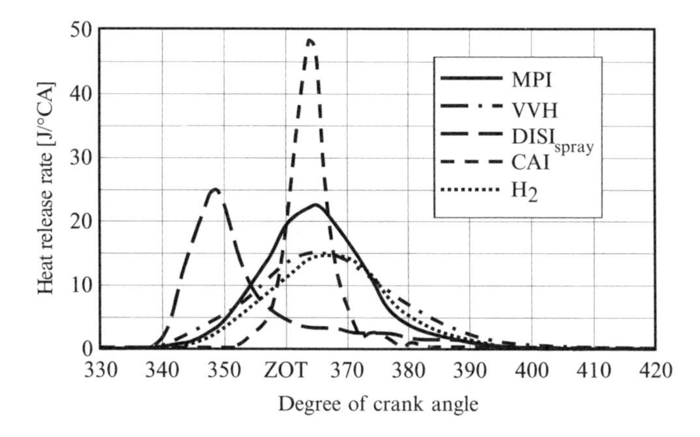

Fig. 3.33 Comparison of the characteristic combustion profiles of various combustion and load control processes

complete combustion with little emission (HC) and a late injection for the purpose of a consumption-optimal center position.

The hydrogen engine possesses a relatively similar burning duration as the multi point injecting SI engine, which results from the high burning speed of hydrogen being mitigated by the very lean mixture ($\lambda > 3$) which slows flame propagation.

3.2.3.2 Comparison of the Loss Distribution of Various Combustion Systems

The loss distributions for the operating conditions described above are represented in Fig. 3.34.

The combustion process with the fully variable intake valve lift control (WH) possesses greater potential for perfect engine efficiency than the throttled engine (MPI) because of higher residual gas capability and gas exchange optimization. However, the losses via imperfect combustion are clearly higher and partially compensate for the significantly smaller charge changing losses.

The spray-guided DI engine ($DISI_{spray}$) possesses the highest potential of the perfect engine at over 57%, since it has a very high global air-fuel equivalence ratio in the described operation point because of its ability to run in stratified-charge mode. The higher compression ratio of 12 as opposed to 10.5 in other combustion processes reinforces the higher basic potential. At this operation point, we recognize the small losses due to imperfect combustion. Through the early center point, the combustion losses are, however, higher than those of multi point injectors, which has a detrimental effect on wall heat loss. These effects temper the potential of DISR, yet they show at the same time a large area for improvements. The reduced charge changing losses, which cause in sum a approx. 7% higher net indicated efficiency, are clearly recognizable. This means a fuel consumption improvement of about 20% in comparison with variable valve lift control at these

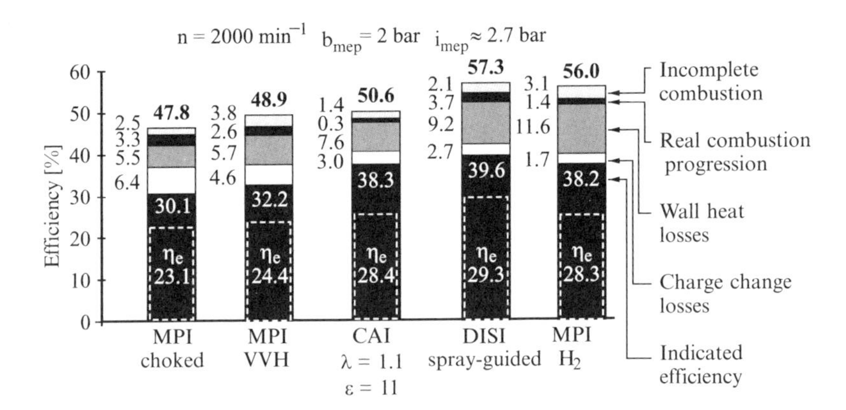

Fig. 3.34 Comparison of loss distribution in various combustion and load control processes

operating conditions. Hereby not under consideration are, however, losses through exhaust gas treatment measures (lean NO_x trap regeneration).

Clearly recognizable is the great potential of controlled autoignition, in which, despite smaller basic potential from the standpoint of the perfect engine, extremely minute combustion losses, relatively small wall heat losses, and very minimal gas exchange losses lead to a high net indicated efficiency. This is obviously higher than that of variable valve lift control and only narrowly below that of direct injection, spray-guided combustion.

In the case of the hydrogen engine, the high compression ratio and above all the high air-fuel equivalence ratio ($\lambda > 3$) due to the extremely wide ignition boundaries of hydrogen have a very favorable effect on the basic potential of the perfect efficiency. The efficiency of the perfect engine amounts to about 56%. Combustion with an optimal position does reduce combustion losses. However, clearly higher wall heat losses also result from the higher combustion temperatures in hydrogen combustion, which destroys a large part of its potential. Nevertheless, a comparatively high indicated efficiency results.

It can be concluded from the above analysis that short combustion duration, resulting small combustion duration loss will cause higher wall heat losses. In this case, a compromise has to be found in order to be able to realize low fuel consumption. The same is valid for the relation between a higher compression ratio for the sake of a higher efficiency of the ideal engine and wall heat losses.

3.3 Optical Diagnostic Techniques

3.3.1 Introduction

Optical techniques have been developed to the point that they can be applied to institutional engine development. The following tables give a listing of various optical techniques and their potential applications in engine development. There are numerous methods applied in engine and combustion research, however, just a few of them have the potential of being applied in practical engine development.

This section gives an overview of optical methods and their application to various aspects of combustion analysis. The focus is on describing optical techniques which were found to be useful in supporting pre- and series development of IC engine combustion systems. The decision for application of a specific optical technique is always based on requested information and the efforts and chances to gain the required results with a given method. Consequently, the methods which have proven their benefits for combustion system development are easy to use and yield specific insight into relevant development tasks.

3.3.2 Optical Methods: An Overview

Tables 3.3–3.7 summarize fields of application, basic features of optical sensors, self-radiating in-cylinder objects which are accessible with passive optics and others which require external light sources for illumination. Some laser based techniques suitable for engine specific analysis tasks are given in Table 3.8.

Table 3.3 In-cylinder optical techniques – the tasks

Field of application	Task	Priority	Optical technique
Research	Understand combustion events	Scientific precision	Specific method and boundary conditions are defined by research task
Methods development	Comparison/ verification with simulation	Precise definition and description of test conditions	
Combustion system development	Testing of specific development variants	Realistic and relevant engine operation	Must comply with required engine operation, effort / benefit economy

Table 3.4 Basic features of optical sensors

Signal	Sensor	Time sequence	Temporal resolution	Spatial resolution
Image	Camera	Single shot and high speed camera	Shutter speed or illumination pulse	Pixel resolution via imaging
Radiation	Single- and multichannel sensors	Continuous	Bandwidth of signal converter and digitalization rate	Integral along aperture cone, or integral surface spot per channel

Table 3.5 Signal sources for passive optical techniques

Signal source	Object	Engine	Application
Radiating gas	Flame front, burned gas	SI engines	Flame kernel formation, flame propagation
Hot particles	Soot radiation	Diesel and GDI engines	Diffusion flame identification flame interactions in diesel engines, soot temperature, soot concentration
Hot surfaces	Surface radiation	Engine components	Component surface temperature

Table 3.6 Illuminated in-cylinder objects

Object	Illumination	Application
Fueal spray	Continuous or flash lamp	Spray propagation
Fuel wall film		Piston, liner, cylinder head surface
Deposits		Injector, valves, any surface
Engine components		Component functionality

Table 3.7 Laser based in-cylinder analysis techniques, Lackner (2008)

Method	Object	Information	Sensors
PIV, particle image velocimetry	Seeding particles, fuel droplets	Flow field	Camera
LDA, laser doppler anemomentry	Particles, droplets	Local flow velocity	Photodiode, multiplier
PDA, phase doppler anemometry	Droplets	Droplet size	Photodiode, multiplier
LIF, laser induced fluorescence	Fluorescent molecules	Species concentration	Camera
LII, laser induced incandescence	Soot particles	Soot distribution	Camera
Raman scattering	Molecules	Concentration, temperature	Multiplier
Light absorption	Molecules, particles	Concentration	Photodiode, multiplier

Table 3.8 Optical analysis methods in transparent research engines and in standard engines

Method	(DI) SI engine	(DI) Diesel engine	Sensors
Single cylinder transparent engine	Mixture formation, flame quality	Spray propagation, flame distribution	Camera
Endoscopic imaging in standard engine	Spray – combustion chamber interaction Diffusion flames	Flame distribution	Endoscope and camera
Two color method in standard engine		Flame temperature, soot concentration	Front optic elements and optical fibers
Flame measurement techniques in standard SI engines	Mixture quality evaluation on basis of flame radiation		Single and multichannel fiber optical sensors

3.3.3 Application Examples of Optical Methods

This article presents the following measurement techniques:

- Imaging techniques to support combustion system development for Diesel and SI engines. Mixture formation and combustion analysis in transparent engines
- Imaging techniques with endoscopic access to the combustion chamber
- Flame radiation measurement in Diesel engines, signal evaluation with two color method
- Flame radiation measurement in SI engines

These methods' applications for SI and Diesel engine development are summarized in Table 3.8.

3.3.4 Diesel Engines

The image arrangement in Fig. 3.35 shows an ideal positioning for fuel spray, vapor cloud and diesel flame within the boundaries of a combustion chamber bowl.

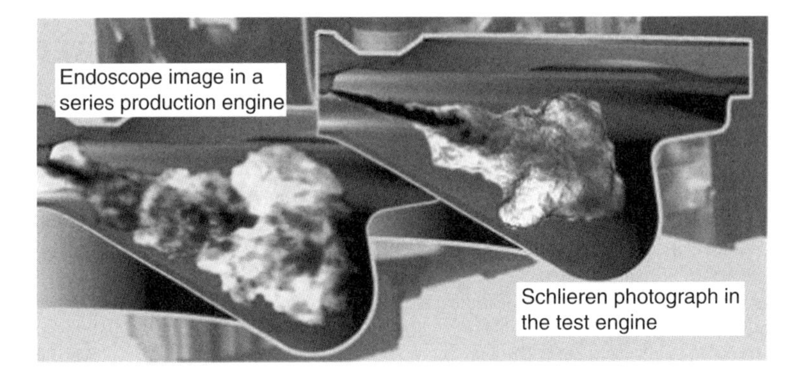

Fig. 3.35 Image arrangements: Diesel spray with fuel vapor cloud and Diesel flame inside the cross section of a combustion chamber bowl

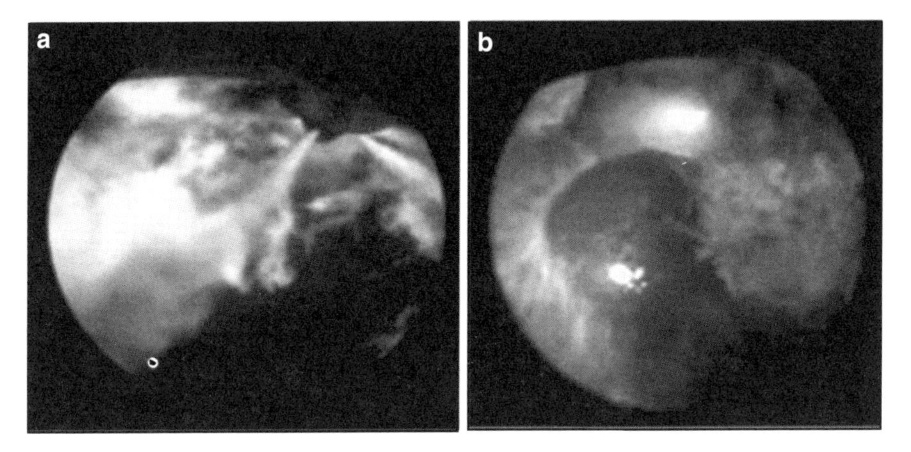

Fig. 3.36 Diesel fuel sprays and flame in warm engine (**a**), fuel wall film and flames in cold engine (**b**)

The Schlieren image shows the spray "core" near the nozzle hole as well as the wide field of the fuel vapor cloud. Such fuel vapor is formed as fuel droplets are heated by the compressed in-cylinder air and ongoing heat transfer finally results in ignition and flame formation, Winklhofer et al. (1992).

The second part of this image arrangement shows the diesel flame. The boundary drawings have been selected to suggest best use of available space without flames touching the head or piston surface.

Such ideal conditions are unattainable in real engine operation. At high load, flames always impinge on combustion chamber walls, in the cold engine at start and in the warm up phase, fuel droplets impact on the piston surface and may result in massive wall film formation. Examples for both cases are given in the endoscope images of Fig. 3.36.

3.3.4.1 Endoscopic Combustion Chamber Inspection

Endoscopes are optical instruments for image transfer by means of long, rod-shaped lenses or by coherent optical fiber bundles. In an engine, the endoscope accesses the combustion chamber via optical windows. Such combustion chamber windows are designed to withstand the pressure and temperature conditions of usual engine operating modes. With adequate design and material selection, window applications include full load operation in both Diesel and gasoline engines. Examples for an endoscope, window design and assembly in an engine are given in Fig. 3.37.

3.3.4.2 Combustion Chamber Imaging via Piston Windows in Research Engines

In combustion engine research and pre-series development there is growing need for combustion chamber inspection via transparent piston configurations. A large piston window allows simultaneous inspection of all fuel sprays as well as observation of the entire flame field as accessed by the window. Spray to spray and flame to flame uniformity are especially of interest in Diesel combustion studies. Figure 3.38 shows examples for Diesel flames, Lindstrom private communication. As optical access in such engines is provided by large windows and mirrors, such configurations are well suited to accommodate various optical and laser optical analysis techniques.

Radiation of a Diesel flame essentially results from thermal radiation of the soot particles formed under heterogeneous diffusion combustion of the fuel sprays. Spectrum and intensity of this radiating cloud of soot particles are used to evaluate soot particle concentration as well as their temperature. In a most simplified method, this evaluation is achieved with measurement of flame radiation intensity

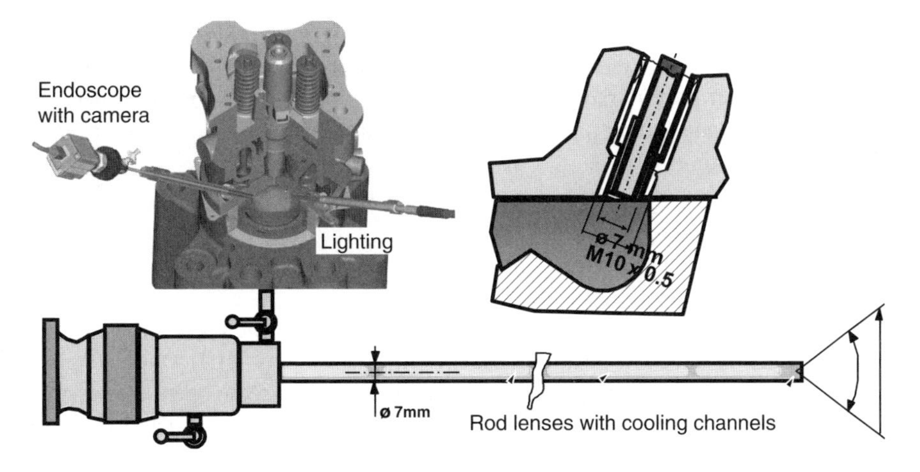

Fig. 3.37 Endoscope and endoscope insertion in the cylinder head of a Diesel engine

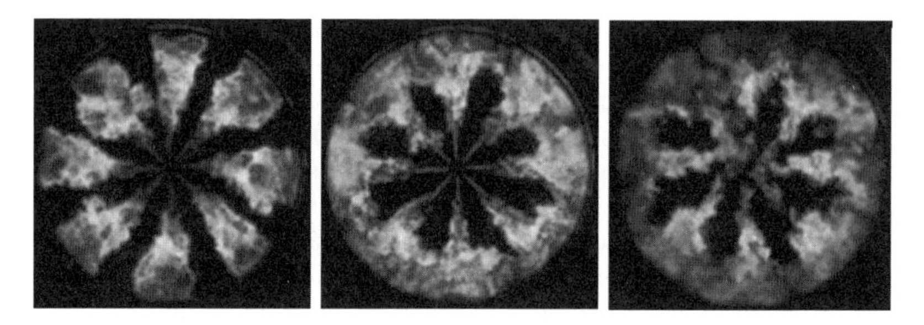

Fig. 3.38 Flame distribution seen through the piston of an optical research engine, Lindstrom private communication

within two narrow band spectral regimes and comparison of the spectral signals with Planck's radiation formula. This "two color" method was applied to gain the results presented in Fig. 3.40.

3.3.4.3 Flame Image Evaluation Examples

Flame position and propagation versus crank angle are the primary information provided by Diesel flame imaging. Evaluation of a Diesel flame's spectral brightness yields flame temperature and concentration of the soot particles contributing the flame radiation. Such spectral image evaluation is achieved with the two color flame analysis method, Gstrein (1987).

Diesel Flame Spectrum

Diesel flames are so called diffusion flames with remarkably high radiation intensity. Their spectral content is dominated by the thermal radiation of the soot particles present in the Diesel flame. A flame spectrum given in Fig. 3.39 shows this broad band thermal radiation together with the low intensity narrow band UV emission lines from OH radiation. In presence of such strong thermal diffusion flame radiation, any further narrow band molecular radiation is scarcely noticeable, Kuwahara and Ando (2000).

Soot Formation: Soot Oxidation Variants Analysis

Figure 3.40 gives a comparison of flame evaluation results which were derived from flame images recorded with an endoscope in a heavy duty (HD) Diesel engine. Combustion modes were under the influence of EGR and needle opening pressure, both of which have influence on engine out soot emission levels. The two color flame evaluation procedure is applied to an area of interest as shown in the flame photograph. Total visible soot concentration within this area is given for an entire

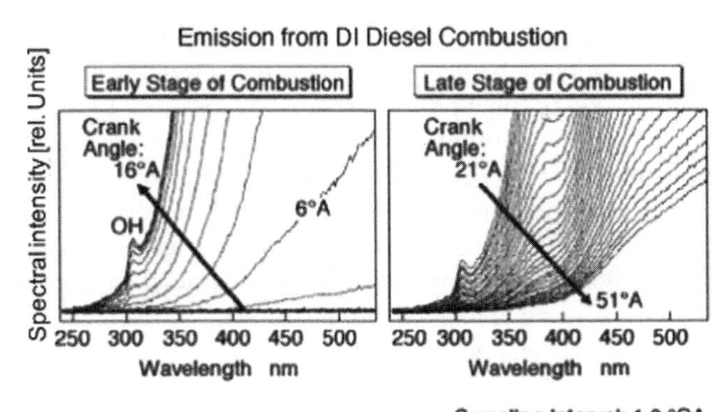

Fig. 3.39 Spectrum of a Diesel flame from early to late combustion. Narrow band OH molecular radiation and intensiv thermal radiation of soot particles, Kuwahara and Ando (2000)

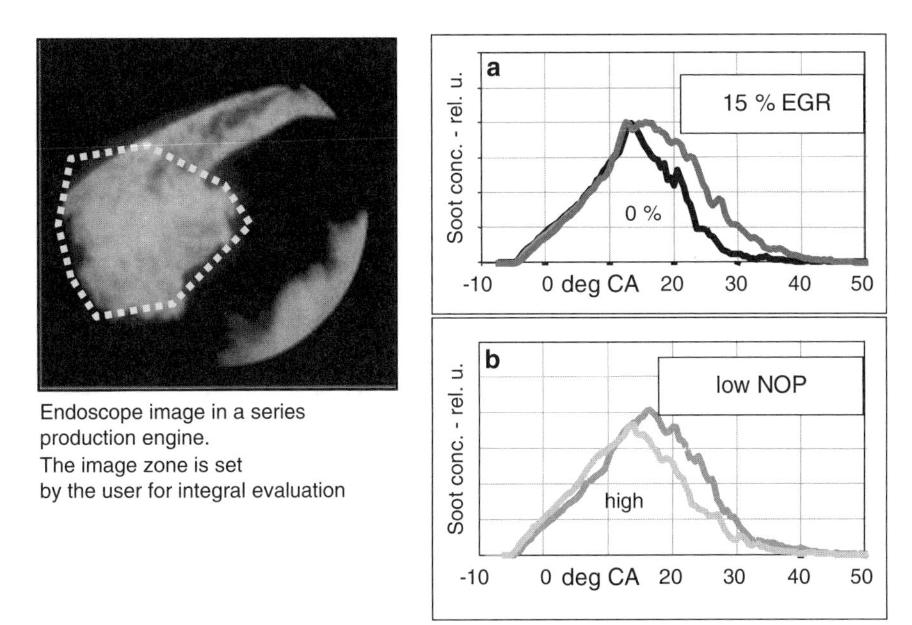

Fig. 3.40 Integral flame evaluation in area of interest: Soot bildup and soot oxidation at EGR variation (**a**) and at variation of needle opening pressure NOP (**b**)

set of crank angle resolved flame images. Consequently, the soot concentration graphs in Fig. 3.40 comprise the input from a few hundred individual flame images taken at consecutive cycles under stationary engine operating conditions.

The results given in the graphs of Fig. 3.40a, b show the steady rise of soot concentration up to a maximum at about the time of end of injection. The following decrease of the soot concentration signals reflects soot oxidation.

This oxidation process reflects the influence of the engine operating variants. With EGR, soot oxidation slows down and results in the well known soot emissions tradeoff. Raising nozzle opening pressure results in faster soot formation, but also in a more effective soot burn-off after end of injection.

In the late expansion stroke, soot particles are either oxidized or they are too cold to maintain sufficient signal intensity for endoscopic imaging. The comparison of soot emission levels with in cylinder soot radiation signals recorded throughout the late expansion stroke shows significant correlation of both signal sets. Such comparisons confirm the suitability of the method relating relative in-cylinder soot signals with absolute engine out emission levels.

Soot Evaluation with the Two Color Method

Diesel engine emissions development requires ever growing attention to transient operation. Engine start and gear change and tip in transients are main contributions to engine out soot and NOx emissions. This defines the task for identification of cylinders and cycles which are predominant in contributing to overall emissions.

Figure 3.41 gives an example for a tip in test with conventional engine out soot opacity measurements. The graphs show a set of six repeated tip in tests (2,000 rpm, IMEP step from 3 to 19 bar) with a typical opacity maximum before the engine gains thermal equilibrium (A). One out of four cylinders was equipped for integral two color flame measurement. Signal evaluation provides a relative soot number for each cycles ["V-soot" signal, Winklhofer et al. (2006)]. At constant load, the signal traces in Fig. 3.41b show noticeably higher fluctuations than the opacity signal, at the tip in it immediately responds to the load change event and it shows peak V-soot numbers slightly ahead of the steady state high load mode.

One out of four cylinders was equipped for integral two color flame measurement. Signal evaluation provides a relative soot number for each cycle ["V-soot" signal, Winklhofer et al. (2006)]. At constant load, the signal traces in Fig. 3.41b show noticeably higher fluctuations than the opacity signal, at the tip in it immediately responds to the load change event and it shows peak V-soot numbers slightly ahead of the steady state high load mode.

Such comparison first of all shows the different features of engine out opacity and in-cylinder flame measurements. In the opacity meter, the sample gas arrives at the sensor position after passing through the exhaust manifold and the instrument tubing and it comprises a gas mix of all cylinders. The opacity instrument time resolution is 0.1 s.

The in-cylinder flame measurement avoids these time delay and cylinder mixing effects. The V-soot (or V-NOx) data are available for each individual cycle of the cylinder of interest. Crank angle resolved flame temperature and soot concentration data are accessible for refined analysis and flame interpretation.

Deriving recommendations for engine development or actuator calibration from such signal traces requires separation of general trends from sporadic combustion events. This is achieved with repetitive test runs and signal evaluation for average trends and sporadic events. The mean cycle test diagram in Fig. 3.41c shows

- The immediate response of enhanced soot formation at tip in
- The peaking of soot formation in the high load cycles before the engine has achieved its high load thermal equilibrium

3.3.5 SI Engines Optical Diagnostics

In a gasoline engine the ideal situation shows fully vaporized fuel which, at stoichiometric concentration, is homogenously mixed with air and residual gas. The spark plug discharge ignites this mixture, it forms a flame kernel that progressively consumes the unburned charge under turbulent diffusion of the flame front

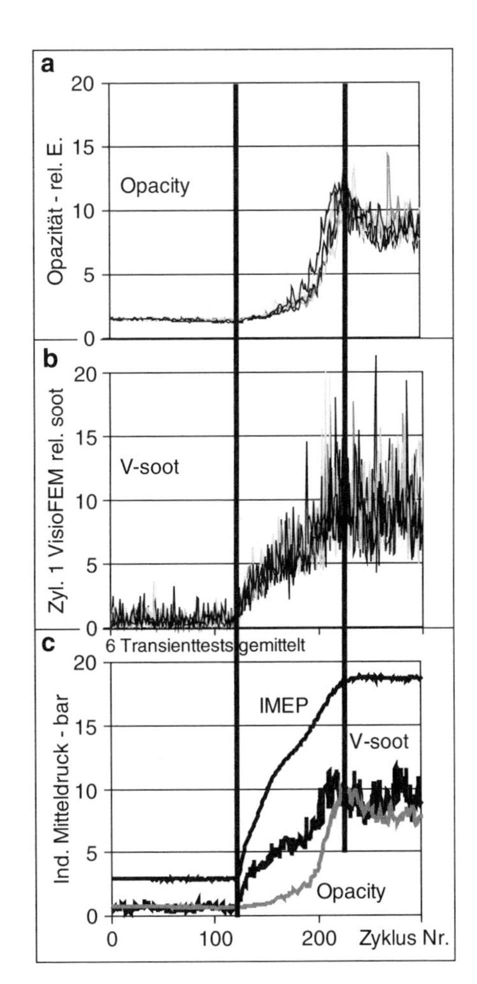

Fig. 3.41 Soot formation in a tip-in test. The two color V-soot signal shows cylinder and cycle specific soot peaks. (**a** and **b**) test repetitions, (**c**) test ensemble average

constituents. Mixture homogenization ensures combustion to be free of soot formation. Stoichiometric conditions yield burned gas temperature near the adiabatic bulk gas limit of around 2,400 K. Such gas conditions, together with three way catalysis are prerequisite for a low emissions engine.

Combustion system development must ensure that mixture formation procedures finally yield such fully vaporized fuel which is homogenously mixed with air or which provides well defined stratified mixture to burn under highest possible thermodynamic efficiency.

Optical analysis techniques support such development work to achieve targets for:

1. Emissions: mixture formation for soot free combustion in port injection and direct injection systems
2. Stability: flame kernel formation, flame front propagation under influence of turbulent in-cylinder flow
3. Efficiency: knock initiation site analysis
4. Irregular combustion: engines with high power density combustion systems are under risk of uncontrolled self-ignition events. Such irregular ignition and combustion is identified with optical diagnostic procedures
5. Non-contact surface temperature measurement: the use of infrared sensitive signal converters extends optical sensing techniques for thermal radiation evaluation, Winklhofer et al. (2009)

Emissions: evaluate mixture formation quality by means of flame radiation analysis

3.3.5.1 Turbulent Flame Propagation in Premixed Charge

In ideally premixed charge, flame propagation is driven by the diffusion of the flame front constituents into the unburned charge. This flame front propagation is under the influence of turbulent and large scale gas flow. Activation energy to support the combustion process is provided by the heat exchange between combustion products and unburned gas, thus the chain reaction to support ongoing combustion is maintained until all available charge is converted into products.

The main part of reaction enthalpy is released by oxidation processes within the flame front. This results in the build up of cylinder pressure. In simultaneous events, the radiative recombination of molecules generated by the oxidation process contributes to flame luminosity.

Consequently, in such premixed flames, heat release due to combustion and flame radiation is pseudo simultaneous events. Comparison of the rate of heat release with flame radiation intensity thus yields information on how well the mixing process achieved premixed charge, see Fig. 3.42a for a typical signal example with good mixture preparation.

The primary mechanism for flame propagation is molecular diffusion. In homogeneous charge engines, this results in isotropic flame propagation. Such isotropic

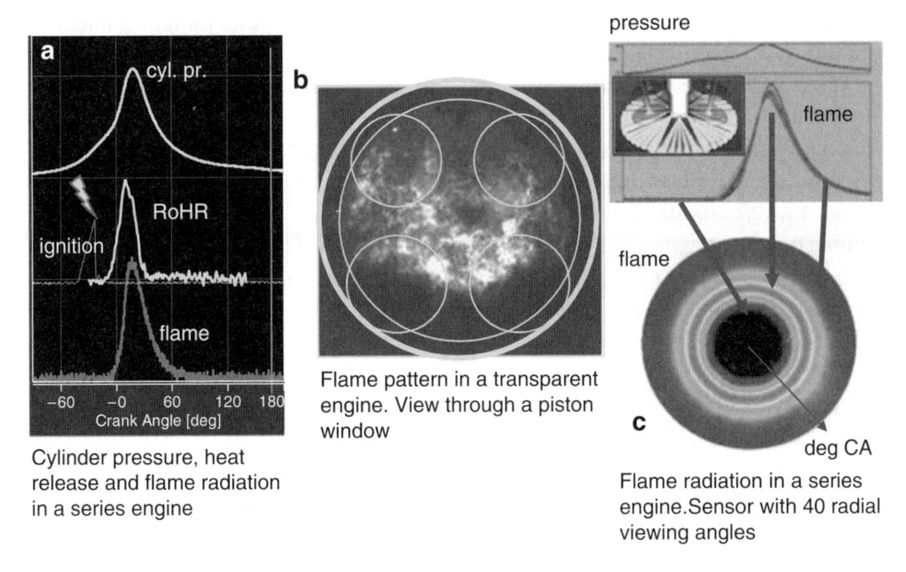

Fig. 3.42 Premixed combustion (**a**): simultaneous heat release and flame radiation. (**b**): flame front under influence of turbulent gas motion. (**c**): isotropic signature of flame pattern signals at premixed combustion

propagation is enhanced by the motion of the fluid. Local flow vortices result in turbulent flame front distortions and any macroscopic flow in the overall convective motion of the expanding flame. The flame photograph in Fig. 3.42b shows such turbulent flame front structure as well as the drift of the flame towards the combustion chamber exhaust side.

Measurement of flame radiation within the narrow apertures of multichannel fiber optic sensors provides a simple and informative technique to check for premixed combustion. An isotropic premixed flame yields highly similar flame radiation signals along individual sensor apertures. The graphics in Fig. 3.42c show the similarity of the multichannel signals. Flame consistency is evident in the polar plot of the flame radiation signals. Highest flame radiation intensity is coincident with the peak rate of heat release.

3.3.5.2 Flame Radiation in Heterogeneous Mixture

Even with incomplete fuel evaporation, combustion essentially starts with a premixed flame kernel which expands into the fuel air mixture near the spark plug. As such premixed flame gets in contact with fuel droplets or with fuel wall films, these rich mixture regimes start to burn under diffusion (sooting) conditions. The flame photograph of Fig. 3.43b shows such heterogeneous flame patterns with fuel droplet combustion (sooting flame).

The resultant flame radiation time traces show the early coincidence of premixed flame radiation and heat release. As fuel rich regimes ignite and burn under sooting conditions, their radiation surpasses the premixed flame in both radiation intensity and duration, see Fig. 3.43a.

The rich mixture is mostly concentrated in specific locations, so diffusion flames are highly localized. This becomes evident in flame images, as well as in multichannel flame radiation measurements. The signal patterns in Fig. 3.43c give evidence of wall film combustion throughout the expansion stroke near one intake valve of a PFI engine. Combustion pressure traces again show that such diffusion combustion has a negligible contribution to heat release despite high light intensity .

3.3.5.3 Applications of Flame Analysis Techniques

The examples in Figs. 3.42 and 3.43 show flame images or flame radiation signals to be well suited to identify combustion going on under ideal premixed conditions, or to understand that rich mixture gives rise to diffusion flames with the risk of incomplete combustion. As such diffusion flames are the source for soot formation, identification of such events is helpful to improve mixture formation for low soot combustion. The benefit of identifying soot formation with in-cylinder measurements is seen in the ability of evaluating individual cylinders and cycles contributing to engine out emissions. This is especially useful under emissions relevant engine start and transient operation.

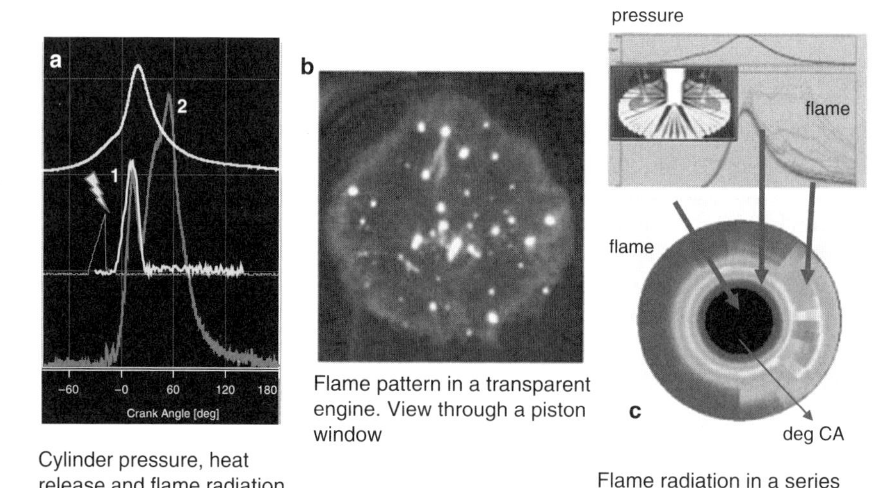

Fig. 3.43 Premixed and diffusion flames (**a**): premixed flame 1: with simultaneous heat release and flame radiation. Diffusion flame 2: high flame radiation intensity at negligible heat release. (**b**): rich mixture (droplets, wall film) is ignited by premixed flame. (**c**): flame pattern signals, diffusion flames result in highly anistropic flame signature

3.3.5.4 Test Examples

Acceleration

Especially in GDI engines, a fast tip-in transient (from part load to full load) can result in pronounced soot emissions as in-cylinder wall temperature is insufficient to evaporate the sudden increase of injected fuel. Calibration of such transient modes requires test iterations to best adjust injection parameters for acceptable soot emissions. Exhaust gas measurements, however, are just capable of evaluating the overall engine emissions behavior. Contributions from individual cylinders or specific cycles are smeared with engine out or tailpipe measurements. Such precise cycle by cycle evaluation for each individual cylinder is achieved with flame radiation measurements.

Combustion signal records for cylinder pressure, rate of heat release and flame radiation intensity are given in Fig. 3.44. Fuel storage effects at a tip-in event become evident in the crank angle resolved flame signals before the combustion chamber temperature becomes sufficiently high to evaporate all injected fuel for premixed combustion.

Adjustment of fuel injection parameters such as injection timing and pressure yields significant reduction of such fuel storage effects, see Fig. 3.44b.

Engine Start

Starting an engine requires injection of excess fuel to compensate for partial evaporation in the low temperature engine environment. The evaporated fuel fraction is expected to result in near stoichiometric, ignitable mixture, whereas the excess fuel forms wall films and just marginally contributes to combustion. Such conditions result in diffusion flames and soot emissions. Suitable selection of fuel injection parameters must avoid misfire cycles and minimize the time required to achieve low emissions premixed combustion. Such mixture and flame conditions are well suited for analysis with the flame radiation measurement techniques of Figs. 3.42–3.44.

In case of misfire cycles, there is the need to understand the root causes in order to compensate fuel injection for either lean or rich misfire events.

The examples of Fig. 3.45 have been recorded in one cylinder of a four-cylinder engine. Pressure and rate of heat release signals show three misfire cycles before the onset of successful ignition combustion. The flame signals show the absence of any flame activity in cycle nr. 1. The following cycle shows flame activity, however without any noticeable heat release. In cycle 3, there is flame activity from the early compression stroke onwards. Here, obviously, combustion residuals from cycle 2 have ignited the fresh charge. These misfire and irregular combustion cycles are then followed by regular combustion. With this signal sequence for cylinder pressure, heat release and flame radiation, it is concluded that overfuelling was the primary cause for the misfire and irregular combustion events.

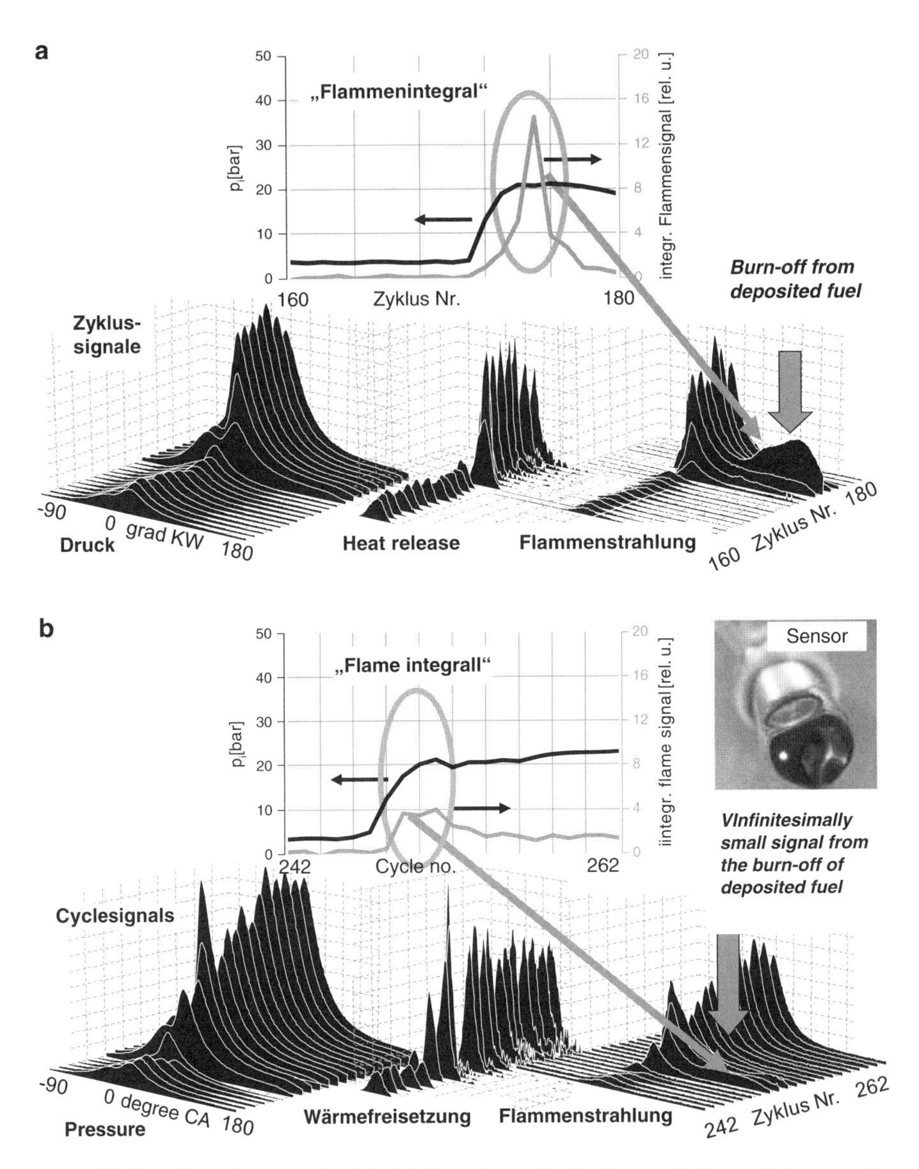

Fig. 3.44 Soot formation at tip-in: (**a**): rich flame is seen by its flame radiation peak in expansion stroke. (**b**): improved fuel evaporation avoids high diffusion flame peaks

Verification of Flame Signals for Soot Evaluation

The above examples have shown that flame intensity signals recorded with a broadband (200–1,100 nm) photodiode provide a simple crank angle trace to identify premixed as well as diffusion flame combustion. Signal comparisons,

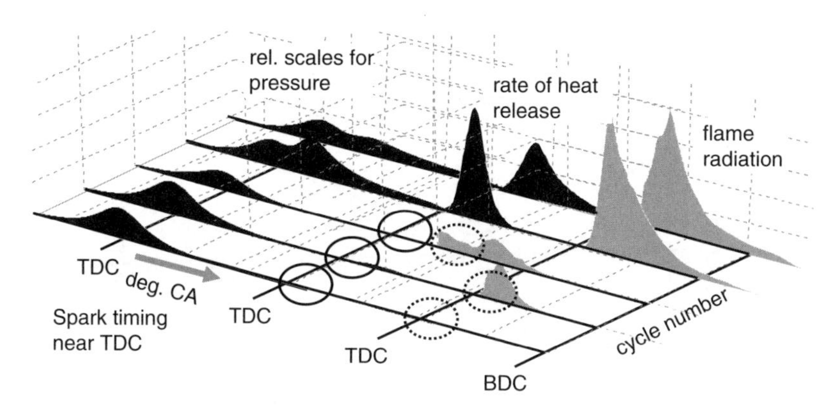

Fig. 3.45 Misfire cycles at engine start. Heat release together with flame radiation signals enable identification of root causes

furthermore, show the large intensity contribution from diffusion combustion. Thus, a simple integration of the flame intensity signal for an entire engine cycle yields a cycle specific number to describe subsequent combustion events. Under stationary operating conditions, this cycle integral flame signal is compared with engine out soot emissions. Emissions trends for particulate mass as well as particulate number counts in Fig. 3.46 confirm the usefulness of this simple cycle integral number to identify combustion cycles with significant contributions to engine out soot emissions. It must be noted that this cycle integral number is to serve as an identifier of combustion cycles with noticeable diffusion flames, it is not capable of substituting any engine out particulate measurement technique.

Local Diffusion Flame Information

Fiber optic spark plugs with sensor channels arranged for circumferential flame detection have already been shown in Figs. 3.42 and 3.43 for the evaluation of symmetric or non-symmetric flame patterns in a PFI engine test. Flame evaluation examples for a GDI engine are given in Fig. 3.47. Injector operating conditions yield specific distribution patterns of resultant diffusion flames.

Combustion Stability: Flow Field, EGR, Lambda

In engine development, combustion stability is defined by the coefficient of variation of $IMEP_H$ (CoV_{IMEPH}) with an evaluation of bulk combustion duration (10–90%) and flame formation period (spark -10%)with crank angle timing of characteristic integral heat release percentages (T10%, T50%, T90%). In many cases, the primary reason for insufficient stability is found in an elongated ignition and flame kernel formation phase. As this flame formation phase is susceptible to

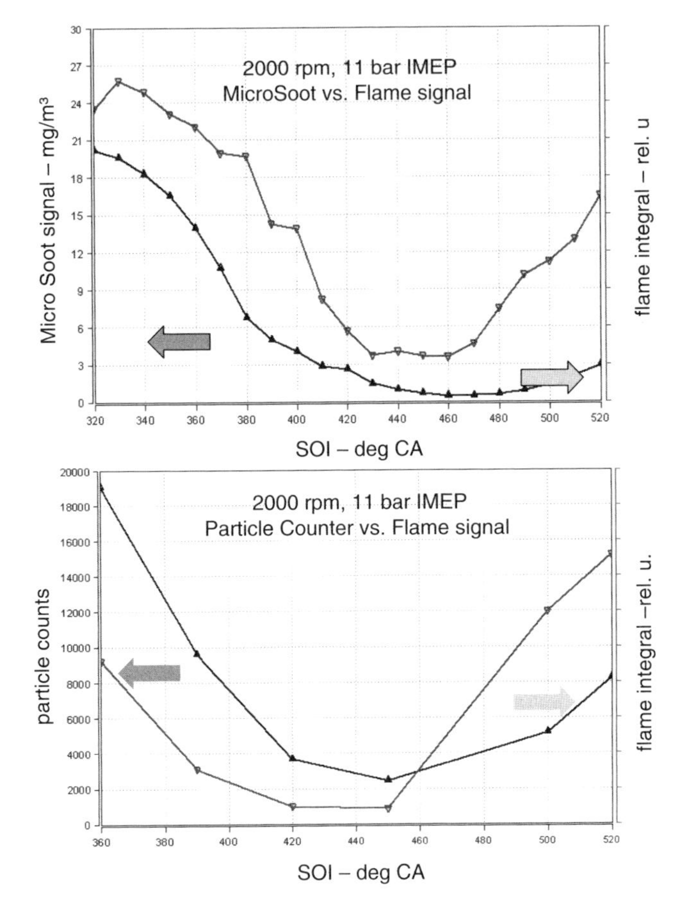

Fig. 3.46 Trend comparison: emissions data at stationary engine operation (particulate mass and number count) compared with cycle integral flame radiation signals

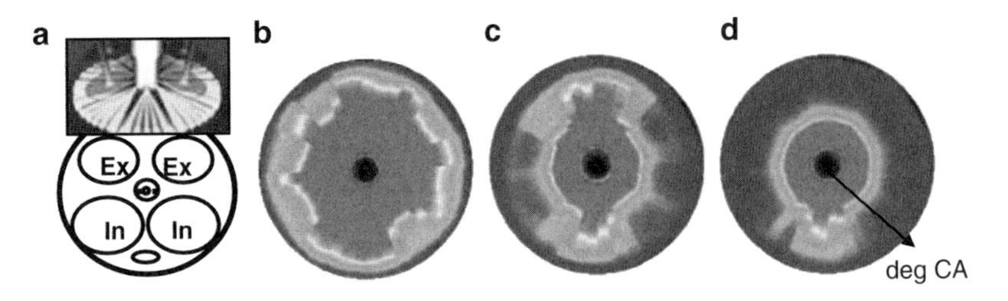

Fig. 3.47 Flame pattern signals in a GDI engine (**a**): sensor configuration, (**b**): injection early, sooting combustion of piston wall film, (**c**): late injection results in diffusion flame near liner, (**d**): best calibration with residual diffusion flames underneath injector

variations of local flow field, and mixture and residual gas concentrations, there is need for measurement techniques suited to identify any of these parameters. Methods and sensors suited for such measurement in either standard or in optical research engines are given in Table 3.9.

Table 3.9 Measurement tasks and methods to evaluate combustion fluctuation root causes

Measurement task	Method	Sensor	Reference
Lambda	HC, narrow band absorption	Spark plug with absorption path, spark emission spectroscopy	Hall and Matthews (2002), Berg et al. (2006), Fansler et al. (2002)
Flow field, turbulence	LDA	Spark plug with LDA optics	Ikeda et al. (2000)
EGR	CO2 absorption	Spark plug with absorption path	Berg et al. (2006)
Flame propagation	Flame radiation, light barriers	Fiber optic spark plug sensor, ion probe head gasket	Witze et al. (1997), Winklhofer and Salzinger (2004)
Heat release	Cylinder pressure	Pressure sensor	
Flame kernel	Flame photography	Transparent engine endoscope	

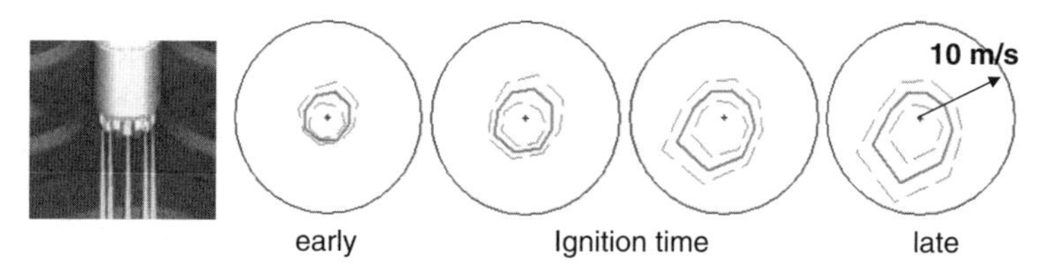

Fig. 3.48 Flame kernel propagation, cycle average (n = 100) and standard deviation at spark advance variation

An example for flame kernel propagation measurements is given in Fig. 3.48. The polar diagrams show the radial flame velocity components as the flame kernel moves through the individual channel apertures of the fiber optic spark plug. As the flame kernel expands under the influence of the local flow and gas concentration, variation of this environment is reflected in the flame kernel velocity components. At constant spark timing, the flame kernel fluctuations are shown by their RMS values. With the variation of spark timing, the polar components give evidence for the flame kernel's response to flow and turbulence.

Combustion Efficiency: Knock Center Detection

An engine's compression ratio (CR) is selected to allow most efficient combustion for a given fuel quality under myriad speeds and loads. This results in full load operation to compromise with spark timing in order to limit knocking pressure amplitudes. As such spark advance limitation has influence on combustion efficiency, power density, combustion noise, engine durability and driveability, there is considerable effort put into the exploitation of an engine's knock limit. Optical flame analysis techniques support these development efforts with detection of knock center distribution and a root cause analysis of spark advance limitations.

Knocking combustion is the result of end gas auto-ignition. After regular ignition, the flame requires a few 10's of deg CA to propagate through the combustion chamber and consume the fuel air mixture. Throughout this combustion phase, the temperature of the unburned gas rises in response to the pressure rise from compression and combustion. This isentropic heating can result in thermal auto-ignition of the unburned mixture. Endgas self-ignition can be avoided if flame propagation is sufficiently fast, and if endgas time-temperature history is limited. Combustion system development must provide measures to enhance flame propagation into critical areas and it must avoid overheating of endgas regimes. Such development efforts are supported with measurement of flame propagation and with identification of self-ignition centers.

Measurement methods for the detection of auto-ignition sites must fulfill requirements as listed in Table 3.10. This is essentially met with techniques which detect in-cylinder gas signals and provide location of the knock signal origin (Mazoyer et al. (2003), Wytrykus, Düsterwald (2001), Philipp et al. (1995), Philipp et al. (2001)). The spontaneous nature of a self-ignition event and the high propagation velocity must be accounted for with adequate signal recording techniques.

For practical measurement applications, multi channel optical spark plug sensors have found wide acceptance for knock center measurement in passenger car engines. They are replacing standard spark plugs without any further need for special engine adaptation. A signal example together with the basics for identifying knock center location is given in Fig. 3.49.

Auto-ignition occurs in endgas areas. Figure 3.50 shows how retarded flame propagation allows formation of endgas areas which eventually give rise to auto-ignition whenever thermochemical reactions have sufficient time to trigger exothermal chemical kinetics.

Table 3.10 Topics and measurement demands for knock center determination

Topic	Feature
Engine operation	Spark advance at knock limit
Knock event at which time?	Spontaneously in a time window close after pressure peak
Gas response to self-ignition	– Pressure wave is started by spontaneous self-ignition – Pressure wave is linked to gas density wave – Density wave is linked to brightness wave in luminous burned gas
Sensor signal pressure	Local pressure acting on surface of sensor
Sensor signal light	Brightness signal recorded within (volume) aperture field of sensor
Signal duration	PRESSURE (density, luminosity) wave transit time across combustion chamber: 50–200 μs
Velocity of knocking pressure wave	Sound velocity in hot compressed gas: up to 1,000 m/s at full load
Required measurement sensitivity	Knock center location at knock amplitudes larger than 0.5 bar
Optical measurement system task	Find origin of propagating brightness wave
Alternative method pressure	Find origin of propagating pressure wave
Acoustic oscillation in combustion chamber	Form after reflection of primary wave front

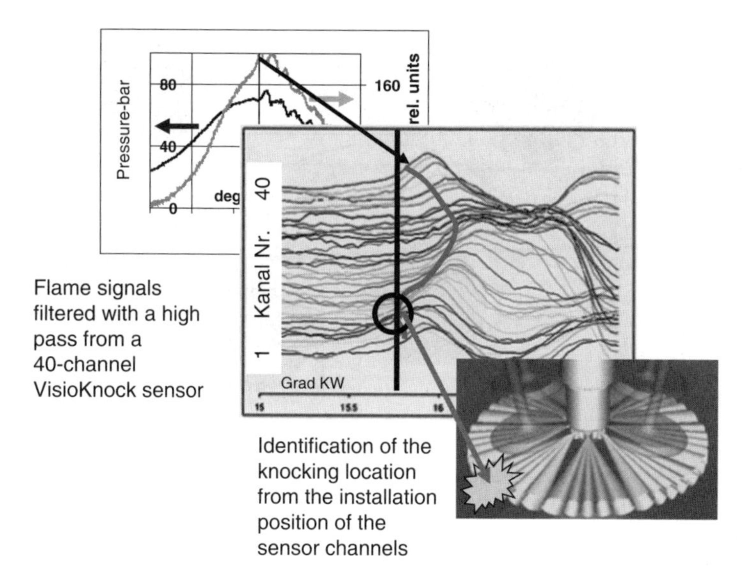

Fig. 3.49 Knock center measurement in a standard multicylinder engine with a VisioKnock spark plug sensor

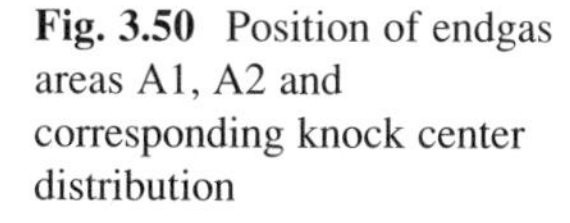

Fig. 3.50 Position of endgas areas A1, A2 and corresponding knock center distribution

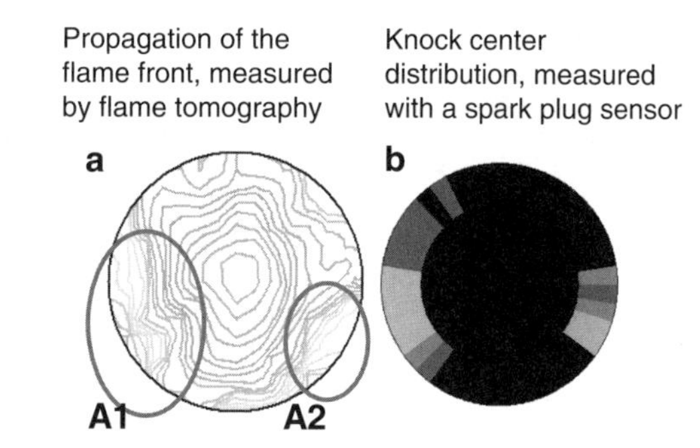

The steps and benefits achieved in a systematic combustion development project are summarized in the engine performance plot of Fig. 3.51. The initial engine configuration had a full load limit with significant knock center agglomeration on the exhaust side. Modification of in-cylinder air motion provided enhanced flame propagation towards the exhaust side and resulted in effective endgas burn-off in those critical areas. This enhanced flame propagation allowed a significant advance of spark timing. The resultant full load line provided BMEP improvements of up to 1 bar.

3.3.5.5 In-cylinder Self-Oscillation

Knocking combustion, first of all, is recognized by its ringing noise. This ringing noise signature is a result of acoustic resonance excited by the auto-ignition event and enhanced by pressure wave reflection at the combustion chamber walls.

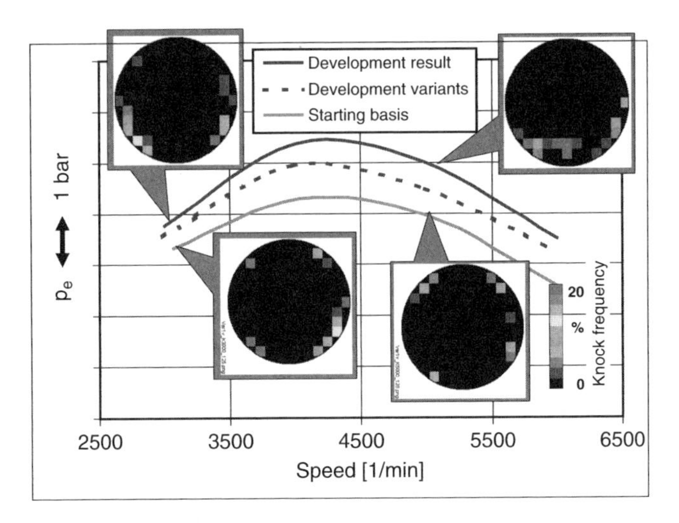

Fig. 3.51 Engine development for knock limit improvement. Basis: knock centers at exhaust side limit spark advance. Final variant: knock centers at intake side enable significant BMEP improvement

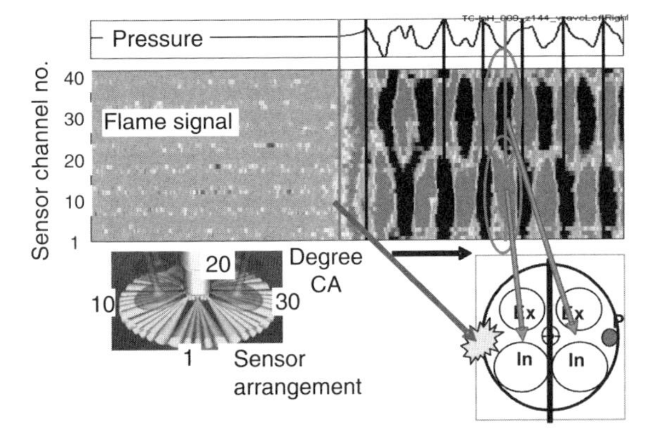

Fig. 3.52 High pass filtered signals of cylinder pressure and flame radiation. The flame signals show the pressure (= brightness) wave with ongoing reflections and excitation of a resonant frequency. P: position of pressure sensor

As optical sensor arrays provide measurement of brightness signals in the entire combustion chamber, the signal patterns show the acoustic field of the in-cylinder gas oscillations. The signal example in Fig. 3.52 shows the modulation of the gas brightness field as started by the self-ignition event and enhanced and filtered by ongoing reflection of the oscillating gas.

Irregular Combustion

Increasing the efficiency and power density of gasoline engines can result in operating conditions which come into the risk of thermochemical activation of uncontrolled pre-ignition. Ignition time and location are no more under the control of engine electronics. In case of pre-ignition at full load fuelling, such "irregular ignition" can result in pressure amplitudes that are large enough to cause mechanical damage.

Reducing this risk requires measurement techniques which are capable of identifying the locations of such irregular ignition and eventually to understand and avoid their root causes. Table 3.11 gives an overview of potential causes and engine conditions which bear the risk to transition into such undesirable behavior.

The task for combustion engineers is to detect cycles and cycle sequences with irregular combustion events and to use flame sensor arrays which enable localization of irregular ignition origins. Sensors and instrument configurations capable of meeting these requirements are shown in Fig. 3.53 and described in more detail in Winklhofer et al. (2005). Figure 3.54 gives an example for signal traces recorded in an irregular combustion cycle and evaluation of the self-ignition center.

Non-contact Temperature Measurement

Optical sensors and endoscopes are as capable of transmitting near infrared thermal radiation. Thus, together with IR sensitive photodiodes or cameras the optical systems can be extended for measurement of in-cylinder or exhaust system components' temperatures. Calibration procedures account for the geometric sensor to object arrangement and provide measurement accuracy of around 10 K in normal

Table 3.11 Topics and metrological requirements for irregular ignition and combustion

Potential root cause	Engine operating mode	Initiating irregular ignition	Engine development action
Hot spot	High load operation	Load increment	Local cooling
Glowing deposits	Part load – full load sequences	Spontaneously after deposits buildup, after normal knocking combustion	Avoid fuel and oil deposit formation
Knock to irregular ignition transition	Engine at knock limit	Increased heat transfer to all surfaces under knocking combustion	Reduce knock limit, improve local cooling
HCCI ignition	Low engine speed, hot engine start	High load limit, engine start calibration, residual gas scavenging	
Residual gas and lambda fluctuations	High load	Following a misfire cycle	Avoid misfire cycles
Regular ignition/ irregular combustion	High load	Regular ignition followed by very fast combustion	Turbulence, mixture composition and temperature

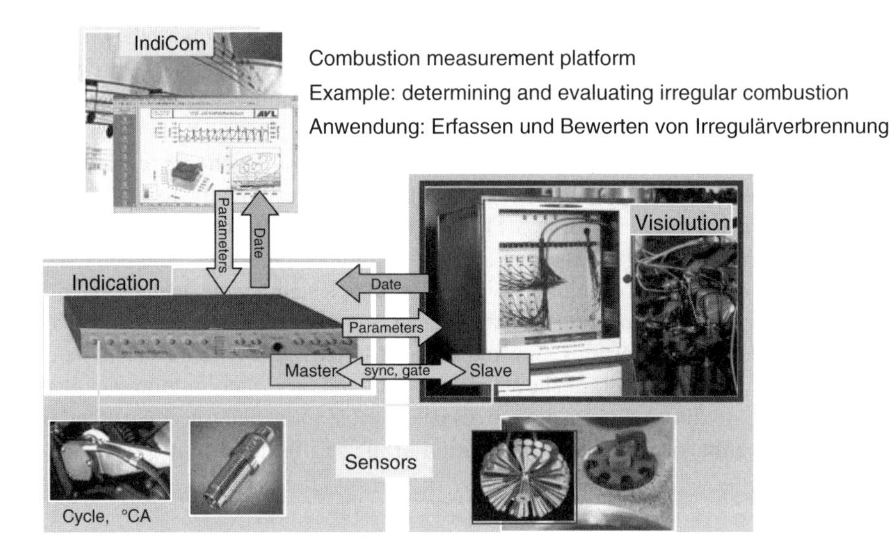

Fig. 3.53 Modules of a "combustion measurement platform" for the recording of spontaneous, irregular combustion events

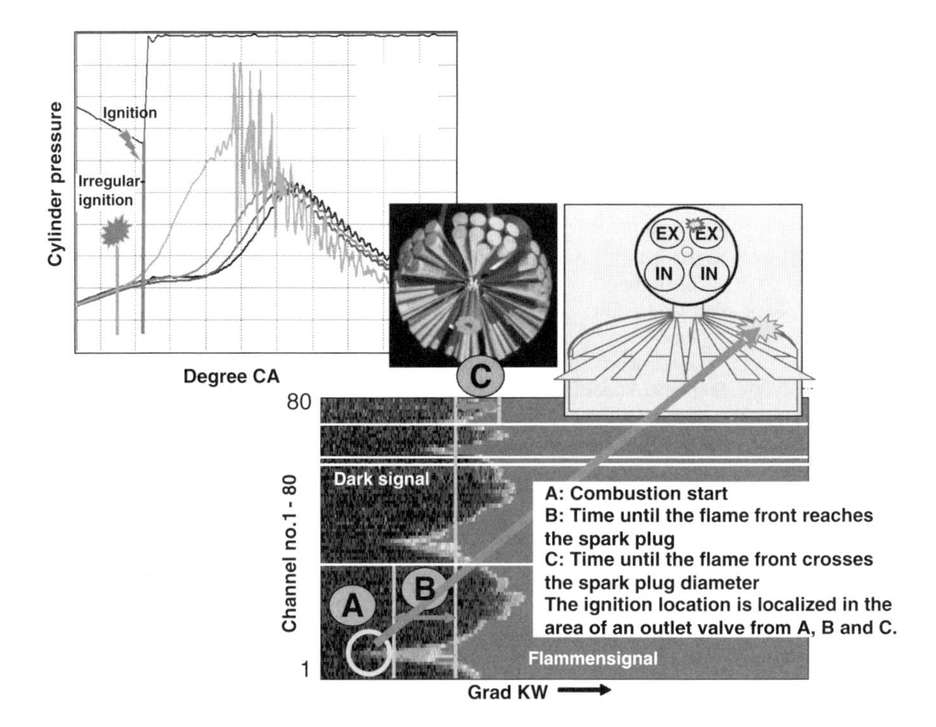

Fig. 3.54 Signal example shows location of irregular ignition

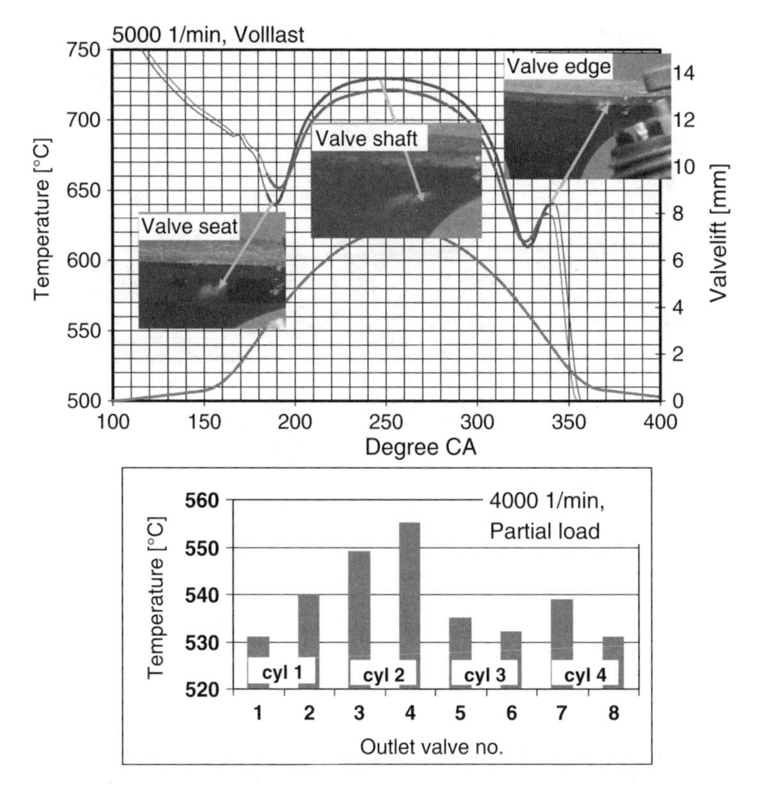

Fig. 3.55 Exhaust valve temperature measurement is accomplished with evaluation of thermal valve radiation

test bed operation, Winklhofer et al. (2009). An example for exhaust valve measurements in a four cylinder engine is given in Fig. 3.55.

3.3.6 Laser Based Measurement Techniques

Table 3.7 has listed a number of laser based measurement techniques to be used for local or planar in-cylinder diagnostics. To enable their application in engines, some of these techniques need large sized optical windows in transparent engine configurations to access the combustion chamber and to enable the use of optical receivers and cameras for signal detection.

In research applications, laser optic effects are exploited for measurement of velocity, concentration and temperature of gases and fuel droplets. Laser absorption techniques are used for soot concentration measurement, Lackner (2008).

An example supporting development of mixture formation processes in direct injection gasoline engines is given in Fig. 3.56. Laser induced visualization of fuel

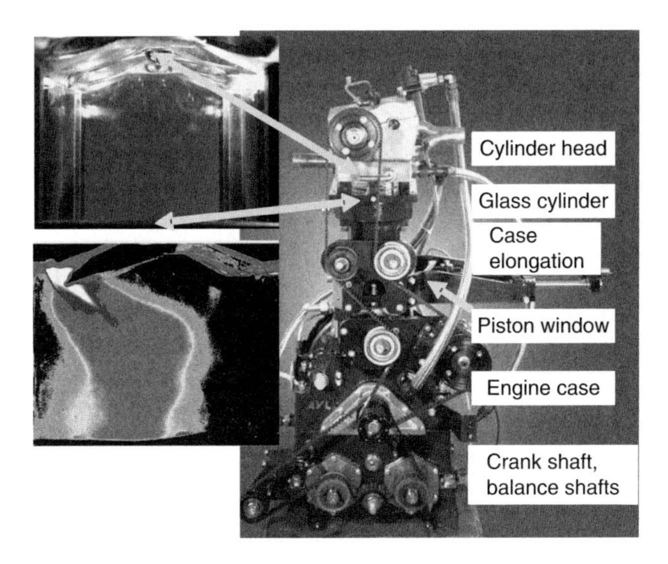

Fig. 3.56 Transparent single cylinder engine with large size windows (glass cylinder and window in piston) provides access for laser diagnostic techniques. Example shows fuel mixture cloud in a GDI engine. Fuel droplets and vapor are visualized by means of laser induced fluorescence (LIF)

spray and vapor distribution enables selection of injectors and their operating modes best suited to meet injection and mixing requirements for premixed combustion. As such imaging is accomplished under real engine flow and geometry conditions, it has become part of industrial engine development procedures, Fuchs et al. (2009).

3.3.7 Optical Combustion Diagnostics: Status and Forecast

Sensor access into the combustion chamber together with straightforward signals interpretation are considered to be the key elements for a successful application of optical diagnostic techniques in IC engine combustion system development. Sensors, signal recorders and results interpretation must interface with the engine test environment and complement conventional analysis techniques while providing otherwise unattainable information. Specific analysis tasks result from requirements to provide relevant insight and decisions to improve mixture formation, combustion as well as exhaust aftertreatment processes.

The features of any diagnostic system must enable standardization of routine analysis procedures. Expanding applications and meeting yet undefined requirements must be provided for with adequate interfaces. This is best accomplished with modularity of sensors, signal converters, recorders and memory. Such modularity is made available within the IC engine "combustion measurement platform".

On this platform, specific measurement tasks are fulfilled with dedicated modules. The platform itself

- Enables synchronized data recording of modular components
- Handles trigger signals between modules
- Provides master-slave signal recording logics
- Provides the user interface to access local data and results

An example of such modular system configuration for the measurement and analysis of irregular combustion events has been shown in Fig. 3.53.

The tasks required from practical combustion diagnostics start with standard thermodynamic analysis, they comprise emissions formation issues related to fuel injection, flame behavior as well as functionality of exhaust aftertreatment systems. Analysis of such combustion and emissions related tasks is further extended into the fields of durability and robustness of combustion related components as fuel efficient engines march the closer exploitation of theoretical engine operation limits while meeting ever more stringent emissions regulations.

References

Berg T, Beushausen V, Thiele O, Voges H (2006) Fiber optics spark plug sensor for the optimization of engine combustion processes. MTZ Motortechnische Zeitschrift

Bertola A et al (2008) Neue Möglichkeiten der Niederdruckindizierung mit piezoresistiven Hochtemperatur-Absolutdrucksensoren. 8. Internationales Symposium für Verbrennungsdiagnostik, Baden-Baden

Davis RS, Patterson GJ (2006) Cylinder pressure data quality checks and procedures to maximize data accuracy. SAE Paper 2006-01-1346

Fansler TD, Stojkovic B, Drake MC, Rosalik ME (2002) Local fuel concentration measurements in internal combustion engines using spark emission spectroscopy. Appl Phys B 75: 577–590

Feßler H (1988) Berechnung des Motorprozesses mit Einpassung wichtiger Parameter. Dissertation, Technische Universität Graz

Frommelt A, Loisch R, Binder S (2008) 15 Jahre Fortschritt in der Messtechnik – Was heisst heute ‚richtig indizieren'?. 8. Internationales Symposium für Verbrennungsdiagnostik, Baden-Baden

Fuchs H, Hopfner W, Kapus P, Winklhofer E (2009) Methods and criteria for fuel injector integration in boosted gasoline direct injection engines. IMECHE C677

Gatowski JA, Balles EN, Chun KM, Nelson FE, Ekchian JA, Heywood JB (1984) Heat release analysis of engine pressure data. SAE paper 841359

Glaser J (1983) Fehler von Quarzdruckaufnehmern und Probleme bei der Druckindizierung von Verbrennungsmotoren. Dissertation, Technische Universität Graz

Gstrein W (1987) Ein Beitrag zur spektroskopischen Flammentemperaturmessung bei Dieselmotoren. Dissertation Technische Universität Graz

Hall M, Matthews R (2002) Mid – IR fiber optic sensors for internal combustion engines. Internationales Symposium für Verbrennungsdiagnostik, Baden-Baden

Heywood JB (1988) Internal combustion engine fundamentals. McGraw-Hill, New York, xxix + 930 pp

Hohenberg G (1983) Experimentelle Erfassung der Wandwärme von Kolbenmotoren. Habilitationsschrift, Technische Universität Graz

Hohenberg G (1994) Indiziertechnik Historischer Überblick und Entwicklung bis heute. 1. Darmstädter Indiziersymposium

Ikeda Y, Nishihara H, Nakajima T (2000) Spark plug-in fiber LDV for turbulent intensity measurement of practical SI engine. 10th International Symposia on applications of laser techniques to fluid mechanics, July 10–13

Karst D (2000) Finite-Elemente-Simulation der zyklischen Temperaturdrift piezoelektrischer Druckaufnehmer in der Motorenmesstechnik. Dissertation, Technische Universität Graz

Krempl P, Schleinzer G, Wallnöfer W (1997) Gallium phosphate, GaPO4. A new piezoelectric crystal material, Sensors and Actuators A61:361–363

Krieger RB, Borman GL (1966) The computation of apparent heat release for internal combustion engines. ASME 66-WA/DGP-4, 16 pp

Kulite (2009) http://www.kulite.com/techinfo.asp

Kuwahara K, Ando H (2000) Time series spectroscopic analysis of the combustion process in a gasoline direct injection engine. 4th international symposium for combustion diagnostics, pp 130–137, Baden-Baden, May 18–19

Lackner M (ed) (2008) Lasers in chemistry, vol 1. Wiley, Weinheim, ISBN 978-3-527-31997-8

Lancaster DR, Krieger RB, Lienesch JH (1975) Measurement and analysis of engine pressure data. SAE Trans., 84, Paper 750026, pp 155–172

Mazoyer Th, Fayet P, Castagne M, Dumas JP (2003) Development of a multi-sensor head gasket for knock localization. SAE 03P-293

Patterson GJ, Davis RS (2009) Geometric and topological considerations to maximize remotely mounted cylinder pressure transducer data quality. SAE Paper 2009-01-0644

Philipp H, Plimon A, Fernitz G, Hirsch A, Fraidl GK, Winklhofer E (1995) A tomographic camera system for combustion diagnostics in Si-engines. SAE 950681

Philipp H, Hirsch A, Baumgartner M, Fernitz G, Beidl Ch, Piock W, Winklhofer E (2001) Localisation of knock events in direct injection gasoline engines. SAE 2001-01-1199

Pischinger R, Sams Th, Klell M (2002) Thermodynamik der Verbrennungskraftmaschine. 2. Aufl., Springer, Wien – New York

Rassweiler GM, Withrow L (1938) Motion pictures of engine flames correlated with pressure cards. SAE Trans., Vol 38, pp 185–204

Vogt R (1975) Beitrag zur rechnerischen Erfassung der Stickoxidbildung im Dieselmotor. Dissertation, Universität Stuttgart

Wimmer A (2000) Analyse und Simulation des Arbeitsprozesses von Verbrennungsmotoren – Modellbildung und meßtechnische Verifizierung. Habilitationsschrift, Technische Universität Graz

Wimmer A, Glaser J (1996) Welche thermodynamischen Aussagen sind mit Miniaturdruckaufnehmern möglich? Vortrag beim 2. Indiziersymposium in Offenbach/Main

Wimmer A, Glaser J (2002) Druckindizierung an Verbrennungsmotoren. Anwender-Handbuch AVL

Wimmer A, Beran R, Figer G, Glaser J, Prenninger P (2000) Möglichkeiten der genauen Messung von Ladungswechseldruckverläufen. 4. Internationales Symposium für Verbrennungsdiagnostik, Baden-Baden

Winklhofer E, Salzinger R (2004) Zylinderdruck und Flammenleuchten – Leitsignale für die praxisnahe Verbrennungsentwicklung. Entwicklungstendenzen im Automobilbau, ISBN 3-9808512-1-4, 61–70, Zwickau 17/18 Juni 2004

Winklhofer E, Ahmadi-Befrui B, Wiesler B, Cresnoverh G (1992) The influence of injection rate shaping on diesel fuel sprays – an experimental study. Proc. IMechE, Vol 206, pp 173–183

Winklhofer E, Kapus P, Knorz C, Moik J (2005) SImotoren im Hochlasttest. Technische Universität Graz, 10. Tagung. DER ARBEITSPROZESS DES VERBRENNUNGSMOTORS

Winklhofer E, Hirsch A, Leifert Th, Rzehorska M (2006) Kurbelwinkelbezogene optische Flammenmessung in HSDI Dieselmotoren. 7. Internationales Symposium für Verbrennungsdiagnostik, Baden-Baden

Winklhofer E, Hirsch A, Kapus P, Kortschak M, Philipp H (2009) TC GDI engines at very high power density – irregular combustion and thermal risk. SAE 09ICE-0216
Witze PO, Hall MJ, Wallace JS (1997) Fiber-optic instrumented spark plug for measuring early flame development in spark ignition engines. Transactions of the SAE 97, 3.813
Witt A (1999) Analyse der thermodynamischen Verluste eines Ottomotors unter den Randbedingungen variabler Steuerzeiten. Dissertation, Technische Universität Graz
Wlodarczyk M (1999) Long-life fiber-optic pressure sensors for harsh environment applications. 9th Trade Fair and Conference Sensor99
Wytrykus F, Düsterwald R (2001) Improving combustion process by using a high speed UV-sensitive camera. SAE 2001–010917

Chapter 4
Engine Combustion

Sebastian Rakowski, Peter Eckert, and Andreas Witt

4.1 Fuels

SI and diesel engine fuels are each mixtures of several hundred different hydrocarbons of various groups ($C_xH_y[O_Z]$). These components differ with reference to molecular size and structure and as a result have sometimes strongly varying properties. Moreover, their composition diverges and with it the properties of gasoline, diesel and potentially alternative fuels, making a separate treatment necessary. The limit values of the most important material properties of engine fuels adhere to standards (DIN, EN, etc.) in order to guarantee consistent quality and composition as well as reliable engine operation (Table 4.1).

In the following, the structure and function of the most important hydrocarbon compounds in engine fuels will be described. In addition, oxygenic components will be shown. These are especially relevant for ignition and combustion processes.

Aliphatic Hydrocarbons. Besides the *alkanes*, which have no double bond and are the most common group in engine fuels, the aliphatic hydrocarbon group also contains *alkenes* (formerly: olefins) with (at least) one double bond as well as *alkines* (formerly: acetylenes) with a triple bond. The alkane group is further subdivided into the *n-alkanes*, which have a straight chain structure, and *iso-alkanes*, which have a branched chain structure. Because of the intermolecular forces which increase with the length of the chain, boiling temperature, viscosity etc. increase along with the chain length. Alkanes (n- and iso-) represent proportionally the largest group in engine fuels.

Figure 4.1 shows examples of the structural formulae of a few aliphatic hydrocarbons.

Alicyclic and Aromatic Hydrocarbons. The alicyclic hydrocarbons comprise the ring-shaped hydrocarbons. The *cycloalkanes* (formerly: napthenes) consist only of simple bonds and are very similar to the alkanes in their properties.

The characteristic structure of the *aromatics* is described by a delocalized charge cloud in the center of the molecule, which is also their distinguishing feature from the cycloalkanes. The benzene ring represents the basic building block for all aromatic compounds. Aromatics with more than one benzene ring are designated

G.P. Merker et al. (eds.), *Combustion Engines Development*,

DOI 10.1007/978-3-642-14094-5_4,

Table 4.1 Properties of engine fuels in the (DIN) EN standards

	Diesel fuel (DIN) EN590 (2008)	SI engine fuel (DIN) EN228 (2004)	Biodiesel (FAME) (DIN) EN 14214 (2008)
Density in kg/m^3 at 15°C (min./max.)	820/845	720/775	860/900
Viscosity in mm^2/s at 40°C (min./max.)	2.0/4.5	Not defined	3.5/5.0
Octane number/Cetane number (min.)	51	ROZ95/MOZ85	51
Sulfur content in mg/kg (max.)	50 10 (from 1/1/2009)	50 10 (from 1/1/2009)	10
Percentage of **alt. fuels** (max.)	7% FAME	5% ethanol	–

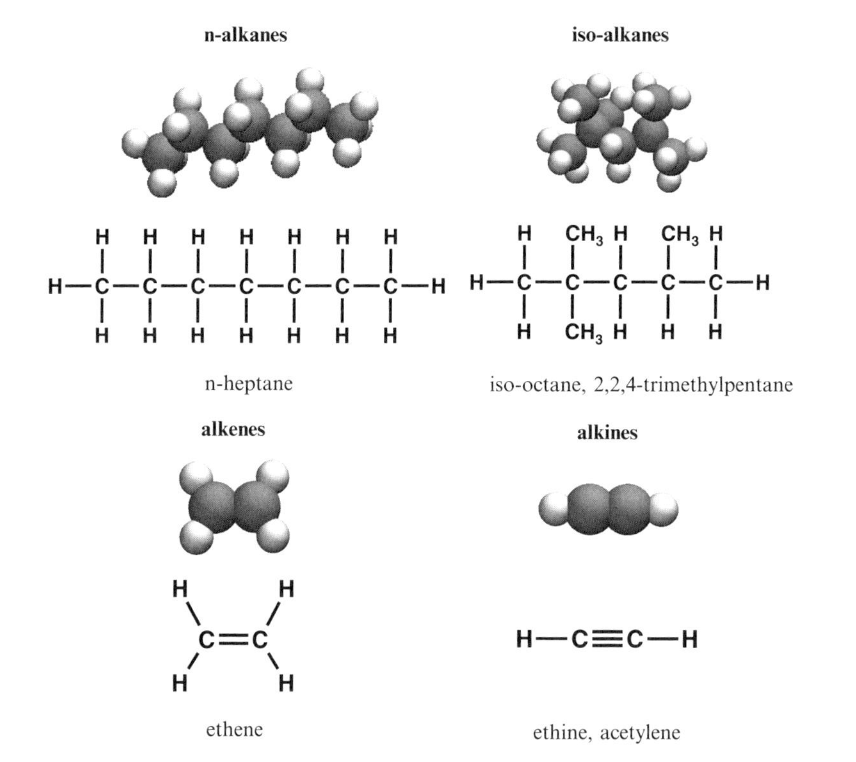

Fig. 4.1 Structural formulae and graphic representation of aliphatic hydrocarbons

correspondingly as di-, tri- or tetraaromatics. By adding alkyl groups (methyl, ethyl, etc.), more complex aromatics are formed.

Besides the alkanes, the aromatics represent the most common fraction in engine fuels and have a central role in the development of soot particles. Special complex structures that build upon polyaromatics and are not counted among the legally

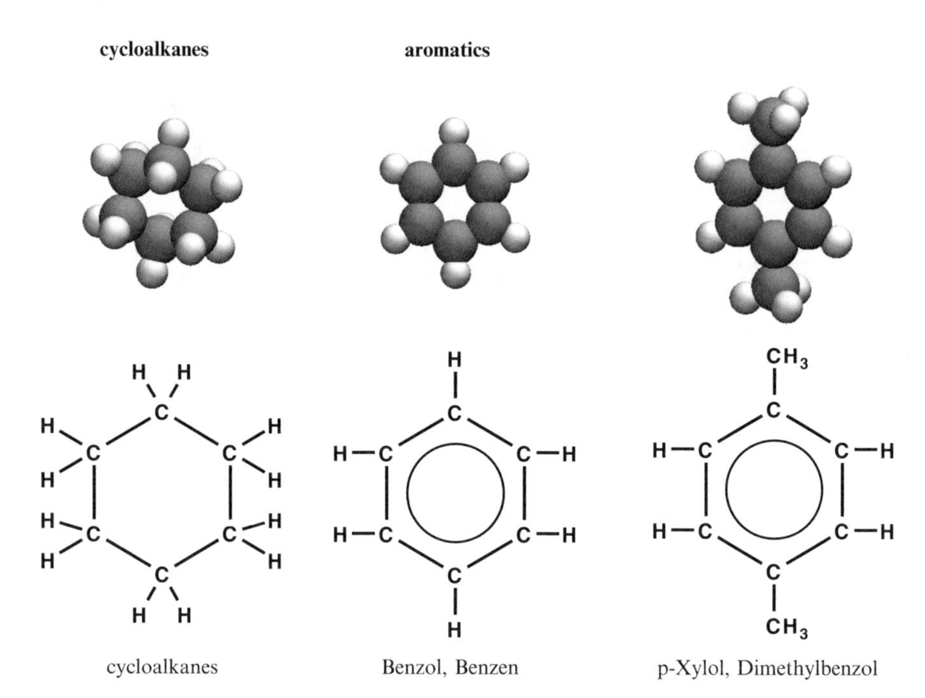

Fig. 4.2 Structural formulae and graphic representation of alicyclic and aromatic hydrocarbons

regimented pollutants are explained in the chapter on pollutants. Figure 4.2 shows examples of cycloalkanes and aromatics.

Oxygenic Hydrocarbons. Oxygenic hydrocarbons are chain-formed compounds subdivided according to their functional groups into alkanoles (alcohols), ketones, ethers, aldehydes, carboxylic acids and (carboxylic acid) esters, see Fig. 4.3.

Alcohols contain a hydroxyl group (R–OH). The simplest (monovalent) alcoholes are methanol (CH_3OH) and ethanol (C_2H_5OH), which are also components of fuels. *Ethers* are alkyl groups bonded with an oxygen bridge (R_1–O–R_2). Although ethers have the same amount of atoms as the corresponding alcohols, their boiling point is much different. *Ketones* are alkyl groups bonded by a carbonyl group (R_1–CO–R_2). *Aldehydes* contain a terminal carbonyl group (aldehyde group, –CHO). Aldehydes derived from the homologous series of alkanes are called alkanals. Aldehydes play an important role as intermediate products in the oxidation of hydrocarbons. The boiling point of alkanals is between that of alkanes and alcohols. *Carboxylic acids* have one or more carboxyl groups (−COOH), which dominate the properties of the molecule; the esters of carboxylic acids are called *carboxylic acid esters* and represent the majority of natural fats. Carboxylic acids and carboxylic acid esters are also added directly into fuels to lower corrosion and to improve their lubricative properties.

Although both SI engine and diesel engine fuels are obtained by refining from the same raw material, crude oil, their individual composition of the various

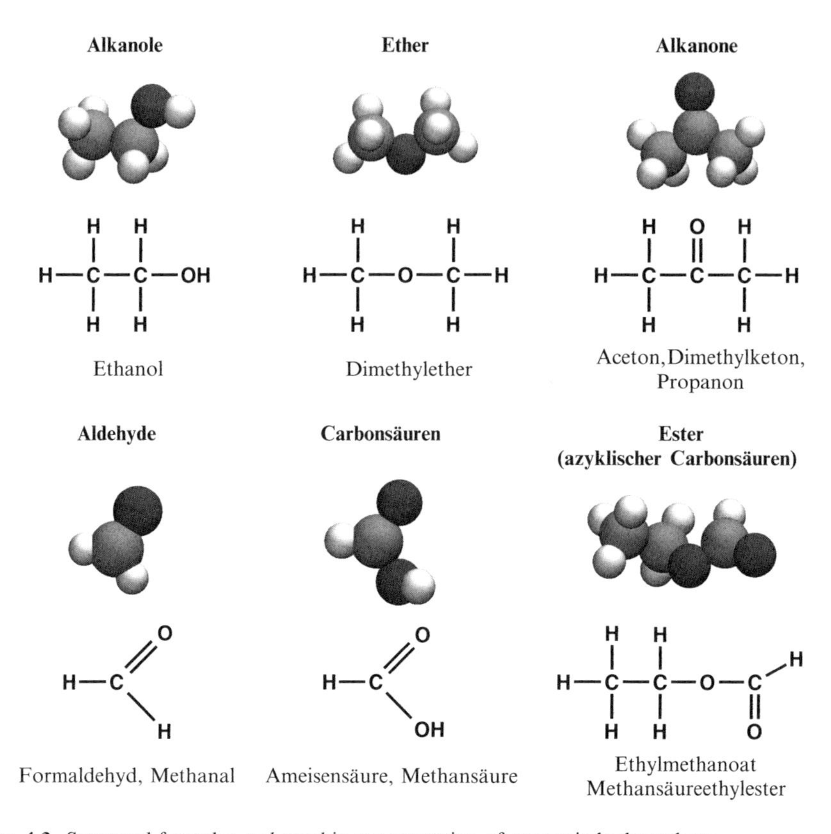

Fig. 4.3 Structural formulae and graphic representation of oxygenic hydrocarbons

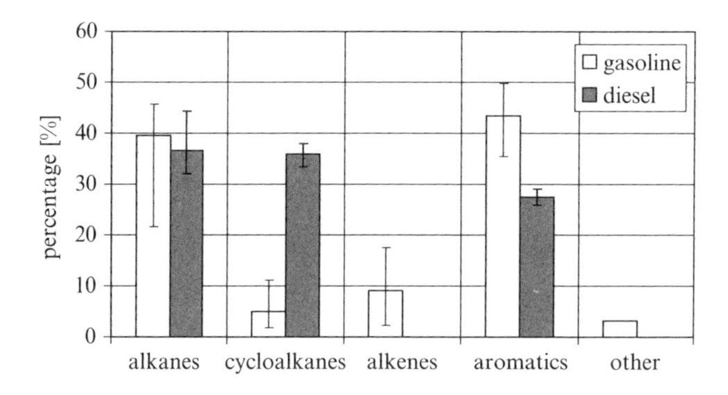

Fig. 4.4 Composition of diesel and gasoline acc. to DGMK (2002, 2003)

hydrocarbon groups is very diverse, as shown in Fig. 4.4. Especially noticeable are the high percentage of cycloalkanes in diesel fuel and the high percentage of aromatics in gasoline. Also important is the large spread in the composition of both fuels in the consideration of different fuel samples (DGMK 2002, 2003). This is the most apparent in the case of gasoline.

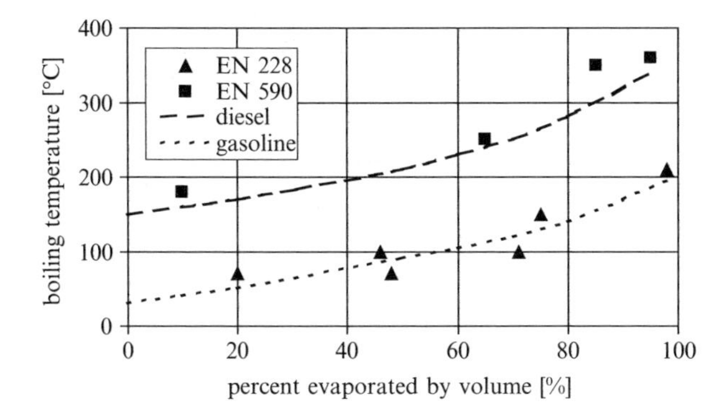

Fig. 4.5 Typical boiling curves for gasoline and diesel

The differing composition of gasoline and diesel is especially clear in their boiling properties. Since both are mixtures, there is no discrete boiling point as with pure substances but rather a boiling curve as a function of the already evaporated portion. Figure 4.5 shows typical boiling lines for gasoline and diesel as well as their limits set by the European standards EN 228 and EN 590.

4.1.1 Gasoline and SI Engine Fuels

Gasoline fuels are used in engines with external and internal mixture formation with spark ignition. Due to the low temperatures and cylinder pressures, the fuel must be highly volatile for the sake of a sufficient mixture formation. Because of the long delay times in the combustion chamber combined with locally hot areas (outlet valve), one must also guarantee at all times against premature inflammation. As a result, only hydrocarbon compounds with short chain lengths are used in gasoline fuels. Only alkanes with four to eight carbon atoms and aromatics built on one benzene ring (monoaromatics) have a mentionable fraction in the gasoline mixture. The characteristic property of a gasoline fuel is described by its octane number. The higher it is, the less ignitable and more knockproof it is. The octane number is detected according to a standardized process (EN ISO 5163 and 5164). In it, a mixture of n-heptane and iso-octane serves as a comparison fuel to determine the octane number. Table 4.2 shows the octane numbers of a few fuels; values for octane numbers >100 are extrapolated.

The octane numbers for the individual hydrocarbon groups vary and are shown in Fig. 4.6. It is noticeable that the octane numbers for alkanes and alkenes fall with increasing chain lengths, while the number increases for aromatics. The reason for this is the increasing amount of branching in the aromatics with increasing chain lengths. This is very distinct for different isomers of heptane: n-heptane

Table 4.2 Research octane numbers RON and engine octane numbers EON for selected fuels

Components	RON	EON
n-heptane	0 (defined)	0 (defined)
Iso-octane	100 (defined)	100 (defined)
Super-gasoline	95 (min)	85 (min)
Liquefied petroleum gas	103–111	–
Natural gas	120–130	–

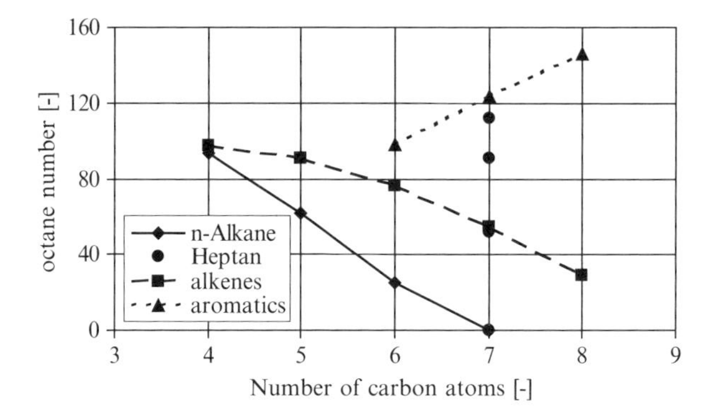

Fig. 4.6 Octane numbers for hydrocarbons of different groups and chain lengths

(RON = 0), 3-methyl-hexane (RON = 52), 2,3-dimethyl-pentane (ROZ = 91.1), and 2,2,3-trimethyl butane (RON = 112).

4.1.2 Diesel Fuels

The process of diesel combustion places different demands on the fuel than in the SI process. Due to the need for autoignition in the diesel engine combustion process, the fuel must have a composition such that autoignition can take place under the predominant temperatures and pressures. The characteristic value for ignitability is the cetane number, which is determined in analogy to the octane number in accordance with a specific process (EN ISO 5165). Since the cetane number behaves reciprocally to the octane number, the ignitability of hydrocarbons increases with the chain length.

A high cetane number corresponds to a high level of ignitability. For a diesel fuel, it should be between 40 and 65, for modern direct-injection diesel engines >50. Because of these requirements, the average molecular mass is higher in the case of diesel fuels than SI engine fuels. The chain lengths with a mentionable percentage are between 9 and 26 carbons atoms, and aromatics exist not only as monoaromatics but also as diaromatics and triaromatics (two or three connected benzene rings). Figure 4.7 shows the cetane numbers for different hydrocarbon groups as a function of the chain length.

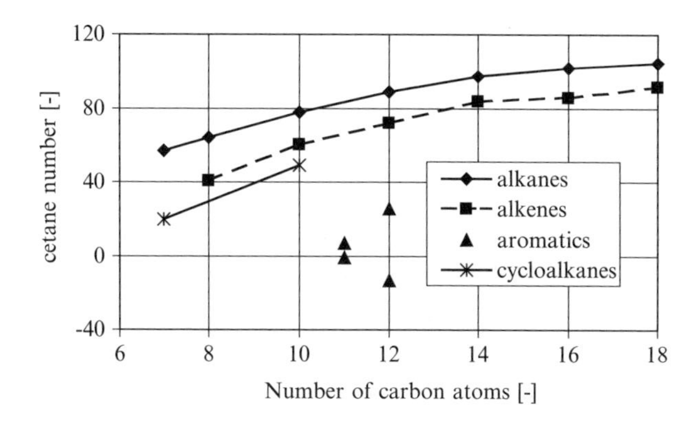

Fig. 4.7 Cetane numbers for hydrocarbons of different groups and chain lengths

4.1.3 Alternative Fuels

The term alternative fuels includes a varied assortment of fuels that are used either as alternatives or supplements to diesel and gasoline. Some of them have been designated as biofuels in guideline 2003/30/EG and are being treated with special consideration by lawmakers (subsidization, tax advantages).

Dimethyl Ether DME, LPG. *Dimethyl ether* is the simplest existing ether (Fig. 4.3) and is usually obtained by dehydrogenation of synthetic gas made of coal and natural gas. Under normal conditions, DME is gaseous and can be liquefied under pressure for efficient storage. DME has a cetane number of 60 and can thus serve as a substitute fuel in diesel engines. In order to ensure the operational reliability of the injection system however, additives must be mixed in to improve lubricity. If DME is obtained from bio-synthetic gas, then it is considered a biofuel according to 2003/30/EG.

LPG (liquefied petroleum gas) is a mixture of different gases obtained in the extraction and refining of raw oil. The main components are propane and butane, and besides them are also their alkenes (butene, propene). Because of the short chain lengths and molecular sizes, LPG has a very high octane number >105 and is therefore very suitable for use in SI engines. To this end, LPG is mostly gaseous and stored at a pressure of 5–10 bar. LPG is especially common in southern and eastern European countries, but it is becoming increasingly important in Germany as well.

Vegetable Oil, Rapeseed Methyl Ester RME, Biodiesel. *Vegetable oils* consist mainly of triglycerides (triple esters of glycerin with fatty acids/carboxylic acids). They are extracted from oily plants; in Europe this is mainly rapeseed oil. Due to its very high viscosity, it can only be used in diesel engines to a limited extent and also has a very high coking tendency.

Its use in winter with low environmental temperatures is especially problematic. In order to facilitate its use in engines, vegetable oils are esterified with methanol making vegetable oil esters/methyl esters of rapeseed (*RME, Biodiesel*) or denatured with small amounts of RME. Aside from operation with pure biodiesel, fatty acid methyl esters (FAME), with which biodiesel is also included, added to diesel to an amount of up to 7%. The properties of biodiesel are monitored by DIN 14214. In order to ensure consistency of quality and composition of different rapeseed oils, the most important properties are regulated in a prestandard (DIN V 51605).

Natural Gas. Natural gas is a combustible gas found in subterranean deposits, frequently along with crude oil. It is used in combustion engines mainly as *CNG* (compressed natural gas). In this case, the gas is stored under a pressure of 200 bar. Liquefaction of natural gas to form *LNG* (liquefied natural gas) does not seem practical for use in engines because of the major cooling required. The main component is always methane (>85%) along with higher alkanes and inert gases (N_2, CO_2). The composition and energy density (H_s = 36–50 MJ/kg) are highly dependent on the deposit/production location; this circumstance must be taken into account for engine operation. Because of the short chain lengths and molecular sizes, CNG has a very high octane number >120 and is thus suited to use in SI engines like LPG. Because of its low carbon/oxygen ratio and resultant reduction of CO_2 emissions, CNG is becoming increasingly significant.

Fuels Containing Alcohol. Ethanol can be added to conventional gasoline to up to 5% according to EN 228. No engine modifications are required at such a percentage. In Scandinavian countries, the US as well as in newly industrializing countries (e.g. Brazil), fuels with up to over 70–100% ethanol are available. These high amounts of ethanol require changes to the engine's electronics due to the lower heat value. These fuels are usually called *flex-fuels* since they contain an arbitrary amount of ethanol the blend with gasoline. Aside from the low heat value, fuels containing ethanol place higher demands on the corrosion resistance of the components conveying the fuel and, because of the high enthalpy of vaporization, on cold starting.

Table 4.3 shows some selected properties of the fuels described.

4.2 Diesel Engines

The conventional diesel engine combustion process is characterized by heterogeneous mixture formation and combustion. In modern diesel engines, fuel is directly injected under high pressure, usually shortly before the top dead center, into the combustion chamber. The fluid fuel entering the combustion chamber is atomized into small droplets, vaporized and is mixed with air, resulting in a heterogeneous mixture of fuel and air. Combustion is initiated by the high temperatures and pressures by an autoignition process, hence the alternative name compression

Table 4.3 Selected properties of engine fuels

	Density	Boiling point	Specific heat value	Ignition temperature
Liquid fuels	kg/m^3 [a]	°C	MJ/kg	°C
Methanol	790	65	19.7	455
Ethanol	790	78	26.8	425
Gasoline	720–775	25–210	43.5	≈400
Diesel	820–845	110–400	42.5	>200
Vegetable oil	900–930	220–320	36	–
Biodiesel RME	860–900	330–350	36	≈150
Gaseous fuels	kg/m^3 [b]			
CNG	0.7–0.84	> −162	≈32–45	≈550
LPG	2.25	> −42	46.1	≈400
DME	0.67	−20	27.6	≈200

[a]At 15°C
[b]At 1,013 mbar

ignition (CI) engine. In the conventional diesel combustion process, usually there is only a very short amount of time available for mixture formation. A fast injection and an as good as possible atomization of the fuel are therefore the prerequisites for a fast and intensive mixing of fuel and air.

The load of the engine is controlled by the amount of injected fuel and combustion start is controlled by the start of injection. Diesel engines are usually operated with a globally lean air-fuel ratio, but direct injection leads to different mixture areas ranging from very lean through stoichiometric to very rich mixture ratios. Such mixture stratification leads inevitably to the formation of pollutant emissions, especially soot particles and nitrogen oxides.

Figure 4.8 qualitatively shows the subprocesses involved in diesel engine mixture formation and combustion. The particular subprocesses proceed largely simultaneously and interact with each other. The modeling of diesel engine combustion is thus extremely complex.

4.2.1 Injection Methods and Systems

As opposed to formerly utilized injection into a prechamber or swirl chamber, direct injection into the combustion chamber is used today almost exclusively. In this case, the combustion chamber is accommodated as a bowl in the piston; the shape of the bowl has a decisive influence on the combustion process. The fuel is injected with a centrally arranged (four-stroke engines) multi-hole nozzle. High injection pressures and many small holes in the injection nozzle result in a good mixture formation, which can be supported by a swirl flow of the combustion chamber gases. The need for increasingly small holes to realize further emission advantages at partial load has in the past led to an increase of the maximum

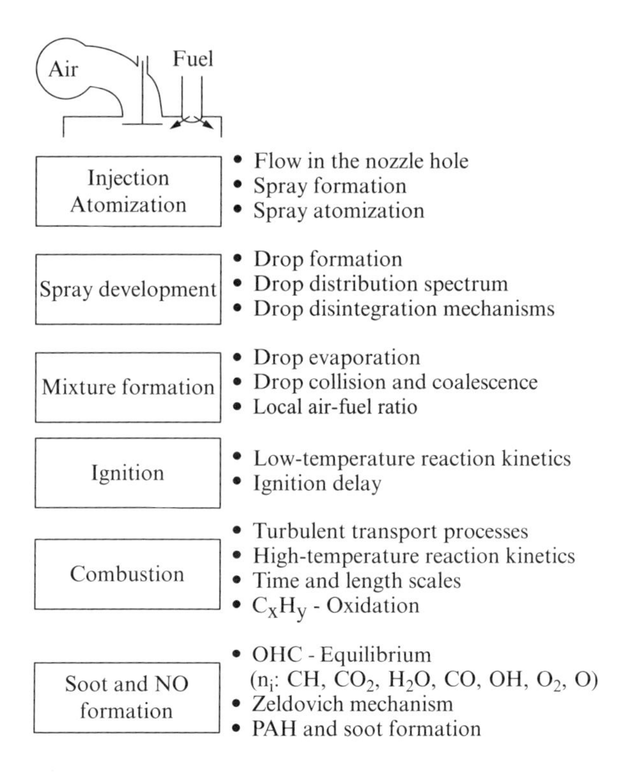

Fig. 4.8 Subprocesses of mixture formation and combustion in diesel engines

injection pressures to currently over 2,000 bar. If possible, the injected fuel should not collide with the relatively cold piston wall, because in this way evaporation and then mixture formation are lagged and the formation of HC emissions is favored.

Direct injection methods in comparison to indirect ones have clearly lower specific fuel consumption. However, because of the high pressure increase gradients at the start of combustion, it also has a much higher noise level. Beyond this, mixture formation is not supported by a fast charge movement in the duct between pre- and main chamber (200–500 m/s) as in pre-chamber engines. The total energy for the mixing of fuel and air is for the most part yielded by the fuel jets injected into the combustion chamber, through which a considerably higher injection pressure is necessary. While in pre-chamber engines injection pressures of approx. 400 bar are sufficient, for direct injection injection pressures are currently between 200 bar at idle and 2,000 bar at full load. For vehicle diesel engines, still higher injection pressures are also continually being discussed.

In the case of injection systems, we distinguish between conventional cam-operated systems as well as the common rail injection systems developed in the last several years. In cam-operated injection systems, pressure increase and fuel metering are coupled mechanically. The cam moves the tappet of the injection pump, which in turn "compresses" the fuel volume. The resulting pressure increase opens a valve and thus releases the feeding pipe for the injection nozzle. In contrast,

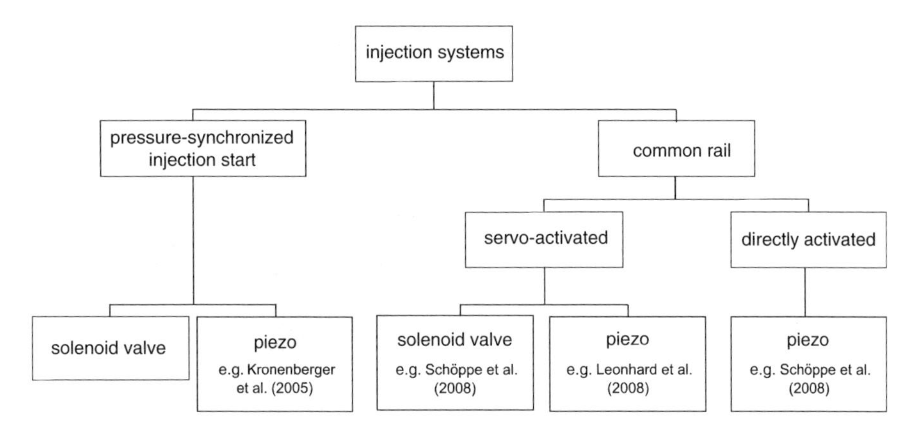

Fig. 4.9 Classification of current injection systems for passenger cars

pressure increase and fuel metering are completely separate in the common rail injection system. Fuel is continuously supplied into a high-pressure fuel reservoir (common rail) by means of a mechanically or electrically driven high-pressure pump. With an electronically controlled injector, fuel is taken from the reservoir and injected into the combustion chamber.

Figure 4.9 gives a graphic representation of the classification of injection systems described above.

Cam-Operated Injection Systems. Cam-operated injection systems – or injection systems that have a pressure-synchronized injection start – can be classified according to the type of actuation. Injection pumps and injection nozzles form a unit in this system, which is installed separately in every cylinder. A quickly switching magnetic or piezo-actuator controls the start and end of injection.

In the pump-nozzle injection system (PN), injection pressures of up to 2,200 bar are possible thanks to the optimization of the dead volumes. This allows for good fuel consumption and emission values due to the fine atomization of the fuel. Figure 4.10 shows the schematic of a PN injection system (Kronenberger et al. 2005). The basic disadvantage of this injection system is in the lack of flexibility of the injection time and the number of injection events, since the pressure build-up is mechanically bound to the cams. This lack of flexibility will make it difficult to implement future laws governing exhaust gas emissions, so injection systems with pressure-synchronized injection starts to no longer play a role in current developments. For this reason, we will not go into further detail about cam-operated systems (distributed fuel injection pumps, in-line injection pumps) except for the PN system.

Common-Rail Systems. In the case of the electronically controlled common rail system, the parameters of injection start and pressure build-up are separated from each other so there is much more freedom with respect to the flexibility of the injection events. Basically, every system consists of individual components which

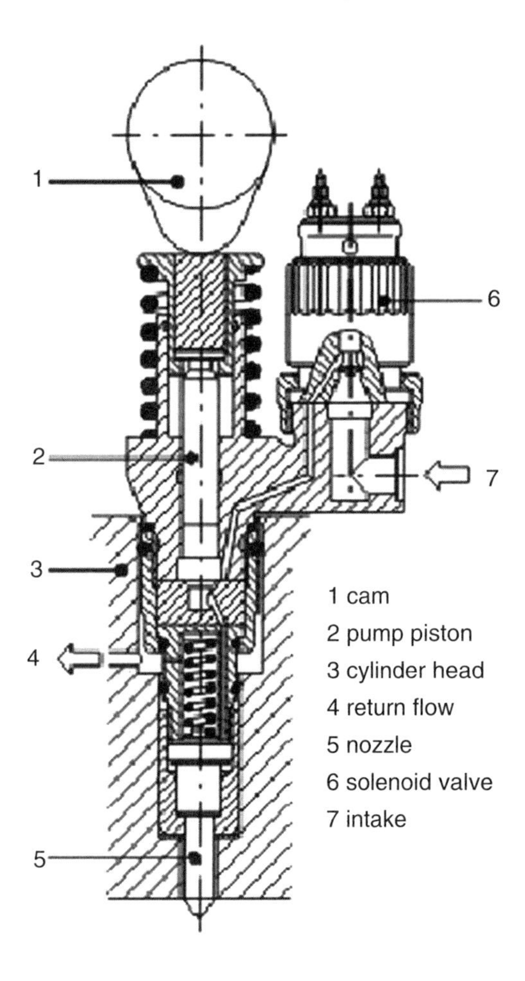

Fig. 4.10 Functional schematic of a pump-nozzle unit

altogether comprise the common rail system (CR system) and will be discussed in more detail in the following. Figure 4.11 provides a schematic representation of the typical structure of a CR system.

A pre-supply pump (not shown) feeds the fuel through a filter and guides it to the high-pressure pump. Before the high-pressure pump there is a suction throttle which restricts flow to the high-pressure pump. This component, also called a metering device, sees to it that only the amount of fuel gets to the pump that is necessary for injection.

The high-pressure pump compresses the fuel to the desired pressure of 200–2,000 bar and guides it to the high-pressure accumulator. The high-pressure pump is usually designed as a single piston pump in current injection systems; one or usually two multistroke cams drive the pump piston. This design has recently replaced the radial piston pump for cost reasons.

The high-pressure accumulator holds available the fuel needed for injection. In this component are also arranged the pressure sensor and pressure control valve.

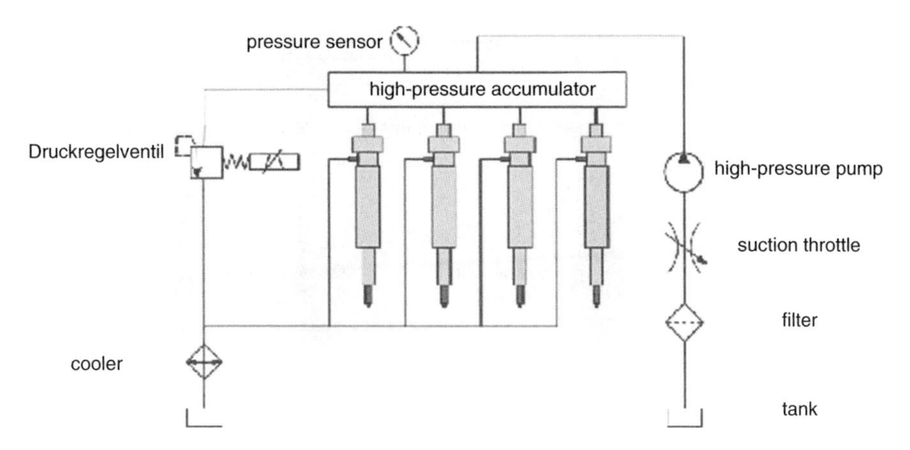

Fig. 4.11 Schematic representation of a common rail system

The volume of the accumulator is designed such that the amount withdrawn for injection does not lead to an excessive loss of pressure.

The pressure control valve diverts fuel under high pressure and guides it back to the tank. In order to prevent heating of the tank, the fuel is cooled on the way. With the pressure control valve (PCV), pressure can be reduced in the rail (e.g. in case of a change in the operating point or when turning off the vehicle) without the necessity for injection. Moreover, with PCV, there is a "fine regulation" of the pressure at critical operating points if the pre-control with the suction throttle is not precise enough (e.g. in idle operation). Since controlled termination of the compressed fuel means a loss of energy, the area of application of the PCV should be as small as possible. Ideally, it is possible to do without PCV altogether (flow control).

The injector is probably the most complex component in a CR system. The injector, or more precisely the nozzle, is the direct interface to the combustion chamber of the engine, so the functionality of the injector always affects the combustion process as well. The most common injector design is the servo-injector, the functional principle of which is shown in Fig. 4.12.

The default state of the injector is closed. The nozzle needle is pressed into the seat due to the pre-stressing of the spring. The outlet valve above the nozzle needle closes the opening and the nozzle needle is pressure-balanced, i.e. the pressure on the pressure level of the needle in the lower part and the pressure above the needle in the control space are identical.

To open the injector, the solenoid valve is lifted and the fuel in the control space above the nozzle needle can flow through the outlet throttle into the return flow. Fuel is also flowing permanently into the control space through the inlet throttle, but the effective cross-section of this throttle is smaller than the outlet throttle so that the overall pressure in the control space is reduced. The ratio of both of these throttles essentially determines the dynamics of the injector. A positive pressure and force difference arises below the needle because of the pressure release in the

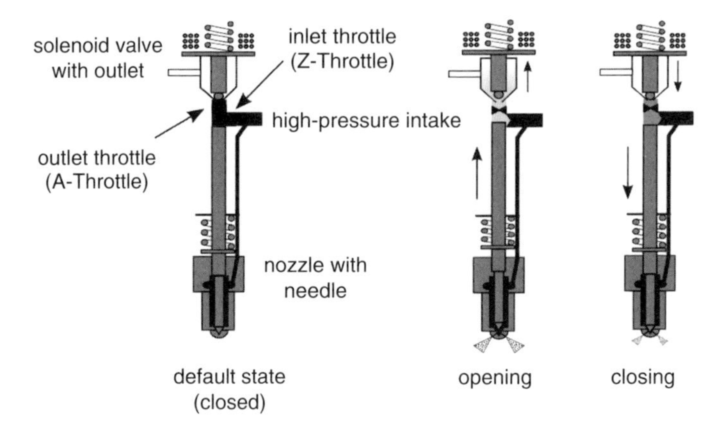

Fig. 4.12 Functionality of a servo-injector

control space, so that the needle is lifted from its seat and the fuel can flow through the nozzle holes.

Closure of the injector is initiated when current is removed from the solenoid valve, and flow from the control space is barred. In this way, pressure is increased again in the control space and the spring presses the needle back into its seat so that the nozzle holes are closed.

The functionality of the example mentioned above was represented with the help of a magnet coil actuator. The basic advantages of robustness as well as the low costs of solenoid valve injectors are accompanied by the disadvantage of delayed dynamics. Due to the merely indirect control of the nozzle needle, there are long delays between activation and injection rate. Figure 4.13 shows an example of the relation between activation and the injection rate. Piezo-actuators are alternatively used instead of servo-valves in order to increase the dynamics of the system. While the piezo-actuator does have the advantages of quick reaction times and a specifically high shift force, a complex control mechanism is necessary to compensate the temperature-dependence of the piezo-properties. Boecking et al. (2005) and Leonhard and Warga (2008) give an example of the functional principle of the servo-injector described here.

In addition to servo-injectors, there also exists the group of directly activated injectors (Schöppe et al. 2008). In the case of this injector type, the piezo-actuator acts directly upon the nozzle needle making it possible to control it directly. The potentials of direct needle control and thus of injection sequence formation with respect to combustion and pollutant formation were already shown early on (Stegemann (2004)), but implementation has in series failed thus far due to the complexity and cost of the system. The advantages of directly activated systems are the very short actuator switching times and the resultantly short distances between the individual injections. A needle motion that is as independent of pressure as

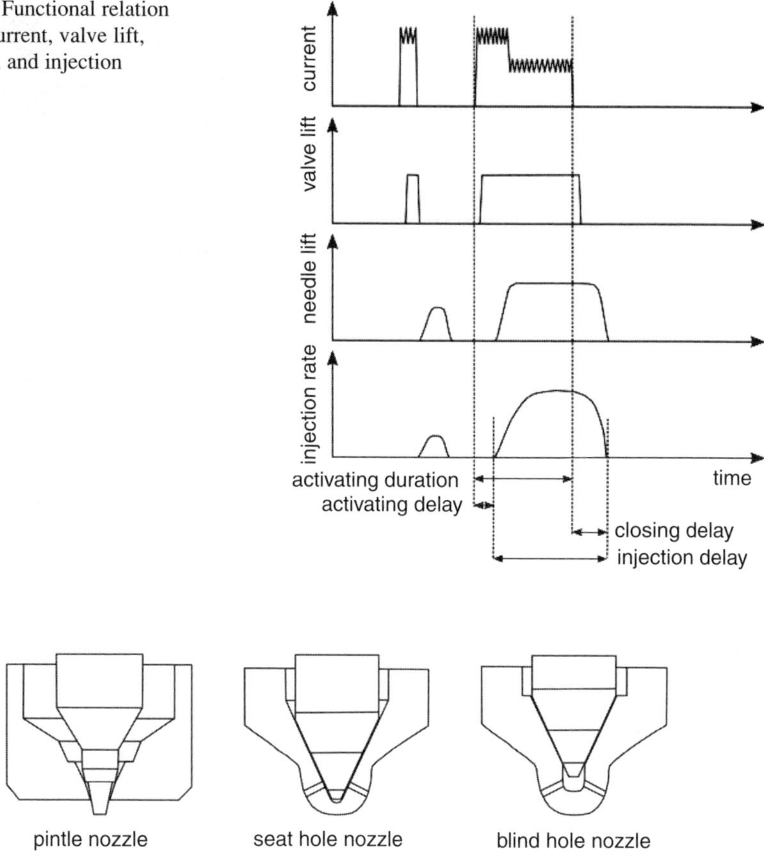

Fig. 4.13 Functional relation between current, valve lift, needle lift, and injection

Fig. 4.14 Injection nozzle designs

possible is realized by the direct control of the needle without the aid of pressure. Further advantages include the lack of leakage lines and thus fewer losses.

Injection Nozzles. Fuel is injected into the combustion chamber through the hole(s) in the injection nozzle. In the injection process, the fuel should be atomized to the highest possible degree (air guided method) or aimed at the surface of the bowl in the piston (wall guided method). However, the latter method no longer plays a role in the current technological standpoint.

It is therefore necessary that different nozzle shapes be used for different combustion methods and fuels (Fig. 4.14).

- Pintle nozzles are employed in pre- and swirl chamber engines. They have a stroke-dependent opening cross-section, are advantageous with respect to combustion noise, tend however towards carbonization. Because of their very large released flow cross-section (annular clearance), they are only suitable for low injection pressures.

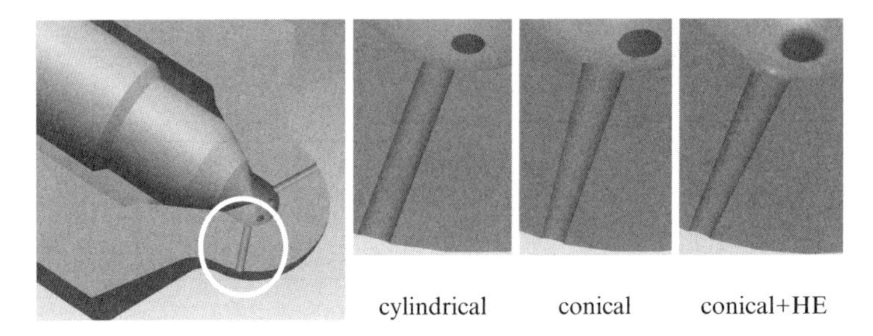

Fig. 4.15 Standard designs of blind hole nozzles

- Orifice nozzles (seat holes and blind holes) are used in diesel engines with direct injection, both in conventional and common rail injection systems.

Pre-chamber combustion processes no longer play a mentionable role in today's diesel engines, so now only seat/orifice nozzles are used. *Seat hole nozzles* have the advantage that the nozzle holes are covered directly by the nozzle needle, and evaporation of residual fuel from the waste space below the needle is not possible. Because of the direct action of the needle on the nozzle hole surface, a stronger coupling of the needle lift/flow cross-section with the injection rate is possible. The design disadvantage of this principle is its elaborate needle guide. If the guide is not precise enough, wobbling of the nozzle holes could create asymmetry in the spray pattern. In *blind hole nozzles*, the nozzle holes empty into a balancing volume that has the task of calming the flow in order to create a homogeneous injection spray pattern. Decoupling the nozzle holes from the sealing surface improves stability, but the volume under the needle after closing leads to evaporation of the fuel and thus to an increase in HC emissions.

Figure 4.15 shows examples of different nozzle hole shapes that exist independently of their position in the nozzle. Cylindrical nozzle holes are the simplest and already have a high level of efficiency and a high flow coefficient. The flow coefficient can be further increased with a positive conicity (inside large and outside small diameter). In order to equate the holes and make the flow consistent, nozzles are rinsed with an abrasive fluid and thereby grounded hydro-erosively. HE grinding minimizes the tendency to cavitation, and the flow coefficient and efficiency are once again increased. Minimizing the tendency to cavitation in flow-optimized nozzle holes has the consequence however that the coking tendency of the nozzle is increased since potential deposits are no longer detached.

4.2.2 Mixture Formation

Diesel engines can be operated both in the two-stroke and four-stroke process. Diesel engines that run in high and average-speeds are utilized, for example, in passenger cars, commercial vehicles, industrial, maritime and stationary applications. These engines are customarily equipped with one or two inlet valves and one or two outlet valves. Part of the combustion chamber is located as a bowl in the piston. In most applications, in every cylinder, a single, centrally positioned injector, combined with a multi-hole nozzle is used. Figure 4.16 shows an example of the spray of such a combustion method. Low-speed two-stroke diesel engines are mostly used to power ships and in stationary power generation. Modern two-stroke diesel engines have uniflow scavenging, intake ports and a centrally arranged outlet valve. In these applications, usually two to four injectors with multi-hole nozzles positioned on the periphery of the combustion chamber are employed. The fuel is introduced to the combustion chamber in a tangential direction.

In addition to the mixing energy induced by the injection system, mixture formation is strongly dependent on the interaction of the fuel spray with the in-cylinder flow. Figure 4.17 gives a schematic representation of the two main in-cylinder flow structures important for mixture formation in a DI-Diesel engine: swirl and squish flow. The swirl flow is a rotating flow around the cylinder axis. It is generated by the geometry of the intake ports and in four-stroke engines additionally by the geometry of the valve seats (masking, phasing). The squish flow is generated when the piston approaches top dead center, displacing the air in the squish area. Both the squish and the swirl flow can improve the mixing between fuel and ambient air. Other directed

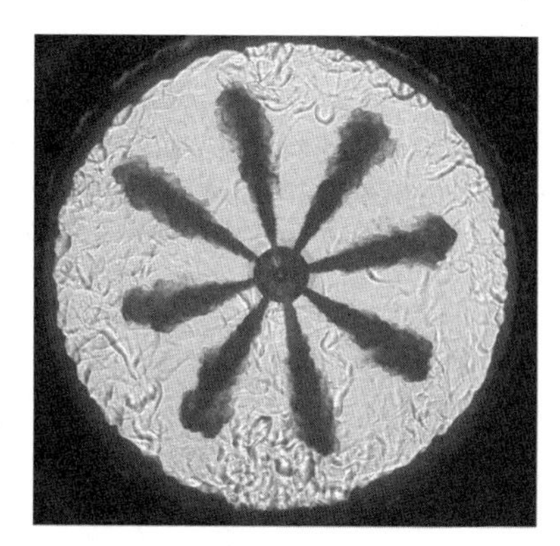

Fig. 4.16 The spray of a direct-injecting four-stroke diesel engine

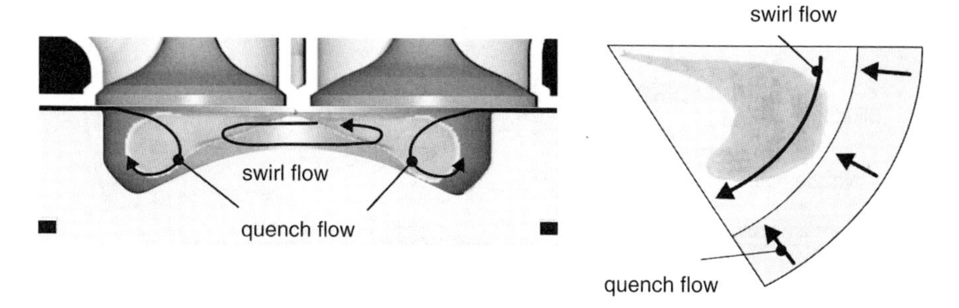

Fig. 4.17 Macroscopic flow structures in the combustion chamber

flow structures generated by the gas exchange, as for example tumble flow, are usually decayed during compression Other aligned flows, such as tumble flows, usually break down during compression (see Sect. 12.5.1).

Especially smaller passenger car and some light-duty truck engines use deep ω-piston bowls with relatively small diameters and high swirl ratios. The swirl ratio is strongly dependent on the engine speed and an optimization of the swirling motion for different points of operation for example by adjustable guiding vanes in the inlet port is necessary. The injection system has to be adjusted to the swirling motion. Combustion processes with high swirl ratios usually use fewer nozzle holes in order to avoid an interaction of adjacent fuel jets. Modern large four-stroke engines as used in trucks, heavy-duty, power generation and marine applications usually use a combustion process with low swirl and a more shallow piston bowl compared to passenger car engines. The advantage of a low-swirl combustion process is the improved volumetric efficiency, since the generation of directed flow structures always increases gas exchange losses. Here, the mixing energy is supplied mainly by the injection system. In contrast, large two-stroke engines have a strong swirling motion due to the scavenging process. However, usually very shallow piston bowls with a diameter equal to the engine bore are used because of the position of the injection system. As a result, two-stroke engines have a weak squish flow.

Phenomenology of Mixture Formation. The injection nozzle represents the link between the injection system and the combustion chamber. The fuel leaves the nozzle at high speed through small holes with diameters in the order of 0.12 mm for passenger car engines up to about 1.5 mm in the case of very large two-stroke diesel engines. Figure 4.18 shows a qualitative sketch of the fuel spray exiting the

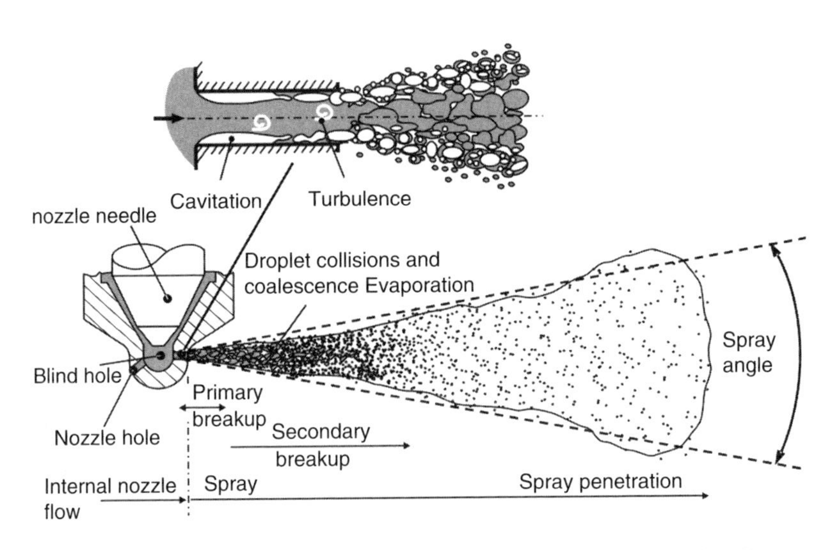

Fig. 4.18 Schematic representation of internal nozzle flow and spray propagation (Baumgarten 2006)

injection nozzle. The spray generated during injection can be roughly subdivided into two regions, one with a dense spray near the nozzle exit and a thin spray region further down the flow. The first decomposition of the cohesive fuel spray into ligaments and droplets is called *primary spray breakup*. In modern high-pressure injection systems, cavitation and turbulence are the most important mechanisms of primary spray breakup (Arcoumanis and Gavaises 1998). In the injection nozzle, the liquid fuel is accelerated into the nozzle holes in the transition from the nozzle blind-hole. The change of flow direction on the edge of the nozzle hole leads to the formation of a "vena contracta", which further lowers the static pressure in the fluid (Baddock (1999)). This reduction is strongly dependent on the geometry of the nozzle and especially on the radius of curvature of the inlet edge of the nozzle hole. If the pressure at the vena contracta falls below the vapor pressure of the fluid, hydrodynamic cavitation is initiated and vapor bubbles are created. Depending on the flow parameters, the cavitation can either be stabilized to reach the nozzle hole outlet or the flow can fully or partly reattach (Kühnsberg-Sarre et al. 1999). Cavitation reduces both the effective flow area of the nozzle as well as friction. In case of small needle lifts, cavitation structures can also arise in needle seat areas which either break up in the blind hole and thus increase turbulence or enter the nozzle holes, thus promoting further cavitation. When they leave the nozzle holes, the cavitation bubbles collapse very quickly due to the high pressures in the combustion chamber, which leads to an increase of turbulence and faster primary spray breakup.

Break-up of already existing droplets into smaller droplets due to aerodynamic forces caused by the relative speed between droplet and the environment is called *secondary spray breakup*. In addition, droplets can collide with each other and coalesce.

The spray momentum leads to air entrainment of the surrounding combustion chamber air into the spray. The droplets are heated up as a result of convective heat transfer and temperature radiation of the hot chamber walls, and the fuel finally begins to evaporate. Besides the physical properties and the combustion chamber conditions (pressure, temperature), the rate of fuel evaporation is determined by the size of the droplet surface formed and thus on primary and secondary breakup as well as on the amount of air entrained into the spray.

In the case of the diesel engine, mixture formation cannot be considered independently of combustion. It is indeed the distinctive feature of diesel engine combustion that spray propagation, mixture formation, and combustion progress in partial simultaneity. Only a small amount of the injected fuel mixes nearly homogeneously with the air in the combustion chamber during ignition delay. After ignition, this amount combusts almost instantly. Afterwards, mixture formation and combustion proceed simultaneously, and combustion is controlled by the mixture formation processes.

Spray propagation and mixture formation are understood quite well at present, at least qualitatively, and can be described approximately with semi-empirical models [see Baumgarten (2006), Ramos (1989), and Stiesch (2003)].

4.2.3 Autoignition and the Combustion Sequence

The period of time between injection start and combustion start is called the ignition delay. The physical and chemical processes occurring during this time are very complex. The essential physical processes are the atomization of the fuel, vaporization, and mixing of fuel vapor with air, forming an ignitable mixture. The chemical processes that lead to autoignition of the hydrocarbons contained in the fuel under typical diesel conditions are characterized by a highly complex, degenerated chain branching mechanism (see Curran et al. (1998)). For a more detailed presentation of chemical processes and modeling approaches, see Sect. 5.2.

In the diesel injection spray, ignition occurs in areas with local air fuel ratios of about $0.25 < \lambda < 0.65$ (Higgins et al. (2000)). The ignition delay can be controlled by means of temperature and pressure at the start of injection, which in turn depend on the inlet temperature and pressure, the compression ratio, injection start and the wall temperatures. In addition, the ignitability of the fuel (cetane number) and further parameters such as injection pressure, the geometry of the nozzle holes and the in-cylinder flow have a major effect on the ignition delay duration and the ignition location.

Figure 4.19 provides a schematic representation of the injection and combustion sequence of a diesel engine with direct injection. As we can see, the sequence of diesel engine combustion can be subdivided into three phases.

Phase 1: Initial Premixed Combustion The first phase follows immediately after ignition. The fuel injected during ignition delay mixes with the air in the combustion chamber and forms a nearly homogeneous and reactive mixture. After the ignition delay, which is physically and chemically controlled, this mixture burns very quickly. Since areas with premixed combustion arise in the main combustion phase as well, this phase is called initial premixed combustion. The rate of heat release is controlled in this combustion phase by the speed of the chemical reactions and by the amount of fuel/air mixture formed during ignition delay. The combustion noise typical of diesel engines is caused by the high speed of pressure rise at the start of combustion. This speed of pressure rise can be influenced by changing the timing of injection, in which case the following rule applies: an early injection start leads to a "hard" combustion and a late start to a "soft" injection (see Fig. 4.20). Moreover, the combustion noise can be considerably reduced by a preinjection. In this case, at first only a small fuel amount of about 2% is injected which leads after the ignition delay to only a small amount of heat release and to a small pressure increase. The increased temperatures lead however to a significant reduction of the ignition delay of the main injection, which leads to a reduction of the amount of premixed combustion with a positive effect on noise.

Phase 2: Main Combustion In the second phase, heat release is controlled by the turbulent mixing processes between the fuel and air and is therefore also called mixture-controlled combustion. In this phase, injection, spray breakup, droplet evaporation, mixing with air, combustion, and pollutant formation all take place

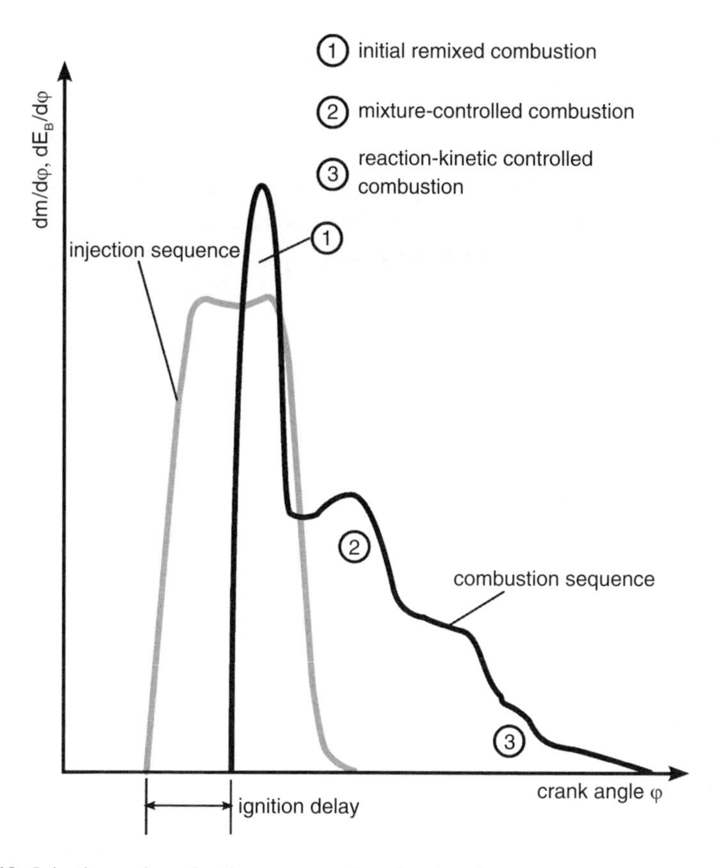

Fig. 4.19 Injection and combustion sequence in a diesel engine

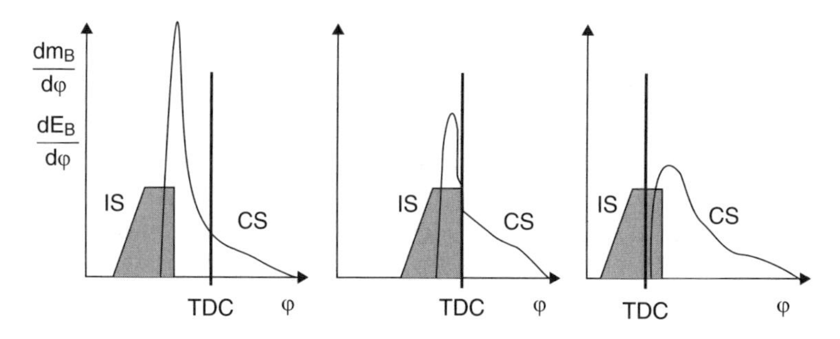

Fig. 4.20 Injection rate (IR) and heat release rate (HRR) with early (*left*) and late (*right*) combustion

simultaneously. Figure 4.21 shows a cross-section through a reacting diesel injection spray following the conceptual model of Dec (1997) and Flynn et al. (1999). The model describes the quasi-steady phase during main combustion and is, strictly speaking, only valid in quiescent conditions. The fluid fuel spray penetrates into the

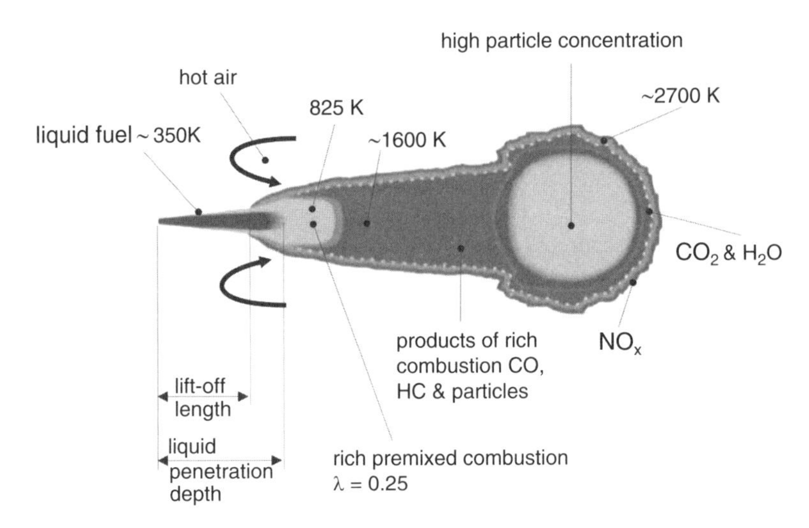

Fig. 4.21 Conceptional model of diesel combustion, acc. to Dec (1997) and Flynn et al. (1999)

combustion chamber, mixes with air and evaporates. The air ratio in the spray increases both with increasing distance to the injection nozzle as well as with the distance to the spray axis. Downstream of the liquid penetration length, a rich mixture zone is formed which leads to partial oxidation of the fuel and temperatures up to 1,600 K. According to Flynn et al. (1999), the air ratio in this zone is in the range of $0.25 < \lambda < 0.5$, and about 15% of the total heat is released in this zone. Among the partially oxidized products of premixed combustion are also found precursor species, which lead to particle formation further downstream in the middle of the flame (Sect. 6.4). A diffusion flame is formed around the injection spray on an iso-surface with a stoichiometric air fuel ratio. The partially oxidized products of the rich premixed combustion and the particles formed move further downstream and are transported into the diffusion flame, where they are completely oxidized into carbon dioxide and water. Temperatures rise to up to 2,700 K. Due to the high temperatures, nitrogen oxides are formed on the lean side of the diffusion flame (see Sect. 6.5). Near the injection nozzle, vaporization processes and chemical reactions in the spray determine the distance from the injection nozzle at which the diffusion flame establishes itself. The axial distance between the injection nozzle and the diffusion flame is called the *lift-off length* and is an important property of a diesel flame with regard to soot formation (see e.g. Siebers and Higgins (2001)).

Phase 3: Post-combustion After the injection process is finished, no additional momentum is added to the spray by the injection and the flame jet structure evolves into a pocket of rich premixed products surrounded by a diffusion flame. The exact properties of this zone depend on the injection system. If the nozzle needle closes very quickly, then the last fuel parcels still have high speed, so that they have a similar combustion sequence as in main combustion. On the other hand, a slow closure of the needle leads to low speeds of the last fuel parcels with low entrainment of oxygen and consequently increased formation of soot. With the expansion of the piston in the direction of the bottom dead center, the temperatures in the combustion chamber are lowered. The reaction rates go down with the temperatures, so that combustion is chemically controlled again. This phase is of extreme importance for the oxidation of the previously formed soot, of which over 90% is decomposed again. As is described in more detail in Chap. 6, the temperatures during this combustion phase should be high, since soot oxidation is very slow below 1,300 K (Glassman 1988).

For the thermodynamic quality of the total combustion process, the released thermal energy is decisive

$$\frac{\mathrm{d}E_B}{\mathrm{d}\phi} = f(\phi). \tag{4.1}$$

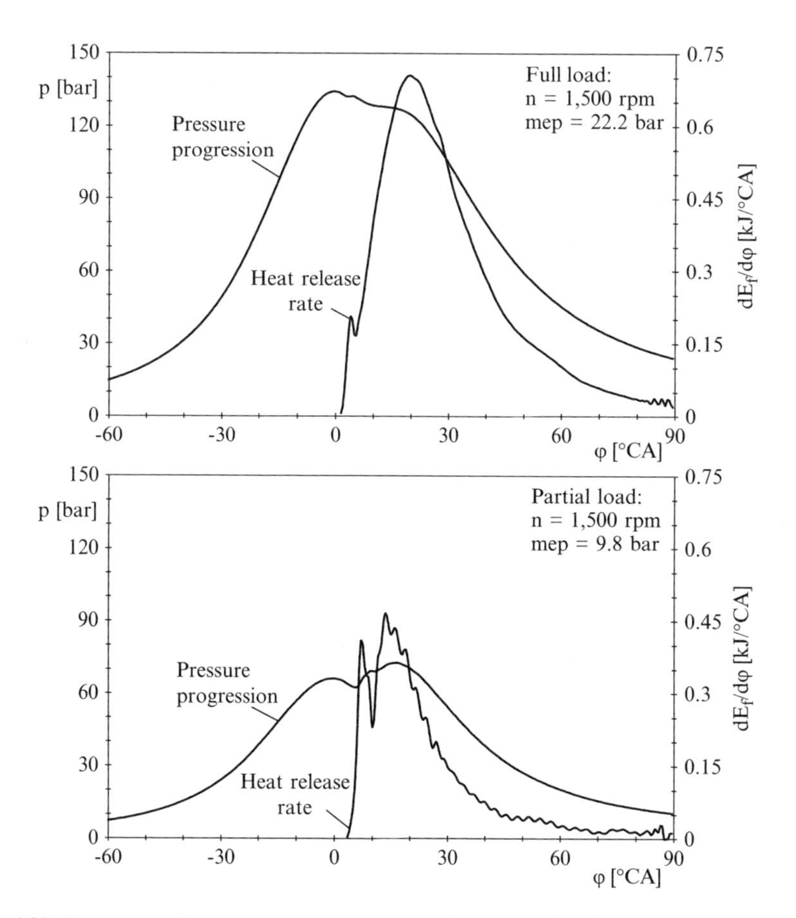

Fig. 4.22 Pressure profile and heat release rate in a high speed diesel engine at full and partial loads

It leads to the heating up of the air fuel mixture in the cylinder and thus to temperature and pressure increase. As an example, Fig. 4.22 shows the pressure progression and heat release rates at full and partial load in a high speed heavy-duty diesel engine with relatively late injection.

4.3 Spark Ignition Engines

4.3.1 Differences Between Premixed Flame and Diffusion Combustion

Mixture formation, ignition and combustion of the SI engine differ fundamentally from the diesel engine.

Classic SI engines with homogeneous mixture formation are characterized by their premixed flame. Fuel and air are mixed long before ignition as completely as possible. This is realized either outside the combustion chamber (intake-manifold fuel injection) or in the combustion chamber itself (homogeneous direct injection in the intake cycle). The mixture exists in a gaseous state in the combustion chamber at the time of ignition. Since SI engine fuel is not a highly ignitable fuel relatively speaking, the mixture must be externally ignited by an ignition spark from the spark plug. Characteristics of new SI engine partial load combustion processes (stratified injection, HCCI) which deviate from these basic principles will be dealt with in their own sections.

As opposed to premixed combustion of the homogeneously operated SI engine, there is the principle of the diffusion flame, which is characteristic of the main combustion phase of the diesel engine. According to Heywood (1988), a diffusion flame is characterized above all by the fact that the reaction partners are not thoroughly mixed in advance, but mixture formation takes place practically directly in the reaction zone. Thus, fuel injection first takes place late in the compression phase in the case of the diesel engine. The fuel droplets of the high-grade ignitable diesel fuel are heated by the highly compressed hot air in the combustion chamber, vaporize, mix with the surrounding air and auto-ignite. These processes must take place in the shortest mixture formation time and parallel to an already initiated ignition, since the further course of fuel injection takes place in the already existing flame front.

A major differentiating factor between premixed flame and diffusion combustion is thus the time available for mixture formation and the question of whether there is a homogeneous mixture at the point of ignition or not. Therefore, signs of diffusion flames can arise in the SI engine under unfavorable mixture formation conditions as well. For example, if the cold combustion chamber walls are wetted after cold start by large fuel droplets and the impinged fuel cannot be completely vaporized toward the end of the compression phase, local mixture inhomogeneities result (e.g. very rich zones) leading to combustion with possible soot formation. A charge movement that has not been optimally designed can lead to similar results, e.g. if fuel droplets wet the cylinder liner due to excessive tumble flow. Since the soot particles have a characteristic radiation, such signs of unintentional diffusion combustion can be detected in the combustion chamber with fiber optical measurement technology (Winklhofer (2007)).

4.3.2 Ignition

In the case of the conventional SI engine, ignition of the air fuel mixture takes place shortly before the top dead center via spark discharge between the electrodes of the spark plug, see Fig. 4.23.

4.3.2.1 Structure of the Ignition System

Modern SI engines are equipped with a resting ignition voltage distribution. Every cylinder has its own ignition coil, which is either mounted directly on the spark plug (e.g. in the form of a rod ignition coil) or conjoined with it via an ignition line.

The ignition coil consists of a primary and secondary coil. During the dwell time shortly before the time of ignition, the primary circuit is closed for a few milliseconds, which builds up a magnetic field in the primary coil. The maximum available ignition energy is thus stored in the magnetic field of the primary coil and can reach values of about 40–100 mJ in present-day ignition coils – in high-energy ignition coil even more. At the time of ignition, the primary circuit is interrupted by the opening of the ignition switch in the ignition output stage of the control unit. The magnetic field in the primary coil breaks down suddenly and thereby induces for a short time a primary voltage of up to 400 V. In accordance with the principle of the transformer, a high-voltage peak value of up to 30 kV is thus induced in the secondary coil.

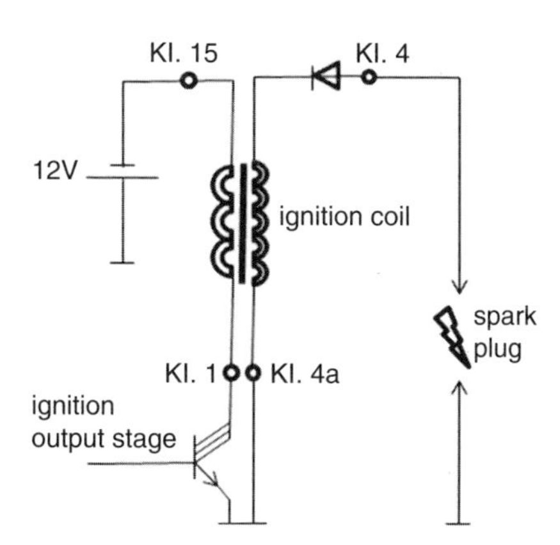

Fig. 4.23 Structure of the primary and secondary ignition systems, acc. to Robert Bosch GmbH (2003)

Due to the high voltage, the mixture between the spark plug electrodes becomes ionized, and an ignition spark can jump between the spark plug electrodes. The plasma duct that then arises reaches temperatures of up to 6,000 K for a short time. For the sake of a stable ignition via the ignition spark, the gas between the electrodes must be of an ignitable composition, and the thermal energy input of the plasma into the mixture must exceed the thermal losses to the surfaces of the electrodes (Pischinger (2001)).

4.3.2.2 The Ignition Sequence

The chronological sequence of spark discharge is subdivided according to Pischinger (2001) into the two extremely short stages of spark initiation and arc discharge, which are completed after a total of about 1 μs, in addition to subsequent glow discharge, which is also called the spark duration (Fig. 4.24).

Spark-over in the initiation phase takes place as soon as the ignition voltage in the secondary circuit has reached the initiation voltage. The necessary amount of ignition voltage depends especially on the mixture density between the electrodes, the electrode distance and the electrode geometry. The ignition system has to be dimensioned such that it can supply this required ignition voltage under all engine operation conditions, in particular under marginal conditions, in which there is an increased ignition voltage requirement, like high load in supercharged operation (high mixture density) as well as increasing spark plug wear (increased electrode distance).

During the short initiation and arc discharge phases, the capacitively stored energy in the spark plug and ignition system is converted. The much larger amount of energy consists however of the inductively stored energy in the ignition coil. This energy is converted in the following glow discharge phase until the coil is discharged and the secondary voltage dies (Fig. 4.25). Depending on the energy supply of the ignition coil, the spark duration in this phase typically lasts about 1 ms, in high-energy ignition coils up to 1.5–2 ms.

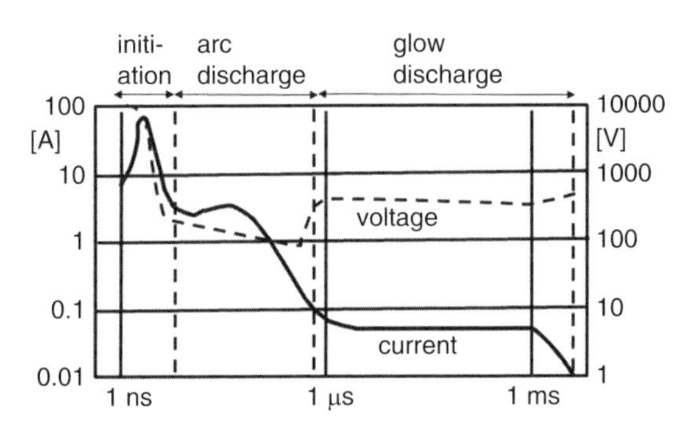

Fig. 4.24 Phases of ignition acc. to Pischinger (2001)

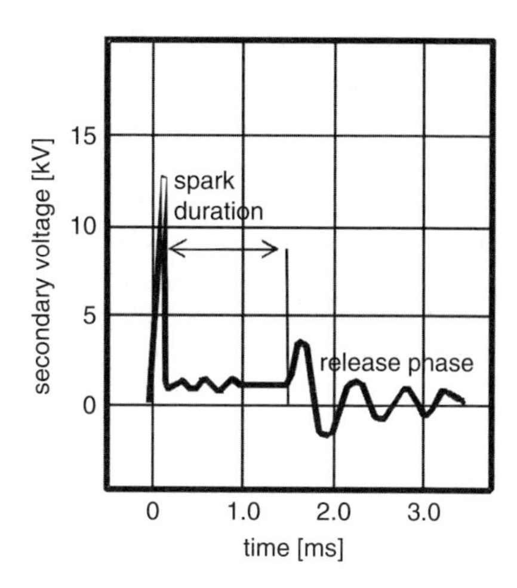

Fig. 4.25 Secondary voltage sequence, acc. to Robert Bosch GmbH (2003)

4.3.2.3 Ignition Spark Deflection and Break, Shunt Circuit

In the case of a normal ignition process, the ignition spark generally does not take the shortest path between the electrodes. Rather, it is more or less strongly deflected under the influence of local charge motion. Since the ignition spark thus covers a larger mixture volume, thereby promoting a more rapid ignition, this effect is quite advantageous as long as the spark does not break off. Burning voltage is increased with increasing spark deflection, which is why one can make inferences about flow velocity indirectly from the burning voltage profile.

However, an excessive local flow speed or (especially in the case of direct-injecting engines) fuel droplets that move through the ignition path, can lead from extreme deflection to repeated breaking of the ignition spark. The spark must then be re-ignited, which manifests itself in the secondary voltage profile as high-frequency voltage changes (Zimmermann et al. 2004). Under these conditions, the ignition energy in the ignition coil can be used up much more quickly, which then shortens the spark duration correspondingly. In case the ignition coil in such situations does not supply enough energy, the probability of a combustion misfire or a highly delayed ignition increases.

Disturbance of ignition can also occur because of a slide spark due to a shunt. One refers to this effect when an electrically conductive path originates due to deposits on the spark plug insulator, and the ignition spark moves on this path along the insulator instead of jumping across the air gap to the mass electrode. The burning voltage profile is characterized in this case by clearly shortened combustion duration. While sometimes ignition can also succeed under such conditions, the probability of a misfire is significantly increased.

4.3.2.4 Heat Rating and Self-Cleaning

In order to prevent shunts caused by depositions, the spark plug must reach sufficiently high temperatures (>400°C) during operation in order that deposits can be burned off. The critical operating condition here is in the lowest load range with the lowest spark plug temperatures. The design of the plug, especially of the length of the insulator base, can influence its heat transfer to the cylinder head and thus the spark plug temperature. This design is characterized by the heat rating, which however is defined differently depending on the spark plug manufacturer. If the spark plug is designed such that it reaches its self-cleaning temperature already in the lowest partial load, its maximum temperature is increased as well. The highest spark plug temperatures should not exceed about 850–950° in the overall engine map, since otherwise the danger of glow ignitions is significantly increased.

4.3.3 Flame Front Propagation After Ignition, the Effect of Turbulence

Ignition of the mixture in the SI engine first requires a supply of activation energy so that the fuel molecules dissociate and oxidize with oxygen in an exothermic reaction (Winklhofer (2007)). At the point of ignition, this activation energy is supplied by the ignition spark. As soon as there is a flame front, it represents the activation energy required to ignite the surrounding fresh gas in the form of available heat. A chain reaction is initiated, in the course of which the flame front moves outwards to the combustion chamber walls and burns the fresh mixture. In the ideal case, the flame front propagates from the spark plug in a spherical shape through the combustion chamber. The best geometrical prerequisites for rapid flame propagation are thus provided by a centrally arranged spark plug, as is standard in present-day four-valve engines. The reaction progresses until the fresh mixture is used up or the flame is prematurely extinguished. This latter effect can come about if the heat released during combustion minus the heat losses can no longer supply the required activation energy to ignite the neighboring fresh gas (Winklhofer (2007)), e.g. in the case of an extremely diluted mixture (large amount of residual gas) or near wall surfaces with high heat flow from the flame front in the direction of the wall (for example, near the fire land).

The speed of flame front propagation depends heavily on the flow condition of the surrounding gas. Here, we differentiate between laminar flame propagation (in the case of a low Reynolds number, i.e. low flow speed) and a flame front in a turbulent flow field (high Reynolds number, high local flow speed in the direct surroundings of the flame front). In a laminar flow, mixing and transport of the reaction partners take place relatively slowly via molecular diffusion processes, while in a turbulent flow field they are facilitated by turbulent eddies within the flow (Heywood (1988)).

As opposed to combustion tests in chambers with resting air, a turbulent flow state builds up in engines near the ignition TDC. It is highly affected by the structure of the inlet duct, the development of the charge motion during the intake process, piston design, the influence of piston movement during the compression stroke and not the least by the engine speed (and thus the intake speed).

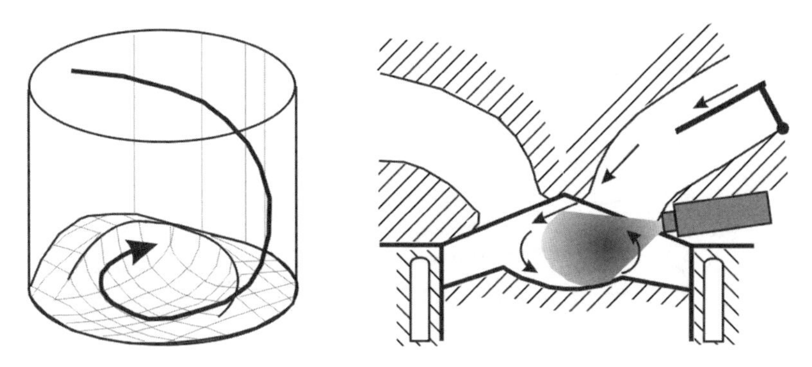

Fig. 4.26 Basic forms of global charge motion: swirl (*left*) as a swirling motion around the vertical axis of the cylinder, here in conjunction with a recessed-head type piston of a wall guided SI/DI process (Holy et al. (1998)). Tumble (*right*) as a swirling motion around the transverse axis of the cylinder, shown here with the help of a tumble valve (Grigo and Wolters 1998)

In order to obtain a distinct state of turbulence, it is important that a global charge motion (Fig. 4.26) such as a tumble vortex, created during the intake stroke, dissipates in the course of the compression and is converted into turbulence. Under these marginal conditions, the flame front in the SI engine proceeds after flame kernel formation quickly from the laminar to the turbulent state, compare Pischinger (2001). The flame front is thereby folded and thus expands its reaction surface, so that fuel dissociation and thus combustion proceed correspondingly faster (Pischinger (2001); Heywood (1988) and Winklhofer (2007)).

Swirl and tumble are the most common forms of charge motion in the combustion engine. Swirl flow is a swirling motion of in-flowing air around the vertical axis of the cylinder, which can be obtained by means of asymmetrically formed inlet ducts or by switching off one of the two inlet valves. However, a swirling flow often remains at least partially in its original form until the time of ignition instead of dissipating completely into turbulence. Tumble flow is defined as a swirling motion of cylinder air around the transverse axis of the cylinder in the direction of the crankshaft axis. If is often created, for example, in supercharged SI engines by relatively flat inlet ducts. Since such a duct shape entails certain charging disadvantages as well, high-performance naturally aspirated engines usually omit such design. As long as the combustion method requires an intensified tumble flow, it can be created as needed by closing a tumble valve in the inlet duct, which closes the lower half of the cross-section.

Combustion accelerated by turbulence is a basic prerequisite for realizing high engine speeds because only in this way can the flame front burn though from the spark plug to the combustion chamber walls completely in the short time available. Values between 0.3 and 1 m/s are quoted in the literature for purely laminar flame speeds of the reference fuel iso-octane (see Pischinger (2001)) depending on pressure and temperature (Fig. 4.27). Specifications for gasoline vary depending on the combustion air ratio in a similar way (Merker et al. (2004)), whereby the maximum speeds are indicated for the case of a rich mixture (about $\lambda = 0.83$).

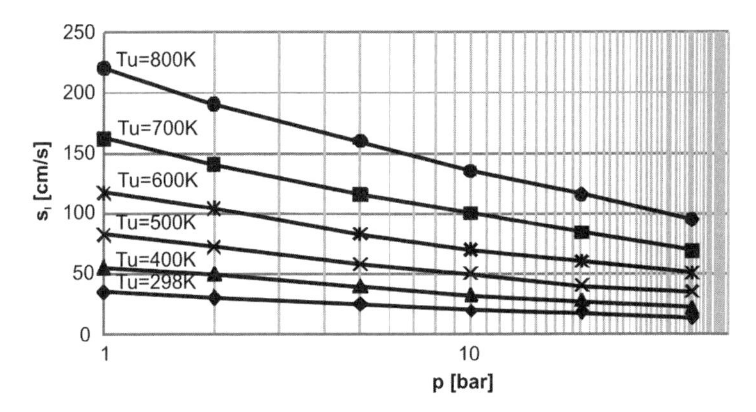

Fig. 4.27 Laminar flame front speed of iso-octane at $\lambda = 1$ acc. to Pischinger (2001)

On the other hand, flame front speeds under real turbulent conditions reach much high values and are increased in particular with the engine speed, since the higher intake speed of fresh air in the intake stroke has a positive effect on turbulence at the time of ignition. Measurements of real flame front speeds near the spark plug shortly after flame kernel formation can be made, for example, with measurement spark plugs, into which optical fibers have been integrated. The optical fibers detect the arrival of the flame front proceeding from the spark plug electrode according to the light barrier principle. By evaluating the time required for these distances, we obtain measurement values such as those in the example measurement along the full load line in Fig. 4.28. In an SI engine with distinct tumble charge motion, flame front speeds of up to 30 m/s could be established shortly after the point of ignition in this way.

4.3.4 Information About Combustion Speed from the Heat Release Rate

The effects of charge motion on flame front propagation speed can also be retraced in the heat release rate.

As explained in Chap. 3 (pressure profile analysis), the heat release rate can be determined by evaluating the first law from the pressure profile measured in the cylinder. It represents the combustion-caused energy release per degree crank angle. The cumulative heat release rate, i.e. the integral representation of the heat release rate between combustion start and end, reaches at the end point the energy chemically bound in the fuel multiplied by the degree of conversion (i.e. the completeness of combustion).

The heat release rate is usually separated into a stage of "ignition duration" (usually defined as 0–5% energy conversion) and "combustion duration" (usually defined as 5–90% energy conversion). The ignition duration thus characterizes the

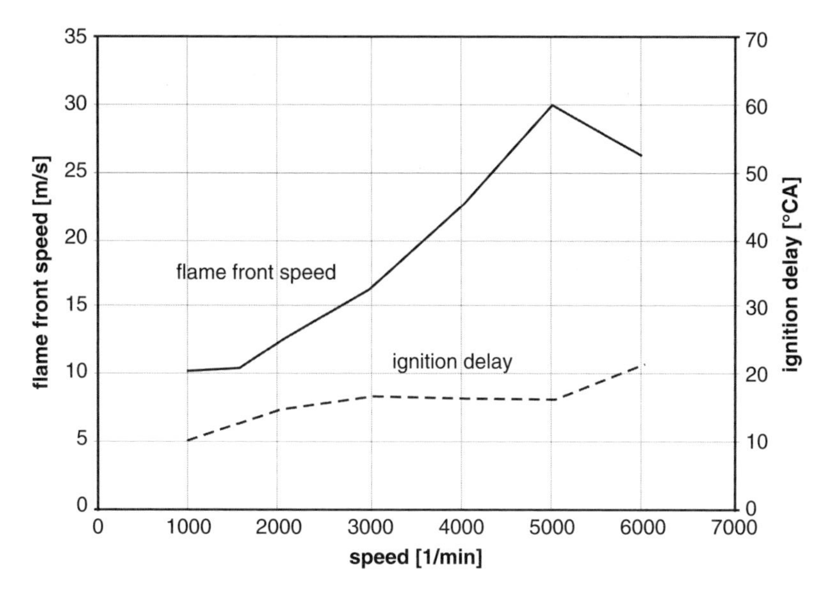

Fig. 4.28 Example measurement of a SI engine with tumble ducts to determine the flame front speed at the spark plug along the full load line

propagation of the flame front between ignition and a first measureable energy release. The combustion duration on the other hand comprises the main combustion phase, in which the flame front moves quickly through the combustion chamber and the largest percentage of the chemical energy of the fuel is released.

All engine marginal conditions that interfere with a rapid propagation of the flame front forming at the spark plug can be traced back especially to extended ignition duration and usually also to longer combustion duration in the heat release profile. Among these are included, for example, a strong charge dilution due to a large amount of residual gas or excess air. Conversely, the effectiveness of measures taken to increase turbulence can be evaluated in the form of shorter ignition and combustion durations. The design of the inlet ducts or the use of a charge motion valve can be taken as examples of ways to boost tumble flow.

Figure 4.29 compares two heat release rate profiles at the same partial load point, differing by the intensity of the tumble created. The shortening of the ignition and combustion duration in the case of increased tumble is clearly identifiable.

4.3.5 *Irregular Combustion*

The term irregular combustion comprises combustion processes in the SI engine that are not (or not exclusively) initiated by the regular ignition spark but rather by autoignition. Of particular importance among these are knocking combustions, preignitions, and glow ignitions.

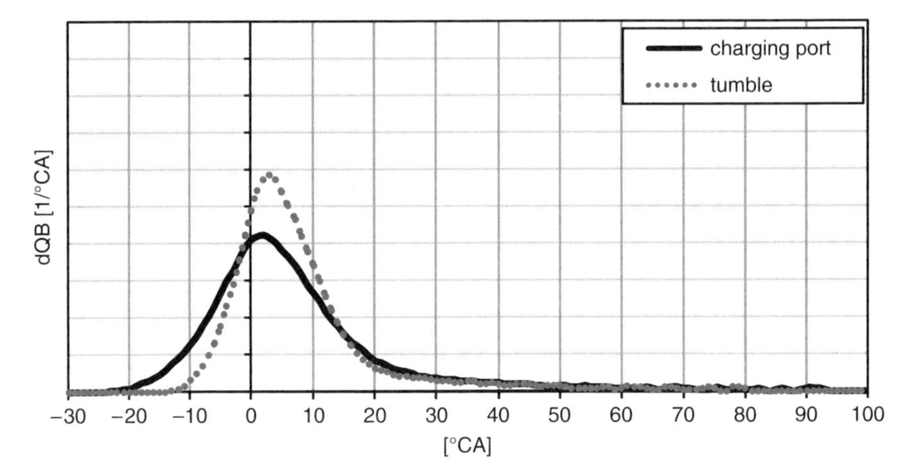

Fig. 4.29 Acceleration of the heat release rate at the homogeneous operating point 2,000/ $w_i = 0.27$ by increased charge motion in the form of tumble (standardized heat release rates)

We should differentiate between

1. Autoignition before the regular ignition time
2. Autoignition after the regular ignition time

Preignitions and glow ignitions belong to the first group, knocking combustion to the second.

4.3.5.1 Knocking

In the case of "knocking", a regular flame front is first initiated by the ignition spark. This increases the pressure and temperature in the entire combustion chamber, i.e. in the unburned mixture as well. Under increased pressure and temperature, pre-reactions take place in the fresh mixture, which in the extreme case can lead to autoignition after a sufficiently long reaction time and provided the necessary chemical and thermal prerequisites are met. In the case of knocking combustion, the regular flame front does not reach the outside area of the combustion chamber quickly enough and autoignition of the unburned mixture found there occurs (the "end gas") before the flame front arrives. The fresh mixture in this area can burn abruptly, thereby generating pressure waves that reflect against the combustion chamber walls and are visible in the measured pressure profile as a high-frequency superimposition during expansion (Fig. 4.30).

This makes it understandable, which boundary conditions increase the danger of knocking combustion:

- High compression ratio, strong charging (high final compression pressure, high temperature)
- Low speed (relatively large amount of time for the formation of prereactions)

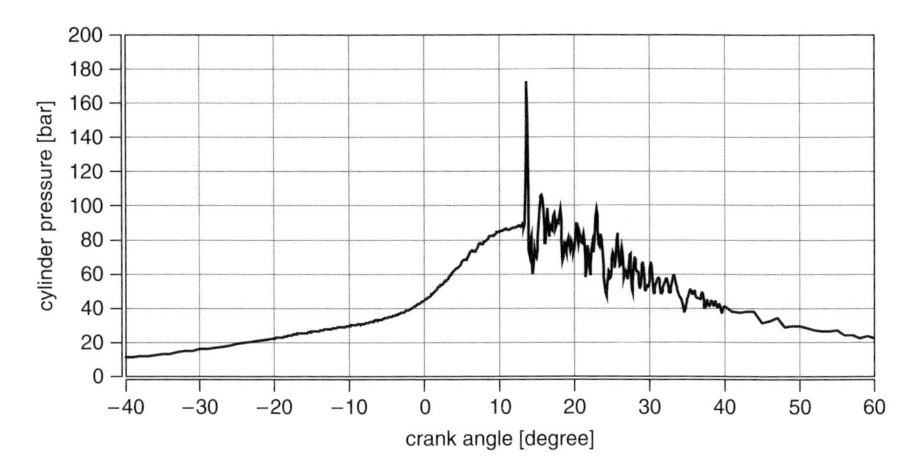

Fig. 4.30 Example of a strongly knocking combustion

- High load/cylinder charging (high final compression pressure)
- Fuel with an excessively low octane number (higher autoignition probability)
- Unfavorable combustion chamber form with long flame paths, slow combustion (flame front reaches the end gas area later)
- Insufficient charge air cooling (high temperature)
- Insufficient cooling of the combustion chamber walls

Since the pressure oscillations in knocking combustion are introduced as structure-borne noise in the engine and radiated from it as airborne noise, strongly knocking combustion is also noticeable acoustically. To assess knocking intensity on the test bench, the cylinder pressure signal can be subjected to high-pass filtering and the peak values of the high-pressure portion examined. In the vehicle, knocking sensors on the engine crankcase detect structure-borne sound excitation. If this signal is above defined threshold values, the knocking controller of the engine control unit can intervene in order to protect the engine from longer knocking operation and this from mechanical and thermal damage. Knocking control first causes a late adjustment of the ignition angle and shifts the engine thereby back into knock-free operation: by means of a later combustion start (in the extreme case even after the top dead center), a softer pressure and temperature increase takes place in the end gas area, and the thermal prerequisites for autoignition of the end gas are no longer there. In the following working cycles, the ignition angle is again shifted back into the "early" direction in small steps until knocking occurs again. In this way, the engine can be operated with a combustion position what is as efficient as possible without risking engine damage due to continuous knocking operation. Knock control thus makes it possible to operate the engine with a fuel that is not very knock-proof, however with a correspondingly less favorable combustion position and thus reduced efficiency.

In the development of combustion processes for full load, a primary focus of interest is on the detection of the points of origin of knocking in the combustion chamber. Depending on the intensity of the charge motion, knocking events could accumulate, for example, on the outlet side in the area of the hot outlet valves. However other areas in the combustion chamber are also possible. The origins of knock also need not appear at all speeds in the same combustion chamber areas. In order to take weak areas into consideration when designing the combustion chamber, a systematic investigation of knocking points of origin, e.g. with measurement spark plugs equipped with optical fibers, is sensible (Winklhofer (2007)).

4.3.5.2 Preignition

Preignitions belong to the group of autoignition that take place before the ignition time. In order to make a clear distinction from glow ignitions, we speak of preignitions in the narrow sense when they arise sporadically in particular working cycles without a noticeable antecedent, but then normal combustion occurs again. The pressure profile in the case of preignition resembles that of a combustion that has been initiated by an ignition point that is too early (Fig. 4.31). The pressure profile and gas temperature thus are significantly increased early during the compression stroke. In the case of a preignition, knocking combustion can also occur in the course of the expansion stroke, but this need not be the case necessarily.

There is still no complete picture in engine research of the possible causes of preignitions. According to experience, preignitions occur especially in the case of supercharged engines in the upper load range, but can also arise in isolated cases in natural aspirated engines. Potential causes include mixture inhomogeneities in the combustion chamber on the one hand, which can lead to local temperature differences and thus to the formation of exothermic zones. Secondly, the chipping off of carbonaceous deposits in the combustion chamber and in the inlet duct can lead to preignitions. They are heated during combustion, remain partially intact during the

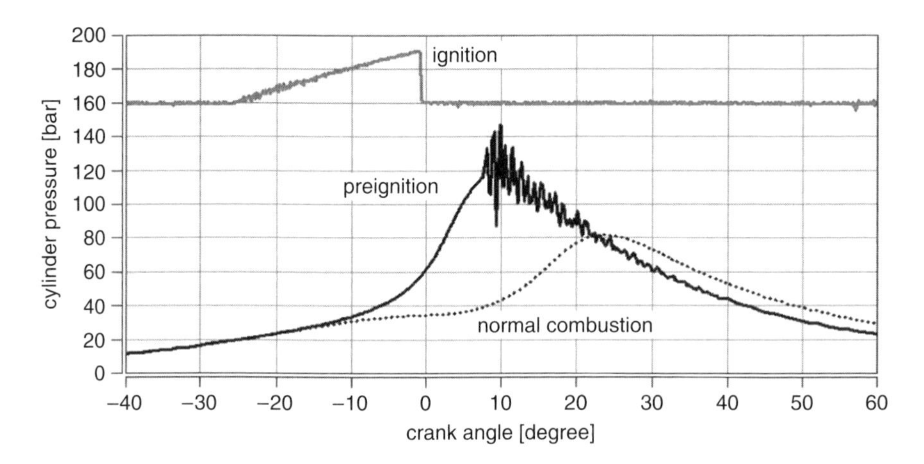

Fig. 4.31 Examples of pressure profiles with preignitions

following charge exchange in the combustion chamber and can lead to premature ignition of the fresh mixture with their high temperature during the next compression phase. The strong pressure increase associated with this can lead to the detachment of further deposits, which are heated in the next cycle and in the subsequent can in turn again initiate preignitions. This explains the characteristic sequence of preignitions and normal combustion cycles which is often recognizable. The causes described here are only some of the entire range of possible triggers however.

4.3.5.3 Glow Ignition

While preignitions were described in the previous section as sporadic events, which disappear again after a certain number of working cycles, glow ignitions involve a self-reinforcing "chain reaction":

- Excessively hot components or surfaces in the combustion chamber lead to autoignition before the ignition time, which is often associated with strong knocking
- Due to the (premature) autoignition, pressure and temperature in the combustion chamber are even further increased, the responsible component or surface is thus even further heated
- In the next cycle, autoignition takes place even earlier, further heating the combustion chamber etc

The typical sequence of a glow ignition (Fig. 4.32) thus extends across several cycles, which are characterized by autoignitions that begin increasingly early and initially strong knocking. At the end of the process, autoignition takes place so early

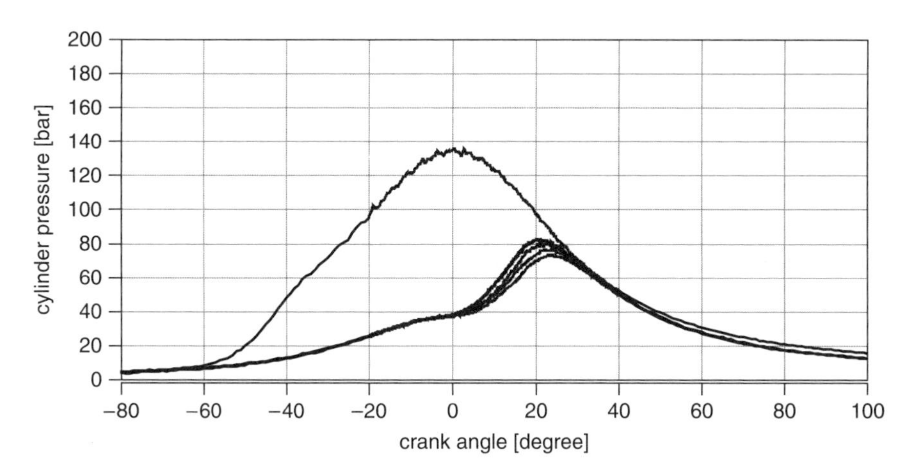

Fig. 4.32 Example of a pressure profile of a fully developed glow ignition without knocking on one cylinder

that no more knocking is identifiable in the pressure profile. The combustion chamber undergoes major thermal overload however, so that usually massive engine damage results at the end of the process (e.g. melted spark plug electrodes, burnt portions of the outlet valves or thermally damaged piston surfaces).

Once a surface in the combustion chamber has become overheated and the glow ignition sequence has begun, late adjustment of the ignition angle as applied in classic knock regulation is of no help: since autoignition already takes place before the ignition spark, a delayed ignition spark has no effect.

Overheating of the combustion chamber must therefore be avoided from the outset. Measures include the classic cooling jacket in the combustion chamber, avoidance of sharp-edged (and therefore more heat-sensitive) surfaces (e.g. on the surface of the piston), careful construction of the spark plug shape and of the heat rating in order to prevent an excessively hot spark plug surface, and a reliable knocking control function.

4.3.6 Combustion Process, Mixture Formation, Modes of Operation

Derivatives of SI engine combustion methods which are presently employed in series or are in development by some manufacturers can be classified as follows:

- *Homogeneous combustion*: The air fuel ratio λ is 1 in almost the entire characteristic map for reasons of optimal exhaust gas treatment with the three-way catalytic converter. Mixture formation takes place at an early time during the working cycle, either by injection of the fuel into the intake manifold or – in most new designs – by direct injection into the combustion chamber during the intake stroke. The mixture thereby attained exists ideally in a homogeneous form with the same λ in the entire combustion chamber.
- *Stratified combustion*: Mixture formation takes place via direct injection into the combustion chamber during the late compression phase. The wall and air guided combustion methods still used a few years ago has in the meantime been replaced by spray guided combustion methods. The spark plug must be positioned on the edge of the spray such that it reaches a locally narrow area with ignitable mixture directly on the edge of the injected fuel spray. As soon as this area is ignited, the flame front can burn even extremely lean zones throughout the combustion chamber in its broad propagation, allowing for global combustion air ratios of $\gg 1$. For this reason, an additional exhaust gas treatment for nitrogen oxide with an excess of air is required in addition to the three-way catalytic converter.
- *HCCI combustion with homogeneous autoignition*: Instead of ignition with a spark plug, this method aims at autoignition of the homogeneous (stoichiometric or slightly lean) mixture. In order to obtain the necessary thermal prerequisites for autoignition of the poorly reactive SI fuel, the combustion chamber must

contain a very large amount of hot residual gas by means of a selection of the valve control times. Because the mixture is considerably diluted, not only is a high level of efficiency obtained, but there are also extremely low amounts of NO_x emissions because of the effect of inert gas and "cold" combustion. As opposed to stratified combustion, the goal is to make a separate NO_x exhaust gas treatment unnecessary.

Stratified operation or the HCCI combustion method can only be applied in a limited range of partial loads. In the operational characteristic map of direct-injecting SI engines, they therefore represent one of several operation modes. In each case, they are supplemented with the operation modes "homogeneous operation" (at higher speeds/loads), "catalytic converter heating" (directly after cold start) and potentially other modes of operation (e.g. "homogeneously lean" or "homogeneous stratified" in the load transition range between stratified operation and stoichiometric operation).

4.3.6.1 Homogeneous Operation with Intake Manifold Injection

Intake manifold injection is implemented today exclusively as multipoint injection. Every cylinder has its own injection valve which is located either in the associated intake runner or directly in the inlet duct of the cylinder head.

The injection valves are supplied with fuel by a common injection rail (Fig. 4.33). Injection pressure is built up by the feed pump and can typically amount to between 3 and 5 bar depending on the system design, sometimes even up to 10 bar. In some cases, the injection pressure is controlled such that a constant pressure difference between the fuel pressure and manifold pressure is established, while other systems work with constant injection pressure (and thus with a variable pressure difference).

For the four-valve cylinder heads common today, there are injection valves with two (usually conical) injection sprays aligned in the direction of both inlet valves. When installed near the valve, wetting of wall surfaces with fuel is reduced, which can minimize potential problems with the formation and detachment of a fuel wall film in case of an unsteady change of manifold pressure.

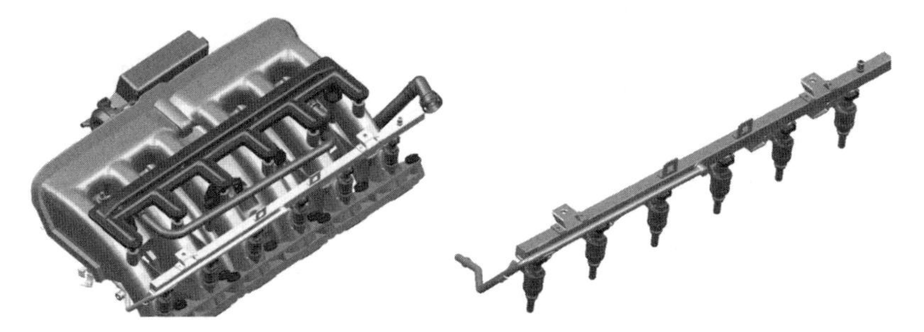

Fig. 4.33 Example of the arrangement of the fuel injection line of a manifold injection system on the intake runners, Albrecht et al. (2000)

The time of injection is determined by the engine control individually for each cylinder. In practice, an early injection time after the end of the intake stroke has become established. In the case of this "earlier" injection, the fuel spray strikes the closed hot intake valves and there is enough time until the next opening of the valves to evaporate as completely as possible. Another possible injection strategy is to inject during the intake stroke while the valves are open and to utilize the speed of the in-flowing fresh air for a best-possible mixture preparation. This method is often used in non-steady operation. It is especially advantageous however in the case of fully variable valve trains with a small intake valve lift in the lower partial load range. In this situation, the in-flowing fresh air in the narrow intake valve gap can reach up to the speed of sound and thus permit a best-possible atomization of the fuel droplets (Fig. 4.34).

The use of variable control times can support mixture formation in still another respect. In the case of internal exhaust gas recirculation via early inlet opening, the hot residual gas arrives in the intake duct and improves evaporation of the injected fuel due to the increase in gas temperature. Alternatively, a similar effect can be realized with the method of late outlet closing, whereby the hot exhaust gas raises the mixture temperature not in the intake duct but later in the combustion chamber.

When the fuel is injected into air, a spectrum of variously sized droplets is formed, which evaporate with different speeds depending on the injection pressure and the intake manifold pressure and temperature.

One characteristic for this spectrum of droplets is the Sauter mean diameter, which describes the diameter of a representative droplet. The ratio between the volume and surface of this droplet has the same value as the ratio of the overall

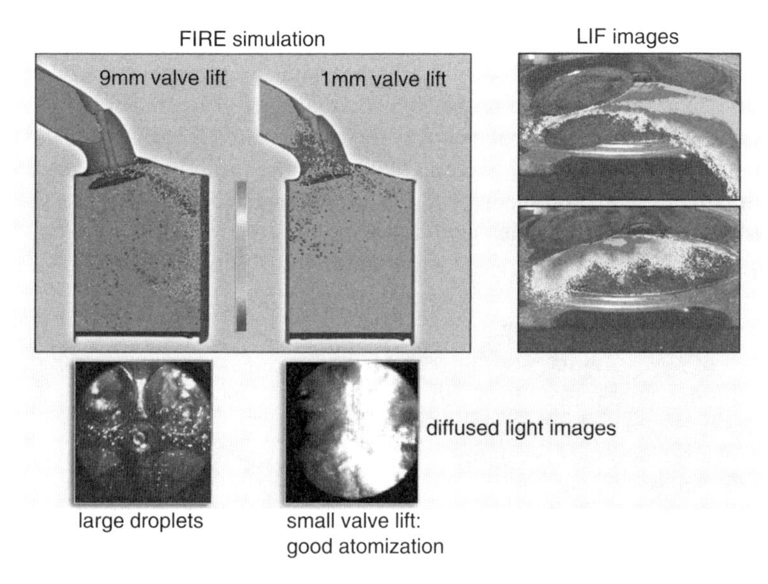

Fig. 4.34 Smaller droplet size and good mixture preparation with manifold injection and reduced intake valve lift, Liebl et al. (2001)

volume of the droplet spectrum to the entire surface of all droplets. A small Sauter mean diameter thus characterizes a droplet spectrum with many small droplets that evaporate with corresponding ease.

With low-priced injection systems, the possibility of further reducing the Sauter mean diameter by increasing the injection pressure is limited. Significantly higher injection pressures can only be obtained with separate high-pressure pumps for direct-injecting engines.

In the case of the engine running under hot conditions, a very good mixture preparation can be obtained with an optimized manifold injection.

Directly after cold start on the other hand, both the intake valve surface and the walls of the intake manifold – and the aspirated air with it – are still at low temperatures. The fuel droplets thus evaporate poorly, and the probability increases that relatively large droplets are injected into the combustion chamber, wetting the chamber walls in a liquid state and participating in combustion either not at all or incompletely. This effect is an important reason for increased HC emissions after cold start and in the warm-up phase. Furthermore, there is an increased risk of oil dilution when the engine is cold if fuel in droplet form wets the oil film on the cylinder liner without evaporating.

4.3.6.2 Direct Injection with Homogeneous Operation

Homogeneous direct injection originally was developed as a secondary mode of operation of the first stratified combustion methods for the upper load range. In the meantime, many newly developed (forced induction in particular) SI engines with direct injection are designed exclusively for homogeneous operation, since the properties and flexibility of direct injection harmonize ideally with charging.

As in manifold injection, load is controlled in accordance with the principle of quantity control in homogeneous direct injection methods as well. The mass of air and the injected fuel mass thus correspond to the air fuel ratio $\lambda = 1$ for the best-possible exhaust gas treatment with the three-way catalytic converter. Fuel is injected under high pressure (50–200 bar) directly into the combustion chamber. The injectors are supplied with fuel in accordance with the common rail concept from a common fuel line after the high-pressure pump. The injectors can be arranged at the side or in the center of the cylinder head. While swirl injectors were still installed (sidewise exclusively) a few years ago, multihole coil injectors (arranged on the side or centrally) or outward-opening piezo-injectors (central) are being applied increasingly in new designs (Fig. 4.35).

The injected fuel vaporizes in the combustion chamber and extracts the heat required for this from the surrounding air in the chamber, the temperature of which drops accordingly. With the same compression ratio ε as in manifold injection, the compression temperature and resultantly the knocking sensitivity would be reduced. This effect of "internal cooling" is however normally used to increase the compression ratio with the same knocking sensitivity by 1 or 2 units to the

Fig. 4.35 Example of a centrally arranged injector with homogeneous direct injection, Mährle et al. (2007)

advantage of efficiency. Typical values of ε for current direct-injecting naturally aspirated engines are around 12, and 10–10.5 for turbo engines.

In comparison to manifold injection, with direct injection, less time is available for mixture formation. Injection start is usually in the first half of the intake stroke (e.g. 340–280° before ITDC). The aim then is, on the one hand, to inject sufficiently early in order to have enough time for mixture formation. On the other, direct contact of the injection spray with the piston or cylinder liner must be avoided as much as possible because wall contact can lead to increased HC emissions, soot and oil dilution. Adequately high piston temperatures (at higher speed and load) and a larger distance between the injector and piston recess can help to minimize the negative effects of wetting the piston. The injection position, spray properties, piston shape, and the interaction of the injection spray with the charge motion play a crucial role in mixture formation effects and must be optimized with CFD methods.

In special cases, multipoint injection can be worthwhile for supporting the described mixture formation effects at high load (and thus with larger fuel amounts). The flexibility in selecting the injection time is increased in the case of very rapid-switching injectors with a short activation dead time before the actual opening of the injector.

With respect to charge motion, one must differentiate between naturally aspirated engines and supercharged engines. Since any kind of charge motion provisions lead to reduced fresh air charging and increased gas exchange losses, in naturally aspirated engines one generally dispenses with them, or switchable systems (e.g. tumble valves) are used that are only activated when needed. Super- or turbocharged engines on the other hand rely on charge motion in the interest of a sufficiently fast combustion and thus a favorable center of combustion at full load. They are thus often designed with distinct tumble ducts, hazarding the consequences of the resultant disadvantages with respect to charge changing work.

4.3.6.3 Direct Injection with Stratified Operation

In the middle of the 90 s, the first direct-injecting SI engines with stratified combustion were introduced. In order to reduce fuel consumption, the principle of quantity regulation (as in conventional $\lambda = 1$-operation) with load-dependent throttling is no longer used for load control, but rather – similar to the diesel engine – quality regulation. Load is shifted primarily by changing the injected fuel mass, while in the theoretical ideal case with a fully opened throttle valve always the maximum air mass is inducted. Charge changing losses can be kept very low in this way. In practical use, the intake manifold pressure must be reduced somewhat in the lowest load range in order to ensure a lowest level of exhaust gas temperature beyond which a sufficiently high catalytic converter degree of conversion is guaranteed.

Operation with an excess of air brings about additional advantages in efficiency in the high-pressure process. The excess air acts as an inert gas. It does not participate actively in combustion, but it must be heated also, thereby reducing the temperature level during and after combustion. This also reduces the temperature-dependent specific heat capacities c_p and c_v of the working gas, and the isentropic coefficient $\kappa = c_p/c_v$ is correspondingly increased. Already from the SI engine idle process, it is known to be a main influencing parameter on high-pressure efficiency in addition to the compression ratio.

If an SI engine is to be operated in partial load with a high global excess of air (λ up to approx. 4), additional measures must be taken to be able to ignite and burn such a lean gasoline/air mixture at all. Different sources in the literature indicate values for a homogeneous mixture in the area of about $0.6\text{–}0.8 < \lambda < 1.5\text{–}1.6$ (e.g. Merker et al. (2004), van Basshuysen and Schäfer 2010). For this reason, one must be sure in the case of the lean-operated direct-injecting SI engine that charge stratification takes place: there should be a cloud of mixture with a locally ignitable composition near the spark plug, while a large amount of excess air is in the outer area of the combustion chamber. To obtain this charge stratification, the fuel is injected very late during the compression phase, in extreme cases very near the ignition time. In the short amount of time available till ignition, the fuel must mix with the air in the combustion chamber and an ignitable mixture cloud form at the spark plug. Investigations concerning the concentration of fuel at the spark plug allow the conclusion that the ignition spark in stratified operation can ignite a richer mixture in the immediate area of the spark plug than in homogeneous operation (see Witt and Kern 2004 and Fischer et al. 2004). Once the flame has formed near the spark plug, it propagates through the combustion chamber though areas with a very high λ gradients proceeding from sometimes extremely rich mixture at the spark plug through stoichiometric zones to very lean zones at the edge of the combustion chamber.

To realize this kind of stratified combustion, different combustion process variants have been developed in the past:

- The wall guided process
- The air guided process
- The spray guided process

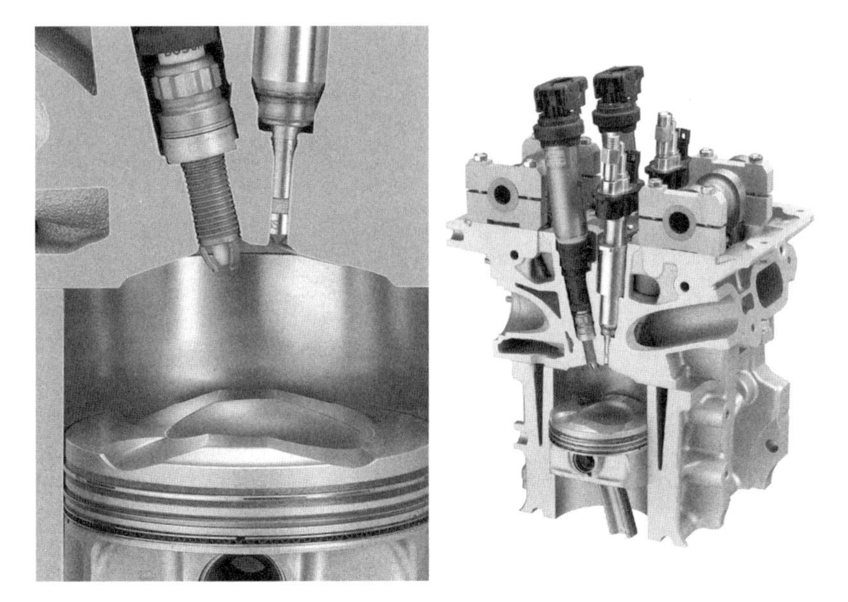

Fig. 4.36 Example of the central injector arrangement of a spray guided combustion process, BMW Presseinformation

These variants are intimately linked with the position of the injector, which can be installed on the side or center of the cylinder head.

In the case of SI/DI engines of the initially mass-produced first generation (wall and air guided), a side injector position was selected. In currently available engines, one can still find injectors positioned on the side, but in the meantime almost exclusively for use with homogeneous direct injection. In modern SI engines with stratified combustion, the injector is installed centrally, and a spray guided combustion process is employed (Fig. 4.36).

Properties of Different stratified Combustion Processes

The *wall guided process* was used in the first SI/DI engines with stratified operation. The characteristic of this process is a distinct piston recess with a deflection function.

The in-flowing air is guided along the combustion chamber walls and the piston recess. The fuel spray is injected during compression from the side in the direction of this recess. The design of the recess and the upwards motion of the piston divert the fuel upward towards the spark plug. The diverted fuel cloud must mix with the surrounding air on its way to the spark plug and form a stable, ignitable mixture at the spark plug at the point of ignition.

This mixture formation process is usually also facilitated by a distinct charge motion in the form of swirl or tumble in both the wall guided and air guided combustion processes (see Fig. 4.37).

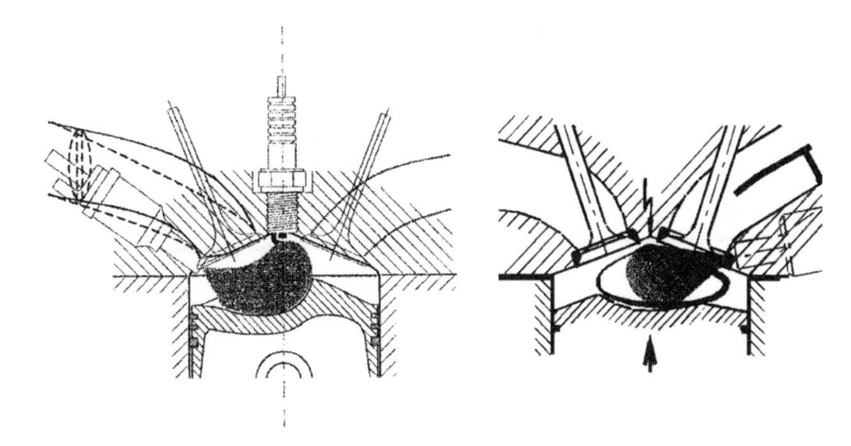

Fig. 4.37 Left: mixture formation in the wall guided process, acc. to Holy et al. (1998). Right: mixture formation in the air guided process with tumble flow, acc. to Grigo and Wolters (1998)

The *air guided combustion process* almost always involves tumble as a charge motion. Demarcation from the wall guided method is more or less fluid. The piston bowl is not as distinct as in the wall guided method and facilitates above all deflection of the side-injected (under a shallower angle) fuel through the entering air, flowing in a tumble pattern in the direction of the spark plug. In the ideal case, this diversion of fuel should not touch the piston. This requirement cannot be realized fully in practice, but at least the piston is not wetted as much.

Both the wall guided and – to a somewhat lesser degree – the air guided process thus must contend with the problem of wetting of the piston with fuel. However, this has the direct result that part of the fuel is no longer evaporated and mixed, but remains in liquid form on the piston, is burned incompletely and is found again in the form of unburned hydrocarbons in the exhaust gas (Fig. 4.38). The resultant loss of efficiency can become considerable in the lower partial load range and thus eliminates part of the advantage of efficiency of stratified operation found in the ideal state.

A further problem of both the wall and air guided combustion processes is the limited characteristic map range in which stratified operation is possible. The upper speed limit is defined by the limited stability of the injection spray and the increasingly poorer mixture preparation. Load is limited upwardly by the clearly increasing soot emissions because the local mixture composition at the spark plug becomes too rich in case of higher injection amounts. In practice, this limits the usefulness of stratified operation to the lower to middle partial load range (typically up to about 3,000 rpm, $w_e = 0.4$ kJ/l).

For these reasons, increased efforts have been made to develop the spray guided combustion method, which in the meanwhile has been introduced to the market in series.

One essential difference to wall or air guided methods is that the spray of a centrally mounted injector is guided directly by the spark plug. The ignitable

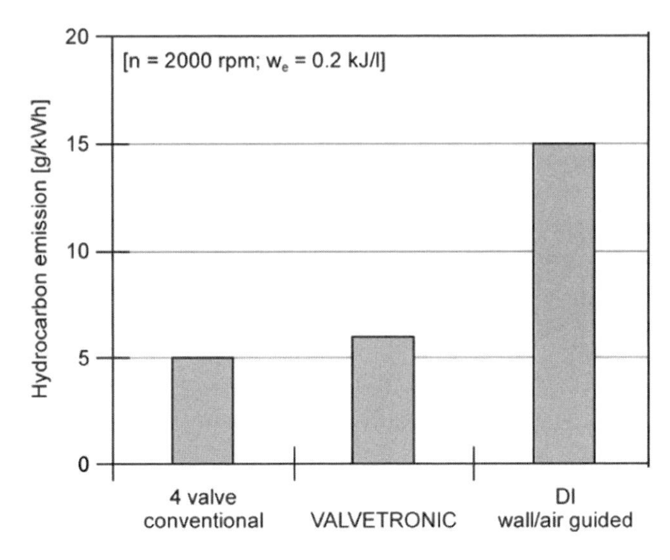

Fig. 4.38 Increasing the HC mass flow at operating point 2,000 rpm, $w_e = 0.2$ kJ/l in the wall-guided process compared to stoichiometrically operated engines, Göschel (2006)

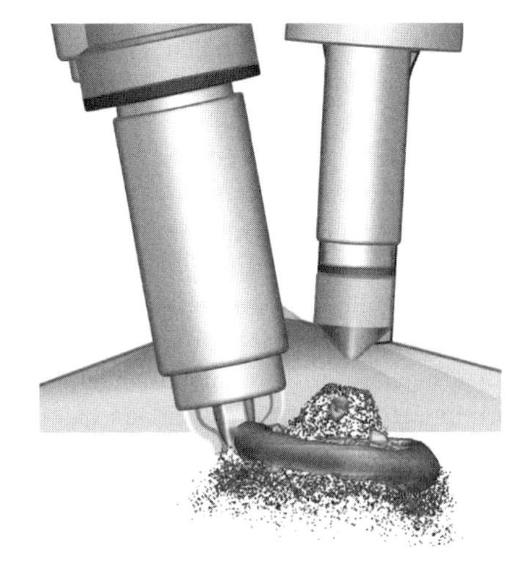

Fig. 4.39 Position of the spark plug at the edge of the spray in spray guided combustion, Fröhlich et al. (2003)

mixture directly at the edge of the spray at the spark plug position is thus a top prerequisite of a reliable ignition. Then, the detour via the piston recess is superfluous, as is most of the efficiency-lowering fuel wetting of the piston. HC emissions can be significantly reduced to the level of good stoichiometrically operated engine (Fig. 4.39).

Due to the spark plug position at the edge of the spray, only a spatially very narrow area is available for mixture formation. The best prerequisites for this are provided by outwards-opening injectors which emit a hollow conical spray with a ring-shaped recirculation vortex at the spray edge. The formation of this swirl area under the conditions of compression injection is characterized by the interaction of the injection spray penetrating at high speed with the nearly resting air in the combustion chamber. The flow state in the surrounding air at the spray edge is much more strongly influenced by the spray itself than by global charge motion. In optimal mixture formation conditions, there is firstly good air access in the vortex area. On the other hand, the fuel spray must consist of a droplet spectrum in which the droplets can vaporize quickly (i.e. they are sufficiently small), while at the same time a stable reproducible outer swirl area is built up from injection to injection. This spray shape must also remain stable in case of a very late injection timing in the compression phase and thus under high cylinder pressure and high gas temperature.

Figure 4.40 shows the spray pattern of an outwards-opening piezo-injector under two extreme conditions. On the left side, fuel is injected into the injection chamber under ambient conditions (i.e. approximately in the conditions of homogeneous operation when injecting in the intake stroke). The right side shows the spray in a state as in the case of a very late injection in the compression stroke under high surrounding pressure and high gas temperature. The increased gas density reduces the penetration length of the spray, but the cone angle and the characteristic boundary vortex remain intact.

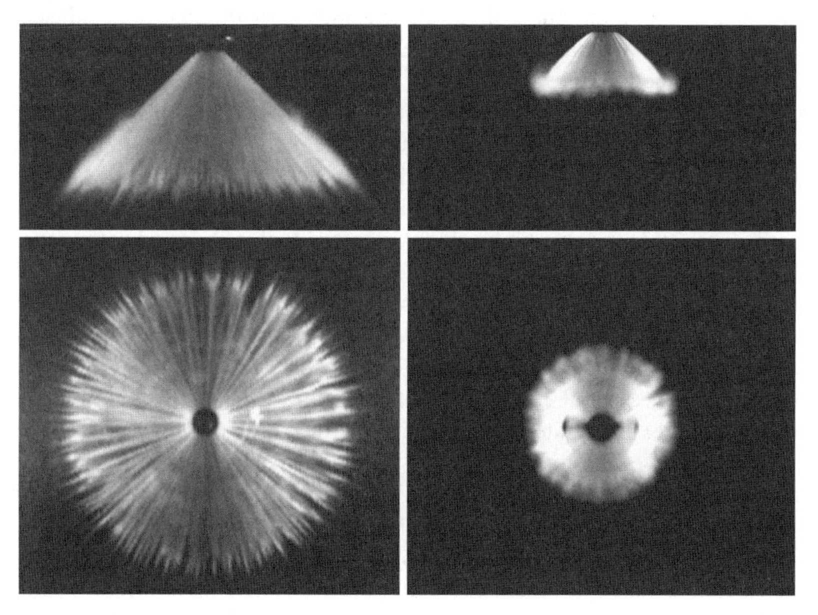

Fig. 4.40 Spray pattern of an outwards-opening piezo-injector under environmental conditions (*left*) and with a late injection time (*right*), Fröhlich et al. (2003)

Injection pressure: 200 bar	Injection pressure: 200 bar
Chamber pressure: 1 bar	Chamber pressure: 18 bar
Chamber temperature: 20°C	Chamber temperature: 400°C
Exposure time: 0.3 ms	Exposure time: 0.3 ms

Injectors with a piezo-drive that open extremely quickly also permit several successive injections. This property allows extensive design freedom with respect to the local air fuel ratio at the spark plug by means of a corresponding injection strategy, with which the useful characteristic map range in stratified operation can be significantly expanded.

The total amount of fuel is customarily split into several short successive partial injections such that the duration of the first injection is changed depending on the load, followed shortly thereafter by a second, very short injection – under higher load often also a third (also very short) injection. The distribution of mass, the dwell between the injections and the choice of ignition time relative to the end of the last injection are of decisive influence on the local air fuel ratio near the spark plug at the time of ignition.

4.3.6.4 Homogeneous Autoignition

The last several years have seen increased activity in both the research and development of a new SI engine partial load combustion process with homogeneous autoignition. The goal is to increase the efficiency of partial load under stoichiometric or lean operation, but without the necessity of additional NO_x exhaust gas treatment such as is required in SI engines with stratified operation.

Thus, to reduce the gas exchange loss characteristics of throttled operation, the aim is to dilute the mixture heavily with residual gas ($\gg$30–80%). At the same time, this entails a drastic reduction of NO_x emissions. A stable ignition by means of a conventional ignition system at such high inert gas rates cannot be ensured however. Instead, autoignition is required.

Commercially obtainable high octane SI engine fuel is designed with respect to its composition to avoid undesired, uncontrolled autoignition (knocking). The thermal boundary conditions for a purposeful, controlled autoignition in the lower partial speed range thus differ fundamentally from the customary spark-ignited SI process. In particular, a sufficiently high compression temperature in the combustion chamber is essential in order to make autoignition of the mixture in the chamber possible. This state can be realized if the combustion chamber contains a large amount of hot residual gas. Fully variable valve trains lay the best foundations for this. The methods of "residual gas storage" (negative valve overlap with residual gas compression in the charge changing TDC with a small valve lift and short opening duration) or "residual gas recycling" (late outlet closing with re-suction of the exhaust gas from the outlet duct, or alternatively a re-opening of the outlet during the intake process) have become established to this end (see Kaufmann et al. (2004)).

The heat release rate in a HCCI combustion process typically shows an extremely rapid energy conversion (combustion duration approx. 10°KW). In optical images of combustion chambers, a classic flame front is not recognizable, but rather the mixture ignites at several locations in the combustion chamber nearly simultaneously.

The local tendency to autoignition at individual locations in the combustion chamber is characterized by the ignition delay and thus by the local temperature and local mixture composition of fresh air, fuel, and residual gas (see Kaufmann (2005) and Maiwald (2005)). Depending on the mixture formation effects and valve control times, one cannot necessarily assume a consistent temperature or completely homogeneous mixture in the entire combustion chamber in practice. As soon as autoignition takes place at those locations with the best prerequisites for ignition, pressure and temperature are increased, which causes other locations in the chamber to reach the conditions of autoignition, and energy conversion then proceeds very rapidly.

The limiting factors for the autoignition range in the characteristic map are determined on the one hand in the lowest load range by the obtainable final compression temperature and on the other by the permissible speed of pressure increase. The latter can take on critical values, especially in the higher load range, due to the nearly simultaneous, abrupt combustion. According to Kaufmann et al. (2004), residual gas storage with intermediate compression is better suited to operation in the lowest engine map range, since the residual gas remaining in the cylinder brings about higher compression temperatures than recycling from the outlet duct. However, this also increases the speed of pressure increase when the characteristic map is extended for higher loads.

In the final analysis, the controllability of the desired thermal state in the combustion chamber by means of a characteristic map range that is as large as possible remains decisive for the prospects of success of the method. Possibilities of control can not only involve valve control times and valve lift curves that are as flexible as possible but also, for example, the injection strategy (Fig. 4.41). Kaufmann (2005) describes a possibility of expanding the HCCI map to idle operation, such that injection takes place near the gas exchange TDC (i.e. into the hot, intermediately compressed residual gas) using the method of residual gas storage. In contrast to a late injection, the physical values of cylinder charging

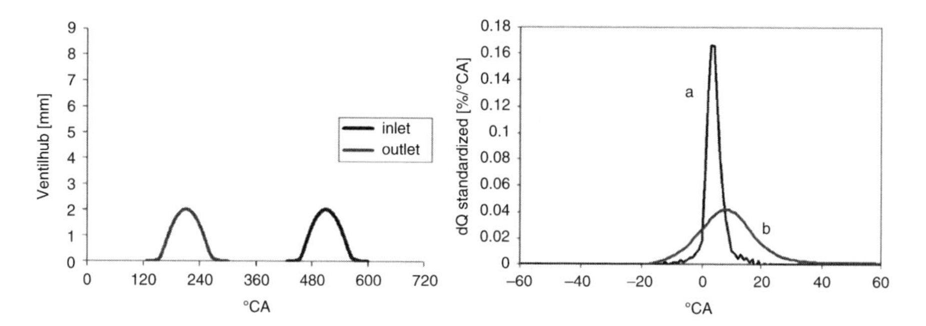

Fig. 4.41 *Left*: valve lift curves for residual gas storage. *Right*: heat release rate for HCCI operation (**a**) and conventional operation (**b**) at operating point 2,000 rpm, $w_i = 0.27$ kJ/lt (Kaufmann et al. 2004)

change thereby during the intake process. The gas temperature can thus reach up to 80 K higher values shortly before ITDC. The time of autoignition can therefore be significantly influenced by a selection of the injection time.

4.3.6.5 Attributes of the Mode of Operation "Catalytic Converter Heating"

The mode of operation "catalytic converter heating" is applied immediately after cold start and should be described in more detail because of its quite specific demands on the combustion method (see also Fig. 4.42).

The constantly tightened exhaust gas regulations require a very rapid, intensive heating of the catalytic converter in the first seconds after cold start. Only after a minimum operating temperature has been obtained is a sufficiently high pollutant conversion rate ensured (especially of unburned hydrocarbons).

To this end, the engine must be operated with an increased idling speed (up to about 1,200 rpm), higher air supply, and extremely late combustion in order to convert as much of the supplied fuel energy into exhaust gas enthalpy as possible. The challenge for practical conversion is to realize this operating state with acceptable running smoothness and low raw HC emissions.

To meet these demands, one possibility is to run the engine with $\lambda < 1$, but simultaneously feeding additional air into the exhaust manifold with a secondary air pump. The products of incomplete combustion, CO and HC react in the manifold exothermically with the secondary air and thereby cause additional heating of the exhaust gas.

In the case of SI engines with direct injection, one can alternately use a suitable multipoint injection strategy. The initial injection takes place during the intake stroke just like in classical homogeneous operation. A late-offset short second injection guarantees local mixture enrichment at the spark plug, stabilizing ignition.

In engines with a sidewise injector position, this second injection takes place usually before the ignition TDC, but ignition first occurs after the ignition TDC.

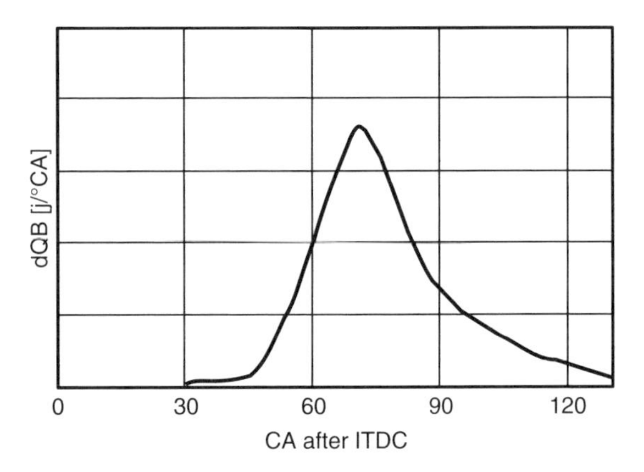

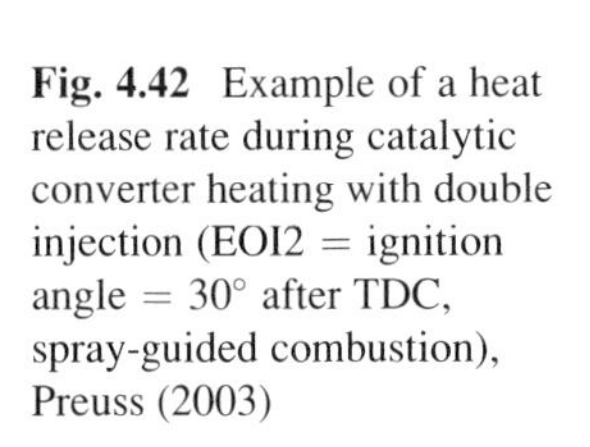

Fig. 4.42 Example of a heat release rate during catalytic converter heating with double injection (EOI2 = ignition angle = 30° after TDC, spray-guided combustion), Preuss (2003)

Suitable design of the piston recess (as in stratified operation) helps divert the fuel in the direction of the spark plug.

In engines with a central injector and spray-guided combustion, the second injection can be offset, in the most favorable case, after ITDC directly before the ignition time. The second injection impulse not only causes local mixture enrichment at the spark plug but also increases the turbulence there. Together, these effects permit a stable, extremely late combustion with relatively little HC emission.

References

Albrecht F, Fischer HD, Kiefer W, Mertl R, Otto E, Griebel CO (2000) Die Technik der neuen BMW Sechszylindermotoren MTZ 61. Vieweg, Wiesbaden

Arcoumanis C, Gavaises M (1998) Linking the nozzle flow with spray characteristics in a diesel fuel injection system. Atomization and Sprays 8:179–197

Baddock C (1999) Untersuchungen zum Einfluss der Kavitation auf den primären Strahlzerfall bei der dieselmotorischen Einspritzung, Dissertation, Universität Darmstadt, Shaker

Baumgarten C (2006) Mixture formation in internal combustion engines, heat and mass transfers in sprays. Springer, Berlin, ISBN 3540308350

Boecking F, Dohle U, Hammer J, Kampmann S (2005) PKW-Common-Rail-Systeme für künftige Emissionsanforderungen, Motorentechnische Zeitschrift MTZ 66, 552–557, Vieweg Verlag, Wiesbaden

BMW Presseinformationen. Internet: http://www.press.bmwgroup.com

Curran HJ, Gaffuri P, Pitz WJ, Westbrook CK (1998) A comprehensive modeling study of n-Heptane oxidation. Comb Flame 114:149–177

Dec JE (1997) A conceptual model of DI diesel combustion based on Laser-Sheet Imaging, SAE paper 970873

Deutsche Wissenschaftliche Gesellschaft für Erdöl, Erdgas und Kohle e.V. (2003) DGMK Forschungsbericht 502–1 Zusammensetzung von Ottokraftstoffen aus deutschen Raffinerien – Winterware 2001/2002, DGMK, Hamburg

Deutsche Wissenschaftliche Gesellschaft für Erdöl, Erdgas und Kohle e.V. (2002) DGMK Forschungsbericht 583, Zusammensetzung von Dieselkraftstoffen aus deutschen Raffinerien, DGMK, Hamburg

DIN EN 228 (2004) Kraftstoffe für Kraftfahrzeuge – Unverbleite Ottokraftstoffe – Anforderungen und Prüfverfahren

DIN EN 590 (2008), (Entwurf), Kraftstoffe für Kraftfahrzeuge – Dieselkraftstoffe – Anforderungen und Prüfverfahren

DIN EN 14214 (2008) (Entwurf), Kraftstoffe für Kraftfahrzeuge – Fettsäure-Methylester (FAME) für Dieselmotoren – Anforderungen und Prüfverfahren

Fischer J, Xander B, Velji A, Spicher U (2004) Zyklusaufgelöste Bestimmung des lokalen Luftverhältnisses an der Zündkerze bei Ottomotoren mit Direkteinspritzung. Internationales Symposium für Verbrennungsdiagnostik, Baden-Baden

Flynn PF, Durrett RP, Hunter GL, zur Loye AO, Akinyemi OC, Dec JE, Westbrook, CK (1999) Diesel combustion: an integrated view combing laser diagnostics, chemical kinetics and empirical validation, SAE paper 1999-01-0509

Fröhlich K, Borgmann K, Liebl J (2003) Potenziale zukünftiger Verbrauchstechnologien, Wiener Motorensymposium

Glassman I (1988) Soot formation in combustion processes, Proc. of the 22nd international symposium on combustion, The Combustion Institute, pp 295–311

Göschel (2006) Die Zukunft des ottomotorischen Antriebs. Vorlesung Technische Universität Graz

Grigo M, Wolters P (1998) Direkteinspritzung im Ottomotor – Chancen, Risiken, Trends. In: Direkteinspritzung im Ottomotor, Expert
Heywood JB (1988) Internal Combustion Engine Fundamentals. McGraw-Hill, Inc.
Higgins B, Siebers D, Aradi A (2000) Diesel-spray ignition and premixed-burn behavior. SAE paper 2000-01-0940
Holy G, Piock W, Wirth M (1998) Ottomotorenkonzepte mit Direkteinspritzung für EURO III/IV. In: Direkteinspritzung im Ottomotor. Expert Verlag
Kaufmann M, Beckmüller M, Hartmann C, Brehm N, Eder A, Schwarz C (2004) Methodeneinsatz bei der Entwicklung des ottomotorischen HCCI-DI Brennverfahrens. Int. Symp. Verbrennungsdiagnostik. Baden-Baden
Kaufmann M (2005) Thermodynamische Analyse des kompressionsgezündeten Benzinmotors. Dissertation, Technische Universität Graz
Kronenberger M, Jovovic D, Pirkl R, Voigt P (2005) Pumpe-Düse-Einspritzelemente mit Piezo-Aktor für EU-4-Dieselmotoren, Motorentechnische Zeitschrift MTZ 66, Vieweg, Wiesbaden, pp 354–360
Kühnsberg-Sarre C von, Kong S-C, Reitz RD (1999) Modeling the effects of injector nozzle geometry on diesel sprays. SAE paper 1999-01-0912
Leonhard R, Warga J (2008) Common-Rail-System von Bosch mit 2000 bar Einspritzdruck für PKW, Motorentechnische Zeitschrift MTZ 69, Vieweg, Wiesbaden, pp 834–840
Liebl J, Poggel J, Klüting M, Missy S (2001) Der neue BMW Vierzylinder-Ottomotor mit Valvetronic. Teil 2: Thermodynamik und funktionale Eigenschaften. MTZ 62, Vieweg Verlag
Mährle W, Krause M, Luttermann C, Klauer N (2007) High Precission Injection in Verbindung mit Aufladung am neuen BMW Twin-Turbo-Ottomotor. MTZ 68, Vieweg Verlag
Maiwald O (2005) Experimentelle Untersuchungen und mathematische Modellierung von Verbrennungsprozessen in Motoren mit homogener Selbstzündung. Dissertation Uni Karlsruhe
Merker GP, Schwarz C, Stiesch G, Otto F (2004) Verbrennungsmotoren. Simulation der Verbrennung und Schadstoffbildung. Teubner, Wiesbaden
Pischinger F (2001) Sonderforschungsbereich 224. In: Motorische Verbrennung. RWTH-Aachen. http://www.vka.rwth-aachen.de/sfb_224/bericht.htm
Preuss (2003) Einfluss der Mehrfacheinspritzung auf Hochdruck-Motorprozess und Abgasnachbehandlung bei Direkteinspritz-Ottomotoren. Dissertation, Technische Universität Graz
Ramos JI (1989) Internal Combustion Engine Modeling. Hemisphere Publ. Corp. New York Washington, Philadelphia, London
Robert Bosch GmbH (Hrsg) (2003) Ottomotor-Management. Vieweg, Wiesbaden
Siebers DL, Higgins B (2001) Flame lift-off on direct-injection diesel sprays under quiscent conditions. SAE paper 2001-01-0530
Schöppe D, Zülch S, Hardy M, Guerts D, Jorach R, Baker N (2008) Common-Rail-Einspritzung mit Direct-Acting-System von Delphi, Motorentechnische Zeitschrift MTZ 69, Vieweg, Wiesbaden, pp 842–851
Stegemann J (2004) Dieselmotorische Einspritzverlaufsformung mit piezoaktuierten Experimentaleinspritzsystemen. Dissertation, Universität Hannover
Stiesch G (2003) Modeling Engine Spray and Combustion Processes. Springer Berlin, Heidelberg, New York
van Basshuysen R, Schäfer F (5th eds) (2010) Handbuch Verbrennungsmotor. Vieweg/Teubner, Wiesbaden
Winklhofer (2007) Flammenmesstechnik für Motorenentwickler. AVL
Witt and Kern (2004) Methoden zur Beurteilung der lokalen Gemischqualität am Zündort bei Otto-DI-Motoren. Internationales Symposium für Verbrennungsdiagnostik, Baden-Baden
Zimmermann K, Wirth G, Grzeszik S, Josefsson S (2004) Entflammung geschichteter Gemische im strahlgeführten Benzin-DI-Motor: Optimierung mit Hilfe von Hochgeschwindigkeits-Visualisierung. Internationales Symposium für Verbrennungsdiagnostik, Baden-Baden

Chapter 5
Reaction Kinetics

Gunnar Stiesch and Peter Eckert

5.1 Foundations

5.1.1 Chemical Equilibrium

A chemical reaction between reactants *Aa*, *Ab*, etc., which form the products *Ac*, *Ad*, etc. can be described in the following form

$$\nu_a A_a + \nu_b A_b + \ldots \rightarrow \nu_c A_c + \nu_d A_d + \ldots . \tag{5.1}$$

The ν_i thereby designate the so-called stoichiometric coefficients of the reaction. Since every chemical reaction can in principle run both forwards as well as backwards, the reaction arrow in (5.1) can be replaced with an equal sign. We thereby obtain the general form of the reaction equation

$$\sum_i \nu_i A_i = 0, \tag{5.2}$$

whereby the stoichiometric coefficients are conventionally negative for all educts and positive for all products.

Every chemical reaction strives towards its state of equilibrium, which is reached under the condition that sufficient time is available. This state of equilibrium can be interpreted as a situation, in which both the forward as well as the reverse reactions progresses with identical speed. The macroscopically visible reaction rate is thus reduced to zero, and the material composition no longer changes. This material composition in the state of equilibrium can, with the help of both laws of thermodynamics, be determined contingent upon the boundary values of temperature and pressure. This method will be demonstrated.

For a closed, compressible system with constant temperature and constant pressure (see Fig. 5.1), the first and second laws of thermodynamics read as follows

$$\mathrm{d}U = \mathrm{d}Q + \mathrm{d}W = \mathrm{d}Q - p\mathrm{d}V, \tag{5.3}$$

G.P. Merker et al. (eds.), *Combustion Engines Development*,
DOI 10.1007/978-3-642-14094-5_5, © Springer-Verlag Berlin Heidelberg 2012

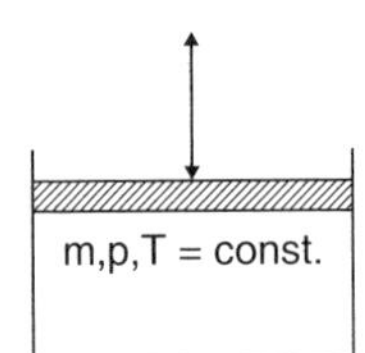

Fig. 5.1 Closed reactor

$$dS = \frac{dQ}{T} + dS_{irr}. \tag{5.4}$$

dS_{irr} thereby designates the increase of entropy due to irreversibility's, which is always greater than or equal to zero. The combination of (5.3) and (5.4) produces therefore the inequality

$$TdS - dU - pdV \geq 0. \tag{5.5}$$

If we introduce to this the free enthalpy G,

$$G = H - TS = U + pV - TS, \tag{5.6}$$

we then obtain, after differentiation and corresponding conversion

$$dG - Vdp + SdT \leq 0. \tag{5.7}$$

For a closed system with constant temperature and constant pressure, the dissipation of free energy is thus always less or equal to zero. That means that every alteration of the composition through chemical reactions reduces the value of G and that the condition

$$dG|_{T,p} = 0 \tag{5.8}$$

is satisfied in chemical equilibrium. Free energy has a minimum in chemical equilibrium.

For a single-phase multicomponent-system, e.g. a combustible gas, the Gibbs free energy is a function of temperature, pressure and composition,

$$G = G(T, p, n_1, n_2, n_3, ...), \tag{5.9}$$

whereby n_i represents the number of moles of the various species i. Now the chemical potential μ_i is introduced. This is defined as the partial derivative of the free energy with respect to the number of moles of i,

$$\mu_i = \left.\frac{\partial G}{\partial n_i}\right|_{T,p,n_j}, \quad j \neq i. \tag{5.10}$$

It can be shown that for an ideal gas – and this is a reasonable assumption for most combustion engine processes – the chemical potential is identical to the molar Gibbs function (Moran and Shapiro 1992).

$$\mu_i = \tilde{g}_i(T, p_i) = \tilde{g}_i^{\mathrm{o}} + \tilde{R}T \ln \frac{p_i}{p^{\mathrm{o}}}. \tag{5.11}$$

The index "o" thereby represents the condition under a reference pressure of 1 atm. The first term on the right side of the (5.11) stands for

$$\tilde{g}_i^{\mathrm{o}} = \tilde{h}_i(T) - T\tilde{s}_i^{\mathrm{o}}(T) \tag{5.12}$$

and can be taken from tabulated property data. It should be pointed out that the molar enthalpy consists of the standard enthalpy of formation and a temperature dependant term

$$\tilde{h}_i(T) = \tilde{h}_{f,i}^{\mathrm{o}} + \Delta\tilde{h}_i(T). \tag{5.13}$$

If the chemical potential introduced in (5.10) is inserted into the condition of equilibrium (5.8), the form

$$\mathrm{d}G|_{T,p} = \sum_i \mu_i \mathrm{d}n_i = 0. \tag{5.14}$$

is obtained.

However, for a general chemical reaction corresponding to (5.2), changes in the material quantities $\mathrm{d}n_i$ are proportional to the corresponding stoichiometric coefficients, so that the equation

$$\mathrm{d}n_i = \nu_i \mathrm{d}\varepsilon, \tag{5.15}$$

with the proportionality factor $\mathrm{d}\varepsilon$ is satisfied for all components i. The equation (5.14) can thus be simplified to the form

$$\sum_i \mu_i \nu_i = 0, \tag{5.16}$$

which contains all the necessary information regarding the equilibrium composition. The solution of (5.16) according to the various species concentrations is only iteratively possible and thus costly in terms of time and effort.

In order to avoid this problem, the concept of equilibrium constant is introduced. Through the insertion of (5.11) into (5.16), we obtain the relation

$$\sum_i \nu_i \tilde{g}_i^{\mathrm{o}} + \tilde{R}T \ln \prod_i \left(\frac{p_i}{p^{\mathrm{o}}}\right)^{\nu_i} = 0, \tag{5.17}$$

where the logarithmic term corresponds to the equilibrium constant

$$K_p = \prod_i \left(\frac{p_i}{p^o} \right)^{\nu_i}. \tag{5.18}$$

This equilibrium constant K_p now contains the information about the equilibrium material composition in terms of the partial pressures p_i of the various species i. Since the first term of (5.17) is exclusively dependant on the temperature [see (5.12)], it is evident that the equilibrium constant K_p is also only a function of T. It can easily be calculated with the help of thermodynamic property data and tabulated for any reaction

$$\ln K_p = \frac{-\sum_i \nu_i \tilde{g}_i^o}{\tilde{R}T} = \frac{-\Delta_R \tilde{g}^o}{\tilde{R}T}. \tag{5.19}$$

The numerator in (5.19) is customarily designated as free molar reaction enthalpy.

With the help of (5.18), the equilibrium composition for a system in which a single chemical reaction elapses, e.g. ($CO + ½\ O_2 = CO_2$), can now be solved for specific temperature and pressure conditions. However, in addition to (5.18), two further conditions are required, since a total of three unknowns, namely the partial pressures of CO, O_2, and CO_2, must be calculated. Both of these conditions result from the atomic balances of both elements involved C and O, i.e. from the fact that the absolute number of atoms of an element does not change during a chemical reaction. Because we usually work with the partial pressures of the components and not with their absolute atomic or molecular number, it is sensible to express the atomic balances as a ratio. This is possible because the ratio of two constants is itself a constant. For the example reaction ($CO + ½\ O_2 = CO_2$), one obtains accordingly for the atomic number ratio $\xi_{C/O}$ of carbon to oxygen atoms before (*) and after the reaction

$$\xi^*_{C/O}\big|_{\text{Edukte}} = \xi_{C/O}\big|_{\text{Produkte}} = \frac{p_{CO} + p_{CO_2}}{p_{CO} + 2p_{O_2} + 2p_{CO_2}}, \tag{5.20}$$

whereby the atomic number ratio before the reaction is known from the material quantities n_i^* of the educts which are to be reacted with each other

$$\xi^*_{C/O}\big|_{\text{Edukte}} = \frac{n^*_{CO}}{n^*_{CO} + 2n^*_{O_2}}. \tag{5.21}$$

Through the ratio formation of both atomic balances, one has however lost an independent equation. This can be surrogated with Dalton's law, which states that the sum of all partial pressures corresponds to the system pressure

$$\sum_i p_i = p_{sys}. \tag{5.22}$$

5.1.2 Reaction Rate

On the micro-scale, i.e. on the molecular level, a chemical reaction always progresses, as is given for example in (5.1), in both forward and reverse directions. The macroscopic reaction direction then results from the simple difference between forward and reverse reactions. Thus, the chemical equilibrium represents a special case, in which the forward and reverse reactions each run equally fast, so that no macroscopically visible material conversion occurs. On the molecular level however, reactions still progress in both directions. While the macroscopic reaction rate is always aimed in the direction of chemical equilibrium, the equilibrium analysis however does not provide any information regarding the absolute reaction rates, i.e. regarding the time necessary for the attainment of chemical equilibrium. This information is supplied by the *reaction kinetics*.

For the chemical reaction specified in (5.1), the temporal change of a species concentration, e.g. for $[A_c]$, can be given with the empirical formulation

$$\frac{d[A_c]}{dt} = \nu_c \left(\underbrace{k_f [A_a]^{\nu_a} [A_b]^{\nu_b}}_{\text{vorwärts}} - \underbrace{k_r [A_c]^{\nu_c} [A_d]^{\nu_d}}_{\text{rückwärts}} \right) \tag{5.23}$$

whereby the first term on the right side describes the reaction rate of the forward direction and the second term the rate of the reverse reaction. k_f and k_r are thereby the so-called rate coefficients of the forward and reverse reactions. They must be experimentally determined for every particular chemical reaction, e.g. in experiments in shock tubes. Since the speed coefficients of most reactions are strongly dependent on temperature, they are customarily represented with an Arrhenius formulation of the form

$$k = A \cdot T^b \cdot \exp\left[-\frac{E_A}{\tilde{R}T} \right] \tag{5.24}$$

The constant A and the exponent b as well as the so-called activation energy E_A are summarized for many chemical reactions in extensive tables, for example Warnatz et al. (2001).

It is sufficient to be familiar with the speed coefficient of either the forward or the reverse reaction. The other respective coefficient can then be determined simply with the inclusion of the corresponding equilibrium constant. This becomes clear if we consider that, in the special case of chemical equilibrium, the integral conversion rate becomes zero, since the reaction in both directions progress equally fast. If we surrogate these conditions in (5.23), we obtain

$$\frac{k_f}{k_r} = \frac{[A_c]^{\nu_c} [A_d]^{\nu_d}}{[A_a]^{\nu_a} [A_b]^{\nu_b}} \equiv K_c, \tag{5.25}$$

whereby K_c is the equilibrium constant defined in reliance on the species concentrations. Through the relation

$$K_c = K_p \cdot \left(\frac{p^{\mathrm{o}}}{\tilde{R}T}\right)^{\sum_i \nu_i} \tag{5.26}$$

it is clearly coupled with the equilibrium constant $K p$, defined in reliance upon the partial pressures and introduced in (5.18) and (5.19). Since both the speed coefficients and the equilibrium constants depend exclusively on the temperature and not on the actual species concentrations, the relation

$$\frac{k_f}{k_r} = K_p \left(\frac{p^{\mathrm{o}}}{\tilde{R}T}\right)^{\sum_i \nu_i} \tag{5.27}$$

is valid not only for the equilibrium state but also in general.

5.1.3 *Partial Equilibrium and Quasi-Steady-State*

In an extensive reaction system with a high number of reactions between participating species, we speak of a *partial equilibrium* if several reactions (not necessarily all) progress fast enough that the assumption of equilibrium between the species appearing in the reaction is justifiable at all times. This does not mean, however, that the absolute concentrations of the species found in partial equilibrium have to be temporally constant. Under altered boundary conditions, the species concentrations can certainly change in time. However, on the assumption of infinitely fast forward and reverse reactions, these concentration changes of all participating species are solidly coupled to each other, such that the determination of species concentrations can be very much simplified: in the case of partial equilibrium, the partial pressures of the corresponding species can be determined in analogy with the method for a single reaction. However, the number of unknowns (i.e. the partial pressures of the species) is now larger, so that additional equations must be set up in order to solve the system. These equations are obtained when we determine an equilibrium constant for every reaction found in the partial equilibrium in accordance with (5.19) and put this according to (5.18) in relation to the corresponding partial pressures.

One example for the appearance of partial equilibrium are the reactions between the species CO, CO_2, H, H_2, H_2O, O, O_2 and OH directly within the flame and also within the hot combustion products in internal combustion engines. The concentrations of these eight species, which contain a total of three different elements (C, O und H), can be calculated via five linearly independent reaction equations each found in partial equilibrium [see Sect. 5.2, reactions (5.38) to (5.42)]. The further three equations necessary for the solution of the eight unknown partial pressures is to be obtained from the atom balances of the three participating atoms. For a more extensive presentation, the reader is referred to Warnatz et al. (2001).

A condition is generally characterized as *quasi-steady*, if in a two-step reaction

$$A \xrightarrow{k_{f,1}} B \xrightarrow{k_{f,2}} C \tag{5.28}$$

the second reaction step progresses much faster than the first, i.e. if $k_{f,1} \ll k_{f,2}$. In this case, it can be assumed that the total mass quantity of B, which is formed in the first partial reaction, is immediately decomposed through the much faster second partial reaction. With this, the temporal rate of change of the concentration of B is approximately equal to zero, i.e. B is quasi-steady

$$\frac{d[B]}{dt} \approx 0. \tag{5.29}$$

Through this assumption, the time dependent determination of the concentrations of the participating species A, B and C can be extremely simplified, as will be shown in the following. In accordance with Sect. 5.1.2, the concentration change rates

$$\frac{d[A]}{dt} = -k_{f,1}[A], \quad \frac{d[B]}{dt} = k_{f,1}[A] - k_{f,2}[B], \quad \frac{d[C]}{dt} = k_{f,2}[B]. \tag{5.30}$$

The integration of the (5.30) under consideration of initial conditions

$$[A]_{t=0} = A_0, \quad [B]_{t=0} = [C]_{t=0} = 0 \tag{5.31}$$

then produce the time dependent course of concentrations of A, B and C , which are schematically presented in Fig. 5.2a for the case $k_{f,1} \ll k_{f,2}$

$$[A] = A_0 \cdot \exp(-k_{f,1} \cdot t), \tag{5.32}$$

$$[B] = A_0 \frac{k_{f,1}}{k_{f,1} - k_{f,2}} \left[\exp(-k_{f,2} \cdot t) - \exp(-k_{f,1} \cdot t)\right], \tag{5.33}$$

$$[C] = A_0 \cdot \left[1 - \frac{k_{f,1}}{k_{f,1} - k_{f,2}} \exp(-k_{f,2} \cdot t) + \frac{k_{f,2}}{k_{f,1} - k_{f,2}} \exp(-k_{f,1} \cdot t)\right]. \tag{5.34}$$

If we surrogate the simplification (5.29) for the rate of change of $[B]$ in (5.30), we obtain instead of (5.33)

$$k_{f,1}[A] = k_{f,2}[B] \Leftrightarrow [B] = \frac{k_{f,1}}{k_{f,2}} A_0 \exp[-k_{f,1} \cdot t]. \tag{5.35}$$

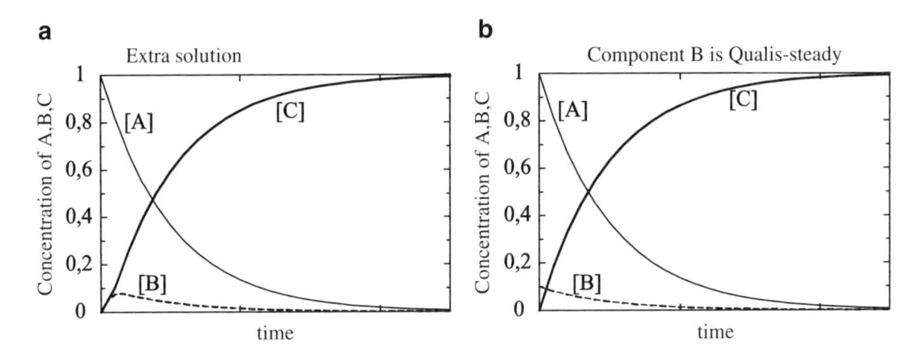

Fig. 5.2 Progression of concentrations of A, B and C under the condition $k_{f,1} \ll k_{f,2}$ in the sequential reaction $A \xrightarrow{k_{f,1}} B \xrightarrow{k_{f,2}} C$ (**a**) exact solution (**b**) assuming that B is quasi-steady

(5.34) thus finally simplifies itself to

$$[C] = A_0 \left[1 - \exp\left(-k_{f,1} \cdot t\right)\right]. \tag{5.36}$$

The concentration progressions of A, B and C for the assumption of quasi-steady-state of B are represented in Fig. 5.2b. In comparison to the exact solution in Fig. 5.1a, it becomes clear that only at the very start of the reaction is the concentration of B incorrectly illustrated. During the most part of the reaction time, the solution determined with the help of the quasi-steady-state assumption agrees however quite exactly with the exact solution. One typical example for a quasi-steady-state in engine combustion is the nitrogen atom N in thermal NO formation. We will go into this example further in Sect. 6.5

5.2 Reaction Kinetics of Hydrocarbons

5.2.1 Oxidation of Hydrocarbons

In the case of complete combustion, hydrocarbon compounds C_xH_y are converted into carbon dioxide CO_2 and water vapor H_2O. This reaction can be described overall by the gross reaction equation

$$C_xH_y + \left(x + \frac{y}{4}\right)O_2 \rightarrow x \cdot CO_2 + \frac{y}{2}H_2O + \Delta H_R. \tag{5.37}$$

Here, the reaction enthalpy ΔH_R represents the heat released by combustion. However, the combustion process does not actually progress in accordance with this gross reaction equation but with a very complex reaction scheme based on elementary reactions.

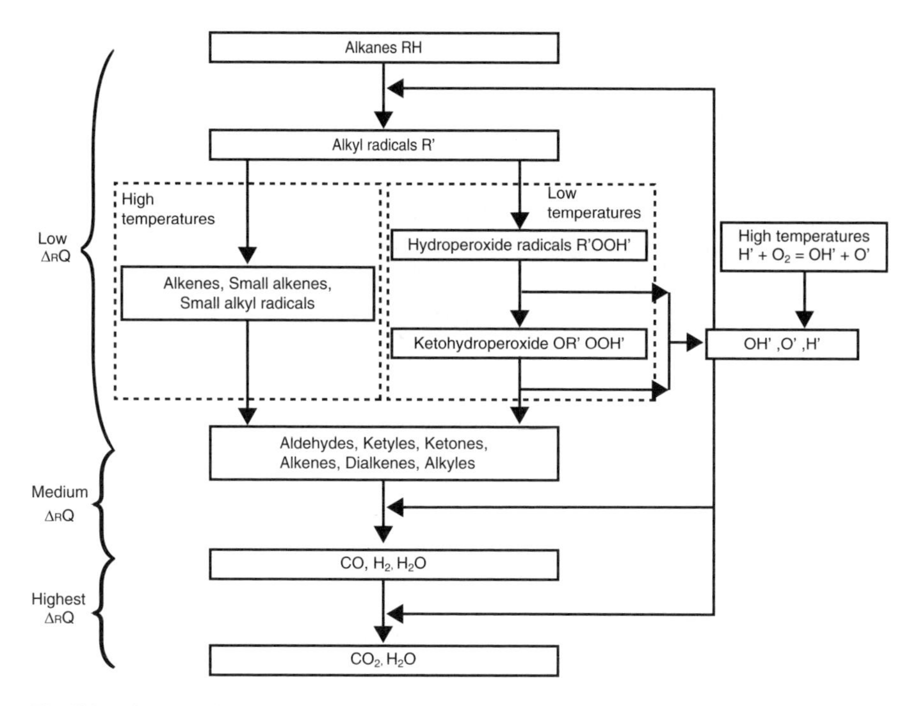

Fig. 5.3 Diagram of the hydrocarbon oxidation process

Figure 5.3 shows a highly simplified reaction diagram of alkanes. At both low and high temperatures, an abstraction takes place, i.e. a hydrogen atom is separated from the hydrocarbon molecule. At low temperatures, hydrocarbon peroxides ($R'OOH^{\bullet}$) are formed, which break down into smaller hydrocarbons due to oxidation and dehydration. These reactions are of decisive importance for engine applications and will be treated extensively in Sect. 5.2.2. At high temperatures, the formation of hydrogen peroxide is circumvented; instead, one alkene and a small alkyl radical each are formed from the larger alkyl radical produced by hydrogen abstraction by means of a β disintegration (Glassmann 1996). Subsequently, alkenes, dialkenes, ketyles, ketones and aldehydes like acetaldehyde (ethanal) CH_3CHO and formaldehyde (methanal) CH_2O are formed. Until the aldehydes are formed, only about 15% of the heat stored in the fuel is released. A major part of heat release occurs therefore only at the end of the reaction sequence during the oxidation of CO to CO_2. Figure 5.4 provides a qualitative overview of the temporal profile of concentration and temperature during hydrocarbon combustion.

To estimate temperature and concentration in the flame front, one can make the simplifying assumption that the eight components $H^{\bullet}$, H_2, $O^{\bullet}$, O_2, $OH^{\bullet}$, CO, CO_2 and H_2O in the flame front are in partial equilibrium due to the high temperatures prevailing there. This "OHC system" is thus described by the five reaction equations

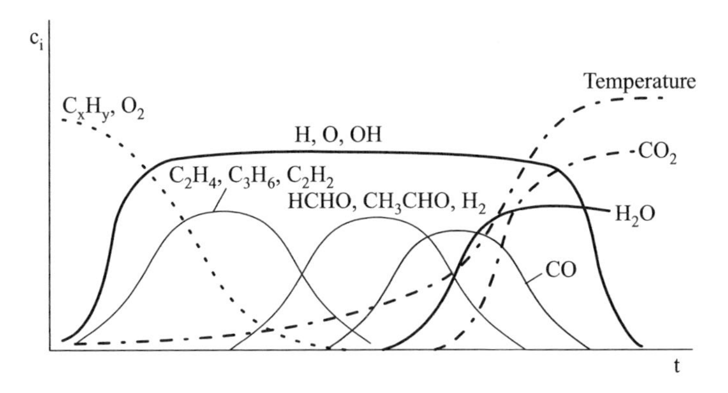

Fig. 5.4 Temporal concentration profile for hydrocarbon combustion

$$H_2 = 2H^\bullet \tag{5.38}$$

$$O_2 = 2O^\bullet \tag{5.39}$$

$$H_2O = \frac{1}{2}\ H_2 + OH^\bullet \tag{5.40}$$

$$H_2O = \frac{1}{2}\ O_2 + H_2 \tag{5.41}$$

$$CO_2 = CO + \frac{1}{2}\ O_2 \tag{5.42}$$

whereby the following is true for the five equilibrium constants:

$$K_{C1} = [H]^2\ [H_2]^{-1} \tag{5.43}$$

$$K_{C2} = [O]^2\ [O_2]^{-1} \tag{5.44}$$

$$K_{C3} = [H_2]^{\frac{1}{2}}\ [OH]\ [H_2O]^{-1} \tag{5.45}$$

$$K_{C4} = [O_2]^{\frac{1}{2}}\ [H_2]\ [H_2O]^{-1} \tag{5.46}$$

$$K_{C5} = [CO]\ [O_2]^{\frac{1}{2}}\ [CO_2]^{-1} \tag{5.47}$$

Together with the atomic balances for the atoms O, H and C (better: CO) and the condition that the sum of the partial pressures of all components must be equal to the total pressure, we finally obtain a non-linear equation system, which is clearly solvable with known numerical integration methods (e.g. the Newton-Kantorowitsch method). The "element potential method" is an example of an

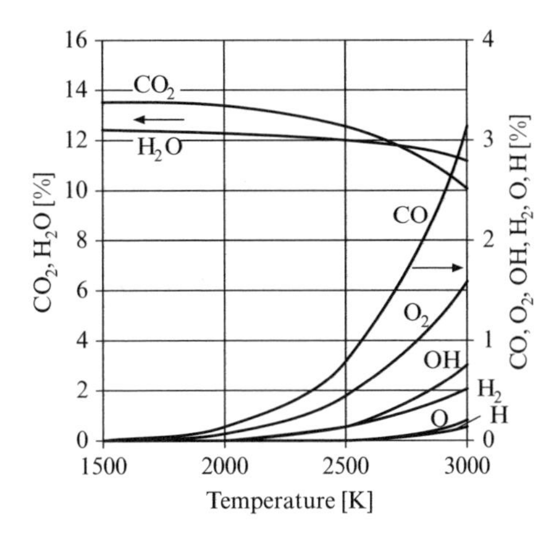

Fig. 5.5 Partial equilibrium of the OHC components as a function of temperature for a total pressure of 1 bar

alternative way to solve the chemical equilibrium of the OHC system (Reynolds 1986).

Figure 5.5 shows an example of the distribution of concentration of the OHC components as a function of temperature for a total pressure of 1 bar.

5.2.2 *Ignition Processes*

Ignition is the transition of a non-reacting fuel-air mixture into a combustion process. Ignition processes can be classified into the categories of thermal reaction and chain reaction. According to Semenov's analysis (Semenov 1935), a thermal reaction occurs when chemical heat production exceeds the heat losses on the combustion chamber walls. In this form of ignition, there is a direct increase of temperature without any delay. In the case of chain reaction on the other hand, usually there is a delay with constant temperature. During this time, the first radicals are formed, which function as chain carriers. Only after a certain amount of these radicals are produced in the system does a sufficient heat release for a temperature increase and subsequent take place. The stages of a chain reaction are subdivided into start, propagation, branching, and termination reactions. Important radicals include the atoms $O^{\bullet}$ and $H^{\bullet}$ as well as the hydroxyl radical ($OH^{\bullet}$), the hydroperoxy radical ($HO_2^{\bullet}$), and the methyl radical ($CH_3^{\bullet}$).

Start reactions form radicals made of stable species, e.g. in the reaction between methane and molecular oxygen:

$$CH_4 + O_2 \longrightarrow CH_3^{\bullet} + HO_2^{\bullet} \tag{5.48}$$

Propagation reactions maintain the number of radical species:

$$CH_4 + OH^{\bullet} \longrightarrow CH_3^{\bullet} + H_2O \tag{5.49}$$

In chain branching reactions, more radicals are formed than are consumed:

$$CH_4 + O^\bullet \longrightarrow CH_3^\bullet + OH^\bullet \quad (5.50)$$

In termination reactions, the number of radical species is reduced, e.g. in the recombination reaction of methyl radicals:

$$CH_3^\bullet + CH_3^\bullet \longrightarrow C_2H_6 \quad (5.51)$$

Chain terminations can also be brought about by the collision of radicals with the combustion chamber walls – a mechanism that is of particular importance at low pressures.

5.2.2.1 The H_2–O_2 System

The H_2–O_2 system has a relatively simple oxidation mechanism and is important both for the investigation of hydrogen combustion and as a subset in reaction mechanisms of more complex fuels. Despite the simple composition of the fuel, hydrogen combustion requires more than 25 reactions between at least eight different species – H_2, O_2, $OH^\bullet$, H_2O, $H^\bullet$, $O^\bullet$, $HO_2^\bullet$, and H_2O_2. The most important reactions connected with ignition are (Glassmann 1996):

$$H_2 + O_2 \rightleftarrows HO_2^\bullet + H^\bullet \quad (5.52)$$

$$H_2 + OH^\bullet \rightleftarrows H_2O + H^\bullet \quad (5.53)$$

$$H^\bullet + O_2 \rightleftarrows O^\bullet + OH^\bullet \quad (5.54)$$

$$O^\bullet + H_2 \rightleftarrows H^\bullet + OH^\bullet \quad (5.55)$$

$$H^\bullet \longrightarrow 0,5H_2 \quad (5.56)$$

$$H^\bullet + O_2 + M \rightleftarrows HO_2^\bullet + M \quad (5.57)$$

Reaction (5.56) is a wall termination. The trimolecular reaction (5.57) may be a propagation reaction formally, but it can be seen as a chain termination, since the $HO_2^\bullet$ radical formed is relatively inert.

The effect of the different reactions on ignition can be explained by means of an explosion diagram (Fig. 5.6). At constant temperature and very low pressures, no ignition occurs, since the radicals formed quickly diffuse to the combustion chamber walls and recombine in reaction (5.56). If the pressure is increased, diffusion slows down so that the chain propagation (5.54) outweighs it and a first explosion limit is reached. If pressure is further increased, the second explosion limit is reached. Here, the highly pressure-dependent reaction (5.57) becomes more

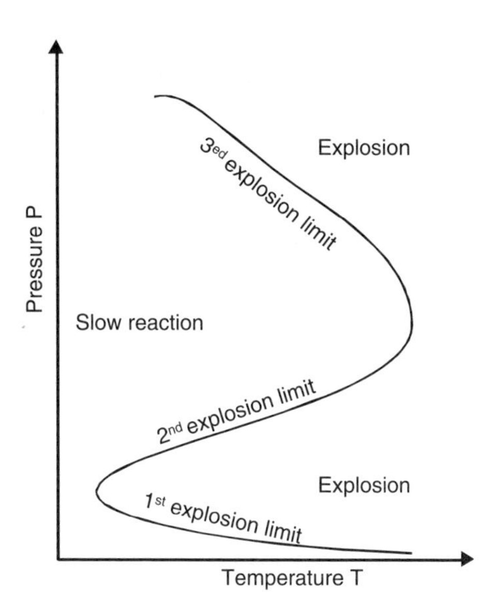

Fig. 5.6 H_2–O_2 thermal explosion diagram

important, and the H_2–O_2 mixture becomes stable again. At the third explosion limit, a further chain propagation by means of the previously inert $HO_2^{\bullet}$ radicals becomes important, and together with the increasing heat release per unit of volume due to the higher pressures, ignition takes place again.

Like hydrogen ignition, the autoignition of hydrocarbons is also a chain process. Hydrocarbons have a much more complex ignition mechanism however, with a much larger number of participating species and reactions. As in hydrogen combustion, there are three explosion limits in the hydrocarbon explosion diagram.

In the case of hydrocarbons, reaction (5.54) is the dominate chain branching at high pressures and temperatures about 1,100 K. Here, oxidation of the fuel proceeds in accordance with the scheme discussed in Sect. 5.2.1. Branching reaction (5.54) however is highly dependent on temperature and becomes quickly insignificant at $T < 1{,}100$ K.

In engine applications, the temperature after compression is usually below 1,000 K. In the case of hydrocarbons, especially alkanes, additional, more complex ignition mechanisms arise in this temperature range.

Ignition in the low and medium temperature range is characterized by the appearance of so-called two-stage ignition. In this case, heat release is increased at first by the first ignition phase and is then reduced again above ca. 900 K. Above about 1,000 K, the second ignition phase begins, leading to complete oxidation of the fuel. The exact temperatures are dependent on pressures.

Two-stage ignition is explained by the negative temperature coefficient (NTC) shown schematically in Fig. 5.7, which describes the fact that the ignition delay becomes larger with increasing output temperature within the NTC regime.

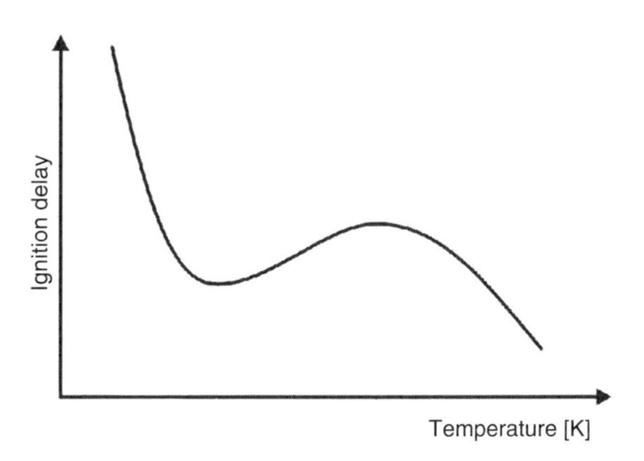

Fig. 5.7 Schematic representation of the negative temperature coefficient (NTC) in the ignition of hydrocarbons

In the low temperature range of T < 900 K, there is a complex chain branching mechanism. In the first step, a hydrogen atom is split from the fuel molecule RH. At the start of ignition, when only a few radicals exist in the system, hydrogen abstraction progresses primarily via the relatively slow reaction with oxygen:

$$RH + O_2 \rightleftharpoons R^{\bullet} + HO_2^{\bullet} \tag{5.58}$$

Subsequently, the hydrocarbon molecules react with the radicals formed in the chain branching process $O^{\bullet}$, $H^{\bullet}$, $OH^{\bullet}$, $HO_2^{\bullet}$, etc. Especially the reaction with the $OH^{\bullet}$ radical has a very low activation energy and is thus very fast:

$$RH + OH^{\bullet} \rightleftharpoons R^{\bullet} + H_2O \tag{5.59}$$

O_2 addition then occurs on the alkyl radical $R^{\bullet}$ that is formed:

$$R^{\bullet} + O_2 \rightleftharpoons RO_2^{\bullet} \tag{5.60}$$

The equilibrium constants of reaction (5.60) are highly dependent on temperature. At low temperatures, equilibrium is on the right side, with increasing temperature the equilibrium shifts to the left. The emerging $RO_2^{\bullet}$ radicals go through an isomerization reaction, in which a hydrogen atom is abstracted in the molecule and added to the O-O group:

$$RO_2^{\bullet} \rightleftharpoons R'OOH^{\bullet} \tag{5.61}$$

A further oxygen molecule is then attached to the generated hydroperoxy-alkyl radical. In the further isomerization reaction, a ketohydroperoxide molecule $OR''OOH$ and an $OH^{\bullet}$ radical are then formed. The ketohydroperoxide molecule continues to break down, forming another $OH^{\bullet}$ radical and a carbonyl radical $OR''O^{\bullet}$:

$$R'OOH^{\bullet} + O_2 \rightleftharpoons OOR'OOH^{\bullet} \tag{5.62}$$

$$OOR'OOH^{\bullet} \rightleftharpoons OR''OOH + OH^{\bullet} \tag{5.63}$$

$$OR''OOH \rightleftharpoons OH^{\bullet} + OR''O^{\bullet} \tag{5.64}$$

In the mechanism described by (5.58)–(5.64) we thus see a chain branching in which, if the process is introduced by reaction (5.59), there is a net production of one $OH^{\bullet}$ radical. This chain branching leads in a low-temperature process to an initial increase in the heat release rate.

Low-temperature oxidation continues until the equilibrium of reaction (5.60) is shifted at a temperature of about 900 K. this shift interrupts the chain branching of the isomerization reaction (5.61), and instead we see the increased formation of alkenes and $HO_2^{\bullet}$ radicals in this middle temperature range. The $HO_2^{\bullet}$ radicals react further, forming hydrogen peroxide, H_2O_2, which is relatively inert at first.

$$R'OOH^{\bullet} \rightleftharpoons \text{Alken} + HO_2^{\bullet} \tag{5.65}$$

$$HO_2^{\bullet} + HO_2^{\bullet} \rightleftharpoons H_2O_2 + O_2 \tag{5.66}$$

Following this, temperature increases slowly until, above ca. 1,000 K, hydrogen peroxide is broken down extremely rapidly and the second ignition phase is introduced:

$$H_2O_2 + M \longrightarrow 2OH^{\bullet} + M \tag{5.67}$$

This process is called degenerated chain branching and is the cause of the negative temperature coefficient. After the equilibrium of reaction (5.60) shifts and reaction (5.65) begins, not enough radicals are formed any more in order to continue the ignition process. Only after the decomposition of hydrogen peroxide large amounts of $OH^{\bullet}$ radicals are produced that accelerate ignition and lead to a second heat release, which introduces a high-temperature oxidation mechanism.

The negative temperature coefficient is the most strongly marked in the case of long-chain alkanes. In contrast, alkenes and aromatics show a weaker or no NTC properties (Leppard 1990). Further details on degenerated chain branching can be found in Curran et al. (1998) among others.

The reaction sequence introduced here is the decisive mechanism for the auto-ignition of hydrocarbons in engine applications. It arises both in (intentional) autoignition in diesel and HCCI engines as well as in (unintentional) autoignition leading to engine knocking in SI engines.

Two-stage ignition can be recognized easily experimentally by means of a fast compression machine. In a compression machine, a homogeneous fuel-air mixture is compressed in a single compression stroke and the piston held at the top dead center. Figure 5.8 shows the pressure profile in such a device over the experiment time. The

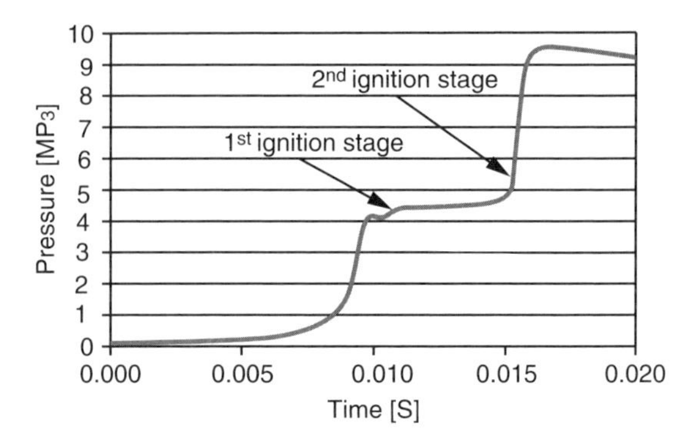

Fig. 5.8 Pressure profile for the ignition of a alkane-air mixture in a fast compression machine

```
                                           H     H
                                           |     |
                                         H-C-H H-C-H
  H   H   H   H   H   H   H   H          H |   H |   H
  |   |   |   |   |   |   |   |          | |   | |   |
H-C---C---C---C---C---C---C---C-H      H-C---C---C---C---C-H
  |   |   |   |   |   |   |   |          | |   |   |   |
  H   H   H   H   H   H   H   H          H |   H   H   H
                                         H-C-H
                                           |
                                           H
```

Fig. 5.9 Molecular structure of n-octane (*left*) and isooctane (2,2,4-trimethylpentane, *right*)

first ignition delay arises after the end of compression at about 9.3 ms. In the first ignition stage; pressure and temperature are increased until the middle temperature range is reached with reactions (5.65) and (5.66). After a second, longer ignition delay, the second ignition stage commences with the following combustion.

The reaction rate and with it the ignition delay time of alkanes depend on the number and kind of existing carbon-hydrogen compounds. Figure 5.9 shows as an example the molecular structure of n-octane and isooctane (2,2,4 trimethylpentane). Although both alkanes have the same amount of carbon and hydrogen atoms, n-octane exhibits much higher ignitability than isooctane. This is due to the different rates of hydrogen abstraction of the differently bonded carbon atoms. Hydrogen atoms, which are bonded to a carbon atom with only one bond to another carbon atom, are called primary hydrogen atoms. Hydrogen atoms bonded with a carbon atom with two neighboring carbon atoms are called secondary hydrogen atoms, and those bonded to a carbon atom with three carbon neighbors are called tertiary hydrogen atoms. The primary bonds are the strongest, i.e. the highest amount of energy must be applied to brake them; tertiary bonds are the weakest. The probability of a reaction of a primary hydrogen abstraction is thus lower than that of a secondary or tertiary bond. In addition, the reaction rate is proportional to the number of bonds existing in the respective molecule. For this reason, larger alkanes are more reactive than smaller ones, and straight-chain alkanes are more reactive than branched ones. Besides the type of the hydrogen atom to be abstracted, the position are plays a role in the subsequent chain branching reactions. Thus, an abstraction of a secondary hydrogen atom at the second position in the n-octane shown in Fig. 5.9 leads to different secondary products than an abstraction of the secondary hydrogen atom at the third or fourth position.

5.2.3 *Reaction Kinetics in Engine Simulation*

Even the ignition and oxidation of simple molecules is, as described in the previous chapter, a complex process. For a complete description of hydrogen combustion for example, we already require over 25 reactions between eight species, and the detailed mechanism GRI-Mech 3.0, which describes the kinetics of methane oxidation, consists of 325 reactions between 53 species (Smith et al. 1999). With increasing length, the reaction mechanisms of hydrocarbons also become more complicated. The complex n-heptane reaction mechanism of Curran et al. (1998) is thus composed of 2,539 reactions between 561 species. The performance of modern computer systems is indeed sufficient to solve such models under homogeneous conditions, but when simulating engine processes, the use of simplified reaction kinetics may be required depending on the application.

5.2.3.1 Single-Step Global Mechanisms

In the simplest case, hydrocarbon oxidation is described by the single-step global reaction (5.37). Westbrook and Dryer (1981) have suggested a relation for the reaction rate of global reactions (5.37) for a selection of hydrocarbons

$$\frac{d[C_xH_y]}{dt} = A \cdot \exp\left(-\frac{E_A}{RT}\right) \cdot [C_xH_y]^m [O_2]^n \tag{5.68}$$

The corresponding parameters A, m and n as well as the activation temperature E_A/R are given in Table 5.1.

Despite its simplicity, single-step global mechanisms are still used today in many combustion models, for example in some phenomenological models for diesel engine combustion (see Sect. 11.1). This is possible because the chemical reactions after ignition are often much faster than physical mechanisms such as turbulence and mixing. However, one must take into consideration the fact that the flame temperature is calculated too high when single-step mechanisms are used. A remedy can be found, for example, in the use of a two-step mechanism, in which the fuel is first oxidized to CO and H_2O, and then CO is oxidized to CO_2 in the second step.

Approaches in the form of (5.68) are sometimes still used to describe ignition as well. With a suitable adjustment of the parameters to the experiment, quite satisfactory results can be obtained so long as one does not deviate too much from the

Table 5.1 Reaction rate parameters for the single-step reaction (5.68), acc. to Westbrook and Dryer (1981)

Fuel	A (mol, cm, s)	Activation temp. E_A/R (K)	m (–)	n (–)
CH_4	8.3×10^5	15.098	–0.30	1.30
C_2H_6	1.1×10^{12}	15.098	0.10	1.65
C_3H_8	8.6×10^{11}	15.098	0.10	1.65
C_4H_{10}	7.4×10^{11}	15.098	0.15	1.60
C_5H_{12}	6.4×10^{11}	15.098	0.25	1.50
C_6H_{14}	5.7×10^{11}	15.098	0.25	1.50
C_7H_{16}	5.1×10^{11}	15.098	0.25	1.50
C_8H_{18}	4.6×10^{11}	15.098	0.25	1.50
C_9H_{20}	4.2×10^{11}	15.098	0.25	1.50
$C_{10}H_{22}$	3.8×10^{11}	15.098	0.25	1.50
C_2H_4	2.0×10^{12}	15.098	0.10	1.65
C_3H_6	4.2×10^{11}	15.098	–0.10	1.85
C_2H_2	6.5×10^{12}	15.098	0.50	1.25
CH_3OH	3.2×10^{12}	15.098	0.25	1.50
C_2H_5OH	1.5×10^{12}	15.098	0.15	1.60
C_6H_6	2.0×10^{11}	15.098	–0.10	1.85
C_7H_8	1.6×10^{11}	15.098	–0.10	1.85

conditions used for calibration. However, it is impossible to represent the negative temperature coefficient as described in the previous chapter.

5.2.3.2 Semi-empirical Multi-Step Models

In order to represent ignition more realistically, a series of semi-empirical multi-step models have been developed. The probably most common model of this kind is the Shell model, originally developed by Halstead et al. (1977) to predict knocking in SI engines and later expanded for modeling diesel ignition by Kong et al. (1995). The Shell model was not formulated by a reduction from a complex mechanism. Instead, it can be seen as a mathematical equation system with which the ignition properties of complex hydrocarbons – including the appearance of the negative temperature coefficient – can be described.

The resultant reaction scheme comprises eight reactions between five generic species:

$$RH + O_2 \longrightarrow 2R^\bullet \tag{5.69}$$

$$R^\bullet \longrightarrow R^\bullet + P + \text{Wärmefreisetzung} \tag{5.70}$$

$$R^\bullet \longrightarrow R^\bullet + B \tag{5.71}$$

$$R^\bullet \longrightarrow R^\bullet + Q \tag{5.72}$$

$$R^{\bullet} + Q \longrightarrow R^{\bullet} + B \quad (5.73)$$

$$B \longrightarrow 2R^{\bullet} \quad (5.74)$$

$$R^{\bullet} \longrightarrow \text{Chain Determination} \quad (5.75)$$

$$2R^{\bullet} \longrightarrow \text{Chain Determination} \quad (5.76)$$

Here, RH again represents a hydrocarbon molecule, $R^{\bullet}$ describes a hydrocarbon radical, Q is an unstable intermediate species, and B a branch species. P designates products of combustion, i.e. CO, CO_2 and H_2O. Reaction (5.69) represents the start reaction in the Shell model, (5.70) to (5.73) are propagation reactions, (5.74) is the chain branching reaction, and (5.75) and (5.76) are termination reactions. The Shell model includes 26 parameters that must be adjusted in order to represent a particular fuel.

Although in the last few years, detailed reaction mechanisms are being increasingly used to calculate ignition and high-temperature oxidation; global, semi-empirical models still have their advantages. For example, complex mechanisms do not necessarily represent all global phenomena well, and an adjustment of complex mechanisms is often required. On the other hand, the parameters of semi-empirical models can be adjusted relatively quickly to the existing problems. Furthermore, fuels used in engines are mixtures of a large number of individual components, which can only be approximately described even with detailed reaction mechanisms of a few components.

5.2.3.3 Detailed Reaction Mechanisms

As opposed to semi-empirical models like the Shell model, detailed reaction mechanisms have the advantage of providing a larger range of application with respect to marginal conditions like temperature, pressure and the fuel-air ratio. With the right mechanism, it is possible to describe the chemistry of low-temperature ignition and high-temperature oxidation in one model, thus avoiding one step between modeling ignition and combustion. Moreover, the concentrations of intermediate species are calculated so that a direct combination with the reactions that are important for pollutant formation is possible.

Highly complex and extensive mechanisms exist for many hydrocarbons. For example, the reaction mechanisms of methane (e.g. Smith et al. 1999), n-heptane (e.g. Curran et al. 2002), and isooctane (2,2,4-trimethylpentane, e.g. Curran et al. 2002) have been investigated in the past especially rigorously. In the recent past, mechanisms for longer hydrocarbons, aromatics and oxygenic molecules have also been increasingly established. In such extensive mechanisms, the reaction rates of the individual reactions are selected partially on the basis of especially well-investigated elementary reactions, partially on the basis of quantum-mechanical theories, but also by adjusting to experiments.

As a rule, the aim is to validate complex mechanisms with experiments that are as varied as possible. Nonetheless, one must take heed that a mechanisms developed, for example, to describe ignition processes, does not necessarily provide good results when calculating laminar flame speeds (Meeks et al. 1988). In addition, some kinetic processes and reaction sequences even today are still not understood in detail.

5.2.3.4 Reduction of Kinetic Mechanisms

Complete, complex reaction mechanisms require a large amount of calculation time, not only due to the large number of simultaneous differential equations. In particular, the differential equations set out for kinetic models are stiff; i.e. they can be solved with explicit solution procedures only with impracticably small time increments, making complex implicit methods necessary for the solution. In complex multidimensional problems, one must solve, in addition to the chemical source terms, transport equations for every species in the system. Complete complex mechanisms should thus be used as a rule only for very detailed problems. For reasons of calculation time, it is necessary to utilize simplified mechanism in most combustion models.

There is a large variety of procedures used to reduce complex mechanisms. The first step in the reduction process is often eliminating unimportant species and reactions. Species that can be eliminated without significantly limiting the predictive accuracy of the mechanism can be determined, for example, via an analysis of the Jacobi matrix, i.e. the derivation of the reaction rates according to the species concentrations:

$$J = \frac{\partial \omega}{\partial c} \tag{5.77}$$

The directed relation graph method is a systematic method for species reduction (Lu and Law 2006). In it, the subject of investigation is the interconnection between individual species. One first proceeds from one or several species that must remain intact in the mechanism, such as the fuel, and species that are important for pollutant formation. One then calculates, for one species A to be sustained, the respective error which arises if one removes another species B from the mechanism:

$$\varepsilon_{AB} = \frac{\sum\limits_{i=1,I} \left| \nu_{A,i} \omega_i \delta_{Bi} \right|}{\sum\limits_{i=1,I} \left| \nu_{A,i} \omega_i \right|}, \quad \delta_{Bi} = \begin{cases} \nu_{B,i} \neq 0 : 1 \\ \nu_{B,i} = 0 : 0 \end{cases} \tag{5.78}$$

Here, the i is the ith reaction in a mechanism with a total of I reactions. If the error ε_{AB} is lower than a limit value, the removal of B leads to a negligible error in the calculation of A. On the other hand, if the error is larger than the limit, the species are linked. If species A must remain intact in the mechanism, species B must as well. In addition, all species must remain intact that are indirectly linked with species A, i.e. that if removed would cause an error in the calculation of species B that is not negligible. The link between the species is then recorded in a graph. One graph can only provide information about one local reaction condition. For this reason, one graph each must be made for several representative points in the reaction sequence of different tests. Only those species that are not linked to the starting species at all points can finally be removed from the mechanism.

Unimportant reactions can be identified by means of a sensitivity analysis. In such an analysis, the reaction rates of individual reactions or of reaction classes are varied and the effect on an result of the mechanism (the ignition delay for example) determined under different marginal conditions. Reactions that exhibit a low sensitivity can be eliminated. Sensitivity analysis can also be used to identify important reactions, the reaction rates of which are especially suitable for adjusting the reaction mechanism to experiments.

To simplify mechanisms, time scale analysis also finds frequent use. Chemical systems often have highly varying time scales, i.e. some species reach an equilibrium concentration extremely quickly during the reaction, while other species are very stable. The goal of time scale analysis is to unlink the fast time scales from the slow time scales, thereby creating a system with fewer variables than before. Two assumptions made in time scale analysis are the quasi-stability of species and the partial equilibrium of reactions. These assumptions were already explained in Sect. 5.1. In addition, more systematic methods exist, for example the ILDM method [intrinsic low-dimensional manifolds (Maas and Pope 1992)]. Tabulation methods, which are being used increasingly in engine simulation, are also based on the assumption that high-dimensional chemical problems can also be described by means of a few variables in a simplifying way.

Lumping, the consolidation of species groups is another reductive measure of interest in the case of long-chain hydrocarbon molecules. As hydrocarbons are oxidized, a large number of intermediate species that have the same molar weight and a similar structure but differ in the location of the radical position or of the oxidation group. These isomers are formed by means of analogous reactions. Depending on the location of the functional group however, the subsequent reaction paths and reaction products are different. For this reason, relations for the percentage of the original species in the consolidated species group must be set up so that one can calculate the following reaction rates. One way of determining this ratio is to assume a partial equilibrium between the isomers of a lumped group (Chaos et al. 2007).

For further details on the reduction of kinematic mechanisms, see the specified literature, Tomlin et al. (1997) and Lu and Law (2009).

5.2.3.5 Surrogate Mixtures for Real Fuels

Fuels used in engine applications are usually mixtures of hundreds of different hydrocarbons (Sect. 4.1). The complex behavior of such a fuel mixture can thus only be insufficiently described by a kinetic mechanism of a single component. For example, the chemical ignition delays arising when diesel fuel is used can only be approximately predicted with n-heptane, while pollutant formation and the physical properties of n-heptane deviate from diesel to a great extent. For this reason, increasing amounts of research is focusing on representing real fuel mixtures by means of several representative components, i.e. by means of simplified surrogate mixtures. For example, one simple surrogate mixture often used for gasoline in the past consists of the components used to determine the octane number, n-heptane and 2,2,4-trimethyl pentane (isooctane). IDEA fuel, a mixture of 70% n-decane and 30% α-methyl naphtaline, is a surrogate fuel used frequently in the simulation of diesel-engine processes (Antoni 1998).

The basic prerequisite for creating a surrogate fuel is that sufficiently detailed and validated reactions mechanisms are available for every component used. A surrogate mixture can be composed on the basis of various criteria. One possibility is to represent every hydrocarbon group – i.e. n-alkanes, iso-alkanes, alkenes, aromatics, oxidized hydrocarbons etc. – with a surrogate component and to weight the average percentage amount of the group in the real fuel accordingly. Alternately, the surrogate species can be composed by determining the functional groups of the real mixture and assembling the surrogate species in such a way that the number of primary, secondary, tertiary, and aromatic carbon corresponds to the actual fuel.

A third possibility consists of selecting the surrogate components such that certain physical and chemical properties are represented by the surrogate mixture. These properties include, for example, the H/C ratio, the boiling curve, the aromatic content, the calorific value, and the octane or cetane number. The decisive point is that the relevant properties of a model mixture can be calculated very quickly. For example, spray and mixture-formation effects are important in the experimental determination of the cetane number, and these effects must be taken into consideration in a simplified manner in the mathematical calculation of the cetane number of a surrogate fuel.

When compiling the reaction mechanisms of the individual components, it is important that the mechanisms are consistent, i.e. reactions that occur in several individual mechanisms have the same reaction rates. In addition, reactions between species that appears in one mechanism in the total set of mechanisms and not in another must be redefined. The resultant set of overall mechanisms is often to large to be used in engine-related inquiries, necessitating a reduction of the model's size.

In summary, detailed reaction mechanisms of surrogate mixtures promise in future a closer approximation of actual fuels than was possible before with global models or detailed models of individual components. However, the use of such surrogate mixtures entails many difficulties with respect to configuration, interaction between the individual components as well as computing time that are still to be solved. For an overview of current investigations into surrogate mixtures for both diesel and gasoline, see Pitz et al. (2007), Farrell et al. (2007) and Meeks et al. (1988).

References

Antoni C (1998) Untersuchung des Verbrennungsvorgangs im direkteinspritzenden Dieselmotor mit zyklusaufgelöster Emissionsspektroskopie. Dissertation, RWTH Aachen

Chaos M, Kazakov A, Zhao Z, Dryer FL (2007) A high-temperature chemical kinetic model for primary reference fuels. Int J Chem Kinet 39:399–414

Curran HJ, Gaffuri P, Pitz WJ, Westbrook CK (1998) A comprehensive modeling study of n-heptane oxidation. Combust Flame 114:149–177

Curran HJ, Gaffuri P, Pitz WJ, Westbrook CK (2002) A comprehensive modeling study of iso-octane oxidation. Combust Flame 129:253–280

Farrell JT, Cernansky NP, Dryer FL, Friend DG, Hergart CA, Law CK, McDavid R, Mueller CJ, Pitsch H (2007) Development of an experimental database and kinetic models for surrogate diesel fuels. SAE Paper 2007-01-0201

Glassmann I (1996) Combustion. Academic, San Diego

Halstead M, Kirsch L, Quinn C (1977) The autoignition of hydrocarbon fuels at high temperatures and pressures – fitting of a mathematical model. Combust Flame 30:45–60

Kong S-C, Han Z, Reitz RD (1995) The developement and application of a diesel ignition and combustion model for multidimensional engine simulations. SAE Paper 950278

Leppard WR (1990) The chemical origin of fuel octane sensitivity. SAE paper 902137

Lu T, Law CK (2006) Linear time reduction of large kinetic mechanisms with directed relation graph: n-Heptane and iso-octane. Combust Flame 144:24–36

Lu T, Law CK (2009) Toward accommodating realistic fuel chemistry in large-scale computations. Progr Energ Combust Sci 35:192–215

Maas U, Pope SB (1992) Simplifying chemical-kinetics – intrinsic low-dimensional manifolds in composition space. Combust Flame 88:239–264

Meeks E, Ando H, Chou C-P, Dean AM, Hodgson D, Koshi M, Lengyel I, Maas U, Naik CV, Puduppakkam KV, Reitz RD, Wang C, Westbrook CK (1988) New modeling approaches using detailed kinetics for advanced engines. 7. International conf. on modeling and diagnostics for advanced engine systems (COMODIA), Sapporo

Moran MJ, Shapiro HN (1992) Fundamentals of engineering thermodynamics, 2nd edn. Wiley, New York, NY

Pitz WJ, Cernansky NP, Dryer FL, Egolfopoulos FN, Farrell JT, Friend DG, Pitsch H (2007) Development of an experimental database and kinetic models for surrogate gasoline fuels. SAE Paper 2007-01-0175

Reynolds WC (1986) The element potential method for chemical equilibrium analysis: implementation in the interactive program STANJAN, Stanford University

Semenov N (1935) Chemical kinetics and chain reactions. Oxford University Press, London

Smith GP, Golden DM, Frenklach M, Moriarty NW, Eiteneer B, Goldenberg M, Bowman CT, Hanson RK, Song S, Gardiner WC Jr, Lissianski VV, Qin Z (1999) http://www.me.berkeley.edu/gri_mech/

Tomlin AS, Turanyi T, Pilling MJ (1997) Mathematical tools for the construction, investigation and reduction of combustion mechanisms. In: Pilling MJ, Hancock G (eds) Low-temperature Combustion and Autoignition, vol 35, Comprehensive Chemical Kinetics., p 293

Warnatz J, Maas U, Dibble RW (2001) Verbrennung: Physikalisch-Chemische Grundlagen. Modellierung und Simulation, Experimente, Schadstoffentstehung. 3. Aufl., Springer, Berlin

Westbrook CK, Dryer FL (1981) Simplified Reaction Mechanism for the Oxidation of Hydrocarbon Fuels in Flames. Combustion Sci. Tech., Vol. 27, 31–48

Chapter 6
Pollutant Formation

Peter Eckert and Sebastian Rakowski

6.1 Exhaust Gas Composition

In the complete combustion of a so-called C_xH_y fuel, consisting only of C and H atoms, the exhaust gas contains the components oxygen (O_2), nitrogen (N_2), carbon dioxide (CO_2), and steam (H_2O).

In real combustion, however carbon monoxide (CO) (CO), unburned hydrocarbons (HC) (HC), hydrogen (H_2), nitrogen oxide (NO_x), and particulates also appear in addition to the above components. As opposed to these substances, which are detrimental to human health, CO_2, which is partially responsible for the greenhouse effect, is not viewed as a pollutant, since it does not pose a direct health hazard and appears as the final product of every complete oxidation of a hydrocarbon. A reduction of CO_2 in the exhaust gas is thus only to be achieved through a reduction in fuel consumption or through an altered fuel having a smaller amount of carbon with reference to its heating value.

A distinction is made between the concepts of complete and incomplete combustion as well as between perfect and imperfect combustion. For air ratios $\lambda \geq 1$, there is theoretically enough oxygen for the fuel to burn completely. In actuality however, combustion progresses under such air ratios at most until chemical equilibrium, i.e. always incompletely, even under ideal conditions. As a result, there is always a certain amount of CO and unburned hydrocarbons after combustion, even if the supply of oxygen is sufficient. In the case of air ratios $\lambda < 1.0$, the fuel cannot burn completely because of the lack of O_2. Under ideal conditions, combustion proceeds at best until chemical equilibrium. At all air ratios, combustion can also be incomplete due to imperfect air fuel mixing or because certain reactions proceed so slowly that chemical equilibrium is not reached.

The formation of CO, HC and NO_x varies air-fuel equivalence ratio λ and the combustion temperature coupled with it, see Fig. 6.1. While CO and HC rise as products of incomplete combustion in a rich mixture ($\lambda < 1.0$), NO_x formation is most rapid at high temperatures with sufficient levels of oxygen ($\lambda \approx 1.1$). With a lean mixture ($\lambda > 1.2$), the combustion temperature drops, so that NO_x emissions fall off and HC emissions increase.

G.P. Merker et al. (eds.), *Combustion Engines Development*,
DOI 10.1007/978-3-642-14094-5_6, © Springer-Verlag Berlin Heidelberg 2012

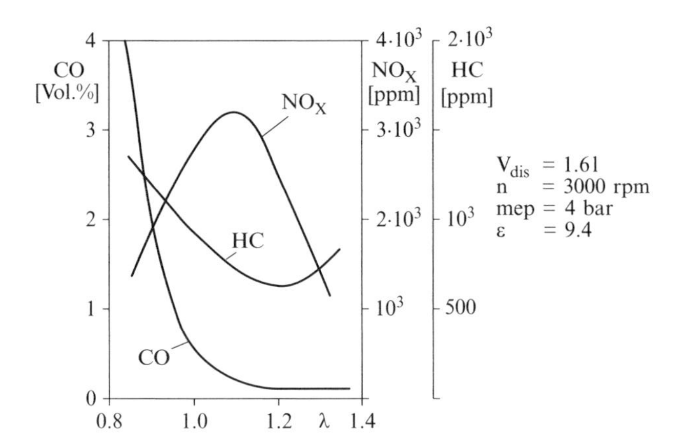

Fig. 6.1 Pollutant formation as a function of the equivalence ratio

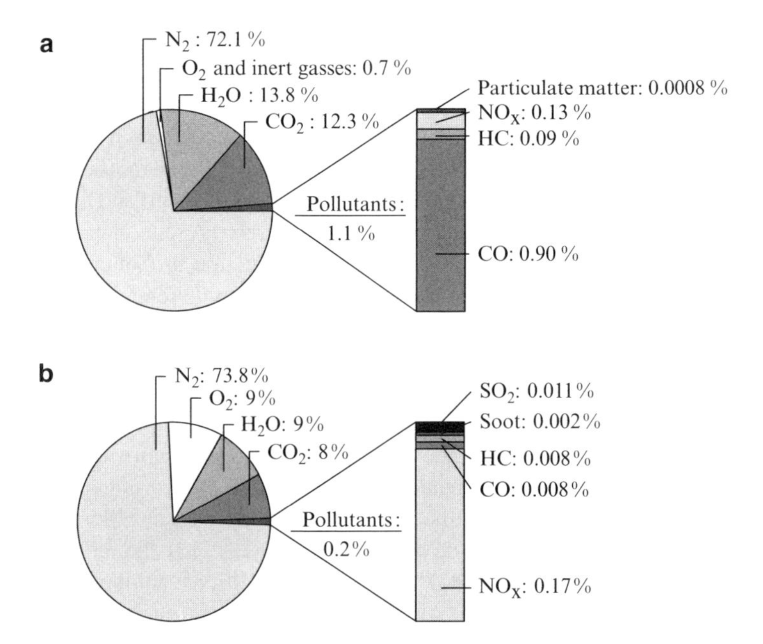

Fig. 6.2 Raw emissions (without catalyst) in percent by volume. (**a**) SI engine and (**b**) diesel engine

In Fig. 6.2, the compositions of the exhaust gasses (without a catalytic converter) of SI and diesel engines are shown. From this we see that the amount of pollutants has, from the point of view of energy, a negligible effect on the engine process, but rather only from the point of view of its potential to jeopardize human health and the environment. Figure 6.2 makes it clear that the amount of pollutant is of no importance for the engine process, but only with respect to the potential danger to human health and the environment. Although the diesel engine emits much less HC and CO than SI engines do, the absolute NO_x concentrations are similar. While in the case of the diesel engine, particulate matter also represent a critical magnitude besides nitrogen oxides, CO is the dominant pollutant component in the SI engine. In Si engines with direct injection however, particle emissions are also important.

6.2 Carbon Monoxide

In the combustion of hydrocarbons, CO arises as an intermediate product of oxidation (Sect. 5.2). Under stoichiometric ($\lambda = 1.0$) and hyperstoichiometric ($\lambda > 1.0$) conditions, CO can theoretically be completely oxidized to CO_2. In case of local air deficiency ($\lambda < 1.0$), CO generally remains intact as a product of incomplete combustion.

The decisive reactions in the oxidation of CO are the reactions with a hydroxyl radical and a hydroperoxyl radical.

$$CO + OH \leftrightarrow CO_2 + H \tag{6.1}$$

$$CO + HO_2 \leftrightarrow CO_2 + OH \tag{6.2}$$

Reaction (6.1) is clearly the dominant one. Reaction (6.2) only plays a role during the autoignition process, since here HO_2 exists in relatively high concentration (Sect. 5.2). Further oxidation reactions, which however are of inferior importance in the absence of hydrogen atoms, include

$$CO + O + M \leftrightarrow CO_2 + M \tag{6.3}$$

and

$$CO + O_2 \leftrightarrow CO_2 + O. \tag{6.4}$$

Since reaction (6.1) is dominant under most condition, CO oxidation is heavily dependent on the concentration of OH radicals. The reaction rate of reaction (6.1) is much slower than that of the reaction between OH radicals and hydrocarbons (Westbrook and Dryer 1984). For this reason, CO oxidation is usually inhibited until the fuel molecules and hydrocarbon intermediate species are oxidized (see also Fig. 5.4 in Chap. 5).

During rich combustion ($\lambda < 1.0$), CO oxidation progresses, due to a lack of O_2, in competition with H_2 oxidation

$$H_2 + OH \leftrightarrow H_2O + H. \tag{6.5}$$

As opposed to the kinetically controlled reaction (6.1), reaction (6.5) is practically in equilibrium at higher temperatures. With an increasing air ratio and increasing temperature, the deviation of the kinetics from the OHC equilibrium is reduced, and as a result the CO concentration is reduced with increasing air ratio λ.

With stoichiometric combustion ($\lambda \approx 1.0$), reactions (6.1) and (6.2) can be described with very good approximation as a gross reaction via the water gas reaction

$$CO + H_2O \leftrightarrow CO_2 + H_2, \tag{6.6}$$

which in this case proceeds near equilibrium, because the surplus concentrations of the chain propagators H and OH are very large.

In the super-stoichiometric range ($\lambda > 1.0$), CO oxidation no longer progresses in competition with H_2 oxidation and is dominated again by reaction (6.1).

During lean combustion ($\lambda > 1.4$), increased CO develops again due to the lower temperatures and incomplete combustion in the area near the wall of the combustion chamber. Generally, CO oxidation is highly contingent on temperature, so that reaction (6.1) becomes increasingly slow during expansion as well. The CO concentration in the exhaust gas thus corresponds approximately to the equilibrium concentration at 1,700 K.

6.3 Unburned Hydrocarbons

In the combustion of C_xH_y fuels, no measurable HC concentrations appear "behind" the flame front at $\lambda > 1$ and the air and fuel are well mixed. HC thus originates in zones that are not completely or not at all involved in combustion, and are composed of a number of different fuel components, which are either completely unburned or already partially oxidized. Most legislation today restricts only the sum of all HC components, which are usually determined with a flame ionization detector (FID). In this way, nothing is said concerning the composition of these unburned hydrocarbons, and the particular hazardous potential of certain components is thus not considered. Included among the unburned hydrocarbons are also the polycyclic aromatic hydrocarbons, or “PAHs”, which are of importance in diesel engine soot formation (Sect. 6.4.2).

6.3.1 Sources of HC Emissions

When considering the sources of HC emissions, a distinction must be made between combustion methods with a homogeneous mixture such as in conventional SI engines with lean stratified intake-manifold fuel injection or HCCI combustion processes, and combustion processes with a heterogeneous mixture, for example in DI diesel engines or SI engines with direct injection. As a rule, engines emit the largest part of unburned hydrocarbons already in the cold start and warm-up phases where the temperature in the combustion chamber is relatively low, so that little re-oxidation occurs.

In conventional SI engines with homogeneous mixture formation, the most important sources of HC emissions are (Cheng et al. 1993):

- Flame extinguishing within crevices as a result of flame quench
- Adsorption and desorption of the fuel in the oil film on the cylinder liner
- Adsorption and desorption of the fuel in deposits on the combustion chamber walls
- Fluid fuel with a high molar weight in the cylinder, which does not sufficiently vaporize before combustion
- Boundary layer quench of the flame while approaching a cold wall
- Flame extinguishing as a result of excessively low flame speed during expansion (rapid temperature drop) or local flame extinguishment in the case of lean mixtures
- Leakage of the fuel-air mixture through the closed outlet valves (mechanical defect)

One must take into consideration the mechanisms mentioned above do indeed prevent the combustion of certain amounts of fuel, but this fuel can however be at least partially reoxidized at a later time. This is shown in Fig. 6.3. At the start of combustion, part of the mixture is compressed into the piston ring crevice and is thus not consumed by the flame front. Part of this unburned mixture remains in the crankcase and is vented back into the induction system. However, much of this unburned mixture returns to the combustion chamber when the cylinder pressure drops. At this late time, the combustion gas chamber temperatures are often already

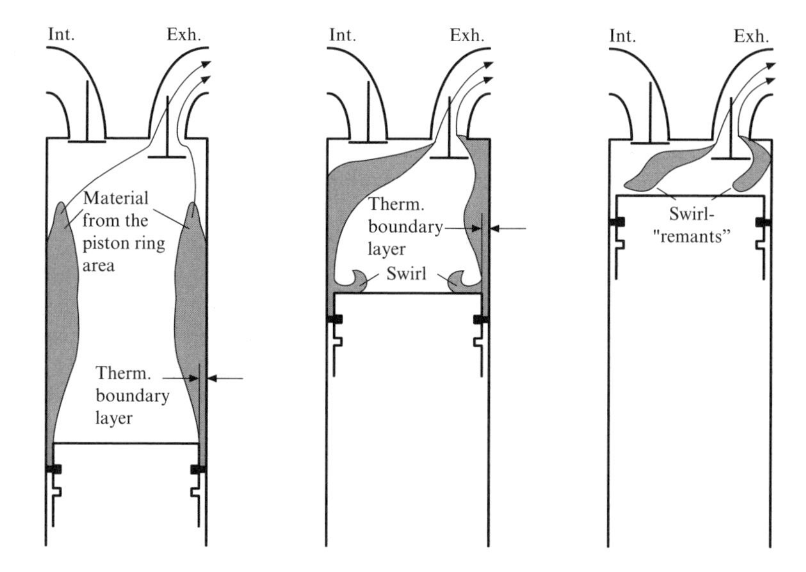

Fig. 6.3 Schematic representation of HC formation

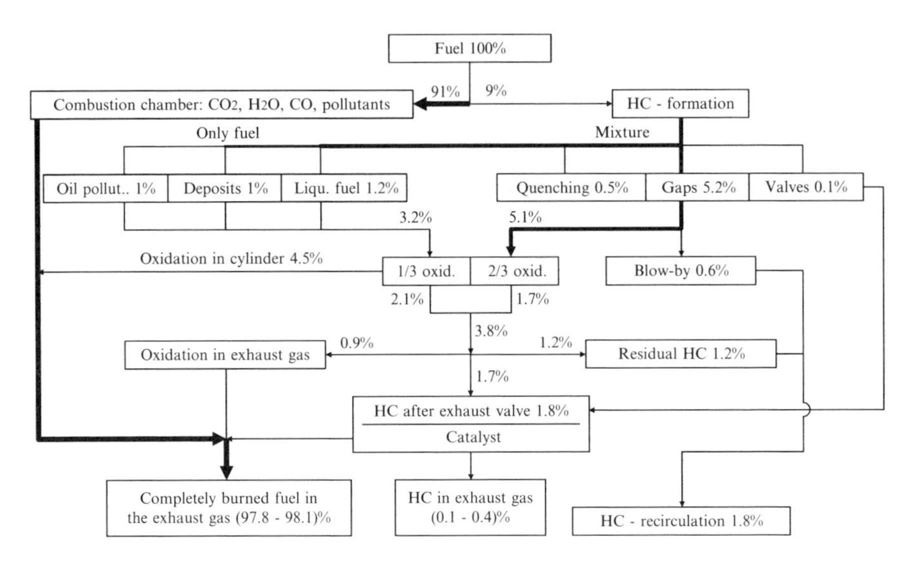

Fig. 6.4 HC-formation mechanisms in SI engine combustion, Cheng et al. (1993)

so low that only a small amount of post flame oxidation is possible. Part of this mixture remains in the cylinder as residual gas and can thus be oxidized in the following combustion cycle. In particular, unburned hydrocarbons that arise by means of flame quench at the combustion chamber walls make up a relatively small part of the hydrocarbons measured in exhaust gas, due to post flame oxidation and diffusion.

The different paths of HC formation mechanisms in older SI engine combustion are represented in Fig. 6.4 (Cheng et al. 1993). One can see that flame extinguishing in cracks are the main source of HC emissions. In total, 8.3% of the fuel does not make contact with the flame front at first. In the exhaust gas before the catalytic convertor however, only about 1.7% of the fuel remains as unburned hydrocarbons.

The combustion chamber architecture has a large influence on the HC emissions of conventionally operated SI engines. For example, ignition close to the bore wall will significantly lower HC emissions by causing burned gas to be compressed into the piston ring crevice (Borrmeister et al. 1997). Ignition timing has a major influence on HC emissions. Adjusting the ignition time late lowers peak cylinder pressure reducing the mass of mixture compressed into the ring pack and increases gas temperatures during expansion, so that fuel emerging from gaps, oil and deposits can be better oxidized. However, very late ignition times lead again to an incrcasc in HC emissions. For further details on the effect of ignition timing, see Eng (2005).

In the case of an ideal HCCI combustion process, there is no flame front in the traditional sense. Instead, distributed ignition takes place, where. the homogeneous mixture ignites in many places simultaneously. Nonetheless, the sources for HC emissions are similar to those of the conventionally operated SI engine.

The mechanisms of HC development mentioned can be complex, and a quantitative prediction of HC emission in SI engines is thus not yet practical.

An approximate calculation of HC emissions requires a very detailed representation of the combustion chamber geometry, including all gaps and the piston ring area, as well as detailed combustion models that can predict flame extinguishing and partial oxidation.

In lean stratified operation only small amounts of fuel-are compressed into the piston ring volumes, so that these mechanisms are of less importance. The most important sources for HC emissions in diesel engines and SI engines with direct injection and very late injection are:

- The periphery of the spray – the mixture composition lies outside flamability limits (too lean)
- The inner spray area – the mixture composition is too rich to fully oxidize
- Extinguishment of the diffusion flame by rapid pressure and temperature decreases during expansion and exhaust valve opening
- Fuel adhering to the wall that is not completely mixed and oxidized due to insufficient temperatures
- "After-injection" due to renewed opening of the nozzle needle after injection finish. From this results extremely large fuel drops, which can only evaporate and combust slowly
- Evaporation of unburned fuel from the blind hole volume of the injection nozzle towards the end of combustion
- In the case of SI engines with direct injection, lean stratified operation can lead to misfires in some working cycles if the fuel-air mixture is too lean at the spark plug
- Gaseous emission of the fuel via injector leakage, especially after stopping the engine and the subsequent cold start

With a view to the last point, Fig. 6.5 provides a schematic representation of the influence of the blind hole volume on HC emissions of diesel engines. The HC

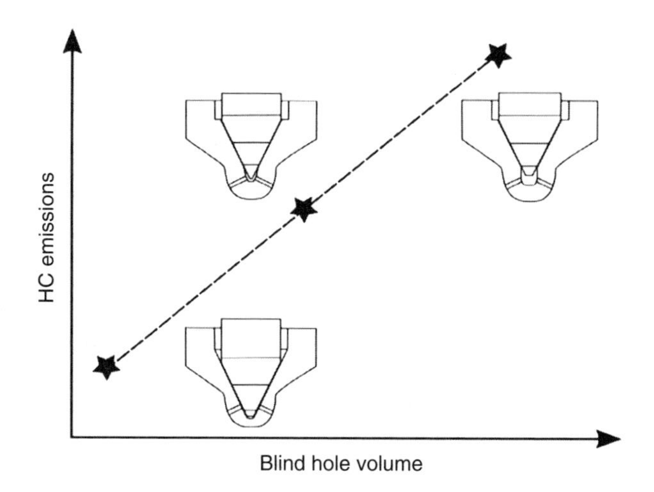

Fig. 6.5 Schematic representation of HC emissions as a function of the blind hole volume

emissions increase approximately linearly with the blind hole volume. The HC emissions for seat nozzles are the lowest due to their functional principle. For further details on HC formation mechanisms in SI engines with direct injection, see Basshuysen (2008).

As in the case of combustion processes with homogeneous mixtures, the HC formation mechanisms of combustion processes with heterogeneous mixtures are very complex and occur in local areas, such as evaporation from a nozzle, so here too only a qualitative calculation is possible.

6.3.2 Non-limited Pollutant Components

Among the total mass of unburned hydrocarbons are some potentially harmful compounds, the amount of which has still today not been explicitly regulated yet, which are particularly important because of potential health hazards.

6.3.2.1 Carbonyl Compounds

Carbonyl compounds can harm the human organism by affecting it directly or via by-products formed in the atmosphere. They contribute, for example, towards the formation of ozone close to the ground (photochemical smog) in concert with nitrogen oxide.

With the carbonyl compounds rank aldehydes and ketones, which each have at least one characteristic carbonyl group. They originate as partially burned combustion components, the complete oxidation of which had prematurely ended for the reasons given previously.

In Fig. 6.6, an extract from the hydrocarbon oxidation system is shown qualitatively with the aldehyde R–CHO, which appears in the final phase of oxidation, as well as formaldehyde HCHO. This visualization also conveys an idea of the complexity of the oxidation system underlying C_xH_y oxidation.

Supplementing this, in Fig. 6.7 are represented the carbonyl compounds detectable in the present day and in Fig. 6.8 the distribution of carbonyl compounds in the exhaust gas of the diesel engine of a commercial vehicle, see Lange (1996).

6.3.2.2 Dioxins and Furans

Dioxins are aromatic hydrocarbons ranging from non-toxic to extremely toxic compounds. The definition frequently also includes the chemically and toxicologically related furan class. Since the chemical accident of Seveso in 1976 however, the extremely toxic 2,3,7,8 tetrachlordibenzo-p-dioxin ("Seveso poison") released then is often representative for all dioxines. For the explanation of the chemical structure, the structural formulae of the benzene ring and of a few chlorinated as

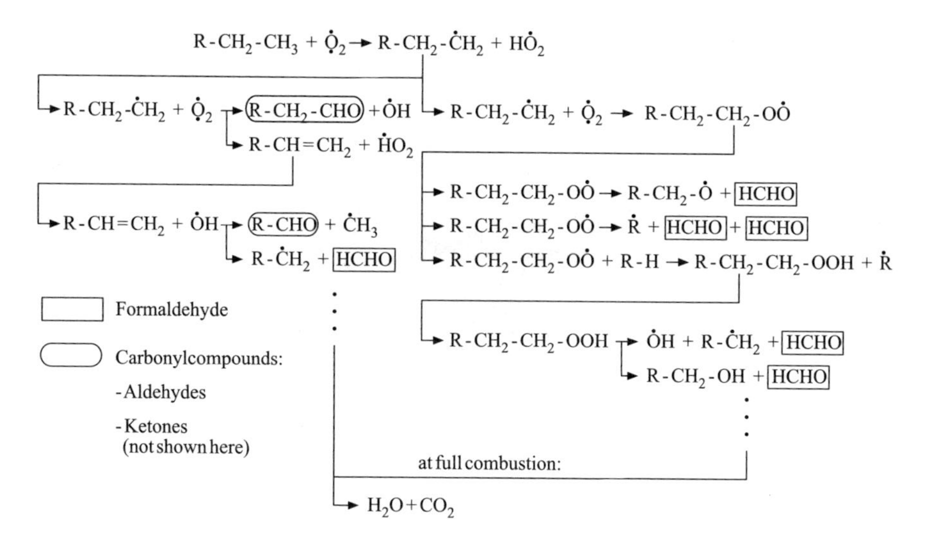

Fig. 6.6 Excerpt from the hydrocarbon oxidation diagram

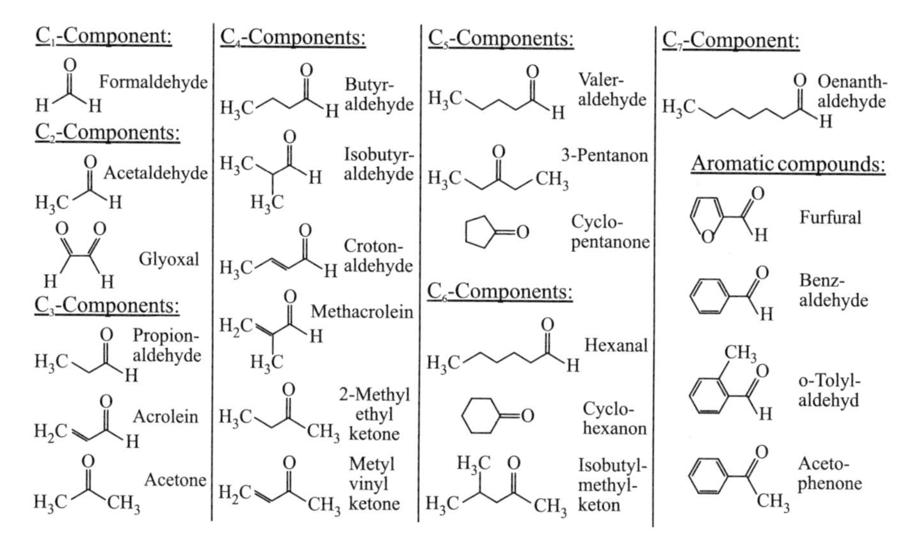

Fig. 6.7 Detectable carbonyl compounds, Lange (1996)

well as polycyclic aromatic hydrocarbons are shown in Fig. 6.9, and in Fig. 6.10 are shown the structural formulae of heterocyclic aromatic compounds pyridine, dioxine, and furane, as well as two substituted compounds.

To be mentioned as precursors of dioxins and furans are the polycyclic aromatic hydrocarbons (PAH), and polycyclic biphenyls (BCB)

Cl—⌬—⌬— Cl

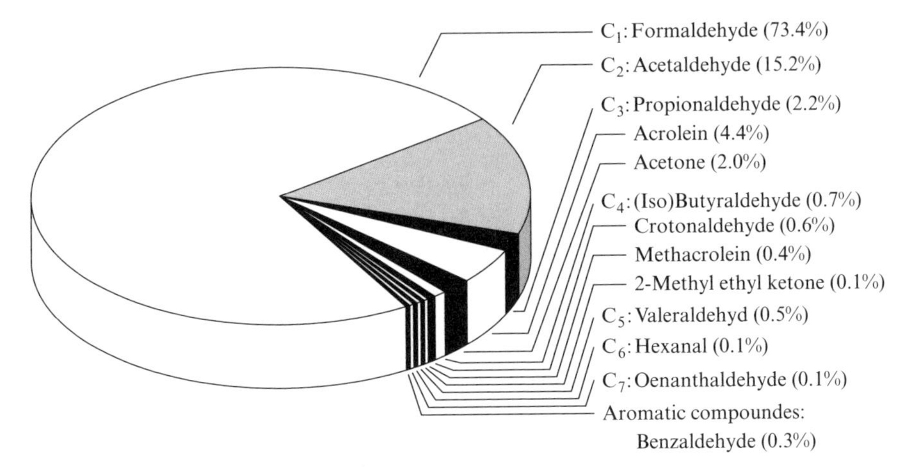

Fig. 6.8 Distribution of carbonyl compounds in the exhaust gas of a truck diesel engine, Lange (1996)

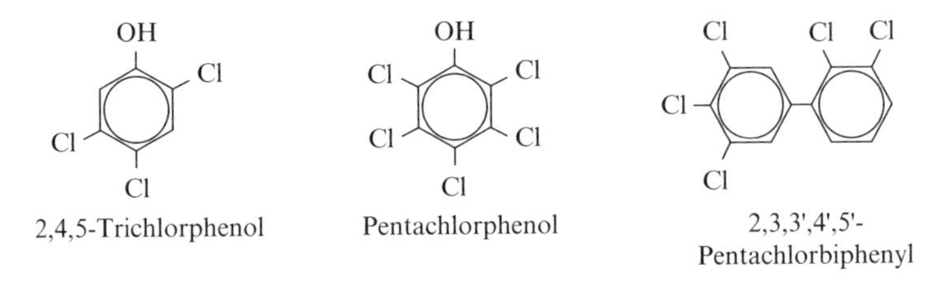

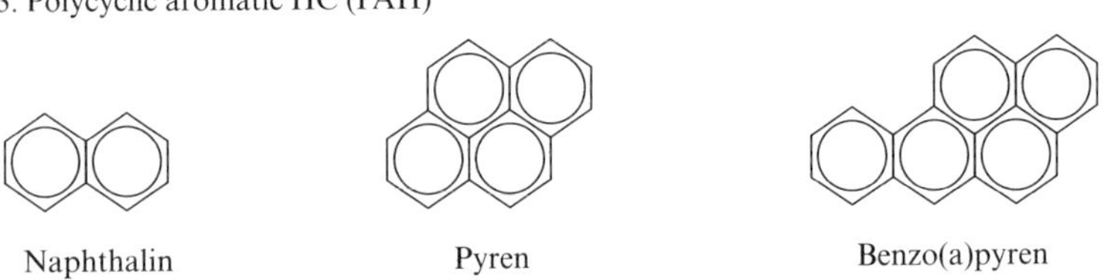

Fig. 6.9 Structure of various aromatic hydrocarbons

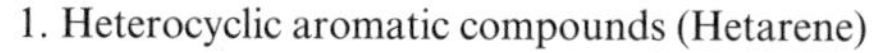

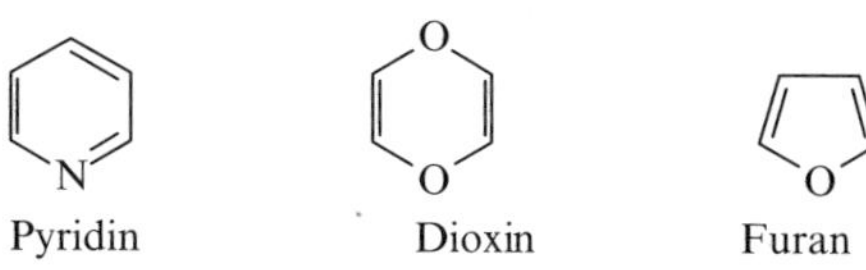

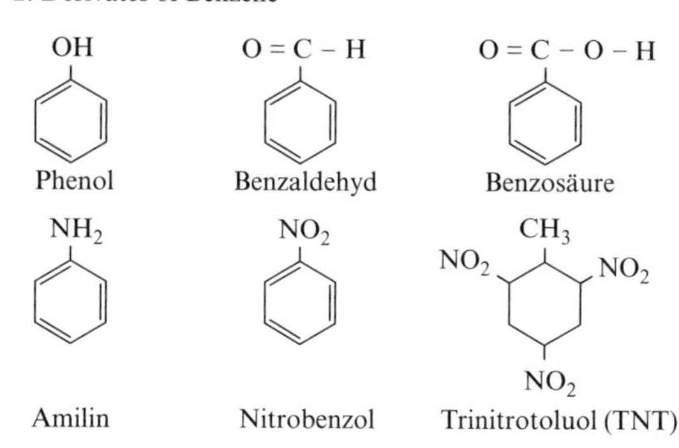

Fig. 6.10 Structure of heterocyclic aromatic hydrocarbons

and polychlorinated terphenyls (PCT).

Cl

Cl Cl

In Fig. 6.11, the structural formulae of dibenzofuran, dibenzodioxin, so-called Seveso poison, as well as the number of possible derivatives are given.

Figure 6.12 shows the orders of magnitude of the concentrations of various pollutant components in the exhaust gas of an internal combustion engine. Different engine construction types as well as gasoline and diesel fuels do not differ as far as the order of magnitude is concerned. We recognize that the concentrations of all dioxines and furanes lie in the order of magnitude of 10^{-9} kg per kg in the exhaust gas and the concentrations of the infamous Seveso poison in the order of magnitude 10^{-14} g per g in the exhaust gas, i.e. far beneath the detection margin of present-day measurement techniques.

For further explanation, see Bühler (1995) and Bühler et al. (1997).

6.4 Particle Emission in the Diesel Engine

6.4.1 Introduction

Exhaust particle content is defined as the amount of all materials that are captured by a certain filter after the exhaust has been diluted and cooled to $T<52°C$ in accordance with a defined process (EN ISO 8178). The different components

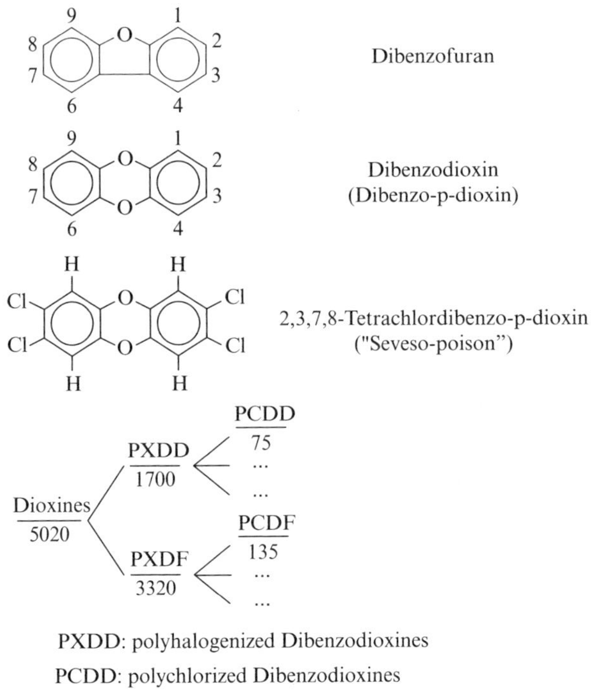

Dibenzofuran

Dibenzodioxin
(Dibenzo-p-dioxin)

2,3,7,8-Tetrachlordibenzo-p-dioxin
("Seveso-poison")

Dioxines 5020
PXDD 1700
PCDD 75
PXDF 3320
PCDF 135

PXDD: polyhalogenized Dibenzodioxines
PCDD: polychlorized Dibenzodioxines

Fig. 6.11 Dioxine structures

existing in engine particles are shown in Fig. 6.13. Elementary carbon, usually called soot, hydrocarbons and sulfates take up the largest amount of the entire particle weight, whereby the exact composition can vary depending on the combustion process and the operating point. Figure 6.14 shows as examples excerpts from the compositions of particle emissions of a commercial vehicle diesel engine at different load points at constant speed determined by Kweon et al. (2002). The total particle weight is about four times larger at the operation points with 75% and 100% load than at 25 and 50%. We can see that the organic components make up the largest portion of the particle weight at lower partial load and thus under a relatively large air-fuel ratio λ, while at higher partial load and full load, i.e. under a low air-fuel ratio λ, elementary carbon is clearly dominant. The percentage amount of sulfates in the total particle weight tends to increase with increasing load, taking up 5% at 75% load. However, one must take into consideration that Kweon et al. (2002) used a diesel fuel with a sulfur content of 350 mg/kg fuel. Diesel fuels used in Europe today, have with a maximum sulfur content of 10 mg/kg, so the amount of sulfur is presumably much lower.

The chemical and physical processes occurring in the origination of soot particles have been understood roughly, but still insufficiently in many details. The modeling of soot formation is thus quite problematic.

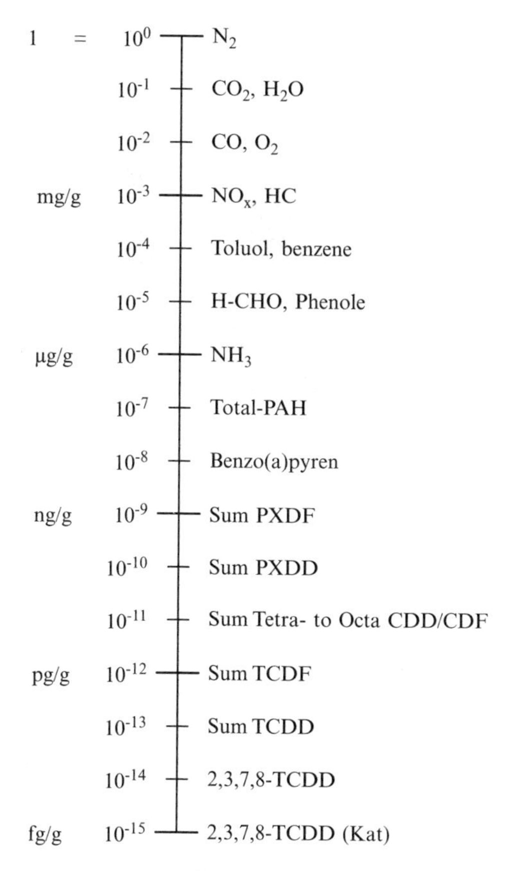

Fig. 6.12 Pollutant concentrations in the exhaust of combustion engines, Bühler (1995)

Particle formation progresses according to present knowledge approximately according to the following scheme, shown schematically in Fig. 6.15:

- Chemical reduction of the fuel molecules under oxygen-deficient conditions to small hydrocarbons, of which small alkenes, dialkenes, alkines and their radicals are of special importance; formation of the first benzene ring
- Formation of polycyclic aromatic hydrocarbons via polymerization of rings and advancing dehydration, also proportional increase of C atoms
- Condensation and formation of soot kernels (nucleation) with dimensions of approx. 1–2 nm
- Surface growth and coalescence of soot kernels to form primary soot particles (surface growth) with diameters of about 20–30 nm and then addition of various substances
- Coalescence of primary soot particles to long, chain structures (agglomeration)
- Breaking down of soot particles and intermediate species via oxidation with O_2 molecules and OH radicals

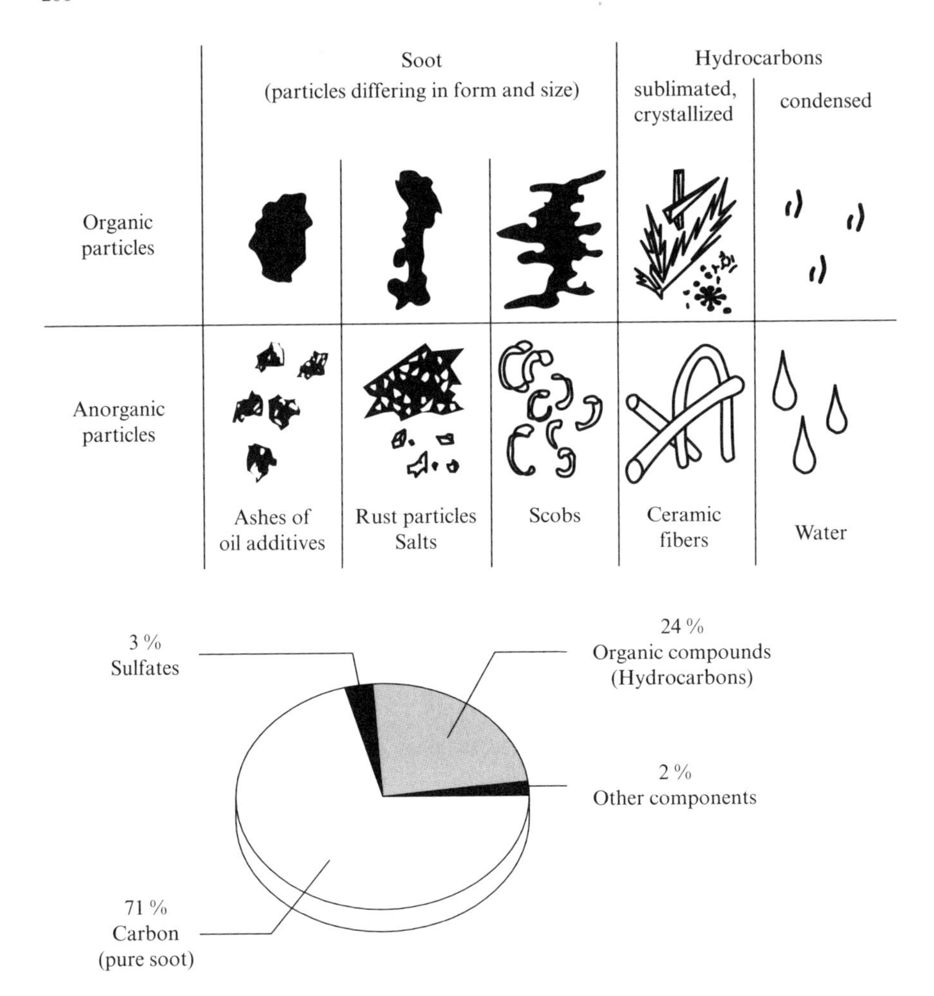

Fig. 6.13 Various components in particle emissions

6.4.2 Polycyclic Aromatic Hydrocarbons

The formation of the first aromatic hydrocarbon (benzene) ring is an important step in soot formation because it often determines that formations speed. For example, soot formation in methane flames not pre-mixed and enriched with different heptane isomers scales in a linear fashion with the concentration of formed benzene (McEnally et al. 2003).

The first benzene ring can be formed by means of different reaction paths, the relative amounts of which depend on the combustion process and the fuel used. The acetylene and ion path will be presented here as an example. For further details, see McEnally (2006).

In the case of the acetylene path, ethine (acetylene, C_2H_2), formed during rich combustion, plays a decisive role. In the initial reaction, an ethine molecule reacts

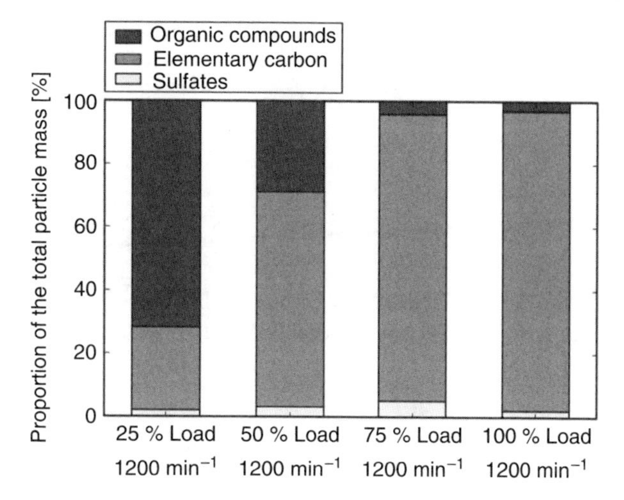

Fig. 6.14 Composition of diesel particles, Kweon et al. (2002)

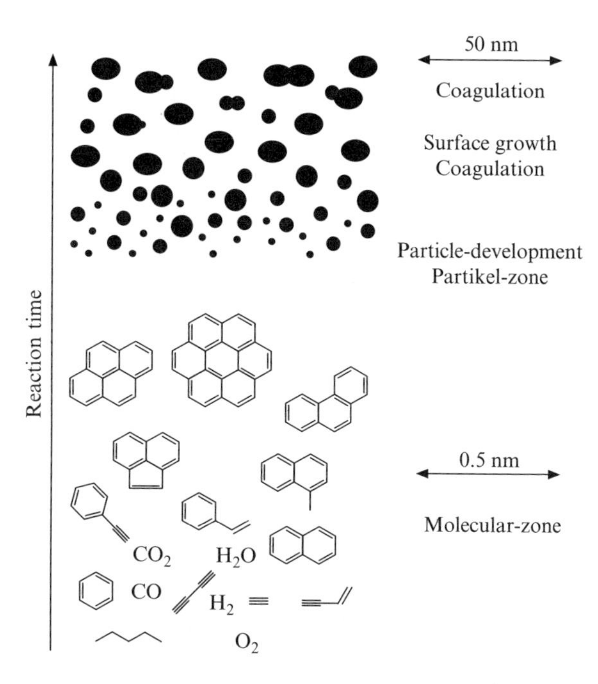

Fig. 6.15 Schematic diagram of soot formation, acc. to Bockhorn (1994)

with a vinyl radical (C_2H_3). As shown in Fig. 6.16, two reaction paths are subsequently possible depending on the local temperature, whereby each time one ethine molecule is used to close the benzene ring (Frenklach and Wang 1994). At high temperatures, a phenyl is formed, at low temperatures a benzene ring.

In the case of the ion path, ethine molecules combine themselves at first with CH- or CH2 groups also found in the fuel-rich mixture to form C_3H_3 ions. Two such C_3H_3 ions can then unite with the rearrangement of two H atoms to form a ring, see Fig. 6.17.

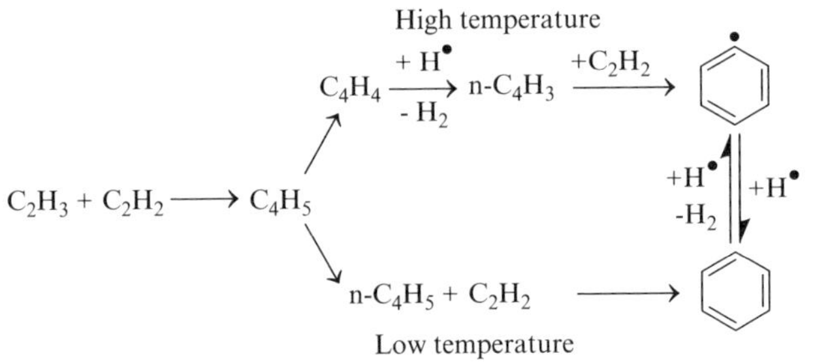

Fig. 6.16 Reaction path in the formation of benzene rings, acc. to Frenklach and Wang (1994)

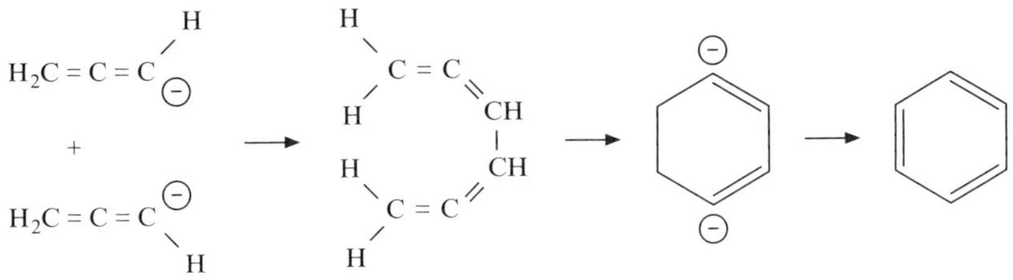

Fig. 6.17 Development of benzene rings, acc. to Warnatz et al. (2001)

Via progressing splitting off of H and C_2H_2 addition, the so-called HACA mechanism (H abstraction, C_2H_2 addition), PAH rings originate, see Fig. 6.18a. However, benzene rings can also directly unite themselves, thus constructing complex ring compounds, see Fig. 6.18b. These days we assume that aromatics contained in the fuel do not initially break down into smaller hydrocarbons under oxygen-deficient conditions but participate directly in the PAH formation process (McEnally et al. 2006).

6.4.3 Soot Formation

The polycyclic aromatic hydrocarbons grow to form larger and larger structures. Customarily, we refer to them as soot particles at the moment in which the PAHs are no longer arranged on a plane but rather represent a spatial structure. One possible mechanism that could lead to the formation of such a structure is the collision between two PSH molecules, which after a certain decisive molecule size can remain adhering to each other via van der Waals forces. Soot kernels formed in this way have diameters of about 1–2 nm. Although the soot kernels only take up a small part of the total particle volume, they are of decisive importance for the further formation of soot. Surface growth then occurs on the soot kernel, a process in which the molecules from the gaseous phase coalesce on the soot particles. We assume that surface growth takes place to a large extent with ethine molecules in a process similar to the HACA mechanism. Surface growth can also take place by means of

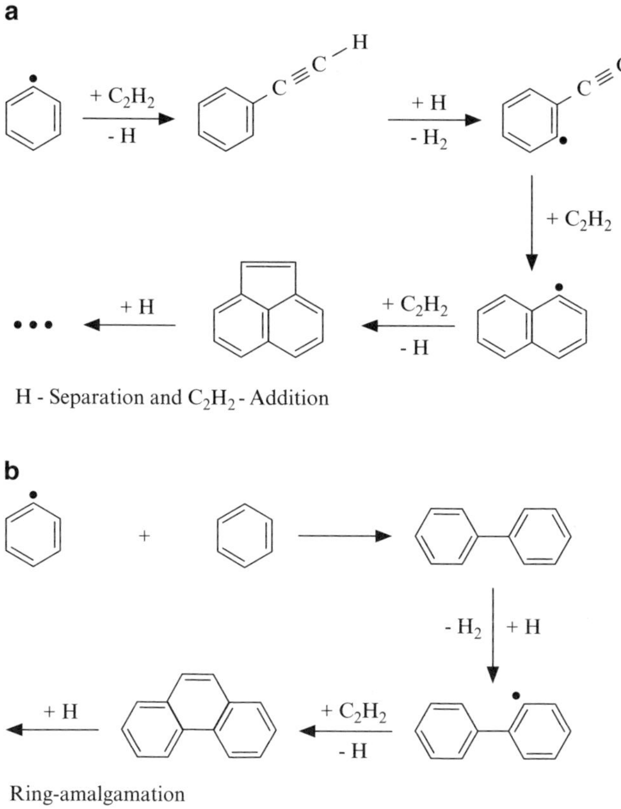

a

H - Separation and C_2H_2 - Addition

b

Ring-amalgamation

Fig. 6.18 PAH growth, acc. to nach Frenklach and Wang (1994)

the condensation of polycyclic aromatic hydrocarbons on the soot surface. Sulfates can also adhere to the surface if present.

In the case of relatively small particles, a collision leads to coagulation, i.e. two colliding, approximately spherical particles form a larger, also spherical particle. Larger particles agglomerate, i.e. the particles taking part in the collision keep their shape and adhere to each other. The individual agglomerated particles are called primary soot particles. In the case of diesel engines, these have a diameter of about 15–35 nm (Mathis et al. 2005). The agglomerated particles often form a branched, chain-like structure.

During the entire soot formation process, soot oxidation can also occur, whereby both molecular oxygen and the hydroxyl radical OH can have the dominant role depending on the marginal conditions.

Individual soot particles in a flame exhibit different rates of surface growth, oxidation, condensation, coagulation, and agglomeration, resulting in a distribution of particles with different diameters that can be described with a probability density function. The size of the particles extends across a wide range of $2 < d_P < 150\,\text{nm}$, though particles with sizes up to 10 μm can also be formed; the maximum of the distribution is however limited to about 100 nm. The density of the soot particles formed is about 2,000 kg/m^3, the surface/mass ratio between 20 and 200 m^2/g.

The cumulated soot volume is given by

$$V_P = \frac{\pi}{6}\, N\ \tilde{d}^3 = \frac{\pi}{6}\, N\left[\sum \frac{1}{N}\ N_i\ d_{P,i}^3\right] \tag{6.7}$$

where N represents the particle number of all soot particle size classes i and $\tilde{d}$ the average diameter of the discrete soot particle size distribution.

The influence of temperature on soot formation is difficult to describe in a consistent way, because high temperatures favor both formation (pyrolysis) and consumption (oxidation). The temperature window 1,500 < T < 1,900 is critical for soot formation under premixed conditions. This is made clear in Fig. 6.19, in which the percentage soot yield is represented as a function of the air-fuel equivalence ratio and temperature. We can recognize the critical temperature range 1,500 < T < 1,900 and an extreme increase in soot emission for equivalence ratios λ < 0.6 (Pischinger et al. 1988). In the case of non-premixed flames such as one finds in diesel engines, soot formation increases with temperature in rich mixture areas. At the same time, high temperatures lead to rapid soot oxidation in lean mixture areas.

The pyrolysis-oxidation problem is clarified in Fig. 6.20. Here we have illustrated the soot mass fraction as a function of the crank angle in a diesel engine.

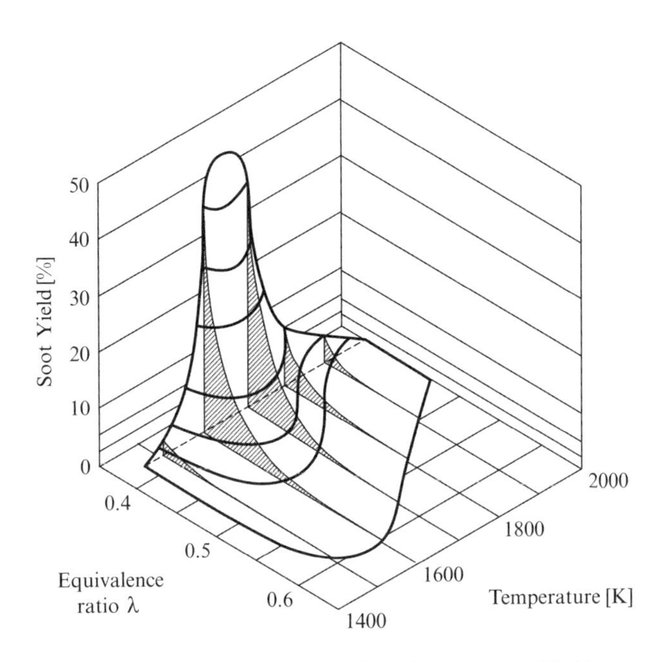

Fig. 6.19 Soot yield as function of equivalence ratio and temperature, Pischinger et al. (1988)

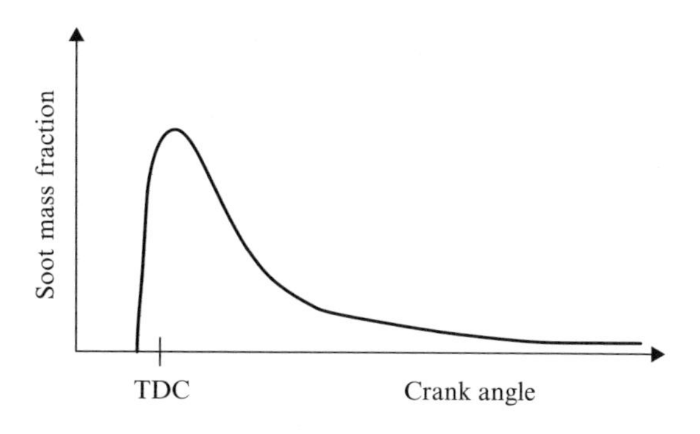

Fig. 6.20 Temporal progression of soot in the DI diesel engine, Stiesch (2003)

One can recognize that, at the beginning of combustion, a relatively large amount of soot is produced, which is however for the most part oxidized again during the main and post-combustion phases. The amount of particles measured in the exhaust gas is thus only a fraction (approx. 1–10%) of the maximum formed particles. For this reason, soot formation and oxidation is very difficult to model, since even exact models for formation and oxidation that exhibit errors in themselves in the low percentage range can lead to a very large error in the absolute soot amount in the exhaust gas (Stiesch 2003). Modeling soot emissions under engine conditions is for this reason most suitable for qualitative inquiries.

6.4.4 *Particle Emission Modeling*

As described in the previous section, soot formation and oxidation is a complex process, the modeling of which under engine conditions is a major challenge and suited primarily to qualitative problems. For this reason, simple empirical models are still often used today, such as the two-equation model of Nishida and Hiroyasu (1989). In this model, formation and oxidation are each described with one empirical equation. The net change of the soot weight is the result of the difference of both of these quantities:

$$\frac{\mathrm{d}m_{P,b}}{\mathrm{d}t} = A_b\ m_{B,g}\ p^{0{,}5}\ \exp\left[-\frac{6.313}{T}\right], \tag{6.8}$$

$$\frac{\mathrm{d}m_{P,ox}}{\mathrm{d}t} = A_{ox}\ m_P\ x_{O_2}\ p^{1{,}8}\ \exp\left[-\frac{7.070}{T}\right], \tag{6.9}$$

$$\frac{\mathrm{d}m_P}{\mathrm{d}t} = \frac{\mathrm{d}m_{P,b}}{\mathrm{d}t} - \frac{\mathrm{d}m_{P,ox}}{\mathrm{d}t}. \tag{6.10}$$

The soot formation in (6.8) is directly proportional to the concentration of fuel. With this simple model, we can gain insight into the general trend of soot formation, but no quantitatively reliable results can be obtained. To describe soot oxidation, often the relation provided by Nagle und Strickland-Constable (1962) is used instead of (6.9). In this semi-empirical model, it is assumed that the surface of the soot particles consist partially of reactive areas A and less reactive areas B – an assumption also used in some detailed soot models. In the Nagle and Strickland-Constable model, the oxidation rate is yielded by

$$\omega = A_{Ruß}\left\{\left(\frac{k_A p_{O_2}}{1 + k_Z p_{O_2}}\right)x + k_B p_{O_2}(1 - x)\right\}, \tag{6.11}$$

where the first term in the brackets is the reaction of the reactive A surface and the second the reaction of the less reactive B surface. The amount of A surface is

$$x = \frac{p_{O_2}}{p_{O_2} + (k_t/k_B)}. \tag{6.12}$$

The reaction rates k_A, k_B, k_t and k_Z can be found in Nagle and Strickland-Constable (1962). Additionally, the influence of the OH radicals on soot oxidation is often calculated based on the gas-kinetic collision rate (Neoh 1976).

In order to prevent the direct dependence of soot formation on fuel concentration, a series of phenomenological soot models were developed in the past, for example Belardini et al. (1994) or Fusco et al. (1994), whose models were later expanded by Kazakow and Foster (1998) and Tao et al. (2005). In these models, intermediate species such as ethine (acetylene, C_2H_2) are also balanced, and the steps of soot formation shown in the previous chapter (PAH-formation, soot kernel formation, surface growth, coagulation and oxidation) are described. The model by Tao et al. (2005) encompasses nine reactions between six species, in addition, the soot particle number density is also balanced. Figure 6.21 provides a schematic representation of the individual steps of the model. An up-to-date comparison between results calculated with the model and experiments for different engines and operation points can be found in Tao et al. (2008).

One alternative to the phenomenological models described above for describing soot formation is the use of the detailed reaction kinetics of hydrocarbon oxidation and the formation of the first aromatic rings in combination with the empirical approach of (6.8)–(6.11) Vishwanathan and Reitz (2009). The idea behind this approach is that the formation of the first aromatic In detailed soot models, the detailed reaction kinetics of hydrocarbon oxidation and PAH formation are combined with a phenomenological description of soot particle dynamics with soot kernel formation, surface growth, condensation, coagulation, agglomeration and surface oxidation. Examples of such detailed methods can be found in Frenklach and Wang (1994) and Mauß (1997).

The core of detailed soot models is, in addition to the reaction kinetics, the description of the dynamics of soot particle size distribution. The Smoluchowski model describes the temporal development of the distribution in a system of colliding and coagulating particles with an infinite number of differential equations for the particle number density N_i (von Smolouchowski 1917; Frenklach 2002):

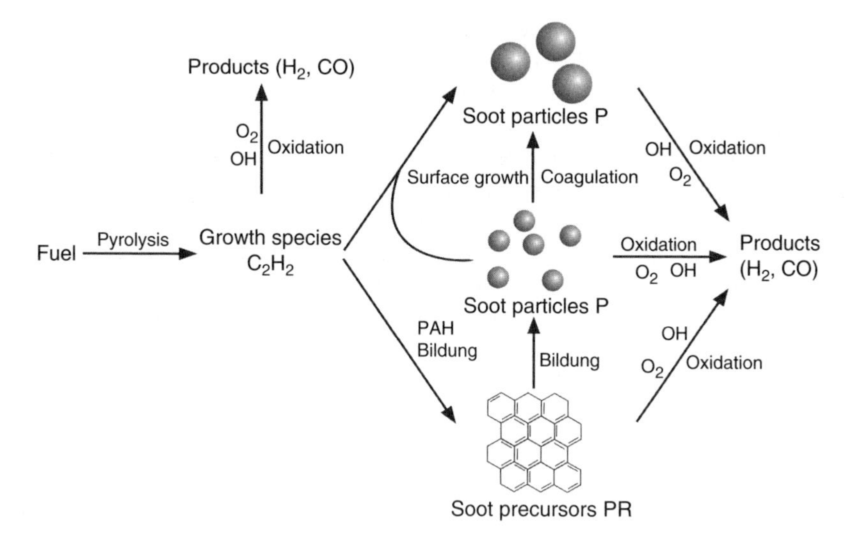

Fig. 6.21 Schematic representation of the phenomenological soot model, acc. to Tao et al. (2005)

$$\frac{dN_i}{dt} = \frac{1}{2}\sum_{j=1}^{i-1}\beta_{j,i-j}N_jN_{i-j} - \sum_{j=1}^{\infty}\beta_{j,i}N_iN_j; i = 2,\ldots,\infty. \tag{6.13}$$

In (6.13), β is a collision coefficient and the index i stands for the particle size class (for example, the number of C atoms in the particle). The first term of the equation describes the formation of particles of size i from two particles of size j and i–j. The second term describes the reduction of particles of size i by collision with other particles. The further steps of soot formation and soot oxidation can be integrated into (6.13) with source terms.

One frequently selected approach for solving the Smoluchowski equation is the description of the probability density function by means of their moments (*Method of Moments*, Frenklach 2002):

$$M_r = \sum_{i=1}^{\infty} m_i^r N_i. \tag{6.14}$$

Here, M_r is the rth moment of the probability density function. The 0th moment describes the total particle number density; the first moment is proportional to the total volume of the particles. To convert the Smoluchowski equation into moment equations, it is necessary to interpolate between the moments in order to obtain broken moments (*Method of Moments with Interpolative Closure*, Frenklach 2002). Theoretically, knowledge of all the moments ($r = 0, 1, \ldots, \infty$) supplies us with the same information as knowledge of all the particle number densities N_i, but in practical applications only the equations for a few moments are solved (Mauß 1997). The main advantage of the "method of moments" is thus its computational efficiency. One disadvantage of this method is that the actual distribution function cannot be recovered from the calculated moments without further assumptions.

Alternative methods for solving the particle size distribution include the discretisation of the probability density function into individual sectors (*Sectional Method*, Netzell et al. 2007), a solution using the Galerkin method (Appel et al. 2001), and approaches based on the Monte-Carlo method (e.g. Mosbach et al. 2009).

Despite the advances made recently in modeling soot formation and soot oxidation, considerable efforts still need to be made. Under engine operating conditions, we must especially take heed of the fact that the quality of the soot formation model strongly depends on the quality of the modeling of preceding processes such as spray and mixture formation, ignition and turbulent combustion. Moreover, further developments are required in the understanding of the oxidation of hydrocarbons, PAH formation, particle reactions and particle dynamics as well as of the numerical description of these processes.

6.5 Nitrogen Oxides

In the troposphere, nitrogen oxides (NO_x) favor the formation of ozone, while close to the ground tend to form photochemical smog. In engine combustion, mainly nitrogen monoxide (NO) is formed, which is eventually converted almost completely into nitrogen dioxide (NO_2) under atmospheric conditions. In combustion, NO can be formed in three different ways. In this case, we distinguish between so-called thermal NO, which is formed among the combustion products at high temperatures according to the Zeldovich mechanism from atmospheric nitrogen, so-called prompt NO, which develops in the flame front via the Fenimore mechanism from atmospheric nitrogen, NO formed from the N_2O mechanism, and finally so-called fuel NO, which is produced by nitrogen portions in the fuel.

The importance of the various formation mechanisms is highly dependent under engine conditions on the operational conditions. Under most conditions, the Zeldovich mechanism is the dominant one. At low temperatures, reached for example in exhaust gas recycling, the prompt NO path increases in importance under fuel-rich conditions, while the N_2O path becomes more important under nitrogen-rich conditions.

6.5.1 *Thermal NO*

The thermal NO formation occurs "behind" the flame front in the hot burned gases. It was first described by Zeldovich (1946). The two-step mechanism formulated by Zeldovich was later extended by Lavoie (1970). This extended Zeldovich mechanism consist of the three elementary reactions

$$O + N_2 \overset{k_1}{\longleftrightarrow} NO + N \tag{6.15}$$

$$N + O_2 \overset{k_2}{\longleftrightarrow} NO + O \tag{6.16}$$

$$\mathrm{N} + \mathrm{OH} \overset{k_3}{\longleftrightarrow} \mathrm{NO} + \mathrm{H} \tag{6.17}$$

with experimentally determined speed constants k_i. Although thermal NO formation according to the Zeldovich mechanism has been one of the most investigated of all reaction mechanisms, there is still uncertainty concerning the choice of speed constants. Deviating values are suggested for this in the literature, of which a few are summarized in Table 6.1.

For the NO formation rate, we obtain with reaction (6.15)–(6.17)

$$\begin{aligned}\frac{\mathrm{d[NO]}}{\mathrm{d}t} &= k_{1,r}[\mathrm{O}][\mathrm{N_2}] + k_{2,r}[\mathrm{N}][\mathrm{O_2}] + k_{3,r}[\mathrm{N}][\mathrm{OH}] \\ &\quad - k_{1,l}[\mathrm{NO}][\mathrm{N}] - k_{2,l}[\mathrm{NO}][\mathrm{O}] - k_{3,l}[\mathrm{NO}][\mathrm{H}],\end{aligned} \tag{6.18}$$

and for the temporal change of nitrogen atom concentrations follows

$$\begin{aligned}\frac{\mathrm{d[N]}}{\mathrm{d}t} &= k_{1,r}[\mathrm{O}][\mathrm{N_2}] - k_{2,r}[\mathrm{N}][\mathrm{O_2}] - k_{3,r}[\mathrm{N}][\mathrm{OH}] \\ &\quad - k_{1,l}[\mathrm{NO}][\mathrm{N}] + k_{2,l}[\mathrm{NO}][\mathrm{O}] + k_{3,l}[\mathrm{NO}][\mathrm{H}].\end{aligned} \tag{6.19}$$

If the instantaneous NO concentration is below the equilibrium concentration at the corresponding temperature, as is the case throughout much of the combustion process, the forward reaction has a decisive influence on the total conversion. Only when the momentary NO is above the equilibrium concentration of the corresponding temperature is the total conversion substantially determined by the reverse reaction. In the engine however, this situation appears at best towards the end of the expansion stroke, when temperature is already quite low.

We know by the speed coefficients for the forward reactions that NO formation via reaction (6.15) progresses much more slowly than via reactions (6.16) and (6.17). For a temperature of $T = 1{,}800$ K, for example, we obtain

Table 6.1 Speed coefficients for the forward reaction of the Zeldoch mechanism

Reaction i	$k_{i,r}$ [cm^3/mol s]	Authors
(6.14)	$1.8 \times 10^{14} \exp\left[-\frac{38.400}{T}\right]$	Baulch et al. (1999)
	$0.544 \times 10^{14} T^{0.1} \exp\left[-\frac{38.020}{T}\right]$	GRI-MECH 3.0 (2000)
	$0.76 \times 10^{14} \exp\left[-\frac{38.000}{T}\right]$	Heywood (1988)
	$4.93 \times 10^{13} T^{0.0472} \exp\left[-\frac{38.061}{T}\right]$	Pattas (1973)
(6.15)	$6.4 \times 10^{9} T \exp\left[-\frac{3.150}{T}\right]$	Baulch et al. (1999)
	$9.0 \times 10^{9} T \exp\left[-\frac{3.280}{T}\right]$	GRI-MECH 3.0 (2000)
	$1.48 \times 10^{8} T^{1.5} \exp\left[-\frac{2.860}{T}\right]$	Pattas (1973)
(6.16)	3.0×10^{13}	Baulch et al. (1999)
	$3.36 \times 10^{13} \exp\left[-\frac{195}{T}\right]$	GRI-MECH 3.0 (2000)
	4.1×10^{13}	Heywood (1988)
	4.22×10^{13}	Pattas (1973)

$$k_{1,r} \approx 1.0 \times 10^2,\ k_{2,r} \approx 2.0 \times 10^9,\ k_{3,r} \approx 2.8 \times 10^{10} \left[\frac{\mathrm{m}^3}{\mathrm{kmol\ s}}\right].$$

The first reaction possesses a high activation energy because of the stable N_2 triple bond and thus proceeds sufficiently fast only at high temperatures; hence the designation "thermal". It is therefore the rate-limiting step. The above numerical values show that at 1,800 K, the first reaction progresses at around seven to eight decimal powers slower than the second and third. Figure 6.22 shows the path of speed coefficients $k_{1,r}$ in [g/(mol s)] as a function of temperature T.

We see that a doubling of the temperature raises thermal NO formation by the factor 10^3, and at a temperature elevation from 2,000 to 2,500 K the thermally formed NO climbs about 50-fold. Because of this considerable temperature-dependency, we refer to a kinetically controlled NO formation. That means that the chemical reaction kinetics at the temperatures in the combustion chamber is slow in comparison to the physical time scales of the flow field and thus that chemical equilibrium cannot be achieved. This should be clarified by Fig. 6.23, in which NO concentrations under assumption of equilibrium and under consideration of kinetics are qualitatively illustrated according to Zeldovich. The kinetically controlled process produces, according to Zeldovich, essentially less NO than under the assumption of equilibrium ($\Delta 1$) at first, but in the late combustion phase, because of the extremely slow process at lower temperatures, produced NO does not regress over the reverse reactions ($\Delta 2$). We speak of a reaction "freeze".

Because the reactions speed of the forward reactions (6.16) and (6.17) is higher than that of reaction (6.15) by several orders of magnitude, the atomic nitrogen

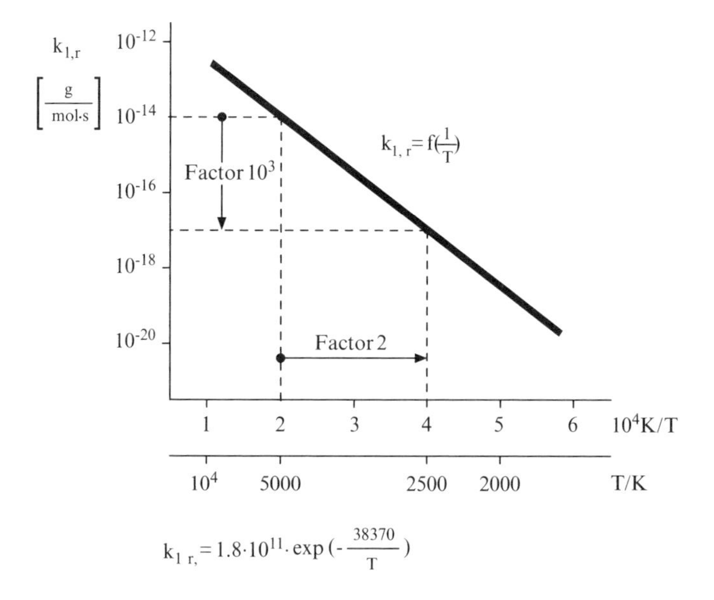

Fig. 6.22 Reaction coefficient of the first Zeldovich reaction

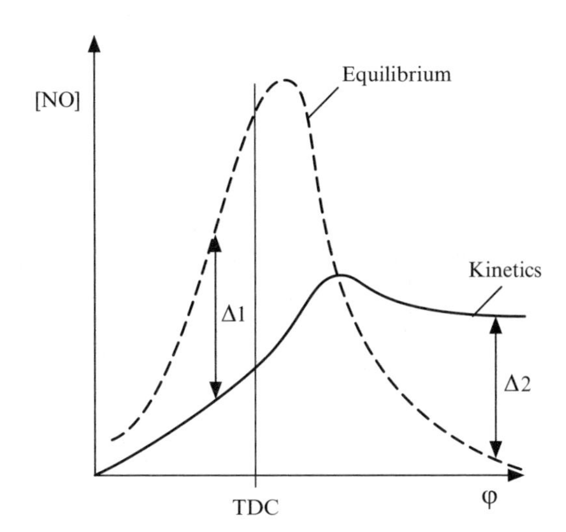

Fig. 6.23 NO-concentrations under condition of equilibrium/kinetically controlled NO-formation

formed in the first reaction step is immediately further converted to NO in the second and third steps. The concentration of atomic nitrogen thus remains constant after a short initial phase. The concentration of [N] can thus be assumed to be quasi-steady, see Sect. 5.1:

$$\frac{\mathrm{d[N]}}{\mathrm{d}t} \approx 0. \tag{6.20}$$

With this follows after adding the relations (6.18) and (6.19):

$$\frac{\mathrm{d[NO]}}{\mathrm{d}t} = 2k_{1,r}[\mathrm{O}][\mathrm{N_2}] - 2k_{1,l}[\mathrm{NO}][\mathrm{N}]. \tag{6.21}$$

For the unknown concentrations of nitrogen atoms [N], we obtain via the transformation of (6.19) under observance of (6.20)

$$[\mathrm{N}] = \frac{k_{1,r}[\mathrm{O}][\mathrm{N_2}] + k_{2,l}[\mathrm{NO}][\mathrm{O}] + k_{3,l}[\mathrm{NO}][\mathrm{H}]}{k_{1,l}[\mathrm{NO}] + k_{2,r}[\mathrm{O_2}] + k_{3,r}[\mathrm{OH}]}. \tag{6.22}$$

With this, (6.21) contains, besides the concentration of NO, only the easily determined concentration of N_2 and the concentrations O, O_2, OH, and H of the OHC system, which can be solved assuming partial equilibrium.

In Fig. 6.24, NO formation and disintegration in a thermal reactor at 60 bar and $\lambda = 1$ is represented for varying temperatures as a function of time. The illustration shows that equilibrium is reached more quickly the higher the temperature; for $T = 2{,}400$ K after about 20 ms and for $T = 2{,}800$ K already after about 3 ms.

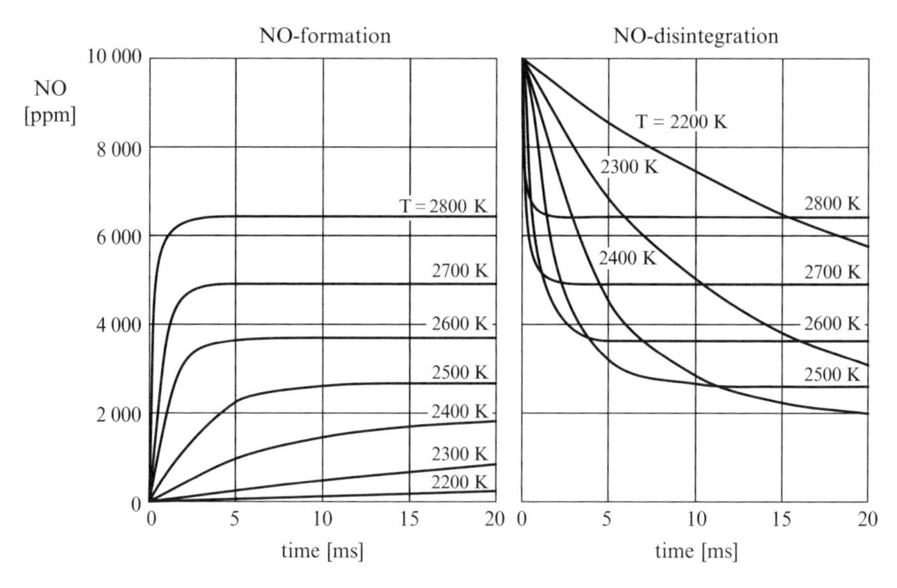

Fig. 6.24 NO-formation and disintegration in a thermal reactor, $p = 60$ bar; $l = 1.0$

Since a temperature of 2,800 K is reached at best during an extremely short time interval directly in the flame front and because the temperature in the burned mixture rapidly drops as a result of the pressure drop during expansion and exhaust blow down, equilibrium cannot be achieved within the time available. NO formation in engine combustion must therefore be calculated with the help of the reaction kinetics.

6.5.2 Prompt NO

The formation of prompt NO in the flame front itself is much more complicated than thermal NO formation, because this process is closely related to the formation of the CH radical, which can react in many ways. "Prompt NO" formation was first described by Fenimore (1979). According to Fenimore, the decisive reaction is that of CH with N_2 to form HCN (hydrocyanic acid), which rapidly reacts further to form NO,

$$CH + N_2 \xrightarrow{k_{fr}} HCN + N. \tag{6.23}$$

GRI-MECH 3.0 (2000) formulates the reaction with

$$k_f = 3.12 \times 10^9 \exp\left(-\frac{10,130}{T}\right) \frac{\mathrm{m}^3}{\mathrm{kmol\,s}} \tag{6.24}$$

Acetylene, (C_2H_2) as a precursor of the CH radical is only formed under fuel-rich conditions in the flame front, hence the concept "prompt NO". Because of the relatively low activation energy of the reaction prompt NO formation proceeds already at temperatures of about 1,000 K.

Subsequently, HCN reacts using various paths to form NCO and NH (Miller and Bowman 1989):

$$HCN + O \leftrightarrow NCO + H, \tag{6.25}$$

$$HCN + O \leftrightarrow NH + CO, \tag{6.26}$$

$$HCN + OH \leftrightarrow CN + H_2O, \tag{6.27}$$

$$CN + O_2 \leftrightarrow NCO + O. \tag{6.28}$$

The following reactions of NH and NCO, which form N atoms, are relatively fast, so that the above-mentioned reactions are rate-determining. The distribution of NO and NO_2 in flame is then dependent on the Zeldovich reactions (6.15) and (6.17) competeing for the nitrogen atom.

Reaction (6.22) had been accepted until a few years ago as the decisive reaction in the prompt NO mechanism. The theoretical reaction rate calculations of Cui et al. (1999) and comparison with experimental research have shown however that the theoretically calculated rate is two orders of magnitude too small. Moskaleva et al. (2000) investigated a prompt NO path with NCN as the intermediate species:

$$CH + N_2 \xrightarrow{k_r} NCN + H. \tag{6.29}$$

Based on their calculations, the authors concluded that this reaction is more dominant than the original reaction (6.22) and that the experimental data can be better reproduced with this reaction. Sutton et al. (2008) have experimentally confirmed the relevance of the NCN path with the help of measurements made with laser-induced fluorescence in methane flames.

The formed NCN can react with different species, forming HCH, CN, NCO and NO (Glarborg et al. 1998).

$$NCN + H \leftrightarrow HCN + N, \tag{6.30}$$

$$NCN + O_2 \leftrightarrow NO + NCO, \tag{6.31}$$

$$NCN + OH \leftrightarrow HCN + NO, \tag{6.32}$$

$$NCN + O \leftrightarrow CN + NO. \tag{6.33}$$

For a comparison between a GRI-MECH 3.0 mechanism, in which the HCN path is replaced by the NCN path, and experimental results, see Sutton and Fleming (2008).

As the various reaction paths shown above prove, much research is still required with respect to the exact mechanism as well as the individual reaction rates of prompt NO formation.

6.5.3 NO Formed via N_2O

This reaction mechanism is only significant when lean air-fuel mixtures repress the formation of CH, and thus little prompt NO is formed, and furthermore if low temperatures stifle the formation of thermal NO.

N_2O (Nitrous Oxide or laughing gas) is formed analogously to the first, rate-limiting reaction of the Zeldovich mechanism,

$$N_2 + O + M \rightarrow N_2O + M. \tag{6.34}$$

The reaction takes place however with a collision partner M, which is not changed by the reaction and considerably lowers the activation energy in comparison with reaction (6.14). NO formation then results trough oxidation of N_2O

$$N_2O + O \ \rightarrow \ NO + NO. \tag{6.35}$$

Because N_2O is only formed in a three-way collision reaction, this reaction path progresses preferably at high pressures. Low temperatures hardly slow down this reaction. NO formed via N_2O is the essential NO source in lean premixed combustion in gas turbines. This mechanism is however also observable in SI engine lean combustion. It is therefore important in the case of lean combustion in SI engines as well as modern diesel-engine combustion processes with high point pressures. It is presumably also the essential NO formation mechanism in lean HCCI combustion (Amnéus et al. 2005).

6.5.4 Fuel Nitrogen

The conversion of nitrogen bound in the fuel to nitrogen oxide does not play a role in engine combustion, because fuels for internal combustion engines contain negligible amounts of nitrogen. It can however play a role in the case of certain low-quality heavy oils (Besio and Nobile 2001). The formation of NO via fuel nitrogen is also of importance in the combustion of coal, since even "clean" coal contains about 1% of nitrogen.

According to common notions, the reaction progresses via hydrocyanic acid HCN and ammonia (Miller and Bowman 1989). The further conversion of HCN takes place in accordance with the mechanisms indicated in (6.25)–(6.27).

6.5.5 Reactions Forming NO_2

The most important reactions for the NO/NO_2 distribution in flames are (Miller and Bowman 1989):

$$NO + HO_2 \leftrightarrow NO_2 + OH, \tag{6.36}$$

$$NO + OH \leftrightarrow NO_2 + H, \tag{6.37}$$

$$NO + O_2 \leftrightarrow NO_2 + O. \tag{6.38}$$

Here, NO_2 is primarily formed by reaction (6.35) at low flame temperatures, in which there are high concentrations of HO_2, and broken up especially by reaction (6.37) at high temperatures. Usually, the ratio of NO_2 to NO in the exhaust gas of internal combustion engines is relatively low. In the case of lean conditions, very high exhaust gas recycling rates or very late injection times however – such as are sometimes used in lean-operated gas engines, HCCI or diesel combustion processes – a much larger amount of NO_2 was detected in the exhaust gas (Liu et al. 2004; Hill und McTaggart-Cowan 2005; Upatnieks et al. 2005). One explanation for this increase is that in these combustion processes, more NO_2 is formed by reaction (6.38) at low temperatures, and this NO_2 can no longer react again to form NO because of poor mixing and a slower combustion process overall (Amnéus et al. 2005).

References

Amnéus P, Mauß F, Kraft M, Vressner A, Johansson B (2005) NO_x and N_2O formation in HCCI engines. SAE paper 2005-01-0126

Appel J, Bockhorn H, Wulkow M (2001) A detailed numerical study of the evolution of soot particle size distributions in laminar premixed flames. Chemosphere 42:635–645

Basshuysen R van (Hrsg.) (2008) Ottomotor mit Direkteinspritzung: Verfahren, Systeme, Entwicklung, Potenzial, Vieweg/Teubner, 2. Aufl

Baulch DL, Cobos CJ, Cox AM, Frank P, Haymann G, Just T, Kerr JA, Murrels T, Pilling MJ, Twe J, Walker Rw, Warnatz J (1999) Compilation of rate data for combustion modeling. Supplement I, J Phys Chem Ref Data 22, 847

Belardini P, Bertori C, Cameretti MC, Del Giacomo N (1994) A coupled diesel combustion and sod formation model for KIVA II code: characteristics and experimental validation. International symposium COMMODIA 94. pp 315–323

Besio G, Nobile M (2001) A challenging fuel for diesel engines: orimulsion, from the concept to the application. CIMAC Congress, Hamburg

Bockhorn H (1994) A short introduction to the problem – structure of the following parts. In: Bockhorn H (ed) Soot formation in combustion. Springer, Berlin

Borrmeister J, Hübner W (1997) Einfluss der Brennraumform auf die HC-Emissionen und den Verbrennungsablauf. Motortechnische Zeitschrift MTZ 58:408–414

Bühler U (1995) Prüfstandsuntersuchungen zur Dioxin-Emission von Verbrennungsmotoren. Dissertation, Universität Stuttgart

Bühler U, Essers U, Greiner R (1997) Dioxin-Emission des Straßenverkehrs. MTZ 58:422–425
Cheng WK, Hamrin D, Heywood JB, Hochgreb S, Min K, Norris M (1993) An overview of hydrocarbon emissions mechanisms in spark-ignition engines. SAE paper 932708
Cui Q, Morokuma K, Bowman JM, Klippenstein SJ (1999) The spin-forbidden reaction CH (2Π) + N2 → HCN + N(4 S) revisited. II. Nonadiabatic transition state theory and application. J Chem Phys 110:19
Eng JA (2005) The effect of spark retard on engine-out hydrocarbon emissions. SAE paper 2005-01-3867
EN-ISO 8178-01 (1996) Hubkolben Verbrennungsmotoren, Abgasmessungen, Teil 1: Messung der gasförmigen Emissionen und der Partikelemissionen auf dem Prüfstand
Fenimore CP (1979) Studies of fuel-nitrgen in rich-flame gases. 17th Symp Comb, The Combustion Institute, Philadelphia, pp 661
Frenklach M (2002) Method of moments with interpolative closure. Chem Eng Sci 57:2229–2239
Frenklach M, Wang H (1994) Detailed mechanism and modeling of soot particle formation. In: Bockhorn H (ed) Soot formation in combustion. Springer, Berlin
Fusco A, Knox-Kelecy AL, Foster DE (1994) Application of a phenomenological soot model to diesel engine combustion. International symposium COMMODIA 94. pp 571–576
Glarborg P, Alzueta MU, Dam-Johansen K, Miller JA (1998) Kinetic modeling of hydrocarbon/ nitric oxide interactions in a flow reactor. Combust Flame 115:1–27
GRI-MECH 3.0 (2000) http://www.me.berkely.edu/gri_mech
Heywood JB (1988) Internal Combustion Engine Fundamentals. McGraw Hill, Inc
Hill PG, McTaggert-Cowan GP (2005) Nitrogen oxide production in a diesel engine fueled by natural gas. SAE paper 2005-01-1727
Kazakow A, Foster DE (1998) Modeling of soot formation during DI diesel combustion using a multi-step phenomenological soot model. SAE paper 982463
Kweon C-B, Foster DE, Schauer JJ, Okada S (2002) Detailed chemical composition and particle size assessment of diesel engine exhaust. SAE paper 2002-01-2670
Lange J (1996) Bestimmung der Carbonylverbindungen im Abgas von schwerölbetriebenen Dieselmotoren, Fortschritt-Berichte VDI. Reihe 15, Nr. 161, VDI Verlag, Düsseldorf
Lavoie GA, Heywood JB, Keck JC (1970) Experimental and theoretical investigation of nitric oxide formation in internal combustion engines. Combust Sci Technol 1:313–326
Liu Y, Amr A, Reitz RD (2004) Simulation of effects of valve pockets and internal residual gas distribution on HSDI diesel combustion and emissions. SAE paper 2004-01-0105
Mathis U, Mohr M, Kaegi R, Bertola A, Boulouchos K (2005) Influence of diesel engine combustion parameters on primary soot particle diameter. Environ Sci Technol 39(5): 1887–1892
Mauß F (1997) Entwicklung eines kinetischen Modells der Rußbildung mit schneller Polymerisation. Dissertation, RWTH Aachen
McEnally CS, Ciuparu DM, Pfefferle LD (2003) Experimental study of fuel decomposition and hydrocarbon growth processes for practical fuel components: heptanes. Combust Flame 134: 339–359
McEnally CS, Pfefferle LD, Atakan B, Kohse-Höinghaus K (2006) Studies of aromatic hydrocarbon formation mechanism: Progress towards closing the fuel gap. Prog Energy and Combustion Sci 32:247–294
Miller JA, Bowman CT (1989) Mechanism and modeling of nitrogen chemistry in combustion. Prog Energy Combust Sci 15:287–338
Mosbach S, Celnik MS, Raj A, Kraft M, Zhang HR, Kubo S, Kim K-O (2009) Towards a detailed soot model for internal combustion engines. Combust Flame 156(6):1156–1165
Moskaleva LV, Xia WS, Lin MC (2000) The CH + N2 reaction over the ground electronic doublet potential energy surface: a detailed transition state search. Chem Phys Lett 331:269–277
Nagle J, Strickland-Constable RF (1962) Oxidation of carbon between 1000–2000°C. Proceedings of 5th conference on carbon, Vol 1. pp 154–164
Neoh KG (1976) Soot burnout in flames. Ph.D. thesis, MIT

Netzell K, Lehtiniemi H, Mauss F (2007) Calculating the soot particle size distribution function in turbulent diffusion flames using a sectional method. Proceedings of the Combustion Institute, Vol 31, pp 667–674

Nishida K, Hiroyasu H (1989) Simplified three-dimensional modeling of mixture formation and combustion in a DI diesel engine. SAE Paper, 890269

Pattas K (1973) Stickoxidbildung bei der ottomotorischen Verbrennung. MTZ 34, 397–404

Pischinger F, Schulte H, Hansen J (1988) Grundlagen und Entwicklungslinien der Dieselmotorischen Brennverfahren, VDI Berichte Nr. 714, VDI Verlag

Stiesch G (2003) Modeling Engine Spray and Combustion Processes. Springer, Berlin Heidelberg, New York

Sutton JA, Fleming JW (2008) Towards accurate kinetic modelling of prompt NO formation in hydrocarbon flames via the NCN pathway. Combust Flame 154:630–636

Sutton JA, Williams BA, Fleming JW (2008) Laser-induced fluorescence measurements of NCN in low-pressure CH4/O2/N2 flames and its role in prompt NO formation. Combust Flame 153: 465–478

Tao F, Liu Y, Rempel-Ewert BH, Foster DE, Reitz RD, Choi D, Miles PC (2005) Modeling the effects of EGR and injection pressure on soot formation in a high-speed direct-injection (HSDI) diesel engine using a multi-step phenomenological soot model. SAE paper 2005-01-0121

Tao F, Reitz, RD, Foster DE, Liu Y (2008) Nine-step phenomenological diesel soot model validated over a wide range of engine conditions. Int J Thermal Sci 48:1223–1234

Upatnieks A, Mueller CJ, Martin GC (2005) The influence of charge-gas dilution and temperature on DI diesel combustion processes using a short-ignition-delay, oxygenated fuel. SAE paper 2005-01-2088

Vishwanathan G, Reitz RD (2009) Modeling soot formation using reduced polycyclic aromatic hydrocarbon chemistry in n-heptane lifted flames with application to low temperature combustion. J Eng Gas Turbines Power Vol 131 pp. 032801-1–032801-7

von Smolouchowski MZ (1917) Versuch einer mathematischen Theorie der Koagulationskinetik koloider Lösungen. Zeitschrift für Physikalisch Chemie 2:129–268

Warnatz J, Maas U, Dibbe RW (2001) Combustion. Physical and Chemical Fundamentals, Modeling and Simulation, Experiments, Pollutant Formation. 3th ed. Springer-Verrlag, Berlin Heidelberg

Westbrook CK, Dryer FL (1984) Chemical kinetic modeling of hydrocarbon combustion. Prog Energy Combust Sci 10:1–57

Zeldovich YB (1946) The Oxidation of Nitrogen in Combustion and Explosions. Acta Physiochimica, USSR, 21:577–628

Part II
Simulation of the Overall Process

Chapter 7
Calculation of the Real Working Process

Christian Schwarz

In the filling and emptying method, a *zero-dimensional* model, in which the process quantities depend only on time, not on location nor particular *subsystem* of the engine, e.g.:

- Combustion chamber
- Intake and exhaust pipes
- Valves and flaps
- Charge system

are physically and mathematically described either via *substitute systems plenums, pipes*, junctions, restrictions or with *characteristic maps.*

Figure 7.1 shows an "engine model" for a turbocharged diesel engine.

The centerpiece of the internal combustion engine, the combustion chamber, is thereby described with the thermodynamic model of the "ideally mixed volume", the turbocharger on the other hand with characteristic maps for the compressor and the turbine.

In the filling and emptying method, only the laws of conservation for *mass* and *energy*, but not the *law of conservation* for impulse is considered (see Sect. 7.4). Because in this way no flow fields are considered, the resulting model is also designated as a zero-dimensional thermodynamic model. For a general examination, see Ramos (1989).

In describing processes in the fresh air and exhaust gas system of intake engines (today almost exclusively SI engines), the process quantities must be described with the help of one-dimensional gas dynamics as a function of time and location. Such complex systems live from the dynamics of pressure waves surging to and for and the interaction of these pressure waves with the valve control times for the purpose of as good a filling as possible. The intake and exhaust system can be represented with a number of pipe components connected over pipe branches, the volumes, orifice plates, the cylinders, and sometimes flow machines (see Sect. 7.4).

Chapter 8 goes into the particular requirements for the simulation of charging aggregates and air cooling.

G.P. Merker et al. (eds.), *Combustion Engines Development*,
DOI 10.1007/978-3-642-14094-5_7, © Springer-Verlag Berlin Heidelberg 2012

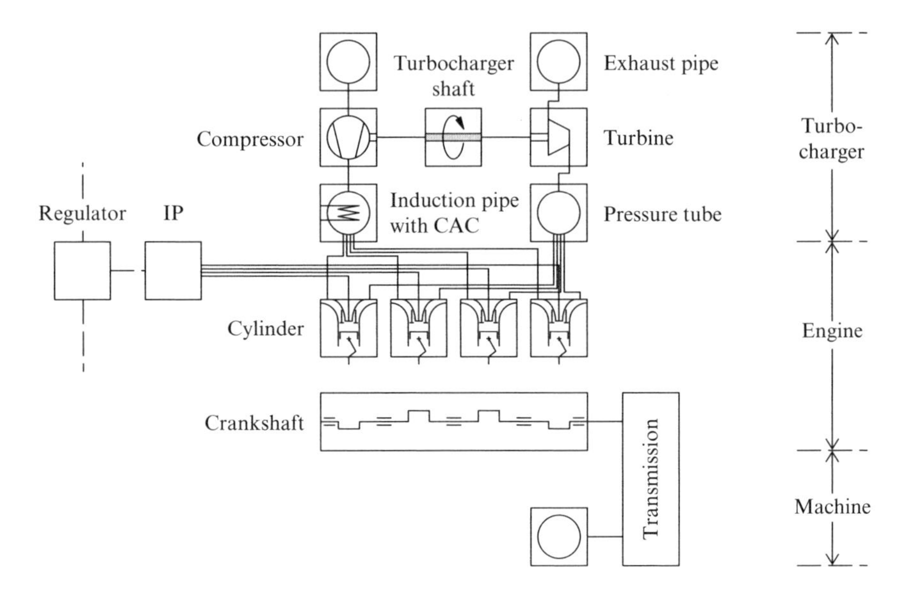

Fig. 7.1 Engine model

7.1 Single-Zone Cylinder Model

7.1.1 Fundamentals

Figure 7.2 shows the combustion system of an internal combustion engine. This is limited by the combustion chamber walls, the piston, and the valves. The combustion chamber walls simultaneously represent the system boundaries. The entire combustion space is viewed as an ideally mixed volume, whereby heat release through combustion is described with the help of a substitute heat release rate later will take a closer look at *multi-zone models*, with which we can describe, for example, NO_x formation in the combustion chamber (Sect. 7.2).

It must thereby be considered that the volume, but not necessarily the mass of the combustion chamber continually changes with time (or crank angle) as a result of piston movement. In considering the balance equations in the combustion chamber, we must distinguish between different concepts of fuel introduction. While the fuel – as long as one does not calculate with a wall film model – is taken in proportionally to the fresh air in the mixture-intaking SI engine, in the direct injection SI engine the fuel is injected either during the opened intake valve (homogeneous operation, injection before TDC) or shortly before ignition (stratified operation). The fuel must then be prepared/evaporated in the combustion chamber. In the diesel engine, the fuel is directly injected. An evaporation of the fuel is generally not considered (exception: multi-zone models with fuel breakup models).

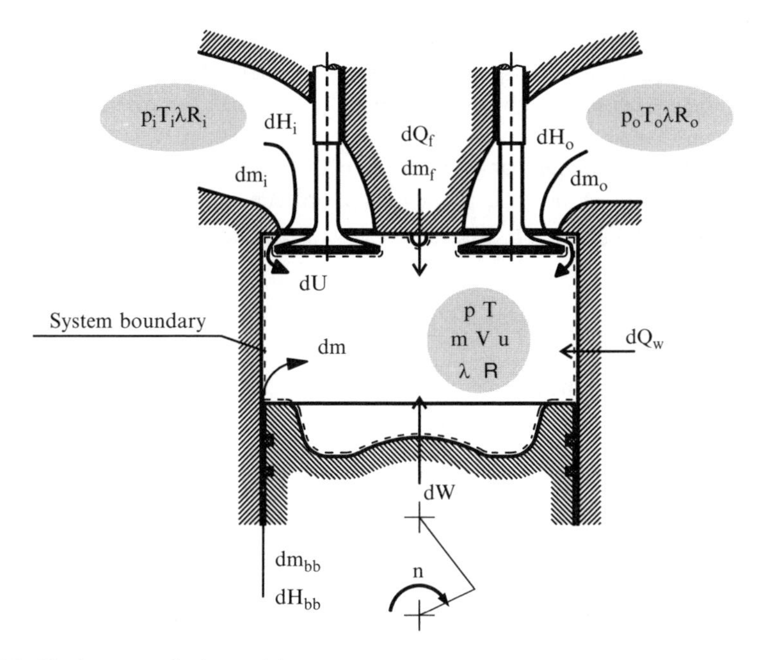

Fig. 7.2 Single-zone cylinder model

The *mass balance* for the cylinder provides for the description of all above mentioned possibilities for fuel insertion

$$\frac{dm_{Sys}}{dt}=\frac{dm_i}{dt}+\frac{dm_o}{dt}+\frac{dm_{bb}}{dt}+\frac{dm_{f,evap.}}{dt}. \tag{7.1}$$

The mass flow entering the engine can be, as already mentioned, pure air, an air-fuel mixture, an air-exhaust gas mixture, or a combination of air, fuel and exhaust gas.

The *energy balance* or the *first law of thermodynamics* provides for the cylinder neglecting the kinetic energy

$$\begin{aligned}\frac{dE_{Sys}}{dt}=\frac{dU}{dt}=\frac{dQ_{heat}}{dt}+\frac{dQ_w}{dt}-p\frac{dV}{dt}+\frac{dm_i}{dt}h_i+\frac{dm_o}{dt}h_a\\+\frac{dm_{bb}}{dt}h_{bb}+\frac{dm_{f,evap.}}{dt}h_{f,evap.}+\frac{dQ_{evap.}}{dt}.\end{aligned} \tag{7.2}$$

As long as the fuel is injected into the combustion chamber but is not yet evaporated, it takes up such a small volume that it is irrelevant to thermodynamic analysis. The introduced fuel only becomes "effective" for the mass and energy balance when it is evaporated and hence in the gaseous phase. Evaporation requires latent energy for which a corresponding amount of heat is taken from the gas. The situation is the same for the heating up of fuel vapor to the gas temperature. Normally, these effects in the lower heating value are considered such that the evaporation enthalpy of the fuel (350–420 kJ/kg) is added to the lower heating value. The same is valid for the re-condensation of fuel and for the water content in the air if the gas cools enough when expanded (e.g. in load control via early "intake closes").

Either time t or the crank angle φ can be chosen as the *independent variable*. In more recent equations, time t is usually given precedence. Valid for the relation of time with the degree of crank angle is

$$\begin{aligned} \varphi &= \omega t \\ \mathrm{d}\varphi &= \omega \mathrm{d}t. \end{aligned} \tag{7.3}$$

For the solution of the mass and energy balance, we need the thermal condition equation already mentioned

$$pV = mRT. \tag{7.4}$$

For the solution of this equation system, we still need relations for energy release via combustion, a so-called "combustion model", a relation for the heat transfer between the gas mixture and the combustion chamber walls, a so-called "heat transfer model", as well as a gas exchange model (e.g. a two-zone model for two-stroke engines) and, under certain circumstances, an evaporation model. The volume path is given by a crankshaft drive model. We will take a closer look at particular partial models in the following.

7.1.2 Mechanical Work

The output at the piston $\mathrm{d}W/\mathrm{d}t$ can be calculated from the cylinder pressure and the change in cylinder volume

$$\frac{\mathrm{d}W}{\mathrm{d}t} = -p\frac{\mathrm{d}V}{\mathrm{d}t} = -p\omega\frac{\mathrm{d}V}{\mathrm{d}\varphi}. \tag{7.5}$$

In Sect. 2.2, the geometrical relations at the crankshaft drive are represented. The crankshaft drive can be described by the geometrical quantities crankshaft radius r, connecting rod length l, eccentricity e, and cylinder diameter D, from which the volume change $\mathrm{d}V/\mathrm{d}\varphi$ can be determined.

7.1.3 Determination of the Mass Flow Through the Valves/Valve Lift Curves

In the valve gap, the flow is constricted. This has the result that the actual cross-section surface is smaller than the geometrical. Because of friction in the ports, the actual mass flow is also smaller than the theoretical. This fact is taken into consideration by the introduction of a flow coefficient

$$\mu \equiv \frac{\dot{m}_{real}}{\dot{m}_{theo.}}. \tag{7.6}$$

For the determination of the mass flow through a valve, we utilize the flow function derived in Sect. 2.3.1. For this, the actual mass flow contingent upon valve lift is determined for the valves on a so-called flow test bench and set in relation to the theoretical mass flow [see equation (2.33)].

$$\dot{m}_{theo} = A_{geo}\sqrt{p_1\rho_1}\,\Psi\left(\frac{p_2}{p_1},\kappa\right).$$

Figure 7.3 shows the conditions at the flow test bench in principle. The flow coefficient is usually related to a circular surface at the port entrance. Since varying cycle calculation programs possess varying definitions of the valve opening surface, usually a corresponding recalculation is necessary. It must thereby be taken into consideration that the actual cross-section surface remains the same independent of the definition of the reference cross-section surface for the respective valve position.

Figure 7.4 qualitatively shows the flow coefficients determined at the flow test bench contingent on the valve lift.

We recognize thereby that varying flow coefficients are adjusted for varying flow directions at the intake and exhaust valves. The main flow direction for the intake valve is "inflow". However, an outflow from the intake valve is also possible. Due to the geometrical relations, this flow case is worse than that of inflow. The situation is the same for the exhaust valve. In this case however, the main flow case is "outflow", from which an, in principle, inferior flow behavior results.

The flow coefficients can be determined either in a stationary test or calculated by means 3D CFD codes in order to make statements about the quality of ports without concrete hardware. In this case, the port geometry and the cylinder are

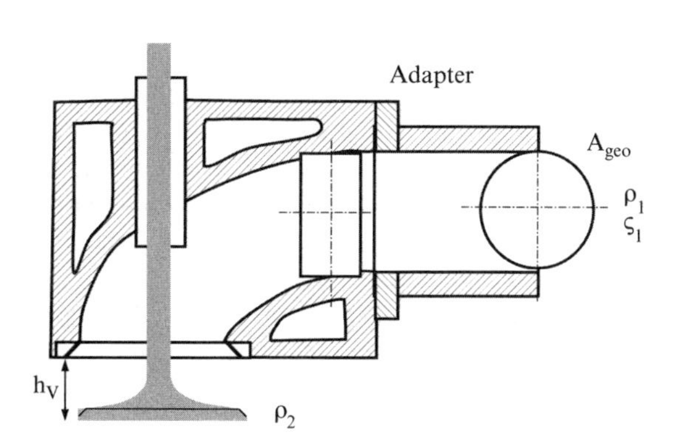

Fig. 7.3 Determining the flow coefficients on the flow bench

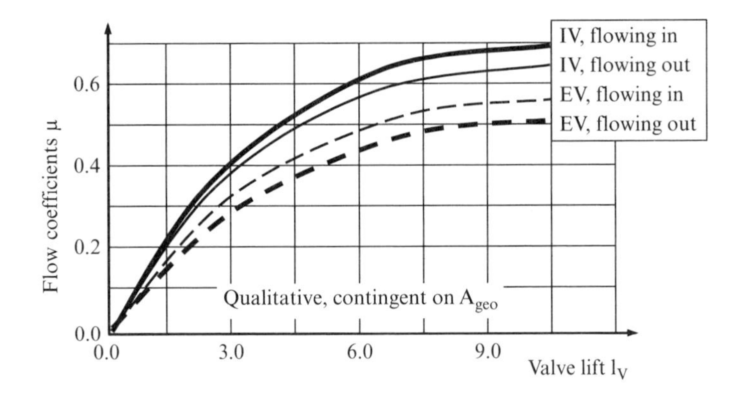

Fig. 7.4 Flow coefficients $\mu = f(h_V)$

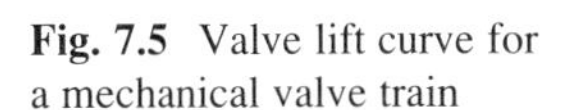
Fig. 7.5 Valve lift curve for a mechanical valve train

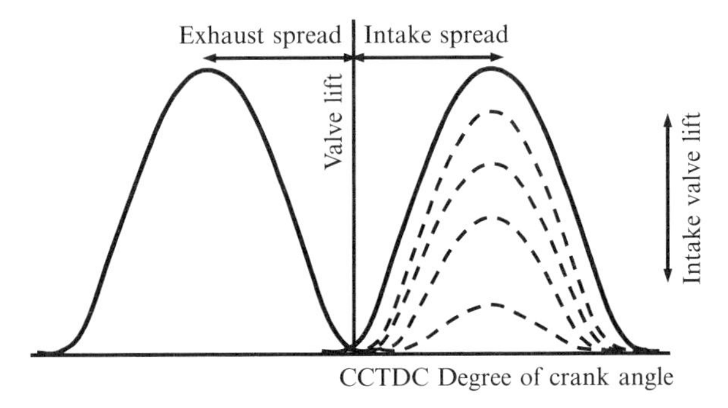

duplicated and a corresponding pressure gradient applied at the margins. In discrete steps, the valve lift is changed and the "actual" mass flow calculated. This can then be compared with the theoretical flow in the same way as in measurement.

Figure 7.5 shows the valve lift graphs for a conventional valve train. For the calculation of the working process, it is sufficient to provide rate of valve lift curves in steps of 1–5°CA and to interpolate between the supporting points. Valve centerline angle is the distance between the maximum of valve lift to top dead center overlap. The exhaust centerline should be negative but can be expressed as a negative or positive number.

Figure 7.5 shows the valve lift graphs for a conventional valve train. For the calculation of the working process, it is sufficient to provide rate of valve lift curves in steps of 1–5°CA and to interpolate between the supporting points. Valve centerline angle is the distance between the maximum of valve lift to top dead center overlap. The exhaust centerline should be negative but can be expressed as a negative or positive number.

In Fig. 7.5, inlet valve lift curves for a fully variable mechanical valve train are included, in which a continuous adjustment of valve lift/duration is possible. With this variability for a quantity regulated SI engine, a load control without a throttle valve is possible, since the flow and thus the fresh gas mass can be adjusted via the

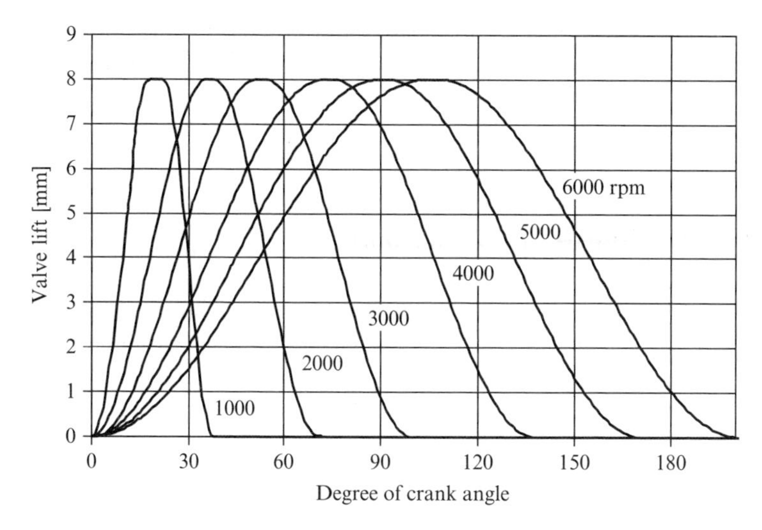

Fig. 7.6 Valve lift curves for an electromechanical valve train

valve lift. For the low load range, the valve lift graduation has to be given in the range of tenths of a millimeter; after approx. 3 mm a range of half to entire millimeter steps is sufficient. Intermediate steps are thereby linearly interpolated. Nothing changes in curves for the flow coefficients for fully variable valve operation, since the flow coefficients only depend on the valve lift. Only in the range of small valve lifts is a finer grid recommendable here as well.

The behavior of so-called electromechanical valve trains should be seen completely differently for simulation. The electromechanical valve train concerns a one mass oscillator, which is usually attracted in a regulated manner and then held at the respective final positions by a magnet. In this case, ideally only the loss energy from swinging from one final position to the other via the magnet is added. In an ideal case, the valve lift curve in electromechanical valve trains is thus only contingent on time and not on the degree of crank angle. For varying speeds, the valve lift curves result as given in Fig. 7.6. In this example, the valve is move upwards again as soon as the lower final position is reached and is not held at this final position. The movement differential equations for the one mass oscillator is

$$m\ddot{x} + d\dot{x} + cx = F_{\text{fric.}}(t) + F_{\text{Magnet}}(t) + F_{\text{valve}}(t) + F_{\text{adhesive}}(t). \tag{7.7}$$

Not under consideration are furthermore the so-called adhesive forces at the actuator, which are contingent on its thermal state and on the presence, for example, of oil. For the values presented in Fig. 7.6, the attenuation is constant and all external forces have been set to zero. Thus results as an idealized solution for the differential equation a cosine function

$$x(t) = x_{\max}\left[\frac{1}{2} - \frac{1}{2}\cos\left(\frac{2t}{\tau}\pi\right)\right] \quad \text{mit } 0 < t < \tau. \tag{7.8}$$

The so-called flight time τ, which describes the process between the opening of the valve until closing without holding the vales in the opened position, serves as a measure for the period duration. We will in this context do without a more extensive modeling.

7.1.4 Heat Transfer in the Cylinder

The description of heat transfer in the internal combustion engine places the highest demands on modeling and is usually based on a global inspection of complex relationships. Heat transfer consists of a convective and a radiation component

$$\frac{dQ_w}{dt} = \frac{dQ_\alpha}{dt} + \frac{dQ_\varepsilon}{dt}. \tag{7.9}$$

Usually the radiation component dQ_ε/dt is added to the convective heat transfer coefficient, although the maxima of dQ_α/dt and dQ_ε/dt actually appear phase-shifted with reference to the crank angle.

Proceeding from the Newtonian method, for the description of the wall heat flow

$$\frac{dQ_w}{dt} = \sum_i \alpha_i A_i (T_{w,i} - T_{gas}). \tag{7.10}$$

is valid.

We usually subdivide the combustion chamber thereby into three areas (see Fig. 7.2):

- The piston
- The cylinder head
- The portion of the liner exposed by the piston including deck height and piston top land

The valves are usually lumped with the cylinder head, or, in very detailed modeling, as their own area. The surfaces for the pistons and the cylinder head are usually larger than the cylinder liner when temperature is right, since these describe, for example, the roof form in a SI engine combustion process or the piston bowl shape in a diesel or SI engine combustion process. The part of the liner released by the piston amounts to

$$A_{\text{liner}} = A_{\text{fire land}} + A_{\text{Kolbenrückstand}} + D\pi s(\varphi). \tag{7.11}$$

Piston motion $s(\varphi)$ versus crankshaft rotation has already been described in Sect. 2.2.

The calculation of heat transfer with the help of the Newtonian approach and the heat transfer coefficients requires an accurate prediction of gas and wall temperatures. The mean gas temperature results from the local averaging of gas temperature in the combustion chamber. Since the combustion chamber system is usually seen as an ideally mixed volume, the mean gas temperature is easy to determine from the condition equation for an ideal gas. At respective wall temperatures, we are dealing with the internal wall temperature averaged over one working cycle. For the piston and the cylinder, usually local constant temperatures are used. In the case of the liner, the wall temperature depends on the engine type and on whether the liner is completely or only partially surrounded by the water jacket. In giving the temperature for the liner, we usually subdivide it into several areas, or we provide a temperature profile over the length of the liner. The temperatures can either be determined by measuring, or we can use a simple, iterative method for the calculation of internal wall temperature for stationary operating points. For this however, knowledge of the temperatures in at least one operating point is required. For non-stationary calculations, neither method is sufficient any more, which is why we use a more concrete heat conduction model, which takes the thermal inertias of the respective wall into consideration. All models are described in Sect. 7.1.6.

For the calculation of heat transfer coefficients, semi-empirical methods are usually used, since many influence factors can only be determined experimentally. Therefore, exterior quantities are used as influence parameters that characterize the operating point. In this section, essentially two approaches will be introduced: The method of Woschni, which was constructed for diesel engines in 1969 and has continually been further developed and that of Bargende, which was introduced in 1990 for SI engines. A multitude of further approaches exist in the literature, e.g. from Hohenberg (1980) and Kleinschmidt (1993), which we will however not go into details here any further.

7.1.4.1 Woschni Heat Transfer Model

Woschni (1970) assumes a stationary, fully turbulent pipe flow. For the dimensionless heat transfer coefficient, the Nusselt number, we obtain from a dimension analysis the semi-empirical power equation

$$\mathrm{Nu} = C\,\mathrm{Re}^{0.8}\mathrm{Pr}^{0.4} \tag{7.12}$$

with the Nusselt number

$$\mathrm{Nu} = \frac{\alpha D}{\lambda}, \tag{7.13}$$

the Reynolds number

$$\mathrm{Re} = \frac{\rho w D}{\eta} \tag{7.14}$$

and the Prandtl number

$$\mathrm{Pr} = \frac{\eta}{\rho\, a}. \tag{7.15}$$

If we view the gas in the combustion chamber as an ideal gas,

$$\rho = \frac{p}{R\,T}, \tag{7.16}$$

thus follows at first

$$\frac{\alpha\, D}{\lambda} = C \left(\frac{p}{R\,T} \frac{w\, D}{\eta} \right)^{0.8} \mathrm{Pr}^{0.4} \tag{7.17}$$

and from that by means of conversion for the convective heat transfer coefficient

$$\alpha = C\, D^{-0.2} p^{0.8} w^{0.8} \frac{\mathrm{Pr}^{0.4}\, \lambda}{(R\, T\, \eta)^{0.8}}. \tag{7.18}$$

With the physical characteristics

$$\mathrm{Pr} = 0.74; \frac{\lambda}{\lambda_0} = \left(\frac{T}{T_0} \right)^x; \frac{\eta}{\eta_0} = \left(\frac{T}{T_0} \right)^y \tag{7.19}$$

and with the assumption that the characteristic speed w is equal to the mean piston speed, i.e. $w \equiv c_m$, we further obtain

$$\alpha = C^*\, D^{-0.2} p^{0.8}\, c_m^{0.8}\, T^{-r} \quad \text{with} \quad r = 0.8(1 + y) - x. \tag{7.20}$$

Through a comparison with the measurement values, the exponent r for the temperature dependence is determined as $r = 0.53$ and the constant as $C^* = 127.93$. For fired engines, a modification of the characteristic speed is additionally introduced, which takes the change in heat transfer as a result of combustion into consideration. We thus obtain

$$\alpha = 127.93 D^{-0.2}\, p^{0.8}\, w^{0.8}\, T^{-0.53} \left[\frac{\mathrm{W}}{\mathrm{m}^2\,\mathrm{K}} \right] \tag{7.21}$$

with

$$w = C_1\, c_m + \underbrace{C_2 \frac{V_{swept}\, T_1}{p_1\, V_1} (p - p_0)}_{\text{combustion term}} \tag{7.22}$$

and p_1, T_1, V_1 at compression start, i.e. at "intake closes". For the constants C_1 and C_2 we obtain via adjustment to measured values

$$C_1 = \begin{cases} 6.18 + 0.417 \dfrac{c_u}{c_m} : \text{gas exchange} \\ 2.28 + 0.308 \dfrac{c_u}{c_m} : \text{compression/expansion} \end{cases} \tag{7.23}$$

$$C_2 = \begin{cases} 6.22 \cdot 10^{-3} \left[\dfrac{\text{m}}{\text{s K}}\right] : \text{pre-chamber-engine} \\ 3.24 \cdot 10^{-3} \left[\dfrac{\text{m}}{\text{s K}}\right] : \text{direct injection-engine} \end{cases} . \tag{7.24}$$

For the intake swirl c_u/c_m, the range of validity is given as $0 \leq c_u/c_m \leq 3$, with swirl determined in the stationary flow experiment on the flow test bench with the Tippelmann or the impeller method. In this case, an impeller of diameter d is arranged at a distance of 100 mm beneath the cylinder head in the cylinder liner. The flow through the intake valve is thereby adjusted such that this impeller receives a flow of mean piston speed c_m. With the speed of the impeller to be measured n_d we correspondingly obtain

$$c_u = D \pi n_d \tag{7.25}$$

the circumference speed and thus the swirl.

Depending on the respective phases of a cycle, several terms or parameters in the heat transfer equation are altered. This leads, for example, in the transition between expansion and gas exchange during the opening of the exhaust valve, to a leap in the constant C_1. In the same way, the term with the constant C_2 is only valid after the start of combustion. However, the transition between the compression phase and combustion is continuous via the term $(p - p_0)$.

With the term $(p - p_0)$, the difference is given between the cylinder pressure during combustion and the cylinder pressure during motored operation. Pressure p_0 can be calculated with a polytrophic relationship from the cylinder volume. The determination of the polytrope exponent ends shortly before combustion, as the polytrope exponents for the, for example, last 10°CA before combustion are averaged. For p_0 then results

$$p_0(\varphi) = p_{before\ comb.} \left(\frac{V_{before\ comb.}}{V_{cyl}(\varphi)} \right)^{\bar{n}} . \tag{7.26}$$

Figure 7.7 demonstrates an example for wall heat flow in a turbocharged diesel engine at a speed of 2,000 rpm and an b_{mep} of 2 bar.

In the lower part of the diagram, the heat flows of the piston, cylinder head, and the liner, as well as the total heat flow are represented. Above them the heat transfer coefficients according to Woschni are shown, as well as the mean mass temperature

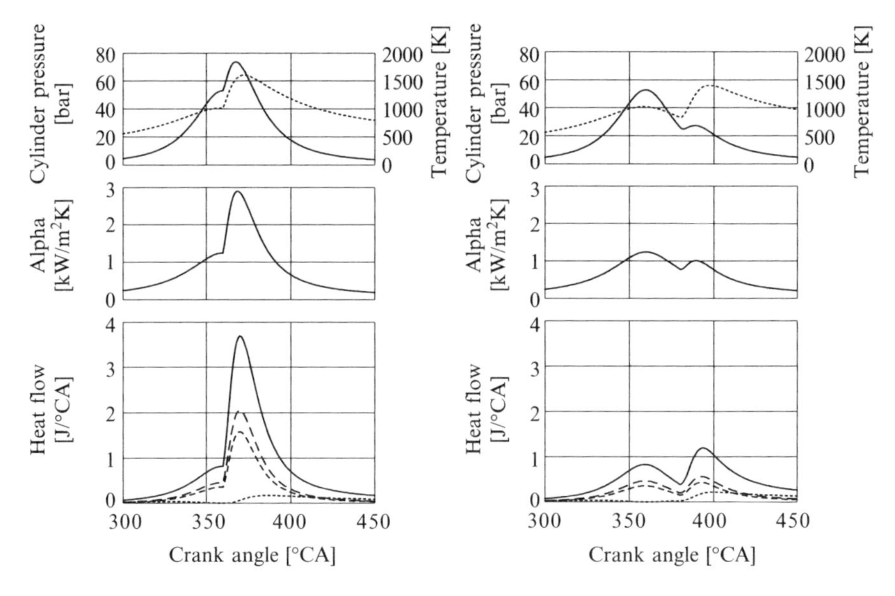

Fig. 7.7 Heat transfer in a turbocharged diesel engine, start of combustion 355°CA (*left*) and 268°CA (*right*)

and the pressure in the cylinder. In the left side, a start of combustion of 355°CA is recorded, in the right a start of combustion of 368°CA. We recognize hereby that the heat transfer coefficient contingent on pressure and temperature is evidently smaller at a later start of combustion.

7.1.4.2 Modifications to the Woschni Heat Transfer Model

Kolesa's investigations on heat transfer with insulated combustion chamber walls (1987) arrived at the result that the wall heat coefficient at wall temperatures over 600 K climbs significantly. For the constant C_2, Schwarz (1993) developed a constant function

$$C_2^* = \begin{cases} C_2 & \text{für } T_W < 525\,\text{K} \\ C_2 + 23 \cdot 10^{-6}(T_W - 525) & \text{für } T_W \geq 525\,\text{K}. \end{cases} \tag{7.27}$$

The speed corrected by Woschni with the combustion term in the equation provides for motored engines and, in the lower load range however, excessively small values, as Huber (1990) has shown. Therefore, the heat transfer equation has been corrected for low loads. For

$$2C_1\, c_m \left[\frac{V_c}{V(\varphi)}\right]^2 i_{mep}^{-0.2} \geq C_2 \frac{V_{swept}\, T_1}{p_1\, V_1}(p - p_0) \tag{7.28}$$

the following is valid

$$w = C_1\, c_m \left[1 + 2\left(\frac{V_c}{V}\right)^2 i_{mep}^{-0.2}\right]. \tag{7.29}$$

Also valid is: $i_{mep} = 1$ for $i_{mep} \leq 1$.

Additional investigations on the heat transfer coefficient – especially on the effect of insulation on combustion chamber wall deposits (soot, oil coke) – have been carried out by Vogel (1995). From these investigations have resulted further changes to the equation altered by Huber. For

$$2\, C_1\, c_m \left[\frac{V_c}{V(\varphi)}\right]^2 C_3 \geq C_2 \frac{V_{swept}\, T_1}{p_1\, V_1} (p - p_0) \tag{7.30}$$

is valid

$$w = C_1\, c_m \left[1 + 2\left(\frac{V_c}{V}\right)^2 C_3\right]. \tag{7.31}$$

The constant C_2 for direct injection diesel engines is extended for SI engines in its range of validity

$C_2 = 3.24 \cdot 10^{-3} \left[\frac{\text{m}}{\text{s K}}\right]$: DI engine, SI engine (gasoline)

new constants,

$C_2 = 4 \cdot 10^{-3} \left[\frac{\text{m}}{\text{s K}}\right]$: SI engine (methanol)

$C_3 = 0.8$: for gasoline

$C_3 = 1.0$: for methanol, and

$C_3 = 1 - 1.2\, e^{-0.65\,\lambda}$: for diesel

are introduced. In the case of large diesel engines running at medium speed, some deviations result in the calculation of the exhaust gas temperature in comparison with the measurement of about approx. 20 K. The exhaust gas temperature calculated too low leads to a low enthalpy at the turbine and thus to a slightly too low charge pressure. For the design of large diesel engines, this is however decisive, since they are mostly optimized at a stationary operating point. For this reason, Gerstle (1999) has modified the heat transfer according to Woschni for the gas exchange. The constant C_1 is thereby valid beyond the point of exhaust opens, until the intake valve opens. Then the constant is raised by the factor 6.5–7.2 in order to take into account the increased charge motion during the influx.

$C_1 = 2.28 + 0.308 \frac{c_u}{c_m}$: compression/expansion/exhaust valve open,

$C_1 = k \left(2,28 + 0,308 \frac{c_u}{c_m}\right)$: intake valve open, and

$k = 6.5$ to 7.2.

Figure 7.8 shows the modified path of the heat transfer coefficient.

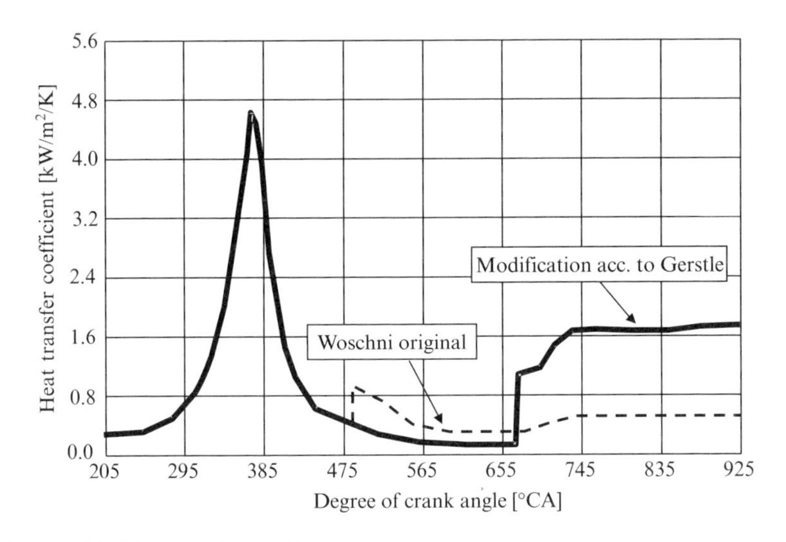

Fig. 7.8 Modified heat transfer coefficient for a medium speed diesel engine

7.1.4.3 Heat Transfer According to Bargende

Bargende (1990) also initially proceeds from Woschni's basic assumptions [see (7.12)] in describing the Nusselt number. The constants in the equation are determined for multi-point injection SI engines

$$\mathrm{Nu} = C\,\mathrm{Re}^m\,\mathrm{Pr}^n \quad \text{mit} \quad m = 0.78 \quad \text{und} \quad n = 0.33.$$

Solved according to the heat transfer coefficient

$$\alpha = C \underbrace{D^{-0.22}}_{\text{characteristic length}} \underbrace{\lambda\left(\frac{\rho}{\eta}\right)}_{\text{material quantities}} \underbrace{w^{-0.78}}_{\text{speed of gas}} \underbrace{\Delta}_{\text{combustion}} \tag{7.32}$$

The characteristic length is expressed over the diameter of a sphere, which has the same volume as the momentary cylinder volume

$$D^{-0.22} \cong 1,11\,V^{-0.073}. \tag{7.33}$$

For the material quantities heat conductivity and dynamic viscosity is valid contingent on the air content r of the gas

$$\lambda = (1.15\,r + 2.02)\,10^{-4}\,T^{0.805}\,\left[\frac{\mathrm{W}}{\mathrm{m\,K}}\right] \tag{7.34}$$

and

$$\eta = (2.57\,r + 3.55)\,10^{-4}\,T^{0.644}\quad\left[\frac{\mathrm{N\,s}}{\mathrm{m}^2}\right]. \tag{7.35}$$

air content r of the gas is defined as

$$r = \frac{\lambda - 1}{\lambda + \dfrac{1}{L_{\min}}}\quad \text{für}\quad 0 \le r \le 1. \tag{7.36}$$

The density of the gas is expressed on the other hand via the ideal gas equation

$$\rho = \frac{p}{R\,T}.$$

In describing the gas conditions on the boundary layer, the average value from the gas temperature and the wall temperature is utilized, since the gas temperature falls off to the wall temperature on the boundary layer

$$T_m = \frac{T_{gas} + T_{wall}}{2}. \tag{7.37}$$

Consideration of material quantities thus results in

$$\lambda\left(\frac{\rho}{\eta}\right)^{0{,}78} \cong \frac{(1.15\,r + 2.02)}{[R\,(2.57\,r + 3.55)]^{0.78}}\,10^{5.36}\,T_m^{-0.477}\,p^{0.78}. \tag{7.38}$$

The speed relevant to heat transfer is calculated by means of an approach from the specific turbulent energy k, which is determined by a simplified k, ε model, see also Sect. 12.1, and the piston speed c_p

$$w = \frac{\sqrt{\dfrac{8\,k}{3} + c_p^2}}{2}. \tag{7.39}$$

Valid for the change in specific turbulent energy is

$$\frac{\mathrm{d}k}{\mathrm{d}t} = \left[-\frac{2}{3}\frac{k}{V}\frac{\mathrm{d}V}{\mathrm{d}t} - \varepsilon\,\frac{k^{1.5}}{L} + \left(\varepsilon_q\,\frac{k_q^{1.5}}{L}\right)_{\phi > CTDC}\right]_{IC \le \varphi \le EO}, \tag{7.40}$$

with $\varepsilon = \varepsilon_s = 2.184$ and the characteristic swirl $L = \sqrt[3]{6/(\pi\,V)}$. For the kinetic energy of the squeeze flow in a bowl-shaped combustion chamber with a bowl diameter of d_{bowl} is valid

$$k_q = \frac{1}{18}\left[w_r\left(1 + \frac{d_{bowl}}{d_{cyl}}\right) + w_a\left(\frac{d_{bowl}}{d_{cyl}}\right)^2\right]^2. \tag{7.41}$$

For the radial speed component w_r and the axial component w_a results

$$w_r = \frac{\mathrm{d}V}{\mathrm{d}t}\frac{1}{V}\frac{V_{bowl}}{V - V_{bowl}}\frac{d_{cyl}^2 - d_{bowl}^2}{4\,d_{bowl}} \tag{7.42}$$

und with bowl volume V_{bowl} and

$$w_a = \frac{\mathrm{d}V}{\mathrm{d}t}\frac{1}{V}s_{bowl}, \tag{7.43}$$

with bowl depth s_{bowl}. The specific kinetic start energy at "intake valve closes" results as

$$k_{IC} = \frac{1}{16}\left[\frac{c_m\,d_{cyl}^2\,\lambda_L}{d_{iv}\,h_{iv}\sin\left(45°\right)}\right]^2. \tag{7.44}$$

In this equation, the mean piston speed c_m, the volumetric efficiency λ_L as well as the intake valve diameter d_{iv} and the intake valve lift h_{iv} are introduced.

The combustion term Δ can be written with the help of the temperature of an intended zone for burned matter T_b and a zone for unburned matter T_{ub}

$$\Delta = \left[X\frac{T_b}{T_{gas}}\frac{T_b - T_{wall}}{T_{gas} - T_{wall}} + (1 - X)\frac{T_{ub}}{T_{gas}}\frac{T_{ub} - T_{wall}}{T_{gas} - T_{wall}}\right]^2 \tag{7.45}$$

With $X = \dfrac{Q_f(\varphi)}{Q_f}$.

The temperature for the unburned zone calculates over a polytrope compression to

$$T_{uv} = T_{gas,IT}\left(\frac{p}{p_{IT}}\right)^{(n-1)/n} \quad \text{mit} \quad 1.34 \leq n \leq 1.37, \tag{7.46}$$

resulting in the temperature for the burned bas,

$$T_b = \frac{1}{X}T_{gas} + \frac{X-1}{X}T_w. \tag{7.47}$$

The heat transfer coefficient according to Bargende is only valid for the high pressure period. Figure 7.9 shows an example for the heat transfer coefficient according to Bargende for a SI engine at 1,500 rpm and $i_{mep} = 7.35$ bar.

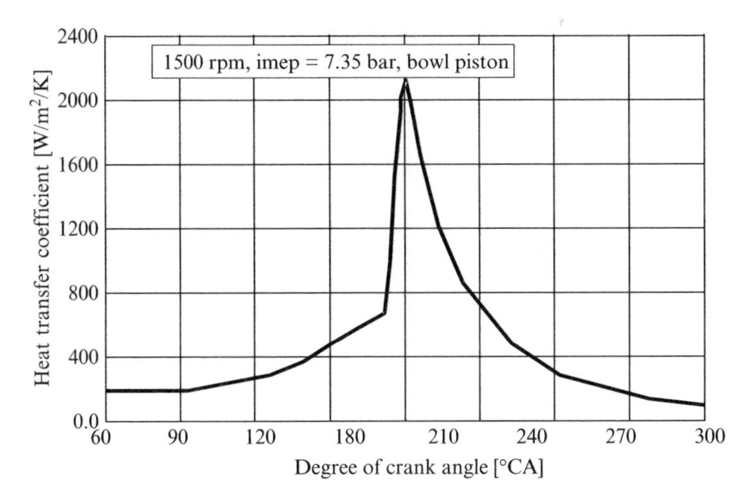

Fig. 7.9 Heat transfer coefficient α for a SI engine, see Bargende (1990)

In the case of the equations for the description of the heat transfer in the cylinder described here, we are dealing with semi-empirical methods, in which the parameters have been determined by means of measured values. We do not thereby distinguish particularly between heat transfer via convection and via radiation. Despite diverse efforts and numerous heat transfer reactions which have been discovered to this day, there is still a need for a relatively simple to use, but physically better founded relation, which physically accurately reproduces especially the portions of heat transfer which result from convection and radiation.

Moreover, the validity of the heat transfer equations should be critically examined for SI engines with unconventional valve timing, as can appear, e.g. in the case of electromechanical valve trains, and for SI engines with direct injection and adjusted with suitable measurements.

The procedure in the determination of the heat transfer coefficient from measured values is described by Merker and Kessen (1999).

7.1.5 Heat Transfer in the Exhaust Manifold

Heat transfer in the exhaust manifold in charged engines plays a decisive role in the exact determination of the exhaust gas temperature and thus in the determination of exhaust gas enthalpy before the turbine. The wall heat losses do not belong to the first law for the cylinder, since its system boundaries end at the valves. Zapf suggested a relation in 1969 for manifold wall heat transfer which is still used today

$$\alpha_{exh.\ man.} = 0.001791\left(1 - 0.797\,\frac{h_{valve}}{d_i}\right)\sqrt{\dot{m}_{exhaust\ gas}}\,T^{-0.41}\,d_{port}^{-1.5}\quad\left[\frac{\mathrm{W}}{\mathrm{m^2\,K}}\right]. \tag{7.48}$$

According to Zapf (1969), the exhaust manifold heat transfer is contingent on the exhaust mass flow $\dot{m}_{exhaustgas}$, the port diameter d_{port}, the gas temperature T – which described the material quantities in a similar way as in cylinder wall heat transfer – as well as on the valve lift h_{valve} and the internal valve diameter d_i.

For the calculation of the wall heat flow, the Newtonian relation should also be used here

$$\frac{dQ_{port}}{\mathrm{d}t} = \alpha_{port} A_{port} \left(T_{w,port} - T\right). \tag{7.49}$$

7.1.6 Wall Temperature Models

7.1.6.1 Stationary Operation

For the experimental determination of internal wall temperatures, there are in principle two possibilities. Since the exact course on the surface is not required as the internal wall temperature (see Fig. 7.10), but rather only the value averaged over a working cycle, we can determine the gradient of the temperature path by means of a differential measurement.

In the combustion chamber wall, a constant temperature gradient appears in the area of the combustion chamber outside wall because of heat conduction according to the adjusted operating point. If we fix two thermocouples offset by a known distance in this area, we can determine the gradient from the difference of the temperatures and the knowledge of the offset distance and with this arrive at a conclusion about the temperature in the inside of the combustion chamber. The other method is the direct measurement of the internal wall temperature, variable over the cycle, by means of a surface thermocouple. Through the use of solutions for the Fourier heat conduction equation, the mean internal wall temperature is easy to determine. This method is used at the same time for the determination of the local heat flows and thus for the

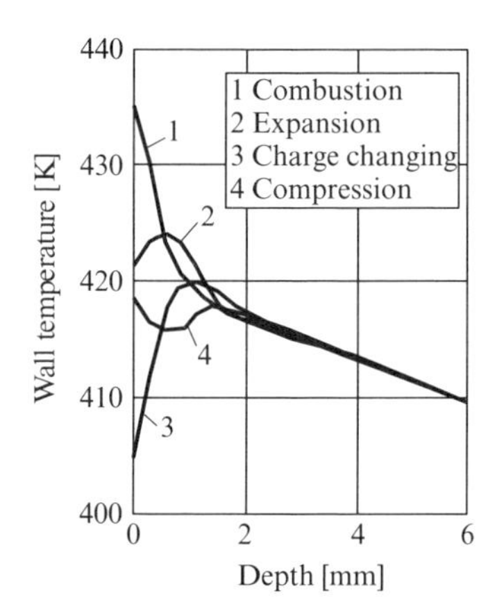

Fig. 7.10 Wall temperature progression during a working cycle

determination of the local heat transfer coefficient in the combustion chamber, see Merker and Kessen (1999), Bargende (1990), or Hohenberg (1980).

If the internal wall temperature is known for an operating point, the thermal substitute conduction coefficient can be determined from this quite easily for stationary operating points. The remaining quantities are known from process calculation (α, Tw) or as a boundary condition (Tc). The equilibrium of the heat flows is hereby valid, which result from the convective heat transfer (Newtonian equation), and the wall heat flow resulting from heat conduction through the combustion chamber wall (discretized Fourier equation).

$$\bar{\alpha} A \left(\bar{T}_{gas} - \bar{T}_w\right) = \frac{\lambda}{d} A \left(\bar{T}_w - \bar{T}_{w,c}\right) = \bar{\alpha}_c A \left(\bar{T}_{w,c} - \bar{T}_c\right). \tag{7.50}$$

If we summarize heat conduction through the wall and the convective heat transfer – to the coolant, for example – we obtain the following relation

$$\bar{\alpha} \left(\bar{T}_{gas} - \bar{T}_w\right) = R_{th} \left(\bar{T}_w - \bar{T}_c\right). \tag{7.51}$$

Solved with the substitute conduction coefficient, we obtain

$$R_{th} = \frac{\bar{\alpha} \left(\bar{T}_{gas} - \bar{T}_w\right)}{\left(\bar{T}_w - \bar{T}_c\right)}. \tag{7.52}$$

The values are averaged over one working cycle, which is indicated by the crossbar. Since the wall thickness in a real engine is not constant everywhere, this effect is also covered in the thermal substitute conduction coefficient. Thus, a local solution of wall temperature is not possible. Upon determining the substitute conduction coefficient, a calculation of the mean wall temperature from the data available at the end of a cycle is possible. On the other hand, since the wall heat flow is dependent on the wall temperature, an iteration loop over several working cycles is necessary. Here is valid

$$\bar{T}_w = \frac{\bar{\alpha} \bar{T}_{gas} + R_{th} \bar{T}_c}{\alpha + R_{th}}. \tag{7.53}$$

The stationary wall temperatures of the piston, cylinder head and liner can be calculated separately from each other according to this method. It is also practical for the determination of exhaust manifold-wall temperature.

7.1.6.2 Unsteady Operation

For the unsteady operation of an internal combustion engine, more expensive models must be considered, which take into consideration the storage capacity of

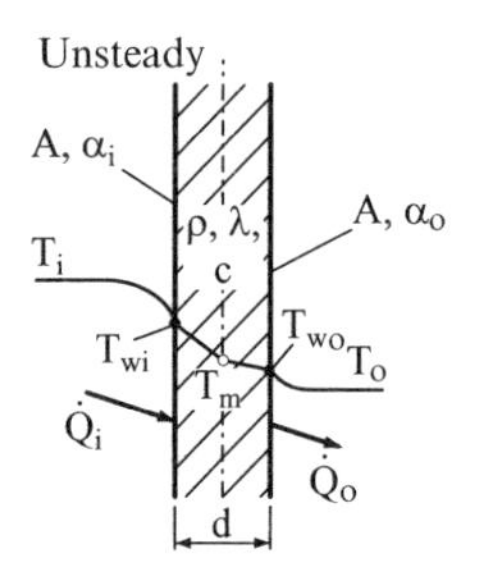

Fig. 7.11 Unsteady wall temperature model

the combustion chamber wall and the heat transfer to the coolant. One simple model for this can be found in Reulein (1998). A level plate as described in Fig. 7.10 serves as a basic model for this approach as well. From the wall, the physical properties heat capacity c, density ρ, and heat conductivity λ are known. The thickness d and the surface A of the wall must also be known. The mean wall temperature T_m can be calculated with the equation

$$\rho\, A\, d\, c\, \frac{\mathrm{d}T_m}{\mathrm{d}t} = \dot{Q}_i + \dot{Q}_o. \tag{7.54}$$

If the wall is divided in the center plane, as indicated in Fig. 7.11, and a quasi-stationary heat conduction is assumed in both wall halves, the following equations also apply:

$$\dot{Q}_i = \frac{2\,\lambda\, A}{d}\,(T_{wi} - T_m) \tag{7.55}$$

and

$$\dot{Q}_o = \frac{2\,\lambda\, A}{d}\,(T_m - T_{wo}). \tag{7.56}$$

The division of the wall into two planes takes place following FEM-models for the calculation of temperature distributions. Through the simulation of the wall with the help of at least two cells, actual temperature distribution within the wall during heating or cooling processes can be approximated much more realistically.

In the case of the piston, we are dealing with a component that is not directly bathed by a coolant. For the stationary case, the substitute conduction coefficient covers the heat conduction over the piston rings and the oil to the liner and to the coolant. In the case of pistons under a full load, an oil injection cooling is employed. For modeling unsteady behavior of the piston, the model of Doll (1989) can be used. In this case, we assume that the heat flow falling in gaseous form into the piston distributes itself to the cylinder liner and the motor oil. The heat flow to the cylinder liner is thereby transferred partially over the piston rings, partially

over the piston shaft. The particular heat flows can be determined with the following equations.
piston ring – cylinder wall:

$$\dot{Q}_{ring} = A_{ring}\,\alpha_{ring}\,(T_{piston} - T_{liner}) \tag{7.57}$$

piston shaft – cylinder wall:

$$\dot{Q}_{shaft} = A_{shaft}\,k_{shaft}\,(T_{piston} - T_{liner}) \tag{7.58}$$

piston underside – motor oil:

$$\dot{Q}_{oil} = A_{piston}\,k_{oil}\,(T_{piston} - T_{oil}). \tag{7.59}$$

The heat transfer coefficients k for the shaft amount to

$$k_{shaft} = \frac{1}{\frac{l_1 + \frac{l_2}{2}}{\lambda_{piston}} + \frac{1}{\alpha_{shaft}}} \tag{7.60}$$

and for the oil to

$$k_{oil} = \frac{1}{\frac{l_1 + l_2}{\lambda_{piston}} + \frac{1}{\alpha_{oil}}}. \tag{7.61}$$

l_1 signifies the free liner length and l_2 the covered liner length. For the heat transfer coefficients, the following assumptions can be made according to Pflaum and Mollenhauer (1977)

$$\alpha_{ring} = 2{,}500\left[\frac{\mathrm{W}}{\mathrm{m^2\,K}}\right]$$
$$\alpha_{shaft} = 1{,}000\left[\frac{\mathrm{W}}{\mathrm{m^2\,K}}\right]$$
$$\alpha_{oil} = 500\left[\frac{\mathrm{W}}{\mathrm{m^2\,K}}\right]. \tag{7.62}$$

The mean temperature of the piston is then calculated with the differential equation

$$m_{piston}\,c_{piston}\,\frac{\mathrm{d}T_{piston}}{\mathrm{d}t} = \dot{Q}_{gas} - \dot{Q}_{ring} - \dot{Q}_{shaft} - \dot{Q}_{oil}. \tag{7.63}$$

7.1.7 The Heat Release Rate

The heat release rate describes the temporal course of energy release in the combustion chamber. The integral of the heat release rate is designated as the total heat release or the burn function. In order to model combustion in the context of process calculation we use various approaches and mathematical modelings, which all have the goal of describing the actual heat release via combustion as exactly as possible as so-called substitute heat release rates. Another possibility for describing heat release are so-called phenomenological models, that pre-calculate the heat release rate, starting from the injection path of the diesel engine for example. In recent times, purely numerical methods are being used because of the increasing complexity of combustion processes (e.g. direct injection with stratified charge in the SI engine, common-rail injection in the diesel engine with multiple injection). In this case we are dealing with so-called neural networks, which have to be trained with results from measured operating points and can be contingent on a number of parameters.

7.1.7.1 The Vibe Substitute Heat Release Rate

Proceeding from the "triangular combustion", Vibe (1970) provided the following relation for the total heat release rate with the help of reaction kinetic considerations

$$\frac{Q_f(\varphi)}{Q_{f,\,total}} = 1 - e^{-a\left(\frac{\varphi-\varphi_{SOC}}{\Delta\varphi_{CD}}\right)^{m+1}} \quad \text{mit } \varphi_{SOC} \leq \varphi \leq (\varphi_{SOC} + \Delta\varphi_{CD}). \tag{7.64}$$

The entire amount of released energy is thereby to be calculated from the product of the fuel mass brought into the combustion chamber and the lower heating value

$$Q_{f,\,total} = m_{fuel}\, l_{hv}. \tag{7.65}$$

Furthermore, φ_{SOC} signifies the start of combustion and $\Delta\varphi_{CD}$ the combustion duration. The so-called Vibe shape parameter is designated with m.

At the end of combustion, i.e. at $\varphi = \varphi_{EOC}$, a certain percentage $\eta_{conv,total}$ of the total energy added with the fuel must be converted (degree of conversion), for which is valid

$$\left.\frac{Q_f(\varphi)}{Q_{f,\,total}}\right|_{\varphi=\varphi_{EOC}} \equiv \eta_{conv.,\,total} = 1 - e^{-a}. \tag{7.66}$$

From this follows for the factor a the relation

$$a = -\ln(1 - \eta_{conv.,total});$$

from which we obtain the following numerical values

$\eta_{conv,total}$	**0.999**	0.990	0.980	0.950
A	**6.908**	4.605	3.912	2.995

If we derive the total heat release rate from the degree of crank angle, we obtain the following for the heat release rate:

$$\frac{dQ_f}{d\varphi} = Q_{f,total}\, a\,(m+1)\left(\frac{\varphi - \varphi_{SOC}}{\Delta\varphi_{CD}}\right)^m e^{-a\left(\frac{\varphi-\varphi_{SOC}}{\Delta\varphi_{CD}}\right)^{m+1}}. \tag{7.67}$$

Figure 7.12 shows heat release rates for various Vibe form factors. One should note here that Vibe form factors smaller than 0 are also possible.

In order to adjust a real combustion by means of a Vibe substitute heat release rate, there are various methods for determining the three Vibe parameters start of combustion, combustion duration, and shape parameter. The parameters can either be adjusted visually or can be determined using mathematical methods (e.g. that of

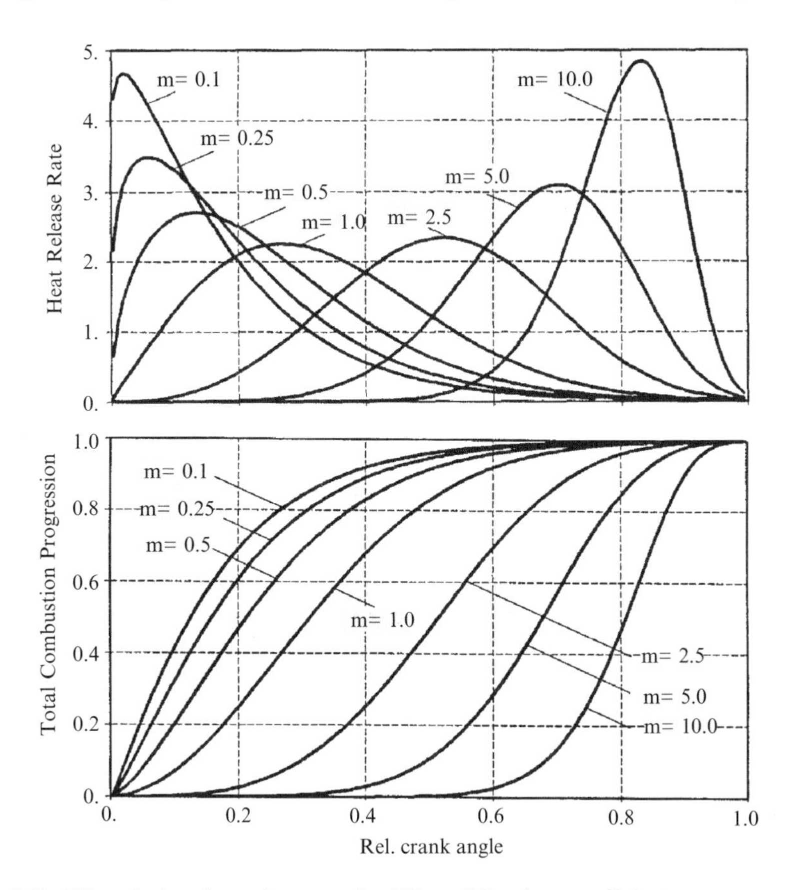

Fig. 7.12 Vibe substitute heat release rate for different Vibe shape coefficients

the least square). It is however important that important process data such as peak pressure, indicated mean pressure and the exhaust gas temperature are calculated with the substitute heat release rate in agreement with reality. The determination of the Vibe parameters is de rigueur in most pressure analysis programs.

7.1.7.2 The Substitute Heat Release Rate as a Double Vibe Function

In engines or operating points with a clearly marked premixed combustion such as in diesel engines, reproduction with a simple Vibe substitute heat release rate is usually too imprecise. We thus often replace the simple Vibe substitute heat release rate with the superimposition of two Vibe functions, the so-called double Vibe function. If one describes an actual heat release rate with a double Vibe function, a subdivision of the energy shares of both functions is also necessary besides the two simple Vibe heat release rates. For a double Vibe heat release rate is valid

$$\frac{dQ_{f,1}}{d\varphi} = Q_{f,1}\, a\,(m_1+1)\left(\frac{\varphi-\varphi_{SOC,1}}{\Delta\varphi_{CD,1}}\right)^{m_1} e^{-a\left(\frac{\phi-\phi_{SOC,1}}{\Delta\phi_{CD,1}}\right)^{m_1+1}} \quad . \quad \text{mit } \varphi_{SOC,1} \le \varphi \le \left(\varphi_{SOC,1}+\Delta\varphi_{CD,1}\right), \tag{7.68}$$

$$\frac{dQ_{f,2}}{d\varphi} = Q_{f,2}\, a\,(m_2+1)\left(\frac{\varphi-\varphi_{SOC,2}}{\Delta\varphi_{CD,2}}\right)^{m_2} e^{-a\left(\frac{\varphi-\varphi_{SOC,2}}{\Delta\varphi_{CD,2}}\right)^{m_2+1}} \quad . \quad \text{mit } \varphi_{SOC,2} \le \varphi \le \left(\varphi_{SOC,2}+\Delta\varphi_{CD,2}\right), \tag{7.69}$$

$$Q_{f,1} = x\,Q_{f,ges} \quad \text{und} \quad Q_{f,2} = (1-x)\,Q_{f,ges}, \tag{7.70}$$

$$\frac{dQ_f}{d\varphi} = \frac{dQ_{f,1}}{d\varphi} + \frac{dQ_{f,2}}{d\varphi}. \tag{7.71}$$

The diagram on the left in Fig. 7.13 shows the reproduction of the heat release rate in the rated performance point of a high speed high performance diesel engine via a double Vibe substitute heat release rate. One can clearly see that the double Vibe function can not exactly reproduce combustion lasting until the exhaust phase. This has to do with the mathematical form of this substitute heat release rate. The diffusion phase is described by the central exponential term of the double Vibe function. This exponential term with its asymptotic course towards burning matter can under certain circumstances still not exactly describe large energy release rates at the burning location in the concrete case, which is why on occasion a threefold Vibe function is used instead of the double Vibe function.

The exact reproduction of premixed combustion is an important prerequisite for a good agreement of the working cycle calculation with measurements, which is why the adjustment and weighing of the first Vibe function should receive careful consideration. One problem in the double Vibe function is – as in the case of the

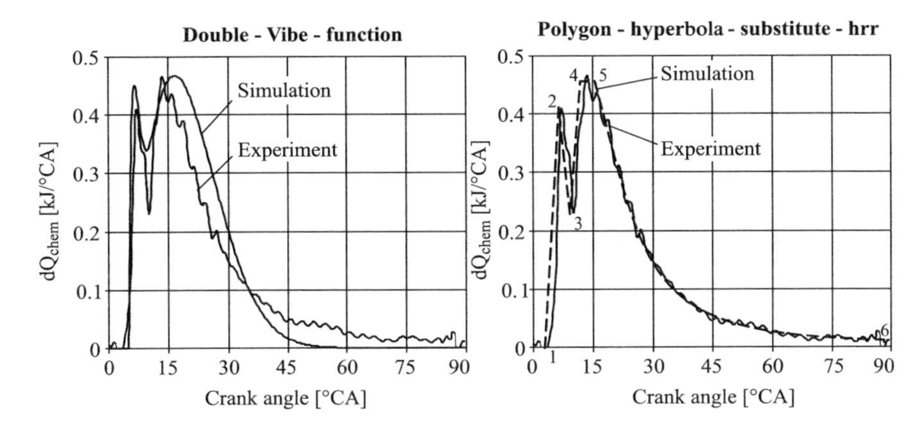

Fig. 7.13 The double-Vibe function and polygon-hyperbola heat release rate

simple Vibe function as well – the exact description of the burn-out phase, which is, especially in charged engines, responsible for the exhaust gas temperature level. Good indicator for this is the total heat release rate, in which deviations are most clearly visible. The method of the least square has proved to be the most suitable adjustment routine in the case of the double Vibe function as well.

7.1.7.3 The Polygon-Hyperbola Substitute Heat Release Rate

In view of the calculation of NO_x formation by means of multi-zone models, a further substitute heat release rate, the polygon-hyperbola substitute heat release rate, was suggested by Schreiner (1993), which is represented in Fig. 7.13 on the right. It consists of a polygonal port (1-4-5) and a hyperbola connected to it (5-6). A triangle (1-2-3) is superimposed over the polygonal port, which serves in the description of premixed combustion. For the mathematical description of this polygon-hyperbola substitute heat release rate, we need nine parameters.

Premixed phase:

$$\frac{dQ_{f,pre}}{d\varphi} = y_2^* \frac{(\varphi - \varphi_1)}{(\varphi_2 - \varphi_1)} \quad \text{with } \varphi_1 \leq \varphi \leq \varphi_2 \tag{7.72}$$

$$\frac{dQ_{f,pre}}{d\varphi} = y_2^* \frac{(\varphi_3 - \varphi)}{(\varphi_3 - \varphi_2)} \quad \text{with } \varphi_2 \leq \varphi \leq \varphi_3 \tag{7.73}$$

Diffusion phase:

$$\frac{dQ_{f,diff}}{d\varphi} = y_4 \frac{(\varphi - \varphi_1)}{(\varphi_4 - \varphi_1)} \quad \text{with } \varphi_1 \leq \varphi \leq \varphi_4 \tag{7.74}$$

$$\frac{dQ_{f,diff}}{d\varphi} = y_4 \text{ with } \varphi_4 \leq \varphi \leq \varphi_5 \tag{7.75}$$

$$\frac{dQ_{f,diff}}{d\varphi} = h_3 + h_1 (\varphi - \varphi_1)^{h_2} = y_4 - \frac{(y_4 - y_6)}{1 - \left[\frac{(\varphi_6 - \varphi_1)}{(\varphi_5 - \varphi_1)}\right]^{h_2}} + \\ + \frac{(y_4 - y_6)}{(\varphi_5 - \varphi_1)^{h_2} - (\varphi_6 - \varphi_1)^{h_2}} (\varphi - \varphi_1)^{h_2} \\ \text{mit } \varphi_5 \leq \varphi \leq \varphi_6 \,. \tag{7.76}$$

For the entire combustion process

$$\frac{dQ_{f,total}}{d\varphi} = \frac{dQ_{f,pre}}{d\varphi} + \frac{dQ_{f,diff}}{d\varphi} \,. \tag{7.77}$$

is valid.

The quantity y_2^* sets the height of the peak in premixed combustion. The three hyperbolic parameters h_1, h_2 and h_3 result from the conditions that the hyperbola go through points 5 and 6 and that the integral under the substitute heat release rate has the value of 1. h_2 thereby fixes the path of the hyperbola. The maximum height of the substitute heat release rate in the diffusion phase y_4 results from the condition that the center of the substitute heat release rate agree with the experimentally determined heat release rate. The quantity y_6 describes the conversion at the end of combustion.

The agreement in Fig. 7.13 (right) is very good, such that there practically no disparity arises in the results of the actual working cycle calculation between the point-for-point provision and the approximation of experimental data with this substitute heat release rate.

Further details and an extensive analysis of the polygon-hyperbola substitute heat release rate can be found in Schreiner (1993).

7.1.7.4 Combinations of Substitute Heat Release Rates

In order to describe the long burn-out phase and the variable pre-injection possible with the common-rail injection system, a combination of the Vibe substitute heat release rate and the hyperbola substitute heat release rate was suggested by Barba et al. (1999). However, the equations described and their conversion regularities are so complex that a general formulation is impossible, which is why we simply refer here to a corresponding source in the literature.

7.1.7.5 Heat Release Rate Precalculation for Diesel Engines

In order to take into account the quantity of fuel energy introduced into the combustion chamber required for the respective operating point and correctly to reproduce the resulting changes in the thermal state quantities of the engine, like exhaust gas temperature, a possibility must be created to pre-calculate the heat release rate contingent upon the respective operating point data in the simulation. This is above all inevitable for non-stationary calculations. The Vibe substitute heat release rate is determined – as described above – by the three parameters start of combustion, combustion duration, and the shape parameter. Conversion regularities for these parameters can be found contingent on the respective operating point. The foundations for the description of these regularities were laid, amongst others, by Woschni and Anisits (1973) and will here be represented in a universally valid form. A known initial operating point of the engine (index A) is used as the basis for the conversion. Partially constant values can be found in the literature for the parameters introduced in the following equations. Here, we will however represent the possible quantities of influence on the particular heat release rate quantities as parameters. The following equations describe the conversion of Vibe parameters for the diesel engine.

For the determination of start of combustion, proceeding from the geometrically set delivery start of the injection pump, the following relation is applicable

$$\varphi_{SOC} = \varphi_{SOD} + \Delta\varphi_{IND} + \Delta\varphi_{IGD}. \tag{7.78}$$

For an injection pump without regulation of the start of injection, the injection delay between the geometrically fixed delivery start of the injection pump and the start of injection – i.e. the opening of the injection nozzle needle – must be considered, see Merker and Schwarz (2001)

$$\Delta\varphi_{IND} = \Delta\varphi_{IND,A}\left(\frac{n}{n_A}\right)^{a_{IND}}. \tag{7.79}$$

In engines with regulation of the start of injection, the geometrical delivery start is delayed until the start of injection, which can be determined via the needle lift signal, agrees with the desired value. The determination of the injection delay time and the geometrical delivery start are thereby left out. The equation for start of combustion is then

$$\varphi_{SOC} = \varphi_{SOI} + \Delta\varphi_{IGD}. \tag{7.80}$$

The ignition delay can be described with different, formulaic connections. According to Woschni and Anisits (1973), for the pre-chamber engine is valid

$$\Delta\varphi_{IGD} = a_{IGD}\, 6\, n\, 10^{-3}\, 1.3\, e^{\frac{990}{T(\varphi_{SOI})}}\, p\left(\varphi_{SOI}\right)^{-0.35}. \tag{7.81}$$

For direct injection diesel engines, the ignition delay can be calculated according to Sitkei (1963)

$$\Delta\varphi_{IGD} = 6\,n\,10^{-3}\left[a_{IGD} + b_{IGD}\,e^{\frac{7.800}{6.9167RT}}\left(1.197\,p^{-0,7}\right)\right.$$
$$\left. + c_{IGD}\,e^{\frac{7.800}{6.9167RT}}\left(1.0197\,p^{-1.8}\right)\right]. \qquad (7.82)$$

For pressure and temperature, the pressure and temperature values averaged from start of injection until start of combustion must be employed. The combustion duration depends on the air-fuel ratio and, in the case of some engines, on the speed as well

$$\Delta\varphi_{CD} = \Delta\varphi_{CD,A}\left(\frac{\lambda_A}{\lambda}\right)^{a_{CD}}\left(\frac{n}{n_A}\right)^{b_{CD}}. \qquad (7.83)$$

For engines with exhaust gas recirculation, the description of the combustion duration via the air-fuel ratio is insufficient, since the air-fuel ratio in exhaust gas recirculation is contingent to a large extent on the mass of the recycled gas. It is thus not a clear indicator of engine load. For this reason, the combustion duration for engines with exhaust gas recirculation is described contingent on the injected fuel mass. The equation then reads

$$\Delta\varphi_{CD} = \Delta\varphi_{CD,A}\left(\frac{\lambda_A}{\lambda}\right)^{a_{CD}}\left(\frac{n}{n_A}\right)^{b_{CD}}\left(\frac{m_f}{m_{f,A}}\right)^{c_{CD}}. \qquad (7.84)$$

The shape parameter describes the course of energy conversion and is thus contingent on the ignition delay (mixture preparation time), on the speed, and on the gas conditions or on the gas mass in the cylinder at "intake closes". In order to calculate shape parameters smaller than 0 as well, the equation is supplemented with an additional part. Shape parameters smaller than 0 describe heat release rates with very high heat release rate increase speeds, which can appear in direct injection diesel engines under weak loads and low speeds, when a load pressure is not yet available.

$$m = (m_A + \Delta m)\left(\frac{\varphi_{IGD,A}}{\varphi_{IGD}}\right)^{a_{VM}}\left(\frac{n_A}{n}\right)^{b_{VM}}\left(\frac{p_{IC}\,V_{IC}\,T_{IC,A}}{p_{IC,A}\,V_{IC,A}\,T_{IC}}\right)^{c_{VM}} - \Delta m. \qquad (7.85)$$

Table 7.1 provides, besides the "basic" parameters known from the literature, an overview of usual parameters as they have been used by the authors for extensive calculations with good agreement with reality.

Conversion regularities for double Vibe heat release rates are much more complex, since another conversion for energy distribution of both individual portions is necessary for the second set of Vibe parameters. Conversion regularities for a high speed diesel engine are described by Oberg (1976).

Table 7.1 Parameters for the pre-calculation of heat release rates

	Original equation	Large diesel engine	Commercial vehicle DI	Passenger car DI
a_{IND}	1.0	1.0	–	–
a_{IGD}	1[a]; 0.5	0.39	0.625[a]	0.1
b_{IGD}	0.135	0.105	–	0.135
c_{IGD}	4.8	3.12	–	4.8
a_{CD}	0.6	0.6	–0.3	0.0
b_{CD}	0.5	0.5	–0.65	0.0
c_{CD}	0.0	0.0	0.0	–0.1
Δm	0.0	0.0	0.4	0.3
a_{VM}	0.5	0.5	0.5	0.2
b_{VM}	0.3	0.3	–0.8	–0.4
c_{VM}	1.0	1.0	1.0	1.0

[a]Anisits

For the polygon-hyperbola substitute heat release rate, the following contingencies are necessary for the conversion of the parameters for high performance diesel engines according to Schreiner (1993)

$$\varphi_{CCD} = \int_{\varphi_1}^{\varphi_6} \frac{\mathrm{d}Q_f}{\mathrm{d}\varphi} \varphi \,\mathrm{d}\varphi, \tag{7.86}$$

$$IND = IND_A \left(\frac{n}{n_A}\right), \tag{7.87}$$

$$\frac{\frac{ID}{ID_A} - e_1}{1 - e_1} = e_2 \left(\frac{n}{n_A}\right) + e_3 \left(\frac{m_f}{m_{f,a}}\right) + (1 - e_2 - e_3) \left(\frac{n}{n_A}\right) \left(\frac{m_f}{m_{f,a}}\right), \tag{7.88}$$

$$\frac{\tau_{IGD}}{\tau_{IGD,A}} = \left(\frac{e^{\frac{a}{T_{IGD}}}}{e^{\frac{a}{T_{IGD,A}}}}\right) \left(\frac{p_{IGD}}{p_{IGD,A}}\right)^b, \tag{7.89}$$

$$\varphi_{SOC} = \varphi_{SOD} + IND + IGD, \tag{7.90}$$

$$x_{IGD} = \frac{m_{f,IGD}}{m_f}, \tag{7.91}$$

$$\frac{x_{pre} - k_1}{x_{pre,A} - k_1} = \frac{x_{IGD} - 1}{x_{IGD,A} - 1}, \tag{7.92}$$

$$\frac{\frac{CD_{SP}}{CD_{SP,A}} - k_3}{1 - k_3} = \left(\frac{ID}{ID_A}\right) \left(\frac{\lambda}{\lambda_A}\right)^{k_4} \left(\frac{n}{n_A}\right)^{k_5}, \tag{7.93}$$

Table 7.2 Standard values

Injection duration	$e_1 = 0$	Center of comb. duration	$k_3 = 0.3$
	$e_2 = 1 - -e_3$		$k_4 = -0.3$
	$e_3 = 0.8$		$k_5 = 0$
Injection delay	$a = 1,500K$	Diffusion combustion	$k_6 = 0$
	$b = -0.8$		$k_7 = 1$
			$k_8 = 1$
Premixed portion	$k_1 = 0.4$		$k_9 = 0$
	$k_2 = 0.125$		

$$\frac{(\varphi_4 - \varphi_1)}{(\varphi_4 - \varphi_1)_A} = \left(\frac{n}{n_A}\right)^{k_6} \left(\frac{m_f}{m_{f,A}}\right)^{k_7} \quad \text{and} \tag{7.94}$$

$$\frac{(\varphi_5 - \varphi_4)}{(\varphi_5 - \varphi_4)_A} = \left(\frac{n}{n_A}\right)^{k_8} \left(\frac{m_f}{m_{f,A}}\right)^{k_9}. \tag{7.95}$$

Table 7.2 provides standard values for the nine parameters.

7.1.7.6 Heat Release Rate Conversion for SI Engines

A conversion of the Vibe parameters for varying operating points has been introduced for the SI engine as well, e.g. in Csallner (1981). Since the cylinder charge is externally ignited with the spark plug, the derivation of the start of combustion via injection/ignition delay is dropped in the case of the SI engine. In thermodynamic evaluations however, a time shift between ignition (ignition time) and a noticeable release of energy (increase of the heat release rate/5% mass fraction burned) is demonstrable in the SI engine as well. This is explained in that at first some time passes because of the point light ignition until a larger volume of the flame front is included. The time period until the rise of the heat release rate is designated as apparent ignition delay.

Csallner (1981) has described the contingency of the Vibe parameters on the process quantities. The investigations were executed such that the individual operating parameters were varied independently of one another. Csallner thus chose a description by means of a multiplicative method. Proceeding from an referenced point (index A), for the ignition delay thus results.

$$IGD = IGD_A f_{IT} f_n f_p f_T f_{x_{RG}} f_\lambda. \tag{7.96}$$

For the combustion duration, it can be described in the same way

$$\Delta\varphi = \Delta\varphi_A \, g_{IT} \, g_n \, g_p \, g_T \, g_{x_{RG}} \, g_\lambda. \tag{7.97}$$

For the form factor we obtain

$$m = m_A \, h_{IT} \, h_n \, h_p \, h_T \, h_{x_{RG}} \, h_\lambda. \tag{7.98}$$

Table 7.3 Functions according to Csallner (1981)

	Ignition delay	Combustion duration	Shape parameter
Ignition time25–50 b. TDC	$f_{IT} = \frac{430 - \phi_{IT}}{430 - \phi_{IT,A}}$	$g_{IT} = 1$	$h_{IT} = 1$
Speed1,000–4,500min^{-1}	$f_n = \frac{1 + \frac{400}{n} - \frac{8 \cdot 10^5}{n^2}}{1 + \frac{400}{n_A} - \frac{8 \cdot 10^5}{n_A^2}}$	$g_n = \frac{1.33 - \frac{660}{n}}{1.33 - \frac{660}{n_A}}$	$h_n = \frac{0.625 + \frac{750}{n}}{0.625 + \frac{750}{n_A}}$
Cylinder pressure at 300°CA	$f_p = \left(\frac{p_{300}}{p_{A,300}}\right)^{-0.47}$	$g_p = \left(\frac{p_{300}}{p_{A,300}}\right)^{-0.28}$	$h_p = 1$
Cylinder temp. at 300°CA	$f_T = 2.16 \frac{T_{A,300}}{T_{300}} - 1.16$	$g_T = 1.33 \frac{T_{A,300}}{T_{300}} - 0.33$	$h_T = 1$
Residual gas port. 0–10%	$f_{x_{RG}} = 0.088 \frac{x_{RG}}{x_{RG,A}} + 0.912$	$g_{x_{RG}} = 0.237 \frac{x_{RG}}{x_{RG,A}} + 0.763$	$h_{x_{RG}} = 1$
Combus. air ratio 0.7–1.2	$f_\lambda = \frac{2.2\,\lambda^2 - 3.74\,\lambda + 2.54}{2.2\,\lambda_A^2 - 3.74\,\lambda_A + 2.54}$	$g_\lambda = \frac{2.0\,\lambda^2 - 3.4\,\lambda + 2.4}{2.0\,\lambda_A^2 - 3.4\,\lambda_A + 2.4}$	$h_\lambda = 1$

Table 7.4 Functions according to Witt (1999)

	Ignition delay	Combustion duration	Shape parameters
Ignition time 17–57°b. TDC	$f_{IT} = \frac{a + b\,\phi_{IT}^2}{a + b\,\phi_{IT,A}^2}$	$g_{IT} = \frac{a + b\,\phi_{IT}^{-0.5}}{a + b\,\phi_{IT,A}^{-0.5}}$	$h_{IT} = \frac{a + b\,\phi_{IT}^{-2}}{a + b\,\phi_{IT,A}^{-2}}$
Residual gas port. 10–26%	$f_{x_{RG}} = \frac{a + b\,x_{RG}^2}{a + b\,x_{RG,A}^2}$	$g_{x_{RG}} = \frac{a + b\,x_{RG}}{a + b\,x_{RG,A}}$	$h_{x_{RG}} = \frac{a + b\,x_{RG}^2}{a + b\,x_{RG,A}^2}$
Speed 1,000–4,000 min^{-1}	$f_n = \frac{a + b\,\ln(n)}{a + b\,\ln(n_A)}$ $f_n = \frac{a + b\,n^{-2}}{a + b\,n_A^{-2}}$	$g_n = \frac{a + b\,n^{-0.5}}{a + b\,n_A^{-0.5}}$	$h_n = \frac{a + b\,n^{1.5}}{a + b\,n_A^{1.5}}$
Indicated work 0.2–0.8 kJ/l	$f_{wi} = \frac{a + b\,wi^{1.5}}{a + b\,wi_A^{1.5}}$	$g_{wi} = \frac{a + b\,wi}{a + b\,wi_A}$	$h_{wi} = \frac{a + b\,\ln(wi)}{a + b\,\ln(wi_A)}$

The particular functions *f*, *g* and *h* are to be found in the following table (Table 7.3).

The range of validity for these conversions is confined to very minimal residual gas amounts (Table 7.4). Modern combustion processes show higher amounts of residual gas, which is why Csallner's influence equations were adjusted for a throttled and an unthrottled SI engine (fully variable valve train) by Witt (1999). In Witt (1999), the following is valid:

$$IGD = IGD_A f_{IT} f_{x_{RG}} f_n f_{imep} \tag{7.99}$$

$$\Delta\varphi = \Delta\varphi_A\, g_{IT}\, g_{x_{RG}}\, g_n\, g_{imep} \tag{7.100}$$

$$m = m_A\, h_{IT}\, h_{x_{RG}}\, h_n\, h_{imep} \tag{7.101}$$

The parameters for this can be taken from the following table (Table 7.5).

Table 7.5 Parameters for the conversion of the heat release rate parameters

			IT	x_{RG}	n	i_{mep}
Ignition delay	Throttled	a	0.678	0.879	0.992	1.112
		b	2.383×10^{-4}	3.648×10^{-4}	-1.246×10^{-4}	−0.545
	Unthrottled	a	0.638	0.914	−1.284	1.162
		b	2.614×10^{-4}	2.795×10^{-4}	0.292	−0.589
Combustion duration	Throttled	a	0.596	0.429	1.355	1.115
		b	2.480	0.031	−18.49	−0.346
	Unthrottled	a	0.477	0.690	1.701	1.295
		b	3.200	0.017	−34.50	−0.699
Shape parameters	Throttled	a	0.964	1.076	1.046	1.007
		b	75.56	-2.534×10^{-4}	-4.075×10^{-7}	0.004
	Unthrottled	a	1.000	1.061	1.016	1.053
		b	19.36	-1.656×10^{-4}	-1.206×10^{-7}	0.065

7.1.7.7 Neural Networks for Vibe Heat Release Rates

Besides the analytic methods of pre-calculation of heat release rates already presented, there is still another possibility in determining the heat release rate contingent upon its influence quantities. For this, the Vibe parameters are determined by means of a neural network, which has previously been trained by a number of measured operating points evaluated with reference to the Vibe parameters. In principle, all types of neural networks are suitable for this task. In comparison to the analytic relations presented in the previous section, clear improvements can be attained by determining the Vibe parameters by means of a neural network, which makes possible a halving of the averaged error and thus positively influences the quality of the calculation results. The disadvantage of neural networks in this case is the impossibility of extrapolation and the lacking transparency with reference to the contingencies of particular influence quantities in comparison to the analytic approach. An advantage is to be seen in the possibility of obtaining statements about the contingencies of the Vibe parameters from a number of unsystematic experimental results. A more exact description of neural networks can be found at the end of this section.

7.1.7.8 Neural Networks for Discrete Heat Release Rates

In the previous section, the description of heat release rate functions via Vibe parameters precalculated with neural networks was introduced. However, because of the increasing complexity of combustion processes both in diesel and in SI engines, it is necessary to describe the discrete heat release rates. In the diesel engine, because of common-rail technology, the injection and thus within certain boundaries the combustion as well become freely formable (pre-/post-injection). In the SI engine also, the combustion is influenced in a lasting way by load control by means of fully variable valve trains or via direct injection. In gasoline direct injection, a clearly lengthened burn-out phase appears for the stratified area. For both cases, an identification of the combustion via a simple substitute heat release rate function is therefore no longer sufficient in describing correctly the details of the combustion process. Moreover, the number of influence quantities is becoming larger and larger in such complex combustion processes, to the extent that an experimental scanning of these quantities is practically impossible technically.

Figure 7.14 shows a heat release rate for a stratified operating point in gasoline direct injection, in which the delayed burn-out phase is clearly recognizable. In the literature, a large number of methods are described for calculating heat release rates discretely with a neural network, see Zellbeck (1997). The method of Reulein et al. (2000) introduced in the following distinguishes itself from known methods and attempts to avoid the disadvantages of these methods. The experimentally determined heat release rates are first filtered, standardized to 1, and centered to their 50 mfb point. Through this, we can exclude significant deviations in the training data and make the network input regular. This process is represented in Fig. 7.15. In this process, it is necessary to train two neural networks, of which the first reproduces the path contingent on crank angle and the other contains the position of the center point. However, precisely this method shows clear advantages, since, on the one hand, the number of input parameters corresponding to the influence quantities to be described can be clearly enlarged as opposed to a functional description, and, on the other hand, the influence quantities for the shape and the 50 mfb point position can be set separately. In this way, the quality of the results is clearly improved. In most professional simulation tools, training algorithms with corresponding network topologies are now offered. In Fig. 7.14 is presented a comparison between a measured heat release rate and one calculated with a neural network. As far as the utilized network is concerned, we are dealing with a dual-layered multi-layer perceptron network with 25 nodes/points of intersection per layer and sigmoid activation functions, which are trained with a back-propagation algorithm. Of course, the heat release rate to be compared was not used for the

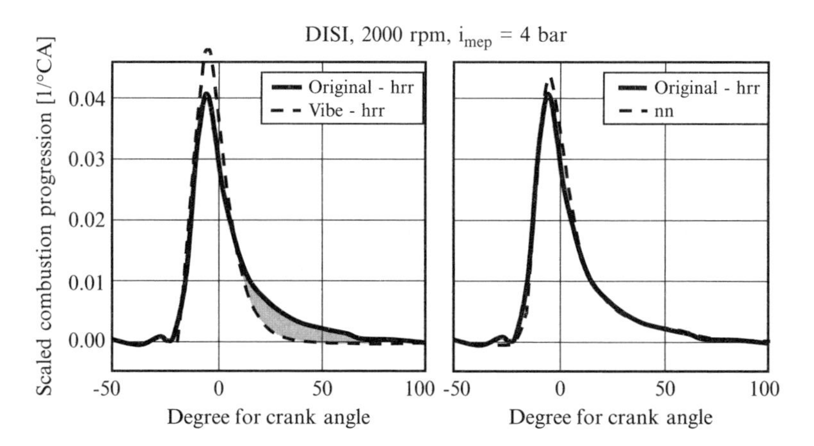

Fig. 7.14 Heat release rate for a DISI engine at $n = 2.000\,\text{min}^{-1}$ and $i_{mep} = 4$ bar. Comparison of the original heat release rate with the Vibe heat release rate (*left*) and with the neural network heat release rate (*right*)

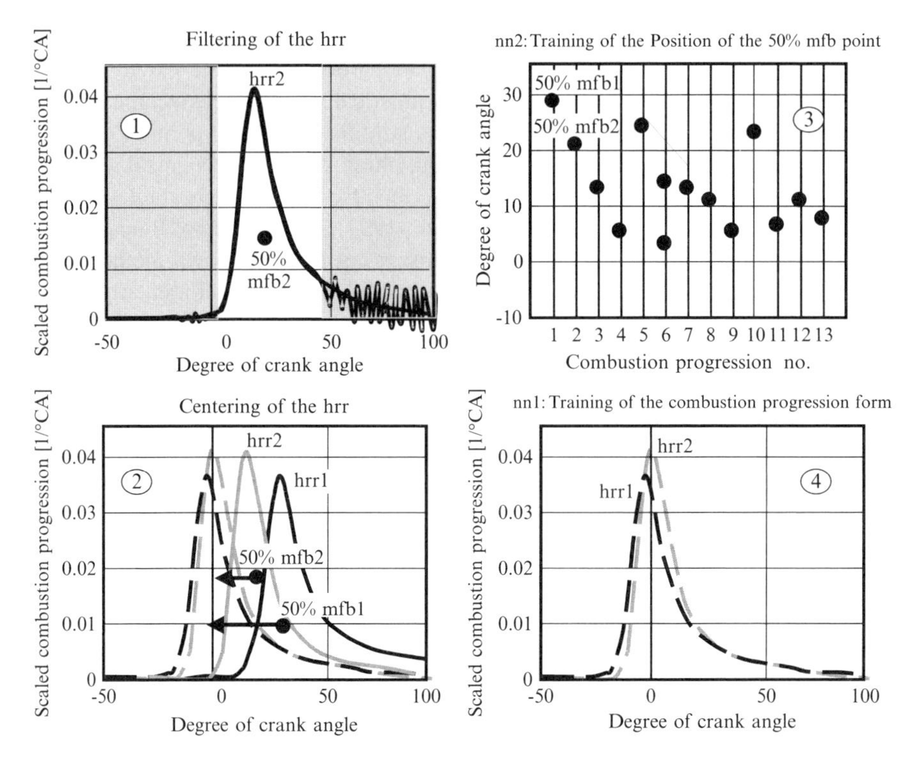

Fig. 7.15 Net training procedure

network training and determined within process calculation with the following input parameters:

- Degree of crank angle
- Speed
- Air mass
- Fuel mass
- Ignition time
- Amount of residual gas

On a critical note, we should note about neural networks that they are only valid within their trained range. An extrapolation is in principle impossible and must be made recognizable to the user, as otherwise the simulation results are considerably falsified.

7.1.7.9 Phenomenological Models

One fundamentally different method for modeling combustion is the use of phenomenological models. In the literature, several phenomenological models for the calculation of the heat release rate of a diesel engine from the injection rate have become known, see Stiesch (1999), Eilts (1993), or Chmela et al. (1998) and Constien (1991). For a description of these models, see Chap. 11.

7.1.8 Knocking Combustion

While in Sect. 7.1.7 methods were described for calculating normal combustion, in this chapter we will go into the simulation of autoignition and into the criteria for the prediction of the appearance of knocking combustion.

Knocking combustion occurs in spark ignited engines (SI engines). Under knocking combustion is understood autoignition of parts of the final gas before it is covered by the flame front, which is initiated by the spark plug, see also Sect. 4.3.5. A superimposition of several pressure waves in the combustion space then occurs, which leads to mechanical damage in the engine. The pressure curve of a knocking combustion is shown in Fig. 4.30.

The goal of the calculation of autoignition behavior is not the statement that a case of knocking combustion in the working cycle is concerned, but the predetermination of the degree of crank angle at which autoignition of the final gas appears.

In the literature, one distinguishes between methods with detailed or reduced reaction kinetics and empirical, phenomenological models. The first describe more or less extensively the reaction mechanisms of hydrocarbon oxidation in the low temperature range of the final gas, whereby so-called chain embranchments reinforce radical formation and chain breakage weakens it. The mechanisms are usually derived from shock wave pipes or rapid compression machines for a particular fuel type. An exponential increase in radical concentrations in the fresh air region over a certain limit value is assessed as an indicator for the commencement of knocking combustion. With this modeling approach there is thus the possibility of a prediction of the degree of crank angle, at which knocking begins. These modeling methods are, however, very complex, require a lot of calculation time, and would exceed the confines of this book. We therefore refer to the corresponding literature, e.g. Halstead et al. (1975, 1977), Li et al. (1992, 1994, 1996), Schreiber et al. (1994), Kleinschmidt (2000) and Klaiß (2003).

The phenomenological methods view the processes in the final gas through an average reaction speed, which can be expressed by means of an Arrhenius relation. We will take a closer look at phenomenological approaches in the following.

7.1.8.1 Knocking Criterion According to Franzke (1981)

Since reaching a certain temperature level in the final gas area is not a sufficient criterion for the setting in of knocking combustion, the pressure-temperature history in the combustion chamber must be considered. For this, a so-called critical prereaction level is calculated as follows

$$I_K = \frac{1}{\omega} \int\limits_{\varphi_{IC}}^{\varphi_K} p^a \, e^{\left(\frac{b}{T_{ub}}\right)} \, d\varphi = \text{const.} \tag{7.102}$$

The parameter φ_K describes thus the degree of crank angle, at which the critical pre-reaction level is reached.

The temperature of the unburned T_{ub} is calculated via a polytrope state change from pressure and temperature at start of combustion and a polytrope exponent, which results from the conditions at compression start and start of combustion

$$T_{ub}(\varphi) = T_{SOC} \left(\frac{p(\varphi)}{p_{SOC}} \right)^{\frac{n_{SOC}-1}{n_{SOC}}}. \tag{7.103}$$

The parameters a and b are provided in Table 7.6.

In addition, Franzke defines the appearance of knocking combustions by means of a constant parameter K specific for the combustion chamber space. This depends on different quantities (e.g. change motion level, combustion chamber space etc.) and represents a quotient of the combustion progress at the start of knocking and the entire combustion duration

$$K = \frac{\varphi_E - \varphi_{SOC}}{\Delta\varphi_{CD}}. \tag{7.104}$$

The difference of the 95% mfb point and start of combustion (1% mfb) is assumed as the combustion duration.

The parameter φ_E describes the degree of crank angle during the combustion progress, at which the critical pre-reaction level in the unburned gas must be reached, such that knocking combustion can appear.

Valid for the appearance of knocking combustion according to this is

$$\varphi_K < \varphi_E. \tag{7.105}$$

With the equation method of Franzke it is thus possible to determine the crank angle at which knocking combustion begins.

Table 7.6 Parameters for the knocking criterion

	Franzke	Spicher/Worret Simple 2-ZM, HTR Bargende	Spicher/Worret Actual 2-ZM, HTR Bargende	Spicher/Worret Simple 2-ZM, HTR Woschni
a [−]	1.5	−1.299	−1.267	−1.262
b [K]	−14,000	4,179	4,080	3,964
c [−]	–	2.370	2.124	2.714
a_{IK} [−]	–	−0.557	−0.449	−0.553
a_K [−]	–	−0.236	−0.241	−0.231
b_K [−]	–	1.292	1.395	1.275
c_K [−]	–	0.251	0.313	0.244
a_{kp} [−]	–	0.211	0.227	0.273
b_{kp} [−]	–	0.288	0.277	0.233

7.1.8.2 Modifications of the Knocking Criterion of Franzke

Modifications of Franzke's approach were carried out by Spicher and Worret (2002). For the pre-reaction state I_K is valid

$$I_K = \frac{1}{6n}\frac{1}{c\ 10^{-3}}\int_{\varphi_{CS}}^{\varphi_K} p^a\, e^{\left(\frac{b}{T}\right)}\,\mathrm{d}\varphi = 1. \tag{7.106}$$

φ_{CS} thereby signifies the degree of crank angle at calculation start, i.e. 90°CA before the TDC. The parameters a, b and c can be taken from Table 7.6.

Table 7.6 describes the parameters adjusted to the respective model. According to Spicher and Worret, a parameter sensitivity contingent on the chosen modeling is given. Methods for a simple and an actual two-zone model and heat transfer relations acc. to Bargende and Woschni were thereby considered. With the method introduced, it is possible to determine the start of knocking combustion in modern SI engines up to ±2°CA.

In order to limit the range of deviations to a tolerable level, the 75% mfb point (φ_{75}) must be taken into consideration according to Spicher and Worret. We thereby utilize – as in the conversion of heat release rates – an additional reference point, for which all parameters must be known. The formal relationship reads

$$I_K = I_{K,ref}\left(\frac{\varphi_{75}+6}{\varphi_{75,ref}+6}\right)^{a_{IK}}. \tag{7.107}$$

In the same way, there are also corrections for the point position of combustion (φ_{CCD}) and the air-fuel ratio λ. The following equation shows the relations.

$$K = K_{ref}\left(\frac{\varphi_{CCD}+8}{\varphi_{CCD,ref}+8}\right)^{a_K}\left(\frac{b_K - c_K\,\lambda}{b_K - c_K\,\lambda_{ref}}\right). \tag{7.108}$$

As opposed to Franzke, Spicher and Worret calculate the gas states in the fresh gas by means of a two-zone model (see Sect. 7.2), which in the simplest case assumes an adiabatic calculation of temperature in the fresh gas zone.

Spicher and Worret also provide an equation method for knocking probability:

$$kp = 100\left(a_{kp} + b_{kp}\,\ln\left(\frac{\varphi_{KBR,\max}}{2}\right)\right)^{\varphi_{KBR}}. \tag{7.109}$$

In it, φ_{KBR} signifies the degree of crank angle, which results from a superimposition of the scatter bands of φ_E and φ_K (see Fig. 7.16).

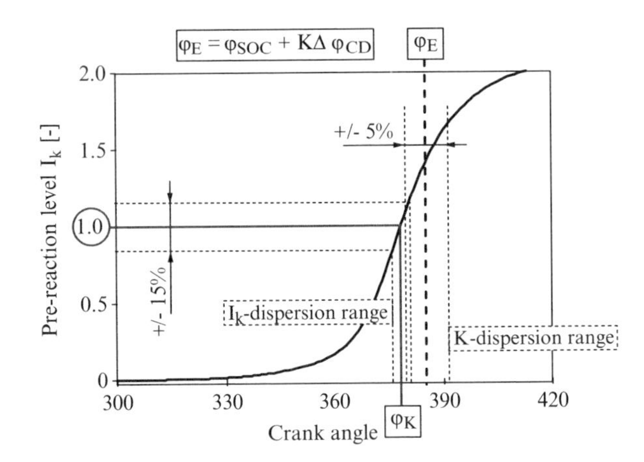

Fig. 7.16 Effect of amount of fluctuation of I_K and K

7.1.9 *Internal Energy*

The internal energy or enthalpy of the gas in the cylinder must be calculated as a further term of the first law of thermodynamics. The gas composition is usually assumed to be homogeneous.

7.1.9.1 Component Model

One possible approach in calculating the internal energy is the description via a mixture of single components of the gas, which are each considered as an ideal gas. The internal energy of each single component of the gas (initial products like O_2, N_2, etc. or reaction products CO_2, H_2O) can be calculated separately, since their standard formation enthalpies, the reaction enthalpies, as well as the molar heat are in tabular reference works, e.g. NIST JANAF (1993). Knowing the respective fraction of these individual components, one can then calculate the total internal energy of the gas. One thereby considers the cylinder charge as an ideal mixture of ideal gasses, for which the thermal state equation is valid

$$p_i V = n_i \tilde{R} T. \tag{7.110}$$

With

$$\left.\begin{aligned} p_i &: \text{ partial pressure} \\ n_i &: \text{ mass amount} \end{aligned}\right\} \text{ of the component } i$$

and the caloric state equation

$$\tilde{u}_i = \tilde{u}_i(\tilde{v}_i, T),$$

$$\tilde{h}_i = \tilde{h}_i(p_i, T).$$

For the total internal energy of the cylinder charge, we thus obtain the expression

$$U = \sum_{i=1}^{k} n_i \, \tilde{u}_i \, (\tilde{v}_i, T). \tag{7.111}$$

The only difficulty is that the fractions of the single components must be determined via the chemical reactions. It is thereby at first irrelevant which fuel one considers and whether the reaction progresses with an excess of air (hyperstoichiometrically) or a shortage of air (hypostoichiometrically).

For the sake of simplicity, we will first limit ourselves to stoichiometric operation. For 1 kmol of any fuel $C_xH_yO_z$ and under the assumption that the combustion air consists of 21% oxygen and 79% nitrogen, the following mass amounts result

$$C_xH_yO_z + \left(x + \frac{y}{4} - \frac{z}{2}\right)\left(O_2 + \frac{79}{21} N_2\right) \rightarrow$$
$$x\,CO_2 + \frac{y}{2} H_2O + \frac{79}{21}\left(x + \frac{y}{4} - \frac{z}{2}\right) N_2, \tag{7.112}$$

$$n_{CO_2} = x \left[\frac{\text{kmol CO}_2}{\text{kmol Fuel}}\right], \tag{7.113}$$

$$n_{H_2O} = \frac{y}{2}\left[\frac{\text{kmol H}_2\text{O}}{\text{kmol Fuel}}\right] \tag{7.114}$$

and

$$n_{N_2} = \frac{79}{21}\left(x + \frac{y}{4} - \frac{z}{2}\right)\left[\frac{\text{kmol N}_2}{\text{kmol Fuel}}\right]. \tag{7.115}$$

In the case of the fuel, for example, the amount of sulfur can also enter into the chemical reaction. For lean operation, pure air must also be considered in addition to the stoichiometric combustion gasses. Contingent on the combustion process under investigation, the evaporated fuel must be calculated as a further component (e.g. during the compression phase or under sub-stoichiometric operation). This is especially the case in the SI engine, since according to the definition of diesel combustion the fuel is usually only added to the system just as it is burning.

Although, compared with polynomial methods, table values from data banks are no longer required considering the performance of contemporary computers, they are still in wide use because of their clarity. In Heywood (1988) we find polynomial methods for the most important species. These refer to the JANAF tables.

Valid for molar enthalpy is

$$\tilde{h}_i = RT\left(a_{i,1} + \frac{a_{i,2}}{2}T + \frac{a_{i,3}}{3}T^2 + \frac{a_{i,4}}{4}T^3 + \frac{a_{i,5}}{5}T^4 + \frac{a_{i,6}}{T}\right)\left[\frac{\text{kJ}}{\text{kmol}}\right]. \tag{7.116}$$

For the molar internal energy results

$$\tilde{u}_i = \tilde{h}_i - \tilde{R}T. \tag{7.117}$$

The specific internal energy is obtained by means of division of the molar mass of the respective component

$$u_i = \frac{\tilde{u}_i}{M_i}. \tag{7.118}$$

The parameters for varying substances are given in Table 7.7.

There is also a description by means of polynomial methods for fuel vapor according to Heywood (1988). A consideration of the components, as in the case of gas components, does not exist in the case of fuels. For this, the evaporation behavior of the particular components is too complex. Nevertheless, most of the usual fuels are included. Valid is

$$\tilde{h}_f = \left(A_{f,1}\vartheta + \frac{A_{f,2}}{2}\vartheta^2 + \frac{A_{f,3}}{3}\vartheta^3 + \frac{A_{f,4}}{4}\vartheta^4 - \frac{A_{f,5}}{\vartheta} + A_{f,6} + A_{f,8}\right) 4186.6 \left[\frac{\text{kJ}}{\text{kmol}}\right]. \tag{7.119}$$

The reference temperature for this amounts to 273.15 K. The molar internal energy and the specific internal energy also amount to

$$\tilde{u}_f = \tilde{h}_f - \tilde{R}T, \tag{7.120}$$

$$u_f = \frac{\tilde{u}_f}{M_f}. \tag{7.121}$$

The parameters for the fuels investigated can be found in Table 7.8.

We should make brief mention at this point of the properties of diesel and gasoline fuels. The data for the molar fraction of carbon x and of hydrogen y can be taken from Table 7.9. From the molar mass of the fuel with the corresponding molar fractions of C and H results a mass fraction c for carbon and h for hydrogen for both fuels. As one can easily see, the mass fractions for both fuels are very close to each other.

The calculated molar mass of the fuel amounts to

$$M_{\text{Kst}} = xM_{\text{C}} + \frac{y}{2}M_{\text{H}_2} + \frac{z}{2}M_{\text{O}_2} \quad \left[\frac{\text{kg Fuel}}{\text{kmol Fuel}}\right]. \tag{7.122}$$

For the oxygen balance results

$$\begin{aligned} n_{\text{O}_2,\min} = c &\left[\frac{\text{kg C}}{\text{kg Fuel}}\right] \frac{1}{M_{\text{C}}} \left[\frac{\text{kmol C}}{\text{kg C}}\right] 1 \left[\frac{\text{kmol O}_2}{\text{kmol C}}\right] \\ + h &\left[\frac{\text{kg H}_2}{\text{kg Fuel}}\right] \frac{1}{M_{\text{H}_2}} \left[\frac{\text{kmol H}_2}{\text{kg H}_2}\right] \frac{1}{2} \left[\frac{\text{kmol O}_2}{\text{kmol H}_2}\right] \\ - o &\left[\frac{\text{kg O}_2}{\text{kg Fuel}}\right] \frac{1}{M_{\text{O}_2}} \left[\frac{\text{kmol O}_2}{\text{kg O}_2}\right] 1 \left[\frac{\text{kmol O}_2}{\text{kmol O}_2}\right]. \end{aligned} \tag{7.123}$$

Table 7.7 Coefficients for the molar enthalpy of gases, acc. to Heywood (1988)

Components	Temp.- range [K]	$a_{i,1}$	$a_{i,2}$	$a_{i,3}$	$a_{i,4}$	$a_{i,5}$	$a_{i,6}$	Molar mass
CO_2	1,000–5,000	0.44608(+1)	0.30982(−2)	−0.12393(−5)	0.22741(−9)	−0.15526(−13)	−0.48961(+5)	−0.98636(0)
	300–1,000	0.24008(+1)	0.87351(−2)	−0.66071(−5)	0.20022(−8)	0.63274(−15)	−0.48373(+5)	0.96951(+1)
H_2O	1,000–5,000	0.27168(+1)	0.29451(−2)	−0.80224(−6)	0.10227(−9)	−0.48472(−14)	−0.29906(+5)	0.66306(+1)
	300–1,000	0.40701(+1)	−0.11084(−2)	0.41521(−5)	−0.29637(−8)	0.80702(−12)	−0.30280(+5)	−0.32270(0)
CO	1,000–5,000	0.29841(+1)	0.14891(−2)	−0.57900(−6)	0.10365(−9)	−0.69354(−14)	−0.14245(+5)	0.63479(+1)
	300–1,000	0.37101(+1)	−0.16191(−2)	0.36924(−5)	−0.20320(−8)	0.23953(−12)	−0.14356(+5)	0.29555(+1)
H_2	1,000–5,000	0.31002(+1)	0.51119(−3)	0.52644(−7)	−0.34910(−10)	0.36945(−14)	−0.87738(+3)	−0.19629(+1)
	300–1,000	0.30574(+1)	0.26765(−2)	−0.58099(−5)	0.55210(−8)	−0.18123(−11)	−0.98890(+3)	−0.22997(+1)
O_2	1,000–5,000	0.36220(+1)	0.73618(−3)	−0.19652(−6)	0.36202(−10)	−0.28946(−14)	−0.12020(+4)	0.36151(+1)
	300–1,000	0.36256(+1)	−0.18782(−2)	0.70555(−5)	−0.67635(−8)	0.21556(−11)	−0.10475(+4)	0.43053(+1)
N_2	1,000–5,000	0.28963(+1)	0.15155(−2)	−0.57235(−6)	0.99807(−10)	−0.65224(−14)	−0.90568(+3)	0.61615(+1)
	300–1,000	0.36748(+1)	0.12082(−2)	0.23240(−5)	−0.63218(−9)	−0.22577(−12)	−0.10612(+14)	0.23580(+1)
OH	1,000–5,000	0.29106(+1)	0.95932(−3)	−0.19442(−6)	0.13757(−10)	0.14225(−15)	0.39354(+4)	0.54423(+1)
NO	1,000–5,000	0.31890(+1)	0.13382(−2)	−0.52899(−6)	0.95919(−10)	−0.64848(−14)	0.98283(+4)	0.67458(+1)
O	1,000–5,000	0.25421(+1)	−0.27551(−4)	−0.31028(−8)	0.45511(−11)	−0.43681(−15)	0.29231(+5)	0.49203(+1)
H	1,000–5,000	0.25(+1)	0.0	0.0	0.0	0.0	0.25472(+5)	−0.46012(0)

Table 7.8 Coefficients for the molar enthalpy of fuels, acc. to Heywood (1988)

Fuel	$A_{f,1}$	$A_{f,2}$	$A_{f,3}$	$A_{f,4}$	$A_{f,5}$	$A_{f,6}$	$A_{f,8}$	Molar mass
Methane	−0.29149	26.327	−10.610	1.5656	0.16573	−18.331	4.3000	16.04
Propane	−1.4867	74.339	−39.065	8.0543	0.01219	−27.313	8.852	44.10
Hexane	−20.777	210.48	−164.125	52.832	0.56635	−39.836	15.611	86.18
Isooctane	−0.55313	181.62	−97.787	20.402	−0.03095	−60.751	20.232	114.2
Methanol	−2.7059	44.168	−27.501	7.2193	0.20299	−48.288	5.3375	32.04
Ethanol	6.990	39.741	−11.926	0	0	−60.214	7.6135	46.07
Normal	−24.078	256.63	−201.68	64.750	0.5808	−27.561	17.792	114.8
Premium	−22.501	227.99	−177.26	56.048	0.4845	−17.578	15.235	106.4
Diesel	−9.1063	246.97	−143.74	32.329	0.0518	−50.128	23.514	148.6

Table 7.9 Properties of diesel and gasoline fuels

	Diesel		Otto (Premium)	
	kmol/kmol fuel	kg/kg fuel	kmol/kmol fuel	kg/kg fuel
C	$x = 10.8$	$c = \mathbf{0.874}$	$x = 7.76$	$c = \mathbf{0.877}$
H	$y = 18.7$	$h = \mathbf{0.126}$	$y = 13.1$	$h = \mathbf{0.123}$
O	$z = 0$	$o = 0$	$z = 0$	$o = 0$
Molar mass	148.3 kg fuel/kmol fuel		106.2 kg fuel/kmol fuel	
Mimimal air required	14.33		14.26	
Lower heating value	42,600 kJ/kg fuel		42,900 kJ/kg fuel	

With the fraction of 21% oxygen in the combustion air and the molar mass for air of 28.85 kg/kmol, the minimal air requirement amounts to

$$L_{\min} = \frac{n_{O_2,\min}}{0.21} M_{air}. \tag{7.124}$$

In rich operation incomplete reactions occur, for the determination of these components the so-called water-gas equilibrium is utilized. It describes, contingent on the temperature, the equilibrium constant of the most important reaction (CO_2, CO) under oxygen deficiency and dissociation

$$CO_2 + H_2 \overset{K_p}{\longleftrightarrow} CO + H_2O \tag{7.125}$$

$$K_p = \frac{p_{CO}\, p_{H_2O}}{p_{CO_2}\, p_{H_2}}. \tag{7.126}$$

Values between 3.5 and 3.7 can be set for the equilibrium constant, which approximately corresponds to a usual combustion temperature of 1,800 K.

7.1.9.2 The Justi Method

In the 1930s, investigations were carried out in order to describe the internal energy and enthalpy of combustion gasses in terms of the quantities of temperature, pressure, and gas composition. The gas composition is thereby expressed as the so-called air-fuel ratio. In describing diesel engine combustion gasses, this procedure is correct as long as one stays within the lean range ($\lambda > 1$). The composition of the combustion air is assumed to be constant; a varying water content cannot be taken into consideration. Fuel mixtures and alternative fuels cannot be represented with these methods. The relation is valid, strictly speaking, only for a fixed H/C-ratio. This is however very similar to diesel and gasoline fuel. The dissociation of the gas can also not be considered. Under dissociation we understand the change in otherwise constant equilibrium constants of chemical reactions at very high temperatures, e.g. over 2,000 K. Since a single-zone combustion chamber model is used as a rule only for calculating the cylinder pressure and the caloric mean temperature and with that for further thermodynamic quantities like performance, efficiency, and heat flows, it is usually totally sufficient to represent the internal energy as a function of temperature and the air-fuel ratio, for which Justi (1938) gives the following empirical function

$$\begin{aligned} u(T,\lambda) = 0.1445 \Bigg[& 1356.8 + \left(489.6 + \frac{46.4}{\lambda^{0.93}}\right)(T - T_{ref})\,10^{-2} + \left(7.768 + \frac{3.36}{\lambda^{0.8}}\right) \\ & \times (T - T_{ref})^2\,10^{-4} - \left(0.0975 + \frac{0.0485}{\lambda^{0.75}}\right)(T - T_{ref})^3\,10^{-6}\Bigg] \text{ in } \left[\frac{\text{kJ}}{\text{kg}}\right]. \end{aligned} \tag{7.127}$$

This is a polynomial method, where $T_{ref} = 273.15$ K is valid for the reference temperature.

7.1.9.3 The Zacharias Method

Zacharias (1966) also suggests a polynomial method, but also takes into consideration the pressure of the combustion gas. Otherwise the above restrictions of Justi's relation (1938) are valid for this as well

$$u(T,p,\lambda) = \left[-A\,\frac{\pi}{\vartheta^2}\,e^{\frac{D}{\vartheta^2}}\left(1 + 2\,\frac{D}{\vartheta}\right) + \sum_{i=0}^{6}\left[FA(i)\,\vartheta^i\right] - 1\right] R_0\,T \text{ in } \left[\frac{\text{kJ}}{\text{kmol}}\right] \tag{7.128}$$

$$r = \frac{\lambda - 1}{\lambda + \frac{1}{L_{\min}}} \tag{7.129}$$

$$R_0 = \frac{\tilde{R}}{28.89758 + 0.06021\,r} \tag{7.130}$$

$$\pi = \frac{p}{0.980665} \tag{7.131}$$

$$\vartheta = \frac{T}{1.000\ \text{K}} \tag{7.132}$$

$$A = 0.000277105 - 0.0000900711\,r \tag{7.133}$$

$$D = 0.008868 - 0.006131\,r \tag{7.134}$$

$$\begin{aligned}
FA\,(0) &= 3.514956 - 0.005026\,r\\
FA\,(1) &= 0.131438 - 0.383504\,r\\
FA\,(2) &= 0.477182 - 0.185214\,r\\
FA\,(3) &= -0.287367 - 0.0694862\,r\\
FA\,(4) &= 0.0742561 + 0.016404110\,r\\
FA\,(5) &= -0.00916344 - 0.00204537\,r\\
FA\,(6) &= 0.000439896 - 0.000101610\,r
\end{aligned} \tag{7.135}$$

Despite the supposed increase in precision via the consideration of the pressure in determining the internal energy, in using the equations of Zacharias, exactly because of this pressure contingency, an iterative calculation of the internal energy becomes necessary, which costs additional time. The contingency of internal energy on the temperature and gas composition for the Justi method is shown in Fig. 7.17. If we elect a very high air-fuel ratio, we obtain the graphs for pure air.

We should briefly mention here the various ways of representing the composition of the combustion gas. In this case, clear differences result between diesel

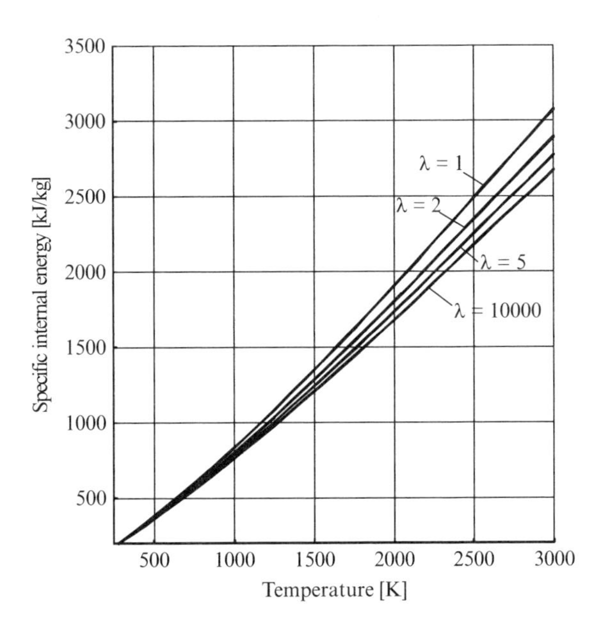

Fig. 7.17 Internal energy, acc. to Justi (1938)

engine and SI engine model representations. While in the case of diesel engine models, the injected fuel normally plays no role, as it is added proportionally to the heat release rate, in the case of SI model representations the fuel must be taken into consideration because of the evaporation heat. The following sketch shows the composition in the case of the diesel engine. Only the states "fuel burned" as well as "air burned" and "air unburned" exist, since in the diesel engine we always assume lean operation.

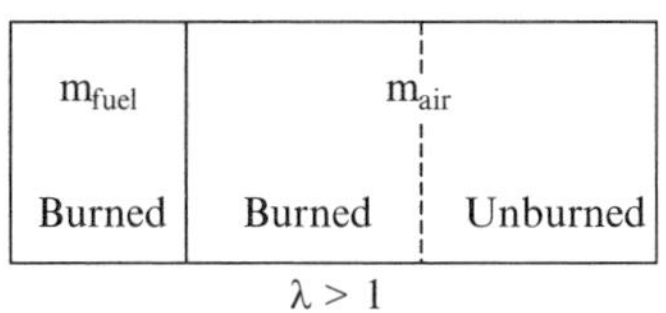

If the internal energy is described according to the methods of Justi or Zacharias, the total differential for the internal energy and the partial differential for the specific internal energy must be formed. From this follows

$$\frac{\mathrm{d}U}{\mathrm{d}t}=\frac{\mathrm{d}(m\,u)}{\mathrm{d}t}=m\,\frac{\mathrm{d}u}{\mathrm{d}t}+u\,\frac{\mathrm{d}m}{\mathrm{d}t}=m\left(\frac{\partial u}{\partial T}\,\frac{\mathrm{d}T}{\mathrm{d}t}+\frac{\partial u}{\partial p}\,\frac{\mathrm{d}p}{\mathrm{d}t}+\frac{\partial u}{\partial \lambda}\,\frac{\mathrm{d}\lambda}{\mathrm{d}t}\right)+u\,\frac{\mathrm{d}m}{\mathrm{d}t}. \quad (7.136)$$

The partial differentials can be calculated with the help of Justi's or Zacharias's relations. The air-fuel ratio is defined as

$$\lambda=\frac{m_{air}}{m_{fuel}\,L_{\min}}. \quad (7.137)$$

The change in the air-fuel ratio amounts to

$$\frac{\mathrm{d}\lambda}{\mathrm{d}t}=\frac{1}{m_{fuel}^{2}\,L_{\min}}\left(m_{fuel}\,\frac{\mathrm{d}m_{air}}{\mathrm{d}t}-m_{air}\,\frac{\mathrm{d}m_{fuel}}{\mathrm{d}t}\right). \quad (7.138)$$

The left side of the first law of thermodynamics is solved from (7.136) and integrated after the change in temperature. The term $\mathrm{d}p/\mathrm{d}t$ can be calculated from the state equation for ideal gas in differential form

$$V\,\frac{\mathrm{d}p}{\mathrm{d}t}+p\,\frac{\mathrm{d}V}{\mathrm{d}t}=m\,R\,\frac{\mathrm{d}T}{\mathrm{d}t}+R\,T\,\frac{\mathrm{d}m}{\mathrm{d}t}+m\,T\,\frac{\mathrm{d}R}{\mathrm{d}t}. \quad (7.139)$$

Because of the fact that in the case of Zacharias the gas constant also depends on the quantities of temperature, pressure, and the air-fuel ratio, under complete differentiation of $\mathrm{d}R/\mathrm{d}t$, terms with $\mathrm{d}T/\mathrm{d}t$ and $\mathrm{d}p/\mathrm{d}t$ arise, which is why an iterative process or setting the differentials equal to zero becomes necessary.

ξ_{fuel} m_{fuel} Evaporated	ξ_{air} m_{air}	$\xi_{exhaust\ gas}$ $m_{exhaust\ gas}$ Burned $\lambda = 1$

$\xi_{air}+\xi_{exhaust\ gas}+\xi_{fuel}=1$

In the case of the SI engine, we subdivide in the same manner three ranges, as the sketch above shows. However, stoichiometrically burned exhaust gas, combustion air, and evaporated fuel exist here as components.

$$\xi_{exhaust\ gas} = \frac{m_{exhaust\ gas}}{m_{total}};\ \xi_{air} = \frac{m_{air}}{m_{total}};\ \xi_{fuel,evap.} = \frac{m_{fuel,evap.}}{m_{total}}, \tag{7.140}$$

$$\xi_{exhaust\ gas} + \xi_{air} + \xi_{fuel,evap.} = 1. \tag{7.141}$$

The total internal energy is to be calculated from the single portions of internal energy of the three areas

$$U = u_{exhaust\ gas}\, m_{exhaust\ gas} + u_{air}\, m_{air} + u_{fuel,evap.}\, m_{fuel,evap.} \tag{7.142}$$

The differential of the internal energy becomes

$$\frac{\mathrm{d}U}{\mathrm{d}t} = \frac{\mathrm{d}(u_{exhaust\ gas}\, m_{exhaust\ gas})}{\mathrm{d}t} + \frac{\mathrm{d}(u_{\mathrm{air}}\, m_{air})}{\mathrm{d}t} + \frac{\mathrm{d}(u_{fuel,evap.}\, m_{fuel,evap.})}{\mathrm{d}t}. \tag{7.143}$$

Since we are concerned in the case of the single shares with "pure" components, the specific internal energies can be calculated either via classical polynomial methods or component for component. For the change in the mass of the single components results

$$\begin{aligned}
\frac{\mathrm{d}m_{fuel,burned}}{\mathrm{d}t} &= \frac{\mathrm{d}Q_{fuel}}{\mathrm{d}t}\frac{1}{I_{hv}} \\
\frac{\mathrm{d}m_{fuel,evap.}}{\mathrm{d}t} &= -\frac{\mathrm{d}m_{fuel,burned}}{\mathrm{d}t} + \frac{\mathrm{d}m_{fuel,evap.,new}}{\mathrm{d}t} \\
\frac{\mathrm{d}m_{air}}{\mathrm{d}t} &= \frac{\mathrm{d}m_{fuel,burned}}{\mathrm{d}t} L_{\min} \\
\frac{\mathrm{d}m_{exhaust\ gas}}{\mathrm{d}t} &= \frac{\mathrm{d}m_{fuel,burned}}{\mathrm{d}t} (L_{\min} + 1)\ .
\end{aligned} \tag{7.144}$$

For the port fuel injection SI engine, the evaporated fuel is already found in the cylinder. In the case of the direct injection SI engine, the fuel is injected during either gas exchange or compression and has to evaporate.

7.2 The Two-Zone Cylinder Model

7.2.1 Modeling the High Pressure Range According to Hohlbaum

In the following, a two-zone model from the zero-dimensional model class will be considered in more detail. This model has gained a certain importance for the calculation of NO_x formation, whereby the heat release rate is given beforehand.

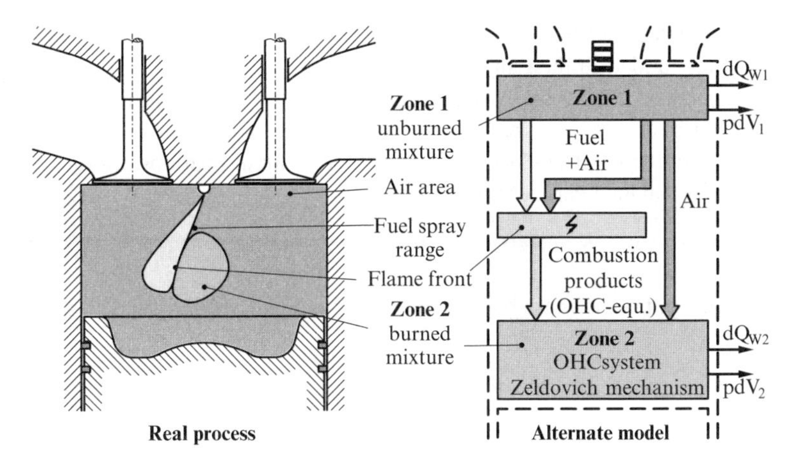

Fig. 7.18 Two-zone model for diesel combustion

In this model, the combustion chamber is divided into two zones, which one should imagine as being divided by the flame front. Strictly speaking, the flame front itself represents a zone onto itself, i.e. the third zone. However, because simple assumptions are made about the reaction kinetics of the flame front and no balance equations are solved, it is usually not considered to be an independent zone, and the designation "two-zone model" has become customary.

An extensive description of this model can be found in Hohlbaum (1992) and in Merker et al. (1993). In the following, only the essential traits of the model will be explained. The basic idea of the model is schematically shown in Fig. 7.18.

Zone 1 contains unburned mixture, i.e. air and fuel, which will in the following be designated as unburned zone and described with the state quantities p, V_1, T_1, $n_{1,i}$, and λ_1.

Zone 2 should contain burned mixture, or more precisely speaking, incompletely oxidized fuel, and will in the following be designated as burned zone and described with the state quantities p, V_2, T_2, $n_{2,i}$, and λ_2. In zone 2, "secondary oxidation" occurs. For this, reaction kinetic models are necessary. Thermal NO_x formation also takes place in zone 2, which is described by the Zeldovich mechanism, see Sect. 6.5.

The flame front separates both of these zones. It is assumed to be infinitely thin and without mass. "Primary oxidation" takes place in the flame front until OHC equilibrium, i.e. the OHC components $O^\bullet$, $H^\bullet$, O_2, H_2, H_2O, CO, CO_2 and $O^\bullet H$ are in chemical equilibrium in the flame front.

By means of a balancing of the relevant mass and energy flows, we obtain equations for the mass and energy in both zones. Valid thereby for the mass balance in zone 1 is

$$\mathrm{d}m_1 = \mathrm{d}m_f - \mathrm{d}m_{1F} - \mathrm{d}m_{12} \tag{7.145}$$

with $\mathrm{d}m_B$: injected fuel mass

$\mathrm{d}m_{1F}$: mass added to the flame front (fuel + air) and

$\mathrm{d}m_{12}$: air mass passing by the flame front and added "directly" to zone 2 (must appear because of $\lambda_1 > 1$); it can also go through the flame front, however without participating in the reactions occurring there and without heat absorption.

The energy balance can be written in this way

$$\mathrm{d}U_1 = \mathrm{d}m_f\, h_f - \mathrm{d}m_{1F}\, h_{1F} - \mathrm{d}m_{12}\, h_{12} + \mathrm{d}Q_1 - p\,\mathrm{d}V_1. \tag{7.146}$$

Analogously, for the mass balance of zone 2 is valid

$$\mathrm{d}m_2 = \mathrm{d}m_{F2} + \mathrm{d}m_{12} \tag{7.147}$$

with $\mathrm{d}m_{F2}$: components of the OHC equilibrium.

The energy law reads:

$$\mathrm{d}U_2 = \mathrm{d}m_{F2}\, h_{F2} + \mathrm{d}m_{12}\, h_{12} + \mathrm{d}Q - p\,\mathrm{d}V_2. \tag{7.148}$$

Because the flame front is assumed to be without mass, it is furthermore valid

$$\mathrm{d}m_{1F} = \mathrm{d}m_{F2} = \mathrm{d}m_F. \tag{7.149}$$

The specific enthalpy of the mass transported from the flame front to zone 2 $\mathrm{d}m_{F2}$ is reaction-enthalpy $\Delta_R\, h$ larger than that which is transported from zone 1 into the flame front, i.e.

$$h_{F2} = h_{1F} + \Delta_R\, h. \tag{7.150}$$

The terms $\mathrm{d}Q_1$ and $\mathrm{d}Q_2$ describe the energy losses of both zones via heat transfer as a result of radiation and convection to the wall limiting the combustion chamber. The total transferred heat

$$\mathrm{d}Q = \mathrm{d}Q_1 + \mathrm{d}Q_2 = \alpha\, A\,(T_w - T)\,\mathrm{d}t \tag{7.151}$$

can, for example, be calculated again according to Woschni's method, whereby T is the energetic mean temperature, which can be determined for the caloric mixture from the relation

$$(m_1 + m_2)\; u\,(T) = m_1\, u_1\,(T_1) + m_2\, u_2\,(T_2) \tag{7.152}$$

However, for the subdivision of the total transferred heat $\mathrm{d}Q$ in $\mathrm{d}Q_1$ and $\mathrm{d}Q_2$, we require a model, because the surface of the flame front and thus the size of the surface of both zones is not defined in the two-zone model. Hohlbaum (1992) proposes for this distribution the following relation

$$\frac{\mathrm{d}\,Q_1}{\mathrm{d}\,Q_2} = \left(\frac{m_1}{m_2}\right)^2 \frac{T_1}{T_2} \tag{7.153}$$

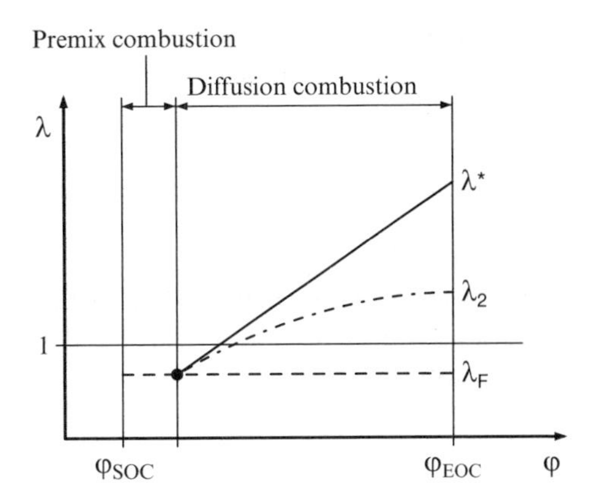

Fig. 7.19 Air ratio for premixed and diffusion combustion

On the one hand, this approach takes into consideration that zone 2 of the burned zone contributes more to the total heat loss because of the higher temperature T_2 than unburned zone 1. On the other hand, the method considers the fact that at the beginning of combustion, the mass of zone 2 and thus its contribution to heat transfer is minimal. Finally, the temporal progression of the bypass air mass flow $\dot{m}_{12}$ must still be determined. The quantity designated as mixture stoichiometry λ^* is defined as

$$\lambda^* = \frac{\mathrm{d}m_{1\,F,air} + \mathrm{d}m_{12}}{L_{\min}\,\mathrm{d}m_{f,\,1\,F}}. \tag{7.154}$$

In Fig. 7.19, the paths of the air-fuel ratios in the flame front and in zone 2, λ_F and λ_2, as well as the mixture stoichiometry λ^* are sketched over crank angle.

Air ratio λ_F in the flame front is assumed to be < 1 and temporally constant, thus $\lambda_F \neq f(\varphi)$, the mixture stoichiometry on the other hand is increasing in linear proportion to the crank angle, such that at the start of diffusion combustion $\lambda^* = \lambda_F$ and the excess air has at the end of diffusion combustion just completely intermixed with the unburned in zone 2.

Especially this assumption for the mixture stoichiometry makes clear the problematic nature of this simple zero-dimensional model; the lack of model depth (lacking physics) has to be substituted with more or less arbitrary assumptions.

7.2.2 *Modeling the High Pressure Phase According to Heider*

Heider (1996) has introduced another way to calculate the conditions in both zones. We distinguish thereby between two zones. In zone 1, which is designated as the reaction zone, energy conversion takes place. In the reaction zone, the air-fuel ratio λ_0 is assumed to be constant throughout the working cycle. The mass in the reaction zone is thus clearly fixed over the heat release rate. Zone 2 describes the unburned and thus sets the remaining volume of the combustion chamber. No combustion occurs.

As opposed to Hohlbaum's model, in the case of Heider, the results are based on the process calculation of the zero-dimensional single-zone model (see Sect. 7.1). These results can be calculated in advance with the working process calculation. The paths of the swept volume, of pressure, and of mass mean temperature as well as the cylinder mass, compression air-fuel ratio, heat release rate, and wall heat losses are assumed to be known. The following assumptions are valid as conditions of compatibility

$$V_1 + V_2 = V(\varphi), \tag{7.155}$$

$$m_1 + m_2 = m(\varphi), \tag{7.156}$$

$$p_1 = p_2 = p(\varphi), \tag{7.157}$$

Provided we know the air-fuel ratio λ_0, the mass in the reaction zone can be calculated

$$\lambda_0 = \frac{m_{air\,1}(\varphi)}{L_{\min}\, m_f(\varphi)} = \text{const.} \tag{7.158}$$

For the fuel mass is valid in the case of a known heat release rate and a known residual gas mass, from which the burned fuel $m_{fuel,0}$ can also be calculated

$$m_f(\varphi) = \frac{1}{I_{hv}} \int \frac{\mathrm{d}Q_f}{\mathrm{d}\varphi}\, d\varphi + m_{f\,0}. \tag{7.159}$$

With this, for the mass of zone 1 we have

$$m_1(\varphi) = m_{air\,1}(\varphi) + m_f(\varphi) = (\lambda_0 L_{\min} + 1)\, m_f(\varphi). \tag{7.160}$$

The ideal state equation is valid for both zones

$$\begin{aligned} p_1 V_1 &= m_1 R_1 T_1 \\ p_2 V_2 &= m_2 R_2 T_2. \end{aligned} \tag{7.161}$$

In the final analysis it must be determined which part of the energy released in the reaction zone is transferred to zone 2. This happens in the model representation essentially via turbulent mixing and less so via radiation and convection. For this, the following boundary conditions must be kept.

At the beginning of combustion, the temperature difference between both zones as a result of the high temperature difference between the flame and the unburned mass is maximal. Furthermore, this temperature difference is contingent on heat release via combustion. The turbulent mixing of both zones leads, with progressing combustion, to a lowering of temperature in the reaction zone and an increase of temperature in the zone with the unburned substance. At the end of the combustion, the temperature difference is around zero, since both zones are then completely intermixed.

These considerations lead to the following empirical method for the temperature difference between both of these zones

$$T_1(\varphi) - T_2(\varphi) = B(\varphi)\, A^*. \tag{7.162}$$

For the function $B(\varphi)$ is valid

$$B(\varphi) = 1 - \frac{\int_{\varphi_{SOC}}^{\varphi} [p(\varphi) - p_0(\varphi)]\, m_1 \, d\varphi}{\int_{\varphi_{SOC}}^{\varphi_{EO}} [p(\varphi) - p_0(\varphi)]\, m_1 \, d\varphi}. \tag{7.163}$$

As in the determination of the heat transfer coefficient according to Woschni (1970), here too, the difference between the cylinder pressure $p(\varphi)$ and the theoretical pressure of the motored engine $p_0(\varphi)$ is utilized for the consideration of the influence of combustion. A^* describes the temperature level in the reaction zone as the start of combustion. Detailed investigations have shown that minimal adjustments of the A^* value and the air-fuel ratio λ_0 are necessary for varying engines and combustion processes. For small to medium-sized diesel engines possessing an intake swirl

$$\lambda_0 = 1.0,$$

$$A^* = A\, \frac{1.2 + (\lambda_{gl} - 1.2)^{C_{gl}}}{2.2\, \lambda_0}. \tag{7.164}$$

is applicable. A is an engine-specific factor, which has to be determined once for the respective engine. For C_{gl} is valid

$C_{gl} = 0.15$ for engines with 4-valve technology and central injection nozzle
$C_{gl} = 0.07$ for engines with 2-valve technology and a side injection nozzle
λ_{gl} describes the global air-fuel ratio.
Valid for large diesel engines without intake swirl is

$$\lambda_0 = 1.03$$

and

$$\lambda_0 = 1.03 - 0.24\, \frac{EGR}{100}$$

in the case of external exhaust gas recycling. In large diesel engines, the A^* value can be assumed to be constant

$$A^* = A = \text{const.}$$

Although this model was first developed only for the diesel engine, it can also be applied to SI engines with favorable results. Then

$$\lambda_0 = \lambda_{gl} \quad \text{and} \quad A^* = \text{const.}$$

is valid.

Figure 7.20 shows a characteristic temperature curve for a high speed diesel engine with approx. 4 l cylinder volume at a speed of 1,400 rpm and an actual load of 8 bar, as can be calculated with this model. In Table 7.10 below, typical A values for various engines are summarized.

Despite the obvious empirical nature of this model, it offers a very good basis for nitrogen oxide calculation described in the following. Moreover, it is convincing in

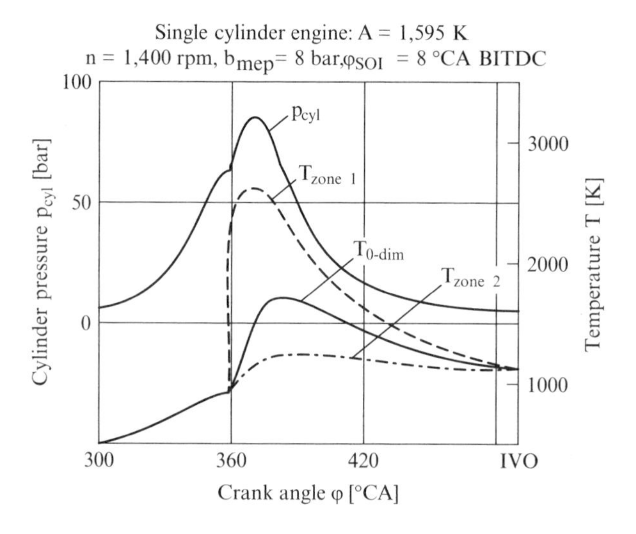

Fig. 7.20 Temperature curve acc. to Heider (1996)

Table 7.10 Typical A values

Engine					
Bore [mm]	Stroke [mm]	Cycle	Compression ratio	rated speed [rpm]	A value [K]
79.5	95.5	4	19.5	4,000	1,650
128	142	4	16	2,100	1,740
160	180	4	14	1,500	1,580
480	600	4	14	450	1,650
580	1,700	2	17	127	1,655

its simplicity. With this model, no assumptions must be made regarding the distribution of the wall heat losses to the two zones, which can only be determined as a whole. The temperatures in both zones can be determined by means of a simple empirical function. The calculation model is appealing because of its short calculation times.

7.2.3 Results of NO_x Calculation with Two-Zone Models

With the two-zone models of Hohlbaum and Heider described in the preceding sections, nitrogen oxide emissions in the zone of the hot combustion products can be calculated. For this, we utilize the description via the so-called Zeldovich mechanism, which is thoroughly described in Sect. 6.5.1.

As an example, Fig. 7.21 shows the effect of the start of delivery timing from "early" to "late" on the rate of formation of NO calculated with the Hohlbaum model. At a start of delivery delay of 5°CA from 20°CA BTDC to 15°CA BTDC the peak temperature drops from about 2,680 K to 2,630 K, and the bulk temperature reaches clearly lower values. This leads to a decrease in the amount of NO formed, which lowers from about 1,500 to 1,200 ppm.

For the operating point described in Fig. 7.20, the temperature curve in the hot zone according to Heider (1996) is represented in Fig. 7.22 on the left. Via this temperature curve, a NO formation rate and the NO concentration in the combustion chamber is adjusted. We recognize that NO formation is over very quickly and that only a minimal reverse reaction occurs. Heider utilizes reaction constants of Pattas, organized in Table 6.1 (Sect. 6.5.1), for the Zeldovich mechanism. For this engine, at a speed of 1,500 rpm, a comparison between measurement and calculation is given for injection time variation (Fig. 7.22, right).

The agreement is very good in this case, as it is in the case of variation of the charge air temperature (Fig. 7.23, left) and the exhaust gas recirculation rate (Fig. 7.23, right), which has a massive influence on NO formation.

We recognize that these very simple models are very much capable of describing reality not only according to tendency, but also quantitatively correctly. It is however of decisive importance that the heat release rate in the cylinder is accurately known. Results of transient calculations of nitrogen oxide emissions in a high speed diesel engine of a passenger car with precalculated heat release rates can be found in Sect. 10.5.4.

7.2.4 Modeling the Gas Exchange for a Two-Stroke Engine

Gas exchange calculation for the two-stroke engine is much more difficult than for the four-stroke engine, since, on the one hand, only a small amount of time is available for gas exchange and, on the other hand, the fresh gas flowing into the

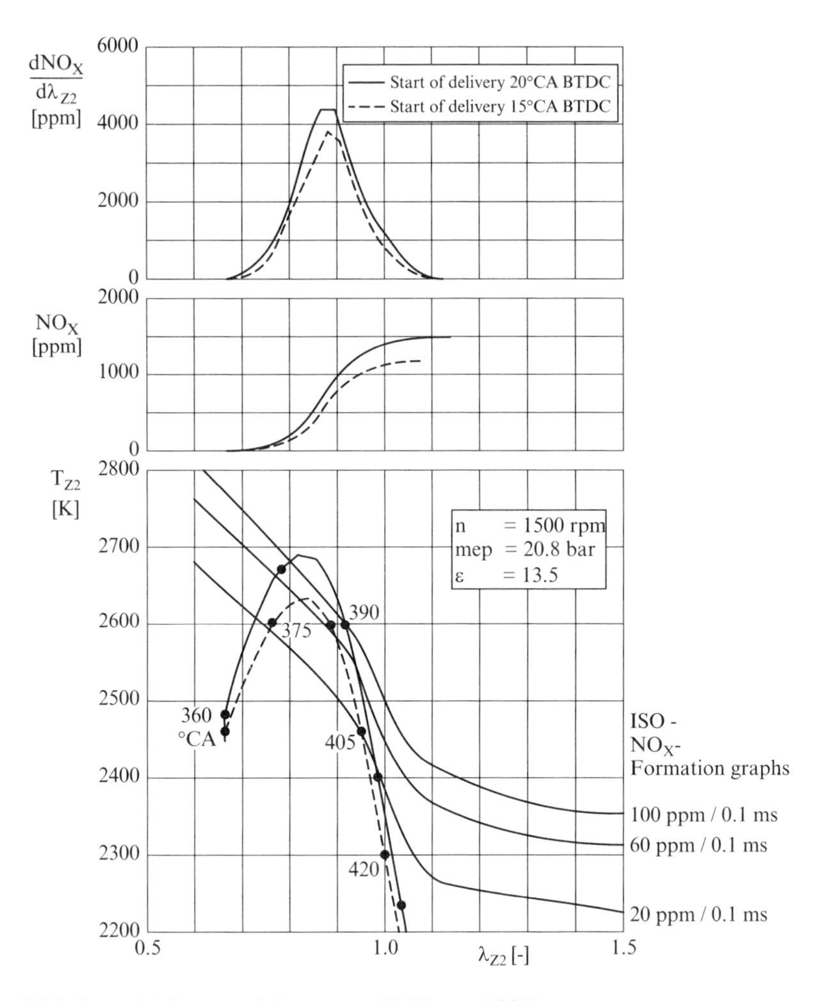

Fig. 7.21 Start of delivery variation, acc. to Hohlbaum (1992)

cylinder has to reduce the exhaust gas found in the cylinder without mixing with it. In literature a large number of models for loop scavenged as well as for longitudinally scavenged two-stroke engines is described. We are hereby dealing with two or three-zone models. Differences arise thereby between zones, in which fresh gas, a mixture between exhaust gas and fresh gas, or pure exhaust gas is found. A more exact list and description of these models can be found in Merker and Gerstle (1997).

The model described here proceeds from a two-zone approach. Since, practically speaking, two components – fresh gas and exhaust gas – are found in the cylinder during gas exchange, this fact must be accounted for with two zones. We will only go into the essential properties and descriptive equations of both zones and their

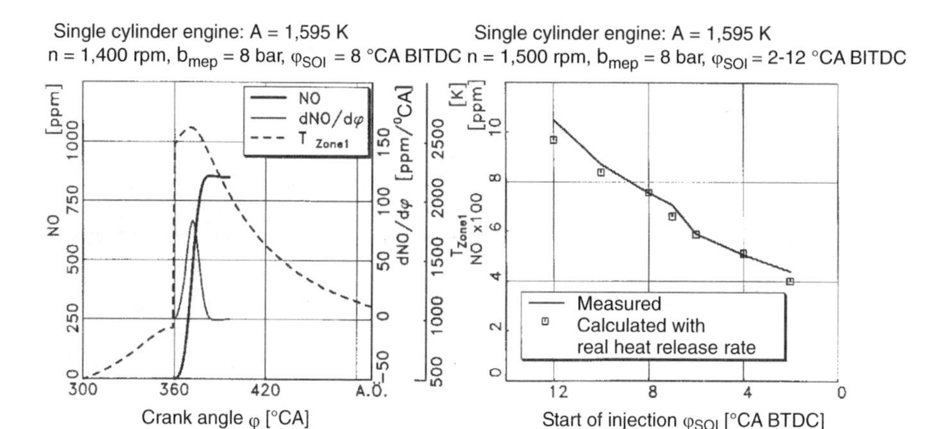

Fig. 7.22 Influence of start of injection on the NO formation rate

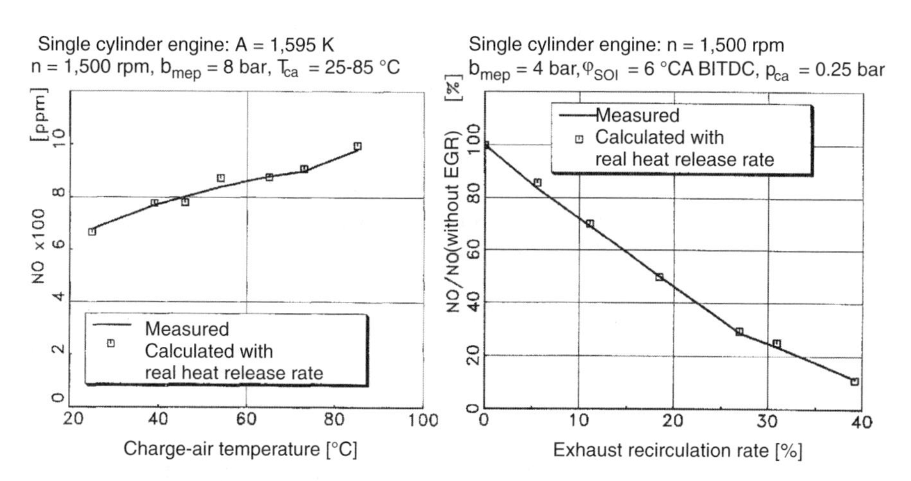

Fig. 7.23 Influence of the charge air temperature (*left*) and the exhaust gas recirculation rate (*right*) on the NO formation rate

interaction. In this model, displacement scavenging with low intake ports and an upper exhaust valve is used as a basis. Figure 7.24 shows a diagram for the two-zone model for the gas exchange calculation for a two-stroke engine, in which the gas mass in the cylinder is subdivided by an infinitely thin, impermeable yet shiftable horizontal membrane. An exchange of gas between the zones is thus impossible, as opposed to the model of Streit and Borman (1971).

At the beginning of gas exchange at exhaust opens – i.e. after combustion and expansion – a homogeneous mixture is in the cylinder. The exhaust gas mass found in the cylinder is subdivided at exhaust opens into a displacement and a mixing zone via a so-called scavenging factor, which is characteristic for the gas exchange properties of the engine. This scavenging factor can have values of

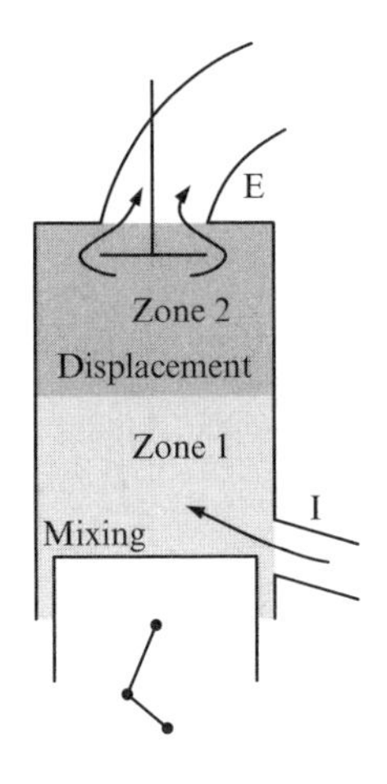

Fig. 7.24 Two-zone model for calculation of gas exchange in a two-stroke engine

0 to 1. A scavenging factor of 0 sets a pure mixture scavenging, the success of which is difficult to grade as a result of the constant ideal mixing of fresh gas and exhaust gas. A scavenging factor of 1 describes pure displacement scavenging, in which both zones cannot mix until the end of the gas exchange process. In this case, the mixing zone has no mass at the start of the calculation, which is why, after the beginning of the intake process, only pure fresh gas is mixed. An ideal mixture is always assumed for this zone, as in the case of the entire gas exchange of the four-stroke engine. The displacement zone in a gas exchange without mixing with the fresh gas flowing in through the intake is the first zone to be pushed out through the exhaust valve. The gas composition thus corresponds at every moment to the composition of the exhaust gas. According to the adjacent scavenging pressure ratio, the displacement zone can be only partially or completely expelled. In addition, part of the mixing zone can also be ejected. Towards the end of gas exchange, both zones, insofar as they still exist and are not completely expelled, must be returned to a homogeneously mixed state. The descriptive equations of the two-zone model for the gas exchange of a two-stroke engine are represented in the following, whereby the index 1 is used for the mixing zone and 2 for the displacement zone. If the mass of the displacement zone is completely expelled and thus only the mixing zone still exists, this is treated like a four-stroke engine. A subdivision of mass or heat flows to the single zones is then no longer necessary. The subdivision of cylinder mass at "exhaust opens" describes the following equation

$$m_1 = m\,(1 - SF), \quad m_2 = m\,SF. \tag{7.165}$$

For both zones, several basic couple conditions are applicable, which must be kept at every moment

$$m_1 + m_2 = m, \tag{7.166}$$

$$V_1 + V_2 = V, \tag{7.167}$$

$$p_1 = p_2 = p = \frac{m_1 R_1 T_1 + m_2 R_2 T_2}{V}. \tag{7.168}$$

The mass balance for both zones reads

$$\frac{dm_1}{d\varphi} = \frac{dm_i}{d\varphi} \tag{7.169}$$

$$\frac{dm_2}{d\varphi} = \frac{dm_o}{d\varphi} \tag{7.170}$$

According to the first law of thermodynamics, for both zones results

$$\frac{dU_1}{d\varphi} = \frac{dW_1}{d\varphi} + \frac{dQ_{w,1}}{d\varphi} + \frac{dH_i}{d\varphi} \tag{7.171}$$

and

$$\frac{dU_2}{d\varphi} = \frac{dW_2}{d\varphi} + \frac{dQ_{w,2}}{d\varphi} + \frac{dH_o}{d\varphi}. \tag{7.172}$$

The enthalpy flow through the valves is formed – as in the case of the four-stroke engine – according to the respective flow direction from the product of the mass flow through the valve and the specific enthalpy of the gas found in the flow direction before the valve.

Via a horizontal splitting of the system, the wall heat flow to the piston can be ascribed to the mixing zone and the wall heat flow to the cylinder lid to the displacement zone. Wall heat flow is distributed to both zones in accordance with the position of the imaginary membrane – i.e. proportional to volume. With that, all equations describing the two-zone model are at our disposal. At the closing of the last control unit, both zones can be ideally mixed again for the following high pressure part, so long as one has at an earlier degree of crank angle not already switched to a single-zone inspection after the complete expelling of a zone. For the mixing temperature in the cylinder is then valid

$$T = \frac{c_{V,1} m_1 T_1 + c_{V,2} m_2 T_2}{c_{V,1} m_1 + c_{V,2} m_2}. \tag{7.173}$$

7.3 Modeling the Gas Path

In order to build a complete engine model, we need besides the cylinder still other components for a description with the filling and emptying method, like the volume, orifice plates, or throttles as well as flow machines for charged engines. An exact description of these components is to be found in the following sections.

7.3.1 Modeling Peripheral Components

7.3.1.1 Volume

A volume (intake manifold, etc.) is usually modeled as a cylindrical solid. Such a model is illustrated in the following sketch.

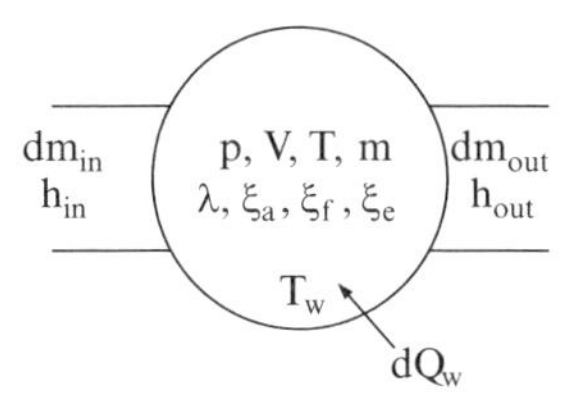

The cross-sectional surface and the volume and the volume itself thus amount to

$$A_{vol.} = d_{vol.}^2 \frac{\pi}{4} \quad \text{and} \quad V_{vol.} = l_{vol.} \, d_{vol.}^2 \frac{\pi}{4}. \tag{7.174}$$

From a thermodynamic perspective, we are thereby dealing with an open, steadily flowed through system. The resulting mass balance is

$$\frac{\mathrm{d}m_{vol.}}{\mathrm{d}t} = \frac{\mathrm{d}m_{in}}{\mathrm{d}t} + \frac{\mathrm{d}m_{out}}{\mathrm{d}t}. \tag{7.175}$$

Furthermore, valid for the energy balance of the volume is

$$\frac{\mathrm{d}U}{\mathrm{d}t} = \frac{\mathrm{d}Q_w}{\mathrm{d}t} + h_{in} \frac{\mathrm{d}m_{in}}{\mathrm{d}t} + h_{out} \frac{\mathrm{d}m_{out}}{\mathrm{d}t}. \tag{7.176}$$

The internal energy can be calculated by means for the relations introduced already in describing the cylinder (see Sect. 7.1.9). Heat transfer is calculated with the Newtonian equation [see (7.10)], whereby we can utilize as heat transfer coefficient the Hausen relation (1976) assuming a turbulent pipe flow

$$\alpha = 0.024 \frac{\lambda_{vol.}}{d_{vol.}} \left[1 + \left(\frac{d_{vol.}}{l_{vol.}} \right)^{\frac{2}{3}} \right] \mathrm{Re}^{0.786} \, \mathrm{Pr}^{0.45}. \tag{7.177}$$

The Prandtl number is set at 0.731. For the Reynolds number and the viscosity we have

$$\mathrm{Re} = \frac{\dot{m}_{mean} \, d_{vol.}}{A_{vol.} \, \eta}; \quad \dot{m}_{mean} = \frac{|\dot{m}_{in}| + |\dot{m}_{out}|}{2}; \quad \eta = 5.17791 \cdot 10^{-7} \, T^{0,62}. \tag{7.178}$$

The heat conductivity according to Woschni amounts to

$$\lambda_{vol.} = 3.65182 \cdot 10^{-4} T^{0.748}. \tag{7.179}$$

It is thus possible to calculate a pipe volume which is switched between two succeeding throttle locations.

7.3.1.2 Orifice Plate (Throttle Valves)

For the simulation of switching flaps, throttle valves, engine bypass flaps, EGR valves, or wastegates in engines, we need flow restriction with constant or variable cross sectional surfaces. The modeling of these components is identical with that of a throttle location at the valves of the cylinder head. For this as well, the so-called flow equation is used.

For the mass flow through a flow restriction is valid (see (2.33))

$$\dot{m} = \alpha A_1 \sqrt{p_0 \rho_0} \sqrt{\frac{2\kappa}{\kappa - 1}\left(\pi^{\frac{2}{\kappa}} - \pi^{\frac{\kappa+1}{\kappa}}\right)}. \tag{7.180}$$

It must also be considered here that at the realization of the critical pressure ratio, the mass flow, as represented in (2.36), is limited. The flow coefficients are shown, contingent on the degree of the opening of the flow restriction, in the form of lines on a characteristic map.

7.3.1.3 Flow Machines

In representing flow machines (compressor, turbine) in charged engines, a consideration with the help of characteristic maps is also possible and sufficient for unsteady processes. Because of the complexity of the presentation of these aggregates, we will take a special look at this in Chap. 8.

7.3.2 Model Building

Figure 7.25 shows a simple example of modeling the gas path of a combustion engine. The model must in principle be built such that a "throttle building block" follows a "storage building block" and then a storage building block again etc. In the storage building block (e.g. the volume), the differential equations for the mass and energy balance are solved. From this results the mass determined for the actual integration step as well as the temperature and, via the general gas equation, the pressure in the storage building block. The mass and enthalpy flows coming in and

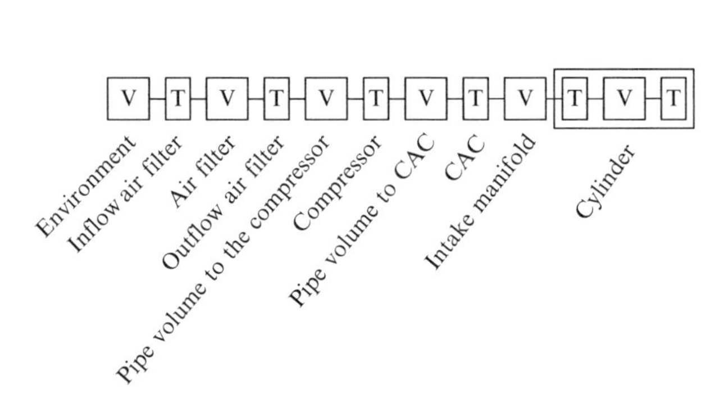

Fig. 7.25 Simple model of the gas path acc. to the filling and emptying method

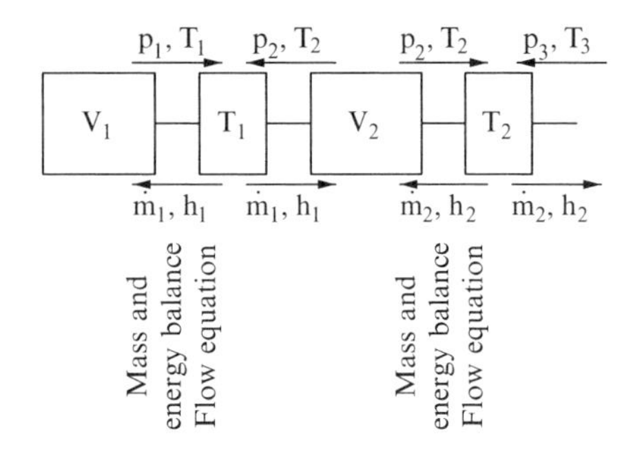

Fig. 7.26 Calculation scheme for the filling and emptying method

out of the system are required as initial quantities for calculating the mass and energy balance. These can be calculated in the throttle building blocks when given the temperature and pressures determined in the previous integration step in the storage building blocks lying before and after the throttle location. In the case of a trigger wheel vane, the calculation of the mass flow takes place, for example, by means of the flow equation. The enthalpy flow is determined upon acquaintance with the gas states (esp. enthalpy) of the storage building block lying in the current flow direction before the throttle location. Figure 7.26 shows this constantly repeating process.

The cylinder in the calculation of the mass and enthalpy flows is usually motored through the valves. Insofar as this is concerned, the cylinder represents a throttle location for the interconnection in the model, although – as shown in Sect. 7.1 – the mass and energy balance is solved and the combustion chamber volume is of course a mass and energy storage space itself.

One additional special position in modeling is occupied by flow machines. While we can designate the flow turbine as a throttle location with energy emission, the flow compressor usually causes an increase in pressure. By means of the use of characteristic maps for describing operating behavior, which is itself dependent on the pressures and temperatures before and after the compressor (see Chap. 8), this does not however play a role for the model described above.

7.3.3 Integration Methods

The so-called Runge-Kutta method of the fourth order has proved to be an adequate method for solving the differential equations of mass and energy in the modules described in this section.

The actual integration time is halved and four gradients are determined – as shown in Fig. 7.27 – which is finally differently weighed. These relations are shown in (7.181).

$$\left(\frac{dT}{d\phi}\right)_0 = f\left(T_0, \phi_0\right)$$
$$T_1 = T_0 + \left(\frac{dT}{d\phi}\right)_0 \frac{\Delta\phi}{2}$$
$$\left(\frac{dT}{d\phi}\right)_1 = f\left(T_1, \phi_0 + \frac{\Delta\phi}{2}\right)$$
$$T3 = T_0 + \left(\frac{dT}{d\phi}\right)_1 \frac{\Delta\phi}{2}$$
$$\left(\frac{dT}{d\phi}\right)_2 = f\left(T_2, \phi_0 + \frac{\Delta\phi}{2}\right)$$
$$T_2 = T_0 + \left(\frac{dT}{d\phi}\right)_2 \Delta\phi$$
$$\left(\frac{dT}{d\phi}\right)_3 = f(T_3, \phi_0 + \Delta\phi)$$

$$\frac{dT}{d\varphi} = \frac{1}{6}\left(\left(\frac{dT}{d\varphi}\right)_0 + 2\left(\frac{dT}{d\varphi}\right)_1 + 2\left(\frac{dT}{d\varphi}\right)_2 + \left(\frac{dT}{d\varphi}\right)_3\right)$$
$$T = T_0 + \frac{dT}{d\varphi}\Delta\varphi\,. \tag{7.181}$$

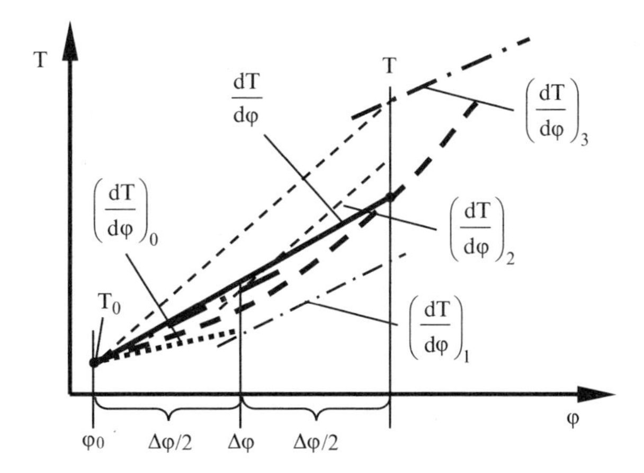

Fig. 7.27 Integration according to the fourth order Runge-Kutta process

The Runge-Kutta method thus distinguishes itself from simple methods like the Euler-Cauchy method, in which only one gradient is formed and the integral thus falls short of reality. We find in the literature a number of other integration methods, but these are usually more complex and not as easy to manage as the Runge-Kutta method.

7.4 Gas Dynamics

7.4.1 Basic Equations of One-Dimensional Gas Dynamics

In principle, the basic equations that describe a one-dimensional flow can be derived from the Navier-Stokes equations introduced in Sect. 12.1 restricting to one dimension and neglecting gravity. Here we will however provide a simple and illustrative "derivative". For this, we will take into consideration the port section sketched in Fig. 7.28 with variable cross section along the x-coordinate.

We assume that the cross sectional alteration of the port along length $\mathrm{d}x$ is small, such that only contingencies on the x-coordinate and on time must be considered.

7.4.1.1 Mass Balance

The port section with volume $\mathrm{d}V$ and a fluid with density $\rho = \rho_x$ contain the mass

$$\mathrm{d}m = \rho_x \,\mathrm{d}V = \rho_x A_x \,\mathrm{d}x. \tag{7.182}$$

Through the cross sections A_x and $A_{x+\mathrm{d}x}$ the medium flows out with a speed of w_x and a speed of $w_{x+\mathrm{d}x}$. Thus we obtain for the mass flows

$$\dot{m}_x = w_x \,\rho_x A_x, \tag{7.183}$$

$$\dot{m}_{x+\mathrm{d}x} = w_{x+\mathrm{d}x} \,\rho_{x+\mathrm{d}x} A_{x+\mathrm{d}x}. \tag{7.184}$$

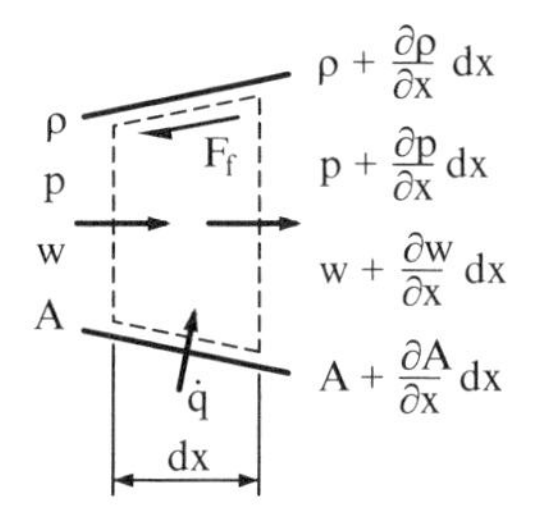

Fig. 7.28 Port length with variable cross section

The mass balance for the port section of length dx then reads

$$\frac{\partial m}{\partial t} = \dot{m}_x - \dot{m}_{x+\mathrm{d}x}, \tag{7.185}$$

$$\frac{\partial}{\partial t}\left(\rho_x A_x \,\mathrm{d}x\right) = w_x \,\rho_x A_x - w_{x+\mathrm{d}x}\,\rho_{x+\mathrm{d}x} A_{x+\mathrm{d}x}. \tag{7.186}$$

With the Taylor series expansion of the state quantities at location $x + \mathrm{d}x$, we obtain

$$\frac{\partial \rho_x}{\partial t}\,\mathrm{d}x = w_x\,\rho_x - \left(w_x + \frac{\partial w_x}{\partial x}\,\mathrm{d}x\right)\left(\rho_x + \frac{\partial \rho_x}{\partial x}\,\mathrm{d}x\right)\left(A_x + \frac{\partial A_x}{\partial x}\,\mathrm{d}x\right)\frac{1}{A_x}. \tag{7.187}$$

Via multiplication and neglecting the terms of higher order, from this follows the continuity equation

$$\frac{\partial \rho}{\partial t} = -\frac{\partial(\rho\,w)}{\partial x} - \rho\,w\,\frac{\mathrm{d}\ln(A)}{\mathrm{d}x}$$

or

$$\frac{\partial \rho}{\partial t} + w\,\frac{\partial \rho}{\partial x} + \rho\,\frac{\partial w}{\partial x} + \rho w\,\frac{\mathrm{d}\ln(A)}{\mathrm{d}x} = 0. \tag{7.188}$$

The index x can be left out to keep things clear.

7.4.1.2 The Law of Conservation of Impulse

Change of the impulse I within the port section under consideration over time is equal to the sum of the impulse flows caused by the mass flow at its cross sections and the external forces that work upon the mass. The impulse is defined as

$$I = m\,w_x = \rho_x A_x \,\mathrm{d}x\, w_x. \tag{7.189}$$

For the impulse flow validity

$$\dot{I}_x = w_x\,w_x\,\rho_x A_x \tag{7.190}$$

or

$$\dot{I}_{x+\mathrm{d}x} = w_{x+\mathrm{d}x}\,w_{x+\mathrm{d}x}\,\rho_{x+\mathrm{d}x} A_{x+\mathrm{d}x}. \tag{7.191}$$

The external forces are composed of the pressure forces F_x, which result from the various cross sectional surfaces, and the friction forces F_f of the fluid on the internal port wall,

$$F_x = p_x A_x, \tag{7.192}$$

$$F_{x+\mathrm{d}x} = p_{x+\mathrm{d}x} A_{x+\mathrm{d}x}, \tag{7.193}$$

$$F_f = k_f \rho_x A_x \,\mathrm{d}x. \tag{7.194}$$

With this, it follows for the impulse balance

$$\frac{\partial}{\partial t}(w_x \rho_x A_x \,\mathrm{d}x) = w_x^2 \rho_x A_x - w_{x+\mathrm{d}x}^2 \rho_{x+\mathrm{d}x} A_{x+\mathrm{d}x} + p_x A_x - p_{x+\mathrm{d}x} A_{x+\mathrm{d}x} - k_f \rho_x A_x \,\mathrm{d}x\,. \tag{7.195}$$

With the Taylor series expansion analogous to the mass balance results

$$\frac{\partial(\rho w)}{\partial t} = -\frac{\partial(\rho w^2 + p)}{\partial x} - \rho w^2 \frac{\partial \ln(A)}{\partial x} - \rho k_f$$

or after transformation under conservation of mass (7.185)

$$\frac{\partial w}{\partial t} + w \frac{\partial w}{\partial x} + \frac{1}{\rho}\frac{\partial p}{\partial x} + k_f = 0. \tag{7.196}$$

The pipe friction coefficient k_f is determined by the pipe friction number λ_f and the internal pipe diameter d contingent on the density of the medium ρ and its speed w

$$k_f = \frac{\lambda_f}{d} \rho \frac{w^2}{2} \frac{w}{|w|}. \tag{7.197}$$

The pipe friction number is determined contingent upon the flow condition and wall roughness (here hydraulically smooth pipes) with the help of the equations of Blasius, and Nikuradse and Prandtl, see Beitz and Grote (1997).

$$\lambda_f = \frac{0.3164}{\sqrt[4]{\mathrm{Re}}} \quad \text{for} \quad 2.320 < \mathrm{Re} < 10^5 \text{ (Blasius)} \tag{7.198}$$

$$\lambda_f = 0.0032 + \frac{0.221}{\mathrm{Re}^{0.237}} \quad \text{for} \quad 10^5 < \mathrm{Re} < 10^8 \text{ (Nikuradse)} \tag{7.199}$$

$$\lambda_f = \frac{1}{\left[2\,\lg\left(\frac{\mathrm{Re}\sqrt{\lambda_f}}{2.51}\right)\right]^2} \quad \text{for} \quad 2.320 < \mathrm{Re} \text{ (Prandtl)} \tag{7.200}$$

7.4.1.3 The Law of Conservation of Energy

On the basis of the first law of thermodynamics, the change in energy in the port section under consideration over time in equal to the sum of the energy flows entering and leaving over the cross sectional surfaces and the heat flow supplied and removed from outside. For the energy and the energy flows is valid

$$E = m\left(u_x + \frac{w_x^2}{2}\right) = \rho_x A_x \, \mathrm{d}x \left(u_x + \frac{w_x^2}{2}\right), \tag{7.201}$$

$$\dot{E}_x = w_x \, \rho_x \, A_x \left(h_x + \frac{w_x^2}{2}\right), \tag{7.202}$$

$$\dot{E}_{x+\mathrm{d}x} = w_{x+\mathrm{d}x} \, \rho_{x+\mathrm{d}x} \, A_{x+\mathrm{d}x} \left(h_{x+\mathrm{d}x} + \frac{w_{x+\mathrm{d}x}^2}{2}\right). \tag{7.203}$$

The heat flow amounts to

$$\dot{Q} = \dot{q} \, A_x \, \mathrm{d}x \, . \tag{7.204}$$

With that follows for the energy balance

$$\begin{aligned} \frac{\partial}{\partial t}\left[\rho_x A_x \, \mathrm{d}x \left(u_x + \frac{w_x^2}{2}\right)\right] &= w_x \, \rho_x \, A_x \left(h_x + \frac{w_x^2}{2}\right) \\ &- w_{x+\mathrm{d}x} \, \rho_{x+\mathrm{d}x} \, A_{x+\mathrm{d}x} \left(h_x + \frac{w_{x+\mathrm{d}x}^2}{2}\right) + \dot{q} \, A_x \, \mathrm{d}x \, . \end{aligned} \tag{7.205}$$

Developed into a Taylor series expansion, we obtain after a quick transformation

$$\frac{\partial}{\partial t}\left[\rho\left(u + \frac{w^2}{2}\right)\right] = -\frac{\partial\left[w\,\rho\left(h + \frac{w^2}{2}\right)\right]}{\partial x} - w\,\rho\left(h + \frac{w^2}{2}\right)\frac{\partial \ln(A)}{\partial x} + \dot{q} \tag{7.206}$$

If we insert the mass and impulse balance [(7.187) and (7.195)] and transform further, the result is

$$\frac{\partial h}{\partial t} + w\,\frac{\partial h}{\partial x} - \frac{1}{\rho}\left(\frac{\partial p}{\partial t} + w\,\frac{\partial p}{\partial x}\right) - \frac{\dot{q}}{\rho} - w\,k_f = 0. \tag{7.207}$$

The heat transfer in the pipe with the help of the Newtonian method becomes:

$$\dot{q} = \frac{\dot{Q}}{A_{circumf.}} = \alpha_w \left(T_w - T_{gas}\right). \tag{7.208}$$

In (7.209) is represented a semi-empirical approach of Gniellinski (see Stephan 1993) based on the Prandt analogy for the mean Nusselt number contingent on the respective range of validity of the Prandtl and Reynolds numbers. The Prandtl analogy assumes a dual-layered model that subdivides the flow into a laminar boundary layer and a turbulent core flow, which is directly joined with the lower laminar stratum. It is assumed that in the fully turbulent flow, the speed, temperature, and concentration profiles depend only on the normal wall coordinate, while in the laminar boundary layer, the total value of thrust tension, heat, and diffusion flow density are independent of the wall normal. The approach is used for components like the charge air cooler or the exhaust manifold, in which heat transfer plays an essential role.

$$\mathrm{Nu}_{m,\,turb} = \frac{\lambda_r}{8} \frac{(\mathrm{Re} - 1{,}000)\ \mathrm{Pr}}{1 + (\mathrm{Pr}^{2/3} - 1)\ 12.7\ \sqrt{\lambda_r/8}} \left[1 + \left(\frac{d}{l} \right)^{2/3} \right]. \tag{7.209}$$

Equation (7.209) is valid in the range $2.300 \leq \mathrm{Re} \leq 5\ 10^5, 0.5 \leq \mathrm{Pr} \leq 2{,}000$ and $l/d > 1$. From the definition of the Nusselt number, the heat transfer coefficient comes to

$$\alpha_w = \frac{\lambda}{d} \frac{\lambda_r}{8} \frac{(\mathrm{Re} - 1.000)\ \mathrm{Pr}}{1 + (\mathrm{Pr}^{2/3} - 1)\ 12.7\ \sqrt{\lambda_r/8}} \left[1 + \left(\frac{d}{l} \right)^{2/3} \right]. \tag{7.210}$$

The determination of the heat transfer for $\mathrm{Re} < 2{,}300$ results via a quadratic averaging of these methods for the turbulent (7.209) and the laminar Nusselt number (7.211) according to (7.212)

$$\mathrm{Nu}_{lam} = 0.664\, \mathrm{Re}^{\frac{1}{2}}\, \mathrm{Pr}^{\frac{1}{3}}, \tag{7.211}$$

$$\mathrm{Nu}_m = \sqrt{\mathrm{Nu}_{m,\,turb}^2 + \mathrm{Nu}_{lam}^2}. \tag{7.212}$$

7.4.2 Numerical Solution Methods

In the previous section, equations were derived that describe one-dimensional gas dynamics. We are dealing here with a partial differential equation system, which is analytically not solvable. For this reason, a large multitude of graphic and numerical solution techniques have been developed, of which numerical solutions in the form of finite differences, supported by the steadily increasing calculation power of computers, has been quite successful, as it offers the necessary flexibility and precision. In this case, a discretization of the location with an approximation of the local gradients takes place. This renders possible the transformation of the partial differential equation system into a row of common differential equations. This process will be described in more detail in the following.

7.4.2.1 The Single-Step Lax-Wendroff Method

The Lax-Wendroff method offers the possibility of describing the gas dynamics in the pipes using finite differences. First the conservation equations are expressed in a form which corresponds to the vector form represented in (7.213)

$$\frac{\partial G(x,\,t)}{\partial t}+\frac{\partial F(x,\,t)}{\partial x}=-C(x,\,t). \tag{7.213}$$

The vectors $G(x,\,t)$, $F(x,\,t)$ and $C(x,\,t)$ are easy to derive from the laws of conservation for mass, impulse, and energy,

$$G(x,\,t)=\begin{bmatrix}\rho\\ \rho\,w\\ \rho\left(u+\frac{w^2}{2}\right)\end{bmatrix},\quad F(x,\,t)=\begin{bmatrix}\rho\,w\\ \rho\,w^2+p\\ w\,\rho\left(h+\frac{w^2}{2}\right)\end{bmatrix}$$

and

$$C(x,\,t)=\begin{bmatrix}\rho\,w\\ \rho\,w^2\\ w\,\rho\left(h+\frac{w^2}{2}\right)\end{bmatrix}\frac{\mathrm{d}\ln(A)}{\mathrm{d}x}+\begin{bmatrix}0\\ \rho\,k_f\\ -\dot{q}\end{bmatrix}. \tag{7.214}$$

In the following, the expression $G_t(x,t)$ will be used for $\partial G(x,\,t)/\partial t$ and the expression $F_x(x,\,t)$ for $\partial F(x,\,t)/\partial x$.

In the next step, the Taylor series expansion of function $G(x,\,t)$ is developed around the point $(x,\,t+\Delta t)$ and a location discretization is carried out. From that results the following equation

$$G(x_i,\,t+\Delta t)=G(x_i,\,t)+\Delta t\,G_t(x_i,\,t)+O(\Delta t^2). \tag{7.215}$$

Then $G_t(x,\,t)$ is substituted with (7.215) and the formation of the local gradient occurs with the central differences. With the approximation

$$G(x_i,\,t_j)\;\approx\;\frac{1}{2}\left[G(x_{i+1},\,t_j)+G(x_{i-1},\,t_j)\right] \tag{7.216}$$

we finally obtain the equation of the single-step Lax-Wendroff process

$$\begin{aligned}G(x_i,\,t_{j+1})=&\frac{1}{2}\left[G(x_{i+1},\,t_j)+G(x_{i-1},\,t_j)\right]\\&-\frac{\Delta t}{2\,\Delta x}\left[F(x_{i+1},\,t_j)-F(x_{i-1},\,t_j)\right]-\Delta t\,C(x_i,\,t_j)\,.\end{aligned} \tag{7.217}$$

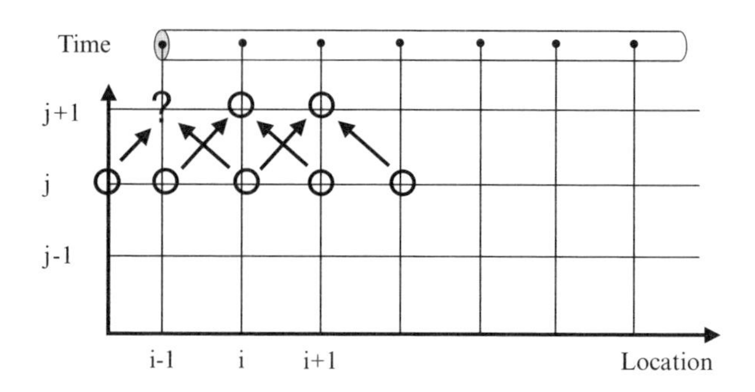

Fig. 7.29 Single-step Lax-Wendroff process with determination of the internal pipe nodes

In Fig. 7.29 we can recognize the method for the determination of the state of the particular internal pipe nodes. The condition of the node at location i at time j is known, as well as the conditions at location i - 1 and $i + 1$ at time j. From the states i - 1 and $i + 1$ at time j, we can determine with the help of the previously described equations the condition of the node at location i at time $j + 1$. However, the conditions at the external pipe nodes cannot be determined with this method, as even though the states at locations $i - 1$ and i at time j are known, the condition at location i - 2 is not. This must take place by means of the pipe-margin coupling, which we look at in more detail in 7.4.3.

7.4.2.2 The Two-Step Lax-Wendroff Method

The two-step Lax-Wendroff method (Fig. 7.30), which follows from the Peyret-Lerat process, describes a method of finite differences, which consists of two steps and can heighten the stability though the use of further coefficients in comparison with the single-step process.

The Peyret-Lerat method can be represented in the following way.

First step:

$$\begin{aligned} G\left(x_{i+\beta},\, t_{j+1}\right) = {} & (1-\beta)\, G\left(x_i,\, t_j\right) + \beta\, G\left(x_{i+1},\, t_j\right) \\ & - \alpha\, \frac{\Delta t}{\Delta x}\left[F\left(x_{i+1},\, t_j\right) - F\left(x_i,\, t_j\right)\right] \\ & - \alpha\, \Delta t\left[(1-\beta)\, C\left(x_i,\, t_j\right) - \beta\, C\left(x_{i+1},\, t_j\right)\right] \end{aligned} \tag{7.218}$$

Second step:

$$\begin{aligned} G(x_i, t_{j+1}) = {} & G(x_i, t_j) - \frac{\Delta t}{2\alpha\Delta x}\left[(\alpha-\beta)F(x_{i+1}, t_j) + (2\beta-1)F(x_i, t_j)\right] - \frac{\Delta t}{2\alpha\Delta x} \\ & \times\left[(1-\alpha-\beta)F(x_{i-1}, t_j) + F(x_{i+\beta}, t_{j+\alpha}) - F(x_{i-1+\beta}, t_{j+\alpha})\right] - \Delta t C(x_i, t_j)\,. \end{aligned} \tag{7.219}$$

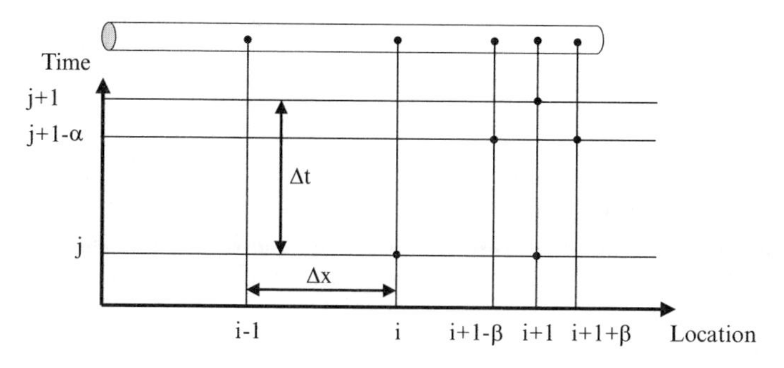

Fig. 7.30 Two-step Lax-Wendroff process

From the Peyret-Lerat method, we obtain via substitution of $\alpha = \beta = 1/2$ the two-step Lax-Wendroff method:

$$G\left(x_{i+1/2}, t_{j+1}\right) = \frac{1}{2}\left[G\left(x_i, t_j\right) + G\left(x_{i+1}, t_j\right)\right] - \frac{\Delta t}{2\,\Delta x}\left[F\left(x_{i+1}, t_j\right) - F\left(x_i, t_j\right)\right] - \frac{\Delta t}{4}\left[C\left(x_{i+1}, t_j\right)\right], \quad (7.220)$$

$$G\left(x_i, t_{j+1}\right) = G\left(x_i, t_j\right) - \frac{\Delta t}{\Delta x}\left[F\left(x_{i+1/2}, t_{j+1/2}\right) - F\left(x_{i-1/2}, t_{j+1/2}\right)\right] - \Delta t\, C\left(x_i, t_j\right). \quad (7.221)$$

7.4.3 Boundary Conditions

Previously, only one port/pipe section has been considered and the conservation equations and their solutions shown. A real system (intake system, exhaust gas system) consists however of a large number of single components (discontinuity points), which are connected with pipes. Only via methods for a coupling of various partial systems does such a complex system become calculable. For this, we must exchange between the partial systems, e.g. the energy and mass flows, thereby setting the boundary conditions of the pipe described above. The structure thereby is always the same and resembles the one described in Sect. 7.3.2: – junction – pipe – junction – pipe – junction.

Consequently, the following components represent junctions from the standpoint of the pipe:

- Pipe end
- Pipe branching
- Flow restriction
- Volume
- Cylinder

- Compressor
- Turbine

The coupling of a pipe with its neighboring component is described by:

- The method of characteristics, which calculates the conditions in the outlet cross section of the pipe (compatibility conditions).
- The general flow equation, which illustrates the influence of the throttle location. It describes the mass flow coming through the throttle location contingent on the conditions in the pipe border and the unsteady location.
- The conservation equations of the respective points of discontinuity, which describe changes in state via common differential equations contingent purely on time. These will be presented in Sects. 7.1–7.3 and Chap. 8.

Since the equations are implicitly contingent upon each other, the solution must be determined iteratively. In the following, the iterative solution of the boundary conditions will be introduced, as it has been developed by Görg (1982) and Stromberg (1977). An exact derivation and arrangement can be found in Miersch (2003). Fundamentally, two cases of flow can be determined at the margin: in the case of elementary flow situation 1, the mass flow flows from the pipe into the margin, while in elementary flow situation 2 the mass flow runs from the border into the pipe.

Figure 7.31 shows the time-location lattice at the pipe border. Point 3 represents the marginal state to be determined in the new point in time. It is determined from the known conditions of the states of rest (index 0) of the border node and from the conditions of the previous time in lattice points 1 and 2. The distance between points 1 and 2 corresponds to the local discretization, the distance between points 2 and 3 the temporal discretization. The location, from which in the former instant the new boundary conditions depart (point 1′ and point 2′), is called the foot of the Mach or particle path. These location coordinates are fixed such that at the new time the conditions starting from point 1′ reach the border (point 3) along the Mach

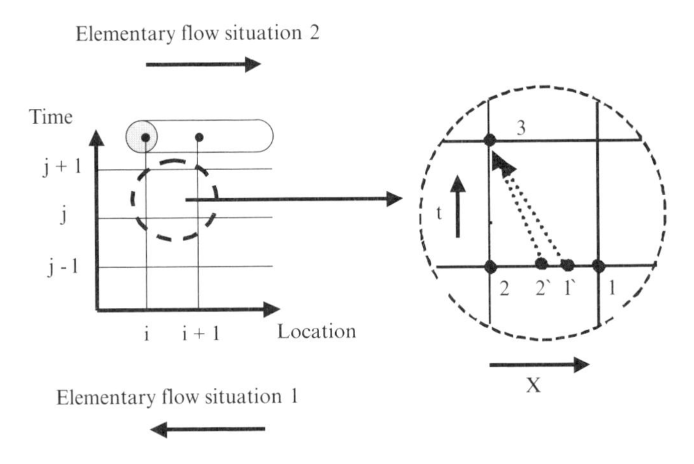

Fig. 7.31 Boundary coupling by of characteristic lines (Miersch 2003)

path and the conditions starting from point 2′ the border (point 3) along the particle path. The foot points are set at the beginning of the iteration to the middle of the distance between points 1 and 2 and then again after every iteration loop as follows:

$$x_{f,\,new} = x_{f,old} - \Delta x_f$$
$$\Delta x_f = \frac{\Delta t}{2}\frac{v_f^* v_2^*}{v_2^* + v_2^*};$$

$v_f = u_f - a_f$; For mach line path (point 1′),
$v_f = u_f$; for particle path (point 2′).

The solution in elementary flow situation 1 is thus determined from the condition of compatibility along the particle path

$$\frac{2\,a}{\kappa - 1}\left(\frac{\partial a}{\partial t} + w\frac{\partial a}{\partial x}\right) - \frac{1}{\rho}\left(\frac{\partial p}{\partial t} + w\frac{\partial p}{\partial x}\right) = 0 \tag{7.222}$$

and the condition of compatibility along the Mach line

$$\frac{\partial w}{\partial t} + (w \pm a)\frac{\partial w}{\partial x} \pm \frac{2}{\kappa - 1}\left[\frac{\partial a}{\partial t} + (w \pm a)\frac{\partial a}{\partial x}\right] = \mp w(w \pm a)\frac{\mathrm{d}}{\mathrm{d}x}[\ln(A)], \tag{7.223}$$

the general flow equation

$$\dot{m} = \alpha A_2\sqrt{2p_{01}\rho_{01}}\sqrt{\frac{\kappa}{\kappa - 1}\left[\left(\frac{p_2}{p_{01}}\right)^{\frac{2}{\kappa}}\left[1 + \frac{(\kappa - 1)}{\kappa R}\frac{\dot{W}}{\dot{m}_{is}T_{01}}\right] - \left(\frac{p_2}{p_{01}}\right)^{\frac{\kappa+1}{\kappa}}\right]} \tag{7.224}$$

and the conditions in the system connected to the pipe.

Figure 7.32 shows the particle and Mach line paths on the flow level. These are described by

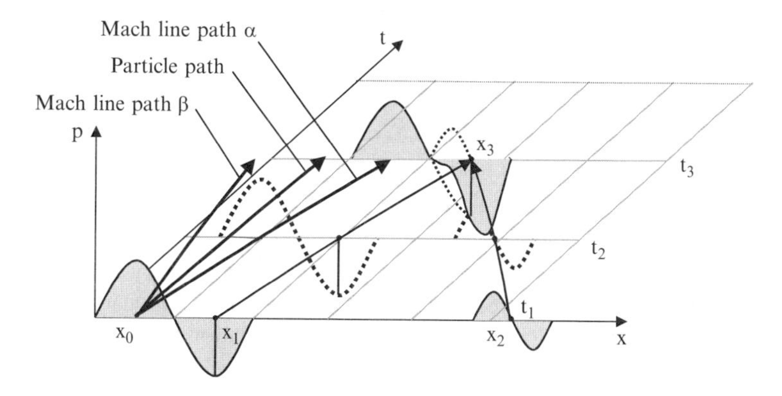

Fig. 7.32 Mach line trace and particle trace in the flow level (Miersch 2003)

$$\frac{\partial x}{\partial t} = w \tag{7.225}$$

and

$$\frac{\partial x}{\partial t} = w \pm a \tag{7.226}$$

The index α describes the propagation of a pressure wave along the Mach line in the flow direction, index β the propagation against the flow direction.

The formulation of the condition of compatibility occurs via transformation of the equation system (7.213) from the representation in the independent variables density ρ, pressure p, speed w to a representation in the independent variables speed of sound a, pressure p, and speed w.

The transformed system consists of coupled common differential equations, the solution of which is found through a purely temporal integration. The detailed derivation is described in Seifert (1962), Stromberg (1977), and Görg (1982) (see Miersch 2003).

The discretization of the conditions of compatibility tracing the Mach lines occurs via

$$\begin{aligned} \mathrm{d}p &= p_3 - p_{1'}\,, \\ \mathrm{d}w &= w_3 - w_{1'}\,, \\ \mathrm{d}t &= \Delta t\,. \end{aligned} \tag{7.227}$$

in the dimensions pressure p and velocity w. In this way, we obtain a linear relation between pressure and speed

$$p_3 = C_1\, w_3 + C_2. \tag{7.228}$$

In elementary flow situation 1, the condition of compatibility along the particle path is discretized according to

$$\begin{aligned} \mathrm{d}p &= p_3 - p_{2'}\,, \\ \mathrm{d}a &= a_3 - a_{2'}\,, \\ \mathrm{d}t &= \Delta t\,, \end{aligned} \tag{7.229}$$

For the solution in elementary flow situation 2, instead of the condition of compatibility along the particle path, that of the quasi-stationary energy balance (7.231) is utilized. Furthermore, one obtains, via the quasi-stationary energy balance between state and state of rest at point 3, a linear relation between pressure and the speed of sound (7.232).

$$\left(\frac{a_2}{a_{01}}\right)^2 = \frac{\hat{A}\left(\frac{p_2}{p_{01}}\right)^{\frac{\kappa-1}{\kappa}}}{\alpha\sqrt{\frac{2}{\kappa-1}\left[1-\left(\frac{p_2}{p_{01}}\right)^{\frac{\kappa-1}{\kappa}}\right]}}\left(\frac{w_2}{a_{01}}\right) \tag{7.230}$$

$$a_0^2 = a_3^2 + \frac{1}{2}(\kappa - 1)\,w_3^2, \tag{7.231}$$

$$a_3 = \frac{p_3}{C_3} + C_4. \tag{7.232}$$

In the context of iteration, the general flow equation (7.224) is solved in the implicit variables pressure or speed. This equation distinguishes itself from the known differential equation of Saint-Venant by the term for the work carried out by the system $\dot{W}$. In this way, there is a possibility of calculating mass flows that flow against a pressure ratio greater than one (e.g. in the case of a compressor).

It is thus possible to describe the boundary conditions of a flow restriction and a flow machine in the same manner. In the case of the flow restriction, the provision of the flow coefficient is necessary, while the flow machine can be seen as a flow restriction for the sake of the solution of the boundary condition.

In the case of pipe branchings with three connections, a total of six flow situations appear (Fig. 7.33), for which the flow coefficients are given contingent on the branching angle and the particular cross sections, mostly in data banks.

For one pipe branching, as a rule three separation situations and their compliments must be taken into consideration, whereby in each case one pipe socket conducts the entire mass flow and the other two only a partial mass flow.

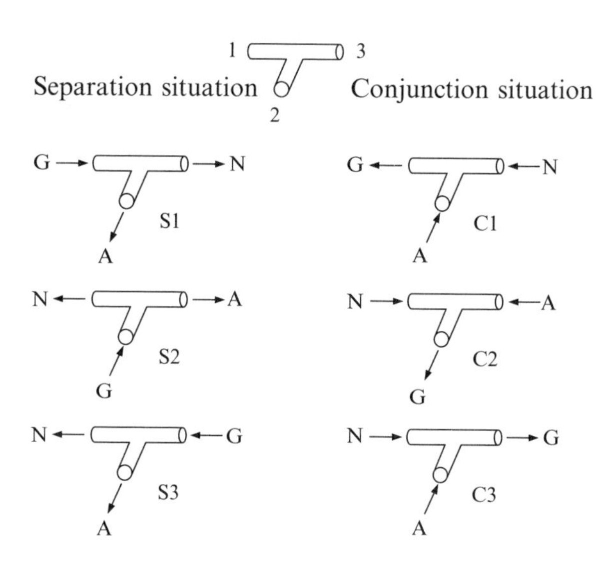

Fig. 7.33 Flow situations at a pipe branching

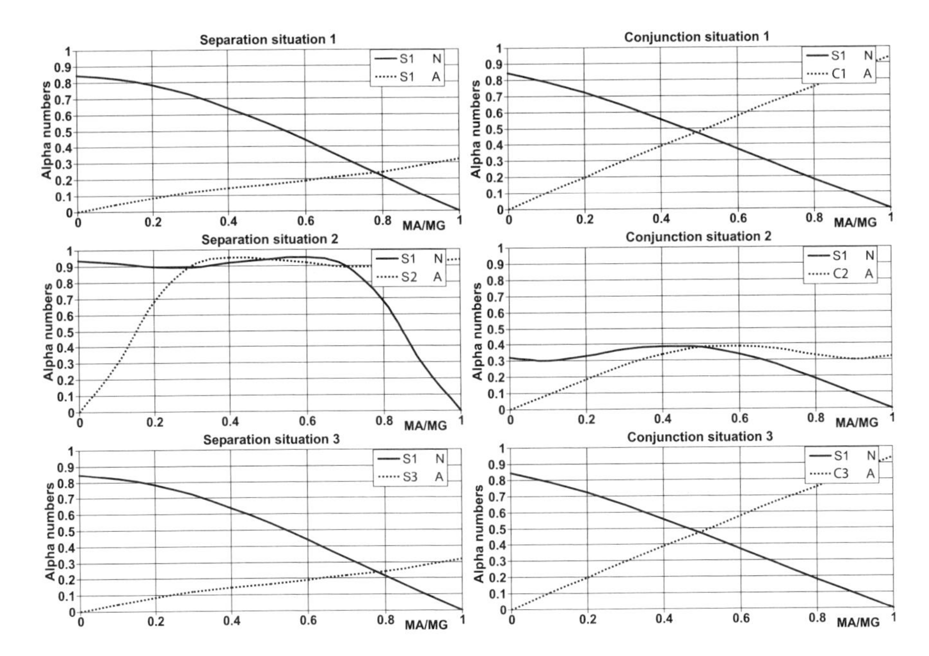

Fig. 7.34 Flow coefficients at a pipe branching

In Fig. 7.34, the flow coefficients from the ratio from the branching mass flow and the total mass flow is applied to the main flow direction and the branching flow direction. The determination of the main flow directions serves only to improve the clarity of the system, having no effect on the calculation result.

The continuity equation must be fulfilled at the pipe branches. This is guaranteed by a corresponding iteration. Figure 7.34 shows an example of the flow coefficient at a pipe embranchment with 120° between each pipe socket. The diameter of the branching pipe socket (2) is twice as large as the cross section of the other two pipe sockets.

Modeling is successfully ended when the internal iteration loop converges over the flow equation and the external iteration loop according to the conditions of compatibility, the marginal equations, and the flow equation with a sufficiently small error.

Also of decisive importance for an exact calculation result is providing the calculation length, which, one the one hand, should of course not be too small, in order to maintain a stable solution and to save calculation time, and should, on the other hand, not be too large, since a secure convergence may not occur. The realization of the time allowance takes place according to the Courant-Friedrichs-Levy criterion of stability

$$\Delta t \leq \frac{\Delta x}{(|w| + a)}. \tag{7.233}$$

These circumstances are shown with the help of Fig. 7.35. The conditions at locations $i - 1$, i and $i + 1$ at time j are known. In order now to draw a conclusion

Fig. 7.35 Courant Friedrichs Levy stability criterium

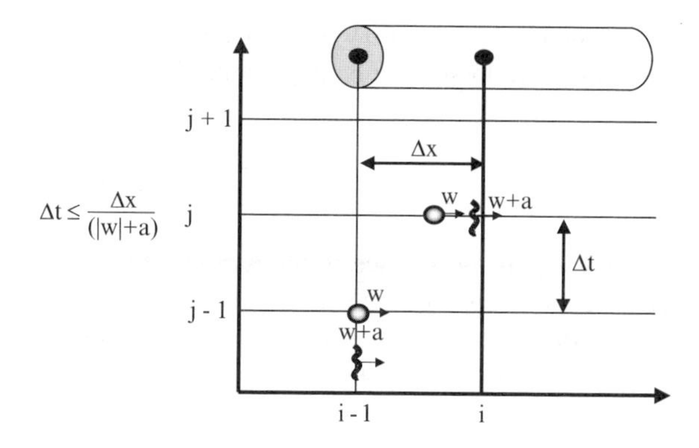

from the conditions ($i-1; j$) and ($i+1; j$) about the condition ($i; i+1$), Δt must be chosen *at least as large*, such that time Δt passes, while the path Δx is traversed. From this results a need of minimal time length in the order of magnitude of 10^{-4} to 10^{-6} seconds.

7.5 Fuel System Simulation

With the increased flexibility of common rail injection systems with respect to number, position and duration of the injection events, the representation of the interactivity of the injection system and the combustion process has also grown in importance. In the context of precalculations and engine process optimization, the effects of injection parameter adjustments, e.g. of spray distance variation (changing the distance between two injections) must be known, making a calculation model indispensible. In the following, hydraulic simulation will be explained on the basis of its use in the calculation of injection with common rail injection systems. The same considerations are valid when calculating oil cycles, for example.

7.5.1 Modeling the Basic Components

Basically, complex hydraulic systems can be analyzed into a sequence of elementary components, similarly to the filling and emptying method. These components essentially include throttles, volumes, and pipes.

7.5.1.1 Throttles

Throttles represent pressure drop for the flowing fluid. The flow is initially constricted to the narrowest location of the throttle and accelerated. Behind the narrow point, the flow is again delayed. Dissipative losses due to acceleration and delay are manifested in a loss of flow pressure across the throttle. The simplest description of these pressure losses can be made with the Bernoulli equation:

$$\Delta p = \frac{\rho}{2}\left(\frac{1}{\zeta \cdot A_T}\right) Q^2 . \tag{7.234}$$

Here, A_D is the surface in the narrowest cross-section of the throttle and ζ the contraction number of the vena contracta. The sign of the flow Q must be taken into consideration for a correct description. (7.235) describes the relation between the contraction number and the flow coefficient C_d.

$$C_d = \frac{1}{\sqrt{\zeta}} . \tag{7.235}$$

The flow coefficient, which depends both on geometric marginal conditions and on the flow velocity, i.e. the Reynolds number, and summarizes the individual loss factors, is a characteristic parameter for calculating flow losses. As an example, Fig. 7.36 shows the profile of the flow coefficient as a function of the flow restriction geometry (Lichtarowicz et al. 1965).

The flow coefficients are usually determined experimentally in an empirical manner for a wide range of geometries and marginal conditions. An extensive compilation can be seen in Idel'chik (1996), for example.

7.5.1.2 Volumes

Hydraulic volumes generally represent connection nodes in hydraulic systems and networks. They are filled with a fluid, which can be compressible or incompressible in accordance with the calculation. Hydraulic volumes usually exhibit a constant volume. If they are connected with another functional component, e.g. a hydraulic cylinder, the volume can also fluctuate. If the selected fluid is compressible, mass

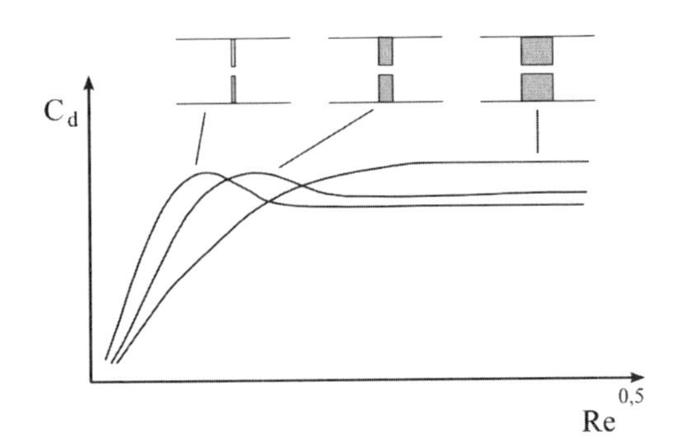

Fig. 7.36 Flow coefficients as a function of throttle geometry and Reynolds #

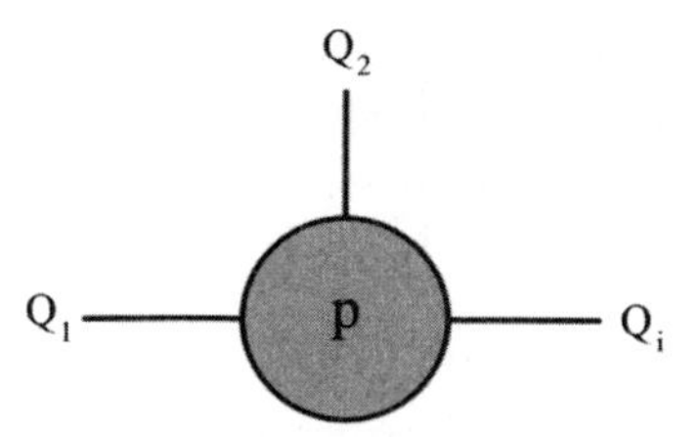

Fig. 7.37 Volume node

conservation in the nodes leads to a differential equation for the pressure in the nodes with the help of the state equations for $p(\rho)$.

If the fluid is compressible, a change of pressure is measured by the bulkelasticity modulus E

$$E = -V\frac{dp}{dV}. \tag{7.236}$$

If the bulk-modulus –of the fluid is known, one can calculate the pressure change for a node of constant volume (Fig. 7.37)

$$\dot{p} = \frac{E}{V}\sum Q_i\,. \tag{7.237}$$

The correct modeling of characteristic fluid properties such as density and the bulk-modulus is extremely important in hydraulic simulation. The consideration of the dissolved gases within the fluid (aeration) has an especially decisive influence on the quality of the calculation.

7.5.1.3 Pipes

Pipes in hydraulic systems are used in order to connect components and volumes. In the simplest case, this is done with a direct connection, which connects only the inlets and outlets of the components without representing the dynamics of a pipe; this is given for example with $l/d < 5$ for very short connections. Since information (pressure waves) move forwards in the hydraulic system with a maximum speed equal to the acoustic velocity of the fluid, pipes must also be modeled with a larger length-to-diameter ratio in order to describe the system correctly. In this case, the following assumptions are made in accordance with Stanciu (2005) due to lacking geometrical resolution:

- Single-dimensional flow in the x-direction
- Constant pressure and density at each pipe cross-section
- Constant average velocity at each pipe cross-section.

Otherwise, pipe models can be arbitrarily complex, for example, by taking into consideration frequency-dependent friction and the mutual influence of a finitely rigid pipe wall. Accordingly, the Navier-Stokes equations are derived for single-dimensional flow under the above-mentioned marginal conditions.

$$\frac{\partial \rho}{\partial t} + \frac{\partial (u\rho)}{\partial x} = 0\,. \tag{7.238}$$

$$\rho\left(\frac{\partial u}{\partial t} + u\frac{\partial u}{\partial x}\right) = -\frac{\partial p}{\partial x} - \mu\frac{\partial^2 u}{\partial x^2}\,. \tag{7.239}$$

(7.238) represents a continuity equation, while (7.239) embodies a single-dimensional impulse equation (neglecting gravity). To solve these partial differential equations, usually a finite difference method is used. The methods of characteristics and the finite element method are also used.

For a detailed description of the simulation of hydraulic systems, see Borchsenius (2003).

7.5.1.4 Cavitation

Cavitation can play a dominant role, particularly in the simulation of injection system components. The reason for this is based on the fact that very high speeds arise at throttle points and curvatures due to the enormous pressure differences in current injection systems. As a result of the drop in static pressure to a value below the vapor pressure of the fluid, the fluid vaporizes and brings about a reduction of the flow cross-section because of the volume expansion of the vapor. This finally causes a reduction of the flow such that a further increase in the pressure drop at constant primary pressure does not bring about a further increase in flow (Fig. 7.38, left). In principle, this phenomenon is comparable in its effect to the choked gas flow, although other physical laws are applied.

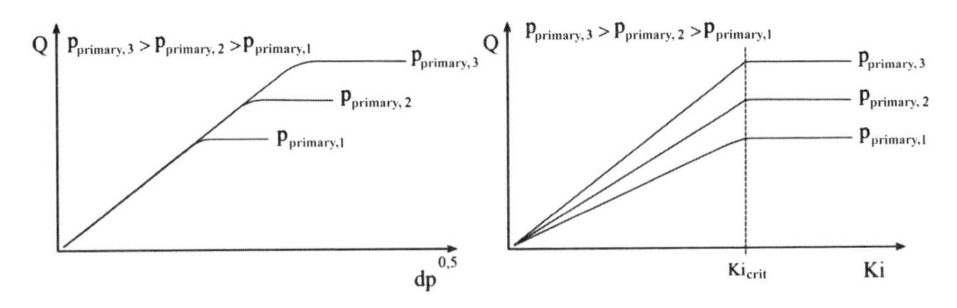

Fig. 7.38 Flow restriction caused by cavitation, plotted over $dp^{0.5}$ (*left*) and the cavitation index *Ki* (*right*)

In order to describe the multidimensional phenomenon in the 1-D environment, a formulation is introduced independently of the adjacent pressure by standardizing the pressure difference

$$K = \frac{p_{primary - p_{back}}}{p_{primary} - p_{evap.}} . \quad (7.240)$$

If the vapor pressure of the fluid p_{dampf} is neglected and instead of the pressure difference its root is inserted, we obtain the cavitation index Ki instead of the cavitation number K

$$Ki = \sqrt{\frac{p_{primary - p_{back}}}{p_{primary}}} = \sqrt{1 - \frac{p_{back}}{p_{primary}}} . \quad (7.241)$$

The right side of Fig. 7.38 clarifies the effect of standardization on the primary pressure. Independent of the adjacent primary pressure, we obtain a value for the cavitation index, beyond which no flow increase takes place despite further increase of the pressure difference. This critical cavitation index Ki_{crit} describes the onset of cavitation independently of marginal conditions and thus characterizes flow behavior through a throttle point.

7.5.2 Application Example

Subcomponents can be formed from the above-mentioned elements by combination. These subcomponents represent customary components of hydraulics; integration into modules makes modeling easier, since one can revert back to modules that are ready-made. Examples of such subcomponents include:

- Valve seats (needle valve, plate valve, ball seat valve, etc.)
- Hydraulic cylinders
- Selector valves
- Leakage points

If these subcomponents are combined with mechanical components and mass inertias and accelerations/forces are introduced, the most complex hydraulic systems can be constructed, such as injectors and high-pressure pumps. These are often called supercomponents. As an example, Fig. 7.39 shows the result of a complex hydraulic model in the form of a piezo-diesel injector consisting of more that 150 (linked) state variables. Illustrated is a triple injection of an emission-relevant partial load operation point at 720 bar rail pressure, consisting of a pre-injection, a main injection and a post-injection. The model was validated with the help of hydraulic test bed examinations (measurement).

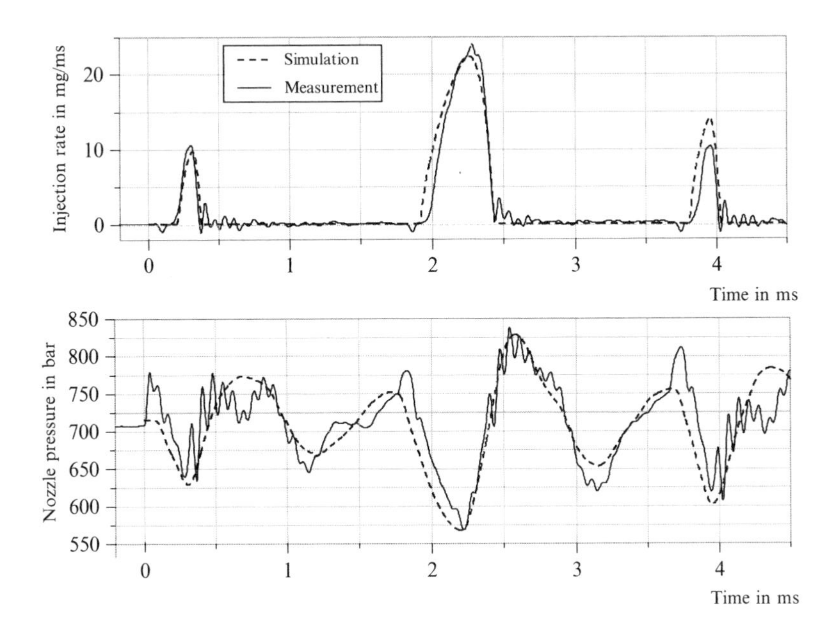

Fig. 7.39 Simulated and measured injection rates and nozzle pressure profiles of a modern piezo-diesel injector

Thanks to their complexity and faithfulness to detail, these models can even be used to pre-calibrate control units and to develop control unit functions.

References

Barba C, Burckhardt C, Boulouchos K, Bargende M (1999) Empirisches Modell zur Vorausberechnung des Brennverlaufes bei Common-Rail-Dieselmotoren. Motortechnische Zeitschrift MTZ 60. Franckh-Kosmos Verlags-GmbH, Stuttgart, pp 262–270

Bargende M (1990) Ein Gleichungsansatz zur Berechnung der instationären Wandwärmeverluste im Hochdruckteil von Ottomotoren. Dissertation, TH Darmstadt

Beitz W, Grote K-H (1997) Dubbel Taschenbuch für den Maschinenbau, Aufl. 19, Springer, Berlin

Borchsenius F (2003) Simulation Ölhydraulischer Systeme. VDI Reihe 8, Nr. 1005, VDI Verlag, Düsseldorf

Chmela F, Orthaber G, Schuster W (1998) Die Vorausberechnung des Brennverlaufs von Dieselmotoren mit direkter Einspritzung auf der Basis des Einspritzverlaufs. Motortechnische Zeitschrift MTZ 59, Heft 7/8, Franckh-Kosmos Verlags-GmbH, Stuttgart, pp 484–492

Constien M (1991) Bestimmung von Einspritz- und Brennverlauf eines direkteinspritzenden Dieselmotors. Dissertation, TU München

Csallner P (1981) Eine Methode zur Vorausberechnung der Änderung des Brennverlaufes von Ottomotoren bei geänderten Betriebsbedingungen. Dissertation, TU München

Doll M (1989) Beitrag zur Berechnung des stationären und transienten Betriebsverhaltens kleiner, schnellaufender Dieselmotoren mit unterteilten Brennräumen. Dissertation, TU München

Eilts P (1993) Modell zur Vorausberechnung des Brenngesetzes mittelschnellaufender Dieselmotoren. Motortechnische Zeitschrift MTZ 54, Heft 3, 134–140, Franckh-Kosmos Verlags-GmbH, Stuttgart

Franzke DE (1981) Beitrag zur Ermittlung eines Klopfkriteriums der ottomotorischen Verbrennung und zur Vorausberechnung der Klopfgrenze. Dissertation, Technische Universität München
Gerstle M (1999) Simulation des instationären Betriebsverhaltens hochaufgeladener Vier- und Zweitakt-Dieselmotoren. Dissertation, Universität Hannover
Görg KA (1982) Berechnung instationärer Strömungsvorgänge in Rohrleitungen an Verbrennungsmotoren unter besonderer Berücksichtigung der Mehrfachverzweigung. Dissertation, Ruhr Universität Bochum
Halstead MP, Kirsch LJ, Prothero A, Quinn CP (1975) A mathematical model for hydrocarbon autoignition at high pressures. Proc R Soc Lond A346:515–538
Halstead MP, Kirsch LJ, Quinn CP (1977) The autoignition of hydrocarbon fuels at high temperatures and pressures – fitting of a mathematical model. Combust Flame 30:45–60
Hausen H (1976) Wärmeübergang im Gegenstrom, Gleichstrom und Kreuzstrom, Aufl. 2. Springer, Berlin
Heider G (1996) Rechenmodell zur Vorausberechnung der NO-Emission von Dieselmotoren. Dissertation, TU München
Heywood JB (1988) Internal combustion engine fundamentals. McGraw-Hill, New York
Hohenberg G (1980) Experimentelle Erfassung der Wandwärme von Kolbenmotoren. Habilitation, TU Graz
Hohlbaum B (1992) Beitrag zur rechnerischen Untersuchung der Stickstoffoxid-Bildung schnellaufender Hochleistungsmotoren. Dissertation, Universität Fridericiana, Karlsruhe
Huber K (1990) Der Wärmeübergang schnellaufender, direkteinspritzender Dieselmotoren. Dissertation, TU München
Idel'chik IE (1996) Handbook of hydraulic resistance, 3rd edn. Begell House, New York
Justi E (1938) Spezifische Wärme, Enthalpie, Entropie und Dissoziation technischer Gase. Springer, Berlin
Klaiß Th (2003) Selbstzündung und Wärmeübergang an der Klopfgrenze von Ottomotoren. Dissertation, Universität Hannover
Kleinschmidt W (1993) Der Wärmeübergang in aufgeladenen Dieselmotoren aus neuerer Sicht. 5. Aufladetechnische Konferenz, Augsburg
Kleinschmidt W (2000) Zur Simulation des Betriebes von Ottomotoren an der Klopfgrenze. Fortschritt-Berichte VDI, Reihe 12, Nr. 422. VDI-Verlag, Düsseldorf
Kolesa K (1987) Einfluß hoher Wandtemperaturen auf das Betriebsverhalten und insbesondere auf den Wärmeübergang direkteinspritzender Dieselmotoren. Dissertation, TU München
Li H, Miller DL, Cernansky NP (1992) A study on the application of a reduced chemical reaction model to motored engines for heat release prediction. SAE Technical Paper 922328, San Francisco
Li H, Prabhu SK, Miller DL, Cernansky NP (1994) Autoignition chemistry studies on primary reference fuels in a motored engine. SAE Technical Paper 942062, Baltimore
Li H, Miller DL, Cernansky NP (1996) Development of a reduced chemical kinetic model for prediction of preignition reactivity and autoignition of primary reference fuels. SAE Technical Paper 960498, Detroit
Lichtarowicz A, Duggins RK, Markland E (1965) Discharge coefficients for imcompressible non cavitating flow through long orifices. J Mech Eng Sci 7(2):210–219
Merker GP, Hohlbaum B, Rauscher M (1993) Two-zone model for calculation of nitrogen-oxide formation in direct-injection diesel engines, SAE-Paper 932454
Merker GP, Gerstle M (1997) Evaluation on two stroke engines scavenging models, SAE-Paper 970358
Merker GP, Kessen U (1999) Technische Verbrennung: Verbrennungsmotoren. B. G Teubner, Stuttgart, Leipzig
Merker GP, Schwarz C (2001) Technische Verbrennung – Simulation verbrennungsmotorischer Prozesse. Teubner, Stuttgart
Miersch J (2003) Transiente Simulation zur Bewertung von ottomotorischen Konzepten. Dissertation, Universität Hannover

NIST JANAF (1993) Thermochemical tables database, Version 1.0
Oberg HJ (1976) Die Darstellung des Brennverlaufes eines schnellaufenden Dieselmotors durch zwei überlagerte Vibe-Funktionen. Dissertation, Braunschweig
Pflaum W, Mollenhauer K (1977) Wärmeübergang in der Verbrennungskraftmaschine. Springer, Berlin
Ramos JI (1989) Internal combustion engine modelling. Hemisphere, New York
Reulein C (1998) Simulation des instationären Warmlaufverhaltens von Verbrennungsmotoren. Dissertation, TU München
Reulein C, Schwarz C, Witt A (2000) Methodeneinsatz bei der Ermittlung des Potentials von Downsizing-Motoren. Tagung "Downsizing von Motoren", Haus der Technik, München
Schreiber M, Sasat Sakak A, Lingens A, Griffiths JF (1994) A Reduced Thermokinetic Model for the Autoignition of Fuels with Variable Octane Ratings. 25th Symposium (Int.) on Combustion, Irvine California, pp 933–940
Schreiner K (1993) Untersuchungen zum Ersatzbrennverlauf und Wärmeübergang bei Hochleistungsdieselmotoren. Motortechnische Zeitschrift MTZ 54, Franckh-Kosmos Verlags-GmbH, Stuttgart, pp 554–563
Schwarz C (1993) Simulation des transienten Betriebsverhaltens von aufgeladenen Dieselmotoren. Dissertation, TU München
Seifert H (1962) Instationäre Strömungsvorgänge in Rohrleitungen an Verbrennungskraftmaschinen. Springer, Berlin
Sitkei G (1963) Über den dieselmotorischen Zündverzug. Motortechnische Zeitschrift MTZ 26, Franckh-Kosmos Verlags-GmbH, Stuttgart, pp 190–194
Spicher U, Worret R (2002) Entwicklung eines Kriteriums zur Vorausberechnung der Klopfgrenze. FVV Forschungsvorhaben Heft 741, Frankfurt/Main
Stanciu AS (2005) Gekoppelter Einsatz von Verfahren zur Berechnung von Einspritzhydraulik, Gemischbildung und Verbrennung von Ottomotoren mit Kraftstoff-Direkteinspritzung. Dissertation, Technische Universität Berlin
Stephan K (1993) Vorlesungsskript Wärme- und Stoffübertragung, Universität Stuttgart
Stiesch G (1999) Phänomenologisches Multizonen-Modell der Verbrennung und Schadstoffbildung im Dieselmotor. Dissertation, Universität Hannover
Streit EE, Borman GL (1971) Mathematical simulation of large turbocharged two-stroke diesel engines. SAE-Paper 710176, International Congress & Exposition
Stromberg H-J (1977) Ein Programmsystem zur Berechnung von Verbrennungsmotorkreisprozessen mit Berücksichtigung der instationären Strömungsvorgänge in den realen Rohrleitungssystemen von Mehrzylinder-Verbrennungsmotoren. Dissertation, Ruhr-Universität Bochum
Vibe II (1970) Brennverlauf und Kreisprozeß von Verbrennungsmotoren. VEB Verlag Technik, Berlin
Vogel C (1995) Einfluß von Wandablagerungen auf den Wärmeübergang im Verbrennungsmotor. Dissertation, TU München
Witt A (1999) Analyse der thermodynamischen Verluste eines Ottomotors unter den Randbedingungen variabler Steuerzeiten. Dissertation, TU Graz
Woschni G (1970) Die Berechnung der Wandwärmeverluste und der thermischen Belastung der Bauteile von Dieselmotoren. Motortechnische Zeitschrift MTZ 31, Franckh-Kosmos Verlags-GmbH, Stuttgart, pp 491–499
Woschni G, Anisits F (1973) Eine Methode zur Vorausberechnung der Änderung des Brennverlaufs mittelschnellaufender Dieselmotoren bei geänderten Betriebsbedingungen. Motortechnische Zeitschrift MTZ 34, Franckh-Kosmos Verlags-GmbH, Stuttgart, 106 ff
Zacharias F (1966) Analytische Darstellung der thermischen Eigenschaften von Verbrennungsgasen. Dissertation, TU Berlin
Zapf H (1969): Beitrag zur Untersuchung des Wärmeübergangs während des Ladungswechsels im Viertakt-Dieselmotor. Motortechnische Zeitschrift MTZ 30, Franck-Kosmos Verlags-GmbH, Stuttgart, 461 ff
Zellbeck H (1997) Neue Methoden zur Vorausberechnung und Onlineoptimierung des Betriebsverhaltens aufgeladener Dieselmotoren. 133. In: Aufladetechnische Konferenz, Dresden

Chapter 8
Charging of Internal Combustion Engines

Claus Reulein

Charging was originally considered a performance improvement method. It has, however, been playing an increasingly large role, whereby fuel consumption and emission questions have stepped more into the foreground. For an exhaustive treatment, the reader is referred to Zinner (1985) and Pischinger et al. (2002). Jenni (1993) has provided an interesting portrait of the historical development of charging.

8.1 Charging Methods

Concerning internal combustion engine charging, we first differentiate between external and internal charging. External charging includes externally powered charge air compressors for single-cylinder trial engines and the scavenge blower for large two-stroke engines. In the case of internal charging, the energy required for the increase of charge pressure is supplied by the internal combustion engine itself without the help from an external energy source. Internal charging methods are classified into those with and without compressors and with and without the exploitation of exhaust gas. Figure 8.1 shows the different combinations. While resonance charging, as used in the naturally aspirated engine, works only by exploiting the pressure waves in cooperation with the valve control times, i.e. without a compressor and direct utilization of exhaust gas, the other methods use either a compressor (mechanical supercharging) both the exhaust gas and a compressor (turbocharging) or only the exhaust gas (pressure wave charging). The following will briefly describe the individual methods. Exhaust gas turbocharging will be treated somewhat more extensively, since this method has become prevalent in nearly all engine types and sizes.

G.P. Merker et al. (eds.), *Combustion Engines Development*,
DOI 10.1007/978-3-642-14094-5_8, © Springer-Verlag Berlin Heidelberg 2012

internal self-charging	without compressor	with compressor
without using exhaust	resonance charging	mech. supercharging
with using exhaust	pressure-wave charging	exhaust turbocharging

Fig. 8.1 Charging types

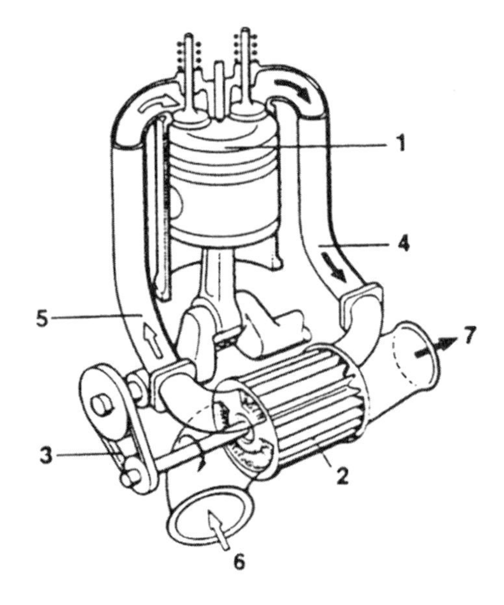

Fig. 8.2 The pressure-wave charger, Hütten (1997)

8.1.1 Pressure-Wave Charging

Pressure-wave charging is based on an invention by Claude Seippel, who first used this principle in 1940/1942 as a high-pressure stage of a gas turbine.

8.1.1.1 Mode of Operation

The pressure-wave charger (Fig. 8.2) consists of a rotor (2), in which open, almost rectangular cells are arranged on both ends in the axial direction, an air casing with an inlet port for fresh air (6) and an outlet port for the compressed air (5) as well a gas casing for the entrance of hot gas (4) and an outlet for the expanded air and so called scavenging air (7). Similar to a turbocharger, the pressure-wave charger has a high and low pressure side on both the fresh gas and hot gas side.

The rotor of the pressure-wave charger can be powered either electrically or with the help of a belt drive (3). As opposed to other charging concepts, the drive does not serve to build up charge pressure but only to control the gas-dynamic processes in the charger.

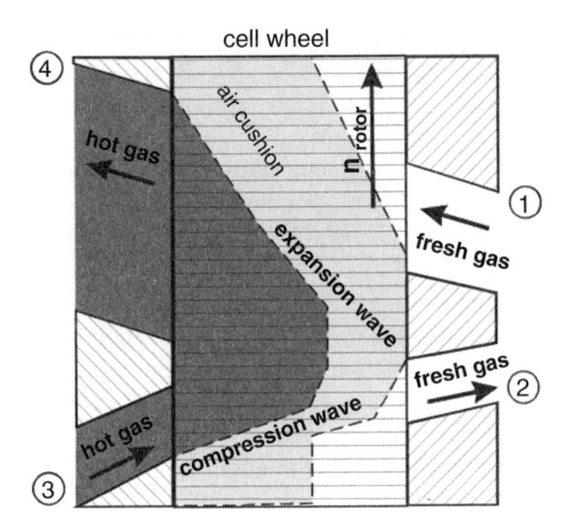

Fig. 8.3 The pressure-wave charger: functional principle

Figure 8.3 shows a view of the cell wheel of a pressure-wave charger. The hot gas (3) enters from the left into the cells of the rotor when these cells move past the inlet corner (from the bottom up). Because of the pressure difference between the hot gas flowing in and the practically stationary fresh gas in the cell, a compression wave is formed, which propagates with the entry velocity of the hot gas, as well as a pressure wave traveling at the speed of sound in the direction of the fresh gas outlet (2). The hot gas thus endows the fresh gas with its pressure by direct exchange of energy and pushes it out of the cell. Before the hot gas can reach the fresh gas outlet, the outlet is closed by the cell movement. Directly after the fresh gas outlet closes, an expansion wave is initiated by the opening of the hot gas outlet (4) and by reflection of the compression wave at the closed cell end. The hot gas flows out of the pressure-wave charger, giving the fresh air the opportunity to flow into the cell of the pressure-wave charger against the lower pressure after the fresh air inlet (1) opens. In the ideal case, the hot gas can be completely displaced.

The quality of the pressure exchange and the purging of the cells depends on the calibration of the pressure-wave charger, which is mainly determined by the penetration depth of the exhaust gas into the cells and the height of the pressure wave. The operating time of the pressure waves (speed of sound) as a ratio of the rotor operating time is responsible for this, which is directly proportional to the speed of the cell rotor. Another decisive prerequisite for proper operation of the pressure exchange process is the purging of the cells, which is significantly determined by the static pressure difference between the fresh gas inlet and the hot gas outlet.

If, as is usual for internal combustion engines, air filters are connected into the inlet pipe before the pressure-wave charger and sound muffling units are connected into the exhaust pipe behind the charger, the static pressure before the fresh gas inlet is always lower than the static pressure after the hot gas outlet because of the associated flow resistances.

The flow direction of the gas from the fresh gas to the hot gas side can thus only be maintained by gas-dynamic effects (expansion waves). For this reason, one must pay heed that the pressure level on the fresh gas side does not drop unnecessarily far and that on the hot gas side is not increased to an unnecessary extent. The pressure-wave charger also requires relatively constant inlet and outlet pressure conditions. To this end, a corresponding stabilizing volume for the exhaust momentum created by the engine must be connected between the engine and the pressure-wave charger.

The pressure-wave charger is normally powered mechanically by a belt drive, which only takes over synchronization to the engine speed and overcoming mechanical friction and must not contribute to the compression of fresh air. Modern designs make use of an electric motor for synchronization, which makes a free arrangement of the pressure-wave charger in the engine space possible and makes it easier to adjust the rotor length to the boundary conditions due to the free choice of speed.

8.1.1.2 Interaction with the Internal Combustion Engine

Figure 8.4 shows the interaction of the pressure-wave charger with the internal combustion combustion engine.

The left diagram shows the high torque potentials of the pressure-wave charger in connection with a very large specific power output in an SI engine in comparison with an engine of the same capacity with turbocharging.

In the right diagram, we see a load transient from partial to full load. We can clearly see the much faster charge pressure increase of the pressure-wave charger compared to turbocharging.

Despite the positive properties of the pressure-wave charger, such as the practically instant energy exchange via the pressure waves and the unlimited operational range

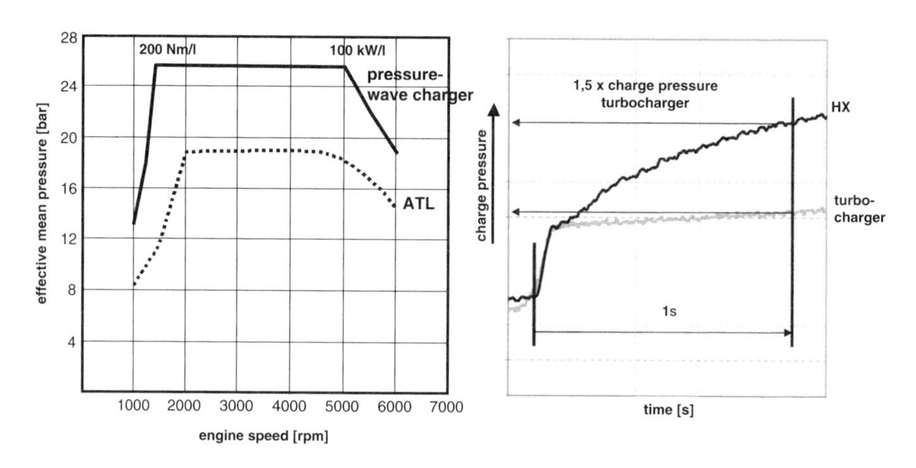

Fig. 8.4 Torque profile and response pressure-wave charger (Flückiger et al. 2006)

(surge limit in the case of the turbocharger), the pressure-wave charger is not subject to mass-production applications. The disadvantageous properties are as follows:

- Larger installation volume and weight than, for example, a turbocharger.
- Intake and exhaust pipes must be dimensioned larger because of their sensitivity to flow losses.
- When powered by a belt drive: arrangement in a direction parallel to the engine's longitudinal axis.
- Crankcase ventilation cannot be conducted before the pressure-wave charger (blockage of the cells by oil residues).
- Both on the intake and exhaust side, there is a very high-frequency noise (>500 Hz), which is difficult to muffle due to the low-pressure sensitivity.
- Compensation volume is necessary on the hot gas high-pressure side in order to reduce the vibrations caused by the exhaust impacts.
- Higher charge air temperatures due to heat exchange in the cells as the scavenge air sinks (calibration).
- High amount of scavenge air lowers the volumetric efficiency of the engine, especially at higher speeds.
- Exhibits weaknesses at start-up.
- Limitations of the power range for vehicle applications by the installation size.

Calculation of the pressure-wave charger in the context of overall process simulation can be achieved as a "pipe bundle" with variable reflection conditions (open and closed tube end) with the help of the one-dimensional gas dynamics described in Sect. 7.4. Since the calculation is very complex and because pressure-wave chargers are not mass-produced, we will dispense with a detailed description here.

8.1.2 Mechanical Supercharging

In the case of mechanical supercharging, a charging aggregate mounted on the engine is usually powered by a belt or gearwheel drive, which is why the engine and compressor speeds are usually rigidly coupled (Fig. 8.5). There is also electromechanical coupling, in which the charger is powered electrically. In the case of a rigid coupling between the internal combustion engine and the compressor, a compressor by pass is necessary to control the excess boost pressure.

In principle, charging aggregates can involve both turbo compressors (see Sect. 8.1.3) and positive displacement chargers.

8.1.2.1 Designs

The displacement chargers suitable for internal combustion engines can be subdivided into different designs that work either with or without internal compression. The speeds of positive displacement chargers are usually about double the engine

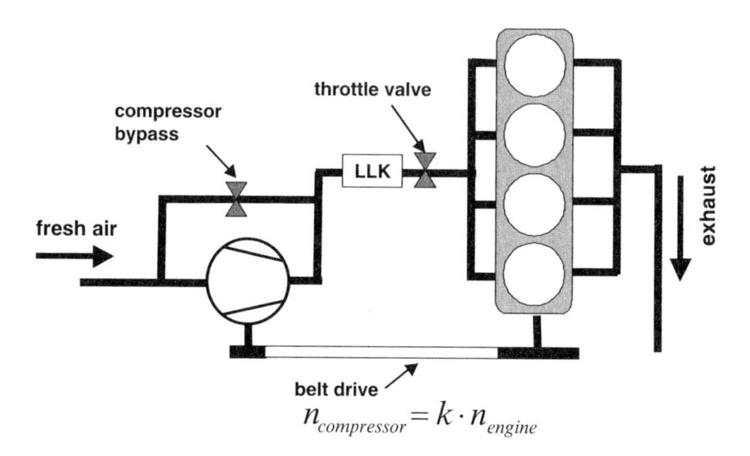

Fig. 8.5 Mechanical supercharging: block diagram

speed. Turbo compressors require a very high transmission ratio, since they are designed for much higher speeds (>100,000 rpm).

The following lists different designs of mechanical supercharging aggregates:

- Oscillating motion
  - Piston compressors (extra aggregate or crankcase for two-stroke engines, internal compression)
- Rotating motion
  - Hier fehlt Strömungsverdichter (siehe Kap. 8.1.3)
  - Twin-cam rotary piston compressors
    - Rotation piston compressors (e.g. roots compressor, no internal compression; internal-axis rotary piston charger, internal compression/expansion depending on the control edge)
    - Screw-type compressors (internal compression)
- Forced rotating compression
  - Spiral chargers (e.g. G-chargers, without internal compression)

Despite the variety of designs of positive displacement compressors, there are some basic common denominators:

- Positive displacement chargers always work between two fixed volume limits.
- Due to their discontinuous functioning, acoustical effects occur in case of an abrupt opening of the conveyor room against the volume lying behind the charging aggregate.
- The volume flow increases approximately in proportion with the speed.
- The performance map is stable in the lower speed and high pressure range despite somewhat increased leakage.
- All designs result in the same characteristic profile of lines of constant speed in the pressure ratio/volume flow map.

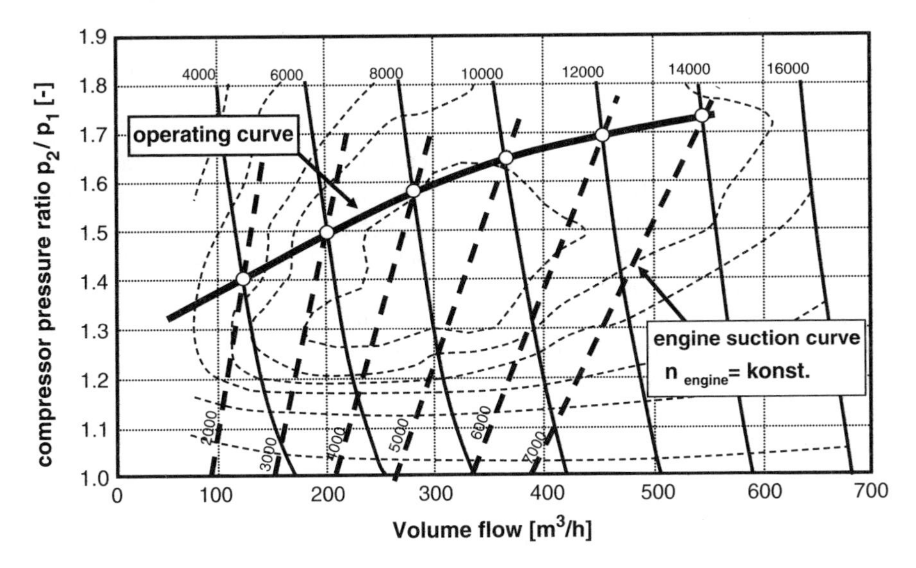

Fig. 8.6 Map and line of operation of a positive displacement charger ---- isentropic efficiency; ——— charger speed [rpm]

Figure 8.6 shows the map of a positive displacement charger showing its characteristic properties. We can see that the lines of constant compressor speeds in the volume flow/pressure ratio map are inclined upwards from the left. With positive displacement chargers, pressure ratios of approx. 2 (exception: reciprocating piston compressors $\gg$2 and screw-type compressors >2) can be reached. The lines of constant compressor efficiency result in a "mussel" characteristic map that is usually at its optimum at average volume flows and higher pressure ratios. Conventional positive displacement chargers have isentropic efficiencies of up to 65%.

8.1.2.2 Interaction with the Internal Combustion Engine

Figure 8.6 shows the engine air mass flow characteristics for a characteristic design in the compressor map. The engine air mass flow characteristics themselves represent the suction capacity of the combustion engine at a constant speed as a function of the pressure ratio before the inlet valve. At a constant transmission (here i = 2) we see the interaction between the suction capacity of the combustion engine and the conveying characteristics of the positive displacement charger, which is plotted into the map as a line of operation. This curve determines the full load of the engine. In the case of a quality-controlled SI engine, the air mass must be adjusted to the load control accordingly. In the list below, the most common measures will be briefly described:

- *Throttle valve after the charger*: common, but not very efficient because the charge pressure previously created under mechanical work is lost again by the throttle valve.
- *Throttle valve in front of the charger*: less common since, because of the low pressure before the charger, motor oil used to lubricate the charger can escape through the charger lining and be taken in by the charge air cooler (an emission disadvantage – clogging of the charge air cooler).
- *Bypass valve on the charger*: common, since this is an efficient performance control measure on the charger.
- *Mechanical shutdown of the charger by a switchable coupling*: common and very efficient, but high mechanical stress on the coupling is required in case of rapid switching as well as high application cost with respect to driving comfort (wear and limited switching range).
- *Variability in the transmission ratio*: not common, since fully variable transmission is costly.
- *Internal charger measures*: not common, since the variability of the control times of the charger is difficult to master mechanically.

Due to the low overall efficiency of the usually low boost pressures compared to turbocharging, mechanical supercharging is almost exclusively limited to SI engines. Here, both a high torque at low speeds and a sufficient surplus torque in unsteady processes are necessary. From this we obtain the selection criteria for mechanically powered charging aggregates:

- Conveyance characteristics adjusted on the engine and maximum mass flow
- Required transmission ratio to maintain the permissible maximum speed of the charging aggregate
- Small moment of inertia and low compressor drive power
- Possibility of operation within a favorable range of efficiency (characteristic map)
- Simple possibility of charge air control for adjustment to the load profile
- High volumetric efficiency at low speeds
- Low drive power in idle operation
- Easy to access and good arrangement of the air inlet and outlet sections
- Low noise emissions
- Small installation space and low weight
- Low charger manufacturing costs
- Low overall engine costs (wiring and control equipment costs)

Figure 8.7 shows the full-load characteristics of a mechanically charged SI engine in comparison to a naturally aspirated engine of the same capacity. We can see that the full load line of the naturally aspirated engine can be shifted almost parallel to higher torques. At low speeds however, the decline of the compressor pressure ratio seen in Fig. 8.6 causes a small increase in torque, which is why the full load curve is slightly inclined downwards to the left.

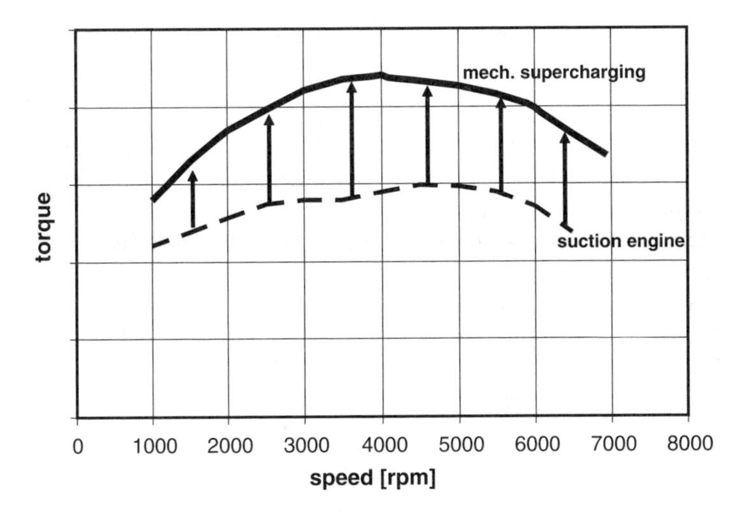

Fig. 8.7 Torque profile of a mechanically charged SI engine

8.1.2.3 Basic Thermodynamic Equations

The actual performance of the engine while running a mechanically powered supercharger comes out to

$$P_e = P_i\,\eta_m - P_{com}\frac{1}{\eta_{tr}} \tag{8.1}$$

with a transmission efficiency η_G. For the compressor performance, we obtain

$$P_{com} = \dot{m}_{com}\,\Delta h_{is,com}\frac{1}{\eta_{is,\,com}}\frac{1}{\eta_{m,\,com}}. \tag{8.2}$$

Should the flow velocities in front of and behind the compressor be about the same, it then follows from the first law of thermodynamics for the isentropic compression

$$\Delta h_{is,\,com} = c_p\,T_{b.com.}\left[\left(\frac{p_{a.com.}}{p_{b.com.}}\right)^{\frac{\kappa-1}{\kappa}} - 1\right] \tag{8.3}$$

and for the isentropic efficiency

$$\eta_{is,\,com} = \frac{\Delta h_{is,\,com}}{\Delta h_{com}} = \frac{T_{a.com.s} - T_{b.com.}}{T_{a.com.} - T_{b.com.}}. \tag{8.4}$$

With these relations, we obtain for the temperature ratio at the compressor

$$\frac{T_{a.com.}}{T_{b.com.}} = 1 + \frac{1}{\eta_{is,\,com}}\left[\left(\frac{p_{a.com.}}{p_{b.com.}}\right)^{\frac{\kappa-1}{\kappa}} - 1\right]. \tag{8.5}$$

The advantages and disadvantages of mechanical supercharging are summarized in the list below.

- Advantages:
  - Spontaneous response behavior thanks to immediately available charge air in case of a sudden load change when a reduction of the compressor power can be quickly eliminated at partial load.
  - High boost pressures already available at low engine speeds (limitation: surge limit in the case of the mechanically powered turbo compressor).
  - Response behavior is hardly affected by exhaust gas aftertreatment systems.
  - No changes are required in the exhaust apparatus since only the inlet side is affected.
  - The exhaust back pressure does not influence the gas exchange/lower gas exchange work.
  - No adverse effects on heating and thus on the response behavior of exhaust gas aftertreatment systems (catalytic converters, soot filters etc.) in case of a cold start or unsteady processes (no heat extraction through the turbines).
  - No heating of the fresh gas via heat transfer from the exhaust (as in the case of the turbocharger).
  - No influence on the bearings or materials of the compressor due to exhaust heat.
  - There is no metal mass in the engine room which can heat up and radiate heat to the environment.
- Disadvantages:
  - Specific fuel consumption at partial load without corresponding control can be higher than that of the natural aspirated engine.
  - High costs for design and control is required to lower the compressor power (coupling, speed shifting and valve control).
  - Under certain conditions, response time is delayed because of shifting or positioning delays in case of a sudden load requirement.
  - High costs due to complex charging aggregates, transmissions with sometimes their own belt drives and difficult controls.
  - The smaller engine installation volume with the same power is compensated by the installation space for the charging aggregate and the drive.
  - Higher noise properties in the case of positive displacement chargers due to the unsteady operation.
  - In the case of the mechanically coupled compressor, the interaction of the engine and the charging aggregate also causes a reduction of the boost pressure at low engine speeds and of the useful engine torque due to the high compressor drive power at full engine load and due to the often high leakage losses of the charging aggregate at low speeds.
  - In the case of positive displacement chargers with abrasive sealing gaskets, there are high friction losses and a resultantly low overall efficiency and often a large amount of wear.
  - To obtain small gap widths, a considerable amount of manufacturing cost is required for positive displacement chargers.

In the case of engines with high efficiency requirements (slow and average-speed large diesel engines), the turbocharger has become preferred to the supercharger due to the high boost ratio and the better overall efficiency. For commercial and diesel passenger applications as well, mechanical supercharging has not become widespread despite its advantages in unsteady conditions due to its complexity and lower efficiency. Only in the field of SI passenger cars has mechanical charging been applied in some niche applications without cost pressure or in combination with turbocharging in order to improve unsteady behavior in the case of extreme downsizing concepts. Here too, the overall efficiency of mechanical charging is at most comparable with turbocharging with moderate boost grades.

8.1.3 Turbocharging

In turbocharging, a turbo compressor is powered with an exhaust turbine using energy from the exhaust gas, i.e. the reciprocating engine is only fluidically connected to the turbo machine "turbocharger" (Fig. 8.8). The exhaust gas of the internal combustion engine slows through the exhaust manifold to the turbine and powers it. A compressor sitting on the main shaft of the turbine, which can in fixed operation convert the drive power of the turbine into compressor power (minus bearing friction), itself compresses fresh air, which must be cooled by a charge air cooler. Via the manifold, the compressed air enters the cylinder. In the case of highly unsteady applications (e.g. vehicle operation), the power of the turbine must be controlled. This is done in the schematic shown here with a so-called wastegate.

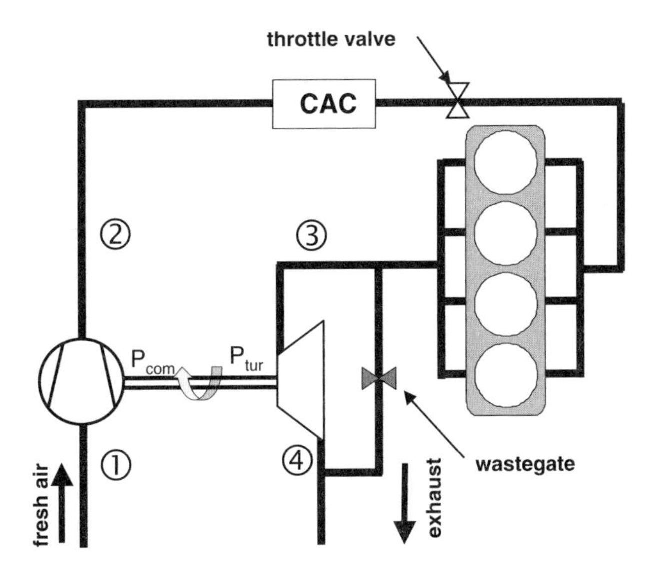

Fig. 8.8 A reciprocating engine and turbo machine. Indexes: 1: b.com, 2: a.com, 3: b.tur, 4: a.tur

8.1.3.1 The Turbo Compressor: Basic Principles

Compressors designed according to the flow principle are classified basically as axial compressors or radial compressors. Figure 8.9 shows a cross-section through a radial compressor.

After entering the compressor (1), kinetic energy is first supplied to the medium to be compressed by means of the mechanical drive energy of the compressor; the medium is accelerated in the rotor (2). The kinetic energy is then converted into pressure energy after passing the very narrow section, similar to a nozzle, and/or after exiting the rotor (3).

Design-engineering variables for turbo machines include the A/R ratio and the trim.

$$Trim = \left(\frac{d_1}{d_2}\right)^2 \tag{8.6}$$

Figure 8.10 shows this process using so-called blade diagrams of a radial compressor.

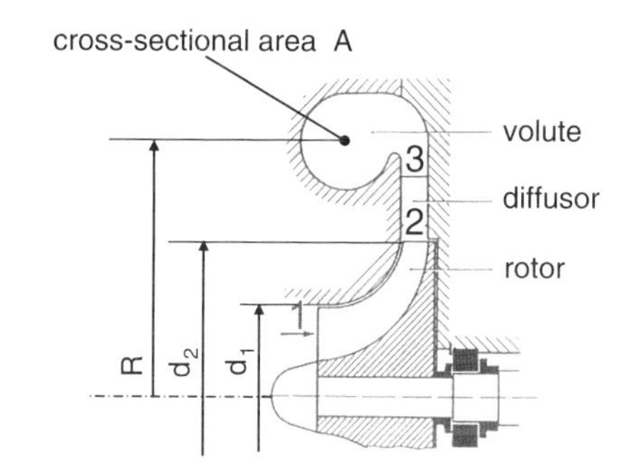

Fig. 8.9 Longitudinal cut through a radial compressor

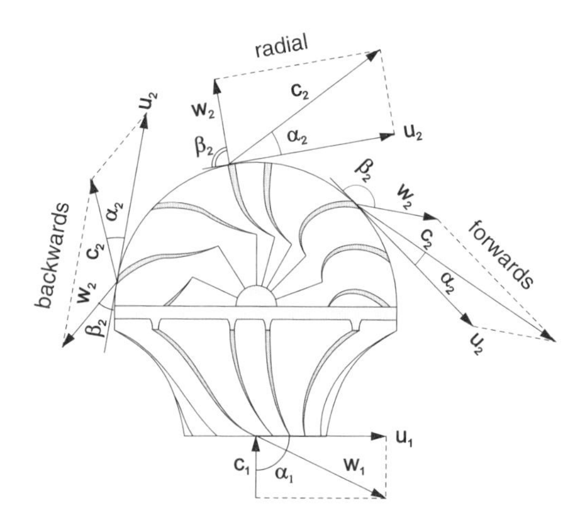

Fig. 8.10 Velocity triangle of a radial compressor

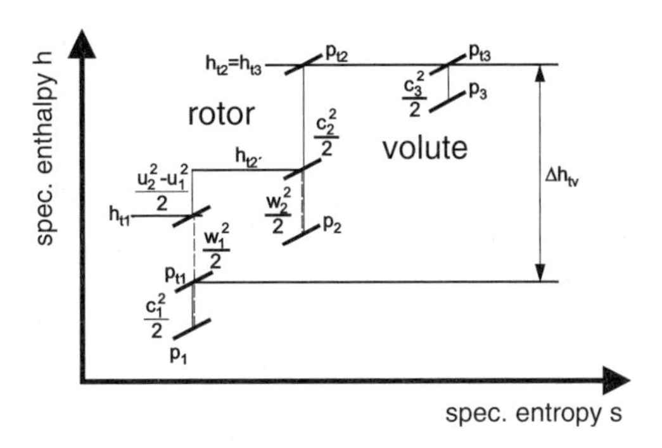

Fig. 8.11 Compression process in the h-s diagram

Here, c means absolute velocity, kursiv the circumferential speed of the compressor rotor and w the relative speed of the medium with respect to the compressor blades in the moved system. The relations between the individual velocities are obtained by means of vector addition as shown in Fig. 8.10.

There are designs with blades bend forwards, terminating radially and bent backwards (Fig. 8.10), whereby these days the backwards-bent design is used in passenger car engines because of the obtainable characteristic map width.

If the focus of the design is placed on good unsteady behavior, the polar moment of inertia of the compressor rotor (proportional to the fifth power of the diameter) should be kept as small as possible. This leads to a reduction of the diameter and thus to an increase of the compressor speed while maintaining the same mass flow.

Figure 8.11 shows the process of compression within a turbo compressor in an h-s diagram. Since the rotor and the volute usually comprise one unit, a global point of view from inlet to outlet is adopted (state 1–3).

From the representation in Fig. 8.11, we can calculate the total enthalpy difference required for compression (Euler equation), from which the compressor power can also be determined.

$$\Delta h_{tV} = -\frac{c_1^2}{2} + \frac{w_1^2}{2} + \frac{u_2^2}{2} - \frac{u_1^2}{2} - \frac{w_2^2}{2} + \frac{c_2^2}{2}, \tag{8.7}$$

The resultant compressor power is then:

$$Pv = \dot{m}_{com} \cdot \Delta h_{tV} \tag{8.8}$$

Otherwise, the relations derived in (8.2) and (8.3) are also valid for a turbo compressor.

8.1.3.2 The Flow Turbine: Basic Principles

Turbines designed according to the flow principle are similarly designated as either axial or radial turbines. While the use of axial turbines is almost exclusively restricted to large diesel engines, radial turbine designs are used comprehensively in all other applications because of their compactness.

In turbines – like turbo compressors, just in the opposite direction – the pressure energy of the gas before the turbines is converted into kinetic energy in order to convert this energy in the rotor of the turbine to the turbine shaft in the form of mechanical energy.

Figure 8.12 shows this process using a blade diagram of a radial turbine. The outer nozzle ring is normally fixed or formed as a short volute slot. In the case of turbines with an adjustable nozzle ring (VTG, see boost pressure control), different absolute velocities c arise at a constant circumferential velocity u depending on the position of the outer guide blade. These absolute velocities then result in different directions (and amounts) of relative velocity due to the vector addition of velocities and lead to a more or less optimal impact of the gas upon the blade geometry. From this no longer optimal incoming flow results the sometimes considerable loss of efficiency characteristic of this turbine design (Fig. 8.13).

Figure 8.14 shows the process of gas expansion in a radial turbine in the h-s diagram. The Euler equation is valid in this case as well [see (8.7)].

With this, we can calculate the power of a turbine according to (8.9).

$$P_{tur} = \dot{m}_{tur} \cdot \Delta h_{ttur} \tag{8.9}$$

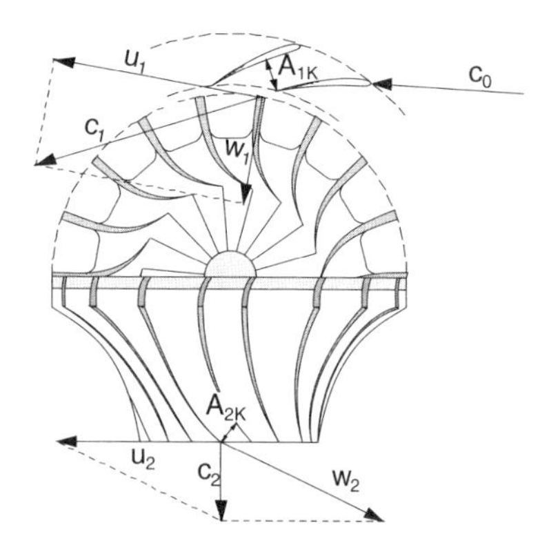

Fig. 8.12 Velocity triangle of a radial turbine

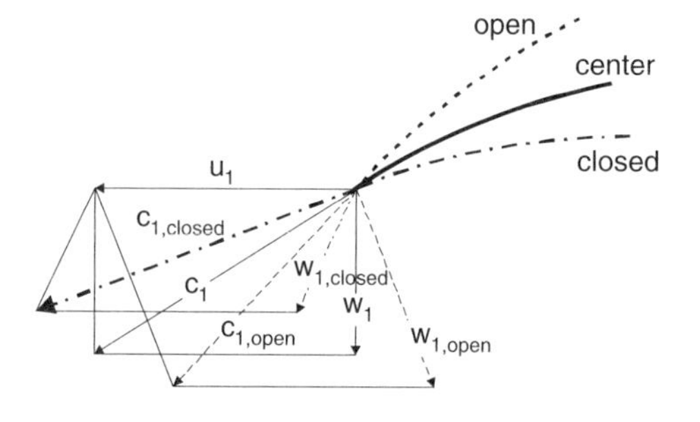

Fig. 8.13 Velocity triangle of an adjustable guide apparatus

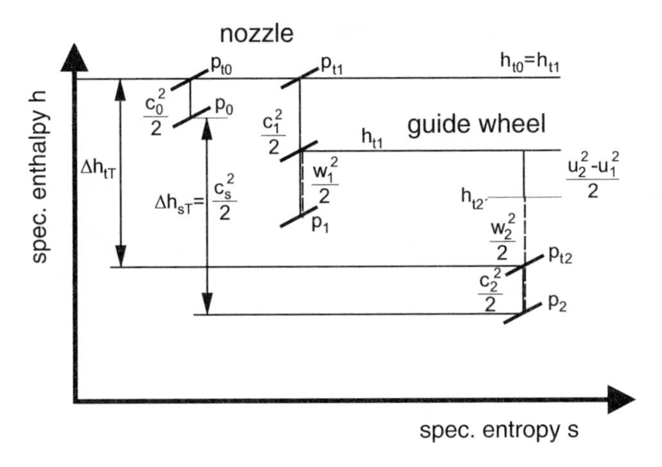

Fig. 8.14 Expansion in a radial turbine in the h-s diagram

$$P_{tur} = \dot{m}_{tur} \cdot \frac{\kappa_{btur}}{\kappa_{btur}-1} R_{btur} T_{btur} \eta_{is,tur} \left(\pi_{tur}^{*\frac{\kappa_{btur}}{\kappa_{btur}-1}} - 1 \right) \quad \text{with} \quad \pi_{tur}^{*} = \frac{p_{btur}}{p_{atur}} \tag{8.10}$$

The mass flow through the turbine, the so-called turbine suction capacity, can be determined approximately with the relation for the flow through a restriction (throttle). With the isentropic substitute section of the turbine, we obtain:

$$\dot{m}_{tur} = A_{is,\,tur}\, \rho_{is}\, v_{is}, \tag{8.11}$$

with

$$\rho_{\text{is}} = \frac{p_{b.tur.}}{R_t\, T_{b.t.}} \left(\frac{p_{a.tur.}}{p_{b.t.}} \right)^{\frac{1}{k_t}} \tag{8.12}$$

b.t. a.t. $\dot{m}_E$

and

$$b_{is} = \sqrt{\frac{2\,\kappa_{tur}}{\kappa_{tur}-1} R_{tur}\, T_{b.tur} \left[1 - \left(\frac{p_{a.tur.}}{p_{b.tur.}} \right)^{\frac{\kappa_{tur}-1}{\kappa_{tur}}} \right]}. \tag{8.13}$$

Insertion of ρ_{is} and v_{is} into the relation for $\dot{m}_T$ and conversion finally provides us with the expression

$$\dot{m}_{tur} \frac{\sqrt{tur_{b.tur.}}}{p_{b.tur.}} = A_{is,tur} \sqrt{\frac{2\,\kappa_{tur}}{R_{tur}(\kappa_{tur}-1)} \left[\left(\frac{p_{a.tur.}}{p_{b.tur.}} \right)^{\frac{2}{\kappa_{tur}}} - \left(\frac{p_{a.tur.}}{p_{b.tur.}} \right)^{\frac{\kappa_{tur}+1}{\kappa_{tur}}} \right]}. \tag{8.14}$$

This is the "second fundamental equation of turbocharging"

$$\dot{m}_{tur} \frac{\sqrt{T_{b.tur.}}}{p_{b.tur.}} = A_{is,tur} f\left(\frac{p_{a.tur.}}{p_{b.tur.}}, \kappa_{tur}\right). \tag{8.15}$$

With the help of this relation, we can form reference or dimensionless variables. This will be dealt with in more detail in Sect. 8.2.

The isentropic turbine substitute cross section is approximately constant and depends only on the geometrical turbine cross section; we thereby ignore its dependence on the speed of the turbocharger and on the pressure ratio $p_{b.tur}/p_{a.tur}$.

8.1.3.3 Interaction with the Exhaust Dynamics of the Internal Combustion Engine

Strictly speaking, turbocharging is classified as either constant-pressure turbocharging or pulse turbocharging.

In constant-pressure turbocharging, only the thermal energy of the exhaust is utilized, since the exhaust of a single cylinder of an internal combustion engine is guided into a large pipe volume between the engine and the turbine, in which it loses its momentum. This momentum results from the discontinuous working process of the internal combustion engine, in which the exhaust abruptly escapes into the manifold when the outlet valves open. There is thus a constant pressure before the turbine in this type of turbocharging, which also has an effect on the individual cylinders. For this reason, a free selection of the ignition sequence of the combustion engine is possible in the case of ideal constant-pressure turbocharging from the thermodynamic perspective. Also, there is no limitation on the cylinder number of the engine joined to the outlet volume. The constant pressure before the turbine has, on the one hand, a positive effect on the average efficiency and suction behavior of the turbine. On the other hand, the velocity contained in the individual impulses of exhaust gas is first converted into thermal energy and thus into static pressure. This energy must then again be converted in the turbine into the more valuable kinetic energy (nozzle, guide apparatus). Since each process of energy conversion is bound with losses however, corresponding losses arise both in the uniformization of the flow and the reacceleration of the flow. Besides the high demands on the package, which can only be met for a stationary application, the exhaust also loses a considerable amount of energy due to the large surface of the volume connected between the engine and the turbine. A large volume with highly unstable engine operation is especially problematic, e.g. as in an airplane, since filling the volume with exhaust in order to obtain a corresponding pressure and thus a corresponding amount of turbine power takes a long time. This is especially true at low speed and loads, since in this case there is already a low pressure level with low mass flows.

In pulse turbocharging on the other hand, the aim is to direct the exhaust to the turbine using as short a path as possible with as little loss as possible. Both the thermal and kinetic energy of the exhaust is exploited. The exhaust system of a pulse-turbocharged engine is therefore characterized by its compactness. The flow of the exhaust is however very inconsistent, since the pressure waves in the exhaust system propagate with sonic speed and, due to the compactness of the pipe system, quickly affect the exhaust counterpressure of the other cylinders joined to the pipe system. This has adverse effects on the turbine on the one hand, which is applied with highly erratic mass and energy flows. The turbine blades are not optimally impacted with flow at every moment in the process. This in turn has a negative effect on efficiency and eliminates part of the benefits. The latter also has a negative effect on the mechanical and thermal strength of the turbine.

On the other hand, the high fluctuations in pressure also have a great effect on the engine gas exchange. Depending on the ignition sequence and the outlet valve control times of the cylinders combined in the exhaust system, it is possible that gas exchange during the push-out phase can be considerably impaired by the previous or subsequent cylinder. The result is that either the pushing-out work becomes very high or that residual gas remains in the cylinder and the combustion of the next working cycle is negatively affected. In the case of a four-stroke engine, three cylinders can be joined assuming a control time of max. 260°CA for one outlet valve. This means that in the case of three or six-cylinder engines (or all cylinder numbers dividable by 3), an aggregation of cylinders that is optimal with respect to gas exchange and turbocharging is possible in principle.

The six-cylinder engine requires however an arrangement of three cylinders each on either two turbines or on one double-flow turbine (twin-scroll), which makes flow separation possible (see below). In the case of the four-cylinder engine, there is an ignition distance of 180°CA, which is why only two cylinders can be joined optimally, which again means two exhaust flows (twin-scroll turbine). Nevertheless, four-cylinder engines in particular also have single-flow aggregates, but to significant disadvantage at low speeds and high loads and in their response behavior. In the V8 engine, the usual ignition sequence necessitates an ignition distance of only 90°CA on every bank, which can only be poorly compensated by flow combination. In this case, usually only combination of four cylinders on one bank remains.

Since complete reduction of the pressure pulse itself is not possible in the case of slowly running engines, while on the other hand an ideal utilization of the exhaust pulses is tied to a real volume of the exhaust system, no pure form of constant-pressure or pulse turbocharging exists.

8.1.3.4 Flow Turbine Designs

In the previous section, the necessity of different flow turbine designs was already pointed out. The basic in this section will be on radial turbines, which can be subdivided into monoscroll – and twin-scroll – or twin turbines. As the name already suggests, monoscroll turbines have only one entry snail which directs the exhaust to the rotor. Figure 8.15 (left) shows a cross-section through a monoscroll

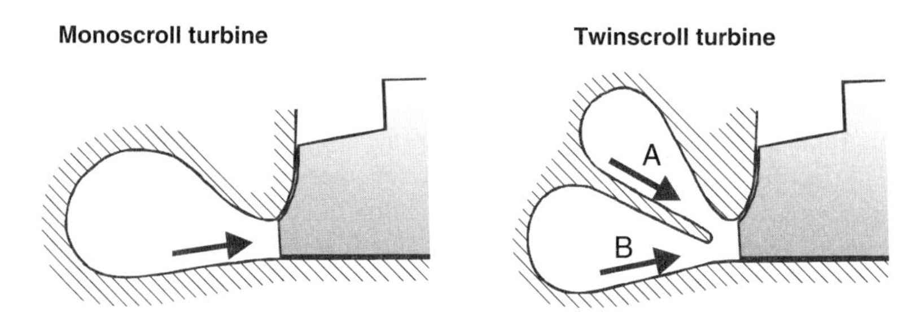

Fig. 8.15 Cross-section through a monoscroll and a twin-scroll turbine

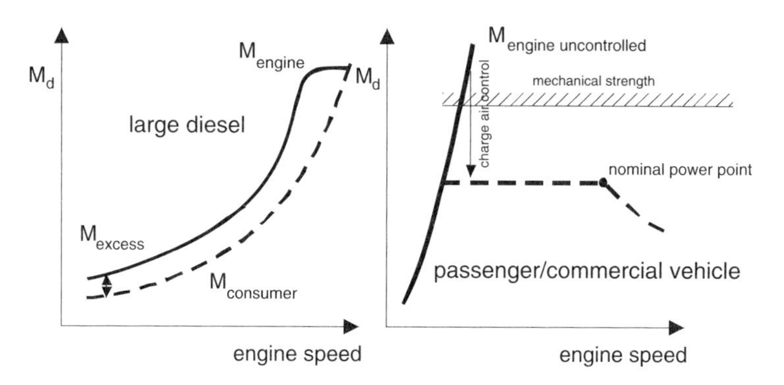

Fig. 8.16 Engine torque and load torque for large diesel and vehicle engines

turbine. On the right is a twin-scroll turbine. We can recognize the two parallel snails that direct the exhaust flows of each cylinder aggregate completely separately to the rotor of the turbine. To fully exploit the functionality of twin-scroll turbines, a complete and consistent separation of the flows until entering the rotor must be assured in order to avoid crosstalk of the gas pulses.

8.1.3.5 Charge Pressure Control Measures

Due to the characteristics of turbo machines, a turbocharger designed for full load without any external control has an approximately quadratic dependence of the boost pressure on the engine speed. At low engine speeds, only a small boost pressure and thus only a small engine torque can be created as a rule. Figure 8.16 (left) clarifies this behavior. This tendency is not as critical for medium speed diesel engines, whose time response is usually not decisive because of the stationary operation near the nominal power point, than for small, high-speed engines, which require high torque already at very low engine speeds due to their dynamic operation. For this reason, measures must be taken in such engines to increase the boost pressure and thus the torque availability at low engine speeds. In the basic design of the turbine, value is placed on high pile-up pressure and thus on high turbine performance already with low mass flows. That means that the size of the turbine is not optimally designed for the nominal power point, but for a much lower engine speed ($<$ 2,000 rpm). With a smaller turbine size, the pressures and turbocharger speeds would become very high with increasing engine speeds due to the almost quadratic increase of mass flow through the compressor and turbine, which would exceed the mechanical strength of the charging system and the combustion engine. Moreover, the charge pressure would increase dramatically, which would lead to serious knocking problems in SI engines due to the resultant high temperatures. For these reasons, boost pressure, exhaust counterpressure and the speed of the turbocharger must be limited by suitable measures in the charging system.

Until now, two methods have become established on the turbine side of the exhaust turbocharger. On the one hand, it is possible to control the exhaust mass flow through the turbine with a wastegate (Fig. 8.8). In this case, part of the exhaust mass flow is guided past the turbine and is thus no longer available for the production of drive power, which is why the compressor conveys a smaller mass flow. In this way, the boost pressure before the engine can be controlled and thus the engine torque as well. For wastegates, the control elements are either pressure sensors that are applied with boost pressure or negative pressure, or electric control elements.

It is also possible to limit the power of the turbine by means of variability in the guide vanes by changing the cross-section by turning the guide vanes. When the cross-section is reduced, this causes a higher pile-up pressure and thus larger turbine power, and if the cross-section is enlarged, it is possible to reduce the pile-up pressure and turbine power.

Nearly all modern turbocharged car diesel engines have charge pressure control using an adjustable guide apparatus. Turbocharged SI engines as a rule require a wastegate because of their much higher exhaust temperatures, since adjustable guide vanes would represent a much larger function risk.

Moreover, SI engines with conventional air-mass load control require, for acoustic reasons, a "diverter valve". By means of reverse flow at the compressor, this valve makes possible a rapid reduction of the charge pressure in case of a sudden intrusion of mass flow due to load reduction. Without a diverter valve, the operating point of the compressor in the characteristic map would shift to the left over the surge limit (see Fig. 8.17) due to the inertia of the exhaust turbocharger. If one keeps the mass flow through the compressor to a higher level with the diverter valve, it is possible not to exceed the suege limit until the speed of the compressor has been reduced.

Figure 8.17 shows the characteristic map of a turbo compressor with its limits (surge, speed and choke limit). The structure of the map is described in more detail in Sect. 8.2.

The full load operating line is drawn in the characteristic map, which runs almost quadratically in the left part. We can clearly see the kink in the operating line, which is cause by the opening of the wastegate due to the charge pressure control described above.

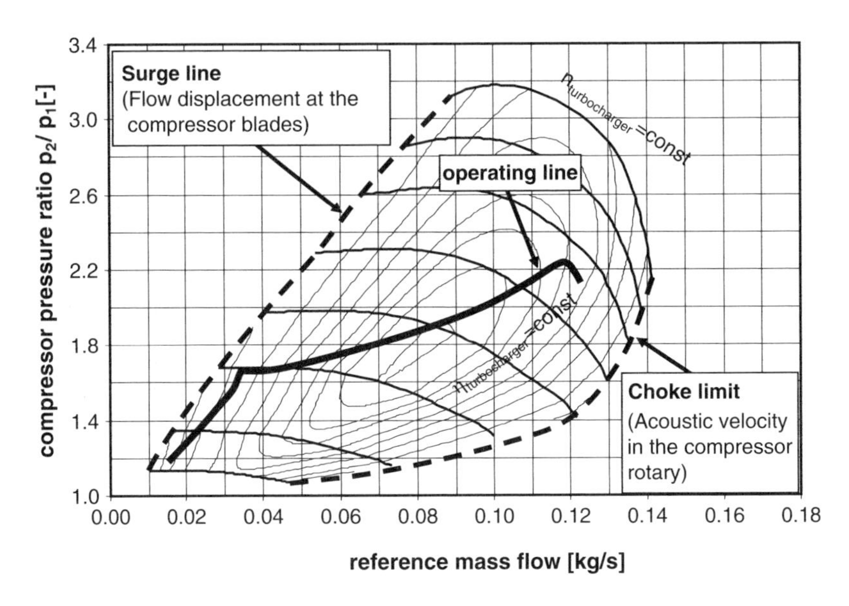

Fig. 8.17 Characteristic map and operating line of a turbo compressor

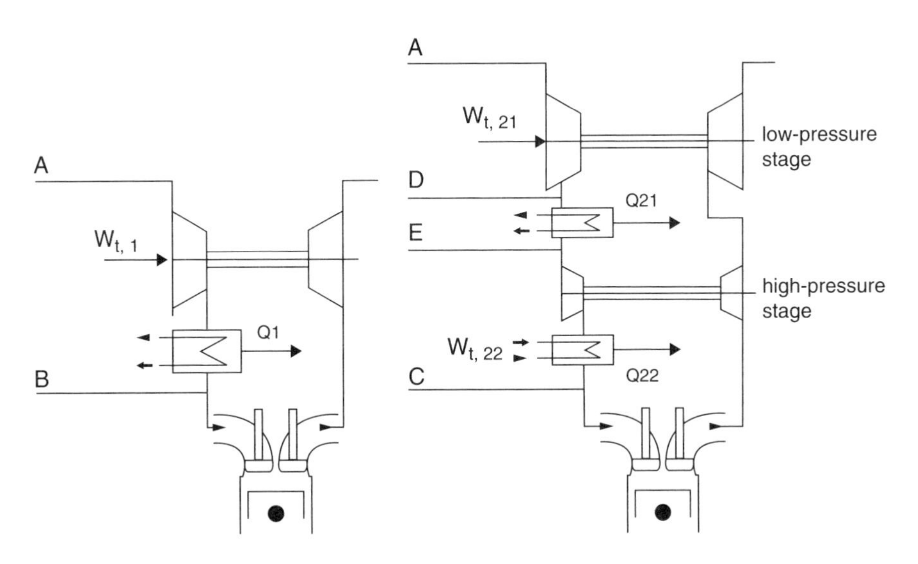

Fig. 8.18 One-stage and two-stage turbocharging

8.1.3.6 Wiring Variants

Figure 8.18 shows a pictorial schematic of

- One-stage
- Two or multi-stage exhaust turbocharging with intermediate cooling

Assuming that the same boost pressure is reached each time, the compressor must work less in the case of two-stage charging than in single-stage charging because of the intermediate cooling (= low-pressure charge air cooler). In accordance with this, the total amount of heat to be removed is also smaller. Theoretically, an isothermal compression would be obtained were an infinite amount of charge stages with corresponding intermediate coolers to be used.

Sequential turbocharging is a one-stage or two-stage charging process in which

- Several equally large turbochargers or
- Several turbochargers of different sizes

are connected sequentially, i.e. with increasing engine loads and speed. It is used in medium-speed and high-speed high-performance engines, while two-stage sequential charging is only used in high-speed high-performance diesel engines.

Figure 8.19 shows the full load characteristic map and operating ranges of a sequential charging with a small and a large turbocharger in a high-performance engine. The small turbocharger ensures a quick response time and good torque at low speeds, while the large turbocharger provides good efficiency at high engine speeds.

A compound process is a combination of different charging processes in one and the same engine, e.g.:

- Mechanical supercharging for operation at low engine speeds in combination with turbocharging
- Turbocharging in combination with an additional power turbine for full load operation

An additional mechanical supercharger can improve the unsteady operation of the engine, since the charge pressure of a mechanically powered compressor is available almost without delay. Design concepts, in which the compressor is electrically powered, have not become established due to the high energy cost involved.

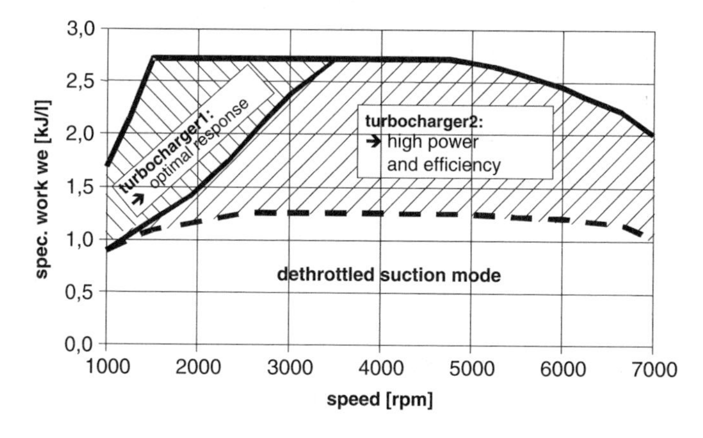

Fig. 8.19 Operating ranges in sequential turbocharging

8.1.3.7 Interaction with the Internal Combustion Engine

Figure 8.17 shows the interaction of a internal combustion engine and a turbocharger with the help of the stationary full load operating line in the compressor characteristic map. This operating line is drawn in Fig. 8.20 in the torque/speed diagram in comparison to a natural aspirated engine of the same capacity as well as a mechanically charged engine. We can clearly see that in the case of the steady-state operation of a turbocharged engine a high torque can be reached already at very low speeds ("low end torque" LET). The torque plateau can be held constant by the wastegate control described above until the desired power is reached (power hyperbola). The early position of LET is made possible by the variable control times in the SI engine caused by the inlet and outlet camshaft phasing. In the case of a positive pressure difference at the turbocharger, in which case the charge pressure is larger than the exhaust counterpressure, in combination with a large overlap of outlet and inlet valve opening during gas exchange, a complete purging of the residual gas from the cylinder and a filling of the cylinder with cold fresh air is possible. In combination withy direct injection, the air fuel ratio in the cylinder can be set slightly rich, which causes very high combustion velocities. The result is a largely knock-free operation, even at low engine speeds along with high torque values.

Figure 8.21 shows the unsteady behavior of a turbocharged SI engine at a constant speed of 1,500 rpm and a load step from partial load to full load, from which we can see that the flow-type coupling of both sub-components results in a significantly delayed torque build-up compared with the steady-state behavior of the engine. Section 10.5 presents some simulation calculations for improving unsteady behavior and for explaining individual effects.

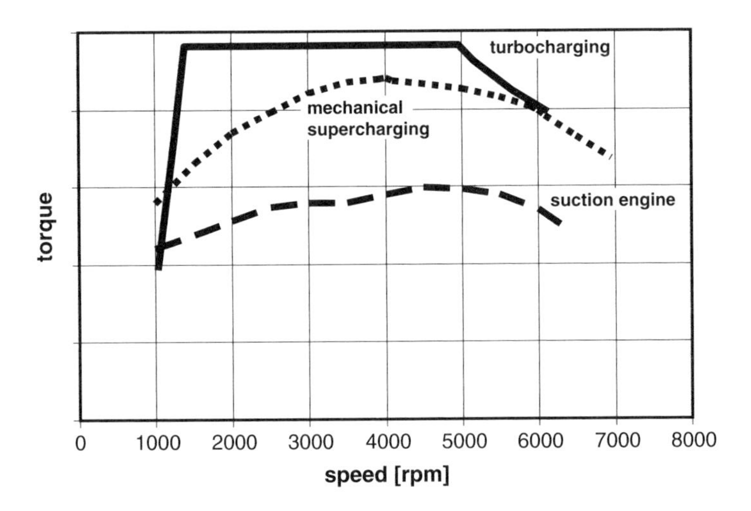

Fig. 8.20 The torque profile of various charging concepts

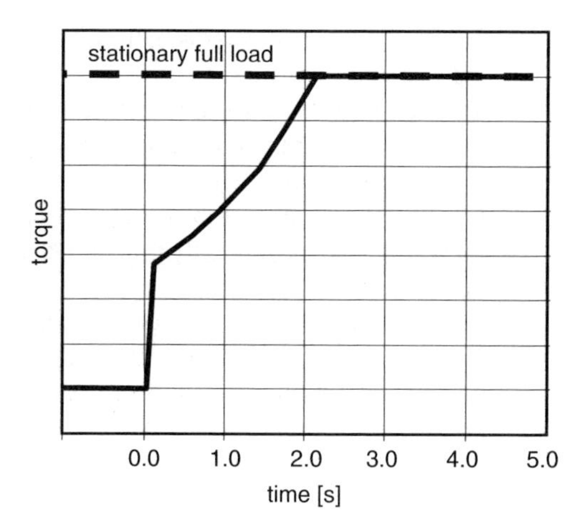

Fig. 8.21 Unsteady torque build-up in a turbocharged engine

The following summarizes the advantages and disadvantages of exhaust turbocharging.

- Advantages:
  - The exhaust counterpressure is reduced in case of partial load. A so-called "self-regulation effect" of the turbocharger: the specific fuel consumption is reduced with it as well.
  - Small turbine cross-sectional surfaces, small polar moments of inertia and high potential turbocharger speeds make for a fast, almost delay-free response time (no noticeable turbo lag).
  - No external drive necessary.
  - Economical manufacture and low manufacturing costs for the turbocharger with high production numbers, since they can be adjusted to the engine on the side of both the turbine and the compressor by means of a change of trim.
  - The small construction size of the turbocharger unit. Integration of the turbine into the exhaust manifold results in additional performance advantages and good response behavior.
  - Utilization of exhaust energy improves engine power and, under some conditions, even the indicated efficiency of the engine.
  - The continuous supply of the turbo compressor.
  - Lower thermal stress on the turbocharging system due to temperature reduction via the turbine.
- Disadvantages:
  - The surge limit restricts the charge pressure at low engine speeds.
  - In the case of large turbine housings (cross-sectional areas), the unsteady response time is clearly delayed (turbo lag).
  - High gas exchange losses and poor gas exchange success of the engine due to the high pile-up pressure before the turbine, especially in the case of a negative purge slope and small turbocharger designs.

- The boost pressure is reduced to low engine speeds in full load, as only low exhaust energy is present.
- Thermal problems due to hot metal masses in the engine room.
- Due to the larger thermal mass of the turbine, the catalytic converter light-off temperature level is delayed.
- Exhaust aftertreatment measures influence the interaction of the engine and the turbocharger more than with other charging concepts, especially in unsteady operation.
- Intervention by injection volume control, which is dependent on charge pressure, magnifies the effect of delayed charge air build-up and has a negative effect on transient operation.
- Exhaust recycling delays transient charge pressure build-up.

8.2 Simulation of Charging

Section 8.1 describes the basics of charging in internal combustion engines. This chapter aims at the understanding of the simulation of charging in the context of zero- and one-dimensional process calculation.

8.2.1 Turbo Compressor

The simulation of charged engines requires, in addition to the process simulation in the cylinder described in Chap. 7 as well as the simulation of pipe and secondary aggregate components, a detailed description of the charging components. The description of these components is based essentially on special, standardized characteristic maps, which, according to the maker, are usually of varying quality. A description of the operating behavior of turbo machines by means of characteristic maps has proved to be sufficiently exact for unstable processes.

8.2.1.1 Reference Quantities

In order to utilize compressor maps measured under more or less arbitrary environmental conditions under altered environmental conditions as well, the quantities that are stored in a characteristic map have to be made independent of the actual environmental conditions. The map quantities are usually converted into standardized reference environmental conditions (e.g. ISA condition: 288 K, 1.013 bar; standard conditions: 293 K, 0.981 bar). This takes place via the introduction of so-called reference quantities with the help of fluid-mechanical laws of similarity. These will now be more thoroughly explicated.

In the case of the speeds for turbo machines, we use the similarity of the dimensionless, so-called Mach number and relate the circumference speed of the rotor, which is directly proportional to the speed, to the sound velocity of the gas at the entry state. The gas constant is assumed to be constant in the following investigations. Furthermore, one must take into consideration in the calculations whether one is dealing with pure air in the compression, for which the isentropic exponent can be set to 1.4 and the gas constant to 287 J/kg K, or whether one has another medium with other physical properties, as for example a mixture or the like.

$$\mathrm{Ma} = \frac{u}{a} = \frac{\omega}{\sqrt{T}} \frac{r}{\sqrt{\kappa R}} = \frac{2n}{\sqrt{T}} \frac{\pi r}{\sqrt{\kappa R} 60 \left[\mathrm{s} \cdot \mathrm{min}^{-1}\right]}. \tag{8.16}$$

One recognizes that the Mach number is proportional to the speed and to the square root from the entry temperature of the gas. According to this is sufficient to define a reference speed or a reference angle speed for which the speed or angle speed is divided by the root of the entry temperature. Using the reference quantities, one can in the case of the turbo compressor simultaneously convert to a certain standard entry condition, in order again to assign the original unit to the reference speed or angle speed.

$$n_{ref} = \frac{n\sqrt{T_{ref}}}{\sqrt{T}}, \quad \omega_{ref} = \frac{\omega\sqrt{T_{ref}}}{\sqrt{T}}. \tag{8.17}$$

For the mass flows as well, independent quantities can be defined from the respective environmental and entry conditions. If we consider at first the flow equation again, the result is

$$\dot{m}_{th} = A \frac{p}{\sqrt{RT}} \sqrt{\frac{2\kappa}{\kappa - 1}\left(\pi^{\frac{2}{\kappa}} - \pi^{\frac{\kappa+1}{\kappa}}\right)} = A \frac{p}{\sqrt{RT}} \Psi. \tag{8.18}$$

With this equation, all components, which are in some way to be seen as throttle locations – i.e. flow turbines or charger aggregates – can be described for the filling and emptying method.

As we can recognize with the help of (8.18), the mass flow through a throttle location is contingent only on the pressure, the square root from the temperature before the throttle location, and the flow function, which for its part is determined practically only by the adjoining pressure ratio. If we rearrange the equation and move the pressure and temperature to the left side of the equal sign, the mass flow referring to these quantities is now practically contingent only on the adjoining pressure ratio. This mass flow is designated as the reference mass flow and can then be referred to for the sake of keeping the corresponding units for the mass flow in the turbo compressor at standard conditions

$$\dot{m}_{ref} = \dot{m}_{th} \frac{\sqrt{T}}{p} \frac{p_{ref}}{\sqrt{T_{ref}}} = \frac{p_{ref}}{\sqrt{T_{ref}}} A \frac{1}{\sqrt{R}} \Psi. \tag{8.19}$$

Analogously to the mass flow, there is also a reference volume flow, which can be determined by division by means of the reference density of the gas and under consideration of the definition of the actual volume flow under environmental conditions over the related density at the entry into the components

$$\dot{V}_{ref} = \frac{\dot{m}_{ref}}{\rho_{ref}} = \frac{\rho \dot{V}_{th}}{\rho_{ref}} \frac{\sqrt{T}}{p} \frac{p_{ref}}{\sqrt{T_{ref}}} = \dot{V}_{th} \frac{\sqrt{T_{ref}}}{\sqrt{T}}. \tag{8.20}$$

The performance admitted by the compressor has already been described in Sect. 8.1. The compressor torque results from a division with the angle speed. Since however quantities are stored in the compressor maps, which were converted according to the laws of similarity to the reference conditions, the compressor torque can also be written with the reference quantities

$$M_{com,ref} = \frac{1}{\omega_{com,ref}} \dot{m}_{com,ref} \frac{\kappa_{b.com.}}{\kappa_{b.com.} - 1} R\, T_{ref} \frac{1}{\eta_{is,com}} \left(\pi_{com}^{\frac{\kappa_{b.com.}-1}{\kappa_{b.com.}}} - 1 \right). \tag{8.21}$$

The torque necessary for the propulsion of the compressor can thus at first be calculated for characteristic map considerations or for the extrapolation of characteristic maps as a reference torque. It can then be converted into the actual torque needed with the help of the definitions of the laws of similarity. This is given by

$$\begin{aligned} M_{com,ref} &= \frac{1}{\omega_{com}} \frac{\sqrt{T_{b.com.}}}{\sqrt{T_{ref}}} \dot{m}_{com} \frac{p_{ref}\sqrt{T_{b.com.}}}{p_{b.com.}\sqrt{T_{ref}}} \frac{\kappa_{b.com.}}{\kappa_{b.com.} - 1} R\, T_{ref} \frac{1}{\eta_{is,com}} \left(\pi_{com}^{\frac{\kappa_{b.com.}-1}{\kappa_{b.com.}}} - 1 \right) \\ &= M_{com} \frac{p_{ref}}{p_{b.com.}} \end{aligned} \tag{8.22}$$

and

$$M_{com} = M_{com,ref} \frac{p_{b.com.}}{p_{ref}} \tag{8.23}$$

We recognize that the reference compressor propulsion torque and the actual compressor propulsion torque differ only by the quotient from the actual pressure adjacent to the compressor entry and the reference pressure. The influence of an altered temperature has already been eliminated because of the theory of similarity via the reference quantities of angle speed and mass flow.

8.2.1.2 Characteristic Map Representation

Figure 8.22 shows the characteristic map of a turbo compressor. The lines of constant reference compressor speeds or constant peripheral speeds and the isentropic efficiency of the compressor are entered in this characteristic map. The so-called surge

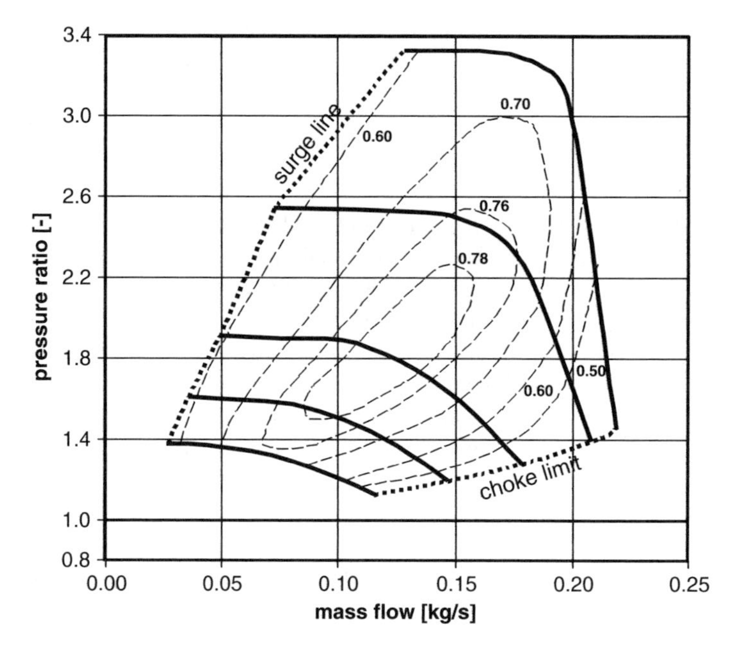

Fig. 8.22 Characteristic map of a turbo compressor

line represents a fictitious boundary line and limits the range of validity of the characteristic map to the left. This limit is however not only set by the compressor type, but results from the interaction of pipe volumes, pipes, and the compressor. Therefore, the characteristic map is at first considered without the surge line and if necessary extrapolated beyond it.

The determination of the isentropic efficiency, which is calculated because of its definition from the temperature difference between compressor intake and exhaust and the compressor pressure ratio, reaches its limits at low pressure ratios and small temperature differences, as is common at low speeds and mass flows. Therefore, a large area below a lowest reference speed is missing in most experimentally determined characteristic maps, see Fig. 8.22. This area is not of interest for a design of the engine under full load, since the interaction of engine and compressor under full load mostly takes place in the area of optimal efficiency from mid to high pressure ratios and speeds. For operating points at lower partial load or for the idle speed of the engine, as they typically appear in the cycles of motor vehicles or, at low speeds, at the propeller line in the diesel engines of large ships, the characteristic map does not provide reliable information. This leads to consequences which are not insignificant for the interaction of engine and the charger aggregate, as the calculations in Chap. 10 shows. While the real engine, even if in area with low efficiency, nonetheless continues to run, in a simulation calculation in these ranges on the other hand, it can come to discontinuations in the program, which makes a calculation of cycles with long partial load or idle phases practically impossible.

For these reasons, one must attempt to extrapolate the characteristic map into areas up to a reference compressor speed of 0 and a pressure ratio of 1.

8.2.1.3 Extrapolation

In this section we will introduce a method, with which the behavior of a turbo compressor can be described beyond the usual characteristic map boundaries, in order to allow the clear representation of all possible operation ranges in simulating the interaction of the engine and the turbo compressor. The characteristic map of the turbo compressor must for these purposes be transported into another representation, which allows us, on the one hand, to extrapolate the map in a simple manner and, on the other, to make it accessible to calculations like the representation previously known. The advantage of the new presentation is a considerably broadened range of validity in areas, in which no more information was attainable in the previous representation. This procedure requires however a high amount of care and constant plausibility control. A reverse calculation into the customary representation after extrapolating is possible with certain assumptions, however, this can not take place in ranges of pressure ratio values smaller than 1.

8.2.1.4 Extrapolation in the Pressure Ratio-Mass Flow Map

The prerequisite for a secure extrapolation is the expansion of the characteristic map in the common and known way of representing it. The lines of constant reference compressor speeds can in most cases be easy expanded to the left and right. One must thereby consider that the lines lightly climb towards the surge line and beyond it, which is required in the case of interpolation in the characteristic map according to Münzberg and Kurzke (1977) as well. In some maps, the lines of a constant reference compressor speed first drop lightly at the surge line or to the left of it and then the rise again. This can be due to the specific events at the test bench during the measurement of the characteristic map (pipe lengths and volumes) and should not be considered in the extrapolation process. Operation left of the supposed surge line has to be indicated in the simulation calculation anyway, in order to leave the decision of whether or not the transgression should be tolerated to the user. For the extrapolation however, the surge line and all actual effects associated with it on the form of the characteristic map are at first ignored.

To the right as well, the lines of constant reference compressor speeds can be led further slightly, since at a constant speed, the maximum mass flow is limited by the speed of sound in the compressor impeller and the speed lines slope down steeply (choke limit).

Since the extrapolation of compressor efficiency based on the existing cannot be executed very easily, the representation of the reference compressor torque (8.21) and as well as the representation of the efficiency of the compressor along the speed line can be helpful here (Fig. 8.23).

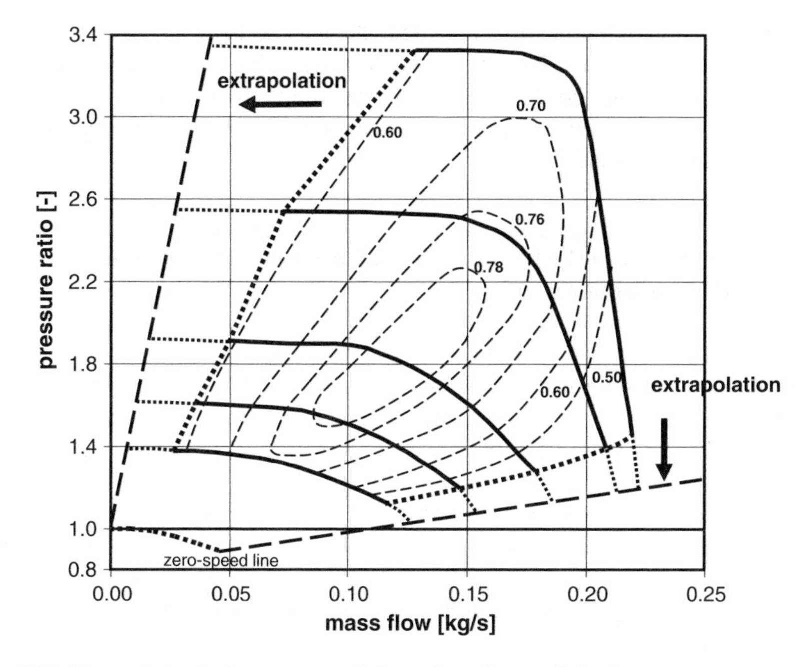

Fig. 8.23 Extrapolation in the pressure ratio/mass flow characteristic map

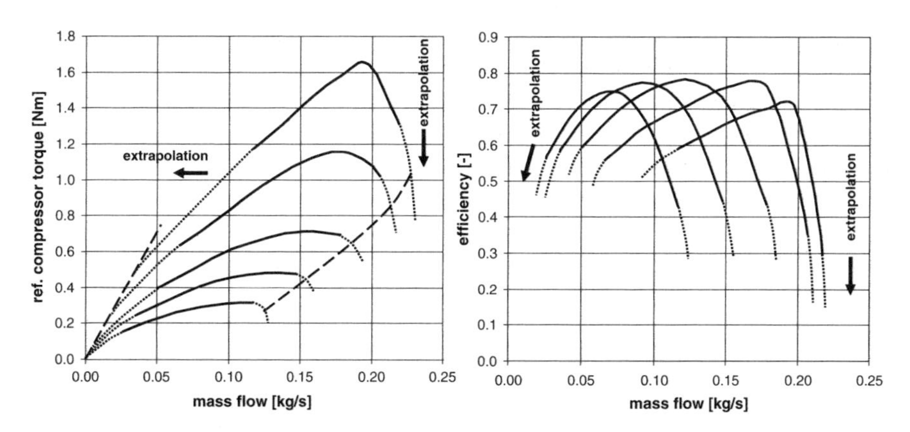

Fig. 8.24 Extrapolation of compressor torque and efficiency along speed lines

Figure 8.24 (left) shows the profile of the reference compressor torque along the lines of constant compressor speed. These lines can be easily extended in both directions. In this context, one can make use of the knowledge that the compressor torque is very small up to mass flow 0 and also at a pressure ratio near 1 will approach 0. At the same time, the extrapolated values can also be checked with the compressor efficiency shown in the right image.

This method makes it possible to extrapolate very securely in the area in which the speed lines are already known via measurement. This extrapolation is not yet suitable for extend the compressor for low and high charger speeds. For this, a lower limit of the compressor map with a stationary charger must first be found by determining the zero-speed line.

8.2.1.5 Determination of the Zero-Speed Line as a Characteristic Map Boundary

A turbo compressor with a speed of 0 rpm represents, from a fluid-mechanical perspective, a simple throttle. If a gas mass should flow through the turbo compressor, a pressure ratio smaller than 1 must be adjacent to it according to the usual definition for the pressure ratio for turbo compressors. In actual engine operation as well, e.g. turbocharged vehicle engines at idle operation, a pressure ratio smaller than 1 at the turbo compressor can occur. At this operating point, the turbine performance available at the compressor is too low to deliver the fresh gas mass flow aspirated by the engine at a corresponding speed and a resulting pressure ratio greater than 1. The speed of the turbo compressor of passenger car diesel engines ranges – according to the respective condition of the engine and the bearing friction contingent on the oil viscosity – between 5,000 and 10,000 rpm. The so-called zero speed line of the compressor is thus an important bottom limitation of the compressor map which one cannot fall beneath.

Figure 8.25 shows the reference volume flow, determined contingently on the pressure ratio via the flow function with a constant actual throttle cross section, in comparison with the zero speed line on the compressor. The good agreement between both graphs permits us to conclude that the zero speed line of a turbo compressor can be determined easily by means of the flow equation if the constant effective throttle cross-sectional surface is known. In the case of the flow

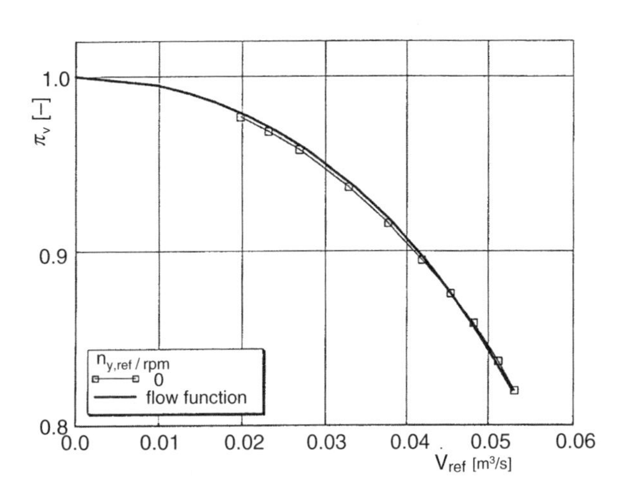

Fig. 8.25 Comparison of the zero speed line and the flow function

equations defined in (7.236), the effective cross-sectional area must be used instead of the cross-sectional surface. In the case of the flow equation defined in (8.18), the effective cross-sectional area μA must be used instead of the cross-sectional area A.

8.2.1.6 Compressor Extrapolation into the Low Speed Range

Proceeding in the next step from the zero-speed line, fan-shaped straight lines will be superimposed over the characteristic map in accordance with Fig. 8.26. The intersections of the fan-lines with the compressor map provide definite support points of the extrapolated compressor map for input into simulation programs. The fan-shaped subdivision guarantees an optimal subdivision of support points in the map area relevant for the simulation, which in turn also optimally supports the previously mentioned interpolation method of Münzberg and Kurzke (1977).

If one plots the reference compressor mass flow and the reference compressor torque over the speed along these fan-lines, the curve sets shown in Fig. 8.27 is obtained, which are already defined over a wide speed range and thus can be easily extrapolated into low and high speed ranges. In the case of the reference volume flow, the extrapolation is also assisted by knowledge of the zero-speed line. The range of extrapolation is shown by the dashes in the illustration.

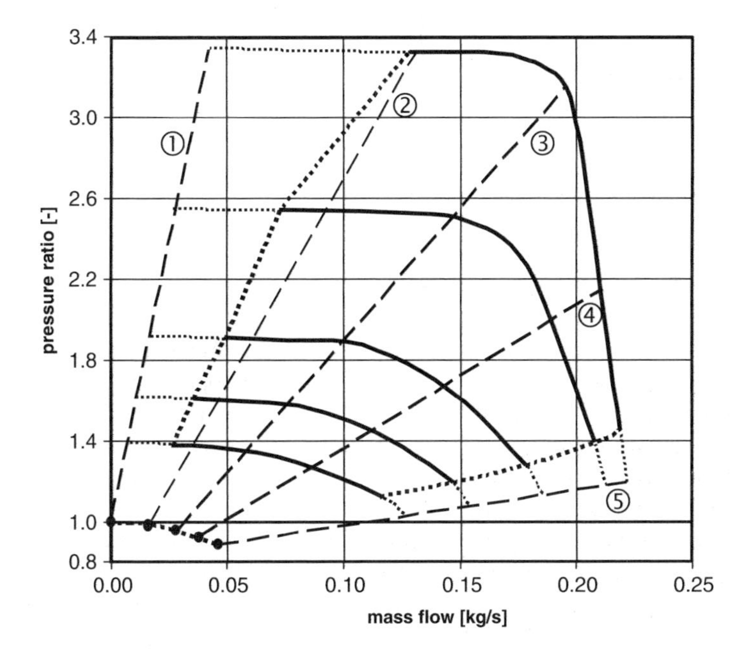

Fig. 8.26 Fan-shaped support lines for compressor extrapolation

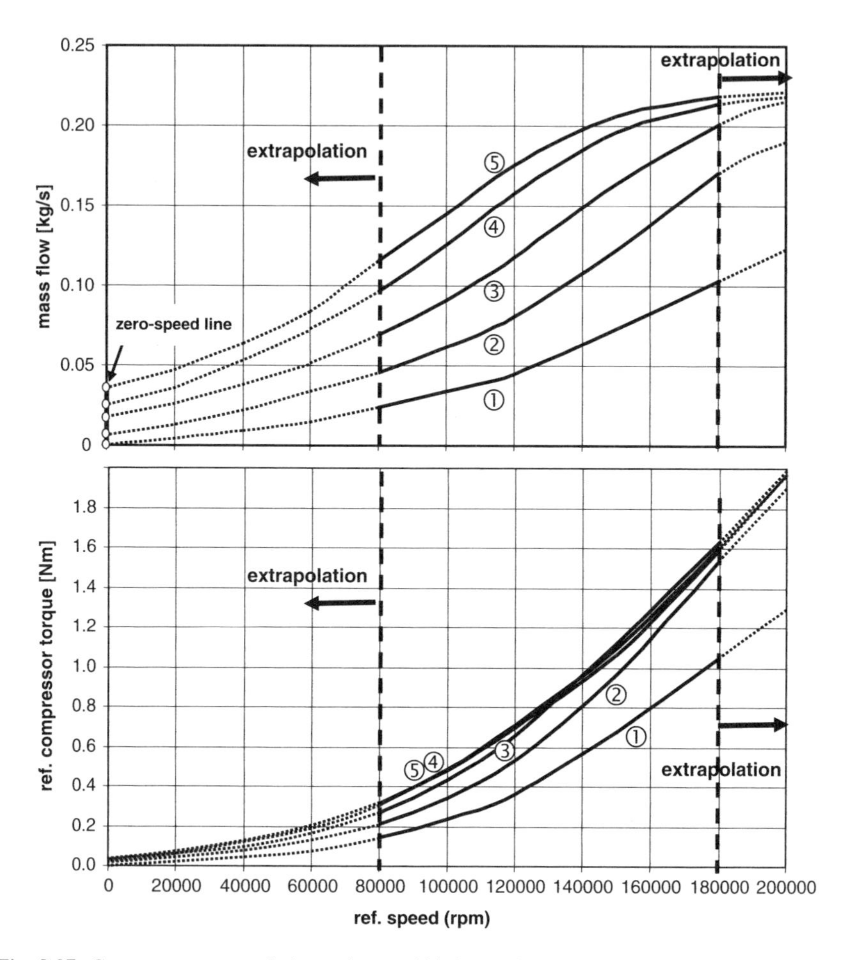

Fig. 8.27 Compressor extrapolation to low and high speeds

8.2.1.7 Behavior at the Surge Limit

The reference mass flow at the surge limit is plotted separately in the characteristic map as a function of the pressure. At every calculated operating point, we check whether the reference mass flow is left of the surge limit. If this is the case, we use instead of mass flow its negative value or 0 for further calculations. With this simple assumption, which should allow for stall of the flow and even a reflux through the compressor, the pressure drops after the compressor due to the reduction of the supplied mass, which is why the pressure ratio is reduced at the compressor and the compressor can again leave the area of the surge. Usually the curve provided in the pressure ratio-volume flow map is used as a profile for the surge limit. Exceeding the surge limit must be indicated by a signal variable in order to prevent misinterpretations of the calculation results.

8.2.2 The Positive Displacement Charger

In the case of positive displacement chargers, the lines of constant charger speeds slope down considerably, similarly to the speed lines near the fill limit in turbo compressors. This slope is grounded however in the working method of positive displacement chargers.

The characteristic map, contingent on the pressure ratio and the reference mass flow, the lines of constant speed and constant isentropic efficiency are represented. Now and then are found presentations of lines of constant temperature differences between the outlet and inlet temperature in the charger, which are however relatively easy to transform into isentropic efficiency or to utilize directly in the determination of the required compressor torque.

$$\eta_{is,com} = \frac{T_1\left(\pi_{com}^{\frac{\kappa-1}{\kappa}} - 1\right)}{\Delta T} \tag{8.24}$$

$$M_{is,com} = \frac{1}{\omega_{com}} \dot{m}\, c_p \Delta T. \tag{8.25}$$

In positive displacement chargers, the friction of the aggregate plays a significant role. While friction in the case of the turbocharger is usually contained in the total efficiency of the turbine, in the case of positive displacement chargers it has to be considered independently. This can take place either by means of the mechanical efficiency or directly with the aid of the lines of constant propulsion performance. These lines can also be extrapolated in areas of pressure ratios smaller than one with good approximation, in order, for example, to make possible a description of the interaction of the engine and the positive displacement charger at a load regulation by means of a change of the transmission ratio also in areas below the full load of the uncharged engine. With that, an extrapolation provides positive displacement charger performances as well, in which a performance is delivered due to the expansion of a gas mass to a pressure below the intake pressure in the charger. Thus the charger possesses thereby, as we would say in the case of the internal combustion engine, a positive gas exchange loop. Figure 8.28 shows such an extrapolation.

8.2.3 The Flow Turbine

The adjustment of flow turbines to the required performance range usually takes place experimentally and extends from an alteration of the spiral housing to an alteration of the nozzle ring or to an alteration of the trim. For these alterations, fixed characteristic maps result, with which the operating behavior of these turbines can be described clearly, also in the context of a simulation computation. However,

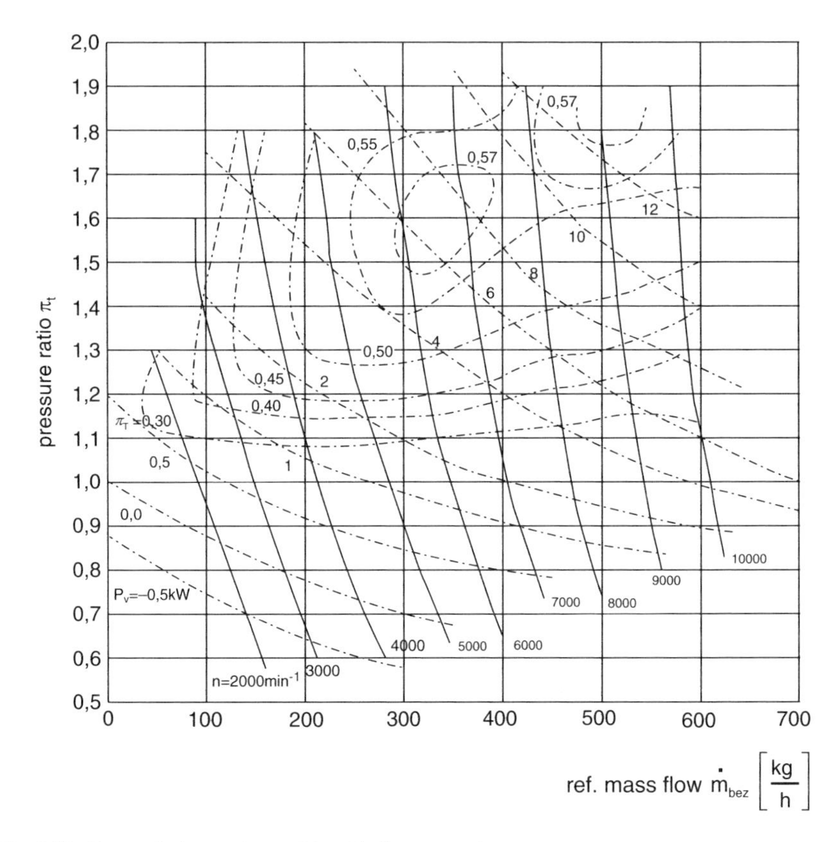

Fig. 8.28 Extrapolation at the positive displacement charger

the influence possibilities on the alteration of turbine performance in running operation (variable turbine geometry) partially cause more or less considerable changes in the form of the characteristic maps and in the efficiency behavior of the turbines. These changes can no longer be covered by the basic maps, which is why several maps – as long as they are available – must be superimposed. It is partially also possible with suitable theories of similarity to describe in sufficient exactness the operating behavior of the turbine being used proceeding from a basic characteristic map. For this and for the basic processing of characteristic maps for the simulation calculation from usually quite limited measurement data for a turbine, laws of similarity for a turbine must first be drawn up.

8.2.3.1 Reference Quantities

For the flow turbine, the laws of similarity described for the turbo compressor are valid as well, in order to prepare the measurement results and with that finally the map characteristic for the entry conditions that deviate from the entry conditions from the determination of the characteristic map. In determining the maps, most turbocharger manufacturers use standardized entry quantities, which are often very helpful in checking the plausibility of extrapolation results. Thus, the reference entry temperature is set in the turbine, for example at 873 K (600°C) by some manufacturers.

Similarity regularities can be derived from the similarity via the Mach number and the fluid mechanical similarity. For the sake of clarity, we have to do without a reference to standard turbine conditions because of the large range of possible entry conditions in the turbine. Only an additional reference to the desired constant reference entry temperature is possible, which is used by some manufacturers. The similarity with respect to the Mach number will be represented here however only with the help of the reference to the actual entry temperature in (8.26)

$$n_{tur,ref} = \frac{n}{\sqrt{tur_{b.tur.}}}; \quad \omega_{tur,ref} = \frac{\omega}{\sqrt{tur_{b.tur.}}}. \tag{8.26}$$

The mass flow through the turbine can be determined by means of the flow equation from the adjoining pressure ratio and a specific flow coefficient, which considers the dependence on a fabricated specific cross sectional surface, which is of course contingent on the turbine speed as opposed to a normal throttle. For this reason, the opportunity presents itself for the turbine to relate the actual mass flow to the entry quantities of pressure and temperature in order to be independent of these quantities at least in the characteristic map representation. Corresponding to (8.18), the following relation is valid for the reference mass flow at the turbine

$$\dot{m}_{ref} = \dot{m}\frac{\sqrt{tur_{b.tur.}}}{p_{b.tur.}}. \tag{8.27}$$

A transformation to standard conditions for the reference mass flow of the turbine is not sensible for the above mentioned reasons.

The reference mass flow through the turbine thus amounts to

$$\dot{m}_{tur,ref} = \mu_{tur} A_{tur} \frac{1}{\sqrt{R}} \sqrt{\frac{2\kappa}{\kappa - 1}\left(\pi_{tur}^{*\frac{2}{\kappa}} - \pi_{tur}^{*\frac{\kappa+1}{\kappa}}\right)}. \tag{8.28}$$

The flow coefficient can be interpolated from the characteristic map described in Fig. 8.30.

In Sect. 8.1, the basic equations of the turbine performance is presented). For the isentropic torque of the turbine results from this

$$M_{tur,is} = \frac{1}{\omega_{tur}} \dot{m}_{tur} \frac{\kappa_{b.tur.}}{\kappa_{b.tur.} - 1} R_{b.tur.}\, tur_{b.tur.}\, \eta_{is.tur}\left(1 - \pi_{tur}^{*\frac{\kappa_{b.tur.}-1}{\kappa_{b.tur.}}}\right). \tag{8.29}$$

The isentropic turbine efficiency can thereby be determined from a characteristic map also described in Fig. 8.30. If we insert (8.28) into (8.29) and convert, we obtain the isentropic turbine torque related to the pressure after the turbine contingent on the reference turbine speed and the reference turbine pressure ratio. We should thereby consider that the reciprocal value of the actual turbine pressure ratio is utilized

$$\frac{M_{tur,is}}{p_{a.tur.}} = \mu_{tur} A_{tur} \eta_{is,tur} \frac{60\left[\frac{s}{min}\right]\sqrt{2R}}{2\pi} \left(\frac{\kappa}{\kappa-1}\right)^{1,5} \frac{1}{a_{tur,ref}} \pi_{tur}^{*\frac{\kappa+1}{\kappa}} \left(1 - \pi_{tur}^{*\frac{\kappa-1}{\kappa}}\right)^{1,5} \tag{8.30}$$

with

$$\pi_T^* = \frac{p_{a.tur.}}{p_{b.tur.}}.$$

Precisely in the context of simulation computation, the turbine cross sectional surface must often be adjusted for the adjustment of the turbine and the fine tuning of possibly existing measurement results. In the case of these alterations, not the size of the turbocharger but only its cross sectional surface is changed, which is why also the reference turbocharger speeds need not be transformed. In an alteration of the cross sectional surface through a change of the nozzle ring, the in-flow of the impeller, but not its size is changed. For this reason, in this case as well, a conversion of the reference turbine speeds can be left out. The changes in efficiency and flow coefficients caused by the varying in-flow of the impeller have to be estimated on an individual case basis. In comparison to the errors made in the preparation of the characteristic maps in the context of extrapolation, they can, however, usually be ignored at minimal changes in quantity.

A further characteristic quantity for a flow turbine is the so-called blade speed ratio, which is expressed in terms of the quotients from the peripheral speed of the impeller and the theoretical flow speed, which can be calculated from the adjoining static enthalpy slope at the turbine. The formal relation for the blade speed ratio reads according to this

$$\frac{u}{c_0} = \frac{u}{\sqrt{\frac{2\kappa R}{\kappa-1} T_{b.tur} \left(1 - \pi_{tur}^{*\frac{\kappa-1}{\kappa}}\right)}}. \tag{8.31}$$

For the peripheral speed at the mean diameter of the turbine d_m is valid

$$u = \frac{\omega_{tur}\, d_m}{2} = \frac{\pi\, n\, d_m}{60[\mathrm{s} \cdot \mathrm{min}^{-1}]}. \tag{8.32}$$

With that, the result for the blade speed ratio of the turbine is the following relationship with the reference speed. Here we already see the advantage of the use of laws of similarity

$$\frac{u}{c_0} = \frac{\frac{\pi\, d_m}{60[\mathrm{s}\cdot\mathrm{min}^{-1}]} \frac{n}{\sqrt{T_{b.tur.}}}}{\sqrt{\frac{2\kappa R}{\kappa-1}} \sqrt{1 - \pi_{tur}^{*\frac{\kappa-1}{\kappa}}}}. \tag{8.33}$$

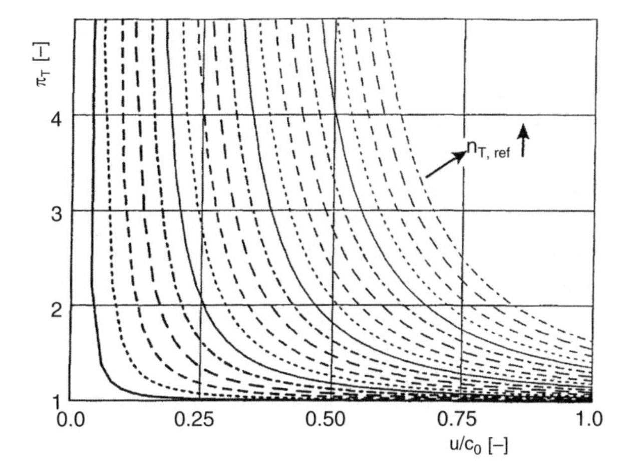

Fig. 8.29 Relation of the referenced turbine speed, blade speed ratio and turbine pressure ratio

Figure 8.29 shows the relationships between the turbine pressure ratio, the blade speed ratio, and the reference turbine speed according to (8.33) in the form of a diagram. The reference speeds increase towards the top right. For the creation of a diagram, a few simplifying assumptions are made, which can also be very helpful in recalculation and extrapolation of turbine characteristic maps without having to hazard all-too large errors. The isentropic exponent κ is set at 1.34 and the gas constant R to 289 J/kg K. Furthermore, a constant mean turbine diameter is taken as a basis.

We recognize in Fig. 8.29 that the blade speed ratio shift to higher values with increasing reference turbine speeds at a realistic turbine pressure ratio (<5). This determination has effects on the extrapolation of turbine maps. The larger the reference speeds, the less important the course of the characteristic map quantities at smaller blade speed ratios and the more important the course at larger blade speed ratios. One further aspect of the diagram analysis is the fact that the blade speed ratios can take on values larger than 1 at larger blade speed ratios at corresponding combinations of turbine pressure ratios and reference speeds, which is also of importance for an extrapolation.

8.2.3.2 Extrapolation

Most turbine maps are measured in the so-called dry cycling of the turbocharger at a combustion chamber test bench. The turbine is thereby admitted with hot gas, which – as mentioned above – usually possesses a constant temperature. The load on the turbine is adjusted via a throttling of the compressor and returned to equilibrium through a heightened mass flow at the turbine. Due to these facts, the working range of the turbine is very limited and the measured characteristic map range of the turbine is very small. For stationary approaches, this range is usually totally sufficient, while in the unsteady consideration of the interaction of engine and turbocharger or in partial or light load ranges, considerable problems result. The same effect makes itself known in this case as in the compressor maps: that precisely the characteristic map range of low reference speeds is lacking in the case of the turbine as well. This range is however of special interest for an exact simulation of the turbocharger.

The characteristic maps measured at the test bench are usually represented with the help of two diagrams. The first diagram contains the reference mass flow as a function of the pressure ratio with the reference turbine speed parameter, and the second diagram the total turbine efficiency again as a function of the pressure ratio with the reference turbine speed parameter. Figure 8.30 (left) shows a characteristic map for a small radial turbine.

In this presentation, the characteristic maps can only be applied in a limited fashion for a sure interpolation of the values necessary for a simulation computation. The characteristic maps must therefore be transferred to a mode of representation, which, on the one hand, allows an extrapolation into the margin areas of the characteristic map, and on the other hand, guarantees a sure interpolation. With the aid of (8.33), the blade speed ratio of the turbine can be calculated from the reference turbine speed and the turbine pressure ratio. The mean diameter of the turbine must be estimated in a suitable way for this, if it is not known. An inaccurate estimation of the mean diameter does not necessarily lead to an error in

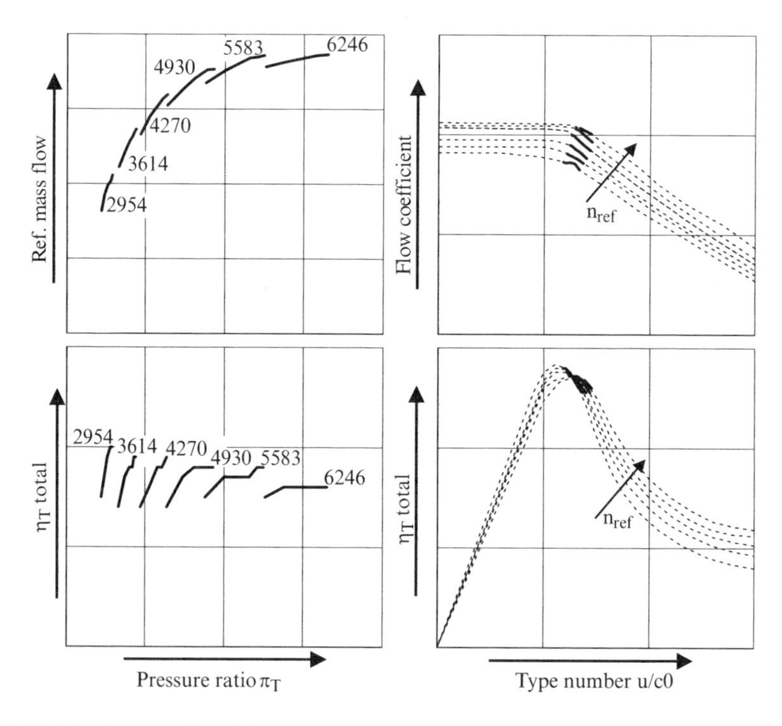

Fig. 8.30 Map for a small radial turbine (*left*), map using the blade speed ratio (*right*)

turbine calculations in the context of simulation calculation, as long as this value remains tightly fixed to the turbine and the calculation of the blade speed ratio always takes place with this value. The isentropic exponent κ and the gas constant R have to be determined with the help of the knowledge of the turbine entry temperature (873 K) and of the air-fuel ratio in the combustion chamber of the turbocharger test bench by means of a corresponding relation between temperature, the air-fuel ratio, and the condition quantities. For a diesel-operated combustion chamber, an air-fuel ratio of approx. 6 can be assumed. With it, the isentropic efficiency can be represented as a function of blade speed ratio with the reference turbine speed parameter. With (8.28), it is possible to determine the flow coefficient of the turbine with the help of the turbine cross sectional surface. Alternatively, the mass flow can be directly plotted here as well. The success of this transformation, as Fig. 8.30 (right) shows by means of the thick solid lines, is unfortunately usually rather disillusioning, since the area of such a transformed characteristic map originates in a narrow range of blade speed ratio.

The characteristic map has to be extrapolated appropriately for unsteady processes, for which the range of the blade speed ratio extends from 0 at a sudden switching-on of the turbine in the context of register charging from a state of rest ($n_{T,bez} = 0$) to very high values for the blade speed ratio at a sudden switching-off of the turbine ($\pi_T = 1$) at high speeds. The blade speed ratio range of the turbine, according to the momentarily adjoining turbine pressure ratio, can deviate strongly also in an impact admission of the turbine at a practically constant turbine speed. The turbine pressure ratio thereby can also take on values of approx. 1, which at least theoretically would cause an infinitely large blade speed ratio.

As has been mentioned, an extrapolation into ranges of small blade speed ratios is not necessary, since this would make necessary more or less infinitely high turbine pressure ratios. Thus at these high speeds, the path of efficiency can be extrapolated from the optimum efficiency towards the left to the coordinate source. At low reference turbine speeds, the blade speed ratios also lie in lower areas. Usually, the optimum efficiency also shifts at these speeds to low blade speed ratios and lower absolute values. This path as well can be extrapolated without much error to the origin of the coordinate system. Naturally, the boundary value problem, as the blade speed ratio approaches the value of 0 at a reference turbine speed other than 0, must be considered here as well, which also allows for an extrapolation to isentropic efficiencies that are minimally larger than 0.

An extrapolation to blade speed ratios right of the optimum efficiency is more difficult. As a basis, one can as a first approximation describe the course of efficiency as a parabola, which has its vertex in the optimum position of efficiency. This extrapolation method provides for most cases satisfactory results, since the range for the blade speed ratio takes on values that are far right of the optimum efficiency – i.e. at higher blade speed ratios – only in a few extremely unsteady cases.

It should be noted in conclusion that there is no patent formula for the extrapolation of turbine maps. Usually, the only thing that helps is a large number of the most varying plotting types for the most varying turbine parameters, the paths of which then, amongst other things, allow for a further extrapolation. These extrapolations must however then be set in relation to each other again and again by means of the equations described above in order to test their plausibility. A greater progress in the describing and extrapolation of turbine maps can be seen, as in the case of the compressor maps, in the use of the zero-speed line, which clearly limits the range of the characteristic map and thus contains valuable information regarding the usually missing intermediate area of the characteristic map.

8.2.3.3 Characteristic Map Representation

Because of the relations between the blade speed ratio, the reference turbine speed, and the turbine pressure ratio, there are differing ways of representing the parameters of the turbine map.

The representational type of contingent quantities of the turbine, like efficiency or the flow coefficient, across the blade speed ratio with the parameter of the reference speed allows for a good interpolation of these quantities, but it is not very effective in describing the interaction of the engine and the turbocharger or exhaust gas turbine. The plotting type shown in Fig. 8.31 has proven itself to be the

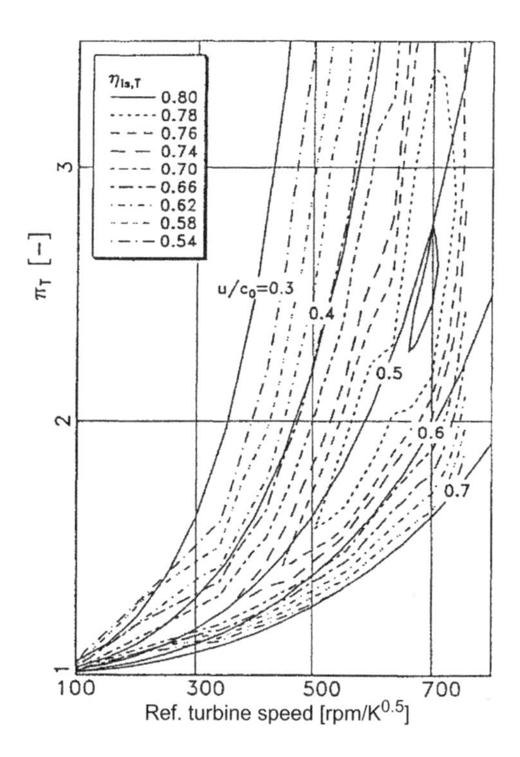

Fig. 8.31 Turbine map for description of operation condition

most effective, whereby instead of the torque of the turbine the turbine efficiency is represented. This is much more common in association with the representation of the turbo compressor in the pressure ratio-mass flow map, in which in the same way the isentropic compressor efficiency is drawn in as a parameter.

The characteristic map bears a close resemblance to a compressor map, but its abscissa is the reference turbine speed and not the mass flow. The pressure ratio-mass flow field is too narrow in the case of turbines to be capable of representing significant alterations. In the case of the chosen mode of representation, the two main influencing quantities of the turbine, i.e. the pressure ratio and the speed, appear directly and thus do not require a transformation.

8.2.3.4 Variable Turbine Geometry

In the case of a turbine with an adjustable diffuser, the guide vane position appears as a further parameter. For discrete guide vane positions between both end positions, a complete separate characteristic map is usually measured, which has to be prepared and extrapolated according to the above regularities. Figure 8.32 shows the characteristic map of such a variable turbine. At a known shifting position, the map quantities for both guide vane positions, between which the actual position lies, must first be interpolated. Finally, the characteristic map values for this shifting position can be determined, for example, by means of a linear interpolation.

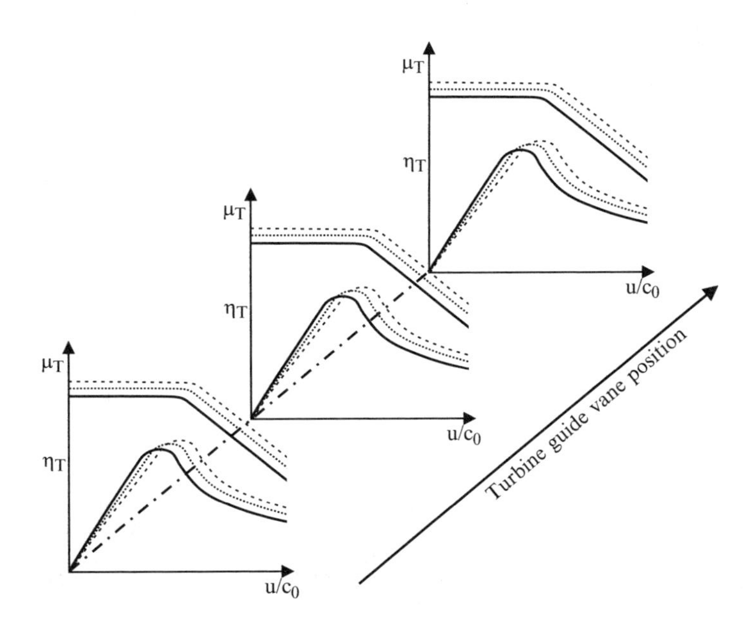

Fig. 8.32 Map of a turbine with an adjustable diffuser

8.2.3.5 Modeling a Twin-Scroll Turbine

Dual-flow turbines – also called twin-scroll turbines – are employed in engine concepts, in which one is trying to achieve a gas exchange which is as free of repercussion as possible for the exploitation of the maximal potential of gas exchange in the internal combustion engine and for the utilization of the exhaust gas pulses at the turbine. Possible areas of use are in-line four or six cylinder engines equipped with a turbine or V8 and V12 engines with a turbocharger for each side. Since in the four-stroke four cylinder engine, for example, the exhaust impulses arrives shifted by 180°CA in the exhaust gas system, the exhaust gas impulse of the following igniting cylinder influences the valve overlap phase of the cylinder proceeding it in the order of ignition considerably at conventional valve timing of approx. 240°CA (see Sect. 10.5). Such a gas exchange leads to increased amounts of residual gas, which cause in the diesel engine a clearly reduced filling and thus losses in performance. In the case of the SI engine, on the other hand, the increased residual gas amounts lead to both losses in filling as well as a considerable increase in knocking potential, since a higher final compression temperature results from higher amounts of residual gas. These effects are shown and briefly described in Sect. 8.2.5 and Fig. 8.39.

For this reason, in the case of the four cylinder engine, we summarize by means of so-called ignition sequence manifold two cylinders with ignition shifted by 360°. With this, one creates two separate flows, the further summary of which can only occur further downstream in order to avoid cross flow. The circumstances for the six cylinder engine are similar considering the other ignition intervals. For emission, response, and package reasons, one must arrange the turbocharger however usually very close to the engine so that one can lead the separate flows without a cross flow location through the turbocharger, so to speak.

Represented in Fig. 8.33 are a sectional view through the turbine scroll (left) and a layout of the turbine scroll (right) for a twin-scroll turbine.

We recognize that two inflow conveyors are arranged beside one another, which admit the turbine wheel directly. The separating wall in the turbine housing reaches almost as far as the turbine wheel, so that because of the high gas speeds there, a cross-flow is avoided similarly to a pulse converter. The flow off after the impeller takes place jointly. An influence upstream the turbine is not to be expected.

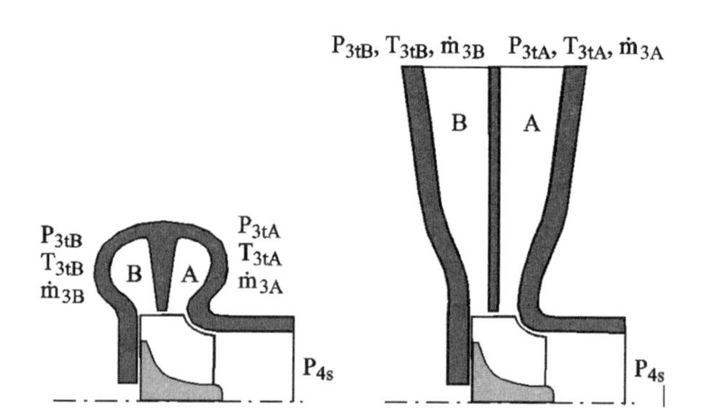

Fig. 8.33 Section through a twin-scroll turbine (Abdel Hamid et al. 2002)

For simulation, several demands result with reference to the modeling of the gas path and the turbine, which we will go into in more detail in the following. Through the twin-scroll control of the exhaust gas pipes, the kinetic energy of the exhaust impulse is used directly at the turbine. A complete conversion into pressure and then into kinetic energy again, complemented with efficiency losses, is not necessary. In order to describe reality exactly, it is essential to model the gas path with the one-dimensional gas dynamics presented in Sect. 7.4. Since there are varying pressures and mass flows in both flows of the turbine, one must already take this into consideration in the experimental determination of the characteristic map. Contingent upon the so-called branch pressure ratio

$$\Pi_{Str.} = \frac{p_{3t,A}}{p_{3t,B}}, \tag{8.34}$$

which describes the ratio between the pressure in branch A to that in branch B, several characteristic maps are recorded.

In this matter, a mass flow emerges in correspondence with the level of pressure in the branch at a given reference turbine speed. Figure 8.34 describes these relations for branch pressure ratios of 0.9 (left) and 1.1 (right). Plotted is the reference mass flow across the pressure ratio for the corresponding branch.

$$\begin{aligned} \pi_{T,A} &= \frac{p_{3t,A}}{p_{4s}} \\ \pi_{T,B} &= \frac{p_{3t,B}}{p_{4s}}. \end{aligned} \tag{8.35}$$

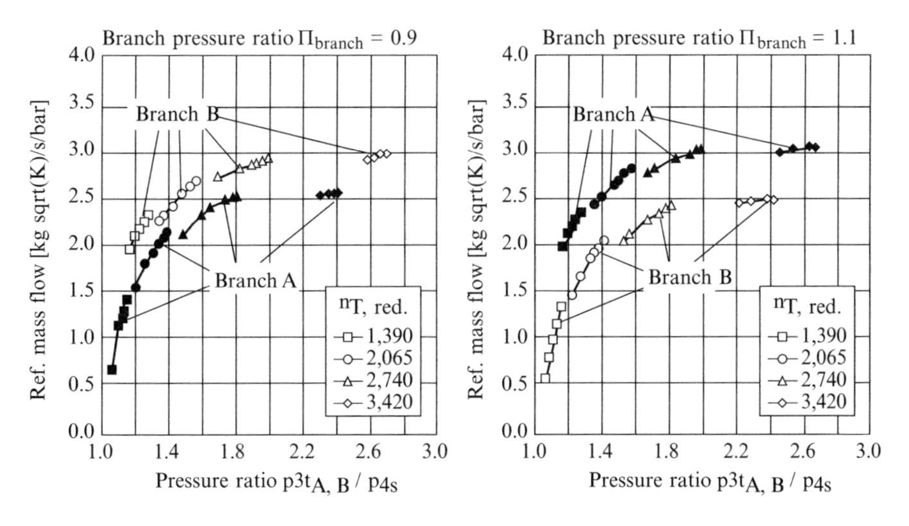

Fig. 8.34 Mass flow map of a twin-scroll turbine for branch pressure ratios 0.9 and 1.1 (Abdel Hamid et al. 2002)

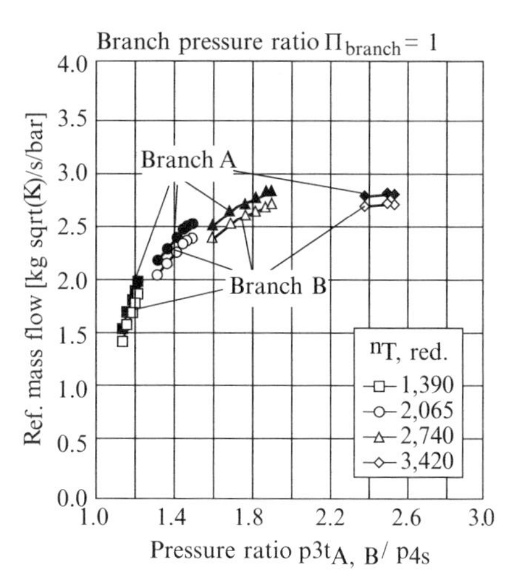

Fig. 8.35 Mass flow map of a twin-scroll turbine for a branch pressure ratio of 1.0 (Abdel Hamid et al. 2002)

For a given branch pressure ratio, the contingency known from the single-flow turbine shows itself, so that in these diagrams, two turbine maps are recognizable. In this example, the branch pressure ratios of 0.9 and 1.1 were chosen such that the mass flow characteristics of branch A in the right diagram corresponded to those in branch B in the left diagram and vice versa. In actuality, this is not the case: in the chosen example, the reference mass flow in branch A is higher than that in branch B. This becomes clear for a branch pressure ratio of 1 also in Fig. 8.35, in which the graphs for both branches should completely match up. This behavior does not have to be characteristic of twin-scroll turbines, but it does represent the difficulty of laying out twin-scroll turbines symmetrically. As one can see in Fig. 8.33, both flows admit different areas of the impeller, from which in principle a different behavior results. Via an adjustment of the cross sectional surface at the separation point, one can achieve a largely symmetrical behavior.

In the simulation of twin-scroll turbines, we can in principle proceed similarly as in the calculation of turbines with variable turbine geometry, for which we interpolate between several characteristic maps. In the case of twin-scroll turbines, separate characteristic maps are created for the particular branches contingently upon the varying branch pressure ratios. According to the engine's cylinder number and the expected height of pressure pulses, a further range for the branch pressure ratio must be covered that can take on orders of magnitude of 0.5–2. Resulting for the efficiency of the turbine is one characteristic map per branch pressure ratio. We do not distinguish hereby between the particular branches, since this is not possible in the experimental determination of the characteristic maps.

We hereby make use of a mean turbine pressure ratio according to

$$\pi_{T,m} = \frac{p_{3t,A} + p_{3t,B}}{2 p_{4s}} \tag{8.36}$$

The maps for efficiency can also be prepared according to the simulation regularities described in this section.

The calculation of a twin-scroll turbine, roughly speaking, follows this pattern: first, the branch pressure ratio $\Pi_{branch.}$ is determined. Then, one interpolates linearly with this branch pressure ratio in the various branch pressure maps. With the reference turbine speeds for both branches, which can differ because of varying temperature levels, the reference mass flow and the efficiency are calculated, from which the turbine performance for the determination of the turbocharger speed can be calculated with the mean turbine pressure ratio. Because of the calculation in the framework of one-dimensional gas dynamics, the process takes place in a relatively iterative fashion and resembles that of a pipe embranchment for a conjunction situation without a crossing of the flows.

As one can see, the calculation of twin-scroll turbines demands the greatest complexity in turbine modeling. Moreover, a plethora of characteristic maps must be prepared for calculation, which implies a cost which is hardly insignificant.

8.2.4 *Turbochargers*

8.2.4.1 The Core Equation for Turbochargers

For the steady turbocharger operation, an equilibrium results between the torque given by the turbine and the torque received by the compressor. The frictional torque should be taken into consideration in turbine efficiency. Solved according to the compressor/turbine pressure ratio, the following relations result

$$\pi_{com} = \left[\frac{\dot{m}_{tur}}{\dot{m}_{com}} \frac{\kappa_{b.com.} - 1}{\kappa_{b.com.}} \frac{\kappa_{b.tur.}}{\kappa_{b.tur.} - 1} \frac{R_{b.tur.} T_{b.tur.}}{R_{b.com.} T_{b.com.}} \eta_{is,com} \eta_{tur} \left(1 - \pi_{tur}^{*\frac{\kappa_{b.tur.} - 1}{\kappa_{b.tur.}}} \right) + 1 \right]^{\frac{\kappa_{b.com.}}{\kappa_{b.com.} - 1}} \tag{8.37}$$

$$\pi_{tur}^{*} = \left[\frac{\dot{m}_{com}}{\dot{m}_{tur}} \frac{\kappa_{b.com.}}{\kappa_{b.com.} - 1} \frac{\kappa_{b.tur.} - 1}{\kappa_{b.tur.}} \frac{R_{b.com.} T_{b.com.}}{R_{b.tur.} T_{b.tur.}} \frac{1}{\eta_{is,com} \eta_{tur}} \left(\pi_{com}^{\frac{\kappa_{b.com.} - 1}{\kappa_{b.com.}}} - 1 \right) + 1 \right]^{\frac{\kappa_{b.tur.}}{\kappa_{b.tur.} - 1}} \tag{8.38}$$

Employed in dynamic propulsion systems, a high measure of unsteady behavior is demanded of engines and thus of the charger aggregates as well. This effect must also be reproducible in the framework of simulation calculations.

8.2.4.2 Angular-Momentum Conservation Law

Whilst in the case of mechanically propelled supercharger aggregates, the propulsion performance comes directly from the engine and the momentum of inertia of the supercharger aggregate is added to the momentum of inertia of the engine corresponding to the square of the transmission ratio, the unsteady simulation of the turbocharger depends primarily on the difference torque of the turbine and the compressor and on the polar momentum of inertia of the mechanism of the turbocharger.

The formal relationship necessary for the simulation of the run-up of the turbocharger for the change in the angle speed of the turbocharger mechanism is

$$\frac{\mathrm{d}\omega}{\mathrm{d}t} = \frac{(M_{tur} - M_{com} - M_{rot} + M_{add})}{\Theta_{tc}}. \tag{8.39}$$

In the case of electrically supported turbochargers, an additional torque, e.g. via an electric motor, is introduced to the turbocharger shaft in order to accelerate the run-up.

For the warm engine, the frictional torque can be considered in the form of a total mechanical efficiency in the case of the turbine torque, since the frictional torque occurs reproducibly depending on the operating point or the turbocharger speed. For the separate determination of the friction torque in the turbocharger, costly measurements of the entire turbocharger are required. Furthermore, an exact attribution and division to both turbo machines is neither possible nor useful, since we are essentially concerned with the frictional torque of the mutual connection shaft of the turbine and the compressor. At high speeds, the frictional torque takes on an order of magnitude of 5–10% of the turbine torque.

The goal of an optimization of the mechanism for unsteady operation must therefore be, on the one hand, to minimize the frictional torque in order to increase the surplus torque between the turbine and the compressor, and, on the other hand, to minimize the polar momentum of inertia in order to obtain a larger gradient of angle speed.

The reduction in size of one component of the turbocharger causes a clear change in the polar torque of inertia. This is clear from the formal relationship in (8.40). The polar momentum of inertia is determined in geometric similarity, on the one hand, by means of the mass of the mechanism, which is proportional to the third power of a geometrical length (e.g. of the diameter). On the other hand, the polar momentum of inertia is determined in addition by square of the diameter

$$\Theta \sim d^3 d^2 = d^5. \tag{8.40}$$

The same relation can be utilized for simulation purposes in the necessary size adjustment of the turbocharger, when the turbocharger as a whole or components thereof are geometrically enlarged or reduced.

8.2.4.3 Dependence of the Frictional Torque on the Oil Temperature

In simulating transient warm-up behavior of engines and turbochargers, knowledge of the dependence of frictional losses on the working medium temperatures is of

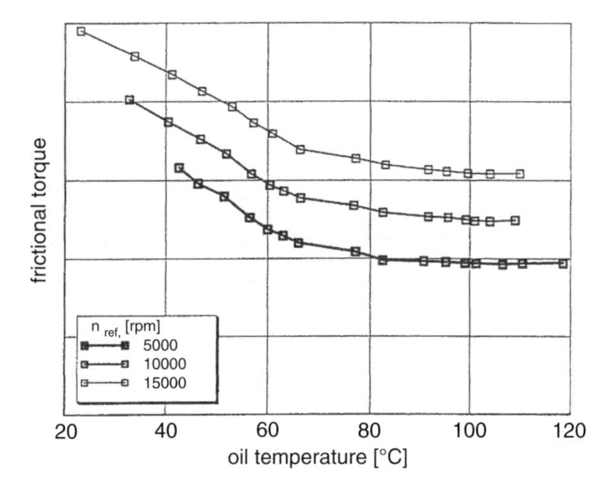

Fig. 8.36 Dependence of the friction torque in a passenger car turbocharger on the oil temperature

decisive importance. Figure 8.36 shows the experimentally determined dependence of the frictional torque on the oil temperature for a turbocharger supported in plain bearings for the passenger car range at low speeds, as they might appear in idle speed operation of the engine.

The graphs show an almost linear fall in the frictional torque up to an oil temperature of approx. 80°C. After this temperature, the frictional torque remains practically constant. This effect remains at higher speeds as well and can be taken into consideration by means of a suitable superimposition of an additional frictional torque contingent on the oil temperature.

8.2.4.4 Overall Thermal Behavior at High Turbine Temperatures

In the case of the turbocharger, the heat release of the turbine influences the efficiency behavior of the compressor enduringly, since between the particular subsystems of turbine, bearing housing, and compressor heat flows are exchanged either via heat conduction or radiation/convection. Since the turbine is usually admitted with a constant gas temperature for the characteristic map measurement and is even occasionally isolated, stationary heat flows between turbine and compressor arise that do not occur in reality. With this, only one aspect of the real behavior of a turbocharger is covered in the adiabatic approach with corresponding isentropic efficiencies of compressor and turbine.

If in real operation a higher gas temperature as in the characteristic map measurement now arises, a greater heat flow occurs from the turbine to the compressor/ into the bearing housing or into the environment. This influences in a sustained way the energy balance of the subsystems and thus also the efficiency behavior. Figure 8.37 shows schematically the heat flows for a water-cooled turbocharger, in

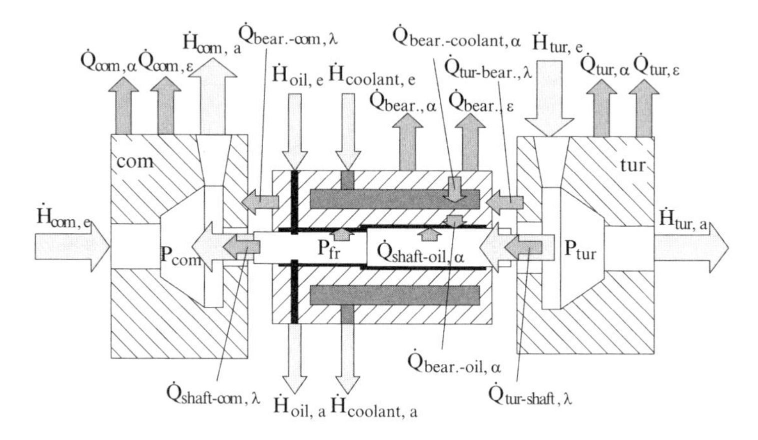

Fig. 8.37 Heat flow model for a water-cooled turbocharger

which the compressor, bearing housing, and turbine are seen as their own subsystems.

At high loads, the turbine releases heat to the environment primarily through radiation, which is partially taken in again by the compressor and bearing housing. According to the current flow around of the turbocharger in the engine package, the three subsystems release heat to the surrounding air via free or forced convection. A large part of the heat is transported via heat conduction from the usually very hot turbine into the bearing housing despite baffle plates and decoupling measures. In the same way, heat conduction takes place in the turbocharger shaft from the turbine into the compressor. A portion of these heat flows is lead away by the lubricant from the bearing housing or the turbocharger shaft. In the case of a water-cooled housing, as is used for thermally heavily burdened turbochargers in SI engines, heat is lead away from the bearing housing by coolant as well.

In order to allow for the description of the individual heat flow terms, we need extensive measurements of the surface temperatures and costly CFD calculations, which leads to unacceptable calculation times in the context of a transient simulation as introduced in Sect. 10.5.

Above all at very high temperatures of up to 1,050°C and the resultantly high wall temperatures, the share of radiation in heat removal becomes more significant. It should be noted that a description of the efficiency behavior of the compressor and above all of the turbine by means of the characteristic maps measured at the stationary turbocharger test bench is results in increased errors, the greater the difference in temperature between the real turbine entry temperature and the gas entry temperature at the turbocharger test bench in the determination of the characteristic map.

As a simple measure for the simulation, a measurement of the maps of the compressor and the turbine with varying turbine entry temperatures is a possibility.

This requires however a very high experimental cost and raises the cost drastically precisely in the case of these turbine models. In the simulation, we must then interpolate between these maps with the real exhaust gas temperature. The thermally slow behavior can however not yet be considered in such an investigation. This requires in addition heat conduction, convection and radiation models for the description of thermal behavior. A detailed formal modeling of these relations would however go beyond the scope of this book.

8.2.5 Charge Air Cooling

8.2.5.1 Foundations

In charged internal combustion engines, the charger aggregate causes an increase in the temperature of the fresh gas as well, due to the change in state occurring in an increase of the pressure of the fresh gas. Via the isentropic state equation and the isentropic efficiency of the charger aggregate, both states are in close relation

$$T_2 = T_1\left[\frac{1}{\eta_{is,\,com}}\left(\pi_{com}^{\frac{\kappa_1-1}{\kappa_1}}-1\right)+1\right]. \tag{8.41}$$

The purpose of charging, i.e. a density increase of the charge mass taken in by the engine via pressure increase, is thereby partially inhibited by the increase in temperature of the charge mass. The increase in density is set in the case of an engine that is not charge air cooled by the isentropic exponent, the compressor pressure ratio, and the isentropic efficiency

$$\rho_2 = \frac{p_2}{R_2 T_2} = \frac{p_1}{R_2 T_1}\frac{\pi_{com}}{\frac{1}{\eta_{is,com}}\left(\pi_{com}^{\frac{\kappa_1-1}{\kappa_1}}-1\right)+1} = \rho_1 \frac{\pi_{com}\,\eta_{is,\,com}}{\pi_{com}^{\frac{\kappa_1-1}{\kappa_1}}-1+\eta_{is,\,com}} \tag{8.42}$$

Figure 8.38 shows a characteristic map, in which the quotients from the density after the charger aggregate and the density before the charger aggregate are represented according to (8.42) depending upon the isentropic efficiency and the pressure ratio at an isentropic compression with an isentropic exponent of 1.4. Even in the case of a possible increase of the isentropic efficiency to the value of 1, which is in reality not achievable, the increase in density is not directly proportional to that of the boost pressure. This effect is all the more pronounced the higher the pressure ratio is, and is grounded in the isentropic change of state. The influence of the isentropic efficiency over the charge air temperature represents an additionally important potential in the increase of density.

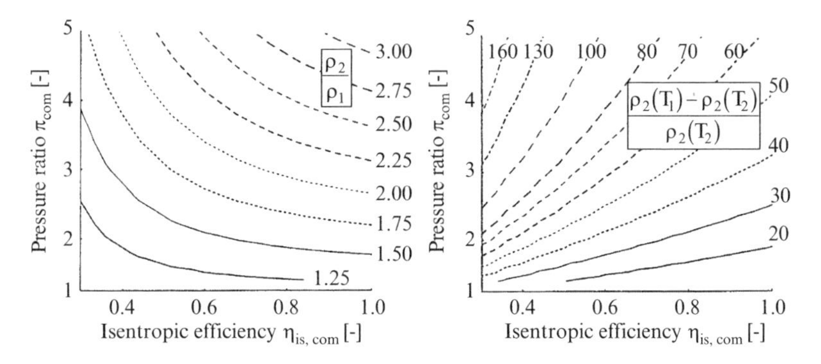

Fig. 8.38 Density ratio due to isentropic compression (left) and percental density factor due to recooling (right)

In Fig. 8.38 (right), the effect is represented of a possible recooling of the charge mass on the temperature before the compressor, again contingent on the isentropic efficiency and the pressure ratio. The theoretically maximum attainable density is thereby directly proportional to the compressor pressure ratio and is thus independent of the isentropic efficiency. The state changes in recooling differ at varying isentropic efficiencies only in the varyingly high amounts of heat that must be removed for the recooling of the charge mass to the state before the charger aggregate. In Fig. 8.38 (right) is represented the percentual increase in density after the charger aggregate at complete recooling as opposed to the density increase shown in Fig. 8.38 (left).

Charge air cooling thus represents a comparatively simple possibility to increase the charge mass and with that the performance of the engine. However, a lowering in temperature has a favorable effect not only with respect to the possibility of performance increase, but also – above all by means of a global lowering of the process temperature level, as well as by means of a lower thermal load on the components – in terms of pollutant emissions (nitrogen oxide) and possible knocking danger in the SI engine, see Dorsch (1982).

Thus, the necessity of charge air cooling also depends on the type of engine. Whilst in diesel engines charge air cooling is only needed for the increase of the charge mass and thus for performance increase or the reduction of emissions, in charged SI- or gas engines, it is absolutely necessary for the avoidance of knocking and for the achievement of a good level of efficiency (optimal position of the center of combustion). Via the reduction of the temperature of the charge mass, we shift the combustion process away from the knocking boundary. The free space thus created can now be made use of either via a further raising of the pressure ratio, of the compression ratio of the engine, or though an efficiency-improving of ignition timing. One can see from this how much charge air cooling, especially in the case of SI- or gas engines, has an effect on the attainable values of performance and efficiency and thus also on fuel consumption, such that the advantages of an

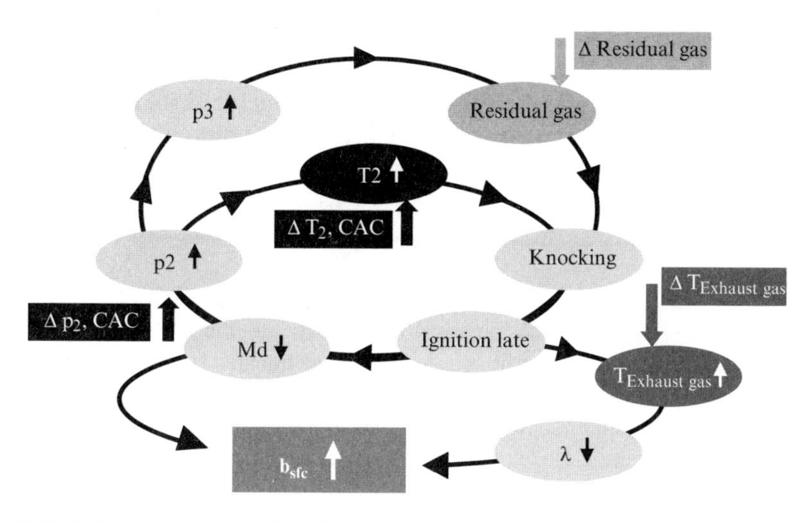

Fig. 8.39 Influence on consumption of a turbocharged SI engine

increase of the charge mass are not spoiled by a retarding ignition delay or a lowering of the compressor ratio.

The relations for a turbocharged SI engine are summarized in Fig. 8.39. A higher loss in pressure in the charge air cooler causes under identical engine performance both a higher exhaust gas counter-pressure as well as a higher inlet manifold temperature. Both cause an increased knocking tendency, which necessities a later ignition as a consequence. Through this, the exhaust gas temperature rises and, at the attainment of a value critical for the turbine or the catalyst, creates an enrichment demand, i.e. higher fuel consumption. At the same time, the torque usually decreases because of the ignition delay, for which can be compensated by a higher charge pressure with the known consequences for exhaust gas pressure and the recooling rate. The spiral is run through once more. We recognize that already a small disturbance of the system can have disastrous consequences. Charge air cooling is thus one of the most decisive keys to efficient turbo engines.

The use and efficiency of charge air cooling depend however primarily on the availability of a corresponding coolant, in order to make possible the reduction of a sufficiently large quantity of heat. The temperature of the charge mass can however be lowered at most to the temperature of the coolant and is moreover still contingent on the heat transfer surface and the mass flow of the coolant. Therefore, various design solutions for charge air cooling result for different areas of use in charged engines. For application in ships or in stationary plants, relatively cold water in partly unlimited amounts is available for the cooling of the charge air, while for application in motor vehicles one can only fall back upon the surrounding air or the coolant of the vehicle. The design of the charge air cooler must in addition be taken into consideration the available mass flow of the coolant and its heat capacity.

8.2.5.2 Numerical Treatment

The most common representation of coolers is shown in Fig. 8.40. In this case, the specific cooling output $\Phi$is plotted in contingence on the mass flow of the cooling medium and the medium to be cooled. The heat flow $\dot{Q}_{LLK}$ removed in the cooler can be calculated with

$$\dot{Q}_{cac} = \Phi(T_{air,in} - T_{cool,in}). \tag{8.43}$$

The influence of the construction form (cross-flow, counter-flow, or equal flow coolers), the heat transfer conditions, and the heat transfer surface are taken into consideration by the varying paths of the specific cooling performance. The effect of cooling is that much greater, the greater the adjoining difference in temperature between the entry temperature of the charge mass $T_{air;in}$ and the entry temperature of the coolant $T_{K,e}$.

Another possibility for describing the charge air cooler is the cooling digit ε, which gives the quotient from the actual temperature lowering and the maximum possible temperature lowering

$$\varepsilon = \frac{T_{air,in} - T_{air,out}}{T_{air,in} - T_{cool,in}}. \tag{8.44}$$

For this as well, a representation in a characteristic map can be selected. With the cooling digit ε results for a rough estimate of the density $\rho_{air;out}$ of the charge mass after the charge air cooler under consideration of (8.42)

$$\rho_{air,out} = \rho_1 \frac{\pi_{com}\,\eta_{is,com}}{\left(\pi_{com}^{\frac{\kappa_1-1}{\kappa_1}} - 1 + \eta_{is,com}\right)(1-\varepsilon) + \varepsilon\,\eta_{is,com}\frac{T_{cool,in}}{T_1}}. \tag{8.45}$$

The heat flow removed in the charge air cooler can thus be calculated contingently on the temperature difference of the charge air at the inlet and outlet of the charge air cooler

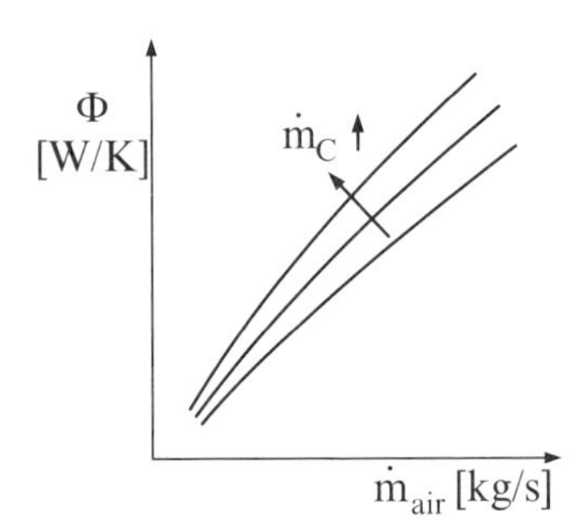

Fig. 8.40 Map of a charge air cooler

$$\dot{Q}_{cac} = \dot{m}_{air}\, c_p\, \Delta T_{cac} = \dot{m}_{air}\, c_p \left(T_{air,in} - T_{air,out}\right). \tag{8.46}$$

With this, we obtain the following relation between the cooling digit and the specific cooling performance:

$$\varepsilon = \frac{\Phi}{\dot{m}_{air}\, c_p}. \tag{8.47}$$

In the case of the filling and emptying method, the heat flow taken from the fresh gas in the charge air cooler is calculated according to (8.43) and removed as wall heat flow in a specific volume module.

In reality, a pressure loss appears in the charge air cooler as in every pipe, which partially thwarts the effect of the increase of density of the charge mass via cooling. Therefore, in the constructive design of the charge air cooler, one must strive for a pressure decrease that is as small as possible. In existing engine vehicle construction types pressure loss values of up to 150 mbar appear under rated power conditions. Such pressure losses can no longer be neglected in the simulation.

The consideration of the pressure losses of the charge air cooler in the simulation is however not possible in a volume module, since the pressure in the volume is calculated from the thermal state equation and is thus set for both volumes exits to an equal amount.

The pressure loss in a charge air cooler can thus only be simulated by means of a specific throttle, which is arranged before or after the charge air cooler volume. Since the description by means of a simple throttle with a constant flow coefficient is insufficient because of the complexity of the charge air cooler, here too we must fall back upon a characteristic map for the description of pressure losses. The losses in the charge air cooler are composed of the losses in the inflow and outflow cases and of the frictional losses in the actual cooling grid. In the characteristic map, these total pressure losses are represented over mass flow. If we know the pressures before and after the charge air cooler throttle, we can very easily determine the mass flow through the throttle from the characteristic map. The charge air cooler throttle thus provides a mass and enthalpy flow, with which the condition quantities in the volumes before and after the throttle can be calculated according to the procedure described in Sect. 7.4. Consequently, a new pressure difference arises, which under stationary conditions corresponds to the pressure loss of the charge air cooler.

The calculation of pressure losses in the charge air cooler using one-dimensional gas dynamics is much more complex. In this case, the charge air cooler is modeled as a pipe-bundle of a number of small tubes corresponding to the cooling shafts, into which the air to be cooled flows. The inflow and outflow cases are represented in this modeling as volumes with the corresponding volumes of the cases. The tubes serve as connections between the volumes. With this method, which is derived from the simulation of catalyst flow, the pressure loss in the charge air cooler can be approximated very well.

However, the heat removal of the charge air cooler is problematic. This has to occur by means of the walls of the pipes.

For this, the heat flow from the charge air to the wall of the tube is determined with the help of methods for the wall heat transfer at the internal pipe wall. The heat exchange of the external pipe wall with the environment is described by the charge air map shown in Fig. 8.40. In order to limit the cost of describing the charge air cooler, we assume that an equal heat flow is removed from each tube. Thus, an equal in-flow of the tubes and an equal distribution of the air to be cooler is assumed for all tubes. With the number i of tubes results

$$\dot{Q}_{pipe} = \frac{\dot{Q}_{cac}}{i}. \tag{8.48}$$

In the case of a cross flow heat exchanger, roughly equal temperatures of the cooling medium exist across the entire pipe course. In the direction of flow however, the temperature of the cooling medium decreases. From this results increasingly small temperature differences in the direction of flow of the cooling medium, which according to Miersch (2003) can be considered by means of the length discretization of the tube, necessary anyway for the calculation, with one segment. These affect the heat flow to be removed per segment as follows

$$.\dot{Q}_{pipe,j} = \frac{\Delta T_j}{\sum\limits_{k=1,..,l} \Delta T_k} \dot{Q}_{pipe;j}. \tag{8.49}$$

In this case, ΔT_k represents the temperature difference between the internal pipe wall and the charge air temperature for one segment. The charge air cooler is viewed as being without mass. A storage effect in the component cannot be considered in this model.

References

Abdel Hamid S, Hagelstein D, Rautenberg H, Seume J (2002) TC-Gesamtkennfeldbestimmung. FVV Forschungsvorhaben

Dorsch H (1982) Ladeluftkühlung bei aufgeladenen Personenwagen-Ottomotoren. Motortechnische Zeitschrift MTZ 43. Franckh-Kosmos Verlags-GmbH, Stuttgart, pp 201–205

Flückiger L, Tafel S, Spring P (2006) Hochaufladung mit Druckwellenlader für Ottomotoren, Motortechnische Zeitschrift MTZ 67. Franckh-Kosmos Verlags GmbH, Stuttgart, pp 946–954

Hütten H (1997) Motoren. Technik, Praxis, Geschichte, 10. Auflage, Motorbuchverlag Stuttgart

Jenni E (1993) Der BBC-Turbolader, Geschichte eines Schweizer Erfolgs. ABB Turbo Systems AG, Baden

Miersch J (2003) Transiente Simulation zur Bewertung von ottomotorischen Konzepten. Dissertation, Universität Hannover

Münzberg HG, Kurzke J (1977) Gasturbinen – Betriebsverhalten und Optimierung. Springer, löschen, Berlin

Pischinger R, Klell M, Sams Th (2002) Thermodynamik der Verbrennungskraftmaschine. 2., überarbeitete Auflage, Springer, Wien, New York

Zinner K (1985) Aufladung von Verbrennungsmotoren. Springer, Berlin

Chapter 9
Exhaust Aftertreatment

Johann Christian Wurzenberger and Reinhard Tatschl

Nomenclature

a_{geo}	Monolith geometric surface
c_g	Gas composition
$c_{j.g}$	Species j gas-bulk concentration
c^L	Species surface concentration
c_j^L	Species j concentration at reactive surface
$c_{p,s}$	Solid phase heat capacity
d_1	Channel diameter
d_{hyd}	Hydraulic channel diameter
h_j	Enthalpy of species j
Δh_i	Reaction enthalpy of i-th reaction
k_w k_{rt} k_{ak} k_{rk}	Permeability of different layers
$m_{russ,in}$	Inlet soot mass flow rate
p_g	Gas-phase pressure
$p_{g,n}$	Channel pressure
Δp_w	Wall induced pressure loss
Δp_{rk}	Pressure loss across soot depth filtration layer
Δp_{ak}	Pressure loss across ash cake
Δp_{rk}	Pressure loss across soot cake
r_i	Species j reaction rate
t	Time
v_g	Gas-phase velocity
$v_{g,n}$	Velocity in channel
$v_{w,n}$	Flow velocity across porous wall
v_w	Wall velocity
$_{wg,1}$	Wall velocity
$v_{w,n}$	Normalized wall velocity
$w_{j,g}$	Species j mass fraction

G.P. Merker et al. (eds.), *Combustion Engines Development*,
DOI 10.1007/978-3-642-14094-5_9,

x	Space coordinate
z	Space coordinate
$A_{F,n}$	Channel free flow area
D_{eff}	Effective diffusion coefficient
F_n	Friction coefficient
Gz_h	Heat transfer Graetz number
Gz_m	Species transfer Graetz number
MG_j	Molar mass of species j
$P_{S,n}$	Channel perimeter
Sc	Schmidt number
S_{rk}	Cake filtration indicator
S_{rt}	Depth filtration indicator
T_g	Gas temperature
T_s	Wall temperature
Z_k	Coverage ratio of k-th stored species
α	Gas/solid-phase heat transfer coefficient
β_j	Species j transition coefficient
δ_{rk}	Soot cake height
δ_{ak}	Ash cake eight
ε_g	Monolith open frontal area
λ_g	Gas-phase heat conductivity
λ_s	Wall heat conductivity
ζ	Friction coefficient
ρ_g	Gas-phase density
$\rho_{g,n}$	Density in channel
ρ_s	Solid-phase density
μ	Dynamic viscosity
$\nu_{i,j}$	Stoichiometric coefficient of species j in reaction i
Γ	Free surface

9.1 Modeling and Simulation

The design and optimization of exhaust aftertreatment systems is increasingly supported by computer modeling and numerical simulations. 1D simulation models support decisions on base designs in the early concept phase by providing virtual investigations of various design variants with respect to overall conversion performance. High detail 3D simulations are applied to investigate individual components with respect to geometry, size, position, coating, flow conditions and thermal behavior. Computational efficient, real-time capable exhaust aftertreatment models, as part of system level engine models, are used to support the development and calibration of control algorithms and functions.

The broad variety of different simulation tasks in aftertreatment system and component optimization typically requires application tailored simulation models suitable for efficient use within the different phases of the development process. By ensuring a maximum of consistency between the physical and chemical models adopted in the 1D/3D simulations and the real-time capable aftertreatment system modeling, model parameterization data can be easily exchanged between the different simulation approaches, ensuring full data compatibility throughout the different product development phases.

Modeling and simulation of exhaust aftertreatment systems can be split – according to the real hardware – into several individual components. Exhaust systems typically comprise of catalytic converters, Diesel particulate filters, dosing units and pipes. The basic modeling assumptions and equations are summarized in the following chapters. Additionally, several selected simulation results are presented and discussed.

9.2 Catalytic Converters

Honeycomb type catalytic converters are applied to reduce engine emissions for many years. The honeycomb structure consists of several hundred parallel channels and provides an as big as possible surface between gas and catalyst wall per unit of overall catalyst volume. At the catalyst wall, within a small wash coat layer, catalytically active centers (e.g. noble metals like platinum, rhodium, palladium, etc.) are located. These active centers support the conversion of pollutants like carbon monoxide, unburned hydrocarbons or nitric oxides. The detailed mechanism of catalytically supported reactions is very complex and only partly understood. One essential effect is that the catalytic support allows reactions taking place at significantly lower temperatures compared to systems in the absence of catalytically active centers. Thus, it is possible to reduce engine emissions at temperature levels which are typical for engines. One key characteristic of simulation models dealing with catalytic converters is to predict its conversion behavior at different operation conditions represented by exhaust temperature, mass flow and species composition.

The development of the theoretical models does not only help to broaden the basic understanding of physical/chemical processes taking place but also supports the design and optimization of real systems. Depending on the computational capacity models with different modeling depth have been developed in the last decade, as summarized and discussed by Depcik and Assanis (2005).

9.2.1 Basis Equations

The physical and chemical effects taking place within one single channel of a honeycomb type catalytic converter are sketched in Fig. 9.1. It can be distinguished between convective and conductive transport in axial direction in the gas and solid

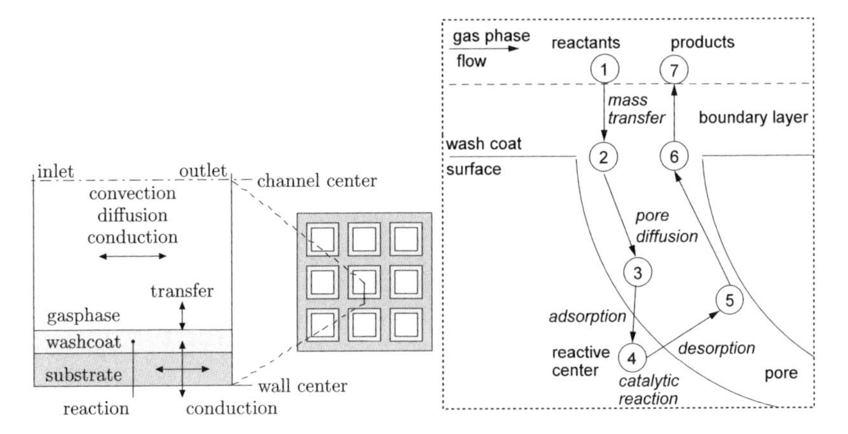

Fig. 9.1 Effects with one single channel of a honeycomb type catalytic converter and schematic of a heterogeneous catalytic reaction (*left figure* taken from Wurzenberger and Peters 2003, *right figure* taken from Hayes and Kolaczkowski 1997)

phase and transfer of heat and mass between the two phases. Besides the intrinsic reaction at catalytic active sites, 1D models typically consider diffusive mass transfer of educts and products through the flow boundary layer, diffusive transport within the porous wash coat and ad-/desorption at the solid surface.

Assuming that radial gradients with the monolith and within the flow cross section of one single channel are negligible, 1D models are sufficient to represent the entire converter. The base model equations of a transient 1D model are summarized in the following section.

The continuity equation of the gas phase is given by

$$\frac{\partial \rho_{\mathrm{g}}}{\partial t}=\frac{\partial \rho_{\mathrm{g}} \cdot v_{\mathrm{g}}}{\partial z}+\frac{a_{\mathrm{geo}}}{\varepsilon_{\mathrm{g}}} \cdot \sum_{j}^{\mathrm{S}} MG_j \sum_{i}^{\mathrm{R}} v_{i,j} \cdot \dot{r}_i\left(c^{\mathrm{L}}, T_{\mathrm{s}}\right). \tag{9.1}$$

The second term on the right side is zero for homogenous gas phase reactions and steady-state approaches of heterogeneous conversion reactions. It is essential to correctly balance transient ad-/desorption processes at the catalyst wall.

The pressure drop within a catalytic converter is described by a momentum balance equation as given by

$$\frac{\partial \rho_{\mathrm{g}} \cdot v_{\mathrm{g}}}{\partial t}=-\frac{\partial\left(\rho_{\mathrm{g}} \cdot v_{\mathrm{g}}^2+p_{\mathrm{g}}\right)}{\partial z}+\zeta(\mathrm{Re}) \cdot \frac{\rho_{\mathrm{g}} \cdot v_{\mathrm{g}}^2}{2 \cdot d_{\mathrm{hyd}}}. \tag{9.2}$$

The energy conservation equation of the gas phase is modeled according to

$$\frac{\partial}{\partial t}\left[\rho_{\mathrm{g}}\cdot\left(\sum_{j}^{\mathrm{S}}w_{j,\mathrm{g}}\cdot h_{j}+\frac{v_{\mathrm{g}}^{2}}{2}\right)\right]=-\frac{\partial}{\partial t}\left[\rho_{\mathrm{g}}\cdot v_{\mathrm{g}}\cdot\left(\sum_{j}^{\mathrm{S}}w_{j,\mathrm{g}}\cdot h_{j}+\frac{v_{\mathrm{g}}^{2}}{2}\right)\right]$$
$$+\frac{\partial}{\partial z}\left(\lambda_{\mathrm{g}}\cdot\frac{\partial T_{\mathrm{g}}}{\partial z}\right)+\sum_{j}^{\mathrm{S}}\frac{\partial}{\partial z}\left(\rho_{\mathrm{g}}\cdot D_{\mathrm{eff}}\cdot\frac{\partial w_{j,\mathrm{g}}}{\partial z}\cdot h_{j}\right)$$
$$+\frac{a_{\mathrm{geo}}}{\varepsilon_{\mathrm{g}}}\cdot\left[\alpha\cdot\left(T_{\mathrm{s}}-T_{\mathrm{g}}\right)-\sum_{i}^{\mathrm{R}}\Delta h_{i}\cdot\dot{r}_{i}\left(c^{\mathrm{L}},T_{\mathrm{s}}\right)\right]. \tag{9.3}$$

The second term (head conduction) and the third term (cross diffusion) on the right hand side are typically small compared to the first term (convection) and only gain significance at very small flow velocities. The last term on the right side represents a heat source due to heat exchange with the wall and chemical reactions. The negative sign at the reaction source is applied to compensate for the implicitly considered heat of reaction given by changed species compositions. Thus, the proper heat of reaction is applied in the energy balance equation of the solid phase where the intrinsic reaction takes place.

The species conservation equation is given by

$$\frac{\partial\rho_{\mathrm{g}}\cdot w_{j,\mathrm{g}}}{\partial t}=\frac{\partial\rho_{\mathrm{g}}\cdot w_{j,\mathrm{g}}\cdot v_{\mathrm{g}}}{\partial z}+\frac{\partial}{\partial z}\left(\rho_{\mathrm{g}}\cdot D_{\mathrm{eff}}\cdot\frac{\partial w_{j,\mathrm{g}}}{\partial z}\right)$$
$$+\frac{a_{\mathrm{geo}}}{\varepsilon_{\mathrm{g}}}\cdot MG_{j}\cdot\sum_{i}^{\mathrm{R}}v_{i,j}\cdot\dot{r}_{i}\left(c^{\mathrm{L}},T_{\mathrm{s}}\right). \tag{9.4}$$

Similar to the energy conservation equation, the conductive transport (second term on the right side) is small compared to the convective transport (first term on the right side). Changes of the species fractions due to chemical reactions depend on the individual reaction rates and are scaled with the stoichiometric coefficients of the individual species in the different reactions.

The energy conservation of the solid phase is

$$\rho_{\mathrm{s}}\cdot c_{\mathrm{p,s}}\frac{\partial T_{\mathrm{s}}}{\partial t}=\frac{\partial}{\partial z}\left(\lambda_{\mathrm{s}}\cdot\frac{\partial T_{\mathrm{s}}}{\partial z}\right)-\frac{a_{\mathrm{geo}}}{1-\varepsilon_{\mathrm{g}}}$$
$$\cdot\left[\alpha\cdot\left(T_{\mathrm{s}}-T_{\mathrm{g}}\right)-\sum_{i}^{\mathrm{R}}\Delta h_{i}\cdot\dot{r}_{i}\left(c^{\mathrm{L}},T_{\mathrm{s}}\right)\right]. \tag{9.5}$$

The second term on the right side is, in analogy to the energy conservation equation of the gas phase, noted with a negative sign. Thus, all heat of reactions are added to the solid phase and exchanged with the gas phase via heat transfer.

In order to account for the two regimes of transport and chemical limitation a steady-state species balance at the surface is applied for species transport and reaction. This is given by

$$a_{\mathrm{geo}} \cdot \beta_j \cdot \left(c_j^{\mathrm{L}} - c_{j,\mathrm{g}}\right) = \sum_i^{\mathrm{R}} \dot{r}_i\left(c^{\mathrm{L}}, T_{\mathrm{s}}\right). \tag{9.6}$$

Besides the conversion reaction, ad-/desorption reactions (e.g. storage of O2, NH3 or HC and different coatings) need to be modeled. Assuming that the number of free surface sites remains constant, the following surface balance equation can be applied

$$a_{\mathrm{geo}} \cdot \Gamma \cdot \frac{\mathrm{d}Z_{\mathrm{k}}}{\mathrm{d}t} = \nu_{i,(\mathrm{S}+k)} \cdot \dot{r}_i\left(c^{\mathrm{L}}, T_{\mathrm{s}}\right). \tag{9.7}$$

The variation of the surface coverage fraction depends on the reaction rates and the corresponding stoichiometric species coefficients. For the case that several free surface sites are given the equation needs to be extended accordingly.

The heat and mass transfer in 1D catalyst model is typically modeled with empirical Nusselt and Sherwood correlations. Sieder/Tate (see Perry and Green 1997) proposed the following Nusselt correlation for heat transfer.

$$\begin{aligned} \mathrm{Nu} &= 1.86 \cdot (\mathrm{Re} \cdot \mathrm{Pr})^{1/3} \\ \mathrm{Sh} &= 1.86 \cdot (\mathrm{Re} \cdot \mathrm{Pr})^{1/3}. \end{aligned} \tag{9.8}$$

Kirchner and Eigenberger (1997) used this approach in analogy also to describe mass transfer.

9.2.2 *Types of Catalytic Converters*

A broad variety of different catalytic converters is used in the automotive application. Four main types can be distinguished. Depending on the engine and exhaust aftertreatment system the catalyst types Diesel oxidation catalysts, three-way catalysts, SCR and NOx storage catalysts are used. The catalytic conversion mechanisms of the single types are comprehensively investigated and documented in literature. Here, only the key features of the particular types are summarized.

9.2.2.1 Diesel Oxidation Catalyst

The diesel oxidation catalyst (DOC) is applied for engines operated under overall lean conditions. The DOC has several conversion tasks. This is the oxidation of carbon monoxide and unburned hydrocarbons, the oxidation of nitrogen monoxide to nitrogen dioxide and the storage of hydrocarbons. Therefore the following global reactions can be noted:

$$\begin{aligned}
&\text{R1:} && CO + \tfrac{1}{2}O_2 \rightarrow CO_2 \\
&\text{R2:} && C_nH_{2m} + \left(n + \tfrac{m}{2}\right)O_2 \rightarrow nCO_2 + mH_2O \\
&\text{R3:} && NO + \tfrac{1}{2}O_2 \leftrightarrow NO_2 \\
&\text{R4:} && C_nH_{2m} + S \rightarrow C_nH_{2m}(S)
\end{aligned}$$

Figure 9.2 shows a typical characterization of a DOC by its light-off curves. These curves represent the pollutant conversion behavior as function of the converter temperature.

Both, the measured and simulated (1D model) curves of CO and HC show first increasing conversion with increasing temperature followed by a plateau at elevated temperatures. The incomplete conversion in this temperature region is caused by transport limitation effects. The significant conversion of HC in the low temperature range indicates storage effects that are not considered in the applied simulation model. The decreasing HC curves up to temperature of about 140°C can be explained by an increasing desorption rate. The light-off curve of NO shows a maximum at about 300°C. At this temperature the reaction changes from a kinetically to a thermodynamically controlled regime. The NO oxidation reaction is of special importance in exhaust systems consisting of DOCs and diesel particulate filters or SCR converters.

9.2.2.2 Three Way Catalyst

The Three Way Catalyst (TWC) is used for lambda controlled engines. Its task is to oxidize carbon monoxide and unburned hydrocarbons and also to support the reduction of nitric oxides. Both, oxidation and reduction is possible by providing

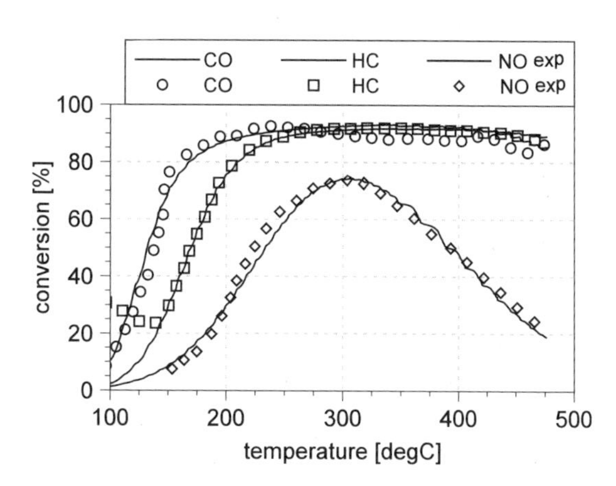

Fig. 9.2 Measured and calculated light-off curves of a diesel oxidation catalyst

engine exhaust gases evolving from stoichiometric combustion conditions. Additionally, in order to support fluctuations of oxygen in the exhaust gas, the TWC apply ceria oxides in the wash coat. Depending on the engine operating point (rich or lean) oxygen is stored at or desorbed from the catalyst surface. The following set of global reactions can be summarized:

$$
\begin{array}{ll}
\text{R1:} & CO + \frac{1}{2}O_2 \rightarrow CO_2 \\
\text{R2:} & C_nH_{2m} + \left(n + \frac{m}{2}\right)O_2 \rightarrow nCO_2 + mH_2O \\
\text{R3:} & 2NO + 2CO \rightarrow 2CO_2 + N_2 \\
\text{R4:} & H_2 + \frac{1}{2}O_2 \rightarrow H_2O \\
\text{R5:} & NO + \frac{1}{2}O_2 \leftrightarrow NO_2 \\
\text{R6:} & CO + H_2O \leftrightarrow CO_2 + H_2 \\
\text{R7:} & C_nH_{2m} + nH_2O \rightarrow nCO + (n + m)H_2 \\
\text{R8:} & 2Ce_2O_3 + O_2 \leftrightarrow 4CeO_2 \\
\text{R9:} & 2CeO_2 + CO \rightarrow Ce_2O_3 + CO_2 \\
\text{R10:} & \left(2n + \frac{5}{4}m\right)CeO_2 + C_nH_{2m} \rightarrow \left(n + \frac{m}{2}\right)Ce_2O_3 + nCO + mH_2O
\end{array}
$$

Figure 9.3 shows a comparison of experimental data and simulation results of the CO and NO conversion in a TWC. The conversion is given as function of temperature (light-off) and as function of exhaust gas composition (lambda-sweep).The simulation is performed using a 1D model. Full conversion can be observed in Fig. 9.3 for CO (and also HC) at lean conditions as soon as the light-off temperature is reached. However, NO shows only a weak conversion. A significant conversion of NO can be seen only if the exhaust gas composition reaches lambda values near one because sufficient CO (R3) is available. The conversion of CO decreases with decreasing lambda values smaller than one because of a lack of oxygen. NO, fully converts in this region because of surplus supply of CO. Out of this analysis one can conclude that the performance of a TWC requires an exhaust composition ranging in a small window around stoichiometric conditions.

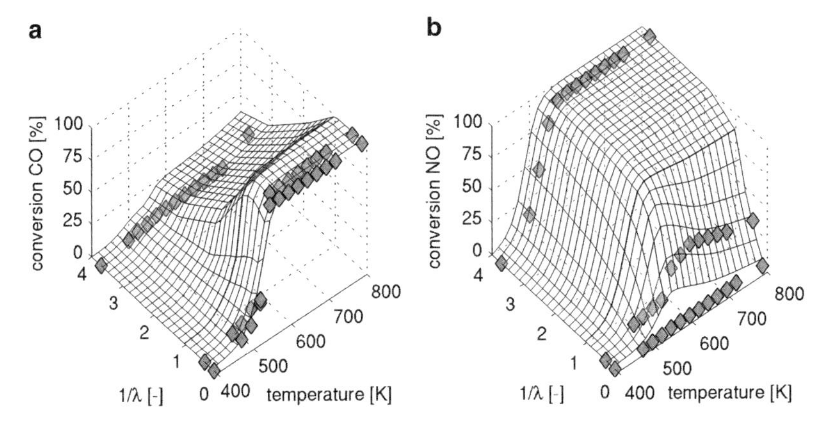

Fig. 9.3 Measured and calculated conversion of CO (*left*) and NO (*right*) in a TWC as function of temperature and gas composition (see Wurzenberger et al. 2006)

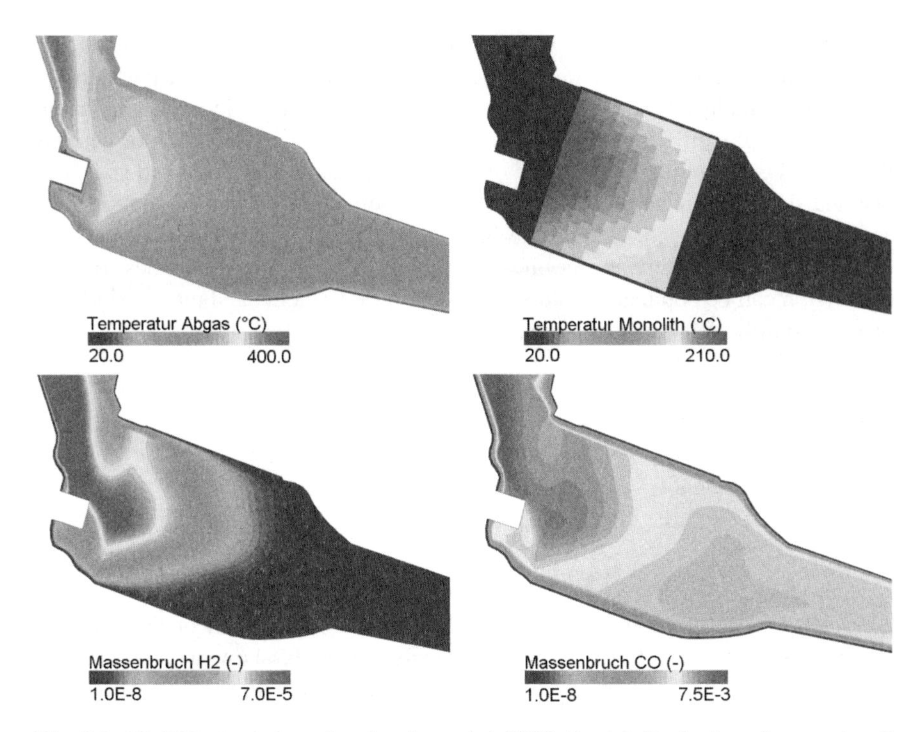

Fig. 9.4 3D CFD simulation of a closed coupled TWC. Spatial distribution of gas and wall temperature (*upper row*) and species mass fractions H2 and CO (*lower row*); (see Missy et al. 2002)

Figure 9.4 shows the temperature and species distributions taking into account the 3D flow situation at the inlet of a closed coupled TWC. The presented results are taken for one selected point of time in a transient heat-up simulation. For the considered geometry the 3D simulation results show significant radial distributions for flow, species fractions and temperature that cannot be neglected during detailed component optimization. The suitable set of reaction parameters is obtained here from an optimization loop adopting 1D simulations to match experimentally obtained conversion rates in steady flow reactor configuration. Prerequisite of such a parameter transfer is that in both the 1D and 3D model identical sub-models for all relevant effects (heat mass transfer, diffusion, reaction mechanism, surface storage etc.) are available. In this case the 1D and 3D models can be combined in an overall workflow without making a systematic break between the adopted models.

9.2.2.3 SCR-Catalyst

The SCR catalyst (selective catalytic reduction) is used for lean operated engines to reduce nitric oxides with the help of ammonia. The introduction of the ammonia in the exhaust systems takes place upstream the SCR catalyst by typically injecting an aqueous urea solution. The injected urea-water droplets convert via the steps of thermolysis and hydrolysis into ammonia, carbon dioxide and vapor. Depending on the operating conditions ammonia adsorbs at the SCR catalyst surface or desorbs from it (R1). The reduction of nitric oxides takes place via the so-called standard, fast and slow SCR reaction (R2–R4) where ammonia gets consumed and nitrogen and water are produced. Here it can be observed that if ratio of nitric monoxide to nitric dioxide is close to one the fast SCR is favored and the highest overall conversion can be obtained. Besides the reaction of ammonia and nitric oxides the first gets oxidized at elevated temperatures.

$$\begin{array}{ll} \text{R1:} & NH_3 + S \leftrightarrow NH_3(S) \\ \text{R2:} & 4NO + 4NH_3 + O_2 \rightarrow 4N_2 + 6H_2O \\ \text{R3:} & NO + NO_2 + 2NH_3 \rightarrow 2N_2 + 3H_2O \\ \text{R4:} & 6NO_2 + 8NH_3 \rightarrow 7N_2 + 12H_2O \\ \text{R5:} & 4NH_3 + 3O_2 \rightarrow 2N_2 + 6H_2O \end{array}$$

In order to get most reliable models suitable for transient drive cycle investigations it is typically required to calibrate the conversion behavior of the catalytic reactions with the help of selected measurements. Figure 9.5 shows a comparison of measured and simulated NO_x-conversion rates given at 20 engine load points characterized by different exhaust mass flows, species compositions and exhaust

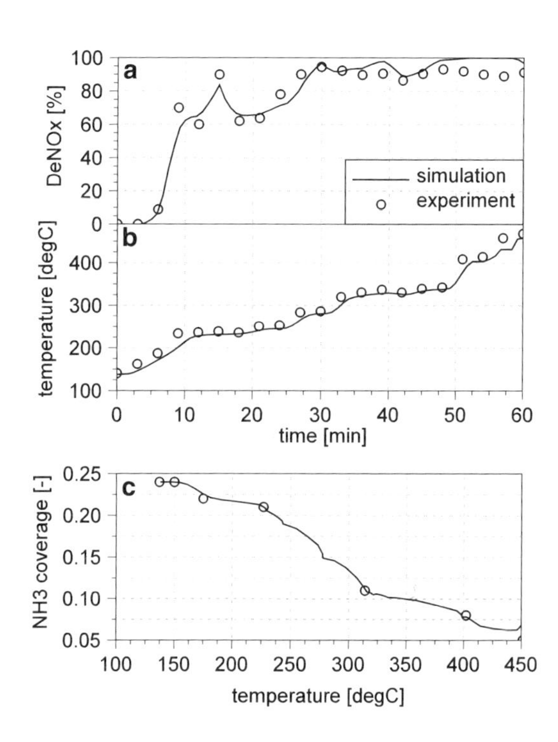

Fig. 9.5 Comparison of experimental data and simulation results of an SCR catalyst

temperatures. A good agreement between simulation and measurement at all steady state points can be observed with the decreasing ammonia surface converge fraction with increasing temperature is correctly reflected by the model. For transient operating conditions the correct description of the ad-/desorption rates and of the surface coverage are the essential values to influence the conversion behavior.

9.2.2.4 NO_x-Storage Catalyst

NO_x-storage catalysts or Lean NO_x-catalysts (LNT) are used as an alternative approach to SCR catalysts for lean burning engines. The basis function of an LNT is provided by barium carbonate particles at the catalyst surface. Barium carbonate converts to barium nitrate in the presence of NO_x as it is given during lean engine operating conditions. During rich operating conditions reductive species like CO react with barium nitrate and build barium carbonate and release NO_x. NO_x is consequently reduced by CO or HC as there is a surplus of the species during rich conditions. The overall reduction of NO_x is obtained by cyclic changes between lean storage and rich regeneration periods.

$$\text{R1:} \quad NO + \tfrac{1}{2}O_2 \leftrightarrow NO_2$$
$$\text{R2:} \quad BaCO_3 + 2NO_2 + \tfrac{1}{2}O_2 \rightarrow Ba(NO_3)_2 + CO_2$$
$$\text{R3:} \quad BaCO_3 + 2NO + \tfrac{3}{2}O_2 \rightarrow Ba(NO_3)_2 + CO_2$$
$$\text{R4:} \quad Ba(NO_3)_2 + 3CO \rightarrow BaCO_3 + 2NO + 2CO_2$$
$$\text{R5:} \quad NO + CO \rightarrow \frac{1}{2}N_2 + CO_2$$

Besides the barium related reactions, conversion reactions and oxygen storage reactions like in three-way catalysts also take place in an LNT. Further effects like intra-particle transport are investigated and modeled by Brinkmeier et al. 2005. One key characteristic of this model the consideration of different pore diameters and transport resistances within barium carbonate and barium nitrate, respectively.

9.3 Diesel Particulate Filter

Wall flow type diesel particulate filters have been used beside other filter types to reduce particulate emissions of diesel engine powered vehicles for several years. The basic structure of the filter monolith is similar to a catalyst, whereas the individual channels are plugged in an alternating way at the inlet and outlet. In a simplifying way the filter can be approximated by a pair of representative channels (see Fig. 9.6), with the inlet channels (seen in flow direction) closed at the downstream end and the outlet channel closed at the inlet side. The channels are separated by a porous wall which filters solid particles from the exhaust gas.

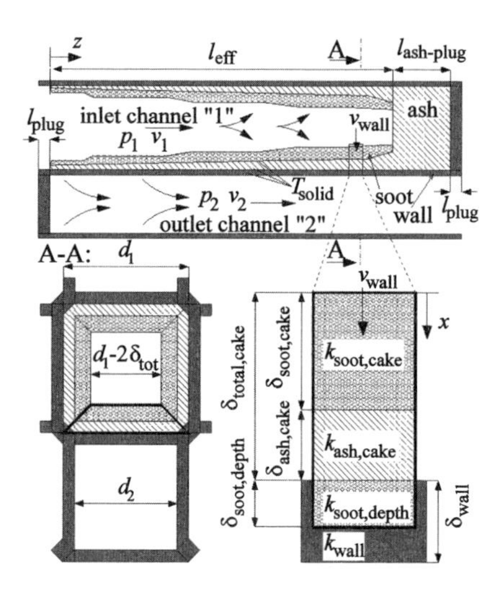

Fig. 9.6 Schematic view of a wall flow diesel particulate filter with asymmetric channel geometry and soot/ash loading in depth and cake filtration layer (Figure taken from Wurzenberger and Kutschi 2007)

These particles, consisting of soot, ash, soluble organic and sulfatic components are referred to in the following as soot for the sake of simplicity. During the filtration the two regimes of depth filtration and cake filtration can be distinguished. The formation of a soot cake and cake filtration starts after the porous wall is completely filled with particles during the depth filtration. Non-regenerating ash in the exhaust gases are accumulated as ash cake, ash plug or as a combination of both, depending on the driving conditions.

9.3.1 Basic Equations

From the modeling point of view the physical and chemical phenomena that occur in diesel particulate filters can be divided into two main effects. First, the filter flow within the channels and the related filter walls and second the loading and regeneration of soot inside the channels. A comparison of the time-scales relevant for both processes reveals that two separated sub-models for the filter flow and the soot layer can be applied. Both sub-models are then combined together in order to form the final modeling approach.

9.3.1.1 DPF Flow

For calculation of the flow and the pressure losses a steady 1D model can be formulated under the assumption that radial gradients in the channels as well as in the entire monolith can be neglected. The continuity equation for the inlet and outlet channels can be written as

$$\frac{\mathrm{d}}{\mathrm{d}z}\left(\rho_{\mathrm{g},1}\cdot v_{\mathrm{g},1}\cdot A_{\mathrm{F},1}\right) = -\rho_{\mathrm{g},1}\cdot v_{\mathrm{w},1}\cdot P_{\mathrm{S},1}$$
$$\frac{\mathrm{d}}{\mathrm{d}z}\left(\rho_{\mathrm{g},2}\cdot v_{\mathrm{g},2}\cdot A_{\mathrm{F},2}\right) = -\rho_{\mathrm{g},2}\cdot v_{\mathrm{w},2}\cdot P_{\mathrm{S},2}. \tag{9.9}$$

The steady version of the momentum conservation equations can be summarized to

$$\frac{\mathrm{d}}{\mathrm{d}z}\left(\rho_{\mathrm{g},1}\cdot v_{\mathrm{g},1}^2\cdot A_{\mathrm{F},1}\right) = -A_{\mathrm{F},1}\cdot\frac{\mathrm{d}p_{\mathrm{g},1}}{dz} - v_{\mathrm{g},1}\cdot\left(F_1\cdot\mu + \rho_{\mathrm{g},1}\cdot v_{\mathrm{w},1}\cdot P_{\mathrm{S},1}\right)$$
$$\frac{\mathrm{d}}{\mathrm{d}z}\left(\rho_{\mathrm{g},2}\cdot v_{\mathrm{g},2}^2\cdot A_{\mathrm{F},2}\right) = -A_{\mathrm{F},2}\cdot\frac{\mathrm{d}p_{\mathrm{g},2}}{dz} - v_{\mathrm{g},2}\cdot F_2\cdot\mu \tag{9.10}$$

The last term of the first equation describes the momentum depletion resulting from a mass depletion caused by the wall flow and ensures that the second law of thermodynamics is fulfilled.

The continuity and momentum conservation equations in the two channels are coupled via the gas flow across the porous wall. The pressure loss of this wall flow can be described by the Darcy law. In order to relate the cake height of both ash and soot with the local velocity, the following equation can be written for quadratic filter inlet channels (see Fig. 9.7)

$$v_{\mathrm{w}}(x) = v_{\mathrm{w},1}\cdot\frac{d_1 - 2\cdot\delta_{\mathrm{rk}} - 2\cdot\delta_{\mathrm{ak}}}{d_1 - 2\cdot\delta_{\mathrm{rk}} - 2\cdot\delta_{\mathrm{ak}} + 2\cdot x}. \tag{9.11}$$

For non-quadratic channel cross-sections the above given wall flow velocity equation can be deduced from the geometrical specifications in an analogous way.

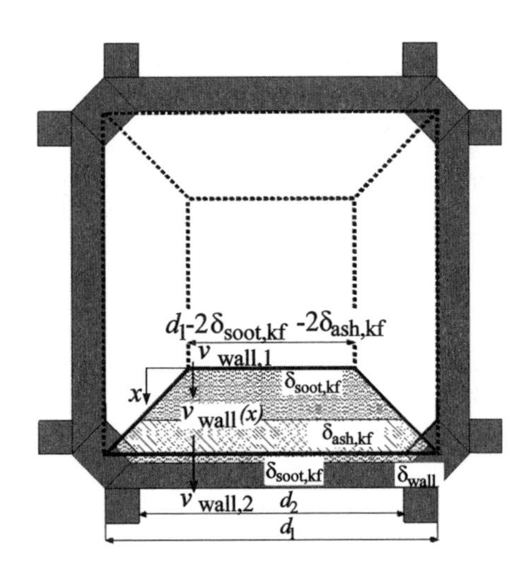

Fig. 9.7 Build up of ash and soot during depth filtration and cake filtration (Figure taken from Wurzenberger and Kutschi 2007)

The relation of the flow velocity and the cake heights is essential for the correct specification of the cake induced pressure losses.

The application of the wall flow velocity within the Darcy law describe the pressure losses for the different sections wall, depth filtration, ash and soot cake. Therefore the following relation between pressure, wall flow velocity and soot surface can be deduced

$$
\begin{aligned}
p_{\mathrm{g},1} - p_{\mathrm{g},2} &= \Delta p_{\mathrm{w}} + \Delta p_{\mathrm{rt}} + \Delta p_{\mathrm{ak}} + \Delta p_{\mathrm{rk}} \\
&= v_{\mathrm{w},1} \cdot \mu \cdot \left(\frac{d_1 - 2 \cdot \delta_{\mathrm{rk}} - 2 \cdot \delta_{\mathrm{ak}}}{d_1} \right) \cdot \left[\begin{array}{l} \dfrac{\rho_{\mathrm{g},1}}{\rho_{\mathrm{g},2}} \cdot \dfrac{\delta_{\mathrm{w}}}{k_{\mathrm{w}}} + \dfrac{\rho_{\mathrm{g},1}}{\rho_{\mathrm{g},2}} \cdot \dfrac{\delta_{\mathrm{rt}}}{k_{\mathrm{rt}}} \\ + \dfrac{d_1}{2 \cdot k_{\mathrm{ak}}} \cdot \ln \left(\dfrac{d_1}{d_1 - 2 \cdot \delta_{\mathrm{ak}}} \right) \\ + \dfrac{d_1}{2 \cdot k_{\mathrm{rk}}} \cdot \ln \left(\dfrac{d_1 - 2 \cdot \delta_{\mathrm{ak}}}{d_1 - 2 \cdot \delta_{\mathrm{rk}} - 2 \cdot \delta_{\mathrm{rk}}} \right) \end{array} \right].
\end{aligned}
\tag{9.12}
$$

The total pressure loss is composed from the partial pressure losses across the wall, the soot depth filtration layer, the ash cake and the soot cake. The partial pressure losses are governed by the geometrical specifications and the permeability of the respective layers.

9.3.1.2 DPF Loading and Regeneration

The soot cake and the soot depth filtration layer can be described as a two-dimensional packed bed. Assuming that the pressure gradients within the layers in axial direction are small compared to the ones in radial direction, the two dimensional packed bed of soot particles can be split into a series of one-dimensional regions (see Fig. 9.8). Each region consists of four layers (wall, soot depth filtration layer, ash cake and soot cake) in which different reactions can take place. These are reactions of soot with oxygen or NO_2 whereby the prior start at temperatures above 850 K and the latter already show significant rates at considerably lower temperature levels (less than 650 K). In the case of catalytically coated filters also

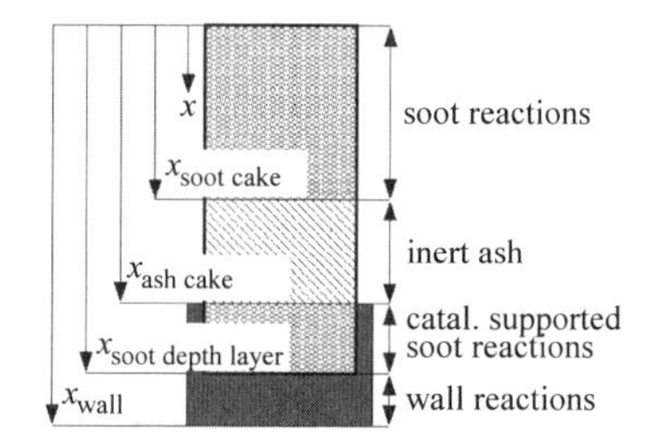

Fig. 9.8 1D region consisting of wall, soot depth filtration layer, ash cake and soot cake (Figure taken from Wurzenberger and Kutschi 2007)

the oxidation of NO to NO_2 can take place within the depth filtration layer. The produced NO_2 then can again react with soot. Due to the catalytic coating additional oxidation reactions can take place within the wall, such as e.g. oxidation reaction of CO and HC, represented by C_3H_6 and C_3H_8.

A steady-state, isothermal and one-dimensional model can be used to calculate the soot regeneration. The continuity equation for the gas flowing through the wall can be written as

$$v_w \cdot \frac{d\rho_g}{dx} = \sum_{j}^{S} MG_j \sum_{i}^{R} v_{i,j} \cdot \dot{r}_i \left(c_g, T_s\right), \tag{9.13}$$

with the spatial change of the mass flow over the soot height being equal to the sum of all reactions at the respective coordinate x. The reactions (function of the local gas composition and the wall temperature) are weighted with the respective stoichiometric coefficient of species j in reaction i, and summed up. This sum is summed up over all species including the molar mass accordingly. The following equation is used to describe the conversion of the individual gas species

$$v_w \cdot \frac{d}{dx}\left(\rho_g \cdot w_{g,j}\right) = MG_j \cdot \sum_{i}^{R} v_{i,j} \cdot \dot{r}_i \left(c_g, T_s\right), \tag{9.14}$$

where the change of the species mass flux is equal to the sum of all reactions where species j is participating.

9.3.1.3 Filter Model

The calculation of flow in the diesel particulate filter and the loading and regeneration delivers – based upon the steady formulation of all equations – solutions for the flow field, the pressure loss distribution, soot loading rates as well as soot regeneration rates for a steady operating condition of the filter. In order to calculate the transient behavior of a diesel particulate filter during loading and regeneration the results of the different sub-models need to be integrated into a transient, non-isothermal two-phase model. This coupling and hence the formulation of the final filter model (see Peters et al. 2004) shows benefits not only with respect to calculation speed but also with respect to its suitability for integration into CFD based simulation software.

9.3.2 Soot Loading and Pressure Loss

The prediction of the pressure loss in diesel particulate filters is one of the relevant calculation tasks in the present context. Typical questions relate to the impact of mass flow, soot distribution, channel geometry, filter porosity, ash, etc. on the pressure loss. For real applications in the vehicle it is important to know how the measured pressure loss correlates with the actual soot loading and whether thermal regeneration needs to be initiated. For this purpose in the following a few selected examples shall be discussed.

As described above the total pressure drop in a diesel particulate filter is composed by a number of individual pressure loss contributions. Depending on the significance of the different pressure loss contributions, benefits for one or the other channel diameter ratios can be observed in different operating points. Figure 9.9a shows the measured and calculated filter pressure loss as a function of the flow velocity for empty filters with different channel diameter ratios. It can be seen that filters with smaller diameters in the outlet channels give the largest pressure losses. This can be explained by the dominance of the friction in the outlet channel that increases strongly with increasing flow velocity. Figure 9.9b shows measured and calculated pressure losses for three channel diameter ratios during loading. It can clearly be seen that for soot loading larger than 1.5 g/l the filter with the largest inlet channel diameter shows the smallest pressure losses. This can be explained by the fact that in this case the pressure loss in the soot cake is predominant and directly correlates with the soot height. That is one and same soot mass

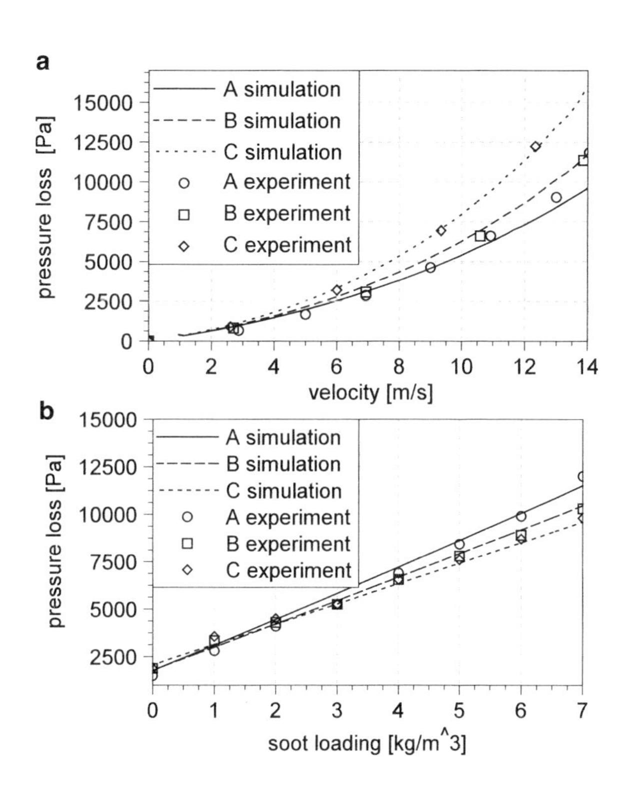

Fig. 9.9 Pressure loss for diesel particulate filters with different channel diameter ratios (inlet/outlet: A > B > C) measured and calculated for empty filters (*top*) and for loaded filters (*bottom*). (calculations from Wurzenberger and Kutschi 2007, measurements according to Ogyu et al. 2004)

shows in channels with larger channel diameters smaller soot heights and hence smaller pressure losses.

9.3.3 Regeneration and Temperature Distribution

The oxidation of soot in the diesel particulate filter is a highly complex heterogeneous chemical reaction process that besides the intrinsic reactions strongly depends on the physical properties of the soot particles. Therefore, it is not possible to use a single global reaction model to characterize the behavior of different filter types. Hence, in the practical application the regeneration model is calibrated based upon measured data and can then be used for calculation of variants of a specific filter type.

Figure 9.10 shows the calibration results of a regeneration model by comparing calculated and measured temperature in three axial positions along the symmetry axis of the filter. For the three selected positions the temporal evolution of the temperature clearly shows the different phases of the DPF regeneration process, namely heat-up, main oxidation and post-oxidation phase. The temperature peaks mark the actual spatial and temporal position of the reaction front that propagates through the filter. Based on such a calibrated model it is then possible to estimate maximum filter temperatures under different driving conditions and regeneration strategies.

Once a 1D diesel particulate filter regeneration model is calibrated properly based upon experimental data, the calibration can be easily transferred to a 3D model, pending the adopted sub-models for pressure loss, heat and mass transfer, chemical reaction, etc. are identical in the 1D and 3D approaches. Figure 9.11 shows an example for a 1D/3D simulation workflow. The conversion model calibrated in 1D is applied to a 3D model of a segmented SiC filter.

In Fig. 9.11 the spatial temperature distribution at one selected time point during the regeneration is shown. A significant radial temperature distribution can be seen.

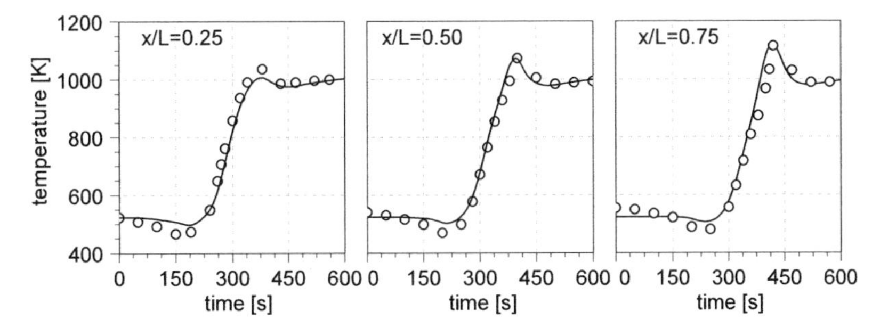

Fig. 9.10 Measured and calculated diesel particulate filter temperatures at three different axial positions along the symmetry line during regeneration

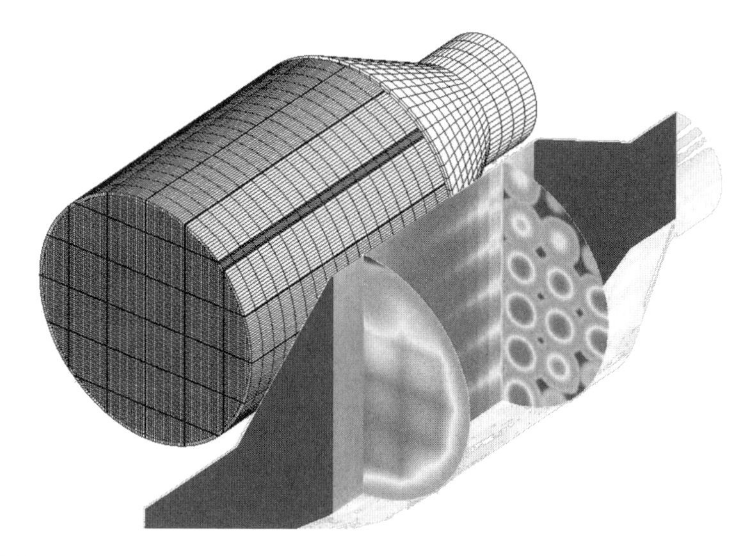

Fig. 9.11 Temperature distribution during the regeneration of a segmented SiC filter

It is caused by the non-uniform inlet flow and by radial heat losses and by the filter segments. The glueing zones of the filter partly block the free flow cross section, influence the flow efficiency of directly attached inlet channels and influence the radial heat transfer. The simulated spatial profiles of temperature and temperature gradients can be further used as input for thermal stress analysis.

9.4 Dosing Units

Within the framework of exhaust aftertreatment systems dosing units are used for injection of liquid into the exhaust line in front of the specific aftertreatment devices. In diesel engine aftertreatment systems the injected liquids are either an urea-water solution or diesel fuel. The injection of urea-water solution is used to enable NO_x reduction in SCR catalysts, diesel fuel is adopted for initiating the DPF regeneration process by elevating the exhaust gas temperature via exothermal fuel conversion in an oxidation catalyst mounted upstream to the DPF.

Simulation is used in this context mainly for analysis and optimization of dosing units with respect to droplet distribution, wall-interaction, evaporation characteristics, the impact of static mixing devices and the related pressure losses, etc. The majority of the above mentioned simulation tasks needs to be handled by 3D CFD based methods (Birkhold et al. 2006; Masoudi 2006) and adequate models for the governing physical and chemical processes. Adopting simplifying assumptions, such as e.g. radial homogeneous distribution and complete evaporation, also 1D models can be adopted in order to analyze dosing strategies and the related control algorithms within the framework of system simulation approaches (see Sect. 9.5).

The most relevant physical and chemical models adopted for simulation of dosing units are presented in the following for the injection of urea-water solution. During the dosing of urea-water solution the following main physical and chemical phenomena can be distinguished.

First, this is the interaction of the droplets with the gas-phase and the related mass, momentum and species exchange between the two phases. Besides the transport of individual droplets in the gas flow it is especially the heat and mass transfer and hence the evaporation characteristics of the droplets that determines the overall temporal evolution characteristics of the liquid and the gaseous phases. The description of the evaporation process requires a multi-component model that allows the proper description of the water evaporation and the thermal break-up of liquid urea (thermolysis) into ammonia and iso-cyanic acid.

The second governing process is the interaction of the liquid droplets with solid walls. Depending on the droplet velocity prior to the wall-impact, on the wall and on the droplet temperature, different interaction regimes can be distinguished, such as rebound, rebound with reduction in droplet size or sticking of droplets to the wall.

The third main phenomenon comprises the formation and transport of a thin liquid film at the wall. Modeling of the liquid film transport is based upon solving conservation equations for the liquid film height, taking into account exchange of mass, momentum, energy and species mass with the droplet and the gas-phase including multi-component evaporation and heat exchange with the walls.

The fourth area comprises the modeling of the chemical reactions. Besides the catalytic SCR-reactions (see Sect. 9.2.2.3), the hydrolysis reaction of iso-cyanic acid to form ammonia deserves attention in this context

$$HNCO + H_2O \rightarrow NH_3 + CO_2$$

Figure 9.12 shows the result of a 3D CFD simulation of an urea-water injection process. The simulation shows the droplet and droplet size distribution, clearly governed by the injection process itself and the flow field, here indicated via representative streamlines.

The influence of a static mixer and the formation of a wall film are shown in Fig. 9.13 for one selected operating point. The impact of the static mixer is

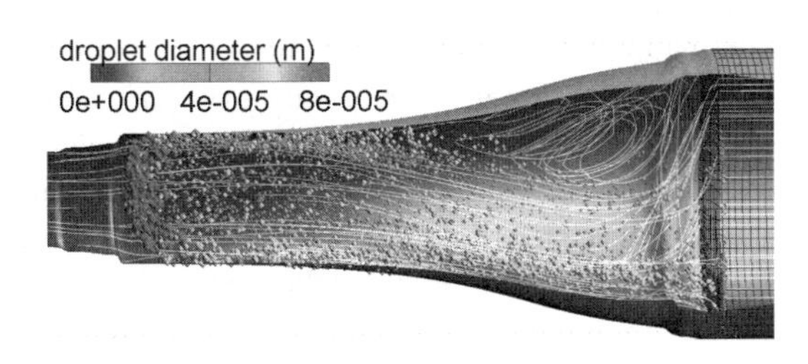

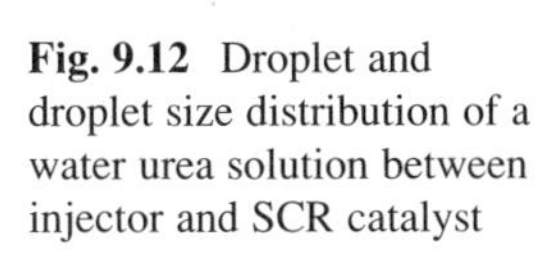

Fig. 9.12 Droplet and droplet size distribution of a water urea solution between injector and SCR catalyst

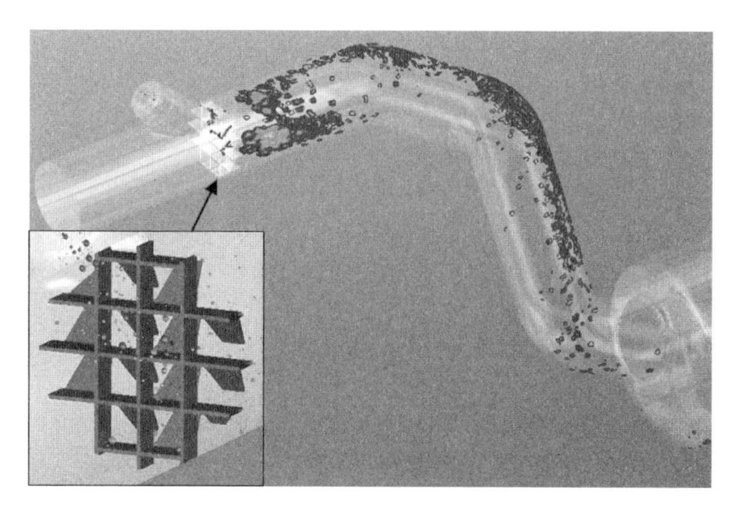

Fig. 9.13 Static mixer downstream of an urea injector and wall film distribution

manifold. If it is applied in the right way it can support break up of droplets by colliding at the mixer blades, it produces turbulence and therefore supports droplet evaporation and thermolysis and it increases the flow and also ammonia uniformity at the catalyst inlet. Due to the broad variety of piping system geometries, injector positions and mixer geometries (if used at all) and of operating conditions relevant for real driving conditions, 1D and 3D simulation models are an essential support in the design and optimization of the dosing units and the respective control function development.

9.5 System Simulation

The simulation of entire exhaust lines aims at the calculation of exhaust systems consisting of different catalysts, diesel particulate filters, dosing systems and pipes on the system level. Typical questions concern the size and arrangement of the different components, the interaction of components, the thermal behavior of the entire system, the application of different control strategies, etc. 1D models have turned out to be very effective to investigate various aspects of the system behavior within reasonable calculation times, important in the context of system simulations of one or several drive cycles. Figure 9.14 shows the results of a drive cycle simulation for a heady duty exhaust system consisting of a DOC, a DPF, an urea dosing unit, an SCR catalyst and an ammonia slip catalyst. The reaction models in the individual components are calibrated based upon experimental data, such as e.g. light-off curves. Measured engine raw emissions are taken as inlet boundary conditions for the system simulation of the European Steady Cycle (ESC) and the European Transient Cycle (ETC).

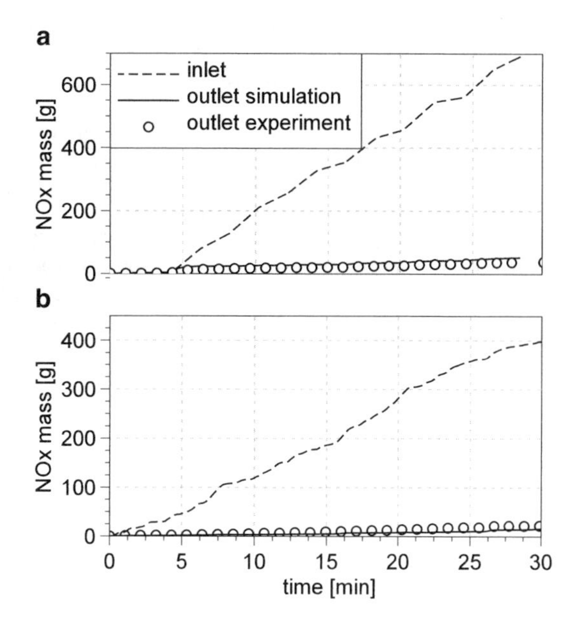

Fig. 9.14 Calculated and measured accumulated NO_x emissions of a HD-truck exhaust system fort the European Steady Cycle (*top*) and the European Transient Cycle (*bottom*)

Figure 9.14 shows a comparison of the calculated and the measured NO_x emissions for the studied drive cycles. Calculations of this kind form the basis for a quantification of emission levels that can be expected during drive cycles and hence allow the specification and optimization of complex exhaust aftertreatment systems already in a very early development phase.

References

Birkhold F, Meingast U, Wassermann P, Deutschmann O (2006) Analysis of the injection of urea-water-solution for automotive SCR DeNOx-systems: modeling of two-phase flow and spray/wall-interaction. SAE Paper 2006-01-0643

Brinkmeier C, Opferkuch F, Tuttlies U, Schmeißer V, Bernnat J, Eigenberger G (2005) Car exhaust fumes purification – a challenge for procedure technology. Chemie Ingenieur Technik 77:1333–1355

Depcik C, Assanis D (2005) One-dimensional automotive catalyst modelling. Prog Energy Combust Sci 31(2):308–369

Hayes RE, Kolaczkowski ST (1997) Mass and heat transfer effects in catalytic monolith converters. Chem Eng Sci 49(21):3587–3599

Kirchner T, Eigenberger G (1997) On the dynamic behaviour of automotive catalysts. Catal Today 38:3–12

Masoudi, M. (2006): Bosch Urea Dosing Approach for Future Emission Legislature for Light and Heavy Duty SCR Applications. 9th DOE Crosscut Workshop on Lean Emissions Reduction Simulation, University of Michigan

Missy S, Thams J, Bollig M, Tatschl R, Wanker R, Bachler G, Ennemoser A, Grantner H (2002) Computer-aided optimisation of the exhaust gas aftertreatment system of the new BMW 1.8-litre valvetronic engine. MTZ 63(2–12):1203–1212

Ogyu K, Ohno K, Sato H, Hong S, Komori T (2004) Ash storage capacity enhancement of diesel particulate filter. SAE Paper 2004-01-0949

Perry RH, Green DW (1997) Perry's Chemical Engineers Handbook. Chemical Engineering Series, 7th Ed., McGraw Hill, Int

Peters B, Wanker R, Muenzer A, Wurzenberger JC (2004) Integrated 1D to 3D simulation workflow of exhaust aftertreatment devices. SAE Paper 2004-01-1132

Wurzenberger JC, Peters BJ (2003) Catalytic converters in a 1D cycle simulation code considering 3D behavior. SAE Paper 2003-01-1002

Wurzenberger JC, Auzinger G, Heinzle R, Wanker R (2006) 1D modelling of reactive fluid dynamics, cold start behavior of exhaust systems. SAE Paper 2006-01-1544

Wurzenberger JC, Kutschi S (2007) Advanced simulation technologies for diesel particulate filters – a modeling study on asymmetric channel geometries. SAE Paper 2007-01-1137

Part III
Simulation of Combustion and Charging

Chapter 10
Total System Analysis

Christian Schwarz

10.1 General Introduction

The term total system analysis refers to the simulation of all engine processes under stationary and transient operating conditions. As a basis, this modeling builds upon the basic building blocks of the engine, like the cylinder, plenum, pipes, flow restrictions or throttle valves, and charger aggregates, which are described in Chap. 7. Since the calculation of the working process in the cylinder only provides indicated quantities, suitable methods for estimating engine friction must be integrated into the total simulation model. In addition, the behavior of the oil and cooling systems must be computed in order to permit a calculation of the frictional torque in thermally unsteady processes, since the influence of oil and coolant temperatures on the frictional torque is considerable. One also needs models, with which a simulation of the loads connected to the internal combustion engine is possible, in order to provide an accurate description of the boundary conditions for the combustion machine.

The behavior of the propulsion system and, for example, the adjustment of the charging unit in reproducible driving cycles can thus be replicated in a fashion which is true to reality. Moreover, the influence of single components within the total system can be optimized by means of an alteration in the model or a change of the parameters or of the characteristic maps for every individual component. This is applicable especially for the selection of charger components and for a possible comparison of varying charging concepts in unsteady processes, for which a description of the individual components as true to reality as possible with a correspondingly large model depth is a requirement. Such a model depth, without particular influences can no longer be exactly worked out, it presupposes considerable calculation times. The description of these engine peripherals as well as examples for stationary and transient calculation results for the internal combustion engine are the object of this chapter.

G.P. Merker et al. (eds.), *Combustion Engines Development*,

DOI 10.1007/978-3-642-14094-5_10,

10.2 Thermal Engine Behavior

10.2.1 Basics

In order to describe the thermal behavior of an engine, e.g. the warm-up, both the oil and cooling systems must be modeled. Moreover, both systems are connected to each other by means of an exchange of heat in the cylinder head, in the engine block, and where applicable, oil-water heat exchangers. The principal heat source is wall heat flow originating from combustion. This heat flow is distributed in the engine block, whereby the largest amount is through by the wall of the cylinder liner and the cylinder head to the coolant. A small amount is conducted via the cooling of the piston to the engine oil. In a turbocharged engine, an additional heat flow is transferred from the turbocharger to the oil. Since the frictional output in the bearings of the engine also heat the oil; yet another heat source joins the oil.

This heat is primarily transferred from the oil to the environment by means of the oil pan and, if present, by the oil cooler and the coolant. Besides that, the heat is emitted by means of the vehicle radiator from the coolant to the surroundings. Considering an unsteady situation, the water and oil masses and the heat storage behavior of the walls must be taken into consideration. Thus, in a cold engine, only a little heat flows from the combustion chamber wall to the coolant system, as the mass of the engine structure must first warm up. The heat removal from the coolant to the environment only begins when the engine itself has reached its normal operation temperature after the thermostat opens (Fig. 10.1).

Reulein (1998) introduced a modeling method for simulating the thermal behavior of coolant and oil systems. The systems were constructed in this case from a combination of pressure vessels and pipelines as well as corresponding modules for additional components such as the water pump, oil pump, heat exchanger, and thermostat.

10.2.2 Coolant System

Figure 10.2 shows, as a flowchart, the coolant system of a turbocharged diesel engine. The coolant is delivered from the water pump (1) through the engine block

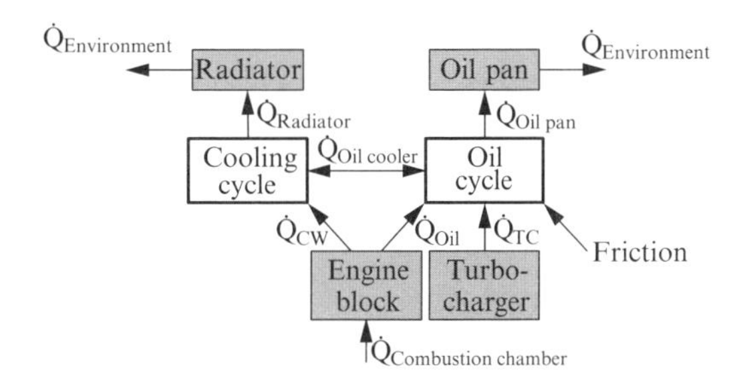

Fig. 10.1 Heat flows in an internal combustion engine

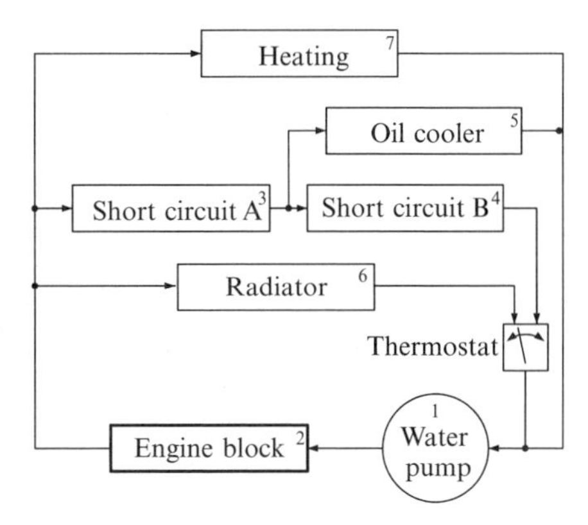

Fig. 10.2 Scheme of the coolant system

(2). Finally, the mass flow divides itself up, contingently on the position of the thermostat, (8) into the system segments of heating (7), radiator (6), and bypass circuit (3). The temperature of the coolant is regulated by the thermostat.

Water Pump. *As a rule, single-stage centrifugal pumps are utilized to circulate the coolant. The delivery flow of the pump depends on its speed and the flow conditions at the entry and exit of the rotor. Corresponding characteristic maps, usually made available by the pump manufacturers, serve as input parameters for the simulation.*

Thermostat. *The control of the coolant temperature takes place by means of a thermostat. Besides two-way thermostats, which only influence the mass flow through the radiator, three-way thermostats are also employed. In the case of elastic material thermostats, the elastic material is surrounded by the flow of the coolant. In modern internal combustion engines, engine manifold pressure controlled thermostats are also used, which controls the coolant temperature as a function of the operational condition of the engine.*

The behavior a thermostat can be described in the coolant system model by a branching of the pipeline system with variable cross-sectional areas in the segments. In this case, the thermostat is controlled by means of characteristic curves.

Heat Exchanger. *For the heat removal to the environment and for vehicle heating, water to air heat exchangers are employed, which are either finned tube radiators or gilled radiators. Engines of higher output and turbocharged engines additionally need oil coolers, which emit their heat either to the coolant or directly into the environment. In this case too, a corresponding characteristic map made available by the manufacturer can be used to describe its behavior.*

10.2.3 The Oil System

Figure 10.3 shows a schematic representation of the oil system of a turbocharged internal combustion engine. Engine lubrication oil is pumped from the oil sump by

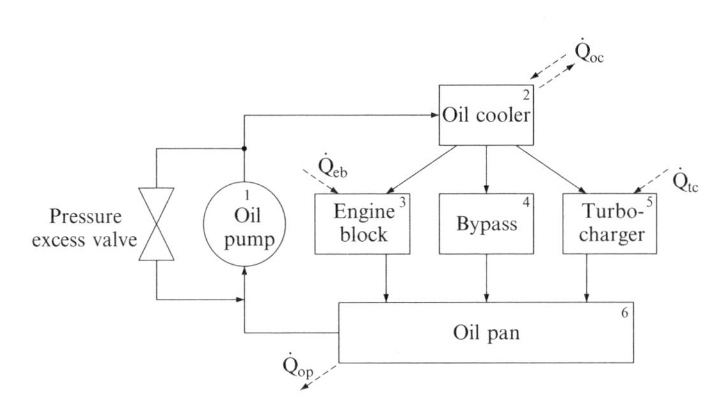

Fig. 10.3 Block diagram of the oil system

the oil pump. It then flows through the oil filter (and oil cooler if applicable). The clean, pressurized oil flows to the main oil gallery in the engine block to branch circuits which then feed each bearing. The lubrication of the turbocharger takes place by means of an additional oil feed.

A bypass line regulates oil pressure in cases where the pump produces more oil than the engine requires w. In addition, a pressure control valve is integrated into the oil pump to protect against excessive oil pressures when the engine is cold.

Heat enters the oil system both from the engine and the turbocharger. The direction of the heat flow in the oil cooler depends on the current operational condition of the engine. In the case of a hot engine, heat is delivered to the coolant; the motor oil receives heat from the coolant during the warm-up phase. Additional oil cooling takes place by means of heat dissipation from the oil pan to the surrounding air.

With the calculation models described by Reulein (1998), user-defined oil systems can be constructed in a modular fashion as simulation models.

Oil Pump. *Gear pumps are generally used to deliver the engine oil. Similar to the procedure when simulating coolant systems, the delivery behavior of the pumps can be determined from characteristic maps as a function of the engine speed and oil temperature.*

10.3 Engine Friction

10.3.1 Friction Method for the Warm Engine

In Schwarzmeier (1992), a method is developed for the friction of an internal combustion engine with the help of experimental investigations of particular tribological systems. The calculation of the friction mean effective pressure of the engine takes place according to the following equation

$$
\begin{aligned}
f_{mep} = & f_{mepx} + C_1 \left(\frac{c_m}{T_{liner}^{1,66}} - \frac{c_{mx}}{T_{linerx}^{1,66}} \right) + C_2 \left(\frac{b_{mep}}{T_{liner}^{1,66}} - \frac{b_{mepx}}{T_{linerx}^{1,66}} \right) \\
& + C_3 \left[\frac{(d \cdot n)^2}{T_{oil}^{1,66}} - \frac{(d \cdot n_x)^2}{T_{oilx}^{1,66}} \right] + C_4 \left[(1 + 0{,}012 c_m) b_{mep}{}^{1,35} - (1 + 0{,}012 c_{mx}) b_{mepx}^{1,35} \right] \\
& + C_5 \left(n^2 - n_x^2 \right)
\end{aligned}
\tag{10.1}
$$

with the constants

$$
\begin{aligned}
C_1 &= \frac{64,0}{z} \\
C_2 &= 12 \\
C_3 &= \frac{30}{z} \cdot 10^{-3} \\
C_4 &= 15 \cdot 10^{-3} \\
C_5 &= \frac{P_{rated}\, x k\, z\, 0,6}{V_{swept}\, n_{rated}^3} + cl\ ila^2\ z\ dl^2 \\
xk &= 0,1 - 0,07\, P_{rated}^{0,04} \\
cl &= 0,14 \cdot 10^{-6}\,.
\end{aligned}
$$

In more detail, the first part of the method (the term with the constant C_1) considers the friction of the piston group in contingence on the glide speed and temperature of the oil film between the piston and the cylinder wall. The second part (C_2) takes the friction amount of the piston group into consideration contingent on engine load and oil film temperature between the piston and the cylinder wall. The third term of the equation (C_3) considers the friction behavior of the main and connecting rod bearings contingently on the oil temperature, engine speed and bearing geometry, the oil temperature-dependent oil pump work, and the ventilation losses of the crankshaft drive. The fourth part of the method reproduces the load- and speed-dependent influence on the injection pump and the last part the output need of the ancillary components, the coolant pump, and the cooling-air fan, which is determined above all be the speed and the geometry of the fan.

For the pre-calculation of the friction mean effective pressure at an arbitrary operating point, knowledge of the friction mean effective pressure f_{mepx}, the mean effective pressure b_{mepx}, the mean piston speed c_{mx}, the friction-relevant cylinder wall temperature T_{linerx}, the engine speed n_x, the lubricant temperature T_{oilx}, and the coolant temperature T_{cx} at a reference point is necessary (index x).

The friction-relevant cylinder liner temperature is calculated according to

$$T_{liner} = T_{linerx} + f_1\left(c_m - c_{mx}\right) + f_2\left(b_{mep} - b_{mepx}\right) + f_3\left(T_c - T_{cx}\right) \qquad (10.2)$$

with $f_1 = 1.6$; $f_2 = 1.5$ and $f_3 = 0.8$. The range of validity of the method is indicated for oil temperatures over 40°C.

10.3.2 Friction Method for the Warm-up

For the calculation of the friction of an engine below 40°C, as is necessary for characterizing warm-up behavior, the method of Reulein (1998) has been expanded.

Friction Method for Low Temperatures. *Schwarzmeier (1992) suggested including the influence of engine pressure on the bearing friction at low oil temperatures by means of*

$$f_{mepB} = \frac{C}{T_{oil}^{a}} b_{mep} \qquad (10.3)$$

With this, we obtain for the friction method under consideration of thermal behavior

$$\begin{aligned} f_{mep} = f_{mepx} &+ C_1\left(\frac{c_m}{T_{liner}^{1,68}} - \frac{c_{mx}}{T_{linerx}^{1,68}}\right) + C_2\left(\frac{b_{mep}}{T_{liner}^{1,68}} - \frac{b_{mepx}}{T_{linerx}^{1,68}}\right) \\ &+ C_3\left[\frac{(d\,n)^2}{T_{oil}^{1,49}} - \frac{(d\,n_x)^2}{T_{oilx}^{1,49}}\right] + C_4\left[(1+0,012c_m)b_{mep}^{1,35} - (1+0,012c_{mx})b_{mepx}^{1,35}\right] \\ &+ C_5\left(n^2 - n_x^2\right) + C_6\left(\frac{b_{mep}}{T_{oil}^{1,49}} - \frac{b_{mepx}}{T_{oilx}^{1,49}}\right). \end{aligned} \qquad (10.4)$$

In accordance with these changes, the coefficients of the remaining terms must be adjusted as well

$$\begin{aligned} C_1 &= \frac{44}{z} \\ C_2 &= 31 \\ C_3 &= \frac{22}{z} \cdot 10^{-3} \\ C_4 &= 6 \cdot 10^{-3} \\ C_5 &= \frac{P_{rated}\, xk\, z\, 0,96}{V_{swept}\, n_{rated}^3} \\ C_6 &= 1,9 \\ xk &= 0,13 - 0,07\, P_e^{0,03} \end{aligned}$$

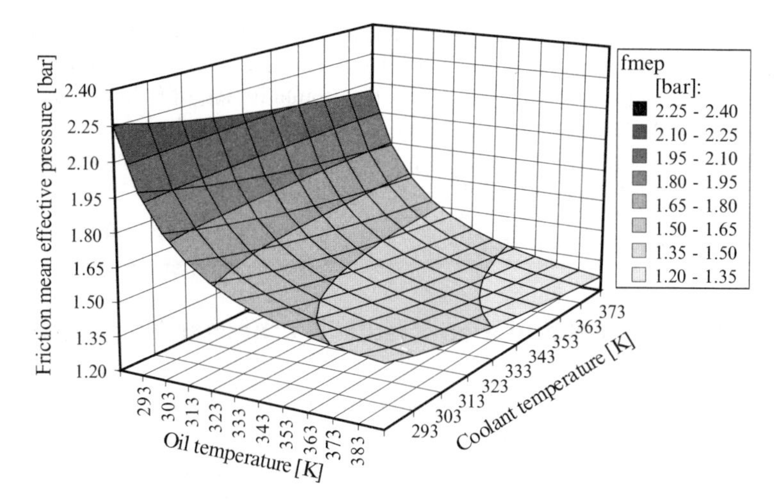

Fig. 10.4 Friction map calculated for a 1.9 l diesel engine at 2,300 rpm, 40 Nm

The friction-relevant cylinder surface temperature T_{Cylw} is calculated according to equation (10.2). For a speed of 2,300 rpm and a torque of 40 Nm is represented in Fig. 10.4 the influence of coolant and oil temperature on the friction mean effective pressure for a 1.9 l turbocharged, direct injection diesel engine. One can see clearly that the influence of the oil temperature is much greater than the coolant temperature.

With the calculation model introduced here it is possible, contingent on the operating point-specific and thermal states of the engine, to pre-calculate the friction mean effective pressure in transient operation. However, the thermal states and the friction are in close connection, since a significant amount of friction is led away to the engine oil as heat. This amount can be determined from the Schwarzmeier (1992) method, if one investigates the single components of the method for their contribution to the warming up of the engine oil.

The friction-relevant oil heat flow originates for the most part through bearing friction at the camshaft and crankshaft, the friction of the piston group and the warming up of the oil due to the inefficiencies through the oil pump. Thus

$$\dot{Q}_{friction} = \frac{V_{swept} n 10^5}{120} \left\{ k_{oil}\, p_{mrx} + C_1 \left(\frac{c_m}{T_{liner}^{1,68}} - \frac{c_{mx}}{T_{linerx}^{1,68}} \right) + C_2 \left(\frac{b_{mep}}{T_{liner}^{1,68}} - \frac{b_{mepx}}{T_{linerx}^{1,68}} \right) + C_3 \left[\frac{(d\,n)^2}{T_{oil}^{1,49}} - \frac{(d\,n_x)^2}{T_{oil}^{1,49}} \right] + C_6 \left(\frac{b_{mep}}{T_{oil}^{1,49}} - \frac{b_{mepx}}{T_{oil}^{1,49}} \right) \right\} \tag{10.5}$$

can be utilized for the calculation of the friction-relevant oil heat flow. In this case, the fraction k_{oil} of the heat flow transferred to the engine oil at the reference point of the total friction output must still be known. A value for k_{oil} of 0.33 can, according to Reulein (1998) be chosen as a first approximation for a high speed diesel engine for a passenger vehicle. With that, it is possible to determine the heat entry resulting from the friction of the engine for a warm-up simulation.

10.4 Stationary Simulation Results (Parameter Variations)

Stationary system calculation is used primarily for characterizing parameter variations. With it, explications for phenomena can be found and understanding about the influence of, and interactions between, particular parameters gleaned, which can be varied only with difficulty in isolation from other parameters under less well controlled conditions. This is one of the most important advantages of the real working process calculation. One should be warned however against all too uncritical parameter studies, since not all entry quantities are physically independent of each other. For example, a change in the compression ratio usually results in a change in the heat release rate, heat transfer losses, and chemical dissociation as well.

In terms of result quantities, we distinguish between those averaged over a working cycle – like peak pressure and temperature, indicated mean effective pressure or the indicated specific fuel consumption – and those represented in degrees of crank angle such as mass fraction burned curves

10.4.1 Load Variation in the Throttled SI Engine

As a first example, a variation of load in the SI engine will be investigated. The engine under scrutiny is a six-cylinder engine with 3 l displacement which was analyzed with the techniques described in Sect. 7.4 for the simulation of gas dynamics. 2,000 rpm was chosen as a speed for this operating point. The valve events for this example are conventional.

Figure 10.5 shows the gas exchange loops in the p, V diagram for a cylinder with an indicated mean effective pressure of 2 bar, of 5 bar, and at full load, with an indicated mean effective pressure of approx. 11 bar. We see that with decreasing load, the gas exchange losses clearly increase, since a sub-atmospheric pressure must be developed for load control in the intake system of the engine via the throttle valve. This pressure is present in the cylinder shortly after the opening of the intake valve.

Depicted in the top portion of Fig. 10.6 below are, for the indicated mean effective pressure of 2 bar, the predicted pressures in the cylinder, and intake and

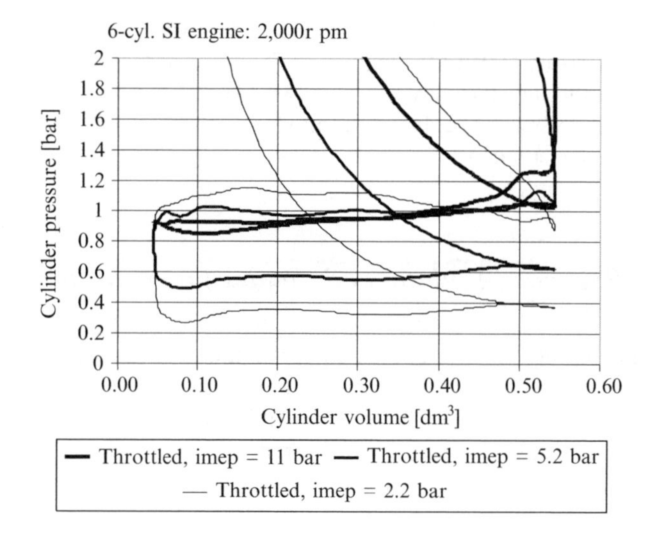

Fig. 10.5 Gas exchange loops under different operating conditions

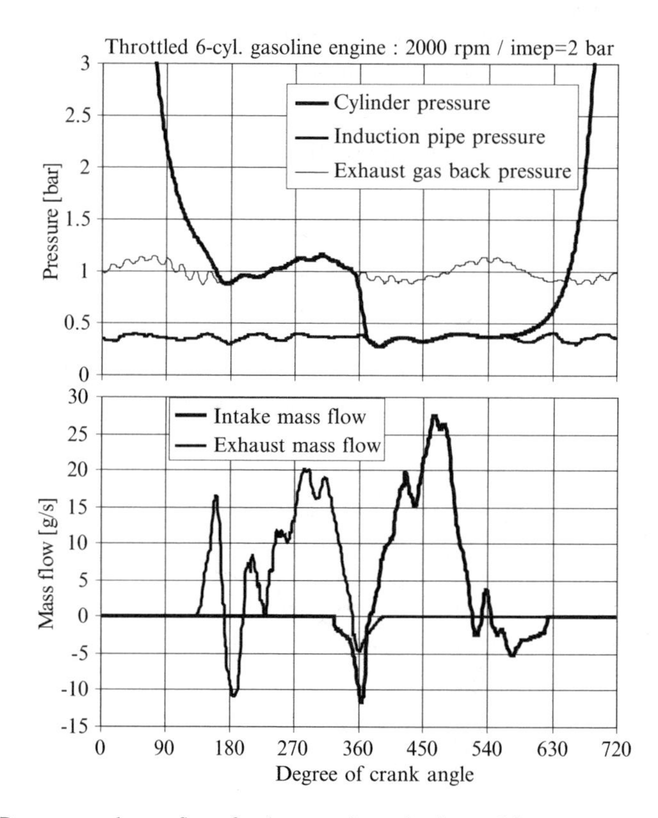

Fig. 10.6 Pressures and mass flows for the operating point $i_{mep} = 2$ bar

exhaust ports at this condition. We can recognize clearly the pressure waveforms in the intake and exhaust ports of the engine. In the lower portion of Fig. 10.6 below, the mass flows through the valves are shown. As we have already seen in the pressure paths, a reversion of exhaust gas into the cylinder occurs shortly before the closing of the exhaust valve.

10.4.2 Influence of Ignition and Combustion Duration

For the SI engine presented in Sect. 10.4.1, variations of the start of combustion (ignition time) and the combustion duration were carried out at an indicated mean effective pressure of about $p_{mi} = 5$ bar. The combustion duration cannot usually be altered in an experiment, but is determined rather by the airflow and combustion system, and valve timing. Nevertheless, a quantitative investigation can supply important information regarding how a possible shortening of the combustion duration affects the process quantities for the combustion process concerned. Figure 10.7 shows the influence of a start of combustion shifted earlier by 9°CA and one shifted later by 9°CA.

The heat release rate and combustion duration were modeled in this case as Vibe substitute heat release rates and held constant (combustion duration: 58°CA, shape

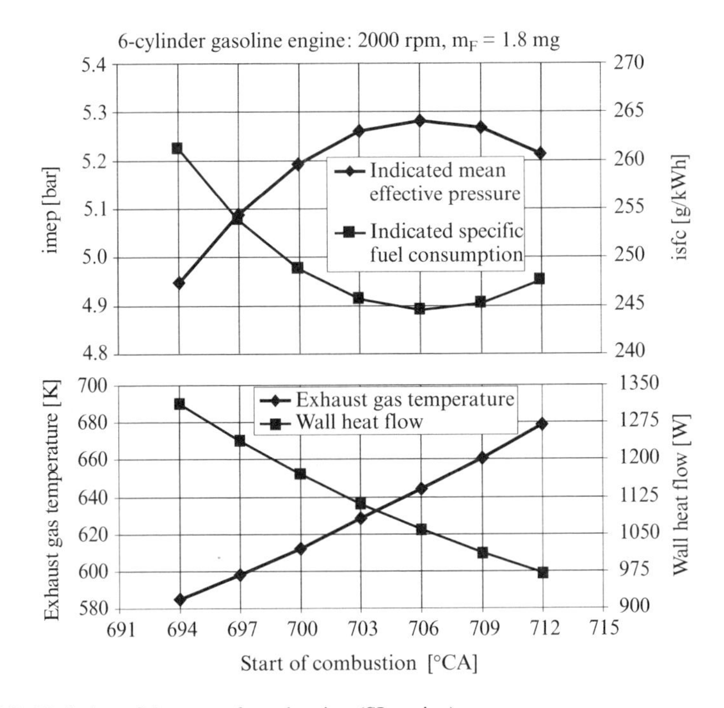

Fig. 10.7 Variation of the start of combustion (SI engine)

factor: 2.3). The indicated mean effective pressure clearly drops on either side of the optimum MBT (Minimum Ignition Advance for Best Efficiency). Because of the fact that the fuel mass for this investigation was left constant, the exact reciprocal behavior is shown under the indicated specific fuel consumption. At an earlier start of combustion, the fuel energy is released so early before the TDC. As Fig. 10.7 further shows that at earlier starts of combustion the peak pressure, and therefore peak temperatures, in the cylinder increase. The result of this is that the heat transfer in the cylinder also increases, leading to higher wall heat losses, as one can also recognize in Fig. 10.7. Accordingly, because of the lower efficiency and reduced heat transfer losses, the exhaust gas temperature decreases with earlier start of combustion and increases at later start of combustion dramatically, since the fuel energy released very late can only make a small contribution to the work of the piston.

On the other hand, the influence of a variation of the combustion duration is much less. Proceeding from the operating point at the consumption-optimal start of combustion, a variation of a combustion duration altered by −10 to +30 degrees was carried out. The 50% MFB point was retained and the shape factor adopted by the start of combustion variation. In Fig. 10.8, one can clearly recognize that the indicated mean effective pressure drops with longer combustion duration and the indicated consumption increases. Because of the retarded combustion, the exhaust gas temperature goes up with increasing combustion duration.

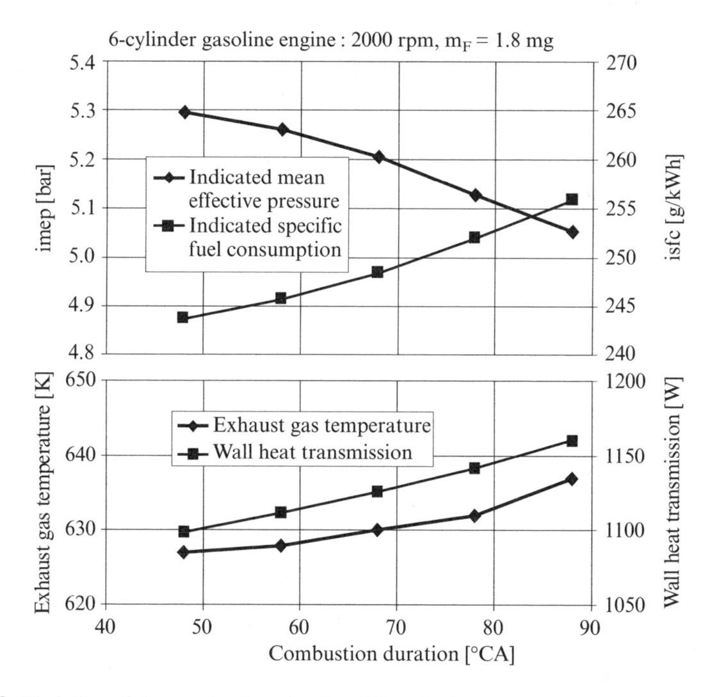

Fig. 10.8 Variation of the combustion duration (SI engine)

These examples make it clear that, as opposed to engine experiments, the influence of particular parameters can be separately investigated in simulation calculations.

10.4.3 Variation of the Compression Ratio, Load, and Peak Pressure in Large Diesel Engine

Figure 10.9 shows a study of a variation of the compression ratio for varying load-peak pressure-ratios for a large diesel engine with a capacity of 113 l per cylinder.

The variation is carried out at the rated speed of the engine of 450 rpm and for combustion duration of 72°CA at an air-fuel ratio of 2.2. The mechanical efficiency is set to a value of 0.92, and the total turbocharger efficiency amounts to 0.65. The description of the turbocharger takes place by means of the so-called first fundamental equation of turbocharger equation, see (8.37). The influence of the compression ratio on the specific fuel consumption is clearly recognizable. For every graph for a constant compression ratio, there is a pronounced minimum for the fuel consumption. This minimum is valid respectively for a quotient from the mean effective pressure and for the peak pressure, which represents a limiting factor with respect to the mechanics. With this, at a pre-set peak pressure, the mean effective pressure for the consumption-optimal operating point can be easily determined. For example, an engine with a peak pressure of 180 bars and a compression of 12 is designed for optimal consumption at a mean effective pressure of approx. 20 bars. Interestingly, most engines are designed more in the direction of higher mean effective pressure and less in the area of optimal consumption. That which envelops the individual curves can be seen as the design graph of these engines, which represents the respective compromise between consumption and output. Especially in the case of diesel engines, simulation is an important aid in designing engines at an early phase.

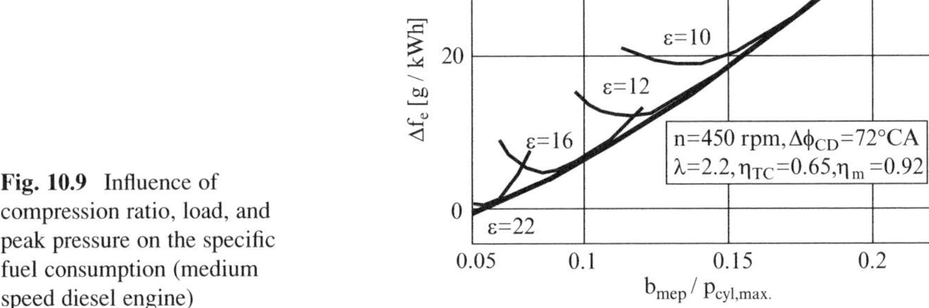

Fig. 10.9 Influence of compression ratio, load, and peak pressure on the specific fuel consumption (medium speed diesel engine)

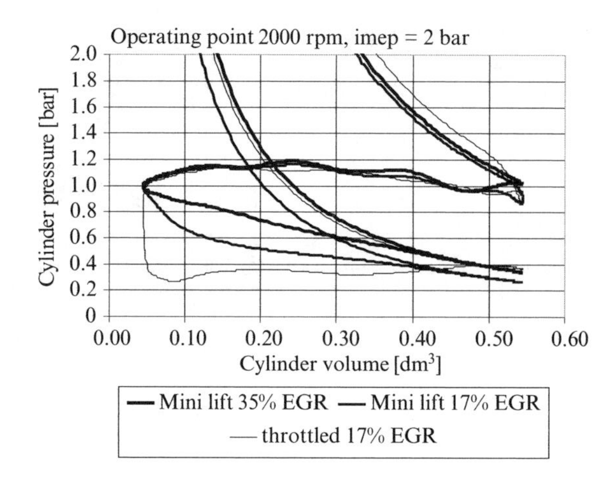

Fig. 10.10 Gas exchange loops of a fully variable mechanical valve train

10.4.4 Investigations of Fully Variable Valve Trains

Figure 10.10 shows two gas exchange loops for the same six-cylinder SI engine in the previous example, now equipped with a fully variable mechanical valve train. At this operating point, the engine has a speed of 2,000 rpm and an indicated mean effective pressure of 2 bar. For reference, the curves for the gas exchange loop for a conventional throttled engine at the same operating point are provided. Since the maximum valve lift for this operating point lies below 1 mm, practically no residual gas can be expelled from the intake side and taken in again later. Control of residual gas is only possible by means of the exhaust valve. Both of the operating points represented differ by a varying amount of residual gas of approx. 17% and 35%. This is possible due to an exhaust spread of 90°CA for the 17% residual gas point and 50°CA for the 35% point. The throttled operating point has a residual gas amount of 17% as well. We clearly see the decrease in gas exchange work at fully variable operation in both of the operating points of equal residual gas. This has the effect of a 6–7% improvement of the indicated consumption in this operating point. With further unthrottling with residual gas, larger fuel consumption gains results. We should in this case notice however that a portion of the gas exchange potential is compensated again by a much retarded combustion phasing.

A completely different kind of load control results in electromechanical valve trains. In this case, at full valve lift, the timing of "intake valve closes" is set such that only the desired filling remains in the cylinder. For this, two kinds of load control are available – the "early intake valve closes" (EIVC) and "late intake valve closes" (LIVC). While an EIVC prevent more charge from arriving in the cylinder, in the case of LIVC, a portion of the cylinder charge taken in is expelled again. Figure 10.11 shows a comparison of these two load control methods with each other and with a throttled engine with a conventional valve train for an operating point of 2,000 rpm and a load of 2 bar. Both methods indicate a residual

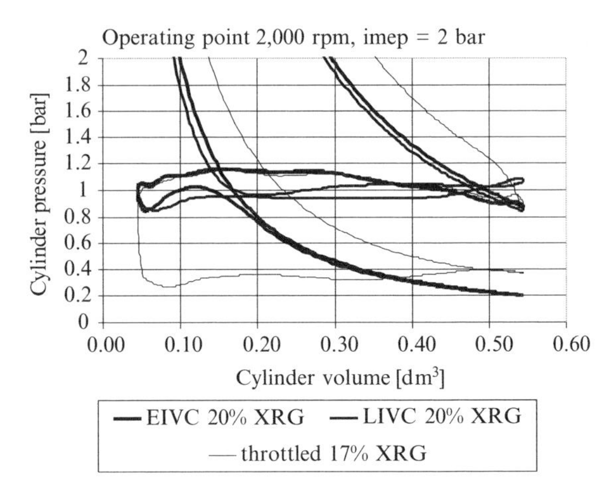

Fig. 10.11 Gas exchange loops for an electromechanical valve train

gas amount of approx. 20%, so that they can be compared with the throttled engine. The improvement of the total indicated specific fuel consumption in comparison with the throttled engine ranges from 7% in LIVC to 9% in EIVC (*these results are not universally applicable – in some well executed studies LIVC was found to have more fuel saving potential due to the stronger mixture motion it inherently develops, which allows more exhaust gas residuals...)*. Part of the existing gas exchange potential is compensated by the system-contingent retarded consumption in the high pressure range. We recognize from this again the necessity of an exact modeling of all components of the working system analysis.

Since electromechanical valve trains provide the highest flexibility with reference to the valve durations and an unlimited number of combination of valve lift and duration, cycle simulation represents a valuable aid in systematic variation and thus in the evaluation of this valve train concept.

10.4.5 Variation of the Intake Pipe Length and the Valve Durations (SI Engine, Full Load)

For a four-cylinder SI engine, in Fig. 10.12 a variation of the intake pipe length is carried out in two steps of 350 mm and 600 mm for the optimization of the full load torque and the output.

Both intake pipes originate in the intake plenum. In the case of the long intake pipe, an additional optimization for the lower speed range of the valve durations for the achievement of lower amounts of residual gas and for the adjustment of the pressure wave timing for the highest possible volumetric efficiency is carried out. The valve durations are thereby varied by the intake spread. This modeling takes place by means of the well understood one-dimensional gas dynamics.

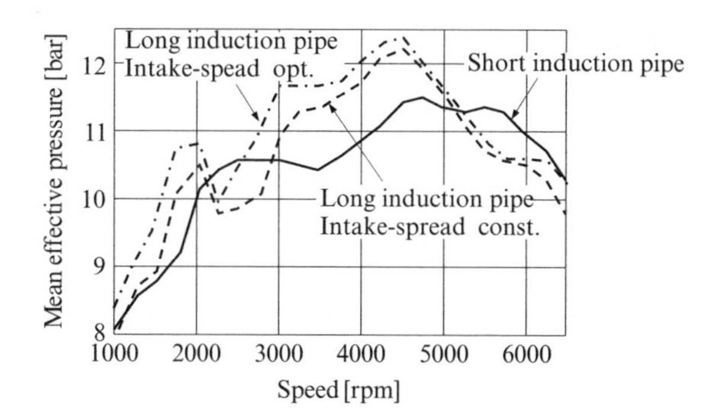

Fig. 10.12 Intake pipe length variation in a SI engine

We clearly recognize the strong effect on specific work short intake pipes have at high speeds (rated performance point) and, for long intake pipes at low speeds, as well as the decline of each curve in the other respective speed ranges. The use of variable valve durations once again heightens the medium pressure for long intake pipe lengths considerably. We see that the quantitative design of an intake system should always be parallel to the calculation process. We will not go further in this context into the exact modeling and consideration of diameters, inlet funnels, valves, etc.

10.4.6 Exhaust Gas Recirculation in the Turbocharged Passenger Car Diesel Engine

Figure 10.13 shows the connection diagram of an engine with charge air cooling and exhaust gas recirculation. In the case of the exhaust gas recirculation – as long as the scavenging pressure is larger than 1 – exhaust gas is directed back to the inlet system by means of an electronically controlled valve connected between the exhaust and inlet systems. The maximum amount of recycled exhaust gas thereby depends on the scavenging pressure ratio of the engine. A recirculation of the exhaust gas to the front of the compressor or the charge air radiator is not possible, since, on the one hand, the entry temperatures in the compressor become very high and, on the other, the oil and soot components carried along in the exhaust gas would foul the compressor or the charge air radiator.

Figure 10.14 shows the engine suction line for an average speed of 3,000 rpm and a temperature of 313 K, which should correspond to the temperature after the charge air cooler and is held constant. This line (roughly dashed) runs between the lines drawn in additionally for constant temperature before the intake valve of 293–393 K. The scavenging pressure ratio is assumed to be a constant 1.25. Below a compressor pressure ratio of 1.6, the exhaust gas recirculation rate increases linearly from 0 to 36% at a compressor pressure ratio of 1. The mass flow taken into the engine decreases however at first due to the increase in mixture temperature with exhaust gas recirculation. The effect of pure temperature increase is depicted

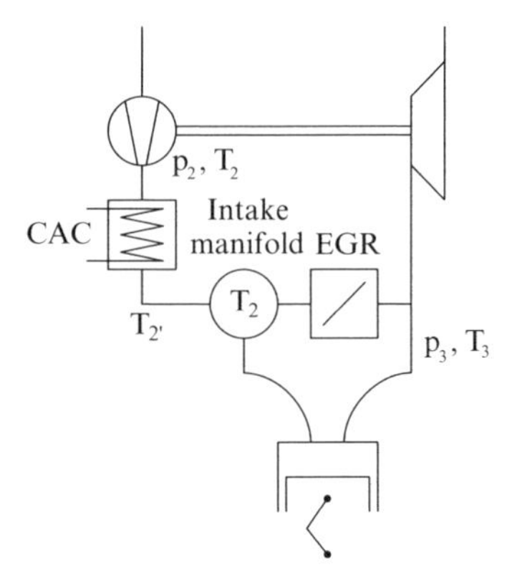

Fig. 10.13 Diagram of an engine with exhaust gas recirculation

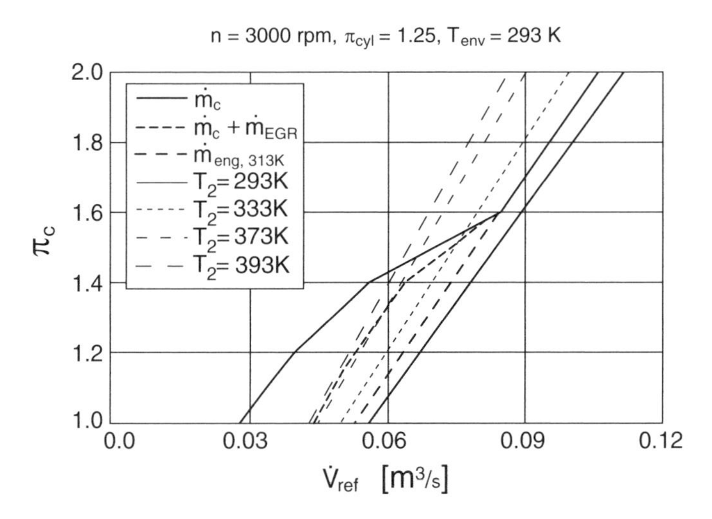

Fig. 10.14 Differences between engine and compressor suction lines under exhaust gas recirculation conditions

by the finely dashed line. In addition however, the mass flow delivered from the compressor is lessened by the recycled exhaust gas mass flow. The solid line describes the mass flow and the reference volume flow through the compressor. The difference between the engine suction line at the respective mixture temperature (fine dashes) and the compressor absorption line corresponds to the recycled exhaust gas mass flow. We recognize a clear shift of the compressor suction line to the left towards the surge line, which is caused on the one hand by the increase in temperature and on the other by the additional recycled exhaust gas mass.

For the clarification of the statements made above, the effect of exhaust gas recirculation is shown in Fig. 10.15 for a four-cylinder diesel engine with 1.9 l capacity and turbo charging. The results are calculated by a simulation of the entire

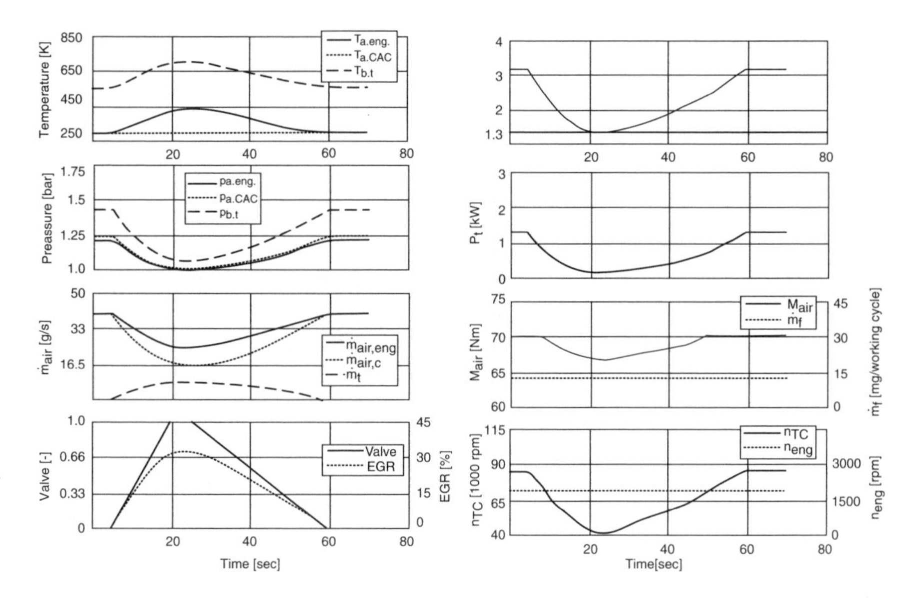

Fig. 10.15 Effect of the lift of the exhaust gas recirculation valve to engine operating conditions

engine. The fuel mass injected per working cycle and the engine speed were held constant by a specially adjusted speed controller. The diagrams show the effects of an adjustment of the exhaust gas valve for the temporal course of engine-specific quantities, proceeding from a stationary engine operation state at a speed of 2,000 rpm and an exhaust gas recirculation rate of 0%. After 5 s, the exhaust gas recirculation valve, as Fig. 10.15 below shows, is linearly opened, until at approx. 20 s it is fully open. After 25 s, the valve is linearly closed until approx. 60 s. This process goes slowly, which is why the conditions can be seen as quasi-stationary. With the increase of the valve opening, the pressures before the turbine and after the compressor conform to each other, but drop as a whole, since the turbine output lowers because of the pressure decrease over proportionally to the increase in exhaust gas temperature. The intake temperature before the engine rises with increasing exhaust gas recirculation, which is why the mass flow into the engine with the known effects on exhaust gas counter-pressure and the turbocharger speed goes back additionally. The air-fuel ratio was limited to 1.3 minimally by the maximum throttle cross-sectional surface of the exhaust gas recirculation valve. Figure 10.15 (left) shows the path of the mass flows through the compressor and into the engine as well as the recycled exhaust gas mass. The indicated mean effective engine torque T_i (Fig. 10.15, right) also collapses in the expected way because of the reduction of the air-fuel ratio caused by exhaust gas recirculation with the known effects on the compression path and the heat release rate as well as on the condition quantities in the cylinder at a constant speed and injected fuel mass. Clearly recognizable as well is the collapse of the compressor speed n_{TC} associated with the collapse of the compressor pressure ratio. Very important in this context is the exact knowledge of the compressor map in the lower speed regions.

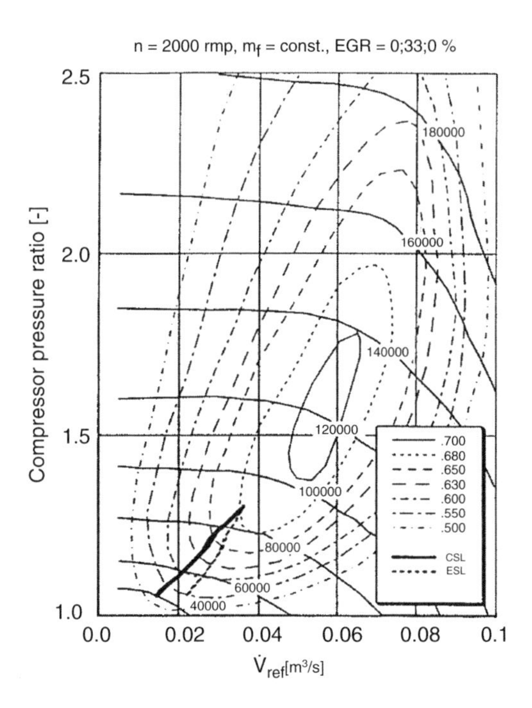

Fig. 10.16 The exhaust gas recirculation process in the compressor map

Figure 10.16 shows the extrapolated map of the compressor used for the engine under investigation. Entered into this characteristic map is the path of the compressor and engine suction lines for the exhaust gas recirculation path shown in Fig. 10.15. We recognize that the compressor absorption line is shifted to the left by the amount of recycled exhaust gas flow and clearly drops with an increasing recirculation rate to lower pressure ratios and compressor speeds. An unsteady acceleration process must therefore begin at a much lower initial boost pressure level, which leads partially to a considerable delay in the transient response.

10.5 Transient Simulation Results

In the following sections, results obtained from the simulation of transient processes in the internal combustion engine will be shown. We will consider processes of power switching for stationary engines, the simulation of unsteady processes in vehicle engines, and the calculation of entire driving cycles.

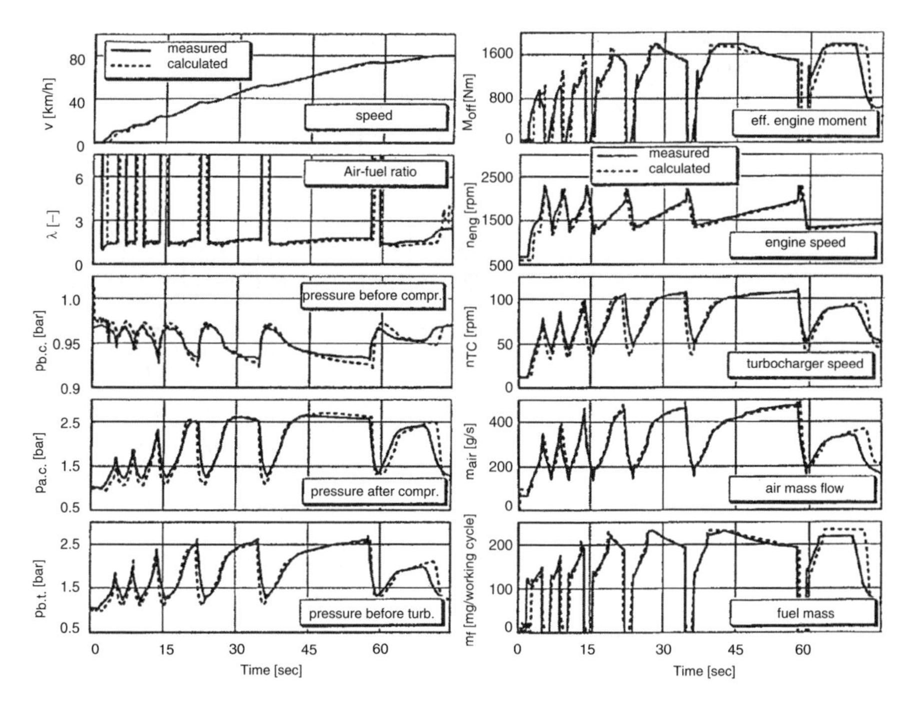

Fig. 10.17 Free acceleration, 40 ton truck

10.5.1 Acceleration of a Commercial Vehicle from 0 to 80 km/h

As a basis for investigating the diesel engine of commercial vehicles, we are using a six-cylinder in-line engine with approx. 12 l capacity. This engine propels a commercial vehicle with a total mass of 40 tons. The measured values forming the basis for the comparison between measurement and calculation were determined on a dynamic engine test bed. Comparisons with the respective measurement (solid) and calculation (dashed) have been illustrated. The propulsion load of the vehicle and the gear shifting transients are supplied to the engine by a transient AC dynamometer. For simulation calculation, the complete power train was illustrated such that the propulsion torque in the engine corresponds to the brake torque for the simulation calculation of the measurement.

Figure 10.17 shows a free acceleration of the truck from standstill with start-up and gear-shifting until a speed of 80 km/h is reached. The speed for the gear-shift is set at exactly 1,900 rpm. For the simulation, at the beginning of the calculation, a target speed of 80 km/h is given. The start and gear shifting at 1,900 rpm are in the further course of the calculation carried out independently by the driver controller. For this reason, the agreement of the measurement and the calculation should receive special attention, since the gear shifting times in the measurement and in the simulation calculation are reached at practically the same time, which testifies

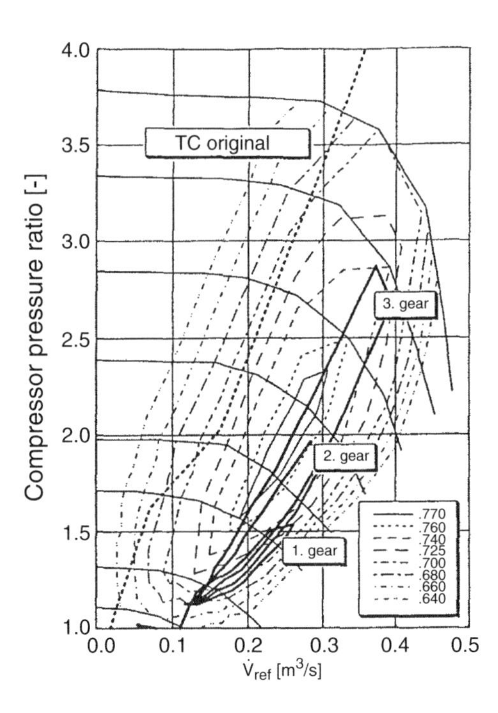

Fig. 10.18 Acceleration process in the compressor map

to a high simulation quality in terms of a firmly pre-set shifting speed and the asymptotic approximation of these speeds at higher gears. The agreement of all quantities of the measurement-calculation comparison in Fig. 10.17, of the fuel mass injected per working cycle, of the air mass flow through the compressor, of the air-fuel ratio, of the effective torque, and of the engine speed already dealt with, confirm the informational value of the simulation. Quantities like exhaust gas temperature, pressure before and after the compressor and before the turbine, turbocharger speed, and the achieved driving speed also show few discrepancies between measurement and calculation, which verifies a correct and realistic procedure in the simulation of the charger aggregate and of the entire engine.

In Fig. 10.18, this acceleration process is depicted in the compressor map. For the sake of clarity, only the compressor operation line is drawn from the start-up to the shift into fourth gear. The circulation direction for the operation lines is clockwise. The low boost pressure build up in the first and second gear and the considerable distance of the operation line from the surge line are clearly recognizable in the characteristic map.

10.5.2 Turbocharger Intervention Possibilities

For a turbocharger diesel engine, different possibilities of improving unsteady behavior at low speeds are being investigated.

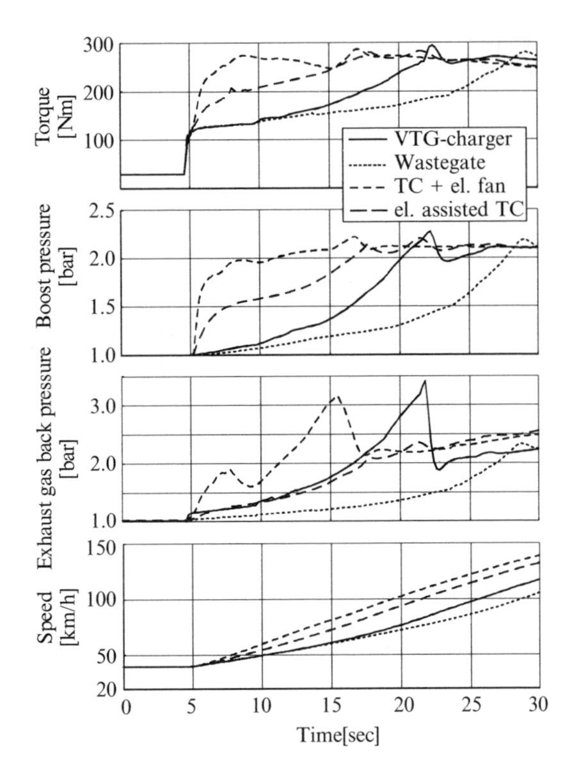

Fig. 10.19 Comparison of different operating strategies of the turbocharger

As an initial basis, a turbocharger configuration with control of the boost pressure by means of a waste gate has been selected. Under investigation are a configuration with variable turbine geometry (VTG), an electrically assisted turbocharger, and an additional compressor switched before the turbocharger, which is electrically driven. Both electric motors have a peak output of approx. 2 kW. The base engine is a 2-l, four-cylinder engine, which is utilized in a 1,350 kg vehicle. In order to workout the effect of the particular measures effectively, a full load acceleration process of approx. 40 km/h in a direct course was selected, in which an engine speed of approx. 1,000 rpm results. This is an extreme acceleration process, since the engine and turbocharger speeds are very low. As is shown in Fig. 10.19, in the case of the conventional engine with a waste gate, it takes approx. 25 s to build up a corresponding boost pressure. With the utilization of a turbine with a variable geometry, the time until complete boost pressure build-up is shorted by approx. 7 s. The electrically supported turbocharger reaches the target pressure approx. 11 s before the engine with the waste gate. The fastest boost pressure build-up is made possible by the additional, electrically propelled compressor. In this case, the boost pressure is reached in a few seconds.

In Fig. 10.19, the engine torque and the speed of the vehicle are additionally shown. The path of the operation lines for the particular concepts in the compressor map is also interesting. This is portrayed in Fig. 10.20.

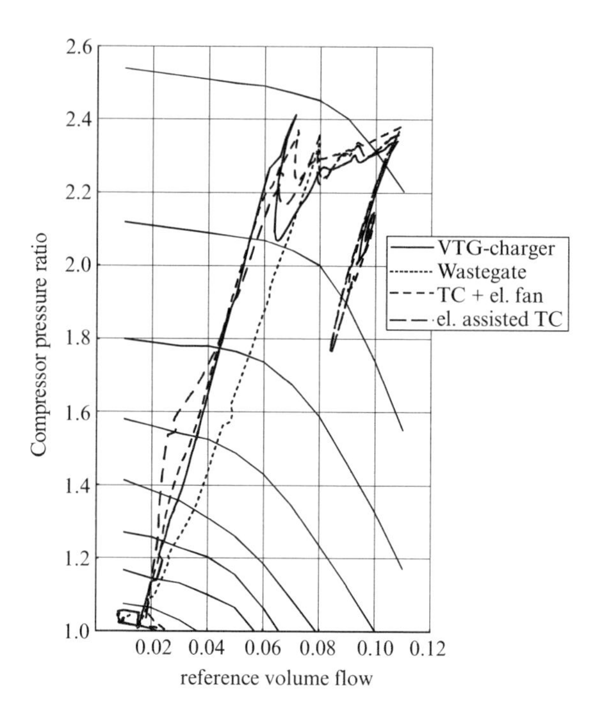

Fig. 10.20 Operating lines in the compressor map

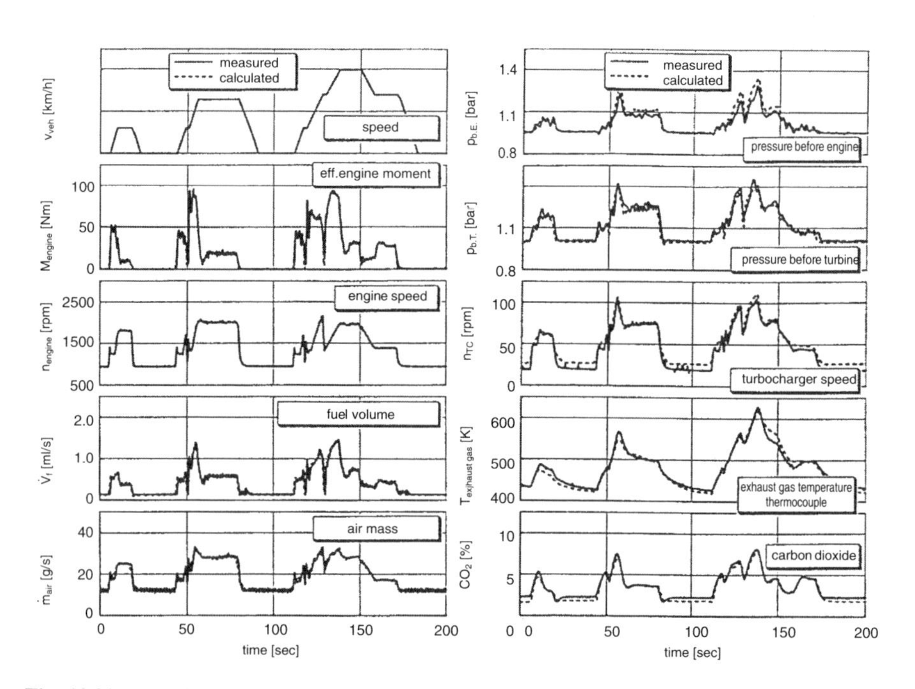

Fig. 10.21 Part of the ECE test cycle

10.5.3 Part Load in the ECE Test Cycle[1]

Part of the ECE test cycle for a turbocharger four-cylinder diesel engine in a medium-class vehicle is represented in Fig. 10.21. In this case, we are dealing with operation in the lower part load of the engine, in which exhaust gas recirculation is activated with the corresponding effects on the unsteady behavior of the turbocharger.

The measured values plotted in the images were taken from an electronically controlled engine test bench, in which both the vehicle with all driving resistance as well as the driver (gas pedal, gears) can be simulated. Since the speed of the ECE cycle is adjusted by the driver controller, the paths for the effective torque and the speed of the engine must correspond exactly to the measured values (Fig. 10.21, left, second and third diagrams). The fuel volume flow and the air mass flow taken in by the engine can adjust themselves freely. Both of these quantities are represented in the fourth and fifth diagrams and display a fine agreement between measurement (solid) and calculation (dashed). The situation is similar in the case of the quantities of boost pressure and exhaust gas counter-pressure as well as the turbocharger speed (Fig. 10.21, right). Thus all quantities characteristic for charging proceed under part load operation congruent to the measurement results. In the case of exhaust gas temperature, represented in the fourth diagram of Fig. 10.21 right, the thermal inertia of the thermocouple that was used for the data acquisition is simulated as well. These paths are also nearly congruent. A decent agreement in the case of the paths for carbon dioxide values results as a necessity from the good agreement of the air and fuel masses, since the values for carbon dioxide emissions can be calculated from the air-fuel ratio.

In Fig. 10.22, the operation line of the compressor for the part of the ECE cycle under consideration is plotted into the compressor map. We recognize that precisely for the quantitative simulation of the part load region, an exact extrapolation into area of lower pressure ratios and speeds in the compressor and turbine maps is essential. Unfortunately, in the case of this configuration, because of exhaust gas recirculation, the operation range is additionally shifted into these difficult ranges.

10.5.4 The Warm-up Phase in the ECE Test Cycle

In Fig. 10.23 results of a simulation calculation of the warm-up process in the ECE test cycle, as calculated by Reulein (1998), are represented with the four-cylinder

[1]Expl.: http://www.dieselnet.com/standards/cycles/ece_eudc.html

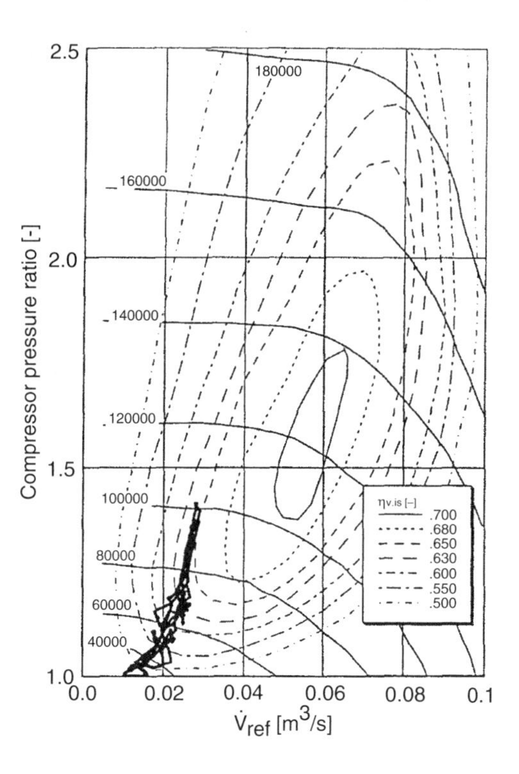

Fig. 10.22 Operating line in the compressor map, part of the ECE test cycle

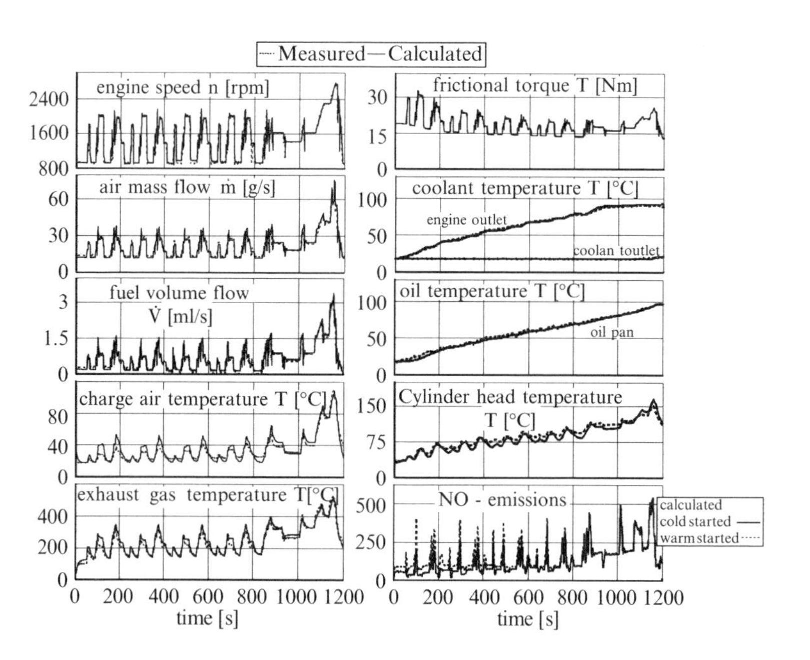

Fig. 10.23 Warming-up in the ECE test cycle

engine described in Sect. 10.5.3. The measured paths are dashed in the diagrams, the calculated paths illustrated with solid lines. All represented curves reveal favorable agreement between measurement and calculations. This makes it clear that the warm-up phase of the internal combustion engine can be realistically simulated with the assumptions made in Sects. 10.2 and 10.3.

A high temperature-dependency is shown by the magnitude of the friction torque, shown in Fig. 10.23 in the top right. Within the first 300 s, a clear increase in engine friction occurs, especially in the higher engine speed range. The paths of the temperatures of the coolant and the oil are shown in the two following diagrams. In this case as well, a very good agreement results between measured and calculated temperature paths. The next diagram shows the paths of the measured and calculated cylinder head temperature. Again, a favorable agreement between calculation and measurement can be seen.

The diagram on the bottom right shows the effect of the engine temperature on the nitrogen oxide concentration in the exhaust gas, which were calculated according to the method of Heider (1996) introduced in Sect. 7.2.2. The path for the engine started at operationally warm temperatures is shown as a dashed line, while the solid line shows in comparison to this the calculated path during the warm-up phase described above. The path of the operationally warm-started engine corresponds very well with measurements, which are however not shown here. Very easy to recognize in the first half of the driving cycle are the initially lower and then gradually increasing nitrogen oxide emissions of the engine in warm-up operation. The causes of these effects are the initially very low combustion chamber wall temperatures, which lower the process temperature and thus control the formation of nitrogen oxide.

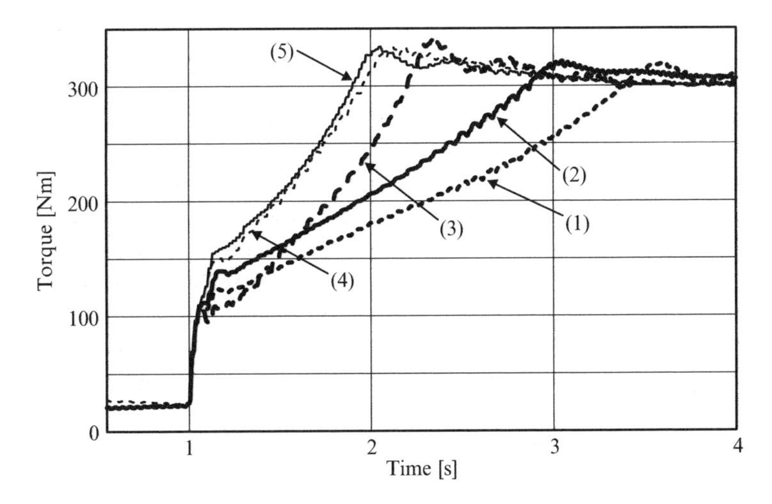

Fig. 10.24 Concept-comparison for a charged four-cylinder engine

10.5.5 Full Load Acceleration in the Turbocharged SI Engine

In the following section, a simulation of the acceleration behavior of a four-cylinder SI engine with a capacity of 2 l, direct injection, and turbocharging in a medium-class vehicle is represented. The engine has high authority cam phasing for both inlet and exhaust camshafts, and can be equipped for modeling both with a “4-in-1 exhaust gas manifold” with a mono-scroll turbine as well as with a “4-in-2 manifold” with a twin-scroll turbine. Since we are concerned with a concept with stoichiometric combustion, the engine has a throttle valve arranged after the charge air cooler.

In Fig. 10.24, acceleration paths in the fourth gear at an initial speed of 1,500 rpm are shown. At approx. 1 s, the full load demand takes place. Clearly recognizable is the relatively fast buildup of the basic torque after the filling of the intake manifold. Five different configurations are represented, in which the reaction behavior is clearly different. In the first, the valve durations optimized for the part load (expressed by the spread – i.e. the position of the maximum of the valve lift curve) are retained. Moreover, the concept is equipped with a 4-in-1 manifold. With this configuration, the torque response is the worst. This has to do with the unfavorable influence of the gas exchange by the exhaust pulses of the respective, previously igniting cylinder and the valve lift curves not adjusted to it. In the second configuration, shortly after the filing of the intake pipe, one switches over to the spreads optimized for full load. In this case, realistic switching times are considered in the modeling. One can clearly recognize of the quick torque response, since filling improves with optimized valve durations and the amount of residual gas remaining in the cylinder can be reduced.

In the case of large diesel engines, the principle of engine bypass is used. This procedure entails shifting the operating point to the right on the compressor map, away from the surge line, so that at an equal engine speed (absorption line), a larger pressure ratio in the compressor and thus a better filling in the cylinder is made possible. At the same time, the compressor efficiency improves significantly. In the SI engine, we can also use this method, since, according to the design, it possesses a positive scavenging pressure gradient in the range up to 3,500 rpm. As a result, both the residual gas is minimized and the cylinder is additionally cooled, which in both cases clearly increases the filling. This is made possible in the SI engine by a cam phaser. With it, the engine absorption line in the compressor map tilts to the right to the higher pressure ratios and efficiencies already described. In the case of the direct injection SI engine, pure air is scavenged thoroughly during the valve overlap phase, which is very favorable with respect to HC emissions. With this, the possibility arises of forming a rich air-fuel ratio in the combustion chamber, which with high burning velocities and cooler temperature makes possible the efficient combustion of the higher filling trapped mass without knocking problems. The unburned fuel reacts in the best case before the turbine with the air mass previously scavenged, thus raising the enthalpy of the exhaust gas before the turbine and increasing its efficiency. In the unfavorable case, the fuel first reacts with the air in the catalyst converter, perhaps raising its temperature catastrophically.

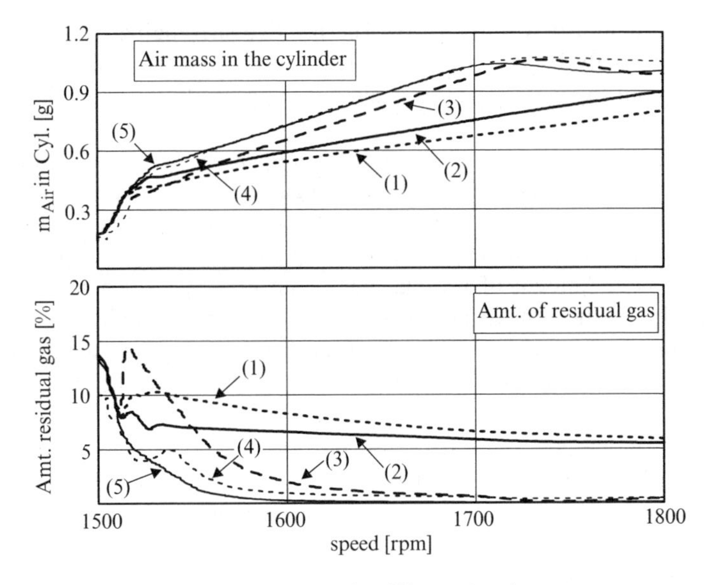

Fig. 10.25 Amounts of fresh and residual gas for different charging concepts

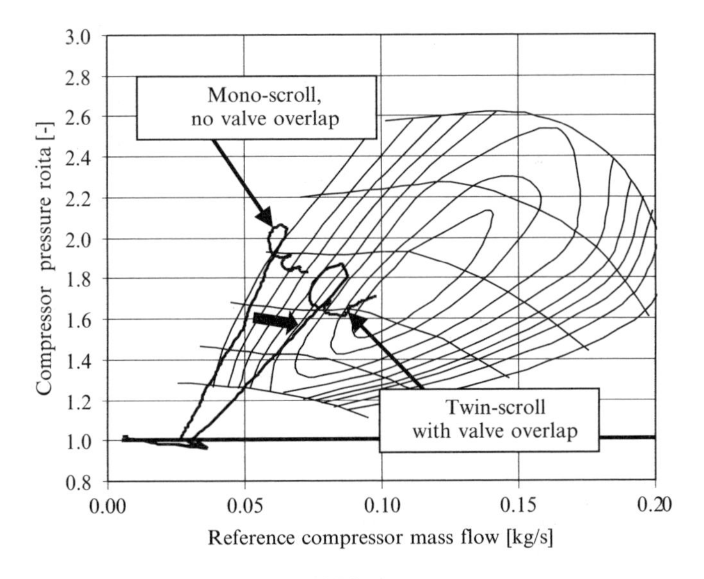

Fig. 10.26 Operating lines for concept 2 and 5 in the compressor map

In the third configuration (3) in Fig. 10.24, we switched to a large valve overlap after the filling of the intake pipe. After a short stagnation of the torque build-up, one can indeed see a clearly larger gradient in the torque build-up, but this behavior is difficult for a driver to administer. At the beginning of the load change process, at first the system aspirates regularly. The positive scavenging pressure gradient breaks down briefly because of the valve durations, and the amount of residual gas spikes, until the system begins again after approx. 1 s to show the expected behavior.

Figure 10.25 shows the fresh gas mass remaining in the cylinder after the closing of the intake valve and the fraction of residual gas in the total cylinder mass for the five configurations. These are, for reasons of comparison, plotted over the engine speed. An exact analysis of the processes during gas exchange for the third concept results in a large amount of residual gas of up to 20%, which is caused by a superimposing of the exhaust pulses of successively fired cylinders and the influence of gas exchange behavior resulting from these pressure pulses. By means of a shortening of the exhaust valve durations to approx. 200° of crank angle, this behavior can be considerably avoided, as curve four in Fig. 10.24 shows. In this case, the interfering pressure wave of the previously fired cylinder is protected by the shorter exhaust valve duration. The residual gas is thus minimized, by means of which the necessary positive scavenging pressure gradient then quickly constructs itself.

The technically cleanest solution is represented by the fifth configuration (5), in which destructive interference during gas exchange of the cylinders can be fundamentally avoided through a 4-in-2 combination in connection with a twin-scroll turbine with a preservation of the longer exhaust valve duration, due to the exhaust processes in each pair of cylinders being 360° apart. This configuration can easily result in an engine with pleasant transient response behavior.

Figure 10.26 the operating lines in the compressor map for the second and fifth concepts – thus an operation optimized with respect to valve durations with and without scavenging of the fresh air. We recognize in the scavenging arrangement the compressor operation line lying further right, which has more safety margin to the surge line. The higher output in the compression of the larger fresh air mass is partially overcompensated by the increased compressor efficiency. With this process, the low end torque speed can be decreased below 1,500 rpm in order to achieve the maximum launch torque.

References

Heider G (1996) Rechenmodell zur Vorausberechnung der NO- Emission von Dieselmotoren. Dissertation, TU München

Reulein C (1998) Simulation des instationären Warmlaufverhaltens von Verbrennungsmotoren. Dissertation, TU München

Schwarzmeier M (1992) Der Einfluß des Arbeitsprozeßverlaufs auf den Reibmitteldruck von Dieselmotoren. Dissertation, TU München

Chapter 11
Phenomenological Combustion Models

Gunnar Stiesch, Peter Eckert, and Sebastian Rakowski

11.1 Introduction

For the calculation of engine combustion processes, various model categories can be employed, which are quite diverse in their level of detail, but also in their calculation time requirements, see Stiesch (2003). Calculation models are customarily designated as phenomenological models that can calculate combustion and pollutant formation contingent upon important physical and chemical phenomena like spray dispersion, mixture formation, ignition, reaction kinetics, etc. Because a spatial subdivision of the combustion space into zones of varying temperature and composition is often necessary, the models are also referred to as quasidimensional models. Phenomenological (or quasidimensional) models differ on the one hand from zero-dimensional (or simplified thermodynamic) models, which simplify the combustion chamber as being ideally mixed at every point in time and are based on empirical approaches for the combustion rate. On the other hand, phenomenological combustion models differ from the CFD codes (CFD = computational fluid dynamics, see Chap. 12 ff.), in that we consciously do without an explicit solution of the turbulent three-dimensional flow field, which reduce the calculation time considerably (Fig. 11.1). The calculation time for one engine revolution lies in the region of seconds in phenomenological models, while in CFD codes it takes hours (Fig. 11.2).

In the following, a few of the most important phenomenological combustion models known in the literature will be introduced. The primary goal of each of these models is to calculate the heat release rate in advance contingent upon characteristic physical and chemical quantities without the need of measured pressure data. Moreover, if statements must be made about pollutant formation, it is necessary to carry out a subdivision of the combustion space into zones of varying temperature and composition. This is because the reaction rates of the chemical reactions decisive for pollutant formation are in general exponentially contingent upon temperature, so that knowledge of the arithmetically averaged cylinder temperature does not suffice on its own, see Chap. 6. Some of the

G.P. Merker et al. (eds.), *Combustion Engines Development*,
DOI 10.1007/978-3-642-14094-5_11, © Springer-Verlag Berlin Heidelberg 2012

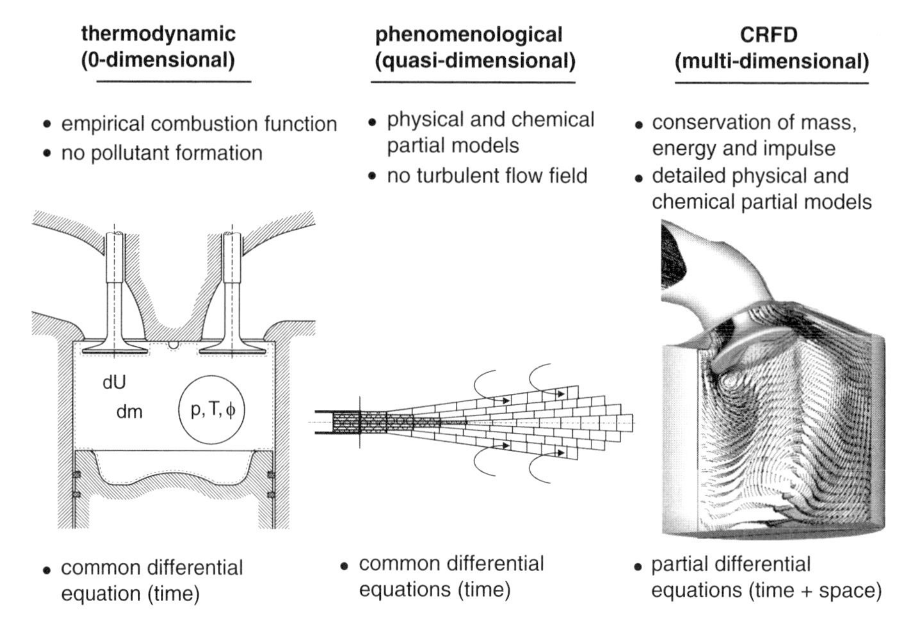

Fig. 11.1 Classification of combustion models

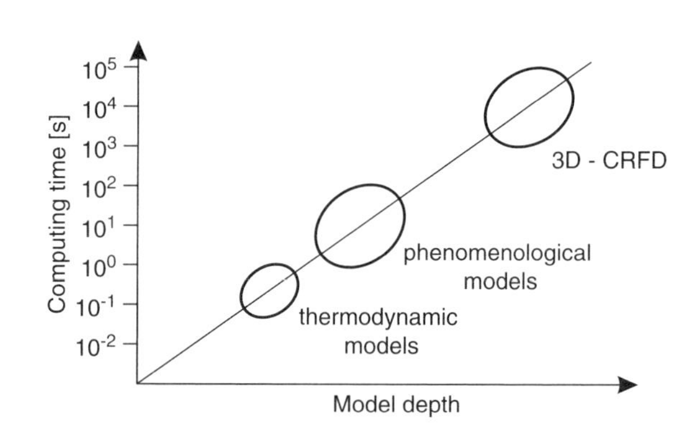

Fig. 11.2 Model depth and calculation time of combustion models

phenomenological combustion models to be described in the following carry out such a subdivision of the combustion chamber into zones of varying temperature and composition automatically, so that the corresponding pollutant formation model can be directly coupled to it. To this belongs, for example, the packet models described in Sect. 11.1.3. In the case of other phenomenological approaches, this zone subdivision is not yet implicitly contained, so that it has to be completed later in order to calculate not only the combustion rate, but also pollutant emissions. For this we can, for example, utilize the two-zone cylinder model explicated in Sect. 7.2.

11.2 Diesel Engine Combustion

11.2.1 Zero-Dimensional Heat Release Function

A relatively simple and thus calculation-time-efficient model for heat release in the diesel engine has been presented by Chmela et al. (1998, 2006). This model borders between zero-dimensional and phenomenological models, as it does not carry out a quasidimensional subdivision of the combustion space into zones of varying composition and temperature. Yet nevertheless gives the heat release rate not empirically, e.g. with a VIBE function, but coupled rather to a few characteristic influence parameters of key importance. These parameters are the fuel mass available at every point in time, thus the difference between injected and burned fuel mass, as well as the specific turbulent kinetic energy, which is taken as representative for the mixing speed of air and fuel.

$$\frac{dQ_B}{d\varphi} = C \cdot f_1(m_B) \cdot f_2(k) = C \cdot \left(m_B - \frac{Q_B}{H_u}\right) \cdot \exp\left(\frac{\sqrt{k}}{\sqrt[3]{V_{cyl}}}\right). \tag{11.1}$$

The temporal progression of the injection rate $dm_{fuel}/d\varphi$ is thereby given as a boundary condition, and the density of the turbulent kinetic energy k is derived only from the injection, since a quantitative estimate shows that the kinetic energy of the injection is about two orders of magnitude more than that of the inlet and quench flow. The production rate of the kinetic energy via injection is

$$\frac{dE_{kin,prod}}{dt} = \frac{1}{2}\dot{m}_B\left(v_{inj}\right)^2 = \frac{1}{2}\left[\frac{1}{\rho_B c_D A_{noz}}\right]^2 \left(\dot{m}_B\right)^3, \tag{11.2}$$

where $c_D A_{noz}$ designates the effective cross-sectional area of the injection nozzle. The dissipation rate of the kinetic energy is treated in a simplified way as proportional to the absolute amount of kinetic energy itself. The result if the differential equation for the change in kinetic energy

$$\frac{dE_{kin}}{dt} = \frac{dE_{kin,prod}}{dt} - C_{diss}E_{kin}. \tag{11.3}$$

It is now assumed that not all the kinetic energy in the combustion chamber is available for the mixing of fuel and air but only a portion of it corresponding to the amount of fuel available for mixture formation

$$E_{kin,mix} = E_{kin}\frac{m_B - Q_B/H_u}{m_B}. \tag{11.4}$$

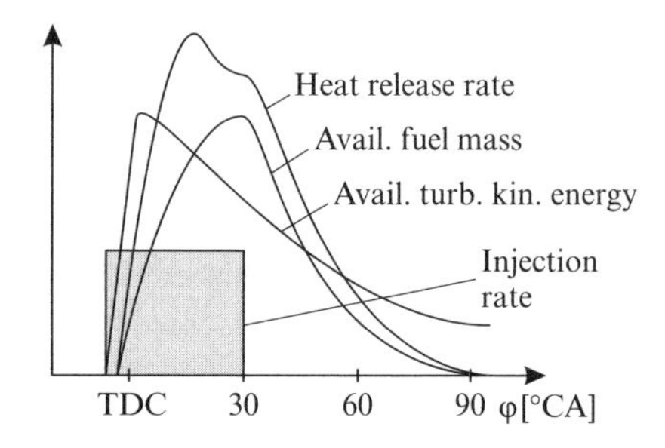

Fig. 11.3 Description of the combustion rate as a function of available fuel mass and turbulent kinetic energy, acc. to Chmela et al. (1998)

The specific turbulent kinetic energy k is finally approximated as the ratio of the kinetic energy available for mixing and the sum of the air and fuel mass in the diffusion flame. The air ratio in the flame is assumed to be stoichiometric:

$$k = C_{turb} \frac{E_{kin,mix}}{m_B(1 + L_{\min})}. \tag{11.5}$$

Figure 11.3 shows a typical temporal development of the available fuel mass and turbulent kinetic energy as well as the combustion rate resulting from the product.

The extremely short calculation time and simple application are advantages of this model approach, as well as the fact that the effects of the injection system (e.g. injection pressure, nozzle orifice cross-section and number) on the heat release rate can, as a rule, be fairly accurately depicted. On the other hand, there is also the restriction that neither the ignition delay nor the typical premixed fraction of diesel engine combustion can be described with this model. Both phenomena are substantially influences by the evaporation speed of the fuel, the additional consideration of which in the model would cause a significantly higher requirement in calculation time.

11.2.2 Stationary Gas Jet

More extensive model approaches, which are each based on the gas jet theory of Abramovich (1963), have been chosen, e.g., by de Neef (1987) and Hohlbaum (1992), in order to calculate heat release in a DI diesel engine.

Under the assumption that evaporation progresses quickly as opposed to mixture formation, injection is described as a quasi-stationary gas jet in an idealized solid body rotational flow, see Fig. 11.4. The combustion rate is then calculated as a direct function of the mixture formation rate, thus of the intermixing of fuel vapor and air.

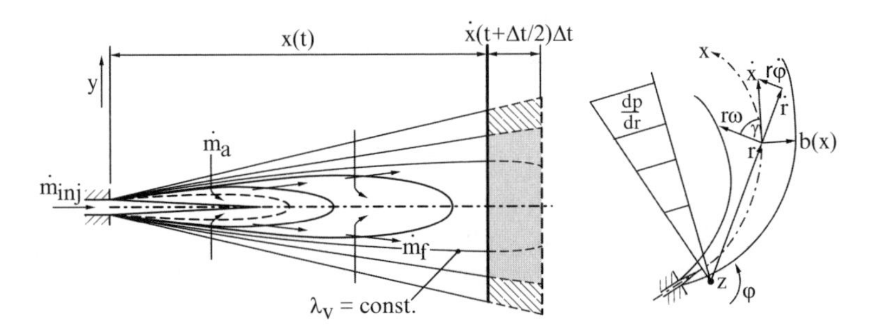

Fig. 11.4 Model of quasi-stationary gas jet in a solid body rotation flow-field

The propagation speed of the spray front as well as its change in direction via the charge movement results analytically from mass and momentum balances of the spray which has been reduced to its central axis. In accordance with Fig. 11.4, the momentum balances in the radial, tangential, and vertical direction of the cylindrical coordinate system are:

$$\frac{\mathrm{d}}{\mathrm{d}t}\left(\mathrm{d}m_{jet}\dot{r}\right) = \mathrm{d}F_r, \tag{11.6}$$

$$\frac{1}{r}\frac{\mathrm{d}}{\mathrm{d}t}\left(\mathrm{d}m_{jet}r^2\dot{\varphi}\right) = \frac{\mathrm{d}}{\mathrm{d}t}(\mathrm{d}m_a)r\omega + \mathrm{d}F_t\ , \tag{11.7}$$

$$\frac{\mathrm{d}}{\mathrm{d}t}\left(\mathrm{d}m_{jet}\dot{z}\right) = 0, \tag{11.8}$$

whereby $\mathrm{d}m_{jet}$ designates the mass of a spray-disc with thickness $\mathrm{d}x$. The magnitudes $\mathrm{d}F_r$ and $\mathrm{d}F_t$ are the radial and tangential forces, which affect the spray-disc, and the index a designates the unburned air surrounding the spray. The radial force is caused by the radial pressure gradient resulting from the rotational movement,

$$\mathrm{d}F_r = -\mathrm{d}V\frac{\mathrm{d}p}{\mathrm{d}r} = -\frac{\mathrm{d}m_{jet}}{\bar{\rho}}\rho_a r\omega^2, \tag{11.9}$$

and the tangential force is roughly

$$\mathrm{d}F_t = 0,1\frac{1}{\bar{c}}\frac{v_{inj}}{b}r(\omega - \dot{\varphi})\mathrm{d}m_f, \tag{11.10}$$

with $b = b(x)$ as the position-contingent radius of the circular spray-disc. The overbar designates the value mass-averaged over the entire spray cross-section.

With the help of the above-mentioned relations, we receive the motion equations of the spray front in the three cylinder coordinates,

$$\ddot{r}+\bar{c}\frac{\mathrm{d}}{\mathrm{d}t}\left(\frac{1}{\bar{c}}\right)\dot{r}=r\left[\dot{\varphi}^2-(1-\bar{c})\omega^2\right], \tag{11.11}$$

$$\ddot{\varphi}+2\frac{\dot{r}}{r}\dot{\varphi}=\left[\bar{c}\frac{\mathrm{d}}{\mathrm{d}t}\left(\frac{1}{\bar{c}}\right)+0.1\frac{v_{inj}}{b}\right](\omega-\dot{\varphi}), \tag{11.12}$$

$$\ddot{z}+\bar{c}\frac{\mathrm{d}}{\mathrm{d}t}\left(\frac{1}{\bar{c}}\right)\dot{z}=0, \tag{11.13}$$

with spray speed $\dot{x}$ and penetration depth S,

$$\dot{x}=\sqrt{\dot{r}^2+(r\dot{\varphi})^2+\dot{z}^2},\ S=x=\int_0^t\dot{x}\mathrm{d}t. \tag{11.14}$$

The spray angle and thus the change in the spray radius along the spray axis has a considerable influence upon the rate of air entrainment to the fuel spray. For combustion processes with no or only a minor amount of air swirl, a standard value of

$$(\mathrm{d}b/\mathrm{d}x)_{\omega=0}=0,16 \tag{11.15}$$

However, this value has to be adjusted if necessary in order to be able to illustrate the real spray angle, which is influenced, for example, by injection pressure, nozzle geometry, or physical characteristics of air and fuel. For combustion processes with marked swirl flow, de Neef (1987) provides the following correction for the spray angle

$$\frac{\mathrm{d}b}{\mathrm{d}x}=\frac{1-C\,(r\omega/v_{inj})}{1+C\,(r\omega/v_{inj})}\cdot\left(\frac{\mathrm{d}b}{\mathrm{d}x}\right)_{\omega=0}, \tag{11.16}$$

with

$$C=\frac{r\dot{\varphi}}{\dot{x}}-\frac{1}{2}\sqrt{2}\frac{\dot{r}}{\dot{x}} \tag{11.17}$$

and

$$v_{inj}=c_D\sqrt{\frac{2\Delta p_{inj}}{\rho_f}}. \tag{11.18}$$

In order to determine the mixture distribution within the spray, the fuel mass fraction averaged over the spray cross-section $\bar{c}$ along the spray coordinate x is at first calculated with the help of mass conservation. Under the presumption that the fuel mass contained in a spray disc of thickness dx is ($\mathrm{d}m_{strahl} \cdot \bar{c} = \text{const.}$) and that the averaged spray density $\bar{\rho}$ within this disc is very small in comparison with the density of the liquid fuel ρ_f, the temporal change of the mean fuel mass fraction contingent upon the spray angle (db/dx) can be expressed as follows

$$\frac{\mathrm{d}}{\mathrm{d}t}\left(\frac{1}{\bar{c}}\right) = \frac{4}{d_{noz}^2 v_{inj}} \frac{\rho_a}{\rho_f}\left[2\left(\frac{\mathrm{d}b}{\mathrm{d}x}\right) b\dot{x}^2 + b^2\ddot{x}\right]. \tag{11.19}$$

With the known fuel mass fraction $\bar{c}(x)$, averaged over the spray cross-section, the local fuel mass fraction $c(x, y)$ can be calculated in a further step. For this, an empirical dependence on the radial position in the spray is assumed

$$c = c_m\left[1 - \left(\frac{y}{b}\right)^{3/2}\right], \tag{11.20}$$

whereby c_m corresponds to the fuel mass fraction on the central axis of the spray.

In the model of de Neef (1987), it is now assumed that the combustion rate is limited by the mass of fuel that is processed per unit of time in stoichiometric relation to air. This quantity is determined as follows. Since the fuel mass fraction is known at every position in the spray, the iso-contours of the air-fuel equivalence ratio λ within the spray represented in Fig. 11.4 can be determined. The dimensionless radius y/b of a definite air-fuel equivalence ratio λv contingent on the axial position in the spray is

$$\frac{y}{b}(\lambda_v, x) = \left[1 - \frac{c(\lambda_v)}{c_m(x)}\right]^{2/3}. \tag{11.21}$$

As the injection spray is assumed to be stationary, the λ distribution within the spray does not change with time. In every numerical advance in time Δt, merely a new disc of thickness Δx is added to the spray, see Fig. 11.4. Due to mass conservation, the fuel mass contained in it is identical to the injected mass during this time increment ($\dot{m}_{inj} \cdot \Delta t$). Therefore, the fuel mass that goes beyond a certain boundary of $\lambda_v = \text{const.}$ within a time step (hatched surface in Fig. 11.4) must be equal to the difference between the injected fuel mass and the fuel which is found within the λ_v boundary (gray surface), thus in the richer mixture.

$$\Delta m_{f,\lambda_v} = \dot{m}_{inj}\Delta t - \pi y^2(\lambda_v)\rho_a c_m\left[1 - \frac{4}{7}\left(\frac{y(\lambda_v)}{b}\right)^{3/2}\right]\dot{x}\Delta t. \tag{11.22}$$

In order to determine the fuel mass being processed in the entire spray in stoichiometric relation with air, (11.22) has to be integrated between the rich ignition limit λ_R and $\lambda = 1$. Since only a fraction of $d\lambda_v$ of the fuel that crosses from $(\lambda = \lambda_v)$ to $(\lambda = \lambda_v + d\lambda_v)$ becomes newly prepared with air (the remaining fraction has already been prepared in previous time steps), we obtain the relation

$$\Delta m_{f,stoic} = \lambda_{v,R} \dot{m}_{f,\lambda_{v,R}} \Delta t + \int_{\lambda_{v,R}}^{\lambda_v = 1} \dot{m}_{f,\lambda_v} \mathrm{d}\lambda_v \Delta t. \tag{11.23}$$

Simplifying, it is assumed after end of injection that the area of spray close to the nozzle is no longer in existence, while the remaining part of the spray located further downstream is still behaving in a stationary manner. This behavior is considered in so far as a second (virtual) spray is calculated that begins to spread out at injection finish and is subtracted from the original spray.

The combustion rate is described with a quasi-kinetic approach which expresses the combusted portion of the stoichiometrically prepared fuel mass

$$X = \frac{m_{f,b}}{m_{f,stoic}} \tag{11.24}$$

with the Arrhenius function

$$\mathrm{d}X = A \rho_{jet} T_{jet}^{\beta} \frac{af_{stoic}(1 - X)^2}{af_{stoic} - 1} \exp\left[-\frac{E_A}{R_m T_{jet}}\right] \mathrm{d}t. \tag{11.25}$$

In this, T_{jet} and ρ_{jet} are the values for temperature and density averaged over the entire spray. The Arrhenius constants A, β and E_A have to be adjusted empirically for a certain engine in order to be able to depict experimentally determined combustion rates.

Since, with the model of stationary gas spray, neither fuel atomization nor drop evaporation are explicitly described, it is hardly possible to model the ignition delay in a detailed fashion. It is instead assumed that combustion begins at the moment in which the air ratio c_m on the spray axis exceeds the lower ignition boundary λ_R for the first time. At this time however, a certain quantity of fuel found in the outer spray regions has already been mixed stoichiometrically with air. This can now be converted very quickly, so that the typical premix-peak of diesel engines results in the heat release rate, see Fig. 11.5.

It should be observed that the spray opening angle in this combustion model, which is to be empirically determined, is a parameter of decisive importance, since it substantially influences the mixing speed of fuel and air and thus also the combustion rate. Beyond this, it is apparent that the assumption of an undisturbed stationary gas spray no longer is applicable if the spray collides with a combustion chamber wall. For this reason, the model appears to be most appropriate for describing large engines with distinct air swirl.

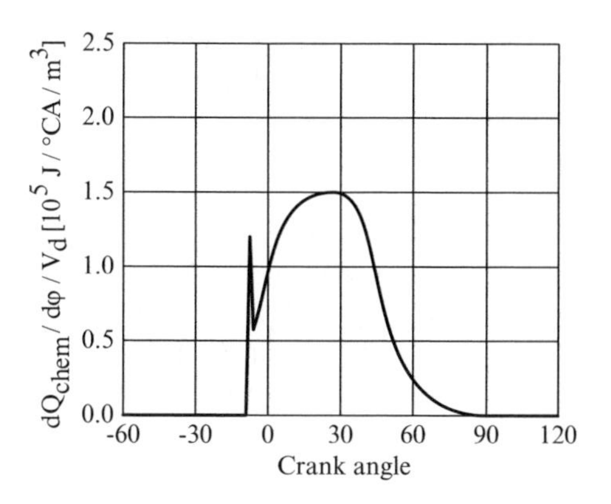

Fig. 11.5 Calculated heat release rate of a high speed high-performance diesel engine at rated power, acc. to Hohlbaum (1992)

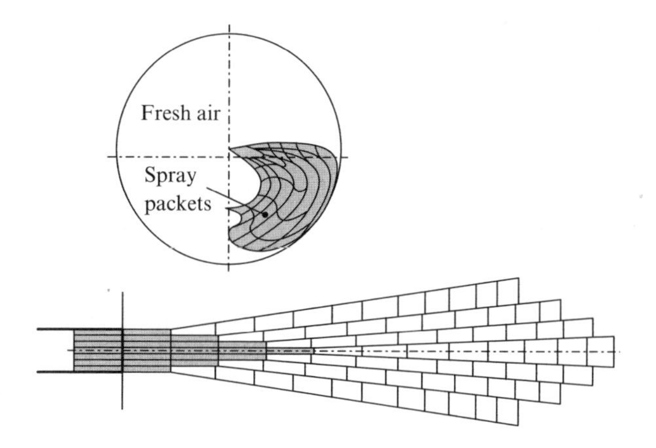

Fig. 11.6 Packet model, see Hiroyasu et al. (1983a, b)

11.2.3 Packet Models

One frequently applied model approach for describing diesel engine combustion is the so-called packet model (Hiroyasu et al. 1983a, b), depicted in Fig. 11.6. The injection jet is subdivided in this approach into many small zones, so-called packets, which illustrate in sum the contour of the entire spray. Usually, only one injection jet is calculated per cylinder, and it is assumed that all the other jets are identical in behavior.

Each of these single spray packets is then considered as a separate thermodynamic control volume, for which the respective mass and energy balances are solved. And, within these limits, the most important sub processes like drop evaporation, air admixture, combustion, and pollutant formation rates are calculated. That means, for every packet, a distinctive composition and temperature history are existed. Via simple addition of burning rates in each packet, we obtain finally the total heat release rate for the cylinder.

The phenomenological combustion model of Stiesch (1999), which will be described in more detail in the following, is based on the fundamental packet approach of Hiroyasu. During the compression stroke, only one zone exists, which extends itself over the entire combustion chamber and is viewed as ideally mixed. Fresh charge air and, in the case of exhaust gas recycling, combustion products are found in this zone. During the injection event, additional so-called spray packets are continuously generated, which reproduce the global form of the injection spray and subdivide it in both axial and radial directions. Independently of the number of nozzle bores, only one single fuel spray is viewed; an interaction of various sprays can thus not be considered. During the injection period, a new axial "disc" of packets is generated at each computational time step, whereby individual packets show a ring form because of their radial subdivision. At the generation instant, only liquid fuel is found in the packet. After the progression of a characteristic time, the liquid fuel is atomized into small drops, and the entrainment of gasses from the surrounding zone of fresh air to the single spray packets begins. The fuel droplets are heated up by the hot gasses which have entered the packets and evaporate. After the end of the ignition delay, the fuel-air-mixture begins to burn, by means of which the packet temperature increases again and pollutant formation (NO and soot) begins as well.

Both atomization and drop evaporation as well as ignition and combustion proceed within the packet boundaries and must therefore be calculated separately for each single packet. After combustion start, the packets can thus contain not only liquid fuel and fresh air, but also fuel vapor and combustion products, see Fig. 11.7. A mixing of various spray packets or an exchange of energy between them does not occur. With exception of air-entrainment into the spray (and thus to the packets) and wall heat transfer, all transport processes thus proceed within packet boundaries.

With the help of mass and energy balances as well as an equation of state, the change rates of composition, temperature, and volume can be calculated for each single packet as well as for the fresh air zone. On the other hand, pressure is viewed as independent of location and only as a function of time. This assumption is justified because of the high speed of sound at high pressures during the combustion phase.

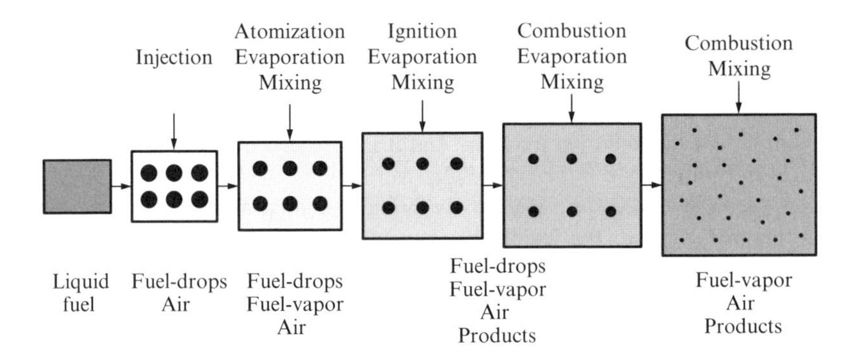

Fig. 11.7 Composition of spray packets

11.2.3.1 Spray Development and Mixture Formation

Immediately after start of injection, one spray packet is viewed as a continuous liquid phase, which moves into the combustion chamber with the constant speed

$$v_{inj} = 0,39\sqrt{\frac{2\Delta p_{inj}}{\rho_{B,fl}}} \tag{11.26}$$

until atomization begins. The liquid fuel mass per packet $m_{B,fl}$ with the momentary injection rate $\dot{m}_{inj}$, the number of packets in the radial direction $k_{\max}$ and the length of time interval Δt amounts to

$$m_{B,P} = \frac{\dot{m}_{inj}\Delta t}{k_{\max}}. \tag{11.27}$$

After a characteristic time span, the liquid phase disintegrates into small drops. This so-called breakup time amounts on the spray axis to

$$t_{bu,c} = 28,65\frac{\rho_{B,fl}D_D}{\sqrt{\rho_L \Delta p_{inj}}}. \tag{11.28}$$

Because the interaction between fuel and air at the spray boundary is more marked than on the spray axis, spray breakup in the outer packets begins earlier corresponding to

$$t_{bu,k} = t_{bu,c}\left(1 - \frac{k-1}{k_{\max}}\right) \tag{11.29}$$

if a linear decrease in breakup time is assumed over the spray radius. Through the entrainment of gasses from the fresh air zone into the spray packet, packet speed is reduced. For packets on the spray axis

$$v_{tip,c} = 1,48\left(\frac{\Delta p_{inj} D_D^2}{\rho_L}\right)^{1/4}\frac{1}{\sqrt{t}} \tag{11.30}$$

is valid, and for packets further outside, it is roughly assumed that a speed profile decreasing exponentially towards the spray border is engaged

$$v_{tip,k} = v_{tip,c}\exp\left(-C_{rad}(k-1)^2\right). \tag{11.31}$$

When five spray packets are viewed radially ($k_{max} = 5$), and it is further assumed that the speed of the outer packet amounts to approx. 55% of the speed on the axis, a value of 0.374 for the constant C_{rad} results. The injection process itself also changes the flow pattern in the combustion chamber decisively. The kinetic energy of the injection jets is approximately two orders of magnitude above the kinetic energy of swirl and squish flows at injection start. As a result of this, the spray packets generated first are much more strongly slowed down by the surrounding gas phase than the ones created towards the end of injection, which move as it were in the "lee of the wind". The packet speed after spray breakup is therefore corrected according to

$$v_{i,k} = C_1 v_{tip,k} \left[1 + \left(\frac{i-1}{i_{\max}-1} \right)^{C_2} \frac{\Delta t_{inj}}{C_3} \right] \tag{11.32}$$

whereby $i = 1$ designates the packet generated first and $i = i_{\max}$ the last. The constant C_1 can be slightly higher that 1, C_2 has the approximate value of 0.5 and C_3 describes the absolute speed difference between the first and the last packet.

The air entrainment rate is calculated with the principle of impulse conservation of the spray packets

$$v_{i,k}(m_{B,P} + m_{L,P}) = \text{const.} \tag{11.33}$$

11.2.3.2 Drop Distribution Spectrum

After the spray breakup time, the liquid fuel of the spray packet disintegrates into many small drops, the integral behavior of which can be described with the Sauter mean diameter. The Sauter mean diameter is thereby the diameter of a representative drop, which has the same volume to surface area ratio as all drops integrated over the entire spray. For this we find the relation

$$SMD = 6156 \cdot 10^{-6} v_{B,fl}^{0,385} \rho_{B,fl}^{0,737} \rho_L^{0,06} \Delta p_{inj}^{-0,54} \tag{11.34}$$

with SMD in m, v in m^2/s, ρ in kg/m^3 and pressure difference Δp_{inj} in kPa. The number of fuel drops in a packet under the assumption that all drops are equally large amounts to

$$N_{Tr,P} = \frac{m_{B,P}}{\frac{\pi}{6} SMD^3 \rho_{B,fl}}. \tag{11.35}$$

For a more detailed description of atomization and thus also of the following evaporation process, the drop size distribution function

$$g(r) = \frac{1}{6} \frac{r^3}{\bar{r}^4} \exp\left(\frac{-r}{\bar{r}} \right) \tag{11.36}$$

can be utilized with the radius

$$\bar{r} = \frac{SMD}{6} \tag{11.37}$$

of the most frequently appearing drop.

11.2.3.3 Drop Evaporation

In describing evaporation, the mixing model is often used, in which the inside of the drop is always assumed to be isothermal. As a comparison fuel, pure tetradekan ($C_{14}H_{30}$) will be used in the following, which has similar physical properties as real diesel fuel. For investigations with two-component comparison fuels, e.g. a mixture of 70 vol.% n-dekan ($C_{10}H_{22}$) and 30 vol.% α-methyl napthaline ($C_{11}H_{20}$), please refer to Stiesch (1999).

With this, we obtain for the convective heat transfer from the gas phase to the drop with the help of the Nusselt number

$$\frac{\mathrm{d}Q_{Tr}}{\mathrm{d}t} = \pi\, SMD\lambda_S(T_P - T_{Tr})\frac{z}{e^z - 1}\mathrm{Nu}, \tag{11.38}$$

whereby z represents a dimensionless correction factor, which diminishes the transferred heat flux under the simultaneous appearance of mass transfer via evaporation corresponding to

$$z = \frac{c_{p,B,g}\frac{\mathrm{d}m_{Tr}}{\mathrm{d}t}}{\pi\, SMD\lambda_S \mathrm{Nu}} \tag{11.39}$$

We calculate the evaporation rate of a drop with the help of the relation for mass transfer as

$$\frac{\mathrm{d}m_{Tr}}{\mathrm{d}t} = -\pi\, SMD\rho_S C_{diff} \ln\left(\frac{p_{cyl}}{p_{cyl} - p_{B,g}}\right)\mathrm{Sh} \tag{11.40}$$

For the Nusselt and Sherwood number is valid

$$\mathrm{Nu} = 2 + 0,6\mathrm{Re}^{1/2}\mathrm{Pr}^{1/3}, \tag{11.41}$$

$$\mathrm{Sh} = 2 + 0,6\mathrm{Re}^{1/2}\mathrm{Sc}^{1/3}, \tag{11.42}$$

whereby the Reynolds number is calculated with a relative speed between drop and gas phase, which is assumed to be up to 30% of the momentary packet speed $v_{i,k}$.

Temperature change of the liquid fuel drops results finally from an energy balance over a drop

$$\frac{dT_{Tr}}{dt} = \frac{1}{m_{Tr} c_{p,Tr}} \left(\frac{dQ_{Tr}}{dt} + \frac{dm_{Tr}}{dt} \Delta h_v \right), \tag{11.43}$$

with drop mass contingent on diameter and drop temperature:

$$m_{Tr} = \frac{\pi}{6} \rho_{Tr} SMD^3. \tag{11.44}$$

11.2.3.4 Ignition Delay

Ignition delay is often described by means of a simple Arrhenius method

$$\tau_{zv} = C_1 \frac{\lambda_P}{p_{cyl}^2} \exp\left(\frac{C_2}{T_P}\right) \tag{11.45}$$

with $C_1 = 18$ and $C_2 = 6{,}000$.

11.2.3.5 Heat Release

As a simplification, we assume that after reaching ignition delay, the fuel is completely converted to CO_2 and H_2O in correspondence with the gross reaction equation. For a detailed consideration of this, the reader is referred to Stiesch (1999).

The maximum combustion rate in the packet is limited by the strictest of the following three criteria. Firstly, only the vaporous fraction of fuel can be burned

$$\dot{m}_{B,Ox,P} \leq \frac{m_{B,g,P}}{\Delta t}. \tag{11.46}$$

Secondly however, the quantity of air in the packet also limits the conversion rate according to

$$\dot{m}_{B,Ox,P} \leq \frac{m_{L,P}}{L_{\min} \Delta t}. \tag{11.47}$$

Thirdly, a maximal chemical conversion rate for premixed flames must still be considered, which is described by the Arrhenius function

$$\dot{m}_{B,Ox,P} \leq 5 \cdot 10^5 \rho_{mix} x_{B,g,P} x_{O_2,P}^5 \exp\left(-\frac{12.000}{T_P}\right) V_P \tag{11.48}$$

and which is important in the late combustion phase, when the temperature in the cylinder has sunk considerably and the chemistry is therefore slow.

The thermodynamic balance equations necessary for further calculation are introduced in Chap. 2 and more exhaustively in Chap. 7. For the determination of thermodynamic state quantities of single components represented in the packets, we refer again to Stiesch (1999).

11.2.3.6 Model Validation

Figures 11.8 and 11.9 show a comparison of a measured and calculated combustion and pressure traces for two operating conditions of a high speed diesel engine with 3.96 l displacement volume per cylinder, 165 mm piston diameter and a speed of 1,500 min^{-1}.

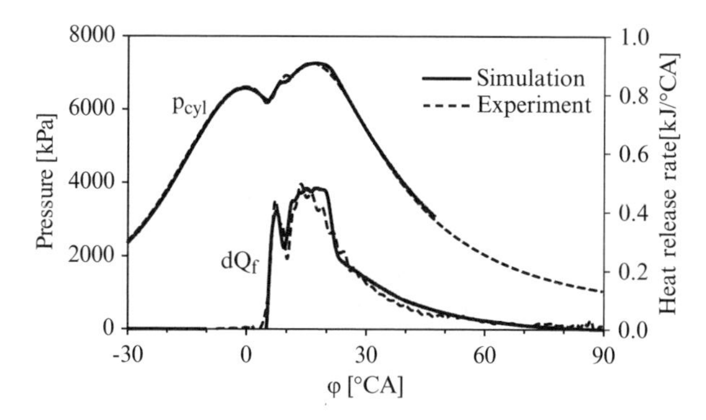

Fig. 11.8 Comparison of measured and calculated pressure progressions and heat release rate for a high speed diesel engine with 3.96 l displacement volume per cylinder at $n = 1{,}500$ rpm and $p_{me} = 9.8$ bar

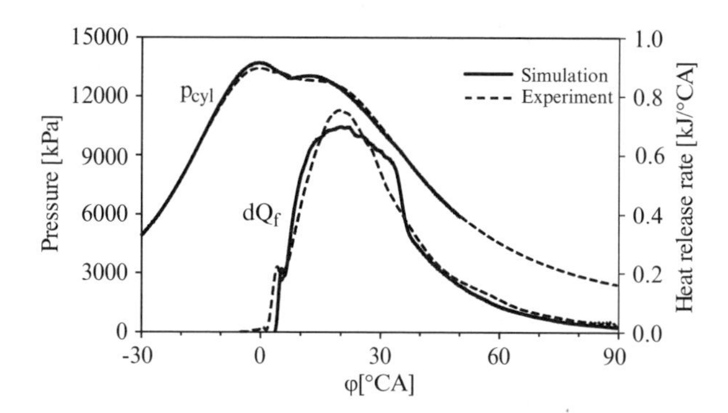

Fig. 11.9 Comparison of measured and calculated pressure progressions and heat release rate for a high speed diesel engine with 3.96 l displacement volume per cylinder at $n = 1{,}500\ min^{-1}$ and $p_{me} = 22.0$ bar

The operation point represented in Fig. 11.8 was chosen as a reference point for the adjustment of the model, whereby more attention was paid to a good agreement of pressure paths than on an exact adjustment of the heat release rate. In total, we can recognize a good agreement between simulation and measurement results. Detailed investigations show however that a more complex heat transfer model, which explicitly considers the influence of soot radiation, is advantageous for a further improvement in agreement.

11.2.3.7 Description of a Pilot Injection

Thoma et al. (2002) expanded upon the packet method in order to describe diesel engine combustion processes with pilot injection as well. However, since the spray penetration curve (11.30) is only valid for continuously injected sprays and not for very small fuel quantities (Stegemann et al. 2002), Thoma et al. (2002) suggest a change of time contingency from $1/\sqrt{t}$ in $1/t$ for the pilot injection packets. Moreover: at the moment, in which the main injection phase begins, the pilot injection packets are compacted into a single so-called pilot injection zone, see Fig. 11.10.

Because the pilot injection slows down so quickly, the packets of main combustion soon penetrate into the pilot injection zone, so that, instead of fresh air, gasses of the pilot injection zone are admixed into the main combustion packets. This entrainment of already hot gasses into the spray packets causes a reduction of the ignition delay of the main injection. From this results the well-known effect of pilot injection, namely, the clear reduction on the premix-peak in the heat release rate. Figure 11.11 shows that this behavior of the model can be depicted very well.

11.2.4 Time Scale Models

Weisser and Boulouchos (1995) have developed a phenomenological model for the heat release rate in the diesel engine, which is based on characteristic time scales, similar to the eddy breakup model which is often used within the CFD code. In this

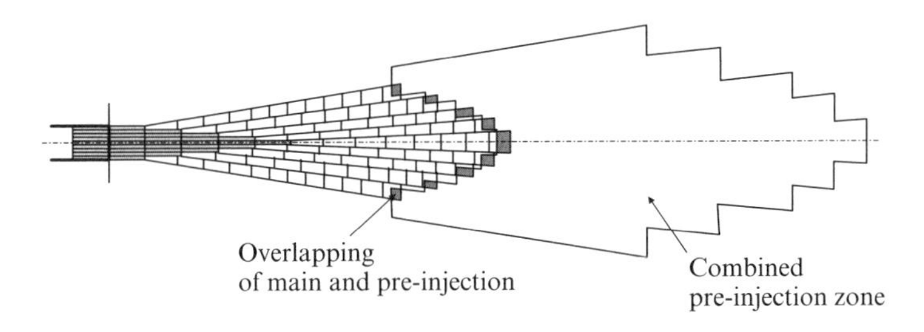

Fig. 11.10 Mixing of the pilot- and main injection pulses, acc. to Thoma et al. (2002)

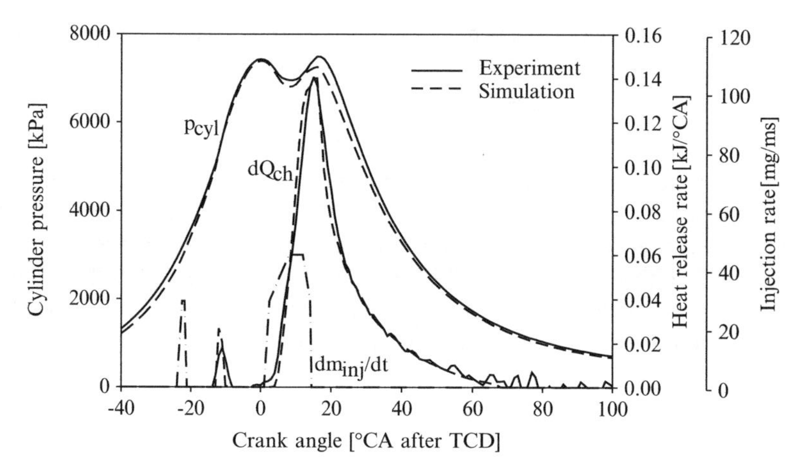

Fig. 11.11 Calculated and measured heat release rate and pressure progressions of a truck diesel engine with pilot injection, Thoma et al. (2002)

case, two different time scales are considered for premixed and diffusion combustion, as it is assumed that premixed combustion is essentially influenced by fuel evaporation and reaction kinetics, while diffusion combustion is above all contingent on the speed of the turbulent mixing of fuel vapor and air.

The atomization and evaporation of fuel is very similarly modeled as in the packet method introduced above. However, the spray is here discretized only in the axial direction, and the spray penetration is calculated with the equation of Dent (1971):

$$S = 3,07\left(\frac{\Delta p_{inj}}{\rho_g}\right)^{1/4}(D_{noz}t)^{1/2}\cdot\left(\frac{294}{T_g}\right)^{1/4}. \tag{11.49}$$

It is now assumed that the fuel fraction that already evaporated before the first ignition is converted as premixed combustion, while the remaining fuel is converted in diffusion combustion, which is controlled by the turbulent mixing. The ignition delay is again similarly determined as in the packet approach on the basis of an Arrhenius equation, see (11.45).

The time scale characteristic for reaction-kinetically controlled premixed combustion is assumed to be proportional to the ignition delay τ_{ID}, so that the conversion rate of the fuel can be given as

$$\frac{\mathrm{d}m_{prem}}{\mathrm{d}t} = C_{prem}\frac{1}{\tau_{zv}}f_{prep}m_{prem,av} \tag{11.50}$$

whereby $m_{prem,av}$ is the total fuel mass added to the premixed combustion. The factor f_{prep} takes into account the fact that only a part of this mass has gone beyond the ignition delay and can actually be converted at the time under consideration.

The conversion rate of diffusion combustion is formulated in analogy to (11.50)

$$\frac{\mathrm{d}m_{diff}}{\mathrm{d}t} = C_{diff}\frac{1}{\tau_{trb}} f_{A,turb} m_{diff,av}. \tag{11.51}$$

We have here the difficulty however that, within the phenomenological approach, the turbulent time scale t_t cannot be determined as is customary in the CFD code directly from the knowledge of the turbulent flow field, but must rather be estimated with the help of a simplified method. For the sake of this, the turbulent mixing frequency – the reciprocal value of the turbulent time scale – is approached as the ratio of turbulent viscosity and of the square of a linear scale characteristic for the problem

$$\frac{1}{\tau_{trb}} = \frac{u' l_I}{(X_{char})^2}. \tag{11.52}$$

In order to estimate the turbulent viscosity $u' l_I$ in the combustion chamber, a simplified method is chosen that assumes two sources of turbulence. The first source is the intake flow of charge-air, for which the turbulence intensity u' is set as proportional to the mean piston speed and the linear scale l_I as proportional to the clearance height. The second source of turbulence is the injection spray itself, for which the magnitudes u' and l_I can be solved with the help of conservation equations, see Heywood (1988). The initial values thereby result from the injection speed and the nozzle orifice diameter.

Correspondingly, for the sum of turbulent viscosity,

$$u' l_I = (u' l_I)_{charge} + (u' l_I)_{inj}. \tag{11.53}$$

is valid. The characteristic linear scale for the process of turbulent diffusion between fuel vapor and air is determined in contingency upon the momentary cylinder volume, the global air ratio, and the number of nozzle orifices

$$X_{char} = \left(\frac{V_{cyl}}{\lambda N_{noz}}\right). \tag{11.54}$$

The factor $f_{A,turb}$ in (11.51) describes the increase of the actual surface of the flame front through turbulent,

$$f_{A,turb} = \frac{u' l_I}{\nu}, \tag{11.55}$$

whereby ν is the viscosity of the combustion gasses. With this, (11.51) can be written as

$$\frac{\mathrm{d}m_{diff}}{\mathrm{d}t} = C_{diff}\frac{u' l_I}{X_{char}^{\;2}}\frac{u' l_I}{\nu} m_{diff,av} \tag{11.56}$$

Finally, the fuel masses $m_{pre,av}$ and $m_{diff,av}$ available to both combustion types are determined through the integration of evaporation rate and combustion rate,

$$m_{i,av} \int_{t_{i,0}}^{t} \left(\frac{\mathrm{d}m_{i,evap}}{\mathrm{d}t} - \frac{\mathrm{d}m_i}{\mathrm{d}t} \right) \mathrm{d}t, \tag{11.57}$$

whereby the index i stands for both combustion types *pre* and *diff*.

11.3 SI Engine Combustion

Phenomenological modeling of combustion in homogeneously operated SI engines is usually based on the following simplifications:

- Fuel, air, and residual gas are mixed homogeneously.
- The volume occupied by the reaction zone is very small in comparison to the volume of the combustion chamber; the flame is usually considered to be infinitely thin (the flamelet model).
- The combustion chamber consists of two zones, the burned and the unburned zone.

The main difficulty in the phenomenological modeling of SI engines is the absence of the most important parameters of the turbulent flow field. Mainly, these are the turbulent fluctuation speed and the associated length scales. Generated by the elements of the air path (e.g. the inlet valve), whirling structures of various sizes are created that have a direct effect on the speed of combustion and heat release. The main task is then to formulate these length scales as a function of the macroscopic engine parameters without the need to integrate additional dimensionalities into the model.

11.3.1 Laminar and Turbulent Flame Front Speed

For the temporal energy release rate and thus also for the pressure increase in the cylinder that results from combustion, the propagation speed of the premixed flame front – the so-called flame speed – is of special importance. We hereby distinguish between laminar flame speed $s_l$and turbulent flame speed s_t. Laminar flame speed designates the propagation speed of a thin premixed flame front in a static air-fuel mixture. Besides the reaction kinetics, it is dependent upon the heat conducting and diffusion processes inside the flame, and it can be assessed as a function of fuel, of the mixture ratio of fuel and air, of residual gas content, and of system pressure and the temperature of the educts. Often, the relation provided by Metghalchi and Keck (1982) is used for flame propagation in hydrocarbon-air mixtures,

$$s_l = s_{l,0} \left(\frac{T_u}{T_0}\right)^{\alpha} \left(\frac{p}{p_0}\right)^{\beta} (1 - c_R f_R) \tag{11.58}$$

whereby f_R designates the mass fraction of the residual gas and T_0 and p_0 the reference conditions at 298 K and 101.3 kPa. The constant c_R defines the influence of the residual gas on the flame speed. Metghalchi and Keck (1982) selected $c_R = 2.1$ from comparative measurements with 85% N_2 and 15% CO_2 as a residual gas. More recent publications with measurements under real engine conditions (e.g. Wallesten 2003) suggest $c_R = 3$. Both the exponents α and β as well as the flame speed under atmospheric conditions $s_{l,0}$ are fuel-dependent quantities. Table 11.1 shows the quantities determined by various authors for isooctane, including the marginal conditions valid for them.

To determine the laminar flame speed for propane, isooctane and methanol, the relation

$$s_{l,0} = B_m + B_\lambda \left(\frac{1}{\lambda} - \frac{1}{\lambda_m}\right)^2, \tag{11.59}$$

is valid, where λ_m is the air-fuel equivalence ratio, at which $s_{l,0}$ reaches its maximal value of B_m. The parameters contained in (11.59) are summarized in Table 11.2.

However, in real engine combustion chambers, the flow field is not laminar, but highly turbulent. Thus, we must additionally consider the influence of turbulence on the propagation speed of the flame front.

Through the interaction with turbulent swirls, the flame front, smooth in the laminar case, becomes wrinkled ("flame wrinkling"), so that its surface A_l becomes larger. The turbulent flame front surface A_t defines the surface with which the turbulent flame propagates into the unburned mixture. The conservation equation is also valid

Table 11.1 Correlations for the laminar flame speed

	α	β	Conditions		
			ϕ	T	p
Metghalchi and Keck (1982)	2.18–0.8 (ϕ–1)	–0.16+0.22(ϕ–1)	0.8–1.2	298–700 K	0.4–50 atm
Gülder (1982)	1.56	–0.22	0.8–1.2	300–500 K	1–8 bar
Bradley et al. (1998)	1.07; 1.01	–0.282; –0.348	0.8–1.0	358–450 K	1–10 bar
Müller et al. (1997)	2.0	–0.27	1	298–800 K	1–40 bar

Table 11.2 Parameters for (11.59)

Fuel	λ_m	B_m [cm/s]	B_λ [cm/s]
Methanol	0.90	36.9	–140.5
Propane	0.93	34.2	–138.7
Isooctane	0.88	26.3	–84.7
Gasoline	0.83	30.5	–54.9

$$A_t s_t = A_l s_l, \tag{11.60}$$

which takes advantage of the insight that the turbulent flame propagates in a laminar fashion in a very small volume, smaller that the surrounding whirl normally to the surface. The geometric expansion of this flame front surface is determined by the position of the spark plug and the shape of the combustion chamber. Correct determination is essential for the exact prediction of the heat release rate. Heywood (1988) provides a representation of the flame front surface as a function of the spark plug position and combustion chamber geometry. The enlargement of the surface via turbulent flame-wrinkling thus actually contributes to the increase of burning velocity. We for this reason introduce the turbulent flame speed s_t, which is calculated from $s_l$and the local turbulence level

$$u' = \sqrt{\frac{2k}{3}}, \tag{11.61}$$

where u' is an average, local speed component due to turbulent fluctuations, see Sect. 12.2. The Damköhler relation (Damköhler 1940) represents a basic approach to this

$$s_t = \left(1 + C\,\frac{u'}{s_l}\right)^n s_l\,. \tag{11.62}$$

The Damköhler constant C depends mainly on the turbulent length scale as well as the thickness of the flame front. The exponent n ranges in the literature between 0.5 and 1. Koch (2002) suggests $C = 2.05$ and $n = 0.7$ for the constants according to his own adjustments. Equation (11.62) makes it clear that the turbulent (effective) flame front speed increases with a higher turbulence level. This is the reason why SI engines can be operated with much higher speeds than, for example, diesel engines. The is only possible because the turbulence level in the combustion chamber caused by the in-flow process increases with rising speeds, so that the flame front speed increases and the mixture can still be completely converted despite the small amount of time available.

11.3.2 Heat Release

In order to calculate heat release in the SI engine, the *entrainment model*, developed by Blizard and Keck (1974) and expanded by Tabaczynski (1980), is often used and will be briefly explained in the following. In this model, heat release or flame front propagation is analyzed into two partial steps. The first step describes the penetration of the flame because of the turbulent propagation mechanism without heat release into the as yet unburned mixture. The penetration speed is additively composed of the turbulent fluctuation speed u' and the laminar flame speed s_l. Alternatively, the turbulent flame front speed can also be used in accordance with (11.62). This amounts under consideration of the continuity condition for the charge mass recorded per time to

$$\frac{\mathrm{d}m_e}{\mathrm{d}t} = \rho_u A_t s_t, \tag{11.63}$$

whereby A_t is the surface of the flame front and ρ_u is the density of unburned mixture.

The second partial step describes heat release through combustion, whereby the fresh gas swirl regions grasped by the flame are converted with laminar flame speed. The dominating swirl size is thereby the Taylor micro-length, which is defined with the integral linear length scale l_I by

$$l_T = \sqrt{\frac{15\, l_I\, \nu}{u'}} \tag{11.64}$$

With that follows for the characteristic combustion time scale

$$\tau = \frac{l_T}{s_l} \tag{11.65}$$

and with this for the conversion rate of the fuel mass found in the flame area

$$\frac{\mathrm{d}m_b}{\mathrm{d}t} = \frac{m_e - m_b}{\tau}. \tag{11.66}$$

While the laminar speed s_l can be determined with (11.58), the integral length scale l_I and the turbulent fluctuation speed u' must be modeled because of lacking flow field resolution. To this end, there are suggestions for u' in the literature, an extract of which is shown in Table 11.3.

The integral length scale l_I of the turbulent flow field and of the flame front structures describes the large-scale whirling in the combustion chamber and must also be taken from measurements. The size of this whirling is in the range of 1 mm $< l_I <$10 mm. Table 11.4 shows some correlations from the literature.

The entrainment model presumes a fully developed flame front. Therefore, an inflamed volume is given as a start value for the combustion, the mass of which

Table 11.3 Suggestions for turbulent fluctuation speed from the literature

Correlation	Source
$u' = \frac{1}{2} c_m$	Wirth (1993)
$u' = 0,08 \cdot \bar{u}_i \left(\frac{\rho_u}{\rho_i}\right)^{\frac{1}{2}}$ with $\bar{u}_i = \eta_v \frac{A_p}{A_{iv}} c_m$	Keck (1982)
$u'_{ZZP} = c_T\, c_m$ with $u' = u'_{ZZP} \left(\frac{\rho_M}{\rho_{M,ZZP}}\right)^{\frac{1}{3}}$	Tabaczynski (1980)

Table 11.4 Suggestions for the integral length scale from the literature

$l_I = 0{,}2 \cdot h_{BR}$	Wirth (1993)
$l_{I,ZZP} = c_L h_{BR}$ with $c_L = 0{,}35$ $l_I = l_{I,ZZP} \left(\frac{\rho_{M,ZZP}}{\rho_M}\right)^{\frac{1}{3}}$	Tabaczynski (1980)

corresponds to 1% of the total charge-mass. In determining the ignition time, the time span between the ignition time and the 1% mass conversion point, the so-called inflammation duration, must be calculated. We set for this

$$\Delta t_{id} = c_{id}\,\tau. \tag{11.67}$$

The modeling approaches introduced here are only valid for the description of homogeneously operated SI engines, which are characterized by an exclusively premixed combustion. Phenomenological models for SI engines with charge stratification hardly exist in the literature, since modeling is very difficult due to the lacking geometrical resolution. Koch (2002) has formulated an approach which subdivides the combustion chamber into two homogeneous zones, the fresh gas and exhaust gas zones. The zones can also themselves contain fresh gas and exhaust gas. In stratification operation, successful injection is followed by the vaporization phase, and the fuel interacts with the fresh gas zone. Vaporization is described by means of a characteristic dead time Δt_{evap} and leads to an enrichment of the fresh gas zone. Combustion is subdivided into premixed and mixture-controlled combustion, whereby the ratio represents a factor that must be adapted. The factor determines which amount of the injected fuel mass burns premixed or diffusively. Premixed combustion is described using an expanded Tabaczynski formulation with

$$\frac{dQ_{vor}}{dt} = A_t \cdot H_U \cdot \rho_u \cdot s_t \cdot Ex. \tag{11.68}$$

Here, the expansion factor Ex describes the superposition of the flame front speed with the propagation speed of the burned zone due to the density difference to the unburned zone (Heywood, 1988). Mixture-controlled heat release is described in analogy to known methods by means of a time scale model:

$$\frac{dQ_{dif}}{dt} = \frac{1}{\tau_{fluid} + \tau_{chem}} \cdot H_u \cdot m_{verf}. \tag{11.69}$$

τ_{chem} represents the time scale of the chemical processes and τ_{fluid} the fluid-dynamic mixing time. Due to the different dependencies of the scales, the chemical time scale is dominant at first. Only after combustion has progressed and the process temperatures have increased does the fluid-dynamic time scale become more important. Jet-controlled combustion in SI engines in stratification operation can be described by means of an adequate adjustment of the subdivision premixed/mixture-controlled.

11.3.3 Ignition

In the SI engine, combustion is introduced by a spark discharge at the spark plug. With the assumption that a constant adiabatic and isobaric temperature exist, which characterizes the ignition limit, the condition

$$h_{AG}\left(T_{ad,ZZP}, p_{ZZP}\right) = \frac{1-\kappa_{RG}}{1+\lambda\, L_{\min}}\; H_u + h_{FG}\left(T_{u,ZZP}, p_{ZZP}\right). \tag{11.70}$$

can be derived by balancing a small volume element in the spark plug area. The thus calculated ignition boundaries agree well in a wide temperature region with measured values, see Scheele (1999).

Figure 11.12 shows a comparison of the characteristic mass conversions determined by pressure course analysis with those calculated by the entrainment model. The constants for the calculation of turbulence intensity is fixed at $c_T = 0.6$ and for the calculation of the integral length scale at $c_L = 0.35$, such that the ratio of the time periods between the mass conversion points agree with the values from the pressure path analysis. The assumption at combustion start of a turbulence intensity

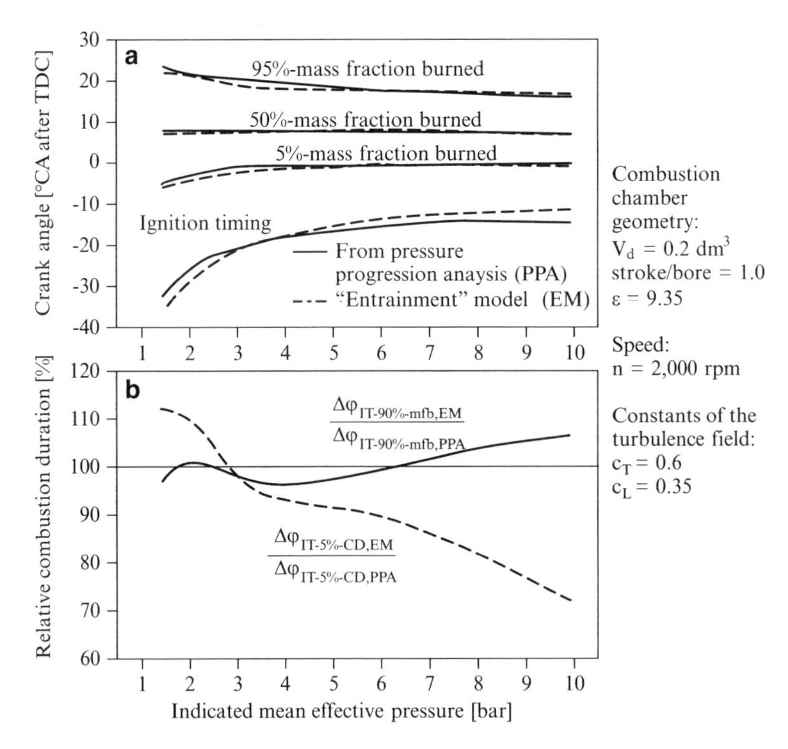

Fig. 11.12 Comparison of the mass conversions determined by the pressure progression analysis and calculated by the entrainment model (from: Scheele 1999)

constant at various loads reproduces the change in burning duration with a sufficient degree of precision.

11.3.4 Knocking

Since irregular combustion or knocking in the SI engine is above all a local phenomenon, phenomenological modeling is correspondingly more difficult without geometrical resolution. Livengood and Wu (1955) postulated that a knocking event occurs when

$$\int_{t=0}^{t_{klopf}} \frac{1}{\tau} dt = 1 \tag{11.71}$$

is satisfied. Here, τ is the ignition delay and t_{knock} the time between compression start and autoignition. The ignition delay is described with an Arrhenius formulation

$$\tau = X_1 \cdot p^{-X_2} \exp\left(\frac{X_3}{T_U}\right). \tag{11.72}$$

Inserting (11.72) into (11.72), the Livengood-Wu integral yields

$$\int_{t=0}^{t_{klopf}} \frac{1}{X_1 \cdot p^{-X_2} \exp\left(\frac{X_3}{T_U}\right)} dt = 1. \tag{11.73}$$

Here, p is the pressure of the combustion chamber and T_U the temperature of the unburned gas. The parameters X_1, X_2 and X_3 have to be adjusted to the experimental data.

Elmqvist et al. (2003) improved with method of Livengood and Wu by adjusting the parameters to extensive knocking measurements for various engine loads and speeds. Minimizing the error squares of the calculation to the measurement results provides the optimized parameters $X_1 = 0.021, X_2 = 1.7$ und $X_3 = 3800\ K$.

The model originally created by Franzke (1981) and further developed by Spicher and Worret (2002) is a comparable approach to representing knocking phenomenologically.

The methods described above have the disadvantage "typical" of phenomenological models that one needs to have extensive measurements at hand (in this case: engine operation at the knocking limit) in order to calibrate the model. Reaction-kinetic approaches do not have this limitation. In these models, hydrocarbon oxidation in the fresh gas zone is described by means of reduced or even complex chemical processes, permitting universally valid pre-calculations of knocking events. The Shell model of Halstead et al. (1975) is a reduced model for calculating the autoignition process, illustrating autoignition with a few elementary reactions between "pseudo species". Li et al. (1996) suggested a more complex model with 29 reactions, which attempts to integrate CO formation into the modeling of autoignition. Sect. 5.2.2 provides a detailed overview of the ignition of hydrocarbons.

References

Abramovich GN (1963) The theory of turbulent jets. MIT Press, Cambridge, MA

Bradley D, Hicks RA, Laws M, Sheppard CGW, Wooley R (1998) The measurement of laminar burning velocities and Markstein numbers for isooctane and isooctane-n-heptane-mixtures at elevated temperatures and pressures in an explosion bomb. Combust Flame 115:126–144

Blizard NC, Keck JC (1974) Experimental and theoretical Investigation of Turbulent Burning Model for Internal Combustion Engines, SAE 740191

Chmela F, Orthaber G, Schuster W (1998) Die Vorausberechnung des Brennverlaufs von Dieselmotoren mit direkter Einspritzung auf der Basis des Einspritzverlaufs. Motortechnische Zeitschrift MTZ 59(8):484–492

Chmela F, Dimitrov D, Pirker G, Wimmer A (2006) Konsistente Methodik zur Vorausrechnung der Verbrennung in Kolbenkraftmaschinen. Motortechnische Zeitschrift MTZ 67:468–474

Damkühler G (1940) Der Einfluß der Turbulenz auf die Flammengeschwindigkeit in Gasgemischen Z Elektrochem 46:601–652

deNeef AT (1987) Untersuchung der Voreinspritzung am schnellaufenden direkteinspritzenden Dieselmotor. Dissertation, ETH Zürich

Dent JC (1971) Basis for the comparison of various experimental methods for studying spray penetration. SAE Paper 710571

Elmqvist C, Lindström F, Angström A, Grandin B, Kalghatgi G (2003) Optimizing engine concepts by using a simple model for knock prediction. SAE Technical Paper 2003-01-3123

Franzke D (1981) Beitrag zur Ermittlung eines Klopfkriteriums der ottomotorischen Verbrennung und zur Vorausberechnung der Klopfgrenze, Dissertation, TU München

Gülder ÖL (1982) Laminar burning velocities of methanol, ethanol and isooctane-air mixtures. 19th Symposium (International) on Combustion, The Combustion Institute

Halstead MP, Kirsch LJ, Prothero A, Quinn CP (1975) A mathematical model for hydrocarbon autoignition at high pressures. Proceedings of the Royal Society, A 346, pp 515–538, London

Heywood JB (1988) Internal combustion engine fundamentals. McGraw-Hill, New York

Hiroyasu H, Kadota T, Arai M (1983a) Development and use of a spray combustion modeling to predict diesel engine efficiency and pollutant emission. Part 1: Combustion Modeling. Bull JSME 26:569–575

Hiroyasu H, Kadota T, Arai M (1983b) Development and use of a spray combustion modeling to predict diesel engine efficiency and pollutant emission. Part 2: Computational procedure and parametric study. Bull JSME 26:576–583

Hohlbaum B (1992) Beitrag zur rechnerischen Untersuchung der Stickstoffoxid-Bildung schnellaufender Hochleistungsdieselmotoren, Dissertation, Universität Karlsruhe

Keck JC (1982) Turbulent flame structure and speed in spark-ignition engines. Proceedings of 19th symposium (international) on combustion. The Combustion Institute, Pittsburgh, PA, pp 1451–1466

Koch T (2002) Numerischer Beitrag zur Charakterisierung und Vorausberechnung der Gemischbildung und Verbrennung in einem direkteingespritzten, strahlgeführten Ottomotor, Dissertation, Eidgenössische Technische Hochschule Zürich

Li H, Miller DL, Cernansky NP (1996) Development of a reduced chemical kinetic model for prediction of preignition reactivity and autoignition of primary reference fuels. SAE Technical Paper 960498

Livengood JC, Wu PC (1955) Correlation of autoignition phenomenon in internal combustion engines and rapid compression machines. Fifth symposium (international) on combustion, pp 347–356

Metghalchi M, Keck JC (1982) Burning velocities of mixtures of air with methanol, isooctane and indolene at high pressure and temperature. Combust Flame 48:191–210

Müller UC, Bolling M, Peters N (1997) Approximations for burning velocities and Markstein numbers foe lean hydrocarbon and methanol flames. Combust Flame 108:349–356

Scheele M (1999) Potentialabschätzung zur Verbesserung des indizierten Wirkungsgrades kleinvolumiger Ottomotoren. Dissertation, Universität Hannover

Spicher U, Worret R (2002) Entwicklung eines Klopfkriteriums zur Vorausberechnung der Klopfgrenze, FVV Abschlussbericht, Heft-Nr. 471

Stegemann J, Seebode J, Baumgarten C, Merker GP (2002) Influence of throttle effects at the needle seat on the spray characteristics of a multihole injection nozzle. Proceedings of 18th ILASS-Europe Conference, Zaragoza, Spain, pp 31–36

Stiesch G (1999) Phänomenologisches Multizonen-Modell der Verbrennung und Schadstoffbildung im Dieselmotor. Dissertation. Universität Hannover

Stiesch G (2003) Modeling engine spray and combustion processes. Springer, Berlin

Tabaczynski RJ (1980) Further refinement and validation of a turbulent flame propagation model for spark ignition engines. Combust Flame 39:111–121

Thoma M, Stiesch G, Merker GP (2002) Phänomenologisches Gemischbildungs- und Verbrennungsmodell zur Berechnung von Dieselmotoren mit Voreinspritzung. 5. Int. Symp. für Verbrennungsdiagnostik, Baden-Baden, pp 91–101

Wallesten J (2003) Modelling of flame propagation in spark ignition engines. Dissertation, Chalmers University of Technology, Göteborg

Weisser G, Boulouchos K (1995) NOEMI – Ein Werkzeug zur Vorabschätzung der Stickoxidemissionen direkteinspritzender Dieselmotoren. 5. Tagung Der Arbeitsprozeß des Verbrennungsmotors, Technische Universität Graz

Wirth M (1993) Die turbulente Flammenausbreitung im Ottomotor und ihre charakteristischen Längenskalen, Dissertation, RWTH Aachen

Chapter 12
Three-Dimensional Flow Fields

Frank Otto and Christian Krüger

Fluid mechanical or CFD simulation (CFD: computational fluid dynamics) is playing an increasingly important role in the simulation of engine processes, as it makes possible the most detailed physical description of the relevant processes. The most diverse processes in the engine field are considered, like charge changing, in-cylinder flow, exhaust gas recirculation, outflow processes, secondary reactions in the exhaust tract, catalyst converter flow, turbine and compressor flow in the turbocharger, internal nozzle flow, or coolant flow.

This technique is still relatively time-consuming process: firstly, computational meshes must be generated, and after the definition of (often extensive) initial and boundary conditions, the actual calculation can finally be started. Evaluation is also typically costly due to the large amount of data. Finally, the theoretical requirements are also quite considerable. We are therefore devoting an entire chapter to the topic of "CFD".

In this chapter, the foundations of numerical fluid mechanics are first explicated. Calculations of internal engine processes like injection and combustion demand a detailed treatment, which we will explore in Chaps. 13 and 14.

12.1 Basic Fluid Mechanical Equations

12.1.1 Mass and Momentum Transport

In the following, the basic equations of fluid mechanics are briefly restated, whereby the components will be expressed in sum convention, i.e. summation over twice-appearing indexes has to be performed. For an exhaustive derivation, see Merker and Baumgarten (2000), Cebeci (2002) and White (1991). The local equation of mass conservation, called the *continuity equation*, reads

$$\frac{\partial}{\partial t}\rho(x,t) + \frac{\partial}{\partial x_i}(\rho(x,t)v_i(x,t)) = 0. \tag{12.1}$$

G.P. Merker et al. (eds.), *Combustion Engines Development*,
DOI 10.1007/978-3-642-14094-5_12, © Springer-Verlag Berlin Heidelberg 2012

The dependency of the field quantities on x or t is usually simply left out. The momentum equation (*Navier–Stokes equation*) then reads

$$\rho\left(\frac{\partial}{\partial t}+v_j\frac{\partial}{\partial x_j}\right)v_i-\frac{\partial}{\partial x_j}\left(\tau_{ij}\left[\frac{\partial v_k}{\partial x_l}\right]\right)=-\frac{\partial p}{\partial x_i}+f_i, \tag{12.2}$$

Whereby

$$\tau_{ij}=\mu\left(\frac{\partial v_i}{\partial x_j}+\frac{\partial v_j}{\partial x_i}\right)+\xi\frac{\partial v_k}{\partial x_k}\delta_{ij} \tag{12.3}$$

μ,ξ: 1st and 2nd viscosity coefficient designates the *stress tensor* and f_i the external force density (e.g. gravity). Often,

$$\xi=-\frac{2}{3}\mu \tag{12.4}$$

is assumed, i.e. τ_{ij} is traceless.

So far the general compressible case was considered, i.e. ρ is variable, a function of location and time. For the case of incompressible flow (typically for liquids), i.e. ρ is constant, the continuity and Navier–Stokes equations are simplified drastically, and we obtain for the incompressible fluid

$$\frac{\partial v_i}{\partial x_i}=0 \tag{12.5}$$

$$\rho\left(\frac{\partial}{\partial t}+v_j\frac{\partial}{\partial x_j}\right)v_i-\mu\Delta v_i=-\frac{\partial p}{\partial x_i}+f_i \tag{12.6}$$

Whereby

$$\Delta=\frac{\partial^2}{\partial x^2}+\frac{\partial^2}{\partial y^2}+\frac{\partial^2}{\partial z^2}$$

designates the *Laplace operator*. Differentiation of the second term in the continuity equation (12.1) according to the product rule provides

$$\left(\frac{\partial}{\partial t}+v_i\frac{\partial}{\partial x_i}\right)\rho+\rho\frac{\partial}{\partial x_i}v_i=0. \tag{12.7}$$

The operator

$$\frac{\partial}{\partial t}+v_i\frac{\partial}{\partial x_i}$$

in (12.7) is called the *convective* or *substantive derivative*. It also appears in the Navier–Stokes equation (12.2) and expresses the temporal change of local fluid quantities in the external, spatially fixed laboratory coordinates. The transition from a local coordinate system carried along with the fluid (*Lagrangian coordinates*) to a global, spatially fixed coordinate system (*Euler coordinates*) thus corresponds to the substitution

$$\frac{\partial}{\partial t_{Lagrange}} \rightarrow \left(\frac{\partial}{\partial t} + v_i \frac{\partial}{\partial x_i}\right)_{Euler}. \qquad (12.8)$$

Assuming Newton's second law:

$$m\frac{\mathrm{d}v_i}{\mathrm{d}t} = F_i$$

is valid in the local coordinate system, carried along with the fluid, whereby force F is composed of an external component and the pressure gradient in the flow, from this follows then *Euler's equation*

$$\rho\left(\frac{\partial}{\partial t} + v_j \frac{\partial}{\partial x_j}\right) v_i = -\frac{\partial p}{\partial x_i} + f_i. \qquad (12.9)$$

This is valid for an ideal, frictionless fluid and distinguishes itself from the Navier–Stokes equation (12.2) by the term of viscosity. That is, in a real flow, an additional force, the force of friction, is in effect, for which commonly the Newtonian formulation

$$f_{i,Reibung} = \frac{\partial}{\partial x_j}\left(\tau_{ij}\left[\frac{\partial v_k}{\partial x_l}\right]\right) \qquad (12.10)$$

is used. This additional term also means that the Navier–Stokes equation is now a 2nd order differential equation in the space coordinates and therefore requires additional boundary conditions. The physical equivalent is the phenomenon of wall friction, because of which a Navier–Stokes fluid rests directly on the wall relative to it and thus forms a *boundary layer*, while an Euler equation-based fluid flowing along a frictionless wall, with finite speed.

Partial differential equations are classified according to their properties. We distinguish between *elliptic*, *parabolic*, and *hyperbolic* equations. The incompressible Euler equation (12.9) is of the hyperbolic type with reference to the variables v.[1] For such a differential equation, *characteristics* exist, i.e. a set of

[1] It is, together with the incompressible continuity equation, elliptic in pressure!

curves, along which temporal development is given by means of a system of ordinary differential equations. Thus, the Euler equation is simplified along a spatial curve $\chi(t)$ with

$$\frac{\mathrm{d}\chi_i}{\mathrm{d}t} = v_i \tag{12.11}$$

to

$$\rho \frac{\mathrm{d}v_i}{\mathrm{d}t} = -\frac{\partial p}{\partial x_i} + f_i, \tag{12.12}$$

i.e. one can imagine, that the solution field "propagates" along the set of curves determined by the characteristics (12.11) and (12.12).[2] Thus, in order to define the boundary conditions, only the initial values $v(t = t_{0,} x_i = x_{0,i})$ and $v(t = t_{0,} x_i = x_{0,i})$ at starting time t_0 must be given for various starting positions $x_{0,i}$. One typical elliptic differential equation is *Poisson equation*

$$\Delta\varphi = \left(\frac{\partial^2}{\partial x^2} + \frac{\partial^2}{\partial y^2} + \frac{\partial^2}{\partial z^2}\right)\varphi = 4\pi\gamma. \tag{12.13}$$

It can be shown, that the solution $\phi(x, y, z)$ depends on all ϕ-values of the boundary of this area; there are neither preferable propagation directions nor characteristics.

The *heat conduction equation* or *Helmholtz equation* (for the description of heat conduction in a solid)

$$\rho\, c_V \frac{\partial}{\partial t} T - \lambda\, \Delta T = 0 \tag{12.14}$$

as well as the Navier–Stokes equation are parabolic equations. They possess a definite propagation direction in time, but not in space. One typical initial condition consists therefore in the specification of all values for the boundary of a certain area at starting time t_0.

The above suggests that these differential equation properties are also decisive for the numerical solution. With reference to the Navier–Stokes equation, we should especially keep in mind that, by adding the viscosity term, an alteration of the type of differential equation is caused; the incompressible Euler equation is hyperbolic, while the Navier–Stokes equation is parabolic with corresponding consequences for the solution method. The propagation behavior of the compressible

[2]This is valid under the assumption of an external pressure and force.

equations is once again different, since now the phenomenon of sound appears as well. Typical engine 3D-flows like cylinder flows are necessarily compressible, but this compressibility is so weak (a measure for this is the Mach number $a = v/c$), that in fact the properties of the incompressible equations are valid.

12.1.2 Transport of Internal Energy and Species

The equation set must still be completed. In the incompressible case, the continuity equation (12.5) and the Navier–Stokes equation (12.6) (a vector equation, i.e. three component equations), four equations in total, are already complete to determine the four unknowns velocities (vectors) and pressure. In the compressible case however, the density must also be determined. For a single-component or a homogeneously mixed gas (which can again be treated in a single-component fashion) we obtain the density from the pressure by means of the thermal equation of state

$$p = \frac{\rho \tilde{R} T}{M} \tag{12.15}$$

This equation now contains the temperature in addition, which by means of the caloric equation of state

$$u = \int_{T_0}^{T} c_V(\vartheta) d\vartheta + u_0 \tag{12.16}$$

is linked with the (specific) internal energy. The internal energy is an extensive quantity, for which a transport equation can be formulated (similarly as for the momentum)

$$\rho \left(\frac{\partial}{\partial t} + v_j \frac{\partial}{\partial x_j} \right) u - \frac{\partial}{\partial x_i} \left(\lambda \frac{\partial T}{\partial x_i} \right) = -p \frac{\partial v_i}{\partial x_i} + \tau_{ij} \frac{\partial v_i}{\partial x_j} + q, \tag{12.17}$$

see Merker and Eiglmeier (1999). The second term on the left side is a diffusion term and corresponds to the viscosity term in the Navier–Stokes equation (λ). The first two terms on the right side represent energy sources and sinks, the first term $-p\partial v_i/\partial x_i$ can assume both signs and corresponds to the reversible mechanical compression work, which is performed on the volume element. The second term $\tau_{ij}\partial v_i/\partial x_j$ describes the heat released by internal friction; this term is always positive (2nd law of thermodynamics). The third term q describes further heat sources, be it via evaporation or via combustion.

With this, our equation system is also complete for the incompressible case – for the seven unknowns, velocity (3), pressure, density, temperature, and internal energy, we have seven equations, (12.1), (12.2), (12.15), (12.16), and (12.17). For the case that the fluid is an inhomogeneous mixture of several components, further equations are required for the determination of the material composition, see below.

We should point out that the energy equation required here is the equation for internal energy and by no means the equation for kinetic energy. The latter results from the Navier–Stokes equation and is therefore not an independent quantity. Of course, instead of the internal energy, the total energy (internal energy + kinetic energy) can be transported, or the thermal enthalpy($w = u + p/\rho$). This is all equivalent, because when one knows the quantities v, p, ρ, all forms of energy can be converted into each other. It is very popular, especially amongst chemists, to transport the total enthalpy (thermal enthalpy + chemical energy). Assuming a known material composition, this is also convertible into the other energy quantities, making the process equivalent.

Finally, we should also consider the case that the fluid under investigation is given as an inhomogeneous mixture of several species. In this case, transport equations for the concentrations $c_{(k)}$ of the particular species must be formulated,

$$\rho\left(\frac{\partial}{\partial t} + v_i \frac{\partial}{\partial x_i}\right) c_{(k)} - \frac{\partial}{\partial x_i}\left(D_{(k)} \rho \frac{\partial}{\partial x_i} c_{(k)}\right) = Q_{(k)}(c_{(j)}, p, T) \tag{12.18}$$

with

$$\sum c_{(k)} = 1,$$

which are constructed analogous to the Navier–Stokes and energy transport equation. On the left side are a convective derivative and the diffusion term and on the right a source term, which is only other than zero when chemical reactions are occurring, which is especially the case in combustion. Also, an additional diffusion term appears in the energy equation (12.17), so that it now reads

$$\rho\left(\frac{\partial}{\partial t} + u_j \frac{\partial}{\partial x_j}\right) u - \frac{\partial}{\partial x_i}\left(\lambda \frac{\partial T}{\partial x_i} + \rho D_{(k)} \sum_{(k)} h_{(k)} \frac{\partial c_{(k)}}{\partial x_i}\right) = -p \frac{\partial v_i}{\partial x_i} + \tau_{ij} \frac{\partial v_i}{\partial x_j}. \tag{12.19}$$

All current CFD codes use the equations of state for ideal gases. Yet this assumption is not particularly suitable for diesel engine conditions (peak pressure of more than 200 bar); in this case, one should rather make use the equations of state for real gases. An implementation is currently unavailable commercially.

12.1.3 Passive Scalars and the Mixture Fraction

Frequently, transport equations of other, formal scalars are defined, like progress variables or the flame surface density. These transport equations follow essentially the pattern of (12.18) (i.e. convection, diffusion, source term). Those quantities however make no contribution to the thermodynamic equations (12.15) and (12.16) and are therefore called *passive scalars* as opposed to the genuine chemical species called *active scalars*.

One important scalar is the *mixture fraction* Z. The mixture fraction field describes the local mixture states of two gases, whereby each of these gasses may be given as a homogeneous mixture of different species; it assumes values between 0 and 1. As long as there is no reaction, one can express it as

$$Z = \frac{\rho_{Gas\,I}}{\rho_{Gas\,I} + \rho_{Gas\,II}}. \tag{12.20}$$

We now establish the fact that Z is a linear function in an arbitrary element mass fraction. The element mass fraction c_X of the element X (e.g. C, O, or H) is defined in this case as

$$c_X = \frac{\rho_X}{\rho_{gesamt}}. \tag{12.21}$$

Where $c_{X,I}$ is the X-mass fraction in gas I, correspondingly $c_{X,II}$ the X-mass fraction in gas II and c_X the X-mass fraction in the local I–II mixture state. We then find the following dependence:

$$Z(c_X) = \frac{c_X - c_{X,II}}{c_{X,I} - c_{X,II}}. \tag{12.22}$$

The basic idea is to use this relation for the definition of the mixture fraction, because this definition is not influenced by chemical reactions, as it is defined on an elemental basis. In this way, we obtain a quantity suitable for describing mixtures independently of reactions (combustion). The mixture fraction is thus an essential concept for the sake of describing diffusion flames. The transport equation of the mixture fraction corresponds to that of a species

$$\rho\left(\frac{\partial}{\partial t} + v_i \frac{\partial}{\partial x_i}\right)Z - \frac{\partial}{\partial x_i}\left(D\rho \frac{\partial}{\partial x_i} Z\right) = 0, \tag{12.23}$$

except that no chemical source term appears.[3] For the diffusion constant D, a "mean" value of the diffusion constant of the species involved must be employed.[4]

[3]Evaporation source terms may however appear.

[4]In the case of turbulent flow, this problem of varying laminar diffusion constant is reduced.

Since the mixture fraction can be used to calculate the mixture composition, it can function as an active scalar.

12.1.4 Conservative Formulation of the Transport Equations

Finally, it shall be mentioned that transport equations (for energy, momentum, and scalars) can also, with the help of the continuity equation, be represented in the *conservative formulation*. For the scalar transport, this reads

$$\frac{\partial}{\partial t}\left(\rho c_{(k)}\right)+\frac{\partial}{\partial x_i}\left(\rho v_i c_{(k)}\right)-\frac{\partial}{\partial x_i}\left(D_{(k)}\rho\frac{\partial}{\partial x_i}c_{(k)}\right)=Q_{(k)}(c_{(j)},p,T), \qquad (12.24)$$

i.e. the convection term (second term on the left) is now also formulated as a divergence of a flow. These formulations are especially relevant for numerical treatment.

12.2 Turbulence and Turbulence Models

12.2.1 The Phenomenology of Turbulence

In the Navier–Stokes equation, the relative order of magnitude of the viscous term has a large influence on the character of the flow. In order to understand this, we will consider a typical flow problem like the flow over of a cylinder[5] (for reasons of simplification we will consider the incompressible case) and introduce characteristic scales for length L (e.g. the cylinder diameter) and velocity v (e.g. in-flow velocity). With this, we obtain in reference to these scales the standardized variables

$$x=x^*L, \quad v=v^*V, \quad t=t^*L/V, \quad p=p^*\rho V^2. \qquad (12.25)$$

With the standardized variables x^*, v^*, t^* and p^*, the problem can be formulated in a scale-invariant manner. We finally obtain the equations

$$\frac{\partial v_i^*}{\partial x_i^*}=0$$

and

$$\left(\frac{\partial}{\partial t^*}+v_j^*\frac{\partial}{\partial x_j^*}\right)v_i^*-\frac{1}{\mathrm{Re}}\frac{\partial^2 v_i^*}{\partial x_i^2}=-\frac{\partial p^*}{\partial x_i^*}, \qquad (12.27)$$

[5]The so-called Kármán vortex street.

Whereby

$$\mathrm{Re} = \frac{\rho V L}{\mu} \tag{12.28}$$

In the Reynolds number, all scale influences are now included. Flows with the same Reynolds number can be formed into each other via rescaling of variables; we say that they are similar to each other. Thus, the Reynolds number classifies flows. In addition, it describes the relative size of the viscous term. If the Reynolds number is small and the viscous term large, then we have the case of a "honey-like", "viscous" flow. In the other limit case, i.e. in the boundary value of an infinitely large Reynolds number, one could at first suppose that the viscous term simply vanishes and that the Navier–Stokes equation would be reduced to the Euler equation. This is however not the case, for, as opposed to the Euler equation, a (indeed increasingly thin) viscous wall boundary layer (property of the Navier–Stokes equation, see above) exists as before, in which the velocity is reduced to zero (relative to the wall velocity). In this way, high velocity gradients exist close to the wall, which lead to vortex formation in the wake of the cylinder.

The flow over of our cylinder resembles for $\mathrm{Re} = 10^{-2}$ approximately as represented in Fig. 12.1a – a laminar, viscous flow. With an increasing Reynolds number, more vortices forms we find behind the cylinder, which also detach, but first still show periodical structures (Fig. 12.1b, c). With further increase of the Reynolds number, the flow finally fluctuates chaotically and three-dimensionally – it is now *turbulent* (Fig. 12.1d, e). Large eddies disintegrate into smaller ones, these in turn into still smaller ones, a eddy spectrum forms down to a very small length scale, the Kolmogorov scale, on which the flow becomes viscous (laminar) again. Such a chaotic process can no longer be calculated deterministically, even with an arbitrarily large computer, because the minutest influence may have the largest effects.

12.2.2 Modeling Turbulence

However, statistical quantities can be provided for the description of turbulent phenomena; the spectrum of turbulent fluctuations is also accessible to theoretical considerations. Typical quantities are the *turbulent length scale* l_t, the *turbulent time scale* τ_t, the *turbulent velocity scale* v_t, as well as the *turbulent viscosity* μ_t. Assuming spatially homogeneous turbulence, there are two independent quantities, most commonly the (specific) *turbulent kinetic energy* k and the *turbulent dissipation* ε are used. Thus we obtain

$$v_t = \sqrt{k}, \quad \mu_t = c_\mu \rho \frac{k^2}{\varepsilon}, \quad l_t = c_l \frac{k^{\frac{3}{2}}}{\varepsilon}, \quad \tau_t = \frac{l_t}{v_t}, \tag{12.29}$$

whereby c_μ and c_l represent proportionality constants. The mathematical method in introducing these turbulence quantities consists in an ensemble averaging of fluid

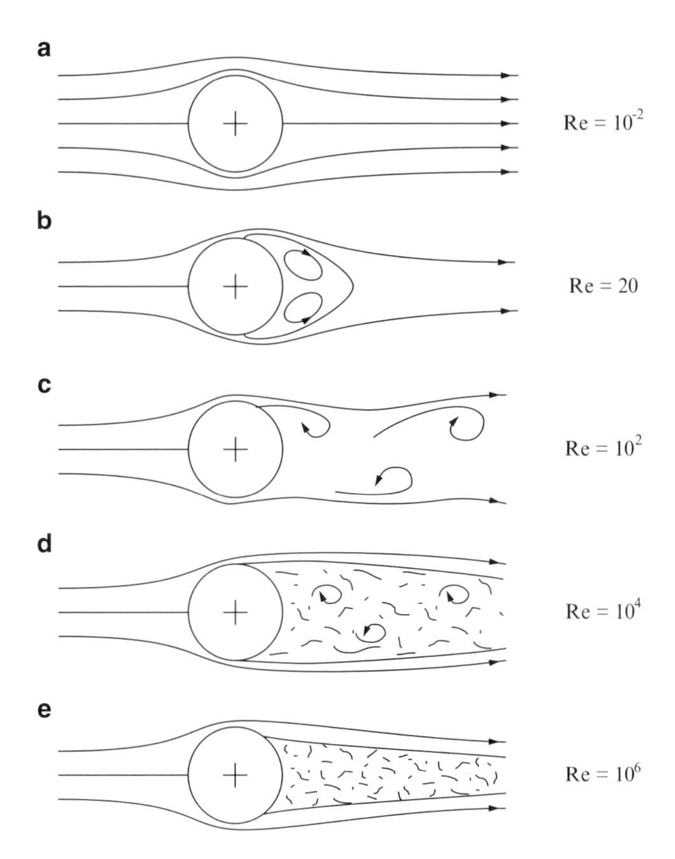

Fig. 12.1 Kármán vortex street (flow over a cylinder) at varying Reynolds' numbers

mechanical equations, i.e. not one single realization is calculated, but the mean value from all possible realizations under identical measurement-technical boundary conditions.[6] This Reynolds averaging will be carried out for the case of a source-free scalar transport equation (12.24).

If $\langle v_i \rangle$ and $\langle c \rangle$ are chosen as the ensemble-averaged quantities, we then obtain for the boundary instantaneous values v_i and c the following reductions

$$v_i = \langle v_i \rangle + v_i' \quad \text{mit} \quad \langle v_i' \rangle = 0 \quad c = \langle c \rangle + c' \quad \text{mit} \quad \langle c' \rangle = 0, \tag{12.30}$$

whereby v_i' and c' now describe the turbulent velocity and scalar fluctuations. The (specific) turbulent kinetic energy introduced above thus amounts to

[6]For engines it is often said that this corresponds to an averaging over several cycles. One should however be wary of such a "descriptive" interpretation. Engine cycle fluctuations are to a large extent caused by for fluctuations of the boundary conditions, e.g. of the injection spray or the mixture motion or preparation, or residual gas content of the previous cycle.

$$k = \frac{1}{2} \sum_i (v_i')^2. \tag{12.31}$$

For all terms linear in v and c the averaging of (12.24) can be simply carried out, but the nonlinear convection term leads to an additional term

$$\frac{\partial}{\partial t}(\rho\langle c\rangle) + \frac{\partial}{\partial x_j}(\rho\langle v_j\rangle\langle c\rangle) + \frac{\partial}{\partial x_j}(\rho\langle v'_j c'\rangle) - \frac{\partial}{\partial x_j}\left(D_c\rho\frac{\partial}{\partial x_j}\langle c\rangle\right) = 0. \tag{12.32}$$

If we now assume that the turbulence behaves like a spatially isotropic fluctuation process uncorrelated at different times ("white noise"), we can show that this additional term takes on the structure of an additional diffusion term with the diffusion constant

$$D_t = \text{const.} \cdot k^2/\varepsilon = \mu_t/(\rho\, Sc_t)$$

This implies the *eddy diffusivity approach*

$$\langle v_i' c' \rangle = -\frac{\mu_t}{\rho Sc_t}\frac{\partial\langle c\rangle}{\partial x_i}. \tag{12.33}$$

Sc_t designates the turbulent Schmidt number, i.e. the ratio of turbulent viscosity and turbulent diffusion. A similar term, $\langle v'_i v'_j\rangle$, appears in the averaged Navier–Stokes equation. It takes over the role of an additional tension stress, the *Reynolds stress* $\tau_{R,ij}$. In analogy to the eddy diffusivity approach, for this the *eddy viscosity approach* is used (we are working as before on an incompressible basis)

$$\left\langle v_i' v_j' \right\rangle = -\frac{\mu_t}{\rho} \cdot \left(\frac{\partial\langle v_j\rangle}{\partial x_i} + \frac{\partial\langle v_i\rangle}{\partial x_j}\right) + \frac{2}{3}\delta_{ij}\, k. \tag{12.34}$$

This means that in addition to the original, laminar diffusion/viscosity in the averaged equations, a usually much larger turbulent diffusion/viscosity appears ($D_t \gg D_c$). One can now easily understand the essential effect of turbulent averaging; the viscous, diffusive character of the differential equations is now more in the foreground and produces a *well posed* problem (alterations in the solution of a given order of magnitude demand a sufficiently large alteration of the initial conditions, there is no longer any chaos). Frequently, because of the dominant turbulent quantities, laminar viscosity and diffusion constants are simply ignored (the same is valid of course for all other transported quantities).

For the compressible case however, we only obtain formally a very similar result, if we carry out the averaging *density-weighed*, i.e. using the *Favre averaging*

$$\langle\Phi\rangle_F = \frac{\langle\rho\,\Phi\rangle}{\langle\rho\rangle}\Phi = \langle\Phi\rangle_F + \Phi'', \tag{12.35}$$

which will be utilized in the following.

For the calculation of k and of ε, we still need additional equations. From the Navier–Stokes equation, a transport equation for the turbulent kinetic energy, the k-equation, can be derived. For this, we only need to assume that pressure correlations are being neglected. The k equation (in the Favre average) reads

$$\langle\rho\rangle\frac{\partial}{\partial t}k+\langle\rho\rangle\langle v_i\rangle_F\frac{\partial}{\partial x_j}k-\frac{\partial}{\partial x_i}\left(\frac{\mu_t}{\mathrm{Pr}_k}\frac{\partial k}{\partial x_i}\right)=\tau_{R,ij}\langle S_{ij}\rangle_F-\langle\rho\rangle\varepsilon-\frac{2}{3}\langle\rho\rangle k\,\nabla\cdot\langle\vec{v}\rangle_F \tag{12.36}$$

with the compressible Reynolds tension stress $\tau_{R,ij}$

$$\tau_{R,ij}=\mu_t\left(\frac{\partial\langle v_j\rangle_F}{\partial x_i}+\frac{\partial\langle v_i\rangle_F}{\partial x_j}-\frac{2}{3}\delta_{ij}\left(\nabla\cdot\langle\vec{v}\rangle_F\right)\right) \tag{12.37}$$

and the shear tensor S_{ij}

$$S_{ij}=\frac{1}{2}\left(\frac{\partial v_j}{\partial x_i}+\frac{\partial v_i}{\partial x_j}\right). \tag{12.38}$$

The fact that the turbulent diffusion constant of the k transport equation can be quite different from μ_t is taken into account by the introduction of a proportionality constant, the turbulent Prandtl number for k and Pr_k. The turbulent dissipation ε amounts to

$$\varepsilon=\nu\left\langle\frac{\partial v''_i}{\partial x_j}\left(\frac{\partial v''_i}{\partial x_j}+\frac{\partial v''_j}{\partial x_i}\right)\right\rangle_F. \tag{12.39}$$

A transport equation for ε can in principle also be derived from the Navier–Stokes equation, however it must be modeled in several terms in accordance with the k-equation. Its usual form reads

$$\begin{aligned}\langle\rho\rangle\frac{\partial}{\partial t}\varepsilon+\langle\rho\rangle\langle v_i\rangle_F\frac{\partial}{\partial x_j}\varepsilon-\frac{\partial}{\partial x_i}\left(\frac{\mu_t}{\mathrm{Pr}_\varepsilon}\frac{\partial\varepsilon}{\partial x_i}\right)&=c_{\varepsilon,1}\frac{\varepsilon}{k}\tau_{R,ij}\langle S_{ij}\rangle_F\\&-c_{\varepsilon,2}\langle\rho\rangle\frac{\varepsilon^2}{k}-\left(\frac{2}{3}c_{\varepsilon,1}-c_{\varepsilon,3}\right)\langle\rho\rangle\varepsilon\,\nabla\cdot\langle\vec{v}\rangle_F.\end{aligned} \tag{12.40}$$

According to the standard, the following coefficient set is used in the k–ε-model.

c_μ	$c_{\varepsilon,1}$	$c_{\varepsilon,2}$	$c_{\varepsilon,3}$	Pr_k	Pr_ε
0.09	1.44	1.92	–0.33	1.0	1.3

Finally, we still require a turbulent transport equation for the internal energy. This differs from the laminar equation only by turbulent transport coefficients and ε as an additional source term.

With that, a closed equation system has been derived. With the help of k and ε, the diffusion constants for the additional turbulent diffusion terms can be calculated in the transport equations. Since we will actually always be looking at Favre averaged equations and quantities in the following, the symbol for Favre averaging $\langle\rangle_F$ will usually be left out.

12.2.3 The Turbulent Law of the Wall

However, there is still a problem concerning the boundary conditions. On the walls, a boundary layer is formed, in which the velocity decreases because of friction to zero, i.e. the flow becomes laminar. Thus in a turbulent flow the boundary layer consists typically of a laminar sub-layer with a turbulent core. The k–ε-model cannot be applied across the entire boundary layer. Moreover, the boundary layers are often so thin (especially in engines) that they are hardly numerically solvable anyway.

The usual way to overcome this problem is by deriving the law of the wall with the help of the boundary layer equations, because the shear stress is constant over the boundary layer, i.e. it describes a tangential momentum flow constant over the wall distance and flowing off onto the wall (this is the definition of the boundary layer). We now need a turbulent law of the wall, i.e. an analytical boundary layer model, in order to calculate shear stress from the local velocities in the cells closest to the wall.

At a given wall shear stress τ_W, only one velocity scale is available in the turbulent boundary layer, the shear stress speed v_τ, with $\tau_W = \rho {v_\tau}^2$. It should be proportional to the turbulent velocity scale, while the turbulent length scale should be proportional to the wall distance y. v_W would designate the speed component parallel to the wall. Under these assumptions, we obtain for the turbulent viscosity

$$\mu_t = \kappa \rho y v_\tau,$$

whereby κ designates a proportionality constant, the *von Kármán constant*. For the shear stress τ_W we now obtain the following relationship

$$\tau_W = \rho v_\tau^2 = \rho \kappa y v_\tau \frac{\partial v_W}{\partial y}, \tag{12.41}$$

from which follows under the assumption of constant density the *logarithmic law of the wall*

$$v_W = \frac{v_\tau}{\kappa} \ln y + C \tag{12.42}$$

with an integration constant C. In the standardized coordinates

$$v^+ = \frac{v_W}{v_\tau} \quad \text{and} \quad y^+ = \frac{\rho v_\tau y}{\mu} \tag{12.43}$$

the logarithmic law of the wall takes on its universal form

$$v^+ = \frac{1}{\kappa} \ln y^+ + \tilde{C} \tag{12.44}$$

with

$$\kappa = 0.4 \quad \text{and} \quad \tilde{C} = 5.5.$$

In this form, a universal range of validity of the law of the wall can now also be given, namely

$$20 < y^+ < 150. \tag{12.45}$$

For the numerical solution, the node nearest to the wall must remain within this boundary layer. The law of the wall is employed in CFD calculation insofar as v_τ and consequently τ_W are calculated from v^+ and y^+ in the cell closest to the wall. This $\tau\ w$ then supplies a momentum source term (negative) as a boundary condition for the Navier–Stokes equation.

A law of the wall can be derived quite analogously for the temperature as well. From the heat flux

$$q_w = \rho \kappa y v_\tau \frac{c_p}{\Pr} \frac{\partial T}{\partial y} \tag{12.46}$$

Follows

$$T - T_W = \frac{q_w \Pr}{\kappa c_p \rho v_t} (\ln y^+ + \text{const.}). \tag{12.47}$$

Finally, one still needs boundary conditions for the turbulence quantities k and ε. Because of the constancy of the shear stress τ_W and the velocity scale v_τ in the boundary layer, a $k_W = $ const. approach is plausible. Such that this can be valid however, production and dissipation of k must be in equilibrium in the boundary layer

$$\mu_t \left(\frac{\partial v_W}{\partial y} \right)^2 = \kappa \rho y\, v_\tau \left(\frac{v_\tau}{\kappa y} \right)^2 = \rho \varepsilon_W. \tag{12.48}$$

Moreover, for the viscosity [see (12.29)]

$$\mu_t = \kappa v_\tau = c_\mu \rho \frac{{k_W}^2}{\varepsilon_W} \tag{12.49}$$

has to be valid. From both of these equations we readily arrive at

$$k_W = \frac{v_\tau^{\,2}}{\sqrt{c_\mu}} \quad \text{and} \quad \varepsilon_W = \frac{v_\tau^{\,3}}{\kappa y}, \tag{12.50}$$

i.e. ε_W diverges towards the wall. For the validity of this relation however, diffusion and source terms of the ε-equation must also be in equilibrium

$$\frac{\partial}{\partial y}\left(\frac{\mu_t}{\mathrm{Pr}_\varepsilon}\frac{\partial \varepsilon_W}{\partial y}\right) + c_{\varepsilon,1}\frac{\varepsilon_W}{k_W}\mu_t\left(\frac{\partial v_W}{\partial y}\right)^2 - c_{\varepsilon,2}\,\rho\frac{\varepsilon_W^{\,2}}{k_W} = 0. \tag{12.51}$$

Insertion of the already obtained results provides a relation between the model constants

$$\kappa = \sqrt{\sqrt{c_\mu}\mathrm{Pr}_\varepsilon\left(c_{\varepsilon,2} - c_{\varepsilon,2}\right)}. \tag{12.52}$$

The derivation of these equations occurred under the assumption of stationary, flows parallel to the wall and constant density. From this it becomes clear that these laws of the wall can only have limited validity. In engines especially, the flows are transient, stagnation points appear (e.g. as the injection jets hit the piston or chamber walls), and the density of the boundary layer is not constant due to higher temperature gradients towards the wall. There are therefore methods for deriving law of the walls for variable density, e.g. the formulation of Han and Reitz (1995). In this case, in the equation for the heat flux (12.46), density ρ is considered as given by the ideal gas equation of state. The integration finally provides the relation

$$T\,\ln\left(\frac{T}{T_w}\right) = \frac{q_w\,\mathrm{Pr}}{\kappa c_p \rho v_t}\left(\ln y^+ + \text{const.}\right) \tag{12.53}$$

instead of (12.47). Of course, the influences of a variable density on k and ε distribution are not hereby considered. Nevertheless, the use of such a formulation deserves strong recommendation especially for combustion calculations.

Finally, it should not be left out that it would be possible to improve models to a significant degree by using methods based on a detailed resolution of the boundary layer, such as the v2f model (e.g. Manceau and Hanjalic 2000). This would be a logical direction for further research.

12.2.4 Modeling the Turbulent Mixture State

The mixture fraction was introduced in Sect. 12.1 as an important reference quantity in describing mixing processes. Its transport equation should also undergo a turbulent averaging process; the result corresponds to the scalar transport equation (12.32) incl. (12.33). However, it must be kept clear that a mean value of Z no longer contains the total information about the local mixture states. Thus, a mean value of $Z = 0.5$ can indicate a fluctuation-free, perfect mixture from equal parts of "flow 1" and "flow 2", but it can also indicate a static superimposing of two totally unmixed states, "only flow 1" and "only flow 2", whereby both states appear with equal probability. Another helpful quantity in describing the local mixture states is the Favre variance of the mixture fraction

$$\left\langle Z''^2 \right\rangle_F = \left\langle \left[Z - \langle Z \rangle_F \right]^2 \right\rangle_F = \langle Z^2 \rangle_F - \langle Z \rangle_F^{\,2}. \tag{12.54}$$

From the mixture fraction transport equation a transport equation for the variance can be derived

$$\begin{aligned} \langle \rho \rangle \frac{\partial}{\partial t} \left\langle Z''^2 \right\rangle_F + \langle \rho \rangle \langle v_j \rangle_F \frac{\partial}{\partial x_j} \left\langle Z''^2 \right\rangle_F - \frac{\partial}{\partial x_j} \left(\frac{\mu_t}{Sc_t} \frac{\partial}{\partial x_j} \left\langle Z''^2 \right\rangle_F \right) \\ = 2 \frac{\mu_t}{Sc_t} \left(\nabla \langle Z \rangle_F \right)^2 - \underbrace{2D \langle \rho \rangle \left\langle (\nabla Z'')^2 \right\rangle_F}_{\chi}. \end{aligned} \tag{12.55}$$

To model the last term χ of this equation, the scalar dissipation rate, the commonly used approach is

$$\chi = c_\chi \frac{\varepsilon}{k} \left\langle Z''^2 \right\rangle_F. \tag{12.56}$$

Based on its character, the term χ corresponds to the turbulent dissipation ε in the k-equation (k is also a variance, that one of velocity). We thus see that gradients of the Z mean value (inhomogeneities) incite Z variance formation, while the latter decays with the turbulent time scale if further production is suppressed. The constant c_χ is usually set to 2 (see Peters 2000).

One can now attempt to reconstruct, by means of structural knowledge of the distribution function, the information about the local mixture states from Z and Z''^2. Such a distribution function f is often named a *pdf* (probability density function). It assigns a rate of occurrence to each mixture state (Z values between 0 and 1). For this reason, it must be standardized to 1

$$\int_0^1 f(Z) \mathrm{d}Z = 1. \tag{12.57}$$

Because of the definition of the mean value and the variance

$$\langle Z \rangle = \int_0^1 Z f(Z) \mathrm{d}Z$$

and

$$\left\langle Z''^2 \right\rangle = \int_0^1 (Z - \langle Z \rangle)^2 f(Z) dZ = \int_0^1 \left(Z^2 - \langle Z \rangle^2 \right) f(Z) dZ$$
$$\leq \int_0^1 \left(Z - \langle Z \rangle^2 \right) f(Z) dZ \qquad (12.58)$$

we obtain the relation

$$\left\langle Z''^2 \right\rangle \leq \langle Z \rangle (1 - \langle Z \rangle). \qquad (12.59)$$

As an example, the statistical superposition of completely unmixed states ($Z = 0$ and $Z = 1$) is described by the linear combination of two Dirac distributions

$$f(Z) = a\,\delta(Z) + b\,\delta(Z - 1) \quad \text{with} \quad a + b = 1. \qquad (12.60)$$

In this case, the sign of equality is valid in (12.59).

In Fig. 12.2, two distribution functions are shown, on the one hand, for the case of minimal mixing, and on the other for a case of considerable mixing. All Z distribution functions should be placed more or less between these limiting cases. In the literature, it has become common to use the beta function[7] for this. It is a simple function with two free parameters, which can be related without difficulty to the mean value and the variance of the distribution.

It holds[8]:

$$f(Z) = \frac{\Gamma(a+b)}{\Gamma(a)\Gamma(b)} Z^{a-1} (1 - Z)^{b-1}, \int_0^1 f(Z) dZ = 1, a, b > 0, \qquad (12.61)$$

$$\langle Z \rangle = \frac{a}{a+b} \quad \text{and} \quad \left\langle Z'^2 \right\rangle = \frac{\langle Z \rangle (1 - \langle Z \rangle)}{1 + a + b}. \qquad (12.62)$$

For a, $b<1$, we obtain the "bath tub form" of Fig. 12.2c, for a, $b>1$ a form like that in Fig. 12.2b. With the parameters $\langle Z \rangle$ and $\left\langle Z'^2 \right\rangle$, which are obtained from

[7]In mathematical literature, the integral of this function as a function of the parameter a and b is called beta function:
$B(a,b) = \int_0^1 (1 - Z)^{a-1} Z^{b-1} dZ = \frac{\Gamma(a)\Gamma(b)}{\Gamma(a+b)}$

[8]The Gamma function is given as $\Gamma(x) = \int_0^\infty e^{-t} t^{x-1} dt$.

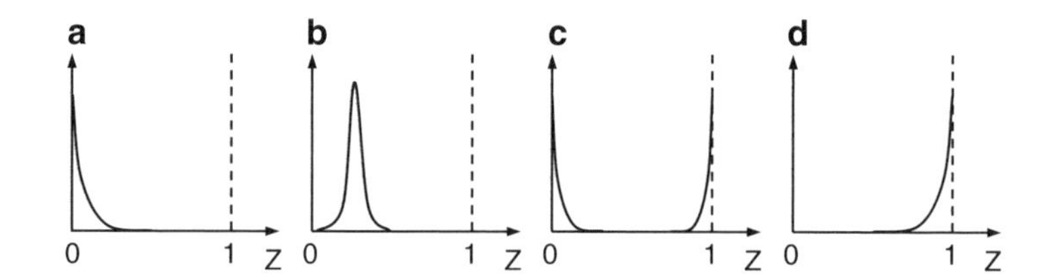

Fig. 12.2 Distribution functions for different mixture conditions: (**c**) describes a slightly mixed and (**b**) a heavily mixed state. (**a**) and (**d**) are the results of a very small/very large average value

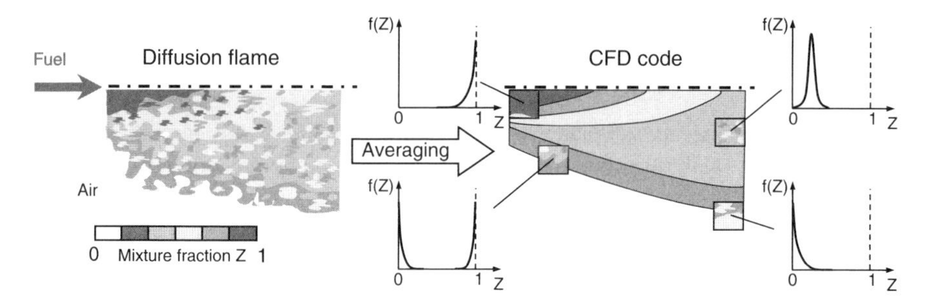

Fig. 12.3 Visual illustration of the effect of ensemble averaging in the case of a turbulent diffusion flame (details of the mixture vanish) and how the statistics of the mixture can be reconstructed by introducing the probability density function

the transport equations, the equation system is now closed, and a and b can be determined from it. Although the total integral over the beta function is normalized to 1, the individual factors of (12.61) are very large for large or very small values of a and b. In the calculation of f, it is therefore recommended to logarithmize (12.61) first:

$$\ln f = \ln \Gamma(a+b) - \ln \Gamma(a) - \ln \Gamma(b) + (a-1)\ln Z + (b-1)\ln(1-Z), \tag{12.63}$$

in order to obtain f by means of exponentiation of the logarithm.

Figure 12.3 provides a visual illustration of both the problem inherent in turbulent mixtures as well as a possible solution with the help of a turbulent diffusion flame. The indicated CFD mesh is not capable of resolving the turbulent mixture, and information is lost by averaging in the cells. Analogous to the Kármán vortex street in Fig. 12.1, the main point of interest is not the resolution of fine spatial and temporal structures but the ensemble-averaged statistics of the mixture. This is achieved by introducing the mixture probability density function.

12.2.5 The Validity of Turbulence Models: Alternative Approaches

The k–ε-turbulence model – a special case of the *Reynolds-averaged turbulence* models *(RANS)* – and the turbulent law of the wall thus far represent the current standard in the calculation of turbulent flow processes, and they have proven to be quite useful in most applications. This is of course also true for engine analysis. Nevertheless, one should always be aware of the fact that this is, in the final analysis, a rather simple model, which can never exactly reproduce the complete Navier–Stokes equations because of its simplified closure approaches, not even in the statistic average. This problem expresses itself in the non-universality of the model constants and the introduction of additional terms for the treatment of particular flow situations.

In certain cases, such modifications can also be of great relevance for engine calculations, for example the Pope correction (Pope 1978) for the calculation of free jets, a type of flow that occurs in direct injection. In many practical cases however, the influence of turbulence model modifications is rather minimal in comparison to the influence of the mesh resolution.

Another RANS model, which far transcends k–ε-models conceptually, is the *Reynolds stress model.* Here, the Reynolds stress $\langle v'_i v'_i \rangle$ is not attributed to the turbulent kinetic energy in accordance with (12.37)

$$k = \frac{1}{2} \sum_{i=1,2,3} \langle v'_i v'_i \rangle$$

but all six independent components of the Reynolds stress are described with their own transport equation. In this way, anisotropies in the turbulence structure, among other things, can be described (unfortunately, one usually works with only one single ε). However, such models demand intricate boundary conditions and a high calculation effort. For this reason, Reynolds stress models are just starting to be introduced to engine-related problems.

One completely different method is the *large-eddy simulation* (LES). In this case, turbulent flow structures larger than a pre-given scale are directly modeled, while smaller ones are modeled with "subgrid models". Such an approach should have a more general validity; and as opposed to RANS models, LES simulation converges to *direct numerical simulation* (DNS) when the mesh is improved (Fig. 12.4).

However, LES models make considerably higher demands on solver and mesh quality compared to RANS. This is easy to understand if one bears in mind the Reynolds numbers and viscosities involved. RANS codes resolve a flow with the effective Reynolds number Re_{eff}, which is about 1,000 times smaller than the actual Reynolds number of the flow.

$$\mathrm{Re}_{eff}^{RANS} = \frac{\rho V L}{\mu_{mol} + \mu_T^{RANS} + \mu_{num}} \tag{12.64}$$

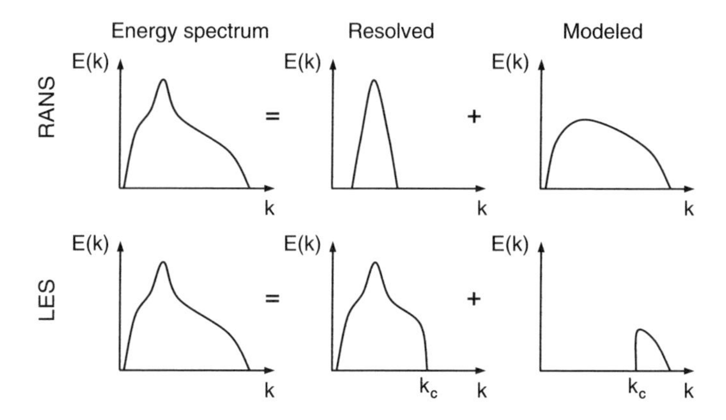

Fig. 12.4 While in Reynolds-averaged turbulence models like the k–ε-model, only a narrow range of the energy spectrum is resolved, LES resolves the entire spectrum until filter width k_c (Angelberger 2007)

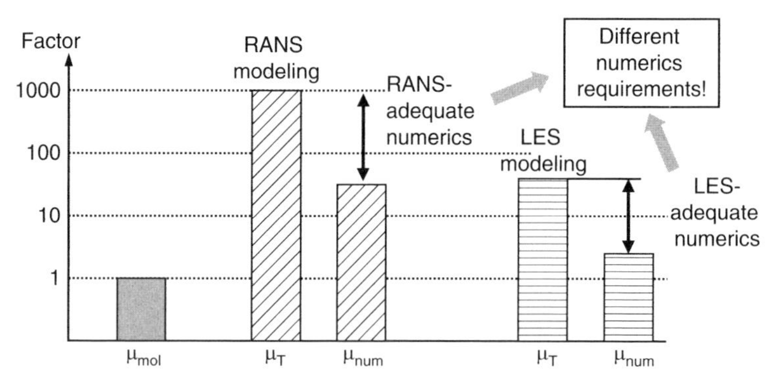

Fig. 12.5 Much higher requirements on the numerics are the result of the much lower turbulent viscosity in LES compared to RANS (see Angelberger 2007)

The reason for this is the much higher turbulent viscosity μ_T^{RANS} compared to μ_{mol}. This eases the requirements on RANS flow resolution considerably, the numeric viscosity μ_{num} of which need only be below μ_T^{RANS}. As discussed in the following sections, even this is not an easy task for engine flows.

Due to the different treatment of turbulence in LES models, edge lengths in the rage of 0.5 mm result for engine flows even on meshes used today.

$$\mu_T^{LES} \approx \left(10^{-1} \text{ bis } 10^{-2}\right)\mu_T^{RANS} \tag{12.65}$$

Now the numeric viscosity of the flow resolution carries much more weight and can affect the result decisively. The various orders of magnitude are depicted again in Fig. 12.5.

Requirements made on solvers are difficult to fulfill with present-day engine RANS codes. On the one hand, the model requires at least 2nd order. This is hardly possible with the currently used codes designed for stability and robustness, especially since these usually utilize "up-wind schemes" (see Sect. 12.1.3), which is more dissipative compared with central differences. One should also consider the discretization of temporal changes. In the case of LES models, an implicit treatment of the time integration is no longer possible, there is thus the requirement of the Courant number $Cou \leq 1$ [see (12.88)]. This cannot be done, at least for the resolution of convection, and in the case of acoustically affected flows in the outlet ports of the engine even the characteristics of the sound must be resolved. This results in time intervals that must be 100 times smaller than is usual in the case of RANS calculations. There are many indications that it would be very difficult to convert RANS codes to LES.

Actually, LES modeling also requires the complete resolution of the turbulent wall boundary layer, i.e. $y^+ \leq 1$. This is absolutely impossible with current computational meshes and possibly not necessary. Recently, a kind of hybrid approach has been developed, *detached eddy simulation* (DES), which continue to describe the boundary layer areas with a Reynolds-averaged law of the wall, while especially areas of detached flow, in which large, inhomogeneous turbulence structures are predominant, are dealt with using the LES method (Spalart 2000).

Figure 12.6 shows velocity fields in the valve cross-section and on the level of the cylinder head gasket simulated both with LES and with RANS for a *flow bench test* (see also Sect. 12.5.2). In the flow bench test, a cylinder head with stationary marginal conditions has a fluid passed through it. Since the resultant flow is of a turbulent character, (stochastically) changing flow field result over time. While one can (steadily) calculate the Reynolds-averaged flow field with RANS, LES calculation must always be calculated in a temporally unsteady fashion. Only after a temporal averaging is a quantitative comparison with the RANS results possible; in this example, the average value images are very similar to the RANS results. The temporal snapshot in LES clearly shows finer flow structures than the RANS result. This example reveals the potential of LES on the one hand, but also the challenges in its use. Transferred to cyclical engine processes, this necessitates the calculation not only one working cycle, but a large number of them, if a meaningful average value image is to be obtained (PIV images in a transparent engine suggest over 50 working cycles). It will therefore require still some time until practical, practicable LES applications will be available, should they be desirable also in the sense of a universal turbulence model.

Finally, we also wish to point out that LES *cannot* supply the resolution of cyclical engine fluctuations, even if it is exactly this that is always given as the main justification for LES in engine-related contexts. Turbulence models can only describe fluctuations that are caused by the turbulent flow itself, which the chaotic, non-deterministic behavior of the Navier–Stokes equations reflects. Variations or fluctuations generated by variations of the marginal conditions cannot be described. These are however the most dominant effects by far (injector, ignition, final cycle, throttle valve, etc.) in an engine. Thus the stochastic case-to-case fluctuations of a

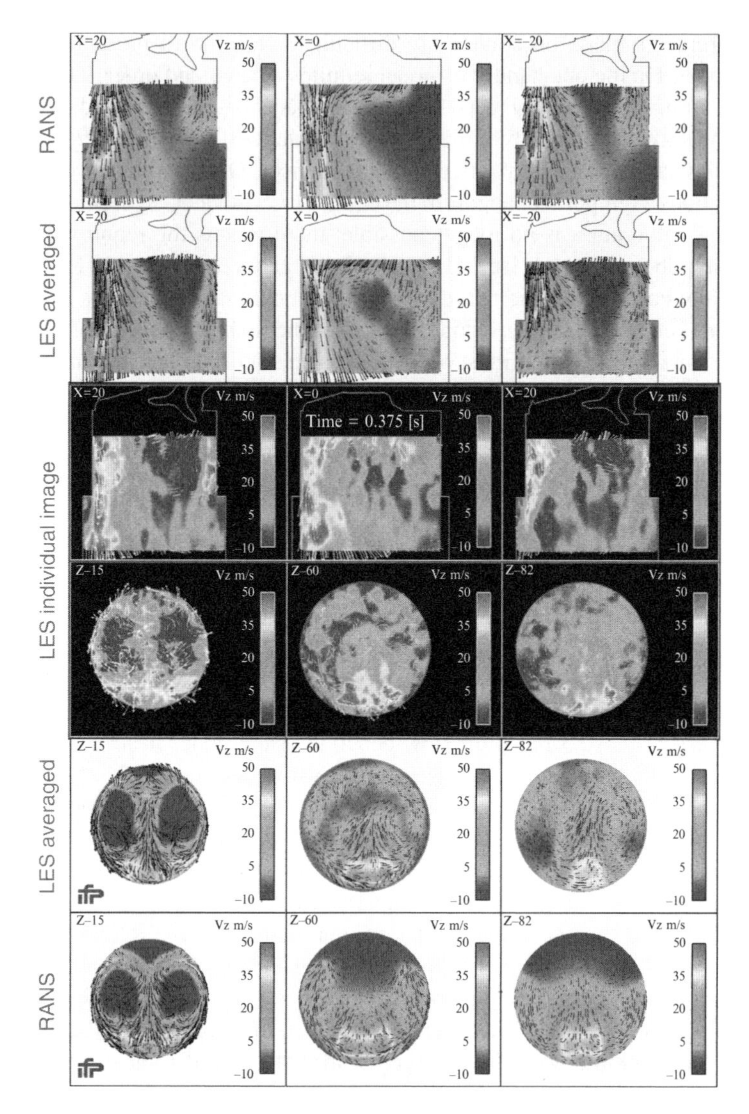

Fig. 12.6 Velocity fields in the valve cross-section and on the level of the cylinder head gasket, simulated for a *flow bench test*, in which characteristic parameters such as the swirl and tumble numbers of a cylinder head are determined (Angelberger 2007)

LES simulation (at unaltered, perfect marginal conditions) will turn out much smaller than the cyclical fluctuations in the actual engine.

Due to its small, directly resolving length and time scales, DNS will not be a candidate for the simulation of engine flows in the foreseeable future. Nevertheless, DNS can be used successfully in the sense of a "numeric experiment" in order to derive submodels for Reynolds-averaged or LES simulations. Examples include not only the verification or calibration of models for turbulent flow processes but also multiphase problems. However, this necessitates particularly simple geometries and low Reynolds numbers. In principle, small, exemplary sections of larger problems are investigated. The derivation of pdf trial functions for turbulent mixing processes (e.g. Reveillon and Vervisch 2000) or the development of splashing models of drops on walls (e.g. Maichle et al. 2003) are two good examples.

12.3 Numerics

In the following, some basic concepts of numerical fluid mechanics will be explained in order to become acquainted with its most essential concepts, which are also of importance for practical work. For a detailed presentation, see Ferziger and Perić (1996) or Patankar (1980).

12.3.1 The Finite Volume Method

Customarily, CFD codes work in the finite volume method. This approach guarantees the numerical preservation of conservative quantities for incompressible flows (unfortunately, this is not obvious). However, the computational mesh should ideally be built hexahedrally. We proceed from a transport equation in the general form

$$\frac{\partial}{\partial t}\Phi + \frac{\partial}{\partial x_i}\Psi_i = \Xi, \tag{12.66}$$

i.e., Ψ_i designates convection and diffusion flows of size Φ, Ξ the corresponding local source. With the help of the Gaussian law follows:

$$\frac{\partial}{\partial t}\int_V \Phi\, dV + \oint_{\partial V} \vec{\Psi}\, d\vec{S} = \int_V \Xi dV, \tag{12.67}$$

or for a hexahedron of the computational mesh

$$[\Phi(t+\Delta t) - \Phi(t)]\Delta V = -\sum_{l=1}^{6} \vec{\Psi}_{(l)}\Delta\vec{S}_{(l)}\Delta t + \Xi\Delta V\,\Delta t, \tag{12.68}$$

whereby the sum runs over the six sides of the hexahedron.

$\vec{\Psi}_{(l)}$ designates the flux vector on the hexahedral side (l), $\vec{\Psi}_{(l)}\Delta\vec{S}_{(l)}\Delta t$ is the (oriented) flux, which in the time Δt leaves the hexahedral volume across side (l) with surface $\Delta\vec{S}_{(l)}$. The same term, but with the opposite sign, appears in the neighboring cell bordering the side (l). Thus insofar as the source Ξ is equal to zero, only fluxes between individual cells of the computational mesh are exchanged. The total amount

$$\sum_{\substack{z:\,Summe\\ \text{über alle Zellen}}} \Phi_z \, \Delta V_z$$

remains constant over time, assuming that the flow over the border of the calculation range is zero.

12.3.2 Discretization of the Diffusion Term: Central Differences

For the discretization of the diffusion term, we will at first look at a (purely elliptic) stationary diffusion equation in one dimension of space

$$\frac{\partial}{\partial x}\left[D\frac{\partial}{\partial x}\Phi\right] = 0. \tag{12.69}$$

The Gaussian law states for a "one-dimensional cell" i

$$D_{i,+}\frac{\partial}{\partial x}\Phi(x_{i,+}) - D_{i,-}\frac{\partial}{\partial x}\Phi(x_{i,-}) = 0, \tag{12.70}$$

whereby $x_{i,+}$ describes the location of the right cell border, $x_{i,-}$ the location of the left cell border (see Fig. 12.7); for equidistant cells of length Δx is valid

$$x_{i,+} = x_i + \frac{\Delta x}{2} \text{ and } x_{i,-} = x_i - \frac{\Delta x}{2}. \tag{12.71}$$

The quantities Φ are only known at the cell centers, i.e. at the locations x_{i-1}, x_i, $x_{i+1}, \ldots$ and we have to attribute the gradients to these. The most obvious solution is the *central difference scheme*, which provides for equidistant cells

$$\frac{\partial}{\partial x}\Phi(x_{i,+}) \cong \frac{\Phi(x_{i+1}) - \Phi(x_i)}{\Delta x} \tag{12.72}$$

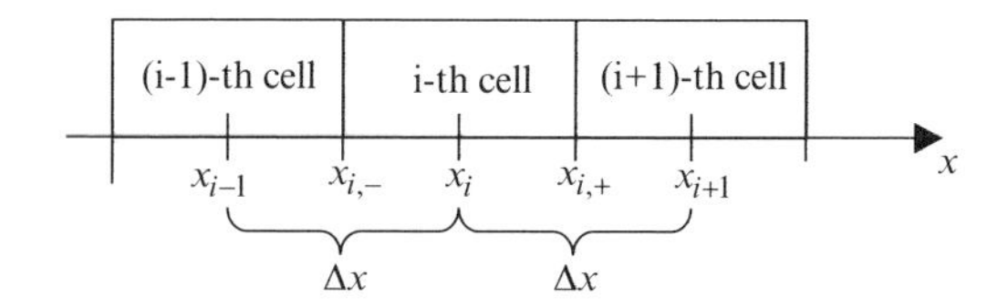

Fig. 12.7 Graphic representation of cells and cell boundaries

which finally (for constant D) leads to the following relation

$$\Phi(x_i) = \frac{1}{2}\Phi(x_{i+1}) + \frac{1}{2}\Phi(x_{i-1}). \tag{12.73}$$

Since in this equation every Φ value is given as the mean value of its neighboring value, this equation system is not directly solvable. Rather, the equations provide, formulated for all cells $i = 1, \ldots, N$, a linear equation system, which should be handled with an appropriate solution technique for linear equations. This equation system corresponds exactly to the character of an elliptic equation, in which all boundary values have to be prescribed for the problem definition.

In general, every source-free transport equation can be attributed to such a pattern, so that the value of a field Φ results from the values of Φ in the temporally and spatially adjacent cells

$$\Phi(x_i, y_j, z_k; t_l) = \sum_{|\tilde{i}-i|+|\tilde{j}-j|+|\tilde{k}-k|+|\tilde{l}-l|=1} \alpha_{ijkl;\tilde{i}\tilde{j}\tilde{k}\tilde{l}} \Phi(x_{\tilde{i}}, y_{\tilde{j}}, z_{\tilde{k}}; t_{\tilde{l}}). \tag{12.74}$$

Obviously (to guarantee that $\Phi =$ constant is also a solution of the discretized transport equation) the condition

$$\sum_{|\tilde{i}-i|+|\tilde{j}-j|+|\tilde{k}-k|+|\tilde{l}-l|=1} \alpha_{ijkl;\tilde{i}\tilde{j}\tilde{k}\tilde{l}} = 1 \tag{12.75}$$

An essential criterion in this formulation is now the positivity of the coefficients

$$\alpha_{ijkl;\tilde{i}\tilde{j}\tilde{k}\tilde{l}} \geq 0. \tag{12.76}$$

Only when this has been achieved does a stable, monotonic numerical scheme exist. Physically, this corresponds to the fact that only via processes of diffusion and convection no local maximum in the value of a field can be generated (which had not already existed).

12.3.3 *Discretization of the Convection Term: The Upwind Scheme*

If we extend the diffusion equation (12.69) by the convection terms, one then obtains

$$\rho \frac{\partial}{\partial x}[v\,\Phi] - \frac{\partial}{\partial x}\left[\rho D \frac{\partial}{\partial x}\Phi\right] = 0. \tag{12.77}$$

For constant ρ and D, we obtain after integration

$$v_{i,+}\Phi(x_{i,+}) - v_{i,-}\Phi(x_{i,-}) - D\frac{\partial}{\partial x}\Phi(x_{i,+}) + D\frac{\partial}{\partial x}\Phi(x_{i,-}) = 0. \tag{12.78}$$

The gradients

$$\frac{\partial}{\partial x}\Phi(x_{i,\pm})$$

can again be described by means of central differences. For $\Phi(x_{i,\pm})$ one could evidently use the difference system

$$\Phi(x_{i,\pm}) = \frac{1}{2}(\Phi(x_i) + \Phi(x_{i\pm 1})). \tag{12.79}$$

This states in all

$$\left(\frac{2D}{\Delta x} + \frac{v_{i,+}}{2} - \frac{v_{i,-}}{2}\right)\Phi(x_i) = \left(\frac{D}{\Delta x} - \frac{v_{i,+}}{2}\right)\Phi(x_{i+1}) + \left(\frac{D}{\Delta x} + \frac{v_{i,-}}{2}\right)\Phi(x_{i-1}). \tag{12.80}$$

For reasons of simplicity, assume from $v_{i,+} = v_{i,-} = v_i$ (this is exactly true in one dimension by the validity of the continuity equation, in several dimensions still approximately true). With this, the positivity condition (12.76) leads to the following inequality[9]:

$$\frac{|v|\Delta x}{D} \le 2. \tag{12.81}$$

If this inequality is violated, which can easily be the case with less fine meshes or high convection velocities, then the difference system (12.79) is no longer stable. Thus, another discretization $\Phi(x_{i,\pm})$ must be introduced. The upwind scheme[10] provides a stable difference scheme

$$\begin{aligned}\Phi(x_{i,+}) &= \theta(v)\Phi(x_i) + \theta(-v)\Phi(x_{i+1}),\ \Phi(x_{i,-})\\ &= \theta(v)\Phi(x_{i-1}) + \theta(-v)\Phi(x_i),\end{aligned} \tag{12.82}$$

i.e. the grid point lying upstream (where the flow is coming from) determines the local Φ value. The difference scheme now leads to

[9]The quantity $Pe = \frac{v\Delta x}{D}$ is also called the Peclet number.

[10]$\theta(x)$ x) designates the Heaviside function:$\theta(x) = \begin{cases} +1 & x \ge 0 \\ -1 & x < 0 \end{cases}$.

$$\left(\frac{2D}{\Delta x}+|v|\right)\Phi(x_i)=\left(\frac{D}{\Delta x}+\theta(-v)|v|\right)\Phi(x_{i+1})+\left(\frac{D}{\Delta x}+\theta(v)|v|\right)\Phi(x_{i-1}), \tag{12.83}$$

i.e. every coefficient is positive. In principle, this unsymmetrical pattern reflects the fact that a pure convection equation (without diffusion) is of the hyperbolic type and thus incompatable with elliptic boundary conditions (the value in a cell results from its bordering e.g. all its neighboring cell values). It would rather have only one initial value (at the "inflow position"), from which it propagates with the flow. This property is passed down to convection-dominated flows.

If one develops $\Phi(x_{i,\pm})$ into a Taylor series, one then recognizes that the central difference scheme (12.79) is more precise than the upwind scheme (namely, of the second order in the discretization parameter Δx, while the upwind scheme is only of the first order). Still, numerical stability is a property one cannot do without for tasks with complex geometries and deformed grids.

Therefore, in numerical fluid mechanics, we cannot avoid an upwind-type scheme (in the literature, there is a large number of modifications, including for higher orders of accuracy), or a hybrid method, which still contains a substantial amount of "upwind".

12.3.4 Discretization of the Time Derivation: Implicit Scheme

If we extend (12.77) additionally by the time derivative, we obtain

$$\frac{\partial}{\partial t}\Phi+\frac{\partial}{\partial x}[v\Phi]-\frac{\partial}{\partial x}\left[D\frac{\partial}{\partial x}\Phi\right]=0. \tag{12.84}$$

After integration and insertion of the previously discussed discretization systems, we obtain

$$\begin{aligned}\frac{\partial\Phi(x_i)}{\partial t}\Delta x&+\left(\frac{2D}{\Delta x}+|v|\right)\Phi(x_i)-\left(\frac{D}{\Delta x}+\theta(-v)|v|\right)\Phi(x_{i+1})\\&-\left(\frac{D}{\Delta x}+\theta(v)|v|\right)\Phi(x_{i-1})=0.\end{aligned} \tag{12.85}$$

In the time variables, the differential equation is clearly hyperbolic, i.e. the state of time $t+\Delta t$ is calculated from the state of time t; one only has to provide a one-sided boundary condition (initial condition). Thus, the following time discretization (the *explicit scheme*) is at first apparent

$$\frac{\Phi(x_i, t + \Delta t) - \Phi(x_i, t)}{\Delta t}\Delta x + \left(\frac{2D}{\Delta x} + |v|\right)\Phi(x_i, t)$$
$$- \left(\frac{D}{\Delta x} + \theta(-v)|v|\right)\Phi(x_{i+1}, t) - \left(\frac{D}{\Delta x} + \theta(v)|v|\right)\Phi(x_{i-1}, t) = 0. \quad (12.86)$$

This leads however to the following equation

$$\Phi(x_i, t + \Delta t) = \left[1 - \left(\frac{2D}{\Delta x} + |v|\right)\frac{\Delta t}{\Delta x}\right]\Phi(x_i, t) + \frac{\Delta t}{\Delta x}\left(\frac{D}{\Delta x} + \theta(-v)|v|\right)\Phi(x_{i+1}, t)$$
$$+ \frac{\Delta t}{\Delta x}\left(\frac{D}{\Delta x} + \theta(v)|v|\right)\Phi(x_{i-1}, t). \quad (12.87)$$

With the Courant number

$$Cou = \frac{|v|\Delta t}{\Delta x}$$

follows the stability condition in accordance with (12.76)

$$\left(\frac{2}{Pe} + 1\right)Cou \leq 1. \quad (12.88)$$

This condition states that the time advance Δt should not be too large, and must be the smaller, the finer the mesh is. In practice, this condition is hardly maintainable for all mesh cells. Alternatively therefore, an *implicit scheme* is formulated, in which the amounts of "spatial" differential operators are set for the "new" time $t + \Delta t$

$$\left[1 + \left(\frac{2D}{\Delta x} + |v|\right)\frac{\Delta t}{\Delta x}\right]\Phi(x_i, t + \Delta t) = \Phi(x_i, t) + \frac{\Delta t}{\Delta x}\left(\frac{D}{\Delta x} + \theta(-v)|v|\right)\Phi(x_{i+1}, t + \Delta t)$$
$$+ \frac{\Delta t}{\Delta x}\left(\frac{D}{\Delta x} + \theta(v)|v|\right)\Phi(x_{i-1}, t + \Delta t). \quad (12.89)$$

This system now has purely positive coefficients again, and is thus monotonic and stable. However, it has the disadvantage that it is implicitly formulated, i.e. it represents a coupled system of linear equations in the $\Phi(xi, t + \Delta t)$ with $i = 1, \ldots, N$ (from (12.87) on the other hand, the quantities $\Phi(xi, t + \Delta t)$ can be directly calculated). Nevertheless, quite all CFD codes use implicit (or semi-implicit, hybridized) systems for reasons of stability.

12.3.5 Discretization of the Source Term

For the complete discretization of the transport equation, we now still need the treatment of the source term. The previously discussed discretization systems lead for the equation

$$\frac{\partial}{\partial t}\Phi + \frac{\partial}{\partial x}[v\Phi] - \frac{\partial}{\partial x}\left[D\frac{\partial}{\partial x}\Phi\right] = Q \tag{12.90}$$

to

$$\left[1 + \left(\frac{2D}{\Delta x} + |v|\right)\frac{\Delta t}{\Delta x}\right]\Phi(x_i, t+\Delta t) = \Phi(x_i, t) + \ldots + Q(x_i, t+\Delta t)\Delta t. \tag{12.91}$$

Insofar as the term $Q(xi, t + \Delta t)$ is unavailable, it has to be attributed to the term $Q(x_i, t)$. A special case exists if Q is directly depending on Φ. In this case, Q can be developed into a Taylor series up to the linear term in Φ

$$Q(\Phi(x_i, t+\Delta t)) \cong Q(\Phi(x_i, t)) + \underbrace{\frac{\partial Q}{\partial \Phi}(\Phi(x_i, t))}_{\alpha}\left[\Phi(x_i, t+\Delta t) - \Phi(x_i, t)\right]. \tag{12.92}$$

Inserted in (12.91), this leads to

$$\left[1 + \left(\frac{2D}{\Delta x} + |v|\right)\frac{\Delta t}{\Delta x} - \alpha\Delta t\right]\Phi(x_i, t+\Delta t) = (1 - \alpha\Delta t)\Phi(x_i, t) + \ldots + Q(\Phi(x_i, t))\Delta t. \tag{12.93}$$

The stability criterion is surely fulfilled for

$$\alpha = \frac{\partial Q}{\partial \Phi}(\Phi(x_i, t)) \leq 0. \tag{12.94}$$

In this case, the employment of (12.93) is indeed highly recommended. If the criterion is offended however, one should rather resort to (12.91) plus substitution $Q(x_i, t+\Delta t) \Rightarrow Q(x_i, t)$, even if this the precision affects (e.g. convergence speed). The precision can namely be raised again, if one iterates (12.91) [or (12.93)]. Proceeding from (12.91), one calculates the approximate values $Q_{(k+1)}(x_i, t+\Delta t)$ from $Q_{(k)}(x_i, t+\Delta t)$ via substitution into the source term, until the value changes only to an inessential extent. One obtains

$$\left[1 + \left(\frac{2D}{\Delta x} + |v|\right)\frac{\Delta t}{\Delta x}\right]\Phi_{(1)}(x_i, t+\Delta t) = \Phi(x_i, t) + \ldots + Q(\Phi(x_i, t))\Delta t \tag{12.95}$$

or

$$\left[1+\left(\frac{2D}{\Delta x}+|v|\right)\frac{\Delta t}{\Delta x}\right]\Phi_{(k+1)}(x_i,t+\Delta t)=\Phi(x_i,t)+\ldots+Q(\Phi_{(k)}(x_i,t+\Delta t))\Delta t$$

12.3.6 *The Operator Split Method*

Especially in internal engine processes, injection and combustion processes appear in addition to pure gas flow, which demand the solving of new transport equations, but also create new source terms in the equations existing already. In this case, a typical transport equation contains terms for convection, diffusion, spray (e.g. evaporation), and combustion. It is recommendable to treat the various effects separately. This takes place by use of the operator split method. $\underline{M}$ and $\underline{N}$ designate two operators in this case, e.g. convection/diffusion and a chemical source term

$$\frac{\partial}{\partial t}\phi = \underbrace{\underline{M}(\phi)}_{\text{Konvektion/Diffusion}} + \underbrace{\underline{N}(\phi)}_{\text{Chemie}}. \tag{12.96}$$

By inserting an intermediate step, the time integration can be split

$$\tilde{\phi}(x_i,t+\Delta t)-\phi(x_i,t)=\int_t^{t+\Delta t} M(\phi)\mathrm{d}t,$$

$$\phi(x_i,t+\Delta t)-\tilde{\phi}(x_i,t+\Delta t)=\int_t^{t+\Delta t} N(\phi)\mathrm{d}t. \tag{12.97}$$

Each of these two steps can then be solved as shown above. The error is of the order $(\Delta t)^2$. In this context, we should point out that the CFD code KIVA calculates for convection and diffusion separately. Because of the different character of the fundamental differential operators (hyperbolic vs. elliptic), this method is not unreasonable.

12.3.7 *Discretization and Numerical Solution of the Momentum Equation*

Finally, the momentum equation for the calculation of velocity and pressure by use of the continuity equation should be considered. For numerical reasons, it is recommendable to resort to *staggered grids*, i.e. pressure and velocity are calculated on computational grids shifted to each other, the pressure for example in the cells (i. e. effectively their centers) and the velocity on the nodes.

The calculation of velocity commonly takes place iteratively, for which several algorithms are known (e.g. SIMPLE, PISO, SIMPISO, . . .). In the final analysis, all have the fact in common that in the first step the momentum equation is solved for the velocities of the momentums kept constant. In the second step, pressure corrections are then calculated with the help of a Poisson equation for pressure. With these pressure corrections, new velocities are then calculated again, and thus again, until a pre-given break off threshold for the convergence is reached. In the incompressible case, the Poisson equation for the pressure shell be given which follows from the velocity equation by divergence formation

$$\Delta p = -\rho \frac{\partial v_i}{\partial x_j} \frac{\partial v_j}{\partial x_i} + 2 \frac{\partial^2}{\partial x_i \partial x_j} \left(\mu_t S_{ij} \right). \tag{12.98}$$

12.4 Computational Meshes

Mesh generation is often the most decisive and strongly limiting factor in CFD calculation accuracy today. It cannot be repeated often enough that a good computation mesh is *the key* to success, much more essential than, for example, the introduction of modified turbulence models. A good computation mesh should consist of hexahedrons, be wall-adapted [i.e. also keep to the y_+ rule (12.45)] and be sufficiently fine/problem-adapted, so that all flow structures (free jets, flames, . . .) can be well resolved (see Fig. 9.4c, d). In practice, these requirements can often not be fulfilled completely satisfactorily.

Another problem is mesh movement, which is solved very CFD code-specifically, so that one has to rely closely in generating moving meshes on the respective code philosophy. Only a few CFD codes exist which provide for the functionality of a mesh movement, as is important in the treatment of in-cylinder problems (moving valves and piston). When calculating engine gas exchange, the quality of the mesh in the area of the valves is of primary importance. Figure 12.8 shows, for a 4-valve SI engine, a qualitatively high-quality mesh structure (primarily cubes, wall-adapted cell layer) around the inlet valve for the closed valve state as well as for partial lift.

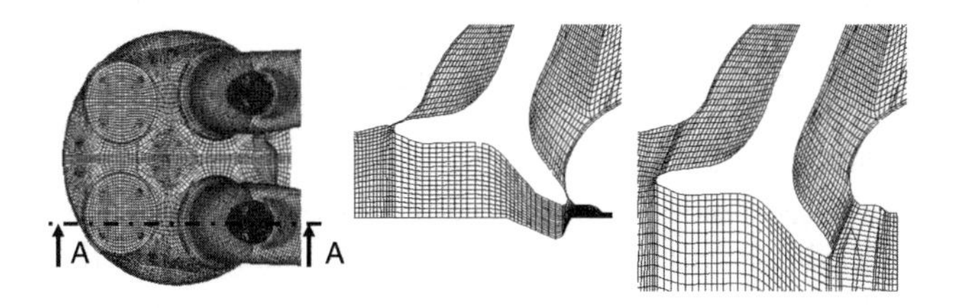

Fig. 12.8 Qualitatively high-grade mesh structure for a 4-valve SI engine for 2 valve lifts

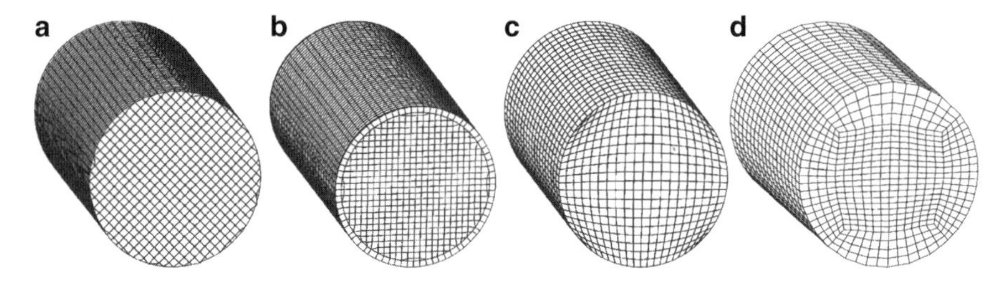

Fig. 12.9 *To the far left* a truncated Cartesian mesh which has a very poor mesh structure. A Cartesian mesh with a surface layer truncated only on the inside (*second to the left*) is already clearly better. *Second to the right* is a representation of a mesh with wall-adapted, hexahedral mesh structure, whereby there are, however, still deformations at the corners of the base meshes. The mesh on the *far right* displays an optimal, wall-adapted mesh structure

In the case of stationary meshes, the situation is less critical, and the meshes can usually be exchanged. One can therefore also resort to independent mesh generation programs. As before, the construction of a qualitatively superior computational mesh still requires high effort – and above all experience.

As opposed to this, in recent times a unfortunate trend is making itself known, also in the case of engine CFD applications "fast" mesh generation with automatic mesh generators, which simply cut off Cartesian meshes on the surface of the object to be modeled (see Fig. 12.9a). What is still acceptable in the case of highly complex geometries like coolant flows, is unsuitable for the calculation of internal engine processes at least, due to the complex, wall-dominated flow structures and the accumulating errors during the transient simulation. In every case, one should pay attention in the case of such truncated meshes that at least one adaptive mesh wall layer exists (y^+ rule) and that the Cartesian mesh is only inscribed into the remaining internal space (see Fig. 12.9b), or else the boundary layer will not be correctly represented. Furthermore, one must check visually whether bad mesh structures have emerged in the automatically produced mesh (an extremely narrow point, e.g. only resolved by a very small amount of cell layers). In addition, the time advantage in automatic mesh generation is usually smaller than one may at first estimate. For on the one hand, an essential part of the mesh generation work consists in the preparation of the surface meshes, and this effort cannot be spared in automatic mesh generation. On the other hand, adaptive, "hand-generated" meshes are often more versatile, because variants can be easily worked in; in the case of automatically generated meshes, a new mesh must be generated for each variant.

One potential way out of the dilemma "high-quality mesh" vs. "complexity of mesh generation" is "surface-based flow resolution", which is currently being introduced in various commercial software packages (e.g. Fire8, StarCD4). Until now, basic cell-based flow resolution have been used, in which case defined cell types (e.g. hexahedrons, tetrahedrons, prisms) are used that are described by the cell centre point and the coordinates of the corner nodes. The flexibility of the mesh generation is naturally limited; in particular, proximity information about the cells, such as are required for higher procedure codes of gradient determination, is unavailable. For example, "hanging nodes", which would have been necessary for adapted mesh refinement, always have a negative effect on the stability of the calculation. In surface-based resolution, the cell center is no longer stored, but rather the relations of the cell surfaces including the node coordinates and information about which cell surfaces belong to a cell. This guarantees that there are no longer any gaps between the cell surfaces. Nodes are always stacked and many polyhedron cells are possible. Implicitly, the clashing of different cells against each other (on the cell surfaces) is known, which has a positive effect on the accuracy and stability of the calculations. These polyhedron meshes are typically generated using the detour of tetrahedrons, in order to be combined later. Pure tetrahedron meshes, such as can be used in cell-based resolution, are not suitable for finite volume methods due to the low precision and the resultantly high required number of tetrahedrons.

12.5 Examples

The CFD applications in the engine sector are, as already mentioned, diverse. In the following, the simulation of internal cylindrical flow structures in SI engines and diesel engines and the simulation of internal nozzle flow in diesel injectors will be looked at as examples.

12.5.1 Simulation of Flow Structures in the Cylinder: The SI Engine

We will now consider the intake and compression flow in a SI engine. The geometry of the calculation mesh is represented in Fig. 12.10 (left).

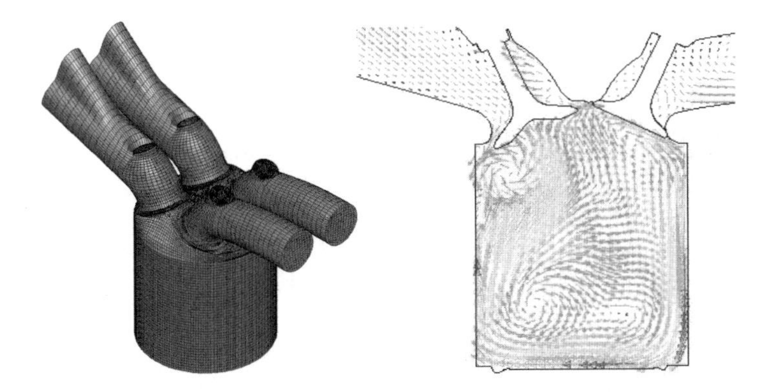

Fig. 12.10 Simulation of the cylinder flow of a SI engine. Basic geometry for the computational mesh (*left*) and the flow image in a section through valves at BDC (*right*)

Because of the mirror-symmetry, only a half-mesh must be taken into consideration, which saves calculation time. Full cylinder meshes of a typical passenger car engine at BDC including intake and exhaust tracts should consist of at least 1,000,000 mesh cells, in order to resolve the relevant flow structures including turbulence sufficiently. In Fig. 12.10 (right), the flow is shown in a section through the valves in the gas exchange BDC. One sees the typical tumble flow (large swirl in the clockwise direction). In three dimensions, a tumble of course does not simply represent a cylinder. We are concerned in the case of the swirl axis (or better: swirl axes) with a complex, unsteady, three-dimensional formation, which is frequently somewhat reminiscent of an "Ω".

Towards Compression TDC (TDCc), the combustion space becomes increasingly flat; the tumble is "squashed" by the piston and disintegrates into turbulence. At first, during the intake phase, the flow is quite directionless. Consequently, many areas of high shear exist, which creates a lot of turbulence. At BDC, the flow has already largely become calm. Only the large swirl structures still exists, while all small structures are already dissipated. Shortly before TDCc the tumble itself decomposes into turbulence, and in the temporal evolution a local maximum develops, which considerably influences the burning velocity (see Sect. 4.1.3). Since the tumble scales with the speed, this is valid approximately for the turbulence "plateau" and thus the burning speed. This is essentially the reason why combustion rates in a SI engine are, to some extent, independent of engine speed.

Figure 12.11 shows with an application example how the cylinder in-flow can be influenced with the geometry around the intake valves. "Masking" forces an overflow of the intake valves in the case of small valve lifts, causing a large-scale tumble motion. Without masking, one obtains two vortexes turning against each other, which weaken each other in the course of the compression phase.

The positive effect of inlet valve masking can be seen clearly both in the tumble and in the turbulence sequence (Fig. 12.11 middle and right). With masking, the maximum tumble is clearly higher, but the dissipation in the compression phase is also stronger. As a consequence, more turbulence energy is available for

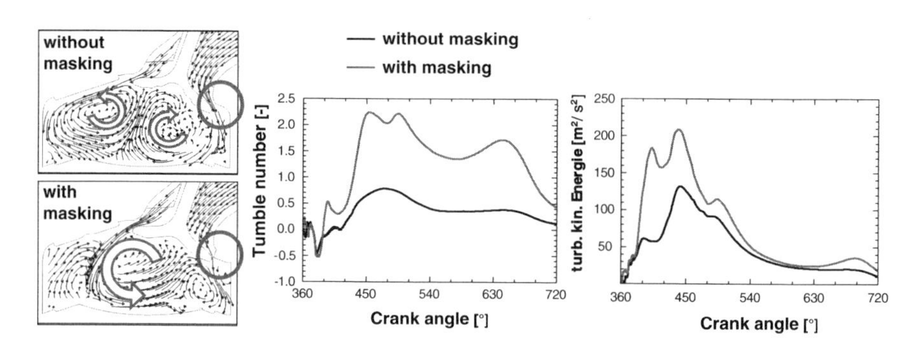

Fig. 12.11 Representation of the flow field during charge changing with and without masking. As a result of the directed tumble motion, a considerably higher level of turbulence results at TDCc (720°) with inlet valve masking TDCc

combustion near TDCc than without masking. This example makes it easy to understand how combustion can be affected by the intake flow – the tumble must be increased. This is the only way in which flow energy can be conserved long enough, i.e. until TDCc, in order to then disintegrate into turbulence. Turbulence sheets or similar installations in the intake ports can however only increase the turbulence level for a short time. This turbulence is long dissipated at TDCc.

Less effective however is an excessive turbulence value at the spark plug. This can lead to ignition arc stretch and failure to successfully ignite. Another important influence parameter for SI combustion is the local residual gas distribution, which can also be determined with CFD.

12.5.2 Simulation of Flow Structures in the Cylinder: Diesel Engines

For the passenger car engine, a high swirl value is very important for a good mixture preparation. Again, swirl generation and evolution are quantities which can be investigated very effectively by means of CFD.

Figure 12.12 shows the intake port geometry of the diesel engine of a passenger car (left) and the time development of the swirl number for two variants (right). To generate swirl, the intake ports are designed as spirals. Moreover, a valve-seat generated swirl, a *valve seat shroud* was introduced as well, i.e. the front intake port has a shroud in the valve seat. As can be seen on the right, this valve seat cover ("variant B") increases the swirl number considerably compared to an equivalent variant without such a valve seat cover ("variant A"). The duct without a valve seat cover (here rear left) functions as a tangential duct. If the left port were also to function as a tangential duct, both flows would compensate for each other, and there would be no resultant swirl. The task of the valve cover seat is to block the tangential component of the flow, thus supporting the functionality of the swirl port.

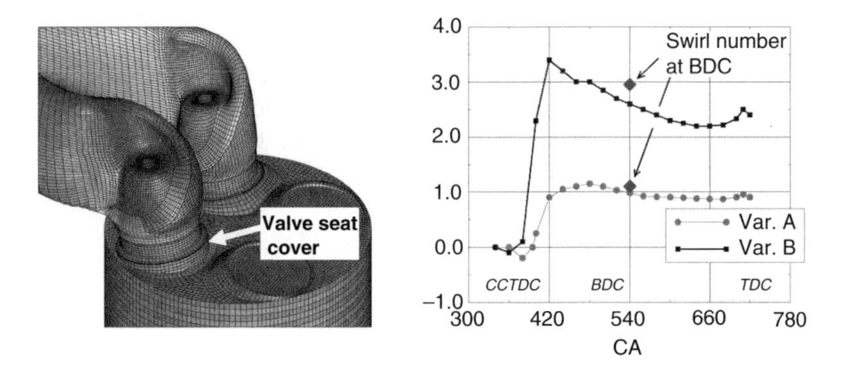

Fig. 12.12 Flow in the intake port of a diesel engine, port geometry (*left*) and time development of the swirl number for two variants (*right*)

Since swirl, geometrically speaking, is much better adapted to the cylinder geometry than tumble, it disintegrates to a much smaller extent during compression; instead, the flow rotating around the cylinder axis is compressed into the piston bowl, by means of which the swirl number can increase because of the conservation of angular momentum as the average mixture radius reduces due to the bowl geometry at the end of the compression stroke.

The swirl number ω is defined as the ratio of the angular momentum L and the momentum of inertia θ related to the engine speed n

$$\omega = \frac{L}{2\pi n\theta} \quad \text{with} \quad L = \int r v_{\tan} \mathrm{d}m \quad \text{and} \quad \theta = \int r^2 \mathrm{d}m. \tag{12.99}$$

In this case, r designates the distance from the cylinder respectively the center of rotation, $v_{\tan}$ the tangential component of the flow velocity. With the calculation at every time step, we obtain transient swirl number evolution across the crank angle. In the flow bench test, a swirl value is determined as a function of the valve lift. Correlating the valve lift graph with the respective crank angle and converting the swirl value into an angular momentum flux, the latter can be integrated over the intake stroke and a swirl number at BDC determined. Because of the small losses in the case of a swirl flow around the cylinder axis, this approximation is usually sufficient, i.e. the real transient swirl number at BDC and the swirl number at BDC do not deviate much from each other, see also Fig. 12.12 (middle). This is assumption is not valid for SI engines because of the much higher flow losses for tumble flow.

During compression, in addition to the swirl in the bowl, another secondary swirl is produced which can be important for the mixture formation processes in the bowl (according to the respective injection). The orientation of this secondary swirl depends however on the swirl strength (Fig. 12.13). However, one must always keep in mind that this describes the flow before injection – the injection event naturally changes the flow structures dramatically.

Similar to the SI engine, all small-scale flow structures disintegrate during compression in the diesel engine as well. Only the large-scale swirl structures (incl. secondary swirls) survive and create, together with the injection sprays, flow structures during the mixture formation phase, whereby the latter is the most dominant by far. For these reasons, it is quite common in the case of diesel engines to start calculation of the in-cylinder processes (mixture formation and combustion) at the closing of the intake valves assuming an ideal "cylinder-shaped" swirl

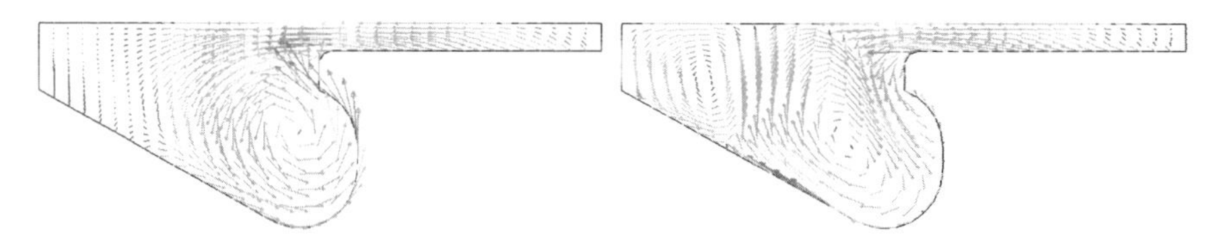

Fig. 12.13 Secondary swirl in the bowl of a diesel engine at 5°CA before ignition TDC with BDC-speed at exactly 0 (*left*) and exactly 2.5 (*right*)

structure. The only unknown, the swirl number, must then be taken from measurements (e.g. from the swirl number BDC, see above). This method is particularly suitable in the case of a combustion process low in swirl (typically e.g. for commercial vehicle engines), because we spare ourselves the generation of a mesh with moving valves on the one hand (as time-consuming as ever) and can restrict the computation time considerably in the case of symmetrical combustion chambers because of the high problem symmetry. For a (symmetrical) 8-hole nozzle for example, only a (360°/8=) 45° combustion chamber sector with cyclical boundary conditions has to be examined.

12.5.3 Internal Nozzle Flow

Another important CFD application is the simulation of internal nozzle flow, since it provides information about the initial conditions of injection spray simulation which will be discussed in the next section. Figure 12.14 introduces a few concepts and explains important phenomena by means of the silhouette image of an optically accessible commercial vehicle blind hole nozzle (see König et al. 2002). Fluid fuel flows from the blind hole into the spray hole. By means of a strong redirection and acceleration of the flow, a cavitation area is formed on the upper surface of the spray hole which extends up to the nozzle exit. This non-symmetry of flow manifests itself in the spray outside the nozzle: the effective exit angle is slightly tilted upwards in comparison to the hole axis.

The appearance of cavitation in the spray hole is a typical event, especially for diesel injectors. It means that, at least at the location of the cavitation's origination, the vapor pressure of the fluid drops locally and cavitation bubbles develop. These bubbles are transported with the fluid, and can grow, shrink, or implode, depending on the environmental conditions. Such a flow can no longer be calculated meaningfully with an incompressible, single-phase fluid model, because negative pressure would inevitably appear at the cavitation origin (the density would however remain constantly high, at the level of the fluid).

Today there are already CFD codes, which offer cavitation modeling on the basis of a turbulent dual-phase flow. In this case, a complete set of transport equations are solved for both phases (fluid on the one hand and gas in vesicular form as a disperse phase on the other), i.e. both phases can possess different velocity fields. However, both phases are coupled by diverse processes; for example bubble growth leads to mass exchange, and the bubble drag creates momentum exchange.

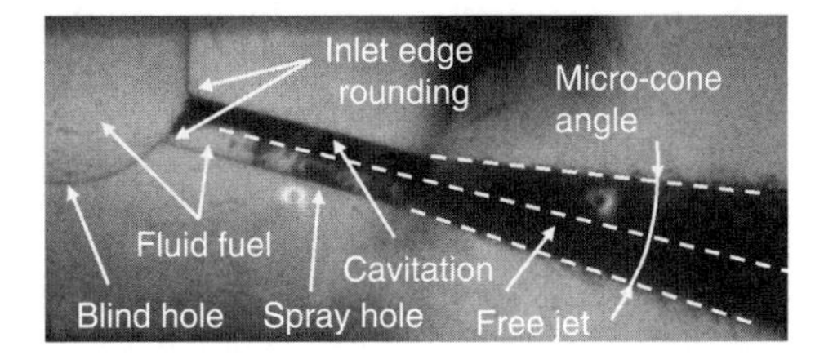

Fig. 12.14 Silhouette image through an optically accessible commercial vehicle injection nozzle

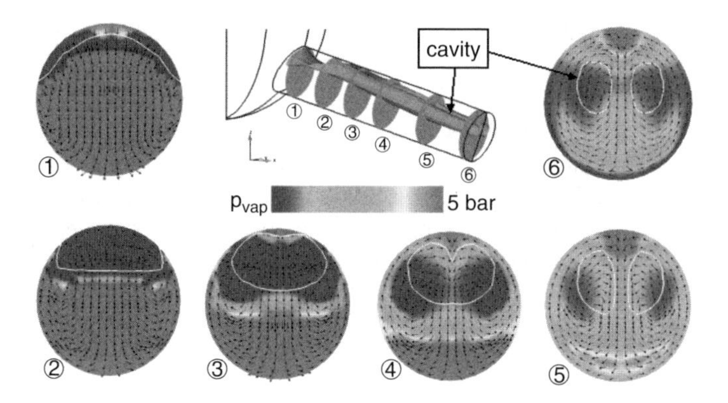

Fig. 12.15 By the strong redirection of the flow at the spray hole entrance, a secondary flow is formed in which cavitation bubbles are transported to the spray hole exit in counter rotating vortices

The simulation result corresponding to Fig. 12.14 in Fig. 12.15 permits a detailed analysis of flow: due to the large curvature of the flow lines at the spray hole entrance, the local pressure falls beneath the vapor pressure and a crescent-shaped cavitation area is formed. As can be seen in cross-section 1, there are extreme pressure differences between the upper surface and lower surface of the spray hole which can be in the order of magnitude of the injection pressure. As a consequence, a secondary flow is generated, in the center of which up to the nozzle exit the pressure remains below the vapor pressure. This is also the reason why cavitation bubbles can reach the nozzle exit at all, despite the high combustion chamber pressure, thus influencing mixture formation.

In the borderline case of highly dominant exchange processes, both phases are firmly coupled, and a single-phase flow with two components arises. For this borderline case, many CFD codes offer cavitation models, and for most practical applications this should be sufficient.

Finally, we will consider the initial conditions for injection spray simulation as the result of internal nozzle flow calculation. If A_{eff} is the effective cross section of the spray and v_{eff} its effective velocity, for the mass flow $\dot{m}$ and the momentum flux $\dot{I}$ at the nozzle hole, we get:

$$\dot{m} = \rho_{fl} v_{eff} A_{eff} \text{ and } , \dot{I} = \rho_{fl} v_{eff}^2 A_{eff}, \tag{12.100}$$

with A_{eff} as the effective cross section limited towards the top by the geometrical hole cross sectionA_{geo}; v_{eff} is limited by the Bernoulli velocity

$$v_{Bern} = \sqrt{\frac{2\,\Delta p}{\rho_{fl}}}$$

whereby Δp designates the difference between injection and combustion chamber pressure. The following loss coefficients

$$C_A = \frac{A_{eff}}{A_{geo}} \leq \quad 1(\text{contraction coefficient}) \tag{12.101}$$

and

$$C_v = \frac{v_{eff}}{v_{Bern}} \leq \quad 1(\text{velocity coefficient}) \tag{12.102}$$

can be defined. With that, the discharge coefficient of the nozzle comes to

$$C_d = \frac{\dot{m}}{\rho_{fl} A_{geo} v_{Bern}} = C_v C_A. \tag{12.103}$$

The discharge coefficient is now usually known or estimable from the injection rate measurements (in the case of rounded conical spray holes it is clearly over 0.8), not however how it is divided into the coefficients C_A and C_v. These can be acquired from the nozzle flow simulation. In our case (Figs. 12.14 and 12.15), cavitation zones exist until the hole exit, which lead to an reduction of the effective flow cross section, on the other hand the velocity in the cavitation-free zones is approximately equal to the Bernoulli speed. Numerical evaluation leads to a value of $C_v \approx 0.9$.

Additional quantities from the internal nozzle flow simulation are the turbulent scales. These lead to the *primary breakup* of the spray. Conversely, imploding cavitation bubbles lead to an increase in turbulence. Jets with increased spray breakup (because of strong atomization of droplets for example) look "bushier" and have other mixture formation properties, see König et al. (2002). This can be seen visually in Fig. 12.16. On the left side is a nozzle with a cylindrical spray hole, on the right a conical spray hole tapering up to the exit. In the conical spray hole there is almost no cavitation, the jet forms a very small cone angle and is very concentrated. In the cylindrical spray hole on the other hand, we can see strong cavitation and the jet is correspondingly wider. These details have a considerable effect on the emission properties of the diesel engine; the developmental trend is pointing towards conical spray holes with strong inlet edge rounding.

But it should not be forgotten that the internal nozzle flow is an extremely transient process, in which the needle lift is of particularly large influence. Figure 12.17 shows that a small needle lift (right) results in a completely different flow image than the

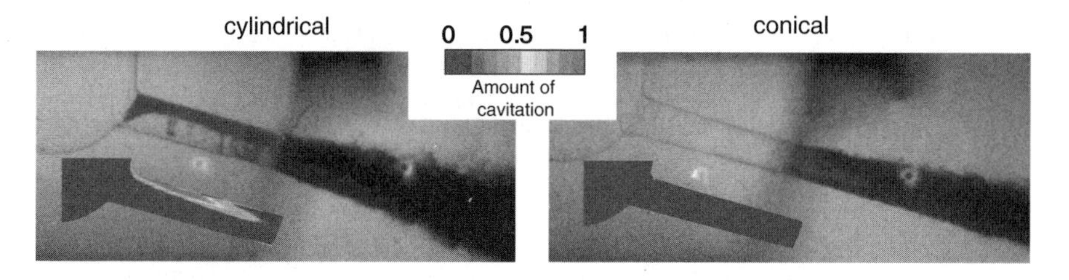

Fig. 12.16 The shape of the nozzle hole has a large effect on the cavitation tendency. The conical nozzle (*right*), almost free of cavitation, leads to a considerably more slender injection jet

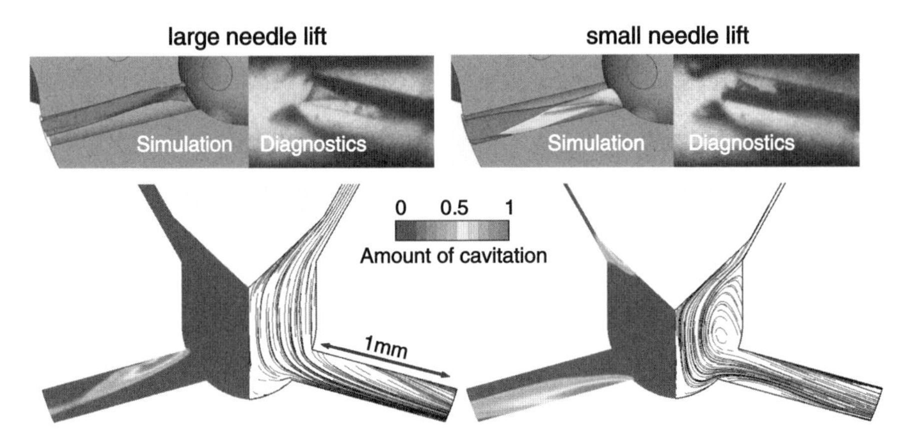

Fig. 12.17 Needle lift affects the in-flow of fuel into the nozzle hole, thus the position of the cavitation tube in the spray hole as well

fully opened needle (left): because of the small lift, the fuel flows in a focused manner into the nozzle hole through the seat area of the nozzle, resulting in a large-scale recirculation area. The spray hole is now flowed from below, the cavitation tube is thus on the bottom of the spray hole as opposed to full needle lift.

References

Angelberger Ch (2007) Interner Bericht

Cebeci T (2002) Convective heat transfer, 2nd revised edn. Springer, Berlin

Ferziger JH, Perić M (1996) Computational methods for fluid dynamics. Springer, Berlin

Han Z, Reitz RD (1995) A temperature wall function formulation for variable-density turbulent flows with application to engine convective heat transfer modeling. Int J Heat Mass Transfer 40:613–625

König G, Blessing M, Krüger C, Michels U, Schwarz V (2002) Analysis of flow and cavitation phenomena in diesel injection nozzles and its effects on spray and mixture formation. In: 5th Internationales Symposium für Verbrennungsdiagnostik der AVL Deutschland, Baden-Baden

Maichle F, Weigand B, Wiesler B, Trackl K (2003) Improving car air conditioning systems by direct numerical simulation by droplet-wall interaction phenomena. ICLASS, Sorrento

Manceau R, Hanjalic K (2000) A new form of the elliptic relaxation equation to account for wall effects in RANS modeling. Phys Fluids 12:2345–2351

Merker GP, Eiglmeier C (1999) Fluid- und Wärmeübertragung. Teubner Verlag, Stuttgart Wiesbaden

Merker GP, Baumgarten C (2000) Wärme- und Fluidtransport – Strömungslehre. B. G. Teubner, Stuttgart

Patankar SV (1980) Numerical heat transfer and fluid flow. Hemisphere, Mc-Graw Hill, New York

Peters N (2000) Turbulent combustion. Cambridge University Press, Cambridge

Pope SB (1978) An explanation of the turbulent round-jet/plane-jet anomaly. Am Inst Aeronautics Astronautics J 16:279–281

Reveillon J, Vervisch L (2000) Accounting for spray vaporization in non-premixed turbulent combustion modeling: a single droplet model (sdm). Combust Flame 121(1/2):75–90

Spalart PR (2000) Strategies for turbulence modelling and simulations. Int J Heat Fluid Flow 21:252–263

White FM (1991) Viscous fluid flow, 2nd edn. McGraw-Hill, New York

Chapter 13
Simulation of Injection Processes

Frank Otto and Christian Krüger

In view of the fact that there are easily operable "spray modules" in all current CFD codes, this topic is much more complex theoretically than one would expect. But if we look at the results gained from such modules critically, the problematic nature of this topic becomes evident. It is still the case that reasonable results can only be produced with a huge amount of computation time, if at all, in the codes for engine applications available in the market today.

In the following, we will first develop the standard spray model, the modeling of the single-droplet processes required for this, and then the stochastical modeling of a droplet ensemble in the Lagrangian formulation. As already mentioned, this approach is associated with many problems and difficulties which will be discussed in detail. Finally, modeling methods in the Euler formulation will be introduced, which may be of help in this context.

13.1 Single-Droplet Processes

Single-droplet processes include the exchange processes of mass, momentum, and heat between a single droplet and the surrounding gaseous phase. The momentum exchange is described in a purely kinetic way in terms of drag, while mass and heat exchange with the environment is generated by diffusion and convection processes in the droplet surroundings.

Droplets are usually described with eight variables: location, velocity (each of them three variables), radius, and temperature. With a modeling of single droplet processes we find the equations of motion of these variables. Sometimes droplet vibration states are introduced as well. Their relevance however has not yet been convincingly demonstrated; we therefore leave them out. Finally, we will introduce two additional model parameters in the context of a multicomponent vaporization

G.P. Merker et al. (eds.), *Combustion Engines Development*,
DOI 10.1007/978-3-642-14094-5_13,

model. In the context of modified models, new statistic parameters such as droplet turbulence will also be introduced, which however no longer describe individual droplets.

13.1.1 Momentum Exchange

If a droplet of radius R, density ρ_{fl} and velocity v_{dr} moves in a gas of density ρ_g and velocity v_g, a decelerating (i.e. directed against the velocity difference to the gaseous phase) force has an effect on the droplet

$$\vec{F} = \rho_{fl} \frac{4\pi}{3} R^3 \dot{\vec{v}}_{tr} = \frac{1}{2} \rho_g C_W \pi R^2 \left| \vec{v}_g - \vec{v}_{tr} \right| (\vec{v}_g - \vec{v}_{tr}). \tag{13.1}$$

Together with the equation

$$\dot{\vec{x}}_{tr} = \vec{v}_{tr} \tag{13.2}$$

the droplet kinetics is determined. The C_W value is usually calculated as follows

$$C_W = \begin{cases} \dfrac{24}{\mathrm{Re}_{tr}} \left(1 + \dfrac{\mathrm{Re}_{tr}^{2/3}}{6} \right) & \text{für} \quad \mathrm{Re}_{tr} \leq 1.000 \\ 0.424 & \text{für} \quad \mathrm{Re}_{tr} > 1.000 \end{cases} \tag{13.3}$$

Whereby

$$\mathrm{Re}_{tr} = \frac{2r\rho_{fl} \left| \vec{v}_{tr} - \vec{v}_g \right|}{\mu_g} \tag{13.4}$$

designates the droplet-based Reynolds number, i.e. for large Reynolds numbers, the drag is quadratically depending on the velocity difference.

It will be important in the following to decompose the flow resistance force of (13.1) into an averaged component and a component dependent on the turbulent fluctuation of the gas velocity[1]:

$$\dot{\vec{v}}_{tr} \approx D_{tr} \cdot \left(\langle \vec{v}_g \rangle - \vec{v}_{tr} \right) + D_{tr} \cdot \vec{v}_g{}'' \quad \text{with} \quad D_{tr} = \frac{3}{8} \frac{\rho_g}{\rho_{fl}} C_W \frac{\left| \langle \vec{v}_g \rangle - \vec{v}_{tr} \right|}{R} \tag{13.5}$$

[1]Equation (13.5) is the formulation common in the literature; strictly speaking, the $\vec{v}_g{}''$-dependence of $C_W \cdot \left| \vec{v}_g - \vec{v}_{tr} \right|$ should also be taken into consideration. For very small velocity differences it approaches zero however [see (13.3) and (13.4)], and for large ones it is small $\left(\left| \vec{v}_g{}'' \right| << \left| \vec{v}_g - \vec{v}_{tr} \right| \right)$.

13.1.2 Mass and Heat Exchange (Single-Component Model)

The continuity equation and the vapor transport equation for the stationary, laminar case read

$$\frac{\partial}{\partial x_i}(\rho v_i) = 0 \,\text{bzw.}\, \frac{\partial}{\partial x_i}(\rho v_i c) - \frac{\partial}{\partial x_i}\left(D\rho \frac{\partial}{\partial x_i} c\right) = 0. \tag{13.6}$$

For reasons of analytical solvability, density, diffusion coefficient and temperature are set as constant. Equation (13.6) is now considered in the environment of a droplet under rotational symmetry; the droplet is thus at rest. The goal is the description of a stationary flow equilibrium for the vapor flux $\tilde{\dot{m}}$ and the heat flux $\tilde{\dot{q}}$ between the droplet surface and infinity. The integration of these equations from the droplet surface to a spherical shell of radius r with the help of the Gaussian law leads to

$$4\pi\rho\, r^2 v(r) = \text{const.}(1), \tag{13.7}$$

$$4\pi\rho\, r^2 v(r) c(r) - 4\pi D\rho\, r^2 \frac{dc(r)}{dr} = \text{const.}(2), \tag{13.8}$$

whereby const.(1) designates the total mass flux and const.(2) the vapor mass flux. Since effectively only vapor is flowing, both should be equal to $\dot{m}$. After solving (13.7) for v, insertion into (13.8), and prescribing boundary conditions $c(R)$ and $c(\infty)$, integration leads to

$$v(r) = \frac{\dot{m}}{4\pi\rho\, r^2} \tag{13.9}$$

and

$$\dot{m} = 4\pi D\rho R \ln\left(\frac{1 - c(\infty)}{1 - c(R)}\right), \tag{13.10}$$

whereby $c(R)$ can be calculated from the droplet temperature by means of the vapor pressure relation (Fig. 13.1).

The equivalent treatment of the heat conduction equation assuming constant specific heat

$$\frac{\partial}{\partial x_i}(\rho v_i c_p T) - \frac{\partial}{\partial x_i}\left(\lambda \frac{\partial}{\partial x_i} T\right) = 0 \tag{13.11}$$

leads first after integration with Gauss's theorem to

$$4\pi\rho\, r^2 v(r) c_p T(r) - 4\pi\lambda r^2 \frac{dT(r)}{dr} = \dot{q}(R), \tag{13.12}$$

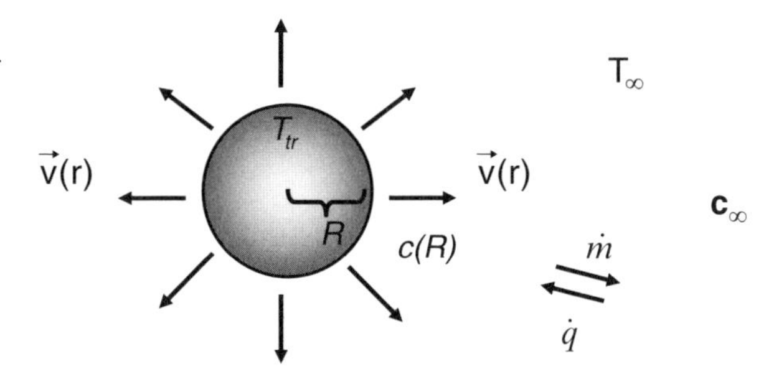

Fig. 13.1 Principle sketch for the representation of vapor and heat fluxes in the droplet environment

finally after inserting (13.9) using the boundary conditions at $T(R) = T_{tr}$ (droplet temperature) and T_∞ auf

$$T(r) = T_\infty + \frac{1 - \exp\left(-\dfrac{\dot{m}\, c_p}{4\pi r \lambda}\right)}{1 - \exp\left(-\dfrac{\dot{m}\, c_p}{4\pi R \lambda}\right)} (T_{tr} - T_\infty). \tag{13.13}$$

For the integration constant $\dot{q}(R)$ (integral heat flow) we obtain

$$\dot{q}(R) = \dot{m}\, c_p T_{tr} + \frac{\dot{m}\, c_p (T_{tr} - T_\infty)}{\exp\left(\dfrac{\dot{m}\, c_p}{4\pi R \lambda}\right) - 1}. \tag{13.14}$$

The (actually non-constant) physical quantities for density, diffusion constant, heat conductivity, and heat capacity are customarily calculated according to the 1/3–2/3 rule, as linear combinations of values at the surface and in infinity, whereby X is representative for the above quantities

$$X = \frac{X_{Tr}}{3} + \frac{2X_\infty}{3}. \tag{13.15}$$

Since however the assumption of a droplet at rest is generally incorrect, in-flow effects are usually considered by means of the following correction according to Ranz-Marschall

$$D \to D \cdot \frac{2 + 0,6\mathrm{Re}^{1/2}\mathrm{Pr}^{1/3}}{2} \text{ und } \lambda \to \lambda \cdot \frac{2 + 0,6\mathrm{Re}^{1/2} Sc^{1/3}}{2}, \tag{13.16}$$

whereby Pr and Sc designate the (laminar) Prandtl and Schmidt number

$$\mathrm{Pr} = \frac{\mu C_P}{\lambda} \text{ und } Sc = \frac{\mu}{\rho D} \tag{13.17}$$

For the correlation of $\dot{m}$ and $\dot{q}$ to the rates of change of the droplet variables T_{tr} and R, the following mass and heat balances can be formulated with the heat capacity of the fluid c_{fl} and the specific evaporation enthalpy $h_V(T)$

$$\dot{q} = \underbrace{\dot{m}\left[h_V(T_{tr}) + c_p T_{tr}\right]}_{\text{Verdampfung}} - \underbrace{\rho_{fl}\frac{4\pi}{3}R^3 c_{fl}\dot{T}_{tr,}A}_{\text{Aufheizung}}, \tag{13.18}$$

$$4\pi R^2 \rho_{fl}\dot{R}_V = -\dot{m}. \tag{13.19}$$

At a given $\dot{m}$ and $\dot{q}$, $\dot{T}_{tr,A}$ and $\dot{R}_V$ can be determined from these equations.

In the literature, one can find several modifications of this modeling. However, these effects appear to be rather unimportant in engine applications. It is still common to use single-component fuel models, although in the meantime interesting multi-component models already exist, which attempt to describe a component spectrum with a small number of form factors as we will look at in the next section. Most CFD codes offer special "synthetic" single-component models for gasoline and diesel fuel. However, n-heptane for gasoline and dodecane for diesel are also reasonable choices. One should always keep in mind, that the properties of a mixture can never be exactly reproduced with a single-component fuel. Moreover, we should point out that we are only concerned with the physical properties, not the chemical properties of the fuels (in this case, for example, n-heptane would be a poor representative for gasoline – to consider only the knocking properties).

What happens now to the fuel in the engine? For this, first consider a GDI engine with late injection. Still relatively cold droplets arrive in a combustion chamber with hot, compressed air. The surface of the droplets must be in a state on the vapor pressure curve (see Fig. 13.2a), it thus takes on the state attributed to the current droplet temperature, such that it "pushes out" a vapor covering with the partial pressure attributed by the vapor pressure curve. The droplets now heat up, i.e. they

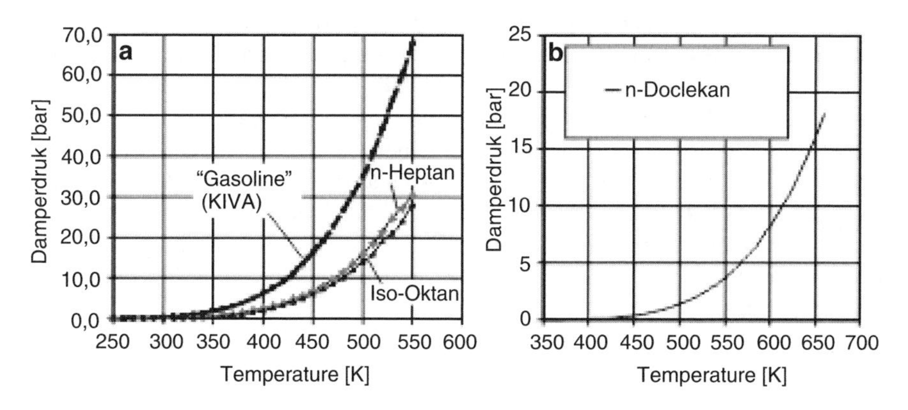

Fig. 13.2 (**a**) Vapor pressure curve for gasoline-class fuels, (**b**) Vapor pressure curve of a component typical of diesel

advance on the vapor pressure curve towards the top right. The vapor covering thickens thereby increasingly. Finally, the point is reached at which the partial pressure is equal to the vapor pressure[2] – the temperature now remains in this state, it boils.

Typically, the droplets are found very close to each other in the combustion chamber (especially in the spray area), all evaporate into a common vapor cloud, the states of which are, consequently, not far away from the vapor pressure curve. This is a confirmation for the reliability of the model of a uniform temperature in the inside of the droplet; because the temperature of the droplet surface cannot move far from the internal temperature, not only because the droplets are small, but also because the temperatures in the environment are limited.

In the diesel engine, a similar process occurs – the droplets proceed on the vapor pressure curve again towards the top right. However, because of the high combustion chamber pressures and the different vapor pressure curves of diesel-typical components, the critical point is now reached, as the end of the vapor pressure curve. After running through the critical point, there is no longer a phase limit. The CFD codes solve this usually by simply eliminating the droplet and attributing it to the vapor phase.

13.1.3 Mass and Heat Exchange in Multicomponent Modeling

Multicomponent models are especially interesting in the case of high residence/ vaporization times, for example in SI engine intake stroke injection or an HCCI process, in which cases a distinctive sequential vaporization and thus demixing of the various components are possible.

The single-component approach shown above can be expanded to 2 or 3 components fairly easily, but the accuracy of the model would not be increased much with respect to the multicomponent-nature of technical fuels. A large number of single components cannot practically be introduced for reasons of computation time as well however. For this reason, we will suggest an approach [see e.g. Lippert et al. (2000), Hallett (2000), and Hermann (2008)], which parameterizes an entire component family. Such a component family can at least characterize to a good extent similar hydrocarbon materials like alkanes as a function of molecular weight (Fig. 13.3). To proceed in a highly correct way, an actual fuel like gasoline would have to be represented by several component families (e.g. one for alkanes, a second for olefins, and a third for aromatics). Since we only want to sketch out the basic method, we will restrict ourselves however to discussing one single component family.

[2]Strictly speaking, droplet temperature never exactly reaches the boiling point, it remains somewhat lower at the wet bulb temperature.

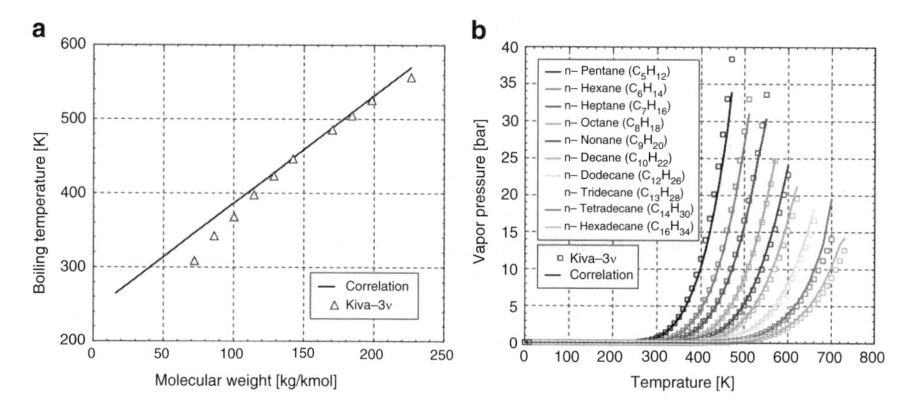

Fig. 13.3 Comparison of model and measurements of boiling temperature (**a**) and vapor pressure curve (**b**) for n-alkanes (from Hermann (2008))

A family is characterized by a molar fraction distribution function n_μ over the molar weight μ. The following suggestion of a four-parameter distribution function is widespread in the literature:

$$n(\mu) = \frac{\delta}{\beta^{\alpha}\Gamma(\alpha)}(\mu - \gamma)^{\alpha-1}\exp\left(-\frac{\mu - \gamma}{\beta}\right) \text{ where } \mu > \gamma. \tag{13.20}$$

The parameter γ is thereby fixed, i.e. also held locally to the predetermined value, while parameters α and β can vary locally as well as between the fluid and gaseous phase. The following moments can be defined:

$$\begin{aligned} n_F &= \int_{\gamma}^{\infty} n_\mu(\mu)d\mu = \delta \\ n_F M_F &= \int_{\gamma}^{\infty} n_\mu(\mu)\mu\, d\mu = \delta(\gamma + \alpha\beta) \\ n_F \Psi_F &= \int_{\gamma}^{\infty} n_\mu(\mu)\mu^2\, d\mu = \delta\left({M_F}^2 + \alpha\beta^2\right). \end{aligned} \tag{13.21}$$

n_F represents the total fuel molar fracture (equal to 1 in the liquid phase), and M_F the average fuel molar weight. If conversely these three moments are known, the distribution can be reconstructed from them (γ is assumed to be fixed). To describe gasoline, the following set of parameters is suggested (Hallett 2000):

$$\alpha = 10.28 \quad \beta = 9.82 \quad \gamma = 0. \tag{13.22}$$

The basic concept now consists in reproducing local differences in the component distributions by local variation of the distribution parameters/moments. To this end, the moments are assumed to be locally distributed, i.e. we set $n_F(x,t), M_F(x,t)$ and $\Psi(x, t)$, for which we must now solve transport equations. Two further transport equations must join the one transport equation for the fuel mass fracture c_F in the gaseous phase, which was already solved. Moreover, every droplet gets its own set of parameters n, M, Ψ (and thus an individual component distribution) for its liquid phase.

It is extremely recommendable to continue working with the mass fraction c as the basis transport parameter, i.e. not to change to a molar-fraction-based formulation (using (13.6) for molar fractions instead of mass fractions), which is unfortunately common in the literature. In particular, the consistency with regard to the single-component vaporization model of the previous section would otherwise be surrendered (this should be reproducible for a δ-function-like distribution n_μ from the multicomponent model). We thus use a mass fraction $c_\mu(\mu)$ with the following conversion to the molar fraction distribution

$$c_\mu(\mu) = n_\mu(\mu)\frac{\mu}{M_{tot}}, \tag{13.23}$$

where M_{tot} designates the average local molar mass across all species (including non-fuel components). With respect to the moments, the following correlations now result:

$$\begin{aligned} \int_\gamma^\infty \frac{c_\mu(\mu)}{\mu}\,d\mu &= \frac{n_F}{M_{tot}} = \frac{c_F}{M_F} \\ \int_\gamma^\infty c_\mu(\mu)\,d\mu &= n_F\frac{M_F}{M_{tot}} = c_F \\ \int_\gamma^\infty c_\mu(\mu)\mu\,dM &= \frac{n_F\Psi}{M_{tot}} = \frac{c_F\Psi}{M_F}. \end{aligned} \tag{13.24}$$

The gaseous phase distribution $n_{\mu,vap,S}$ on the droplet surface is not identical to the distribution $n_{\mu,fl}$ in the liquid phase. The vapor pressure with molar weight μ across the droplet surface is obtained according to the Raoult law from the vapor pressure curve of the μth component $p_{vap}(\mu, T)$ in the following way:

$$n_{\mu,vap,S}(\mu)\,p_{gas} = n_{\mu,fl}(\mu)\,p_{vap}(\mu,T). \tag{13.25}$$

The following approach is taken for the vapor pressure curves:

$$\begin{aligned} &p_{vap}(\mu,T) = p_{atm}\exp\left[\frac{\Delta S}{\tilde{R}}\left(1-\frac{a_S+b_S\mu}{T}\right)\right] \quad \text{with} \quad \Delta S = 87.9kJ/(kmol\cdot K) \\ &p_{atm} = 1{,}013\,bar \quad a_S = 241.4K \quad b_S = 1.45K\cdot kmol/kg. \end{aligned} \tag{13.26}$$

Elegantly, a distribution of type (13.20) is form-invariant under transformation (13.25), i.e. multiplication with a factor $p_{vap}(\mu, T)/p_{gas}$. If the distribution is of form (13.20) with parameters α, β, γ and $\delta = 1$, then $n_{\mu,vap,S}$ also has the form (13.20) with the parameters:

$$\begin{aligned}
\tilde{\alpha} &= \alpha \\
\tilde{\beta} &= \frac{\beta}{\left(\frac{b_S \Delta S}{\tilde{R} T_{tr}}\right) + 1} \\
\tilde{\gamma} &= \gamma \\
\tilde{\delta} &= \frac{p_{atm}}{p_{gas}} \left[\frac{1}{\left(\frac{b_S \Delta S}{\tilde{R} T}\right) + 1} \right]^{\alpha} \exp\left(\frac{\Delta S}{\tilde{R}} \left(1 - \frac{a_S}{T_{tr}}\right) - \gamma \frac{b_S \Delta S}{\tilde{R} T_{tr}} \right).
\end{aligned} \tag{13.27}$$

From this results by means of formulae (13.24) the gas phase distribution moments $n_{F,vap,S}$, $M_{F,vap,S}$ and $\Psi_{vap,S}$ on the surface. The identity $\tilde{\gamma} = \gamma$ confirms the approach of keeping γ constant.

In the droplet environment, one must now solve in addition the transport equations for the distribution moments. From (13.6) follows by multiplication with μ or $1/\mu$ and subsequent integration (here the independence of the diffusion coefficient D from the molar weight μ is again assumed):

$$\begin{aligned}
\frac{\partial}{\partial x_i}\left(\rho v_i \frac{c_{F,vap}}{M_{F,vap}}\right) - \frac{\partial}{\partial x_i}\left(D\rho \frac{\partial}{\partial x_i} \frac{c_{F,vap}}{M_{F,vap}}\right) &= 0 \\
\frac{\partial}{\partial x_i}\left(\rho v_i \frac{c_{F,vap}\Psi_{vap}}{M_{F,vap}}\right) - \frac{\partial}{\partial x_i}\left(D\rho \frac{\partial}{\partial x_i} \frac{c_{F,vap}\Psi_{vap}}{M_{F,vap}}\right) &= 0.
\end{aligned} \tag{13.28}$$

Equations (13.6) and (13.11) and their solutions (13.10) and (13.14) continue to be valid. Equation (13.28) is solved in analogy to (13.11), after integration by means of Gauss's law we obtain:

$$\begin{aligned}
4\pi\rho\, r^2 v(r) \frac{c_{F,vap}}{M_{F,vap}} - 4\pi r^2 D\rho \frac{d}{dr} \frac{c_{F,vap}}{M_{F,vap}} &= \Theta_0 \\
4\pi\rho\, r^2 v(r) \frac{c_{F,vap}\Psi_{vap}}{M_{F,vap}} - 4\pi r^2 D\rho \frac{d}{dr} \frac{c_{F,vap}\Psi_{vap}}{M_{F,vap}} &= \Theta_1.
\end{aligned} \tag{13.29}$$

After inserting (13.9), integration, and inserting the boundary conditions, we obtain for the integration constant (the global moment flows), completely analogously to (13.14):

$$\Theta_1 = \dot{m}\frac{c_{F,vap}(R)}{M_{F,vap}(R)} + \frac{\dot{m}}{\exp\left(\dfrac{\dot{m}}{4\pi D\rho R}\right) - 1}\left[\frac{c_{F,vap}(R)}{M_{F,vap}(R)} - \frac{c_{F,vap,\infty}}{M_{F,vap,\infty}}\right]$$

$$\Theta_2 = \dot{m}\frac{c_{F,vap}(R)\Psi_{vap}(R)}{M_{F,vap}(R)} + \frac{\dot{m}}{\exp\left(\dfrac{\dot{m}}{4\pi D\rho R}\right) - 1}\left[\frac{c_{F,vap}(R)\Psi_{vap}(R)}{M_{F,vap}(R)} - \frac{c_{F,vap,\infty}\Psi_{\infty}}{M_{F,vap,\infty}}\right] \tag{13.30}$$

From this results the temporal development for the moments of the liquid distribution $M_{F,fl}$ and Ψ_{fl} in the droplet [analogously to (13.19)]:

$$\begin{aligned} \frac{d}{dt}\frac{\frac{4}{3}\pi R^3 \rho_{fl}}{M_{F,fl}} &= -\frac{\dot{m}}{M_{F,fl}} + \frac{4}{3}\pi R^3 \rho_{fl}\frac{d}{dt}\frac{1}{M_{F,fl}} = -\Theta_1 \\ \frac{d}{dt}\frac{\frac{4}{3}\pi R^3 \rho_{fl}\Psi_{fl}}{M_{F,fl}} &= -\frac{\dot{m}\Psi_{fl}}{M_{F,fl}} + \frac{4}{3}\pi R^3 \rho_{fl}\frac{d}{dt}\frac{\Psi_{fl}}{M_{F,fl}} = -\Theta_2 , \\ X &= \frac{X_{Tr}}{3} + \frac{2X_{\infty}}{3} \end{aligned} \tag{13.31}$$

where the flows Θ_i from (13.30) are to be inserted. Parameters like D or ρ should again be calculated in accordance with the 1/3–2/3 rule (13.15).

The transport equations in the gaseous phase including source terms now read:

$$\begin{aligned} \rho\left(\frac{\partial}{\partial t} + v_i\frac{\partial}{\partial x_i}\right)\frac{c_F}{M_F} - \frac{\partial}{\partial x_i}\left(\rho D_t\frac{\partial}{\partial x_i}\frac{c_F}{M_F}\right) &= \frac{1}{\Delta V}\sum_{droplets\ in\ volume\ \Delta V}\Theta_1 \\ \rho\left(\frac{\partial}{\partial t} + v_i\frac{\partial}{\partial x_i}\right)c_F - \frac{\partial}{\partial x_i}\left(\rho D_t\frac{\partial}{\partial x_i}c_F\right) &= \frac{1}{\Delta V}\sum_{droplets\ in\ volume\ \Delta V}\dot{m} \\ \rho\left(\frac{\partial}{\partial t} + v_i\frac{\partial}{\partial x_i}\right)\frac{c_F\Psi}{M_F} - \frac{\partial}{\partial x_i}\left(\rho D_t\frac{\partial}{\partial x_i}\frac{c_F\Psi}{M_F}\right) &= \frac{1}{\Delta V}\sum_{droplets\ in\ volume\ \Delta V}\Theta_2 \end{aligned} \tag{13.32}$$

In addition, all required material parameters (diffusion coefficient of the vapor phase, critical temperature, liquid density, vaporization enthalpy, heat conductivity, heat capacity of the liquid and gas phase, viscosity of the liquid phase, surface tension) should now be described as a function of the component distribution; see the relevant literature (Tanim and Hallett 1995).

13.1.4 Flashboiling

So-called "flashboiling" is a special case that sets in when warm fuel is injected into an intake pipe or combustion chamber with a surrounding pressure lying below the vapor pressure corresponding to the droplet temperature. In this case, the spray immediately "explodes", so to speak. This phenomenon cannot be described with the models described above.

Let us consider a mixture of air and fuel with the corresponding molar densities n_A and n_F and the average molar weights M_A and M_F. Let p_F be the fuel partial pressure and p the total pressure. We then have for the fuel mass fracture c_F:

$$\begin{aligned} c_F &= \frac{n_F M_F}{n_F M_F + n_A M_A} = \frac{\frac{p_F}{p} M_F}{\frac{p_F}{p} M_F + \left(1 - \frac{p_F}{p}\right) M_A} \\ &= 1 + \left(\frac{p_F}{p} - 1\right) \frac{M_A}{M_A + \frac{p_F}{p}(M_F - M_A)} \geq 1 \quad \textit{für} \quad p_F \geq p. \end{aligned} \tag{13.33}$$

As soon as the vapor pressure exceeds the surrounding pressure, c_F becomes larger than 1 arithmetically, and the argument of the logarithm in (13.10) is now undefined. The vaporization rate is also very high in reality, i.e. not describable by means of a process of physical equilibrium. However, this "non-equilibrium" vaporization comes to a rest (i.e. the process makes a transition to the previously described equilibrium process) when so much heat is removed by the high vaporization rate that the associated vapor pressure is below the surrounding pressure. Essential that matters then is that the lack of definition in (13.10) be intercepted and instead a high, yet finite vaporization rate be calculated until (13.10) is defined again. One possibility is to calculate the mass flow with a relation

$$\dot{m} = 4\pi D\rho R \ln\left(\frac{1 - c(\infty)}{\max(1 - c(R), \varepsilon)}\right), \tag{13.34}$$

where $\varepsilon > 0$ should be a very small, but finite number, which essentially should be determined by the numerical stability of the code.

13.2 Spray Statistics

An injection spray typically consists of hundreds of millions of droplets. Such an ensemble is (similar to turbulent gas flow) no longer deterministically calculable (i.e. on the basis of each single droplet). For the spray as droplet ensemble, it is therefore appropriate to introduce a statistical description, a droplet distribution density $p(\vec{x}, \vec{v}, R, T)$. It describes the probability of finding at location $\vec{x}$ a droplet with velocity $\vec{v}$, radius R, and temperature T (further parameters should be added if necessary, but we will restrict ourselves to these eight parameters in the following).

Already 120 years ago, Ludwig Boltzmann was confronted with a similar problem as he sought to derive thermodynamics on the basis of statistics from the mechanics of atomic and molecular processes. For this he developed the *Boltzmann equation* which takes his name and with the help of that he could ground the 2nd law of thermodynamics on an atomistic level for the first time in history [see, for example, Landau and Lifschitz (1981)]. This equation is the basis of the kinetic gas theory today. It was applied to the special dynamics of droplets by Williams (1958), in which case it is simply called *spray equation*, the basis for all spray models. However, this equation can no longer be integrated in a closed way. Instead, stochastic methods are used in the standard model – the solution is derived by "throwing dice" (the "Monte Carlo simulation"). This method must also be considered critical from the standpoint of accuracy, which will be discussed in more detail in the next chapter.

Following this, the description of multi-particle processes such as collision or breakup will be introduced. Turbulent gas flow induces turbulent structures even in the spray itself, and the modeling of this – only statistically describable – phenomenon will then be discussed. One essential step will be to simplify the complex source term of the Boltzmann equation by careful discussion such that its solvability can be significantly increased. In principle, this will involve the replacement of the Boltzmann equation with a *Fokker–Planck equation.*

13.2.1 The Boltzmann–Williams Equation

For every single point of the eight-dimensional space of the variables $\vec{x}, \vec{v}, R, T$, the dynamics $\dot{\vec{x}}, \dot{\vec{v}}, \dot{R}, \dot{T}$ is described by (13.1), (13.2), (13.18) and (13.19). One can now summarize the eight variables in the 8-tupel $\alpha = (\vec{x}, \vec{v}, R, T)$ and write the equations of motion for it as

$$\dot{\alpha}_i = A_i(\alpha) \quad \text{with} \quad i = 1, ..., 8. \tag{13.35}$$

The goal is now to find an equation of motion for the distribution function $p(\alpha)$, of a "point cloud" in this phase space, so to speak. The solution of this problem is the so-called *Liouville equation*[3]

$$\begin{aligned}\frac{\partial}{\partial t}p(\alpha, t) &= -\sum_{i=1}^{8} \frac{\partial}{\partial \alpha_i}(A_i(\alpha)p(\alpha, t)) \\ &= -\sum_{i=1}^{8} A_i(\alpha)\frac{\partial}{\partial \alpha_i}p(\alpha, t) - \left(\sum_{i=1}^{8} \frac{\partial A_i(\alpha)}{\partial \alpha_i}\right)p(\alpha, t).\end{aligned} \tag{13.36}$$

[3]In classical mechanics, the Liouville equation is normally formulated for conservative or Hamiltonian systems. The second term of the second line of (13.36) then vanishes because of the canonical Hamiltonian equations, i.e. the phase space volume is incompressible in this case.

The Liouville equation is a hyperbolic equation. Its characteristics are the equations of motion (13.35), as one sees, with the help of the formulation in the lower line of (13.36). The formulation in the upper line of (13.36), on the other hand, shows that the probability is maintained, i.e.

$$\frac{\partial}{\partial t}\int_{Phasenraum} d\alpha\, p(\alpha,t) = 0. \tag{13.37}$$

is valid. In the Liouville equation, all (discontinuous) multi-droplet processes like collision and breakup are not taken into consideration. This was also, applied to atoms and molecules, the problem of Boltzmann. He extended the Liouville equation by one source term, the so-called *collision integral*. First, one extracts the location $\vec{x}$ from the 8-tupel α.

$$\alpha = (\vec{x},\beta)\,\text{und}\,\beta = (\vec{v},R,T). \tag{13.38}$$

For a general dual-particle collision term

$$I_{Stoß} = \int p(\vec{x},\beta_1)p(\vec{x},\beta_2)\sigma(\beta_1\,\beta_2 \to \beta + \ldots;\vec{x})d\beta_1 d\beta_2 - \int p(\vec{x},\beta_1)p(\vec{x},\beta)\sigma(\beta_1\,\beta \to \ldots;\vec{x})d\beta_1 \tag{13.39}$$

The term $\sigma(\beta_1\,\beta_2 \to \beta + \ldots)$ describes the conditional probability that with the presence of droplets β_1 and β_2 at location $\vec{x}$, a droplet of property β emerges through collision; $\sigma(\beta_1\,\beta \to \ldots)$ stands for the conditional probability that droplet β collides with droplet β_1, the consequence of which is the disappearance of droplet β, hence the minus sign of the complete term. Three-particle collision processes (and processes of still higher orders of particles) are usually ignored, "single-particle processes" are nothing but disintegrations. With the above logic, one can write for them

$$I_{Zerfall} = \int p(\vec{x},\beta_1)\sigma(\beta_1 \to \beta + \ldots;\vec{x})d\beta_1 - p(\vec{x},\beta)\sigma(\beta \to \ldots;\vec{x}). \tag{13.40}$$

In summa, we obtain the Boltzmann–Williams equation

$$\begin{aligned}
&\frac{\partial}{\partial t}p(\alpha,t) + \sum_{i=1}^{8}\frac{\partial}{\partial\alpha_i}(A_i(\alpha)p(\alpha,t)) \\
&= \int p(\vec{x},\beta_1)p(\vec{x},\beta_2)\sigma(\beta_1\,\beta_2 \to \beta + \ldots;\vec{x})d\beta_1 d\beta_2 \\
&\quad - \int p(\vec{x},\beta_1)p(\vec{x},\beta)\sigma(\beta_1\,\beta \to \ldots;\vec{x})d\beta_1 \\
&\quad + \int p(\vec{x},\beta_1)\sigma(\beta_1 \to \beta + \ldots;\vec{x})d\beta_1 - p(\vec{x},\beta)\sigma(\beta \to \ldots;\vec{x}).
\end{aligned} \tag{13.41}$$

This equation already completely describes the spray dynamics to be calculated. However, turbulent dispersion, i.e. the interaction of droplets with the flow turbulence, is only implicitly contained in (13.41), specifically in the second term of the left side, as long as the function $A_i(\alpha)$ contains the instantaneous and not the averaged gas velocity. However, we require a formulation with the averaged gas velocity. After conversion, we acquire an additional collision term on the right side which contains the interaction with the turbulent fluctuation of the gas velocity

$$\begin{aligned}
&\frac{\partial}{\partial t}p(\alpha,t)+\frac{\partial}{\partial x_i}\left[v_{tr,i}p(\alpha,t)\right]+\frac{\partial}{\partial v_{tr,i}}\left[D_{tr}\left(\langle v_{g,i}\rangle - v_{tr,i}\right)p(\alpha,t)\right]+\frac{\partial}{\partial R}\left[\dot{R}_V p(\alpha,t)\right]\\
&+\frac{\partial}{\partial T_{tr}}\left[\dot{T}_A p(\alpha,t)\right]=\int p(\vec{x},\beta_1,t)p(\vec{x},\beta_2,t)\sigma(\beta_1\,\beta_2\rightarrow\beta+\ldots;\vec{x})d\beta_1 d\beta_2\\
&-\int p(\vec{x},\beta_1,t)p(\vec{x},\beta,t)\sigma(\beta_1\,\beta\rightarrow\ldots;\vec{x})d\beta_1\\
&+\int p(\vec{x},\beta_1,t)\sigma(\beta_1\;\rightarrow\beta+\ldots;\vec{x})d\beta_1 - p(\vec{x},\beta,t)\sigma(\,\beta\rightarrow\ldots;\vec{x})\\
&+\int \sigma(\vec{x},\vec{v}''_{gas})\frac{\partial}{\partial v_{tr,i}}\left(D_R v''^{i}_{g}p(\vec{x},\beta_1,t)\right)\mathrm{d}^3 v''_{gas}\\
&mit D_{tr}=\frac{3}{8}\frac{\rho_g}{\rho_{fl}}c_w\frac{\left|\langle\vec{v}_g\rangle-\vec{v}_{tr}\right|}{R}.
\end{aligned} \tag{13.42}$$

where $\sigma(\vec{x},\vec{v}''_{gas})$ designates the probability of the appearance of the corresponding velocity fluctuation at location x and the concrete terms already developed in the previous sections have already been inserted on the left side; only the collision and breakup process terms of the right side were left in the formal notation.

13.2.2 The Numerical Solution of the Boltzmann–Williams Equation; the Standard Model (Lagrange Formulation)

Equation (13.42) is a high-dimensional, integro-differential equation. With direct methods, it is not solvable in the foreseeable future in eight-dimensional space. One only has to think of every dimension discretized into ten levels (this is surely still much too rough for many of the eight dimensions, e.g. for the spatial coordinates). In that case, a computational mesh of 10^8 cells has already been produced! For the solution of the equation therefore, another way must be taken. Considering the fact that the underlying Liouville equation is of hyperbolic character, it can be solved along its characteristics (which correspond to the droplet trajectories), i.e. by means of ordinary differential equations. For the case that the relevant area, in which the final droplet density is clearly other than zero, describes only a "low-dimensional surface" in eight-dimensional space, one possibly only needs a "few" trajectories.

With that, the following strategy is chosen: we introduce sufficiently many representative particles, so-called "parcels", which "follow" the characteristics of the Liouville equation (13.36). All continuous processes are thus already handled. The treatment of the collision processes, i.e. the terms on the right side of (13.41) is, however, equally obvious. The representative parcels experience these as a stochastic process, and they are formulated in the collision integrals exactly in this way. For all parcels found at a particular time in a volume element (practically speaking, "in a computational cell"), collision probabilities are calculated in accordance with the formulae (13.39) and (13.40), and, in accordance with these probabilities, the concrete actions are "diced out" (hence the notion "Monte Carlo simulation").

The actual continuous statistical description is thus modified into a discontinuous, equivalent stochastic description; the dynamics of the droplet distribution function is described by means of the dynamics of an ensemble of representative stochastic parcels. But now these parcels look very similar to the original droplets, all continuous processes progress similarly (momentum exchange with flow, evaporation, heating-up); one must only work stochastically for discontinuous processes (formation, breakup, collision). From this originates the name "discrete droplet model" (DDM). This is precisely the charm and intuitive comprehensibility of the stochastic model: one has the "feeling" of working with the individual drops, although one is ultimately only operating with stochastic parcels. All visualizations of injection spray simulations show these stochastic parcels, never droplets. Practically speaking, a parcel has the properties of a droplet (thus in the present case the eight variables $\vec{x}, \vec{v}, R, T$) and also a statistical weight, which can be interpreted, for example, as the number of actual droplets that it represents. We can also interpret it as a parcel mass; the "number of represented droplets" is then the ratio of the parcel mass to the droplet mass.

The individual parcels represent elements of droplet flow, i.e. a description by means of parcels has a Lagrangian character, because it describes the flow by means of moving reference points. The number of parcels is completely independent of the number of droplets and should only be determined from considerations of statistical convergence.

The danger of naive or casual use of spray simulations in the standard Lagrangian model cannot be stressed enough. Due to the use of common differential equations, the model is quite harmless in its mathematical structure, and it is not difficult to produce results – indeed, even original modeling approaches are typically not difficult to implement. The greatest challenge is to generate correct, converged results!

To give an idea of the full extent of the problem, it is practical to review the original gas-dynamic Boltzmann equation for atoms and molecules. It is possible to derive the fluid-mechanical equation set (12.1)–(12.18), thus including the Navier–Stokes equations, from it [with the Chapman–Enskog approach, e.g. see Landau and Lifschitz (1981)], but this is not trivial. The Boltzmann equation thus describes complex collective behavior, clearly different from the individual-particle behavior of atoms and molecules. Naturally, the Boltzmann equation contains much more, including gas-dynamic effects which are not contained in fluid-mechanical equations such as the Navier–Stokes equation. If one is interested only in fluid-mechanical topics however, it is not recommended to use a computer program for solving the Boltzmann equation; it is incomparably more difficult and requires much more processing power to find an correct solution in this way than using the Navier–Stokes equation set, in which the collective behavior of the fluid-mechanical continuum is already been made explicit. It was also probably never Boltzmann's goal to solve his equation for real technical problems.

On the level of the spray, the situation is not simpler. Droplets have a higher degree of freedom including radius and temperature, and other processes are involved such as heating, vaporizing and collision. An injection spray also shows much different behavior to the individual droplet, and it can as collective behavior again only be reconstructed with the Boltzmann–Williams equation with a high amount of processing power.

This also explains the hope to make significant progress with an alternative "Euler modeling". This does not mean that a description in fixed coordinates is less expensive than one in co-moving coordinates (which in the narrower sense describes the difference between "Euler formulation" and "Lagrangian formulation"). Since a direct "Eulerian" solution of the Boltzmann–Williams equation is hopeless, "Euler model" in this context means the extraction of moment equations such as the transport equations for the droplet mass, the locally averaged droplet momentum, the averaged momentum variance etc. from the Boltzmann equation. That is, it is a question of developing a Navier–Stokes equation set for the liquid phase. This equation set would then be solved with the Eulerian method for obvious reasons. We will return to this point later.

For now, let us return to the standard approach. To complete the modeling process, another model-technical description of multi-particle processes will be derived following a mathematical digression concerning the determination of random numbers.

13.2.3 Excursus: The Numerical Determination of Random Numbers

Let us first discuss a concrete problem in programming realization. In the stochastic modeling for solving the spray equation, the determination of random numbers following a pre-given probability distribution is an important building element. According to the standard, computers only make rand m number generators available for the generation of equally distributed random numbers between 0 and 1. The basic task now reads as follows

$$x \in X \subset \Re^n, f : X \to \Re, f(x) > 0, \int_X f(x)dx = 1. \tag{13.43}$$

Drawing randomly an element $x \in X$ corresponding to the distribution function f means that if we repeat multiply the process, the elements $x_1, x_2, x_3, \ldots$ should be distributed in accordance with the function f. To fulfill this task, two varying methods will be discussed:

Method I – "integrate and invert": this approach only works for one-dimensional distributions, i.e. $X = [a, b] \subset \Re$. We first calculate the distribution function $F(x)$

$$F(x) = \int_a^x f(\xi) d\xi \quad F : [a, b] \to [0, 1]. \tag{13.44}$$

Because of the positivity of f, F is strictly monotonous and thus invertible.

Secondly, we determine the inverse function $F^{-1} : [0, 1] \to [a, b]$ (if necessary, we have to numerically integrate and tabularize) and draw a random number $z \in [0, 1]$. The value $x = F^{-1}(z)$ is then our desired random variable. To found this: with the probability dF , the random number lands in the interval $[x, x + \mathrm{d}x]$, whereby

$$\mathrm{d}x = \frac{\mathrm{d}F}{f(x)}$$

The density of probability p is given as the ratio of probability to interval length

$$p = \frac{\mathrm{d}F}{\mathrm{d}x} = f.$$

Method II – "draw and evaluate": this method is suitable for multi-dimensional spaces X as well.

First step: we determine an element $x \in X$ on the basis of equal distribution. With complicated quantities (e.g. the inside of a calculation area with complicated margins), one can proceed as follows: one inscribes $X \subset \Re^n$ into a "n-dimensional rectangle (quader)"

$$X \subset \tilde{X} = [a_1, b_1] \times [a_2, b_2] \times \cdots \times [a_n, b_n] \tag{13.45}$$

and, in an equally distributed manner, draws an element from $x \in \tilde{X}$ with the help of n random numbers $z_1, z_2, \ldots, z_n \in [0, 1]$

$$x = (a_1 + z_1(b_1 - a_1), a_2 + z_2(b_2 - a_2), \ldots, a_n + z_n(b_n - a_n)).$$

There are two possibilities: x lies in X, in which case it is our chosen element. Or x does not lie in X, then it is rejected and a new selection process started. In this way we are sure that all elements from X are chosen with equal probability.

Second step: the variable x chosen in the first step is evaluated. For this, $f_{\max} = \max(f(\xi), \xi \in X)$. We draw a further random number $\tilde{z} \in [0, 1]$ and compare it with the ratio

$$\zeta = \frac{f(x)}{f_{\max}}.$$

If $\tilde{z} \leq \zeta$, x is accepted, otherwise it is rejected and, the process is commenced again from step one until an element is found and accepted. The reason for this: the probability that $\tilde{z} \leq \zeta$ is proportional to $f(x)$. In this way, every element x is chosen with a correct relative probability. Through the process of rejection and repetition in the case of a non-acceptable element x, we safeguard that the normalization of the probability density is fulfilled, and finally an element is selected with probability 1.

Example: one typical task is to determine equally distributed spatial points within the calculation area. The calculation volume is however discretized into mesh cells, and thus we are actually dealing with the determination of mesh cells, under the boundary condition of equal distribution with reference to the volume. One should never in this situation think of simply drawing cell numbers at equal distribution! For the mesh cells possess, in general, (very) different volumes, and one must take this into consideration in the selection process. In this case, method II is used ideally, such that we first draw random numbers equally distributed among the cell numbers and then evaluate the selected cell Z according to its volume V_Z, i.e.

$$f(Z) = V_Z.$$

13.2.4 Parcel Start Conditions at the Nozzle Exit

In the standard model, parcels have to be generated at the nozzle, and logically this occurs stochastically. Typically, one determines the injection direction per parcel in a pre-given solid angle or spray cone area, and possibly an initial droplet size in accordance with a drop size distribution as well. With the toolbox we developed just now, more complex initial conditions like correlations between injection direction and droplet size would also be realizable. Practically speaking however, the experimental data are usually lacking for the derivation of such complex constraints. Simulations of internal nozzle flow can be helpful here.

It is recommendable to give every parcel at the nozzle (independent of the drop size) the same mass (i.e. a parcel "consists" of many small or a few large droplets). This corresponds to the approach, according to which the fuel mass is the quantity that is of actual interest to be discretized by parcels.

As an example, a method for the simulation of uniformly distributed injection into a spray cone area of angle 2φ will be derived. Two angles should be randomly chosen, the azimuth angle $\theta \in [0, 2\pi]$ and the polar angle $\gamma \in [0, \varphi]$. The azimuth angle θ may be selected in an equally distributed fashion, but γ may not. We remind the reader that the spatial angle measure has the form $\sin\gamma \, \mathrm{d}\gamma \mathrm{d}\theta$ in integration, and exactly this distribution must be selected. Because of the equal distribution of θ, we restrict ourselves to the choice of γ. This is solely a one-dimensional problem, and one can therefore use method I. The distribution function of γ has the form

$$F(\gamma) = \frac{\int_0^{\gamma} \sin\tilde{\gamma}\, d\tilde{\gamma}}{\int_0^{\varphi} \sin\tilde{\gamma}\, d\tilde{\gamma}} = \frac{1-\cos\gamma}{1-\cos\varphi}. \tag{13.46}$$

From a random number $z \in [0, 1]$, we then obtain the following γ value

$$\gamma = arc\ \cos(1 - z + z\cos\varphi).$$

13.2.5 Modeling Breakup Processes

Breakup processes influence the spray especially in its early phase close to the nozzle, and thus, practically speaking, form a unit together with the nozzle model. For example, a breakup model, which already provides small drops very fast, can be replaced by a nozzle model with small drops.

Because of the different mechanism, two kinds of spray breakup can be distinguished, primary breakup and secondary breakup. Primary breakup results from properties given to the spray already by the internal nozzle flow, like turbulence and cavitation (which creates turbulence again through cavitation-bubble implosion). For secondary breakup, aerodynamic processes are relevant that are not a result of internal nozzle flow.

For primary breakup modeling, we need information about the internal nozzle flow, about its turbulence and cavitation distribution. Then, breakup time and length can be derived from the turbulent scales and cavitation bubble densities, see e.g. Tatschl et al. (2000). For turbulence parameters k and ε of the fluid inside the droplet or ligament, the following temporal development is assumed:

$$\frac{dk_{fl}}{dt} = -\varepsilon_{fl} + S_K \qquad \frac{d\varepsilon_{fl}}{dt} = -1{,}92\frac{\varepsilon_{fl}}{k_{fl}}\left(\varepsilon_{fl} - S_K\right) \tag{13.47}$$

where S_K is a source term from a cavitation bubble implosion. From these parameters, we can determine the parameters for a breakup model – breakup time τ_{prim} and the stable droplet radius $R_{S,\ prim}$:

$$\tau_{prim} = B_{prim}\frac{k_{fl}}{\varepsilon_{fl}} \quad \text{and} \quad R_{S,prim} = A_{prim} C_{\mu}{}^{\frac{3}{4}} \frac{k_{fl}{}^{\frac{3}{2}}}{\varepsilon_{fl}}. \tag{13.48}$$

However, information on internal nozzle flow is not always available, even today. Particularly in the case of modern diesel injection systems, a very strong

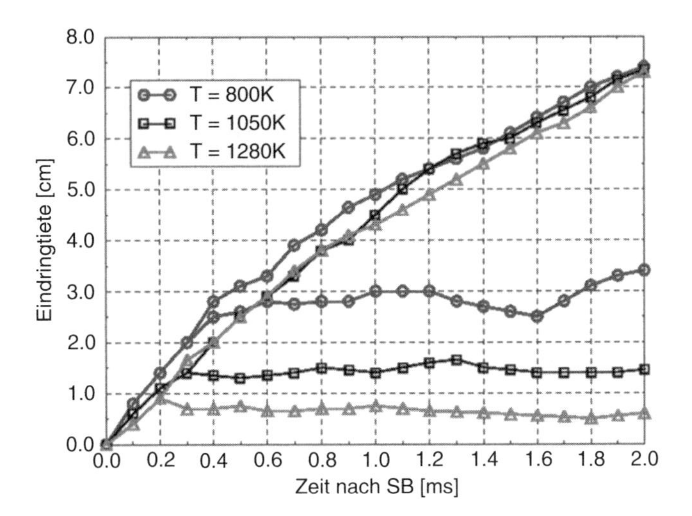

Fig. 13.4 Measurement of the penetration length of a diesel spray in a hot chamber at varying chamber temperatures but equal density (Krüger (2001))

primary breakup occurs; the spray leaves the nozzle in a "foamy" state so to speak. This offers the chance to start already with small droplets at the nozzle, i.e. no longer considering primary breakup explicitly. A typical drop size would be about 5 μm diameter. One sensible method of obtaining information through measurements about initial droplet sizes is described by Krüger (2001). It consists in injecting the spray into a hot chamber at varying chamber temperatures and, e.g. with sliding and Mie scattering techniques, to determine the penetration depth of the liquid and gaseous phases (direct droplet size measurements are hopeless because of the density of diesel injection sprays, at least highly unreliable). Such an experimental result is shown in Fig. 13.4. At equal chamber densities, the gaseous phase penetrates approximately to the same extent, but the penetration depth of the liquid phase is much different, it reduces with increasing temperature. Our task is now to synchronize the spray breakup parameter and initial droplet size such that all of these penetration graphs can be reproduced with one set of parameters. This method led in the previous case to an initial Sauter diameter[4] of 5 μm at small further breakup rates. However, the prerequisite of this procedure is a functioning spray model.

If we diminish the initial droplet size of a spray continuously, then *local homogeneous flow* is initiated as a limiting case. In it, gaseous and liquid phases are in kinetic and thermodynamic equilibrium, because, on the one hand, very small drops have a relatively large flow resistance related to their mass, and as a result, no speed difference between the two phases is possible anymore; on the other hand,

[4]The Sauter diameter d_S of a droplet distribution is defined as the mean value of d^3divided by the mean value of d^2, $d_S = \langle d^3 \rangle / \langle d^2 \rangle$.

because of the high surface rate per volume unit, the gaseous phase must persist in a vapor pressure curve state. In fact, we now have a single phase flow. Experimental investigations show that at least typical diesel injections can be relatively well described as a local homogeneous flow, see e.g. Siebers (1998). The experimental expression for "locally homogeneous flow" is "mixture-controlled". For the spray model, this means that the droplet size (and thus the spray breakup) is no longer a decisive factor if the droplet size is selected sufficiently low!

Secondary breakup processes have aerodynamic causes and thus show longer breakup lengths. They occur in competition with primary breakup. In dense diesel injections with strong primary breakup, they thus play a rather small role. In the case of gasoline direct injection on the other hand, there is hardly any or no primary breakup because of the minimal turbulence and cavitation rates of the internal nozzle flow (varying according to the injector type), and therefore secondary breakup can even play the dominant role. Especially in the case of the complex flow swirl structures of cone sprays, droplet sizes appear to have a real influence on the spray structure, and thus we are farther removed from the limiting case of local homogeneous flow.

In describing secondary breakup, we usually use an analysis of instability; the main effect is the so-called Kelvin–Helmholtz instability. Probably the best-known modeling approach is the WAVE model, see Reitz (1987). The wave length Λ growing the most and its rate of growth Ω read

$$\frac{\Lambda}{R} = 9.02 \frac{\left(1 + 0.45\, Oh^{0,5}\right)\left(1 + 0.4\, \Theta\right)^{0.7}}{\left(1 + 0.865\, We_{tr}^{1.67}\right)^{0.6}} \tag{13.49}$$

and

$$\Omega\left(\frac{\rho_{Tr}\, R^3}{\sigma_{Tr}}\right)^{0.5} = \frac{\left(0.34 + 0.38\, We_g^{1.5}\right)}{\left(1 + Oh\right)\left(1 + 1.4\, \Theta^{0.6}\right)}, \tag{13.50}$$

whereby for the Weber numbers for liquid and gaseous phases We_{tr} and We_g and the Ohnesorge number Oh as well as the Taylor number Θ is valid:

$$We_{g/tr} = \frac{\rho_{g/tr}\, R\, v_{rel}^2}{\sigma_{fl}}, Oh = \frac{\sqrt{We_{tr}}}{\mathrm{Re}_{tr}}, \Theta = Oh\sqrt{We_g}, \tag{13.51}$$

where σ_{fl} designates the surface tension. With these quantities, a breakup time $\tau_{\sec}$ and a stable radius $R_{s,\sec}$ can be defined

$$\tau_{\sec} = \frac{3.788\, B_1\, R}{\Lambda\, \Omega}, \tag{13.52}$$

$$R_{s,\mathrm{sec}} = \begin{cases} B_0\,\Lambda & \text{fur} \quad B_0\,\Lambda \le R \\ \min\left(\sqrt[3]{\dfrac{3\pi R^2 v_{rel}}{2\Omega}}, \sqrt[3]{\dfrac{3R^2\Lambda}{4}}\right) & \text{fur} \quad B_0\,\Lambda > R \end{cases}, \tag{13.53}$$

where B_0 and B_1 are model constants.

There are different options for modeling spray breakup: in the "naive" variant, we emulate droplet disintegration by means of parcels as much as possible by drawing a probability of breakup at random in a time interval per droplet and then creating several daughter parcels from one parcel in accordance with the assumed breakup process from the thereby "diced" breakups. If the average droplet breakup time amounts to τ, then the probability $W(n)$ for n breakups within a time interval Δt is Poisson-distributed, i.e. given by the relation

$$W(n) = \frac{\left(\dfrac{\Delta t}{\tau}\right)^n}{n!} \exp\left(-\frac{\Delta t}{\tau}\right). \tag{13.54}$$

What is problematic here is that the breakup processes cause the number of parcels to increase dramatically, but this number should be determined, as previously mentioned, for reasons of statistical convergence, not due to physical processes. If for example the amount of parcels is statistically sufficient after several breakups, there must be too few before it, i.e. particularly in the critical zone near the nozzle! It is therefore practical to take a path that does not change the number of parcels. We can then assign a possible daughter droplet radius determined by chance to the particle after breakup; the conservation of mass is guaranteed because the parcel mass does not change (i.e. many small droplets come from a few large ones). And in the statistic average (if we have enough parcels), the desired daughter-droplet radius distribution should be obtained.

We can however go still another step and emulate breakup as a continuous average process, i.e. describe how the radius changes on average. To this end, we typically use the relation

$$\frac{\mathrm{d}R}{\mathrm{d}t} = -\frac{R - R_s}{\tau_B} = \dot{R}_Z \tag{13.55}$$

whereby we have already made reference to the parameters of (13.48) or (13.52) and (13.53). The correlation to the elementary process on the level of the droplet can be made quickly. If during breakup m daughter droplets are created on average with breakup time τ, radius change in the time interval Δt on average amounts to

$$R = \frac{R_0}{m^{\frac{\Delta t}{3\tau}}} \text{ i.e. } \frac{dR}{dt} = -\frac{1}{3\tau} \cdot \ln m \cdot R \text{ and thus } \quad \tau_B = \frac{3\tau}{\ln m}. \tag{13.56}$$

In the case of two competing breakup processes (e.g. primary and secondary breakup), the breakup rates must be added:

$$\frac{dR}{dt} = -\frac{R - R_{S,prim}}{\tau_{prim}} - \frac{R - R_{S,\sec}}{\tau_{\sec}} = -\frac{R - R_S}{\tau_B} = \dot{R}_Z \quad \text{where}$$
$$\tau_B = \frac{\tau_{prim} \cdot \tau_{\sec}}{\tau_{prim} + \tau_{\sec}} \qquad R_S = \frac{\tau_{\sec} R_{S,prim} + \tau_{prim} R_{S,\sec}}{\tau_{prim} + \tau_{\sec}} \tag{13.57}$$

While in the first ("naive") case, the breakup term was taken into consideration via (13.40) on the right side of (13.42) in the collision integral, it is described with (13.57) as a continuous process for every particle on the left side of (13.42). The second variant is recommended, so we will restrict ourselves to it in the following. One should however take care that the term $\dot{R}_Z$ is not simply added in (13.42) to the vaporization term $\dot{R}_V$. The reason for this is that $\dot{R}_Z$ only reproduces a "repacking" into smaller droplets, but not a mass loss such as $\dot{R}_V$. If the function $p(\vec{x}, \vec{v}, R, T; t)$ describes the droplet number distribution, the liquid mass distribution $\rho(\vec{x}, \vec{v}, R, T; t)$ is given as

$$\rho(x, v, R, T; t) = \frac{4\pi\rho_{fl} R^3}{3} p(\rho(x, v, R, T; t). \tag{13.58}$$

In order that the breakup term does not lead to a mass defect in the liquid mass, the following must be valid:

$$\frac{\partial \rho(x, v, R, T; t)}{\partial t} = \ldots + \frac{\partial}{\partial R}\left[\dot{R}_Z \rho(x, v, R, T; t)\right] + \ldots . \tag{13.59}$$

Applied to the number distribution function $p(\alpha; t)$ (with $\alpha = (\vec{x}, \vec{v}, R, T)$), this means

$$\frac{\partial}{\partial t} p(\alpha, t) + \frac{\partial}{\partial x_i}\left[v_{tr,i} p(\alpha, t)\right] + \frac{\partial}{\partial v_{tr,i}}\left[D_{tr}\left(\langle v_{g,i}\rangle - v_{tr,i}\right) p(\alpha, t)\right] + \frac{\partial}{\partial R}\left[\dot{R}_V p(\alpha, t)\right]$$
$$+ \frac{1}{R^3}\frac{\partial}{\partial R}\left[\dot{R}_Z R^3 p(\alpha, t)\right] + \frac{\partial}{\partial T_{tr}}\left[\dot{T}_A p(\alpha, t)\right] = I_{Stoßterme}. \tag{13.60}$$

In addition to the Kelvin–Helmholtz instabilities, one can also consider the Rayleigh–Taylor instabilities, see e.g. Patterson (1997) and Patterson and Reitz (1998); however, with realistic droplet sizes, these should not play such a large role. There are also completely different modeling ideas concerning secondary breakup, e.g. vibration-based models like the TAB (Taylor analogy breakup) model, which assume, that drop vibrations lead to breakup. It is however experimentally known, that these breakup types are no longer dominant at high Weber numbers.

However, these methods lead to quantitatively comparable breakup time scales for reasons of dimensional analysis. In Fig. 13.5 a summary of the various aerodynamic breakup models is provided.

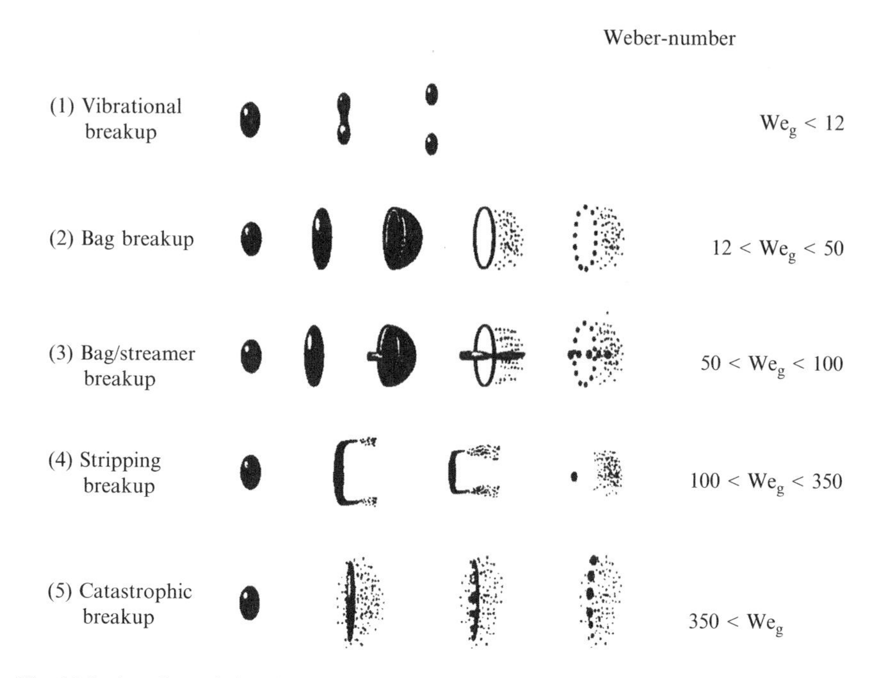

Fig. 13.5 Aerodynamic breakup mechanisms, acc. to Pilch and Erdman (1987)

Quite often in the literature, the all-dominating role is attributed to spray breakup even in the propagation of the gaseous phase, which is physically simply wrong. Unfortunately however, this corresponds to the function that spray breakup models perform in CFD simulations, or with other words, spray breakup models are adjusted to "compensate" for mathematical/numerical model deficiencies (which of course does not really work).

13.2.6 Modeling Collision Processes

Collision processes are subdivided into various subprocesses, including "collision with subsequent separation of droplets" (or "grazing collision") and "coalescence" (KIVA-II Manual, Amsden et al. (1989)). The most important suggestion in this context is to leave out collision processes when calculating with the standard model. This is because collision processes play an especially critical role numerically in the context of the Lagrangian standard model. Their calculation is based on knowledge of two-point correlations. These are, however, not easy to resolve numerically. For example, if we only have one parcel per cell, collisions are impossible, since all droplets in this cell are represented by this one parcel and thus have the same velocity (vectorially as well). But this excludes the possibility of collision. In the case of two parcels per cell, only one collision of droplet "from parcel 1" with droplet "from parcel 2" is possible, etc. In order to obtain a convergent result, the parcels in a cell must resolve in a detailed way the locally existing and generally correlated velocity-size spectrum. This is impossible practically speaking.

Leaving out collision processes is a largely established process today. On the one hand, droplet coalescence is a process that counteracts droplet breakup and is thus is taken into consideration by means of an effective breakup model. The same is true of collision processes that bring about breakup. This is done more or less automatically when the breakup model constants are calibrated. As long as the spray is not too dense, momentum diffusion caused by collision gives way to the effect of turbulent dispersion, in which case a local fluctuation equilibrium of the liquid and gaseous phase momentum arises. Elastic collision processes could maybe become noticeable by disturbing this equilibrium significantly – however, this is to be expected if at all in the immediate vicinity of the nozzle and could then be taken into account once again in an effective nozzle model.

With the elimination of the collision terms in the collision integral on the right side of (13.42), the spray equation (13.42) has become linear in the probability distribution p. It can therefore be understood as a linear superimposition of solutions for individual liquid parcels arising at the nozzle (e.g. in correspondence to individual parcels). There are of course interactions between different droplets or parcels via the gaseous phase.

A further category of collision processes that cannot be eliminated from the calculation are those that take place in droplet–wall contact. In the final analysis, we are concerned with formulating reflection and atomization laws for the individual droplets. There are a few approaches for this suggested in the literature, for example Bai and Gosman (1995) or Stanton and Rutland (1996). Furthermore, there is the matter of the formation of wall film. Wall film dynamics require their own set of equations and solvers. CFD codes offer approaches to this as well, for example KIVA-3V (Amsden (1999)).

13.2.7 Modeling Turbulent Dispersion in the Standard Model

We designate as turbulent dispersion the interaction of droplets with the turbulence of the gas flow. Most CFD codes simply utilize the formulation for flow resistance, which subdivides the active force into a constant and a fluctuating component, whereby the velocity fluctuations $\vec{v''}_g$ that a parcel "experiences" are drawn randomly with the distribution function

$$G(\vec{v''}_g) = \frac{1}{\sqrt{2\pi}\sigma} \exp\left(-\frac{\vec{v''}_g{}^2}{2\sigma^2}\right) \quad \text{with} \quad \sigma = \sqrt{\frac{2}{3}k} \tag{13.61}$$

The lifespan of a turbulent fluctuation, τ_{corr}, as the minimum of the turbulent time scale and the time, which a drop needs to traverse a turbulent swirl (turbulent length scale divided by the relative velocity of droplet to gas), is calculated with:

$$\tau_{corr} = \min\left(c_\mu \frac{k}{\varepsilon}, \frac{l_t}{\left|\langle v_g \rangle - v_{tr}\right|}\right) \quad \text{with} \quad l_t = c_\mu^{3/4} \frac{k^{3/2}}{\varepsilon}. \tag{13.62}$$

This corresponds to the formulation in the last term of the Boltzmann–Williams equation (13.42). Such a motion equation with stochastic force is also called a *Langevin equation* (see e.g. Risken 1989).

We already mentioned in the discussion on collision processes in the last section that turbulent dispersion is an essential process which in no case should be neglected in simulations. Due to a local kinematic equilibrium between the droplet and gas motion, an equilibrium is formed between both which also contains an equilibrium of the velocity fluctuations, i.e. a turbulent velocity spectrum develops in the droplet phase ("droplet turbulence"). In the boundary case of small droplets (droplet radius almost zero), turbulent dispersion should give way to turbulent diffusion as in (12.32)/(12.33).

As opposed to the gas phase however, no averaging via turbulent velocity fluctuations is carried out in the droplet phase in the standard approach discussed here, i.e. no turbulence is introduced, but rather the velocity distribution function is directly simulated by the parcels. If one recalls however that small parcels should show a different velocity spectrum than large ones, it quickly becomes apparent that this makes very large local numbers of parcels necessary. Even 50 parcels per cell can still be too few. Long interaction times also bring about convergence problems because they mean that a single stochastic gas velocity component affects a parcel for a long time. In principle, this should be compensated by other parcels with other stochastic gas velocities of equally long life but not synchronized start and end times. Practically speaking, there will always insufficiency in the case of such parcels. In the case of short interactions, this is less critical because an averaging already occurs in the successive influence of different, short-lived turbulent fluctuations on a parcel.

13.2.8 Describing Turbulent Dispersion with the Fokker–Planck Equation

Based on the previous discussion, it seems sensible to formulate turbulent dispersion in a context that is more mathematically useful. One important step in this direction has already been made by O'Rourke (1989) in the KIVA code. This can be comprehended in a more general sense if we consider that a formulation with a Langevin equation with stochastic force can be converted into a so-called *Fokker–Planck equation*.

To this end, we return to (13.42) [or (13.60)], whereby in the collision integral on the right side now is only found the turbulent dispersion term as discussed earlier:

$$\begin{aligned}&\frac{\partial}{\partial t}p(\vec{x},\vec{v}_{tr},R,T;t)+\frac{\partial}{\partial x_i}\left[v_{tr,i}\,p(\vec{x},\vec{v}_{tr},R,T;t)\right]+\\&\frac{\partial}{\partial v_{tr,i}}\left[D_{tr}\left(\langle v_{g,i}\rangle-v_{tr,i}\right)p(\vec{x},\vec{v}_{tr},R,T;t)\right]+\\&\frac{\partial}{\partial R}\left[\dot{R}_V p(\vec{x},\vec{v}_{tr},R,T;t)\right]+\frac{1}{R^3}\frac{\partial}{\partial R}\left[\dot{R}_Z R^3\,p(\vec{x},\vec{v}_{tr},R,T;t)\right]+\\&\frac{\partial}{\partial T}\left[\dot{T}_{tr,A}p(\vec{x},\vec{v}_{tr},R,T;t)\right]=-\frac{\partial}{\partial v_{tr,i}}\left[D_{tr}v''_{g,i}\,p(\vec{x},\vec{v}_{tr},R,T;t)\right].\end{aligned}\tag{13.63}$$

$\dot{T}_{tr,A}$, $\dot{R}_V$ and $\dot{R}_Z$ describe the heating, vaporization and breakup source terms according to (13.18), (13.19) and (13.57). The goal is to execute a turbulent averaging in the gaseous phase (á la Reynolds) in order to eliminate the term v''_g on the right side. One must take heed that the probability density p also depends on v''_g. In principle, we need to carry out the following averaging operation:

$$\begin{aligned}&\frac{\partial}{\partial t}\langle p\rangle+\frac{\partial}{\partial x_i}\left[v_{tr,i}\langle p\rangle\right]+\frac{\partial}{\partial v_{tr,i}}\left[D_{tr}\left(\langle v_{g,i}\rangle-v_{tr,i}\right)\langle p\rangle\right]+\frac{\partial}{\partial R}\left[\dot{R}_V\langle p\rangle\right]\\&+\frac{1}{R^3}\frac{\partial}{\partial R}\left[\dot{R}_Z R^3\langle p\rangle\right]+\frac{\partial}{\partial T}\left[\dot{T}_{tr,A}\langle p\rangle\right]\\&=-\frac{\partial}{\partial v_{tr,i}}\left[D_{tr}\left\langle v''_{g,i}(t)\int_0^t d\vartheta\, v''_{g,i}(\vartheta)\right\rangle\left\langle\frac{\delta p(t)}{\delta v''_{g,j}(\vartheta)}\right\rangle\right],\end{aligned}\tag{13.64}$$

where $\delta p(t)/\delta v''_g(\vartheta)$ is the *functional* or *Fréchet derivation*. Now it is a matter of giving concrete meaning to the just formal right side.

A formal solution of (13.64) with the starting condition

$$p(\vec{x},\vec{v}_{tr},R,T;t)=P(\vec{x}_0,\vec{v}_{tr,0},R_0,T_0)$$

reads:

$$\begin{aligned}&p(\vec{x},\vec{v}_{tr},R,T;t)=P[\vec{x}_0-\vec{\mathcal{X}}_{tr}(t),\vec{v}_0-\vec{\mathcal{V}}_{tr}(t),R_0-\mathcal{R}(t),T_0-\mathcal{T}(t))]\cdot\\&\exp\int_0^t d\tau\left[D_{tr}\underbrace{\frac{\partial v_{tr,i}}{\partial v_{tr,i}}}_{=3}-\frac{\partial\dot{R}_V}{\partial R}-\frac{1}{R^3}\frac{\partial(R^3\dot{R}_Z)}{\partial R}-\frac{\partial(\dot{T}_{tr,A})}{\partial T}\right]\end{aligned}\tag{13.65}$$

where *x(t)*, *v(t)*, *R(t)* and *T(t)*designate the trajectories corresponding to (13.35). From the motion equations follows for *x(t)* and *v(t)*

$$\frac{d\vec{v}_{tr}}{dt} = D_{tr}\left(\langle\vec{v}_g\rangle + \vec{v''}_g - \vec{v}_{tr}\right)$$
$$\frac{d\vec{x}}{dt} = \vec{v}_{tr} \tag{13.66}$$

the v''_g dependence:

$$\vec{v}_{tr}(t, \vec{v''}_g) = \vec{v}_{tr}(t, \vec{v''}_g = 0) + D_{tr}\exp(-D_{tr}t)\int_0^t d\vartheta\ \exp(D_{tr}\vartheta)\ \vec{v}''_g(\vartheta)$$

$$\vec{x}_{tr}(t, \vec{v}''_g) = \vec{x}_{tr}(t, \vec{v''}_g = 0) + D_{tr}\int_0^t d\vartheta \exp(-D_{tr}\vartheta)\int_0^{\vartheta} d\tilde{\vartheta}\ \exp(D_{tr}\tilde{\vartheta})\ \vec{v}''_g(\tilde{\vartheta}). \tag{13.67}$$

Neglecting the v''_g dependence of *R(t)* and *T(t)*, we finally find the following v''_g dependence of function *p* up to the first order in v''_g:

$$\left\langle\vec{v''}_g(t)p(t, \vec{v''}_g)\right\rangle = -\int_0^t d\vartheta \frac{\partial\langle p\rangle}{\partial v_{tr,i}}(\vartheta)\left\langle v_{tr}(\vartheta, \vec{v''}_g)\ \vec{v''}_g(t)\right\rangle$$
$$-\int_0^t d\vartheta \frac{\partial\langle p\rangle}{\partial x_i}(\vartheta)\left\langle x(\vartheta, \vec{v''}_g)\ \vec{v''}_g(t)\right\rangle$$
$$\approx -\frac{\partial p}{\partial v_{tr,i}}(t)\cdot D_{tr}\ \exp(-D_{tr}t)\int_0^t d\vartheta\ \exp(D_{tr}\vartheta)\left\langle v''_g i(\vartheta)\ \vec{v''}_g(t)\right\rangle$$
$$-\frac{\partial p}{\partial x_i}(t)\cdot D_{tr}\int_0^t d\vartheta \exp(-D_{tr}\vartheta)\int_0^{\vartheta} d\tilde{\vartheta}\ \exp(D_{tr}\tilde{\vartheta})\left\langle\ v''_g i(\tilde{\vartheta})\ \vec{v''}_g(t)\right\rangle, \tag{13.68}$$

where we assume that the derivations $\partial p/\partial x$ and $\partial p/\partial v$ during time $(D_{tr})^{-1}$ are approximately constant temporally. In order to execute an ensemble averaging, a correlation function of the velocity fluctuation at different times is required. We take the approach that is usual here:

$$\left\langle v''_g i(t_1) v''_g j(t_2)\right\rangle = \frac{2}{3}k\ \delta_{ij}\exp\left(-2\frac{|t_1 - t_2|}{\tau_{corr}}\right) \tag{13.69}$$

We then obtain from (13.68):

$$\left\langle\vec{v''}_g(t)p(t, \vec{v''}_g)\right\rangle \xrightarrow[t>>\tau_{corr}]{} -\underbrace{\frac{2}{3}k\delta^{ij}\frac{D_{tr}\tau_{corr}}{D_{tr}\tau_{corr}+2}}_{D_{vv}/D_{tr}\ \delta^{ij}}\frac{\partial p}{\partial v_{tr,i}} - \underbrace{\frac{1}{3}k\delta^{ij}\frac{D_{tr}\tau^2{}_{corr}}{D_{tr}\tau_{corr}+2}}_{D_{xv}/D_{tr}\ \delta^{ij}}\frac{\partial p}{\partial x_i} \tag{13.70}$$

and finally from (13.63):

$$\frac{\partial}{\partial t}\langle p\rangle+\frac{\partial}{\partial x_i}\left[v_{tr,i}\langle p\rangle\right]+\frac{\partial}{\partial v_{tr,i}}\left[D_{tr}\left(\langle v_{g,i}\rangle-v_{tr,i}\right)\langle p\rangle\right]$$
$$+\frac{\partial}{\partial R}\left[\dot{R}_V\langle p\rangle\right]+\frac{1}{R^3}\frac{\partial}{\partial R}\left[R^3\dot{R}_Z\langle p\rangle\right]+\frac{\partial}{\partial T}\left[\dot{T}_{tr,A}\langle p\rangle\right]$$
$$=\frac{\partial}{\partial v_{tr,i}}\left[D_{xv}\frac{\partial\langle p\rangle}{\partial x_i}\right]+\frac{\partial}{\partial v_{tr,i}}\left[D_{vv}\frac{\partial\langle p\rangle}{\partial v_{tr,i}}\right] \tag{13.71}$$

A diffusion term has thus been developed in the velocity-position space. This equation, linear in $\langle p\rangle$, is a variant of the *Fokker–Planck equation* [see Landau and Lifschitz (1981), Risken (1989)]. The original Fokker–Planck equation (in it, there is only one pure velocity diffusion term, i.e. $D_{xv}=0, D_{vv}\neq 0$) was derived in order to describe *Brownian motion* (i.e. the movement of a particle in a liquid under the influence of thermal fluctuations). Yet equation (13.71) describes the motion of a droplet in a gas under the influence of turbulent fluctuations!

From this equation follows – after integration with x, v, R, T and exploitation of the rules for partial integration for the average values, variances and co-variances – the correlations:

$$\begin{aligned}
&\frac{d\overline{x_i}}{dt}=\overline{v_{tr,i}}\\
&\frac{d\overline{v_{tr,i}}}{dt}=D_{tr}\left(\langle v_{g,i}\rangle-\overline{v_{tr,i}}\right)+\overline{\frac{\partial D_{xv}}{\partial x_i}}+\overline{\frac{\partial D_{vv}}{\partial v_{tr,i}}}\\
&\frac{d\left[\overline{v_{tr,i}v_{tr,j}}-\overline{v_{tr,i}}\cdot\overline{v_{tr,j}}\right]}{dt}=2\left(D_{vv}\delta_{ij}-D_{tr}\left[\overline{v_{tr,i}v_{tr,j}}-\overline{v_{tr,i}}\cdot\overline{v_{tr,j}}\right]\right)\\
&\frac{d\left[\overline{v_{tr,i}x_{tr,j}}-\overline{v_{tr,i}}\cdot\overline{x_{tr,j}}\right]}{dt}=\left[\overline{v_{tr,i}v_{tr,j}}-\overline{v_{tr,i}}\cdot\overline{v_{tr,j}}\right]+D_{xv}\delta_{ij}-D_{tr}\left[\overline{v_{tr,i}x_j}-\overline{v_{tr,i}}\cdot\overline{x_j}\right]\\
&\frac{d\left[\overline{x_ix_j}-\overline{x_i}\cdot\overline{x_j}\right]}{dt}=2\delta_{ij}\left[\overline{v_{tr,i}x_j}-\overline{v_{tr,i}}\cdot\overline{x_j}\right]
\end{aligned} \tag{13.72}$$

where was set:

$$\overline{A(x,v)}=\int d^3x\,d^3v\,dR\,dT\,A(x,v)\text{ and}$$
$$D_{xv},D_{vv},D_{tr},\frac{\partial D_{xv}}{\partial x},\frac{\partial D_{vv}}{\partial v}=const \tag{13.73}$$

The drift terms $\partial D_{xv}/\partial x$ and $\partial D_{vv}/\partial v$ in the second equation right were considered because they generate terms of the first order in t for the temporal development of v_{tr} (but this is somewhat arbitrary under certain conditions if one considers large time scales). These relations can also be directly derived from the solution of the Langevin equation (13.67), i.e. the Langevin equation is equivalent to the Fokker–Planck equation. In the following the definitions for variances and co-variances will be formulated:

$$\begin{aligned}\sigma_{vv,ij} &= \overline{v_{tr,i}v_{tr,j}} - \overline{v_{tr,i}} \cdot \overline{v_{tr,j}} \\ \sigma_{xv,ij} &= \overline{x_i v_{tr,j}} - \overline{x_i} \cdot \overline{v_{tr,j}} \\ \sigma_{xx,ij} &= \overline{x_i x_j} - \overline{x_i} \cdot \overline{x_j}.\end{aligned} \tag{13.74}$$

How is the set of equations (13.72) to be interpreted? The velocity variance σ_{vv} ("droplet turbulence") strives against the equilibrium value D_{vv}/D_{tr}. The covariance σ_{xv} strives against $D_{xv}/D_{tr} + D_{vv}/D_{tr}{}^2 = k \cdot \tau_{corr}/3$ and the variance of the position variables σ_{xx} finally increases after a break-in phase in the sense of a diffusion process: $D_{xx} \rightarrow (2/3) \cdot k \cdot \tau_{corr} \cdot t$. In other words, the Fokker–Planck finally results in a spatial diffusion process after the velocity distribution has reached equilibrium. The ratio of droplet relaxation time to the fluid-mechanical time scale is called the Stokes number $\mathrm{St} = 1/D_{tr}\tau_{corr}$. This ratio essentially determines whether and to what extent droplets follow the turbulent flow. For small droplets: $\mathrm{St} \rightarrow 0$.

How should the phenomenon of turbulent dispersion now be integrated corresponding to this equation into the Lagrangian approach? To this end, we will consider the solution of (13.72) for a spatially concentrated liquid parcel for a time interval Δt; the initial condition are $x_i = x_{0,i},\ v_i = v_{0,i}$; variances and co-variances are equal to zero:

$$\begin{aligned}
\overline{v_{tr,i}} &= \langle v_{g,i}\rangle + \left(v_{0,i} - \langle v_{g,i}\rangle\right)\exp(-D_{tr}\Delta t) + \left(\frac{\partial D_{xv}}{\partial x} + \frac{\partial D_{xv}}{\partial x}\right)\Delta t \\
\overline{x_{tr,i}} &= x_{0,i} + \langle v_{g,i}\rangle \cdot \Delta t + \frac{v_{0,i} - \langle v_{g,i}\rangle}{D_{tr}}(1 - \exp(-D_{tr}\Delta t)) + \frac{1}{2}\left(\frac{\partial D_{xv}}{\partial x} + \frac{\partial D_{xv}}{\partial x}\right)\Delta t^2 \\
\sigma_{vv,ij} &= \delta_{ij}\frac{D_{vv}}{D_R}(1 - \exp(-2D_{tr}\Delta t) \\
\sigma_{vx,ij} &= \delta_{ij}\left[\frac{k}{3}\tau_{corr} - \left(\frac{k}{3}\tau_{corr} + \frac{D_{vv}}{D_{tr}^2}\right)\exp(-D_{tr}\Delta t) + \frac{D_{vv}}{D_{tr}^2}\exp(-2D_{tr}\Delta t)\right] \\
\sigma_{xx,ij} &= \delta_{ij}\left[\frac{2k}{3}\tau_{corr}\Delta t - \frac{2}{D_{tr}}\left(\frac{k}{3}\tau_{corr} + \frac{D_{vv}}{D_{tr}^2}\right)[1 - \exp(-D_{tr}\Delta t)] + \frac{D_{vv}}{D_{tr}^3}[1 - \exp(-2D_{tr}\Delta t)]\right]
\end{aligned} \tag{13.75}$$

The associated distribution function $\langle p\rangle$ is a correlated Gauss distribution of x and v_{tr}:

$$\langle p(\vec{x}, \vec{v}_{tr}, R, T)\rangle = \frac{1}{N}\exp(-q_1\left(\vec{x} - \bar{\vec{x}}_{tr}\right)^2 - q_2\left(\vec{x} - \bar{\vec{x}}_{tr}\right)\left(\vec{v}_{tr} - \bar{\vec{v}}_{tr}\right) - q_3\left(\vec{v}_{tr} - \bar{\vec{v}}_{tr}\right)^2). \tag{13.76}$$

The reader may determine the coefficients N and q are as an exercise. The droplet dynamics are first calculated in interaction only with the averaged gas velocity, i.e. for every parcel the first two equations of (13.75) are solved.

The turbulent fluctuation provide in every time step stochastically determined offsets in location and velocity, which are to be selected in agreement with the last three equations of (13.75). The following is true:

$$\begin{aligned} v_{tr,i}(t+\Delta t) - v_{tr,i}(t) &= \overline{v_{tr,i}(t+\Delta t)} - v_{tr,i}(t) + \delta v_i \quad \text{mit} \quad \overline{v_{tr,i}(t)} = v_{tr,i}(t) \\ x_{tr,i}(t+\Delta t) - x_{tr,i}(t) &= \overline{x_{tr,i}(t+\Delta t)} - x_{tr,i}(t) + \delta x_i \quad \text{mit} \quad \overline{x_{tr,i}(t)} = x_{tr,i}(t) \end{aligned} \tag{13.77}$$

The variables $\delta x_i = x_{tr,i} - \bar{x}_{tr,i}$ and $\delta v_i = v_{tr,i} - \bar{v}_{tr,i}$ are normally distributed as determined above, but their distribution is correlated. For random determination, we require two independent variables. If δx_i is the first, then

$$\delta z_i = \delta v_i - \frac{\sigma_{vx}}{\sigma_{xx}} \delta x_i \tag{13.78}$$

provides the other variable, because their covariance $\langle \delta z_i \, \delta x_j \rangle$ disappears (we set $\sigma_{xx,ij} = \sigma_{xx}\delta_{ij}$ etc.). For the variance of these variables is valid[5]:

$$\begin{aligned} \langle \delta x_i \, \delta x_j \rangle &= \sigma_{xx}\delta_{ij} \\ \langle \delta z_i \, \delta z_j \rangle &= \left(\sigma_{vv} - \frac{{\sigma_{vx}}^2}{\sigma_{xx}} \right) \delta_{ij} \end{aligned} \tag{13.79}$$

The solutions from (13.75) should be inserted into these formulae on the right side. Now we can make a random determination to establish δx and δz component-wise with respect to a normal distribution (for this, the inverse error function should be inverted and tabulated as a primitive of the normal distribution a single time and the result be multiplied with the respective standard deviation (root from the variance)). δx and δv, taken from (13.78), can finally flow into (13.77).

As we already mentioned, the CFD code KIVA used a similar approach (Amsden et al. (1989) or O'Rourke (1989)), but only for the case that the correlation time τ_{corr} is smaller than the calculation interval. But it is exactly the other case of long correlation times that is especially critical for reasons of statistical convergence. Then especially should preference be given to the approach presented here (better still: its further development in the next section). The modeling in KIVA is, however, not difficult to alter in the sense suggested above. Comparable approaches in other CFD codes are unfortunately unknown to us.

Turbulent dispersion also affects the thermal behavior of droplets. In droplet heating and vaporization, velocity influences are considered in the pre-factors according to Ranz-Marschall, since in the droplet-related Reynolds number, the amount of velocity difference between droplet and gas appears. These should be

[5]Due to the Cauchy–Schwarz inequality, ${\sigma_{xv}}^2 \leq \sigma_{xx} \cdot \sigma_{vv}$ and thus $\langle \delta z_i \delta z_i \rangle \geq 0$ are valid.

corrected, i.e. averaged. A simple approach to doing this will be shown. First, we analyze the velocity difference and calculate it in accordance with (13.67)–(13.70).

$$\begin{aligned}\left\langle\left(\vec{v}_{tr}-\vec{v}_{g}\right)^{2}\right\rangle &= \left(\vec{v}_{tr}-\langle\vec{v}_{g}\rangle\right)^{2}-2\left\langle\vec{v}_{tr}\cdot\vec{v}'g\right\rangle+\left\langle\left(\vec{v}'_{g}\right)^{2}\right\rangle \\ &\xrightarrow[t>>D^{-1}]{} \left(\vec{v}_{tr}-\langle\vec{v}_{g}\rangle\right)^{2}-4k\frac{D_{tr}\tau_{corr}}{D_{tr}\tau_{corr}+2}+2k \\ &= \left(\vec{v}_{tr}-\langle\vec{v}_{g}\rangle\right)^{2}+2k\frac{2-D_{tr}\tau_{corr}}{D_{tr}\tau_{corr}+2}. \end{aligned} \tag{13.80}$$

This finally leads to an averaged Reynolds number:

$$\left\langle \mathrm{Re}_{tr}\left(\left|\vec{v}_{tr}-\vec{v}_{g}\right|\right)\right\rangle = \mathrm{Re}_{tr}\left(\left|\vec{v}_{tr}-\langle\vec{v}_{g}\rangle\right|\right)\cdot\sqrt{1+2k\frac{2-D_{tr}\tau_{corr}}{2+D_{tr}\tau_{corr}}\left(\frac{1}{\left|\vec{v}_{tr}-\langle\vec{v}_{g}\rangle\right|}\right)^{2}}. \tag{13.81}$$

Since this expression does not directly guarantee that the term under the root will not be negative [as it should be due to (13.80)], this must be explicitly ensured.

The Fokker–Planck equation no longer contains turbulent gas velocity, which must be determined randomly. However, in this approach too, the parcels must reproduce the local statistical velocity distributions of the droplet phase as before, and this a also still a major challenge typically requiring very large amounts of parcels.

13.2.9 Representing Diffusion with the Fokker–Planck Equation

We can make yet another step by explicitly developing the fact that the Fokker–-Planck also describes a diffusion process in space. For the derivation, we proceed from (13.71), which we will imagine as limited to single droplet parcels (see the discussion in the last paragraph of Sect. 12.2.6) so that the variances do not become too large and integrate the base equation as well as the equation multiplied with v over v:

$$\begin{aligned}&\frac{\partial}{\partial t}\eta+\frac{\partial}{\partial x_i}\left[\langle v_{tr,i}\rangle\eta\right]+\frac{\partial}{\partial R}\left[\dot{R}_V\eta\right]+\frac{1}{R^3}\frac{\partial}{\partial R}\left[R^3\dot{R}_Z\,\eta\right]+\frac{\partial}{\partial T}\left[\dot{T}_{tr,A}\,\eta\right]=0\\ &\frac{\partial}{\partial t}\left[\langle v_{tr,j}\rangle\eta\right]+\frac{\partial}{\partial x_i}\left[\langle v_{tr,j}\rangle\langle v_{tr,i}\rangle\eta\right]+\frac{\partial}{\partial x_i}\left[\underbrace{\left(\langle v_{tr,j}v_{tr,i}\rangle-\langle v_{tr,j}\rangle\langle v_{tr,i}\rangle\right)}_{\sigma_{vv,ij}}\eta\right]\\ &-D_{tr}\left(\langle v_{g,i}\rangle-\langle v_{tr,i}\rangle\right)\eta+\frac{\partial}{\partial R}\left[\dot{R}_V\langle v_{tr,j}\rangle\eta\right]+\frac{1}{R^3}\frac{\partial}{\partial R}\left[R^3\,\dot{R}_Z\langle v_{tr,j}\rangle\eta\right]\\ &+\frac{\partial}{\partial T}\left[\dot{T}_{tr,A}\langle v_{tr,j}\rangle\eta\right]+D_{xv}\frac{\partial\eta}{\partial x_j}-\frac{\partial D_{vv}}{\partial v_{tr,j}}\eta=0\end{aligned} \tag{13.82}$$

where

$$\int d^3v \langle p(x,v,R,T;t)\rangle =: \eta(x,R,T;t)$$
$$\int d^3v\, v_i\, \langle p(x,v,R,T;t)\rangle =: \langle v_i\rangle \eta(x,R,T;t)$$
$$etc. \tag{13.83}$$

was set (i.e. we are using the same averaging symbol as in the turbulence-induced ensemble averaging, since here too we are averaging via velocity fluctuations). If we subtract the first equation multiplied with $\langle v_{tr,j}\rangle$ from the second (13.82), we obtain:

$$\underbrace{\eta\frac{\partial}{\partial t}\langle v_{tr,j}\rangle + \langle v_{tr,i}\rangle \eta \frac{\partial}{\partial x_i}\langle v_{tr,j}\rangle}_{\eta\frac{d}{dt}\langle v_{tr,j}\rangle} + \frac{\partial}{\partial x_i}\left[\sigma_{vv,ij}\eta\right] - D_{tr}\left(\langle v_{g,i}\rangle - \langle v_{tr,i}\rangle\right)\eta$$
$$+ D_{xv}\frac{\partial \eta}{\partial x_j} - \frac{\partial D_{vv}}{\partial v_{tr,j}}\eta = 0 \tag{13.84}$$

Solved for $\langle v_{tr,j}\rangle$, we find

$$\langle v_{tr,j}\rangle\eta = \underbrace{\exp(-D_{tr}t)\int_0^t d\tau \exp(D_{tr}\tau)\left[D_{tr}\langle v_{g,j}\rangle + \frac{\partial D_{xv}}{\partial x_j} + \frac{\partial D_{vv}}{\partial v_{tr,j}}\right]\eta}_{\langle \tilde{v}_{tr,j}\rangle \eta}$$
$$- \delta_{ij}\left[\exp(-D_{tr}t)\int_0^t d\tau \exp(D_{tr}\tau)\frac{\partial}{\partial x_i}\left[\sigma_{vv}\eta + D_{xv}\eta\right]\right]. \tag{13.85}$$

Inserted in the first of (13.82), we thus obtain

$$\frac{\partial}{\partial t}\eta + \frac{\partial}{\partial x_{tr,i}}\left[\langle v_{tr,i}\rangle\eta\right] - \frac{\partial}{\partial x_{tr,i}}\frac{\partial}{\partial x_{tr,i}}\left[\underbrace{\exp(-D_{tr}t)\int_0^t d\tau\, \exp(D_{tr}\tau)(\sigma_{vv} + D_{xv})\,\eta}_{\frac{1}{2}\frac{d}{dt}\sigma_{xx,ij}}\right]$$
$$+ \frac{\partial}{\partial R}\left[\dot{R}_v\eta\right] + \frac{1}{R^3}\frac{\partial}{\partial R}\left[R^3\dot{R}_z\eta\right] + \frac{\partial}{\partial T}\left[\dot{T}_{tr,A}\eta\right] = 0 \tag{13.86}$$

This means we have now derived a generalized diffusion equation!

Until now, we have not used an approximation. Let us now however consider the limit case of small droplets $D_{tr} \to \infty$. Then the following is true:

$$\begin{aligned}
\langle v_{tr,i} \rangle &= \langle v_{g,i} \rangle + \frac{\partial}{\partial x_i}\left(\frac{k \cdot \tau_{corr}}{3}\right) \\
\frac{\partial D_{vv}}{\partial v_{tr,i}} &\approx 0 \quad (\text{fur } \mathrm{Re}_{tr} \ll 1) \\
\sigma_{vx} &= \frac{k}{3}\tau_{corr} \\
\sigma_{xx} &= \frac{2k}{3}\tau_{corr}\, t
\end{aligned} \tag{13.87}$$

and the following is developed from (13.86)

$$\begin{aligned}
&\frac{\partial}{\partial t}\eta + \frac{\partial}{\partial x_{tr,i}}\left(\langle v_{g,i} \rangle \eta\right) - \frac{\partial}{\partial x_{tr,i}}\left(\frac{k}{3}\tau_{corr}\frac{\partial \eta}{\partial x_{tr,i}}\right) + \frac{\partial}{\partial R}\left[\dot{R}_V\, \eta\right] \\
&\quad + \frac{1}{R^3}\frac{\partial}{\partial R}\left[R^3 \dot{R}_Z\, \eta\right] + \frac{\partial}{\partial T}\left[\dot{T}_{tr,A}\eta\right] = 0
\end{aligned} \tag{13.88}$$

which corresponds to a classical diffusion equation (diffusion in location x).

How is the approach in (13.86) to be represented arithmetically? We introduce three new variables with the following dynamics for every particle in accordance with (13.72):

$$\begin{aligned}
\frac{d\sigma_{vv}}{dt} &= 2(D_{vv} - D_{tr}\sigma_{vv}) \\
\frac{d\sigma_{xv}}{dt} &= \sigma_{vv} + D_{xv} - D_{tr}\sigma_{xv} \\
\frac{d\sigma_{xx}}{dt} &= 2\sigma_{xv}.
\end{aligned} \tag{13.89}$$

Now the variables δx_i must be determined randomly in time interval Δt with respect to a normal distribution with variance $\sigma_{xx}(\Delta t) \approx 2\sigma_{xv}\Delta t$ and added to the parcel location in accordance with (13.77). If the droplets are sufficiently small (i.e. the diffusion approximation (13.87) is now valid), one need not solve equation (13.89) but simply state

$$\sigma_{xx} = \frac{2k}{3}\tau_{corr}\,\Delta t. \tag{13.90}$$

The Reynolds number correction of (13.80)/(13.81) must be modified since the difference of the average velocities is now the basis. Recalculation results in

$$\begin{aligned}\left\langle\left(\vec{v}_{tr}-\vec{v}_g\right)^2\right\rangle &= \left(\langle\vec{v}_{tr}\rangle-\langle\vec{v}_g\rangle\right)^2+\langle\vec{v}'_{tr}\rangle^2-2\left\langle\vec{v}_{tr}\cdot\vec{v}'_g\right\rangle+\left\langle\left(\vec{v}'_g\right)^2\right\rangle \\ &= \left(\langle\vec{v}_{tr}\rangle-\langle\vec{v}_g\rangle\right)^2+3\sigma_{vv}+2k\frac{2-D_{tr}\tau_{corr}}{2+D_{tr}\tau_{corr}}.\end{aligned} \tag{13.91}$$

This finally leads to an average Reynolds number:

$$\begin{aligned}\left\langle \mathrm{Re}_{tr}\left(\left|\vec{v}_{tr}-\vec{v}_g\right|\right)\right\rangle &= \mathrm{Re}_{tr}\left(\left|\langle\vec{v}_{tr}\rangle-\langle\vec{v}_g\rangle\right|\right) \\ &\times\sqrt{1+\left(2k\frac{2-D_{tr}\tau_{corr}}{2+D_{tr}\tau_{corr}}+3\sigma_{vv}\right)\left(\frac{1}{\left|\langle\vec{v}_{tr}\rangle-\langle\vec{v}_g\rangle\right|}\right)^2}\end{aligned} \tag{13.92}$$

In the limit case of small droplets, the correction term disappears, but the Reynolds number becomes small anyhow due to its dependence on radius.

These modeling approaches are also easy to implement in the KIVA code. There is no longer a stochastic velocity component; the parcels must no longer resolve the local velocity fluctuations. Instead, these fluctuations are described by the correlation functions transported by the parcel. For these reasons, these methods (13.86)/(13.89) or (13.88)/(13.90) should be preferred for the implementation of turbulent dispersion.

In the limit case of small drops, the liquid phase switches into a species of gaseous phase and *turbulent dispersion* becomes *turbulent diffusion*. The turbulent diffusion constants must also be in agreement, and τ_{corr} can be determined from this congruence requirement:

$$\frac{k}{3}\tau_{corr}=\frac{c_\mu}{Sc_t}\frac{k^2}{\varepsilon}=>\tau_{corr}=\frac{3c_\mu}{Sc_t}\frac{k}{\varepsilon} \tag{13.93}$$

13.2.10 Problems in the Standard Spray Model

As already mentioned, the standard spray model suffers under considerable deficiencies in practical applications. These will first be represented with a few examples in order then to explain the causes in more detail. If we compare the relative simplicity of modeling with stochastic parcels to the high complexity of the problem at hand, namely, the solution of an integro-differential equation in eight-dimensional space, then the failure of the spray model is actually not surprising at all. Indeed, the problems are obvious. On the one hand, there is the problem of the small length scales to be resolved at the nozzle hole, and on the other that of lacking statistical convergence.

In the case of typical diesel injection sprays with injection occurring shortly before top dead center, the situation is not that critical since we are dealing with simple spray geometries (hole type nozzle!) and small liquid lengths. The situation becomes much more difficult when investigating conical sprays such as are produced by modern piezo-actuated, outward-opening nozzles for direct gasoline injection. These exhibit highly complex swirl dynamics that are furthermore extremely contingent on boundary conditions such as ambient gas temperature, nozzle and combustion chamber geometry and droplet size. It is only possible to calculate these phenomena correctly with a high-performance spray model. In general however, all applications involving a long fluid lifetime, such as injection processes in the intake stroke or early compression stroke, are to be considered as challenging.

13.2.10.1 Problem: Spatial Resolution of the Spray

One problem in spray modeling is immediately obvious and has also been quickly perceived: typical nozzle orifices are so fine that they normally are not numerically resolved. The consequence of poor mesh resolution is an erroneous calculation of the exchange processes between gaseous and liquid phase.

First however, we should point out that lacking resolution of the spray dimensions also leads to inacceptable results in the case of the open gaseous jet calculated with classical fluid mechanics. Spatial resolution problems are therefore not at all a specific characteristic of the Lagrangian spray model.

A nozzle hole of a passenger car diesel engine now ranges within the order of magnitude of 100 μm. One should still resolve it with 10 mesh cells, which would lead to a mesh cell edge size of 10 μm. Applied to a compression volume of 20 cubic centimeters, this would imply 20 billion mesh cells! Surely, one could still reduce the mesh cell number by meshes refined adaptively to the nozzle opening, and perhaps one could get by with less than 100 mesh cells per nozzle orifice (10×10). But an extremely high number of mesh cells and complexly designed meshes cannot be avoided.

The most common practice differs from this unfortunately. In order to acquire results similar to available measurements despite insufficient mesh resolution, a number of modifications are introduced to the spray model, e.g. to the spray breakup model (a very efficient tuning!). The (unphysical) mode of operation of breakup mechanisms in such simulations can be understood in the following manner: with large droplets, the zone is skipped over with false momentum exchange, and in some approaches this is even explicitly eliminated near the nozzle hole ("intact core length"). However, then small droplets must be created quickly in order to furnish the necessary quality of mixture formation.

As has already been shown in Fig. 13.4, the penetration behavior of a hole type nozzle injection spray is almost independent of the droplet size. With sufficiently small droplets, it takes in a lot of gas and behaves exactly like an open gas jet. In this case, essential influential factors include the momentum flow I as well as the spray angle α (which in turn is dependent on the ambient density ρ_g). The momentum flow in a cross-section of a stationary open jet with opening angle α at distance x from the nozzle amounts to

$$I = \rho_g v^2 \pi x^2 \tan^2 \frac{\alpha}{2}. \qquad (13.94)$$

and is preserved along the spray. A mass point in the open jet then moves according to the relation:

$$x\frac{dx}{dt} = \sqrt{\frac{I}{\rho_g\pi, tan^2\frac{\alpha}{2}}} => x = \sqrt{\frac{2I}{\rho_g\pi\tan^2\frac{\alpha}{2}}}\cdot\sqrt{t}. \tag{13.95}$$

The penetration behavior of the, in principle, unsteady spray peak is also well represented by this relation (at constant momentum flow) – even in the case of complete vaporization. In the latter case, the formula describes the penetration behavior of the vapor phase. But from this immediately follows that the spray behavior is primarily independent on droplet size and thus on spray breakup. The prerequisite for this are sufficiently small droplets, yet this is actually always the case in modern diesel injection sprays. If a penetration depth tuning by means of spray breakup is possible at all, these speaks for non-physically large droplets or a spray model with insufficient statistical convergence (see the next section).

The currently most practical solution consists in using meshes that are adaptively refined in the spray area (Fig. 13.6), whereby the so-called *length scale limiter* (Johnson et al. (1995)) is still extremely useful. In the case of the length scale limiter, the turbulent length scale in the spray area is limited to the spray cross-section radius l_{Str}. With this parameter, we obtain the following constraint for ε at a distance x from the nozzle

$$\varepsilon \geq c_\mu^{3/4}\frac{k^{3/2}}{l_{Str}} = c_\mu^{3/4}\frac{k^{3/2}}{x\tan\frac{\alpha}{2}}. \tag{13.96}$$

In the case this relation is violated, ε is defined in accordance with the equal sign. Such a procedure is surely on the borderline of impermissibility in the sense of fluid mechanics; on the other hand, this method is both physically meaningful and mathematically well-defined. In addition to the turbulent length scale, turbulent diffusion and viscosity ($\propto k^2/\varepsilon$) are also limited, thus a false momentum flow is eliminated, which leads to excessively small penetration properties. An essential aspect of the length scale limiter: it is noneffective for fine, spray-resolving meshes (in which case relation (13.96) is automatically fulfilled).

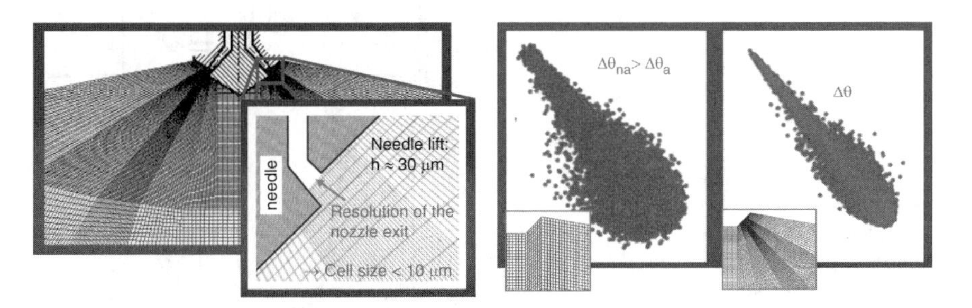

Fig. 13.6 Adaptively refined mesh with nozzle hole resolution for an SI engine with piezo-activated A-nozzle. The spray calculated with it is shown right. An excessively coarse mesh structure creates sprays that are much too bushy (center image)

We should finally point out that even a round open gas jet calculated with a correct mesh resolution in the standard k–ε model (from an orifice nozzle) turns out too bushy and thus exhibits an insufficient penetration depth according to (13.95). This problem can be solved by introducing an additional source term in the ε equation, the *Pope correction* (Pope 1978):

$$\Delta Q = 0.2\frac{k^2}{\varepsilon}\left[S_{ij}\left(\nabla\times\vec{v}\right)_i\left(\nabla\times\vec{v}\right)_j - \left(\nabla\times\vec{v}\right)^2 Tr(S_{ij})\right]. \tag{13.97}$$

A selection of $\varepsilon_1 \approx 1.55$ is a simple constant modification, which also provides a correct open jet penetration depth.

13.2.10.2 Problem: Statistical Convergence

Even if we try to carry out calculations with the standard spray model with a high mesh resolution, we can still encounter difficulties. There is even the (false) opinion that the spray model does not work for fine meshes at all!

But the actual problem is in the lacking statistical convergence per computational cell. The number of parcels per cell is relevant to this; if the number of mesh cells now increases because of mesh refinement, this ratio is naturally poorer (consider Fig. 13.7). To reach convergence, one must not only seek the limit "edge length of the mesh cells approaches zero" but also the limit "parcel number per mesh cell approaches infinity". At the least, this ratio must be held at a high level. This implies however an immensely high number of parcels.

As a rule of thumb for parcel numbers, one possible approach would be to consider that 50 parcels per cell is already a very good number. If N cells of edge length Δl (in the spray direction) are found at the nozzle opening and the injection

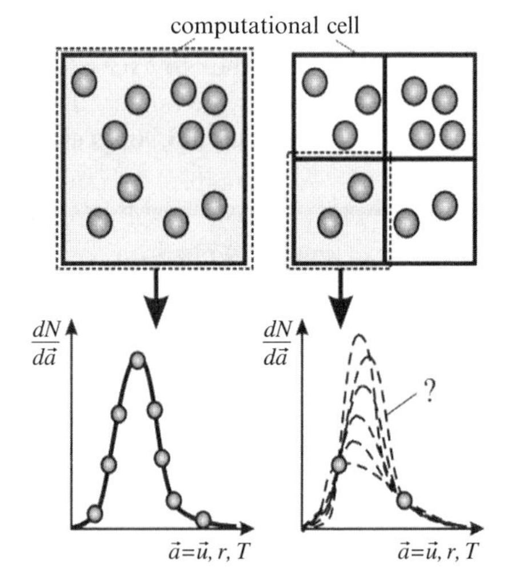

Fig. 13.7 Mesh refinement (parcel number remaining equal) leads to a diminishing number of parcels per cell and thus to poorer statistical resolution of the local droplet properties

speed is v_E, then about 50 N parcels in time $\Delta t = \Delta l / v_E$ should be injected. This is true of course only under the assumption that the number of cells that resolve the spray cross-section along the spray direction does not increase.

Moreover, the number of parcels required is extremely dependent on the applied models for reasons of statistical convergence. This point has already been treated in detail in the discussion regarding individual models. Particularly critical is the calculation of collisions, such as they are still encountered in the naive formulations of the spray model; these require unrealistically high numbers of parcels per cell. Also extremely critical is the naive modeling of turbulent dispersion in accordance with Sect. 12.2.7: the turbulent fluctuations of the gaseous and liquid phases are calculated explicitly. This means higher demands with respect to resolution than are made on the gaseous phase. For good reasons, a turbulence model is utilized there. It must finally be stressed once again that the use of average value equations, the Fokker–Planck equation and their diffusion approximation is extremely important.

Finally, we should remember that the failure of the spray model cannot be attributed to the non-existence of a continuum limit. Specifically, the assertion is often made that, in mesh refinement, eventually a drop is larger than a mesh cell, a case which is undefined. Yet we must always keep in mind that a solution method for the calculation of the probability distribution function equation is applied, which always has a well-defined (continuous) solution. The parcels do indeed have a radius, but this has the character of an "inner" degree of freedom. Certainly, the "paradox" under consideration is typical: practically speaking, many mesh refinements fail due to the problem of "large droplets", since they do not go along with an increase in parcels. Yet this is not a problem of the stochastic approach in principle, but rather one of its incorrect application.

13.2.11 Application Example: Direct Gasoline Injection for Stratified Charge with Centrally Arranged Piezo-Actuated Outward-Opening Injector

Finally, we will provide another example of an injection spray calculation executed according to the principles introduced. As mentioned previously, direct gasoline injection with a piezo-injector opening outwards is one of the most interesting and at the same time most challenging applications of spray simulation. The following rules of quality were taken into consideration:

- The computational mesh is that of Fig. 13.6, i.e. we are using an adaptive, nozzle-resolving mesh.
- We are dealing with a rotation-symmetrical calculation, i.e. only one "disc" of 0.5° sector angle is being calculated. In this sector angle, 100,000 parcels are injected, i.e. extrapolated to 360°, this means 1,440,000 parcels.
- A model of turbulent dispersion in the diffusion approximation of the Fokker–Planck equation (as described in Sect. 13.2.9) is being used.

- No collision model was used.
- The distribution of droplet size at the nozzle was taken from the spectrum of the turbulent length scales of a corresponding internal nozzle flow calculation.

The CFD code used is KIVA-3V. Figure 13.8 shows such an injection process (late injection to produce a stratified charge). The spray cone angle is just under 90°.

The spray cone takes in air inside and outside; an air flow inside the spray cone arises that is directed into it. If the spray cone is sufficiently thin-walled and the opening angle sufficiently large, this flow aimed into the spray cone is stable; there is no negative pressure and the spray cone does not collapse, irrespective of the environmental conditions. This implies that the relative position spray – spark plug does not change even at different engine operation points. A nozzle opening outwards thus has a high level of spray stability with the highest mixture formation quality – an essential prerequisite for spray-guided stratified charge combustion. This is not so in the case of swirl nozzles for example; the spray cone is not sufficiently hollow and thus collapses, especially in the case of higher counter pressures. The turbulence field is shown on the right side of Fig. 13.8; due to the very fine mesh structure, the shear layers on the inner and outer surfaces of the spray are resolved.

Figure 13.9 compares the simulation of an injection in a chamber at different times with spray-diagnostic investigations [spray photography and particle imaging velocimetry (PIV)]. There is very good agreement on the one hand; on the other, fluid-mechanical structures, especially the developing swirl, can analyze and explain the experimental result. The effects of the most diverse combustion chamber conditions or details of nozzle geometry could also analysed with this spray model. Hermann (2008) provides a detailed description of these high-resolution injection spray calculations for direct gasoline injection.

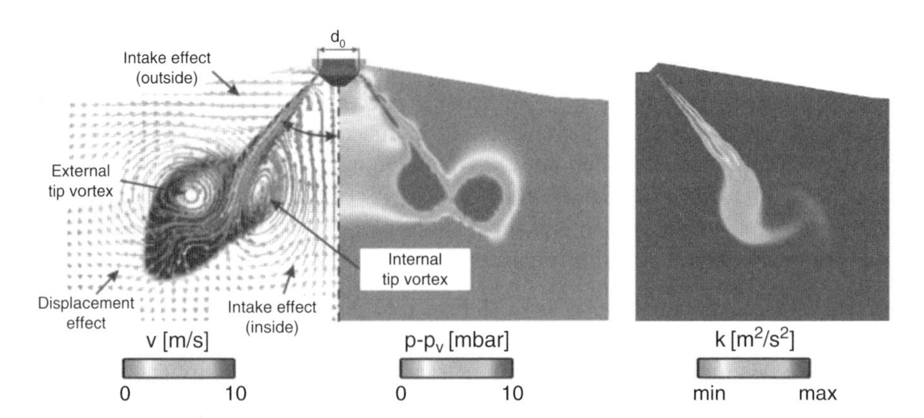

Fig. 13.8 Spray propagation in the case of direct gasoline injection with an outward-opening nozzle at 6 bar combustion chamber pressure. Left: the velocity and pressure field, right: gaseous phase turbulence

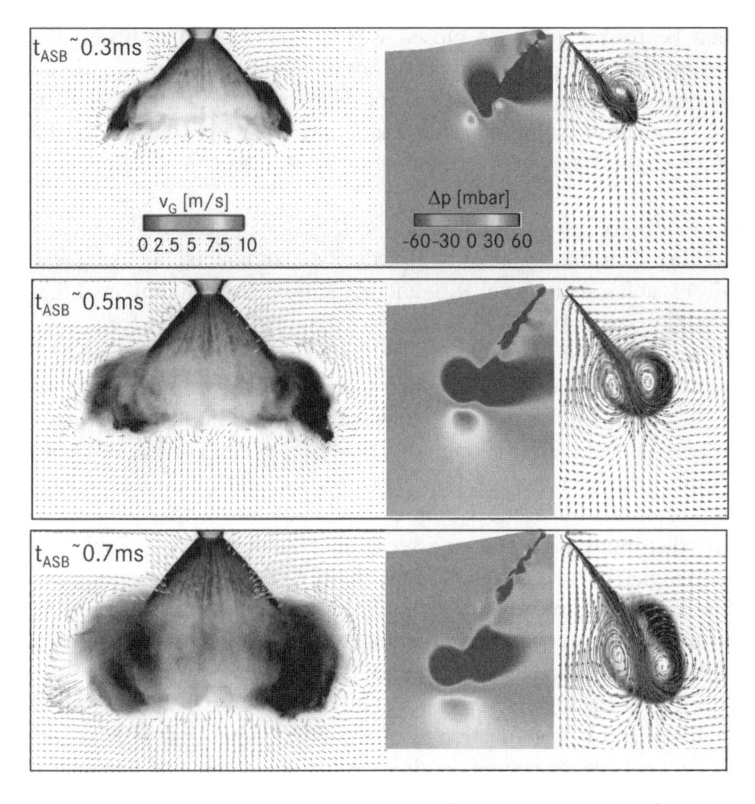

Fig. 13.9 Comparison of parcel image velocimetry (PIV) and simulation (left: static pressure distribution, right: velocity vector field)

It must also be noted that the stable behavior shown in Figs. 13.8 and 13.9 seems to be typical for liquids. Gas sprays from an outward-opening nozzle on the other hand apparently tend to collapse much more (Baratta et al. 2008).

13.3 Euler Spray Models: Formulation of Spray Dynamics with Observable Averages

"Euler spray models" offer the chance to avoid the problems of the Lagrangian model with respect to statistical convergence and turbulent averaging. "Tossing the dice" is always inefficient if one has the chance of already taking averaging processes into consideration in the model. The naive standard model (i.e. the Boltzmann–Williams equation with its complex, non-closed formulation of the source terms) was gradually modified in this sense in the above section by introducing effective average-value processes (breakup) and the Fokker–Planck equation, which makes stochastic behavior statistically comprehensible. The source terms could then be formulated in a closed way as generalized diffusion terms. The dynamic equation for probability distribution itself however is still defined in 8-dimensional state space and is thus not directly solvable, which (still) leads to the quasi-"simulation solution" with Lagrangian parcels (but on the basis of a clearly more "good-natured", effective dynamics than in the "naive" standard model).

Euler models go a step further insofar as they solve, instead of such an equation for probability distribution, a set of equations for moments, i.e. average values of special observables such as velocity and velocity variance (similar to the Navier–Stokes equation for example) directly, i.e. without the use of parcels.

Juxtaposition of the concepts "Euler model" and "Lagrange model" may be common in this context, but it is not particularly appropriate. The difference between both formulations does not consist exclusively in the use of fixed coordinates vs. coordinates carried along with the flow, which is what this concept pair refers to in the narrower sense. Rather, the more fundamental difference is based on the different formulation of the dynamics depending on whether it is defined using a probability distribution function or a set of observables. Only as a solution method do the Lagrangian parcels come into play in the first case. In the physical literature (including statistical physics), the terms *Schrödinger picture* (probability distribution) and *Heisenberg picture* (observable) are used following quantum mechanics to describe this difference in the formulation of the dynamics.

However, the Lagrange model is not at all defined with the help of the dynamics of a probability distribution function by many authors and in most CFD code manuals (KIVA is a laudable exception in this case!), but rather the Lagrangian parcel equations are, so to speak, "directly derived" from the droplet equations. We can only warn against such a casual procedure, as in this way the process of physical model formation is confused with that of numerical solution and the discretization required for it; we can then no longer be certain at all of whether we have a well-defined model. Moreover, abstractly motivated parcel dynamics, such as we obtained with the help of the Fokker–Planck equation, cannot be produced in this way.

Let us now return to the derivation of the dynamics in the observable image. We will start from the following probability distribution dynamics:

$$\frac{\partial}{\partial t}p(x,\beta,t)+\frac{\partial}{\partial x_i}\left[A_i^{(x)}(x,\beta)p(x,\beta,t)\right]+\frac{\partial}{\partial \beta_\mu}\left[A_\mu^{(\beta)}(x,\beta)p(x,\beta,t)\right]$$
$$-\frac{\partial}{\partial x_i}\left[D_{i\mu}^{(x,\beta)}(x,\beta)\frac{\partial}{\partial \beta_\mu}p(x,\beta,t)\right]-\frac{\partial}{\partial \beta_\mu}\left[D_{\mu\nu}^{(\beta\beta)}(x,\beta)\frac{\partial}{\partial \beta_\nu}p(x,\beta,t)\right]=0, \tag{13.98}$$

where $(\beta_\mu)=(\vec{v},R,T)$. Let $O(\beta;x,t)$ be an observable, e.g. momentum, or the square of velocity. In these cases, the observable O is not explicitly dependent on space and time. However, if one considers the square of velocity fluctuation (velocity minus the average of velocity – the average value of the square is then the variance), the observable O can also be explicitly dependent on space and time (in this case by the velocity average). By forming an average value (over β), we obtain a variable $\langle O\rangle(x,t)$ which now depends only on space and time (time-dependent also via the averaging, not only in the explicit sense explained above). For the dynamics we obtain the following relation by means of partial integration:

$$\frac{\partial\langle O\rangle(x,t)}{\partial t}=\frac{\partial}{\partial t}\int(p(x,\beta,t)O(\beta;x,t))\,d\beta=\int\left(p(x,\beta,t)\frac{\partial O(\beta;x,t)}{\partial t}\right)d\beta$$
$$-\frac{\partial}{\partial x_i}\left\langle A_i^{(x)}(x,\beta)\mathrm{O}(\beta;x,t)\right\rangle+\left\langle\frac{\partial\mathrm{O}(\beta;x,t)}{\partial\beta_\mu}A_\mu^{(\beta)}(x,\beta)\right\rangle$$
$$-\frac{\partial}{\partial x_i}\left\langle\frac{\partial}{\partial\beta_\mu}\left(D_{i\mu}^{(x,\beta)}(x,\beta)\mathrm{O}(\beta;x,t)\right)\right\rangle+\left\langle\frac{\partial}{\partial\beta_\mu}\left(D_{\mu\nu}^{(\beta\beta)}(x,\beta)\frac{\partial\mathrm{O}(\beta;x,t)}{\partial\beta_\nu}\right)\right\rangle \tag{13.99}$$

This formulation tends however to lead to closure problems, since new observable averages arise on the right side, the dynamics of which must either be described with a separate transport equation or otherwise modeled.

13.3.1 Locally Homogeneous Flow

In the limiting case of strong coupling between fluid and gas phases, which, for example, is the case with sufficiently small droplets, the special case of locally homogeneous flow arises; the liquid phase can now be described as a species in the context of a single-phase treatment, while the turbulent dispersion changes into turbulent diffusion of a species. To do this, one need only transform the dynamic equation (13.88), which corresponded exactly to this special case, according to the method in (13.99). If one applies this method to the observable R^3 (i.e. the mass up to a constant) and integrates via R and T (we have already integrated via v), we obtain using

$$\rho_{fl}(x;t)=\int dRdT\;\eta(x,R,T;t)\frac{4\pi}{3}R^3\rho_{fl} \tag{13.100}$$

(ρ_{fl} designates the density of the fluid) from (13.88) the following fluid density transport equation:

$$\frac{\partial}{\partial t}\rho_{fl}+\frac{\partial}{\partial x_i}\left(\rho_{fl}\left\langle v_{g,i}\right\rangle\right)-\frac{\partial}{\partial x_i}\left(D_t\frac{\partial}{\partial x_i}\rho_{fl}\right)=\frac{3\rho_{fl}\dot{R}_V}{R}. \tag{13.101}$$

The transport equation for the gaseous phase possesses the inverse vaporization source term:

$$\frac{\partial}{\partial t}\rho_g+\frac{\partial}{\partial x_i}\left(\rho_g\left\langle v_{g,i}\right\rangle\right)-\frac{\partial}{\partial x_i}\left(D_t\frac{\partial}{\partial x_i}\rho_g\right)=-\frac{3\rho_{fl}\dot{R}_V}{R}. \tag{13.102}$$

The continuity equation of the overall system then has the standard form again:

$$\begin{aligned}&\frac{\partial}{\partial t}\rho_{ges}+\frac{\partial}{\partial x_i}\left(\rho_{ges}\left\langle v_{ges,i}\right\rangle\right)=0\\&\rho_{ges}=\rho_g+\rho_{fl}\\&\rho_{ges}\left\langle v_{ges,i}\right\rangle=\rho_g\left\langle v_{g,i}\right\rangle+\rho_{fl}\left\langle v_{g,i}\right\rangle-D_t\frac{\partial}{\partial x_i}\rho_{ges}\,.\end{aligned}\tag{13.103}$$

With the fluid mass fracture $c_{fl}=\rho_{fl}/\rho_{ges}$ we obtain from (13.101):

$$\frac{\partial}{\partial t}\rho_{ges}c_{fl}+\frac{\partial}{\partial x_i}\left(\rho_{ges}c_{fl}\left\langle v_{ges,i}\right\rangle\right)-\frac{\partial}{\partial x_i}\left(\rho_{ges}D_t\frac{\partial}{\partial x_i}c_{fl}\right)=\frac{3\rho_{ges}c_{fl}\dot{R}_V}{R},\tag{13.104}$$

i.e. we have found a species diffusion equation like (12.32). In this approximation, we can therefore transport the liquid phase like a species of a single-phase flow. Under consideration of the observables R^4 and R^3T we obtain analogously the transport equations

$$\begin{aligned}&\frac{\partial}{\partial t}(\rho_{ges}C_{fl}R)+\frac{\partial}{\partial x_i}(\rho_{ges}\left\langle v_{ges,i}\right\rangle C_{fl}R)-\frac{\partial}{\partial x_i}(\rho_{ges}D_t\frac{\partial}{\partial x_i}(C_{fl}R))\\&\qquad=\rho_{ges}C_{fl}(4\dot{R}_v+\dot{R}_z)\end{aligned}\tag{13.105}$$

$$\begin{aligned}&\frac{\partial}{\partial t}(\rho_{ges}C_{fl}T_{tr})+\frac{\partial}{\partial x_i}(\rho_{ges}\left\langle v_{ges,i}\right\rangle C_{fl}T_{tr})-\frac{\partial}{\partial x_i}\left(\rho_{ges}D_t\frac{\partial}{\partial x_i}(C_{fl}T_{tr})\right)\\&\qquad=\rho_{ges}C_{fl}\dot{T}_A+\frac{3\rho_{ges}c_{fi}\dot{R}_v}{R}T_{tr}.\end{aligned}\tag{13.106}$$

Locally, there is thus exactly one average droplet radius and one average droplet temperature, but both quantities are generally varied in space and time (typically, the droplet radius decreases with increasing distance from the nozzle due to vaporization).

If we assume that the liquid and gaseous phases are not only kinematically but also thermodynamically in equilibrium (i.e. truly "mixture controlled"), we can in principle do without equations (13.105) and (13.106); the local amount of vapor corresponds exactly to that demanded by the local temperature (which is the same for the liquid and gaseous phases). If this equilibrium should be shifted by transport processes, is would be immediately restored by local vaporization or condensation. This defines source terms for the transport equations of vapor, liquid phase and internal energy. However, this assumption is rather uncommon. It is quite sensible only to assume kinematic equilibrium which tends to develop faster than thermodynamic equilibrium. Moreover, the latter is easier to calculate, on the one hand because the droplet radius represents a kind of relaxation factor for the thermodynamic exchange process between the phases, on the other because the phase equilibrium in the form required here is not usually available in the CFD codes by default but must first be implemented.

The opposite step is more advisable, that one does not settle for a single droplet radius but rather has the goal of resolving more complex radius distributions (and radius-temperature spectra with them). To this end, it is sensible to introduce "droplet classes". Each of these droplet classes is its own species and describes droplets that are to be assigned at a respective spatial point to a narrowly defined radius and time interval (these intervals are dependent on the point in space). Every droplet class is represented by its own set of equations (13.104) to (13.106).

Equations (13.104) to (13.106) (whether for one or several droplet classes) are common transport equations that can be easily implemented in a standard CFD code. The liquid phase should contribute to the local density of the "gas" mixture, but not to the specific heat or to pressure (as these are the parameters of the gaseous phase, the thermal energy of the liquid phase is represented by the transport parameter "droplet temperature"). This can be achieved within a CFD code by selecting an extremely high molecular weight for the species "fluid" (in principle, a droplet now corresponds to a molecule). The scalars "droplet radius" and "droplet temperature" are passive. Finally, in the transport equation for the species "steam" and in that for the internal energy of the gaseous phase source terms inverse to the ones of the equations (13.104) and (13.106) have to be added.

The source terms on the right side depend on the relative velocity of droplet to gas. Already in the previous section, an averaged Reynolds number was introduced that only required the difference of the mean velocities. In the context of the approximation of small droplets being undertaken here, the Reynolds number could be set to zero.

The model introduced here is especially good for simulating diesel sprays because of the assumption of small droplets. It can be implemented in most CFD codes with little modeling effort. However, nozzle-resolved meshes must still be used; the use of the Pope correction is advisable in the case of hole type nozzle sprays.

13.3.2 The Embedding of 1D-Euler Methods and Other Approaches

All previously described methods require the numerical resolution of the nozzle orifice. As obvious as this requirement may be, it is quite difficult to fulfill in praxis. Thus, so-called embedding methods will be discussed, with the help of which the demands on mesh resolution in the engine CFD code can be minimized. In such a method, the spray, i.e. the liquid and gaseous phases, are calculated in a nozzle-proximate area (ideally in the zone, in which the liquid phase appears) with an independent spray code on a special computational mesh (typically one- or two-dimensional). The exchange terms of both phases (with respect to momentum,

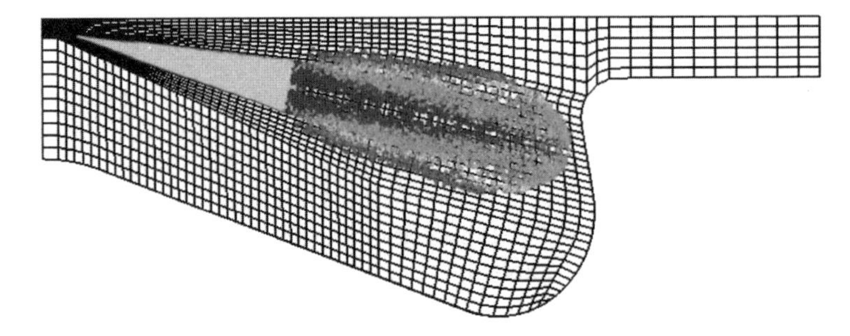

Fig. 13.10 Spray simulation for direct fuel injection in the ICAS embedding method: in the yellow disc 1D calculation, then the standard model

mass, and energy) are calculated then in the engine CFD code coupled in at the corresponding location. In this engine CFD code, only the gaseous phase is calculated. For the thermodynamic boundary conditions, a re-coupling of the engine to the spray code is sensible.

The method should only be applied near to the nozzle, where effects like crossflow still play an insignificant role. Farther down the spray, one can, for example, switch over to the standard Lagrangian model at a defined location, see Fig. 13.10.

The effectiveness of the embedding approach is based on the fact that a sufficiently high resolution can be represented in the spray code and that the spray propagation including all exchange processes between the phases are correctly calculated there. In the engine code, the resolution requirements are now reduced, since an interference of the resolution errors between both phases is avoided. The correct source terms are mapped into the engine calculation, an erroneous calculation of the gaseous phase induces no resulting errors in the source term. However, the resolution requirements on the engine CFD code is still high, and it cannot be recommended enough that one works with spray-adapted meshes.

A comparison of the propagation of the gaseous phase in the engine and in the spray code (it is calculated in both codes!) provides additional certainty about the correctness of the calculation. The models discussed until now (the modified Lagrange method as well as the locally homogeneous 2-phase flow) are suitable as spray codes. However, a 1D Euler code, the ICAS model (integrated cross-averaged spray model) is the most common approach. In this case, genuine 2-phase Euler equations for droplet classes are averaged across the spray cross-section (i.e. within the spray cone). With this averaging, diffusion terms are also largely eliminated; the remaining are neglected compared with the dominant convection. The essential effect of diffusion is implicit in the spray cone angle, which represents an entry parameter. For derivation, one should return to (13.71). The equations for the liquid phase are (we will do without all averaging symbols such as $\langle\rangle$; in these equations all quantities should be seen as averaged/effective):

$$\frac{\partial}{\partial t}\left(r^2 \rho_{ges}\, c_{fl}\right) + \frac{\partial}{\partial r}\left(r^2 \rho_{ges}\, v_{fl} c_{fl}\right) = r^2 \frac{3\rho_{ges} c_{fl} \dot{R}_V}{R}, \tag{13.107}$$

$$\frac{\partial}{\partial t}\left(r^2\rho_{ges}\, c_{fl}v_{tr}\right)+\frac{\partial}{\partial r}\left(r^2\rho_{ges}\, c_{fl}v_{tr}^2\right)=r^2\frac{3\rho_{ges}c_{fl}\dot{R}_V}{R}v_{tr} \\ -r^2\rho_{ges}c_{fl}D_{tr}(v_{tr}-v_g), \tag{13.108}$$

$$\frac{\partial}{\partial t}\left(r^2\rho_{ges}\, c_{fl}R\right)+\frac{\partial}{\partial r}\left(r^2\rho_{ges}\, v_{trl}c_{fl}R\right)=r^2\rho_{ges}\, c_{fl}\left(4\dot{R}_V+\dot{R}_Z\right), \tag{13.109}$$

$$\frac{\partial}{\partial t}\left(r^2\rho_{ges}c_{fl}T_{tr}\right)+\frac{\partial}{\partial r}\left(r^2\rho_{ges}v_{tr}c_{fl}T_{tr}\right)=r^2\rho_{ges}c_{fl}\dot{T}_A+r^2\frac{3\rho_{ges}c_{fl}\dot{R}_V}{R}T_{tr}. \tag{13.110}$$

Equations for the gaseous phase can be formulated analogously

$$\frac{\partial}{\partial t}\left(r^2\rho_{ges}\, c_g\right)+\frac{\partial}{\partial r}\left(r^2\rho_{ges}\, v_g c_g\right)=-r^2\frac{3\rho_{ges}c_{fl}\dot{R}_V}{R}+\mathrm{E}, \tag{13.111}$$

$$\frac{\partial}{\partial t}\left(r^2\rho_{ges}c_g v_g\right)+\frac{\partial}{\partial r}\left(r^2\rho_{ges}v_g^2c_g\right)=-r^2\frac{3\rho_{ges}c_{fl}\dot{R}_V}{R}v_{tr}-r^2\rho_{ges}c_{fl}D_R(v_g-v_{tr}). \tag{13.112}$$

In this case, E designates the entrainment, i.e. the intake of air as a source term of the spray gas mass. Equation (13.111) needs not be a solved together with the other equations, since the remaining equations (13.107)–(13.110) and (13.112) form a 5-dimensional equation system for five variables $c_{fl}, v_g, v_{tr}, R, T_{tr}$); they can be used to calculate the entrainment source term instead. The r^2 terms are dimensional factors, that are induced by the geometry of the problem (cone).

This model has a hyperbolic character, is well solvable and already contains, despite its extreme simplicity, many effects of spray dynamics. In particular, typically "shock wave"-like structures appear because of its hyperbolic character, for example when the liquid phase (of a cold, non-vaporizing spray) propagates faster than the gaseous phase in the spray until it hits the "shock front" of quiescent gas at the spray peak and is thereby slowed down. In effect, a collection of liquid is formed on the spray peak and the spray peak moves in accordance with the law of the stationary free gas jet (13.95). This shock front can be characterized by Hugoniot-type jump conditions (Fig. 13.11, from Krüger (2001)). As in the three-dimensional case, many droplet classes can again be introduced, whereby each can now be described by a set of equations (13.107)–(13.110).

The detailed description of the practical employment of such a method (ICAS) in engine CFD codes can be found in Otto et al. (1999) or Krüger (2001). The method has also been well established in practice (see also Fig. 13.12), especially since it also mitigates the nozzle resolution problem (at least to some extent). Unfortunately, neither a one-dimensional Euler model nor the standard embedment of another spray code has been represented as yet in any commercially available engine CFD codes. The programming complexity required (1D code, embedment

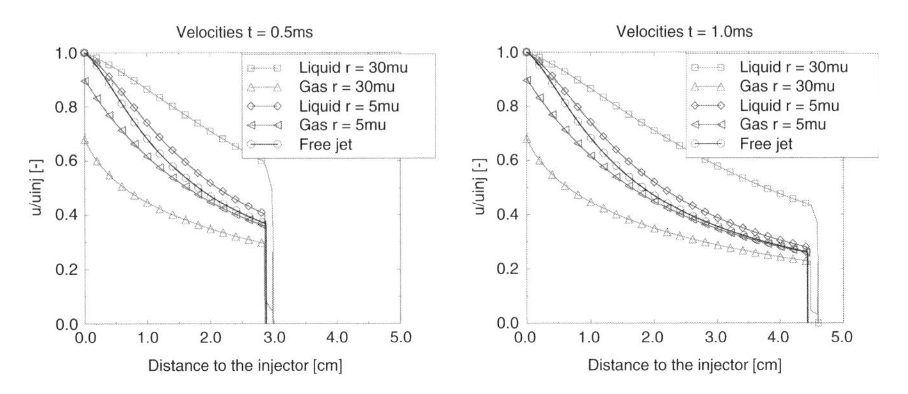

Fig. 13.11 Spray simulation with the ICAS method (Krüger 2001): Propagation behavior of different (non-vaporizing) sprays at various droplet sizes. The spray peak always spreads equally like a stationary free gas jet

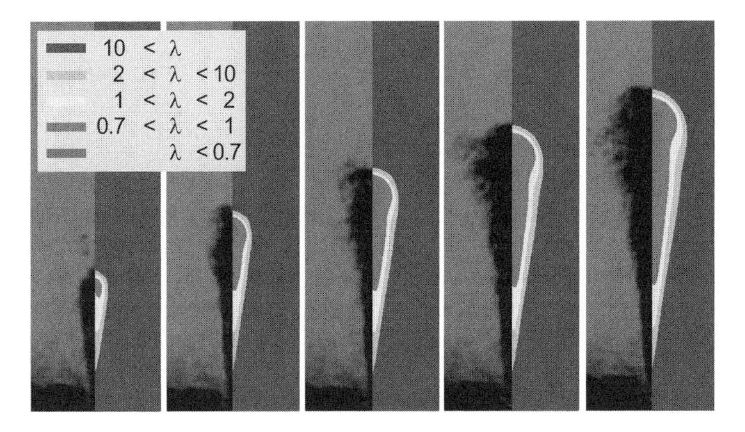

Fig. 13.12 Spray simulation with the ICAS approach (Krüger 2001): in comparison with the schlieren images, the spray form and propagation could be reproduced well

interface, transition to the standard model in the engine code) is very high. For this reason, this method is unfortunately not really recommendable for or available to most users.

13.3.3 The 3D Euler Method

The final option is a complete, three-dimensional, genuinely multi-phase description of the injection spray. Some initial variants are already available in the literature and in different CFD codes. Typically, general multi-phase are used and the model constants adjusted "empirically" for the injection spray application. If one bears in mind however how much and how often this "numerical empiricism" has already gone wrong, this cannot be considered the more effective way. Instead, the goal should be to transfer the standard Lagrangian model, in many ways time-tested (especially after carrying out the modification suggested above) and further developed, into a Eulerian model by theoretical means without modifying its physical content (see also Michels 2008).

We have already developed out concepts and models to such an extent that this step should no longer present large difficulties. We have identified Lagrangian models as methods of solving for the dynamic development of a probability distribution. And a dynamics of observables is hidden behind the concept of "Euler model". Therefore, we need only to transform the spray equation (13.71), which had been formulated as a Fokker–Planck equation, into a dynamic of observables. The second matter is then the closure of unknown terms.

The transformation algorithm is shown in (13.99). We will again use droplet classes that should first resolve a droplet size spectrum at the nozzle. Due to the varying dynamics of different sized droplets, a v–R–T spectrum is formed from this. We work again with the mass (or R^3) weighted observable O and produce after integration with v, R and T per class (index "k") a transport equation for this observable in space and time. For this purpose we again posit

$$\int dv_{tr}\, dR\, dT_{tr} \frac{4\pi}{3} R^3 \rho_{fl} p(x, v_{tr}, R, T_{tr}; t) = \rho(x,t)$$

$$\int dv_{tr} dR\, dT_{tr} \frac{4\pi}{3} R^3 \rho_{fl} p(x, v_{tr}, R, T_{tr}; t) O(x, v, R, T_{tr}) = \langle O(x;t\rangle \rho(x,t) \quad (13.113)$$

and for an arbitrary function f of the observables R and T

$$\langle f(R, T_{tr})\rangle = f(\langle R\rangle, \langle T_{tr}\rangle), \quad (13.114)$$

which is reconcilable with (13.71) and the absence of variance or covariance terms in these quantities. In the following we simply posit $\langle T_{tr}\rangle = T_{tr}$ and $\langle R\rangle = R$.

The transport equation for the mass of the kth class reads:

$$\frac{\partial}{\partial t}\rho_{(k)} + \frac{\partial}{\partial x_i}\left(\rho_{(k)} \left\langle v_{(k),i}\right\rangle\right) = \frac{3\rho_{(k)}\dot{R}_{(k),V}}{R_{(k)}}. \quad (13.115)$$

The transport equations for temperature and radius of the kth class are:

$$\frac{\partial}{\partial t}\left(\rho_{(k)} R_{(k)}\right) + \frac{\partial}{\partial x_i}\left(\rho_{(k)}\left\langle v_{(k),i}\right\rangle R_{(k)}\right) = \rho_{(k)}\left(4\dot{R}_{(k),V} + \dot{R}_{(k),Z}\right), \quad (13.116)$$

$$\frac{\partial}{\partial t}\left(\rho_{(k)}\, T_{(k)}\right) + \frac{\partial}{\partial x_i}\left(\rho_{(k)}\left\langle v_{(k),i}\right\rangle T_{(k)}\right) = \rho_{(k)}\dot{T}_{(k),A} + \frac{3\rho_{(k)}\dot{R}_{(k),V}}{R_{(k)}} T_{(k)}. \quad (13.117)$$

The transport equation for momentum is:

$$\frac{\partial}{\partial t}\rho_{(k)}\langle v_{(k),i}\rangle+\frac{\partial}{\partial x_j}\left(\rho_{(k)}\langle v_{(k),i}\rangle\langle v_{(k),j}\rangle\right)+D_{tr}\rho_{(k)}\left(\langle v_{(k),i}\rangle-\langle v_{g,i}\rangle\right)$$
$$+\frac{\partial}{\partial x_j}\left(\rho_{(k)}\tau_{(k)ij}\right)+D_{xv}\frac{\partial}{\partial x_i}\rho_{(k)}-\left(\frac{\partial}{\partial x_i}D_{vv}\right)\rho_{(k)}=\frac{3\rho_{(k)}\dot{R}_{(k),V}}{R_{(k)}}\langle v_{(k),i}\rangle, \tag{13.118}$$

where, as in (13.72), the coefficients D_{tr}, v_g, D_{xv} and $\partial D_{vv}/\partial x$ and again assumed to be constant and $\dot{R}_V$as independent of velocity to derive the equation. A turbulent stress tensor was defined in the following way:

$$\tau_{(k)ij}=\left\langle\left(v_{(k),i}-\langle v_{(k),i}\rangle\right)\left(v_{(k),j}-\langle v_{(k),j}\rangle\right)\right\rangle. \tag{13.119}$$

We now calculate another transport equation for droplet turbulence, i.e. the velocity variance, which is also given as the trace of the stress tensor:

$$k_{(k)}=\frac{1}{2}\tau_{(k),ii}. \tag{13.120}$$

The equation reads:

$$\frac{\partial}{\partial t}\left[\rho_{(k)}k_{(k)}\right]+\frac{\partial}{\partial x_i}\left[\rho_{(k)}\langle v_{(k),i}\rangle k_{(k)}\right]+\frac{1}{2}\frac{\partial}{\partial x_i}\left[\rho_{(k)}\tau_{(k),ijj}\right]$$
$$-\rho_{(k)}\tau_{(k),ij}\frac{\partial v_{(k),j}}{\partial x_i}+2D_{tr}\rho_{(k)}k_{(k)}+D_{xv}\rho_{(k)}\frac{\partial v_{(k),i}}{\partial x_i}-3D_{vv}\rho_{(k)}$$
$$=\frac{3\rho_{(k)}\dot{R}_{(k),V}}{R} \tag{13.121}$$

with

$$\tau_{(k)ijj}=\left\langle\left(v_{(k),i}-\langle v_{(k),i}\rangle\right)\left(v_{(k),j}-\langle v_{(k),j}\rangle\right)\left(v_{(k),j}-\langle v_{(k),j}\rangle\right)\right\rangle. \tag{13.122}$$

As we have already indicated, we are confronted with a closure problem: we know neither the non-diagonal term of the dual correlation function $\tau_{(k),ij}$ nor the triple correlation function $\tau_{(k),ijj}$.

We will close these terms by putting the source terms of (13.71) in equilibrium:

$$\frac{\partial}{\partial v_{tr,i}}\left[D_{tr}\left(\langle v_{g,i}\rangle-v_{tr,i}\right)p\right]+\frac{\partial}{\partial R}\left[\dot{R}_V p\right]+\frac{1}{R^3}\frac{\partial}{\partial R}\left[R^3\dot{R}_Z p\right]+\frac{\partial}{\partial T}\left[\dot{T}_{tr,A}p\right]$$
$$=\frac{\partial}{\partial v_{tr,i}}\left[D_{xv}\frac{\partial p}{\partial x_i}\right]+\frac{\partial}{\partial v_{tr,i}}\left[D_{vv}\frac{\partial p}{\partial v_{tr,i}}\right]. \tag{13.123}$$

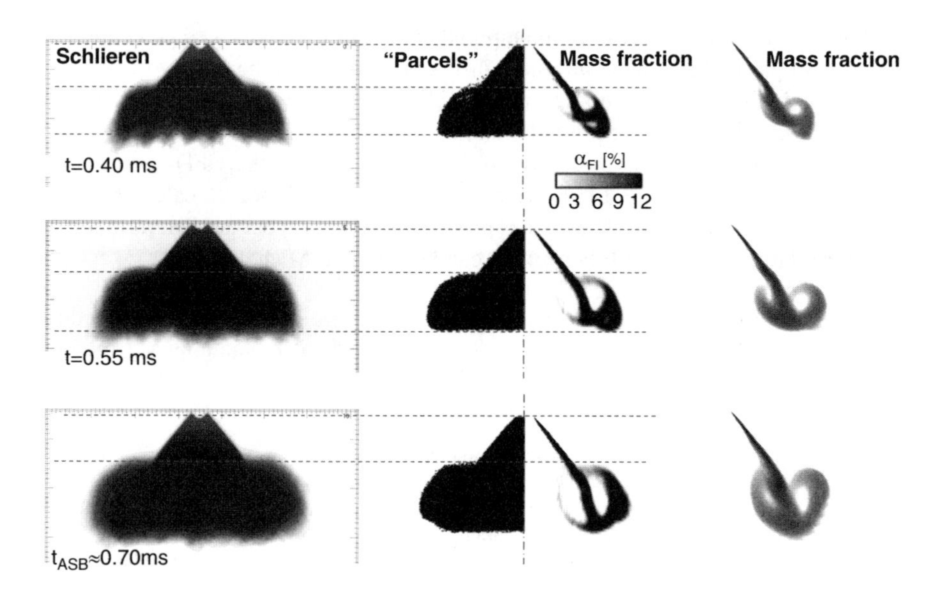

Fig. 13.13 Comparison between diagnostics and simulation for a gasoline direct injection with a outward-opening nozzle. Left: diagnostics; center: simulation in the further developed Lagrangian concept (on the basis of the Fokker–Planck equation) (3D image and 0.5° sector); right: simulation in the 3D Euler model (0.5° sector)

If we multiply this equation with $R^3\left(\tau_{(k),ij} - \delta_{ij}\tau_{(k)ll}/3\right)$ and integrate over v, R and T, we obtain:

$$\tau^{sf}_{(k),ij} = \tau_{(k),ij} - \delta_{ij}\underbrace{\frac{\tau_{(k),ll}}{3}}_{\frac{2}{3}k_{(k)}} = \frac{D_{xv}}{2D_{tr}}\left[\frac{\partial\langle v_{(k),i}\rangle}{\partial x_j} + \frac{\partial\langle v_{(k),j}\rangle}{\partial x_i} - 2\delta_{ij}\frac{\partial\langle v_{(k),l}\rangle}{\partial x_l}\right]. \tag{13.124}$$

An analogous procedure with respect to $\tau_{(k),ijj}$ leads to:

$$\tau_{(k),ijj} = -\frac{2D_{xv}}{3D_{tr}}\left[\frac{\partial k_{(k)}}{\partial x_i} + \frac{\partial \tau_{(k),ij}}{\partial x_j}\right]. \tag{13.125}$$

Inserting into (10.121), we finally obtain:

$$\begin{aligned}
&\frac{\partial}{\partial t}\left[\rho_{(k)}k_{(k)}\right] + \frac{\partial}{\partial x_i}\left[\rho_{(k)}\left\langle v_{(k),i}\right\rangle k_{(k)}\right] - \frac{1}{3}\frac{\partial}{\partial x_i}\left[\frac{D_{xv}}{D_{tr}}\rho_{(k)}\left(2\frac{\partial k_{(k)}}{\partial x_i} + \frac{\partial \tau^{sf}_{(k),ij}}{\partial x_j}\right)\right] \\
&+ \rho_{(k)}\left(\tau_{(k),ij} + \frac{2}{3}k_{(k)}\delta_{ij}\right)\frac{\partial v_{(k),j}}{\partial x_i} + 2D_{tr}\rho_{(k)}k_{(k)} + D_{xv}\rho_{(k)}\frac{\partial v_{(k),i}}{\partial x_i} - 3D_{vv}\rho_{(k)} \\
&= \frac{3\rho_{(k)}\dot{R}_{(k),V}}{R}.
\end{aligned} \tag{13.126}$$

The set of equations exhibits some parallels to the Reynolds-averaged equation (12.36). In particular, typical turbulent diffusion terms are generated.

In addition to these equations, corresponding source terms in the gaseous phase equations must still be considered (mass, momentum, turbulent kinetic energy, heat). This equation set is presently being implemented in the CFD code STAR CD (version 4). Figure 13.13 shows a first result of an A-nozzle calculation for gasoline direct injection. The exact procedure with respect to derivation and implementation of this 3D Euler model can be found in Michels (2008).

References

Amsden AA (1999) KIVA-3V: a block-structured KIVA program for engines with vertical or canted valves; KIVA-3V, Release2, Improvements to KIVA-3V, LA-13608-MS

Amsden AA, O'Rourke PO, Buttler TD (1989) KIVA-II: a computer program for chemically reactive flows. Los Alamos National Laboratory Report, LA-11560-MS

Bai C, Gosman AD (1995) Development of methodology for spray impingement simulation, SAE 950283

Baratta M, Catania AE, Spessa E, Herrmann L, Rößler K (2008) Multi-dimensional modeling of direct natural-gas injection and mixture formation in a stratified-charge SI engine with centrally mounted injector, SAE 2008-01-0975

Hallett WLH (2000) A simple model for the vaporization of droplets with large number of components. Combust Flame 121:334–344

Hermann A (2008) Modellbildung für die Simulation der Gemischbildung und Verbrennung in Ottomotoren mit Benzin-Direkteinspritzung. PhD thesis, Universität Karlsruhe

Johnson N, Amsden A, Naber J, Siebers D (1995) Three-dimensional computer modeling of hydrogen injection and combustion. In: '95 SMC Simulation Multiconference, Phoenix Arizona

Krüger Ch (2001) Validierung eines 1D-Spraymodells zur Simulation der Gemischbildung in direkteinspritzenden Dieselmotoren. PhD thesis, RWTH Aachen

Landau LD, Lifschitz EM (1981) Course of theoretical physics, vol 10 Physical kinetics. Pergamon, Oxford

Lippert AM, Stanton DW, Reitz RD, Rutland CJ, Hallett WLH (2000) Investigating the effect of spray targeting and impingement on diesel cold start, SAE 2000-01-0269

Michels U (2008) Euler-Strahlmodell für Düseninnenströmung und Strahlausbreitung. PhD thesis, Universität Hannover

O'Rourke PJ (1989) Statistical properties and numerical implementation of a model for droplet dispersion in a turbulent gas. J Comput Phys 83(2):345–360

Otto F, Krüger Ch, Wirbeleit F, Willand J (1999) Probleme und Lösungsansätze bei der Simulation der dieselmotorischen Einspritzung. Mess- und Versuchstechnik für die Entwicklung von Verbrennungsmotoren. Haus der Technik, Essen

Patterson MA (1997) Modeling the effects of fuel injection characteristics on diesel combustion and emission. Ph D Thesis, University of Wisconsin-Madison

Patterson MA, Reitz RD (1998) Modelling the effects of fuel spray characteristics on diesel engine combustion and emissions. SAE-Paper 980131

Pilch M, Erdman CA (1987) Use of breakup time data and velocity history data to predict the maximum size of stable fragments for acceleration-induced breakup of a liquid drop. Int J Multiphase Flow 13:741–757

Pope SB (1978) An explanation of the turbulent round-jet/plane-jet anomaly. Am Inst Aeronautics Astronautics J 16:279–281

Pope SB (1985) PDF methods for turbulent reactive flows. Prog Energy Combust Sci 19:119–197
Reitz RD (1987) Modeling atomization processes in high-pressure vaporizing sprays. Atomization Spray Technol 3:309–337
Risken H (1989) The Fokker-Planck equation: methods of solutions and applications. Springer, Berlin
Siebers DL (1998) Liquid-phase fuel penetration in diesel sprays. SAE Paper 980809
Stanton D, Rutland Ch (1996) Modeling fuel film formation and wall interaction in diesel engines, SAE 960628
Tanim J, Hallett WLH (1995) Continous thermodynamics model for multicomponent vaporization. Chem Eng Sci 50(18):2933–2942
Tatschl R, von Künsberg Sarre C, Alajbegovic A, Winklhofer E (2000) Diesel spray break-up modelling including multidimensional cavitation nozzle flow effects. In: 16th ILASS-Europe 2000, Valencia, Spain
Williams FA (1958) Spray combustion and atomization. Phys Fluids 1:541–544

Chapter 14
Simulation of Combustion

Frank Otto and Christian Krüger

This section is devoted to the fluid-mechanical simulation of turbulent combustion, for Diesel and SI engines. In essence, we are "only" dealing with turbulent averaging of the source term of the species transport equations (12.18); however, it is immediately apparent that this is a difficult endeavor, as reaction kinetics are typically exponentially depending on the temperature. The necessary modeling effort for this is not insignificant – with the pure application of commercially standardized models, we (unfortunately) do not get very far.

We should point out that we are only concerned with engine combustion, i.e. with unsteady, turbulent combustion processes in complex, moving geometries, as a consequence of or in accompaniment with complex mixture processes. It thus becomes quickly clear that many combustion models developed for much simpler boundary conditions are adaptable to engines. Another big problem for the enduring establishment of generally recognized advances in modeling continues to be the lack of a reliable spray model, making it difficult to evaluate the quality of a combustion model.

14.1 Excursus: Combustion Regimes

In internal combustion engines, different combustion regimes exist, as well as hybrid forms of these. Figure 14.1 shows examples of these in a triangular diagram. At the bottom left is classical diffusion combustion: fuel and oxidants are first unmixed. They mix in a first step (i.e. a physical process) and react in the second (chemical) step, whereby the second step proceeds much more quickly than the first. This can be expressed succinctly in the formula: "mixed = burned". If the chemical step is the dominant one however, e.g. because the fuel and oxidants are already perfectly mixed in the combustion chamber, we find ourselves in the upper corner of the triangle, homogeneous autoignition. SI compression ignition combustion comes very close to this ideal case. Conversely, a classical Diesel combustion process with small premixed amounts (e.g. in the case one or several preinjections are introduced, these reduce the ignition delay of the main injection dramatically, i.e. they reduce the chemical time scale) comes close to the idealized diffusion

G.P. Merker et al. (eds.), *Combustion Engines Development*,
DOI 10.1007/978-3-642-14094-5_14, © Springer-Verlag Berlin Heidelberg 2012

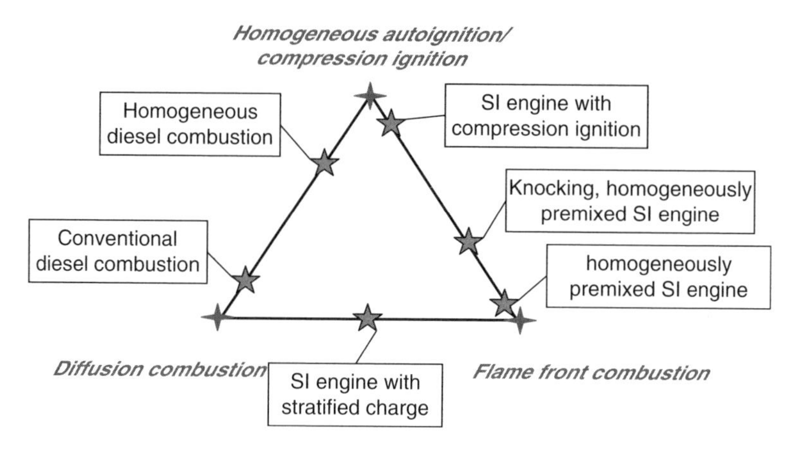

Fig. 14.1 Representation of different engine combustion regimes

combustion process on the bottom left of the triangle. The larger the premixed amount, the further up on the left side of the triangle is the combustion.

The bottom right of the triangle is a classical flame front combustion, which should be strictly differentiated from homogeneous autoignition. In this case, there is no autoignitable mixture locally despite ideal fuel-oxidant mixing (e.g. due to insufficient temperatures and, as a result, excessive ignition delay). Combustion occurs in this case when (starting from an ignition source) a flame front advances spatially through the mixture. The hot burned gases behind the flame front heat up the still unburned gases in front of it, as a result of which ignition of the unburned mixture is initiated. If a flame front is formed in a mixture that is close to the autoignition limit, temperature increase created by the compression of the unburned mixture by the expansion of the burning mixture can already lead to autoignition; this results in flame front speeds in the order of magnitude of the speed of sound and higher, and deflagration becomes detonation. In the context of the engine, we then experience the phenomenon of knocking.

Similar conditions exist in a Diesel engine. Flame front propagation does not tend to occur; a more or less independent autoignition of the premixed fuel occurs basically "simultaneously". If flame front phenomena were to play a larger role (i.e. the already burning premixed fuel part induces ignition of the not yet burning part), undesirable knocking would again immediately come into play.

In SI engines with stratified charge finally, there is a mixture of diffusion flame and flame front premixed combustion. A flame front acts on the inhomogeneous mixture, but behind the flame front there is still fuel (in the locally rich zones), even if it is chemically certainly no longer in its original form, and oxidants (in the locally lean zones), which then are converted in a subsequent diffusion combustion.

Combustion sequences within the triangle (Fig. 14.1) are also possible, if autoignition phenomena occur in an SI engine with stratified charge at sufficiently high load for example. But these conditions appear technically not to be very relevant, at least not yet.

In addition, it should be noted that turbulent flow conditions are predominant in a combustion chamber. The combustion regimes of Fig. 14.1 should therefore all be interpreted in their turbulent variants. In the case of both bottom corners – turbulent diffusion combustion and turbulent flame front combustion – this is also uncritical because we are basically dealing with a scale separation: the chemical time scale is much shorter than the physical (in these cases the turbulent) time scale, which is why the latter is the dominant one. In both cases, turbulence-driven processes determine the sequence, turbulent mixing in the first case and turbulent heat conduction in the second. Here, the turbulent Damköhler number is introduced as a characteristic parameter describing the ratio of the turbulent and chemical time scales:

$$Da = \frac{\tau_t}{\tau_{chem}}. \tag{14.1}$$

Thus, for the previously described cases based on the triangle in Fig. 14.1, Da $\gg$ 1 is valid.

The situation is more complicated for smaller Damköhler numbers, i.e. as soon as the reaction kinetics begin to determine the process, we will experience an upwards movement on the triangle in Fig. 14.1. A time scale separation turbulence-chemistry is no longer possible then, and highly complex ensemble averaging is required for an accurate calculation.

With respect to modeling, one should expect that the premixed SI engine, the stratified charge SI engine, and the conventional Diesel engine highly dominated by diffusion combustion would be represented most advantageously. This is indeed exactly the case, as we find in practice.

14.2 General Procedure

First, the general procedure will be presented. The thermodynamic accuracy of the model and of the boundary conditions must first be secured. This takes place logically by using standardized 1D- and 0D-codes. Especially in Diesel engine applications, one must take into consideration that the real gas behavior[1] (i.e. the behavior that manifests itself by a deviation from the ideal gas equation) already plays a considerable role. However, component elasticity (piston, cylinder head screws) are also relevant, and both effects have (at least in their effect upon the

[1]In the context of engines in particular, the concept of "real gas" is frequently used incorrectly. "Real gas behavior" refers to the case in which the inner energy becomes dependent on volume or pressure (i.e. to a Van der Waals gas, see also Landau and Lifschitz 1980). A gas for which only the temperature dependence of the specific heat is considered is an ideal gas as before. Even an inert mixture of ideal gases is itself ideal.

pressure path) varying signs; ideally, they may compensate for each other. Real gas behavior is unfortunately unavailable in engine CFD codes at this time.

In the second step, one should turn one's attention to the pressure path in the expansion phase (i.e. after the end of combustion). Here exist again two types of sources of error, on the one hand a wall heat transfer calculated too low, which results in pressure values in the simulation which are too high, on the other hand a poor mixture formation (typically in Diesel engine full load calculations), which is responsible for pressure values which are too low. However, even if the pressure graph should "fit" exactly, this does not yet necessarily indicate a correct description of combustion. It is not at all uncommon that both errors compensate for each other (of course only in the pressure graph). For the correct calculation of wall heat transfer, we refer to Sect. 12.2.3: meshes with correct y^+- values must be utilized, and one should apply the Han-Reitz formulation (12.53). Nevertheless, one will still calculate wall heat transfer rates which are too low. An important reason for this is the lacking heat transfer because of soot radiation. As long as nothing better is available, the simplest way of "correcting" this is a scaling of the wall heat transfer, so that the desired value results globally. This target value can be obtained from a heat release rate analysis.

The other case is more critical, namely, when in Diesel or SI engines with stratified charge the pressure path after the end of combustion is too low (i.e. lower than the experimental graph or one calculated with a 0D-program). In that case, there is presumably a calculational mixture formation deficiency. It must be stressed that here, typically, the 0D-simulation is more reliable in the comparison between 0D- and 3D-simulation. A 0D-program might possess an unsuitable (or poorly adjusted) combustion model for the concrete application case at hand, but after the end of combustion, all pressure graphs should lie close to each other, provided only that an equal amount of fuel has been converted. 0D-programs work however mostly with experimentally well adjusted conversion rates. Not so the 3D-simulation. In this case the global conversion rate is not an input parameter, but results from the CFD calculation of flow, mixture formation, and combustion. If we now have, as the consequence of an erroneous mixture-formation calculation, a locally rich mixture zone with $\lambda < 1$, unburned fuel or combustible intermediary species (like H_2, CO, see below) must in this case continue to exist. And no 3D-combustion model in the world (which as such is always locally formulated) can solve this problem. It is therefore not worthwhile to look out for better combustion models, as the spray model is the culprit. The problem with the latter has already been discussed enough in the last section. An ad hoc measure, which indeed "helps", is increasing the injection velocity, usually to completely unphysical values beyond the Bernoulli velocity. This procedure is explicitly not recommended.

If the thermodynamics and mixture formation simulation are largely under control, i.e. the pressure path in compression and expansion is correct/plausible (according to the available occasions for comparison or validation), we can concern ourselves with the analysis of the actual combustion. Typically, the pressure path is not used for this, but rather the gross heat release rate. Experimentally, this one

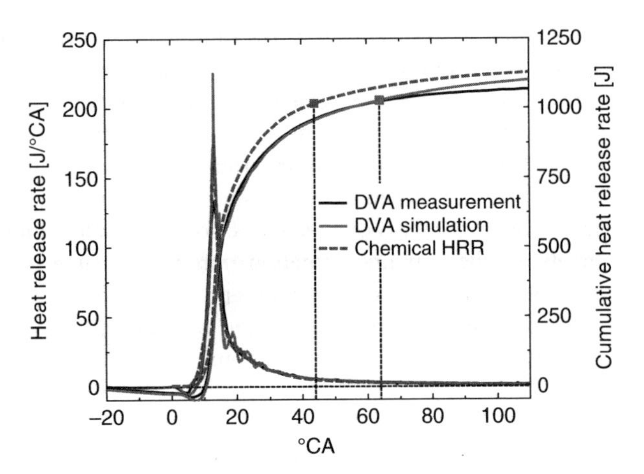

Fig. 14.2 Comparison of two heat release rates from 3D simulation, created by spatial integration and indication evaluation. The heat release rate taken from spatial integration is somewhat earlier

results from the pressure path by indication analysis; the heat release rate follows from this by means of a heat transfer model. In the 3D-simulation it seems to be obvious to obtain the gross heat release rate by summation of the heat release rates in all cells. But this procedure is not compatible with the experimental procedure. If gases of different temperature in pressure equilibrium are mixed in a closed space, an increase in pressure occurs (this is essentially an effect of temperature-dependent heat capacities – we will leave the calculative derivation to the reader as an exercise). Let us now consider two cases: in the first case, heat release occurs homogeneously in the chamber, in the second inhomogeneously and mixing only occurs later. The overall system is closed, the integral heat release identical. Since pressure and inner energy are state variables, the final states must be identical in both cases, but because in the second case a pressure increase is associated with the mixing process, the pressure increase in the previous process of inhomogeneous combustion must turn out smaller than in the homogeneous combustion process of the first case. Applied to the engine problem however, this means that the heat release rate taken from the 3D simulation as a space integral is shifted earlier in comparison with the heat release rate taken from indication evaluation (see Fig. 14.2). Therefore it is best to subject both pressure curves to an indication evaluation in order to compare heat release rates from the measurement and the calculation.

14.3 Diesel Combustion

In the main phase of Diesel combustion, it is a case of diffusion combustion, i.e. the turbulent time scale is dominant ($Da >> 1$). Autoignition and premixed combustion (which dies not correspond to SI engine premixed combustion) are considerably more affected by the reaction kinetics, i.e. the Damköhler number is much smaller. The same is valid for pollutant formation.

The question of course arises of whether combustion simulation is necessary at all. Isn't mixture formation the dominant process in the Diesel engine, is thus a simulation of mixture formation not sufficient to analyze engine behavior? In principle, this approach is, depending on the formulation of the question, not altogether wrong, but mixture formation is heavily influenced by local expansion effects that are caused by combustion (locally burning zones become very hot and thus expand). Thus combustion is not negligible. Combustion also helps to find errors in the mixture-formation simulation, such as when larger amounts of unburned fuel, CO, or H_2 remain at the end of the calculation.

In the following, the simulation of heat release will first be treated and then the more complex phenomena of ignition and pollutant formation.

14.3.1 Simulation of Heat Release

The main phase of Diesel combustion can be seen as a turbulent diffusion flame, i.e. it progresses mixture-controlled, according to the formula "mixed = burned". With diffusion, turbulent diffusion is meant here.

14.3.1.1 Eddy Breakup Models

The simplest method of modeling a turbulent diffusion flame is the so-called "eddy breakup model". In such a model, into the species transport equations of type (14.14) source terms Q are introduced that scale with the species concentrations and the inverse turbulent length scale, i.e. they describe decay and formation processes that run with the turbulent breakup time

$$\tau_t \propto \frac{k}{\varepsilon}$$

e.g.

$$Q \propto \frac{c_A c_B}{\tau_t}, \tag{14.2}$$

A heat release is calculated corresponding to the conversion rate.

To describe Diesel combustion, the combustion progress must still be modeled. The most well-known and used method for this is the mixing time-scale model, see Patterson and Reitz (1998). In it, from the turbulent and a chemical time scale (τ_t and τ_{chem}), an effective time scale τ_{eff} is formed, with which the combustion processes proceed

$$\tau_{eff} = \tau_{chem} + f\,\tau_t, \quad f = \frac{1-\exp(-r)}{1-\exp(-1)}, \quad \tau_{chem} << \tau_t = c_\mu \frac{k}{\varepsilon}, \tag{14.3}$$

whereby r designates the mass fraction of all reaction products. At the start of the reaction, $f = 0$, i.e. the effective time scale corresponds to the (small) chemical time scale, the reaction progresses very quickly, and a "premixed peak" occurs. With the increase in reaction products, f increases (to 1 maximally), and now $\tau_{eff} \approx f \cdot \tau_t$ is valid, i.e. we now have a diffusion combustion. The mixing time-scale model uses seven species N_2, O_2, fuel, H_2O, CO_2, CO and H_2. From a given distribution of concentration ($c^*_{(k)}$, $k = 1\ldots7$), the corresponding equilibrium distribution ($c^*_{(k)}$, $k = 1\ldots7$) is calculated. We assume that every species tends towards local equilibrium with the time scale τ_{eff}

$$\rho\left(\frac{\partial}{\partial t} + v_i \frac{\partial}{\partial x_i}\right) c_k - \frac{\partial}{\partial x_i}\left(D_t \rho \frac{\partial}{\partial x_i} c_k\right) = \rho \frac{c^*_{(k)} - c_k}{\tau_{eff}}. \tag{14.4}$$

The chemical equilibrium code for the mixing time-scale model works with two lambda regimes. The limit between both of these regimes is given by the air-fuel ratio at which the fuel can be completely converted with the existing oxygen to CO and H_2. In the "rich" regime, the lambda values of which are lower than this limiting value, the equilibrium is determined such that the total available oxygen is utilized to create CO and H_2 from the fuel. In addition however, unburned fuel remains. In the "lean" regime, the lambda values of which are higher than the limiting value, we assume that no fuel remains. Besides the non-reactive N_2, five reactive species thus still remain, H_2O, H_2, O_2, CO und CO_2. Their equilibrium concentration is calculated by means of three elemental mass fraction conservation equations (for C, O, and H) as well as two relations, which follow from the law of mass action

$$\frac{[\mathrm{CO_2}]}{[\mathrm{CO}]\ [\mathrm{O_2}]^{0,5}} = K_{\mathrm{C}}(p, T)\,,$$
$$\frac{[\mathrm{H_2O}]}{[\mathrm{H_2}]\ [\mathrm{O_2}]^{0,5}} = K_{\mathrm{H}}(p, T)\,. \tag{14.5}$$

From these five equations, we obtain a polynomial of fourth order, which can be solved analytically with the corresponding solution formula. The results of this equilibrium solver are shown in Fig. 14.3. Corresponding to the reaction rates of the species, a source term for the enthalpy equation is determined by means of their specific reaction enthalpies $h_{(k)}$

$$q = \rho \sum_k h_{(k)} \dot{c}_{(k)}. \tag{14.6}$$

With this seven-species method, local conditions can be very well prescribed, even in the rich area. Both in the modeling of reaction-kinetic influences and in turbulence interaction, the mixing time-scale model is however still very simple. Nevertheless, one can already analyze basic dependencies and mechanisms of Diesel engine combustion with this modeling approach.

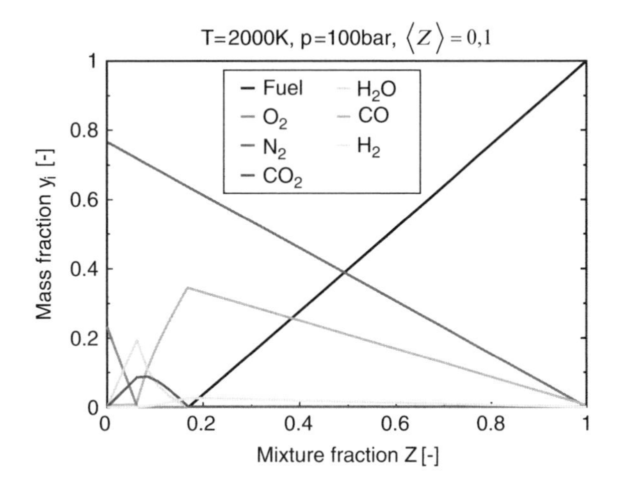

Fig. 14.3 Distributions of the seven species over the fuel mass fraction or the mixture fraction

instead of seven species, we can also work with fewer, the minimum being three:

- Oxidants (air + recirculated residual gas)
- Fuel
- Combustion products

These three species are of a formal nature. Typically, the following reaction equation is set up (the sum formula of the fuel is C_mH_n):

$$\frac{\left(m+\frac{n}{2}\right)}{0,21}\left[0,21 \cdot O_2 + 0,79 \cdot N_2\right] + C_mH_n \rightarrow mCO_2 + nH_2O + \frac{0,79}{0,21}\left(m+\frac{n}{2}\right)N_2 \tag{14.7}$$

The right side represents the species "product". Any residual gas should be added to both sides.

14.3.1.2 β-pdf-Based Models

The modeling of the diffusion part can be improved with an approach corresponding to the mixing time-scale model, the so-called *pdf time-scale model*, see Rao and Rutland (2002). In this model, in addition to the species transport equations, transport equations for the mixture fraction and the mixture fraction variance are solved, from which the local mean values of the relevant species $c_{(k)}^{(pdf)}$ are determined using the *β-pdf* (see Sect. 12.2.4). The species transport equations now read

$$\rho\left(\frac{\partial}{\partial t} + v_i\frac{\partial}{\partial x_i}\right)c_{(k)} - \frac{\partial}{\partial x_i}\left(D_t\rho\frac{\partial}{\partial x_i}c_{(k)}\right) = \rho\,\frac{c_{(k)}^{(pdf)} - c_{(k)}}{\tau_{chem}}. \tag{14.8}$$

The turbulent mixing process, which progresses with time scale τ_t, is described in the term $c_{(k)}^{(pdf)}$, since it tends in a time of the order of magnitude τ_t against the local equilibrium value in homogeneous mixing $c_{(k)}^*$

$$\dot{c}_{(k)}^{(pdf)} \approx \frac{c_{(k)}^* - c_{(k)}^{(pdf)}}{\tau_t}. \tag{14.9}$$

The premixed part is indeed still illustrated with the same "phenomenological" approach, but the turbulence interaction is now better described. This is helpful especially for full-load combustions, in which the premixed part is of only minor importance.

How is $c_{(k)}^{(pdf)}$ calculated? In the original formulation of the model, an averaging for this is carried out over the mixture fraction Z (with β distribution) and the scalar dissipation rate χ (with Gaussian distribution)

$$c_{(k)}^{(pdf)} = \int dZ \int d\chi \, p_\beta(Z) \, pdf_{Gauss}(\chi) \, c_{(k)}(Z, \chi). \tag{14.10}$$

In order to calculate this integral however, knowledge of the functions $c_{(k)}(Z, \chi)$ is necessary. The determination of these functions, so-called *flamelets*, demands costly reaction-kinetic calculations on laminar counter-flow flames. The scalar dissipation rate has the significance of an effective diffusion in the mixture fraction space and is responsible for so-called flame stretching effects, i.e. laminar flow equilibriums between diffusion and reaction. In principle, we thus leave the diffusion flame approach "mixed = burned", and the chemistry is no longer considered as "infinitely fast". Yet the reaction kinetics is very fast especially in the Diesel engine diffusion flame phase because of high temperatures, and effects of finitely fast reaction kinetics are indeed already considered in the chemical time scale τ_{chem} in a highly phenomenological manner. It is thus to be recommended that one neglects the flame stretching effects and applies the seven-species equilibrium kinetics as a reaction-kinetic basis as described above, see Steiner et al. (2004). The calculation and modeling cost is enormously reduced in this way. $c_{(k)}^{(pdf)}$ now amounts to

$$c_{(k)}^{(pdf)} = \int_0^1 dZ \, p_\beta\left(Z; \langle Z \rangle, \left\langle Z''^2 \right\rangle\right) c_{(k)}^*(Z). \tag{14.11}$$

The distribution function p_β is the β function with the mean value$\langle Z \rangle$ and the variance Z''^2. The functions $c_{(k)}^*(Z)$ are the ones shown in Fig. 14.3 (ideally, one should also take into consideration that the temperature is not constant over the mixture fraction).

The functions $c_{(k)}^*(Z)$ are partially quite linear. With this, a very efficient integration scheme is possible for the integral (14.11), if one discretizes the

Z-axis into intervals, in which the functions $c^*_{(k)}(Z)$ can be linearly (or quadratically) approximated. The product of a β-distribution function with a function linear in Z produces however exactly the linear combination of two β-distribution functions

$$\begin{aligned} &N(a,b)^{-1}(1-Z)^{a-1}Z^{b-1}\cdot(A+B\cdot Z)\\ &\quad = AN(a,b)^{-1}(1-Z)^{a-1}Z^{b-1}+BN(a,b)^{-1}(1-Z)^{a-1}Z^{b}. \end{aligned} \tag{14.12}$$

Thus, only integrals of the form

$$B(a,b;x)=N(a,b)^{-1}\int_0^x dZ\,Z^{a-1}(1-Z)^{b-1} \qquad 0\le x\le 1 \tag{14.13}$$

must be solved, which are designated in the literature as incomplete beta-functions. To calculate them, very efficient algorithms are available. However, an efficient method of solving integrals (14.11) is decisive, as these have to be calculated in every time interval in every computational cell. And β-distribution functions may assume very unpleasant forms (at every Z-value, an arbitrarily sharp peak is possible).

With the pdf time-scale model (in conjunction with an accurate spray model), very good full-load results can be obtained. An example of this is depicted in Fig. 14.4.

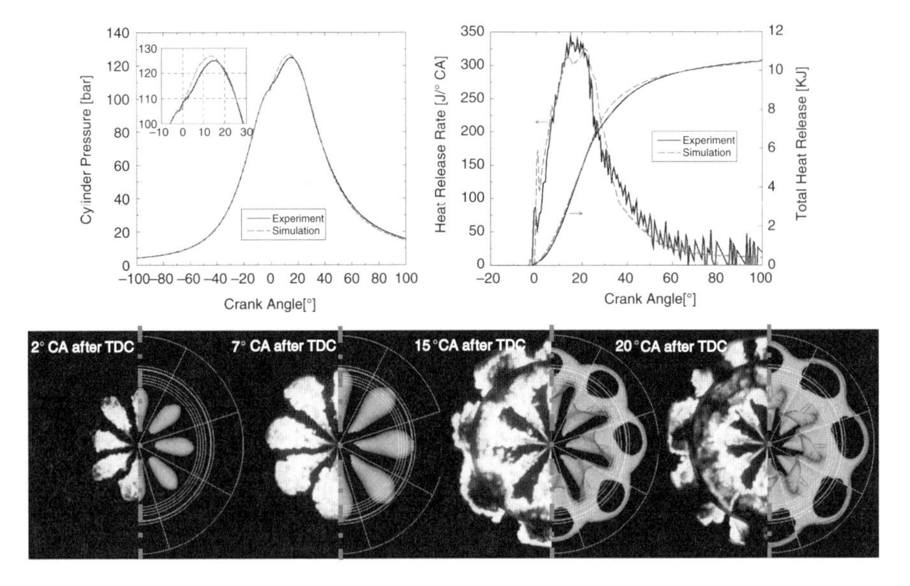

Fig. 14.4 Simulation of the combustion of a commercial vehicle Diesel engine with the pdf time-scale model. (**a**) Comparison of experimental and calculated pressure paths (**b**) Comparison of temperature-isosurface (simulation) with film shot from transparent engine (*left*)

14.3.1.3 Extended Coherent Flame Models

This class of models is based on SI engine CFM (coherent flame) models describing a flame front combustion (see also Sect. 14.4.3). These have been modified for application to Diesel engines however. There in now no longer a propagation of the flame front, in the combustion triangle (Fig. 14.1) we are no longer on the base line but on the left corner; there is an averaging between autoignition (very fast reaction kinetics) and diffusion combustion, but not between flame front propagation and diffusion combustion. Typically, several zones are introduced locally:

- Air zone; unmixed
- Fuel zone; unmixed
- Mixed zone (air + residual gas) + fuel + combustion product

Now species transport equations are formulated for the individual zones; a transfer of "unmixed" to "mixed" takes place as a turbulent mixing process with the turbulent time scale; the reaction in the mixed zone takes place as i a homogeneous reactor. $c_{K,u}$ designates the concentration of unmixed fuel, $c_{Ox,u}$ the concentration of unmixed oxidant (air and residual gas), $c_{K,g}$ the concentration of mixed fuel, $c_{Ox,g}$ the concentration of mixed air, and c_P the concentration of combustion products. A typical model could then look something like this:

$$\begin{aligned}
&\frac{\partial}{\partial t}(\rho c_{K,u})+\frac{\partial}{\partial x_j}(\rho\langle v_j\rangle c_{K,u})-\frac{\partial}{\partial x_j}\left(D_t\rho\frac{\partial}{\partial x_j}c_{K,u}\right)=-A\frac{\varepsilon}{k}c_{K,u}+q_{verd}\\
&\frac{\partial}{\partial t}(\rho c_{Ox,u})+\frac{\partial}{\partial x_j}(\rho\langle v_j\rangle c_{Ox,u})-\frac{\partial}{\partial x_j}\left(D_t\rho\frac{\partial}{\partial x_j}c_{Ox,u}\right)=-\tilde{A}\frac{\varepsilon}{k}c_{Ox,u}\\
&\frac{\partial}{\partial t}(\rho c_{K,g})+\frac{\partial}{\partial x_j}(\rho\langle v_j\rangle c_{K,g})-\frac{\partial}{\partial x_j}\left(D_t\rho\frac{\partial}{\partial x_j}c_{K,g}\right)=A\frac{\varepsilon}{k}c_{K,u}-Bc_{K,g}\exp\left(-\frac{T}{T_{akt}}\right)\\
&\frac{\partial}{\partial t}(\rho c_{Ox,g})+\frac{\partial}{\partial x_j}(\rho\langle v_j\rangle c_{Ox,g})-\frac{\partial}{\partial x_j}\left(D_t\rho\frac{\partial}{\partial x_j}c_{Ox,g}\right)=\tilde{A}\frac{\varepsilon}{k}c_{Ox,u}-\lambda Bc_{K,g}\exp\left(-\frac{T}{T_{akt}}\right)\\
&\frac{\partial}{\partial t}(\rho c_P)+\frac{\partial}{\partial x_j}(\rho\langle v_j\rangle c_P)-\frac{\partial}{\partial x_j}\left(D_t\rho\frac{\partial}{\partial x_j}c_P\right)=(1+\lambda)Bc_{K,g}\exp\left(-\frac{T}{T_{akt}}\right)
\end{aligned}\tag{14.14}$$

where λ designates the stoichiometric air-fuel ratio, q_{Verd} the vaporization source term, T_{akt} an activation temperature, and A, $\tilde{A}$, B and $\tilde{B}$ model constants.

In principle, the subdivision into two zones (among others, "mixed" cs. "unmixed") is very simple, where the β-pdf method is more up to date. For this reason, an extended coherent flame (ECF) model is perhaps a small improvement relative to the mixing-time scale model since it permits an improved treatment of premixed and diffusion combustion by considering these as series-connected processes (first fuel arrives in the mixed zone, then it burns). But it describes combustion in a less detailed way than a pdf time-scale model, which models a continuum of mixture states and not only two. Sometimes β-pdfs are also introduced in ECF models therefore (e.g. to describe the mixed zone), but this makes the model more complex, unwieldy and unnatural. There is no apparent advantage to the pdf time scale model. However, since ECF models are very popular in present-day CFD codes, one is sometimes not completely free when choosing a combustion model and must be satisfied with an ECF approach.

14.3.1.4 The Problem of Distributing the Progress Variables

In the discussion on combustion models, it could be seen that there are very practical approaches to describing diffusion combustion. The premixed, chemistry-dominated part of combustion causes some difficulty. The reason for this is that we have no information about how the reaction progresses as a function of the local mixture conditions (i.e. primarily as a function of the mixture fraction), or stated differently, how a progress variable should be distributed in the flamelet. There is a whole series of approaches to this in the literature, of which none has achieved acceptance thus far (see e.g. Steiner et al. (2004) or Lehtiniemi et al. (2005)). Unsteady flamelets (*representative interactive flamelets, RIF*) also belong to this series (Peters (2000)).

So-called *conditional moment closure (CMC) models* could possibly become accepted eventually as the theoretically best-founded approach (Bilger (1993) and Klimenko and Bilger (1999)). In this case, one solves transport equations for average species concentrations $\langle c_i|Z\rangle$ existing (i.e. are conditioned) at a certain mixture fraction Z:

$$\langle\rho|Z\rangle\frac{\partial Q_\alpha}{\partial t}+\langle\rho|Z\rangle v_i\frac{\partial Q_\alpha}{\partial x_i}-\langle\rho\chi|Z\rangle\frac{\partial^2 Q_\alpha}{\partial Z^2}=\langle\rho\bar{\omega}|Z\rangle, \tag{14.15}$$

whereby the symbol $\langle X|Z\rangle$ always designates the variable X conditioned by the mixture fraction value Z (this is thus also true for the progress variable, since it is typically a case of selected species concentrations). The local ensemble averages are obtained after integration of the conditioned variables over Z. The term on the right side of (14.15) represents the conditioned reaction rate and usually is obtained without any difficulty from the reaction kinetics. The third term left is the most critical; it arises from the diffusion term of a standard transport equation by transformation of the spatial coordinate x on the mixture fraction Z and thus represents the mixture or diffusion term in Z-space. It can, for example, be taken from the temporal development of the mixture fraction-distribution function (e.g. the β-pdf from Sect. 12.2.4). In addition, this distribution function must fulfill a continuity equation:

$$\frac{\partial}{\partial t}\left[\langle\rho|Z\rangle\cdot p_\beta(Z)\right]+\frac{\partial}{\partial x_i}\left[\langle\rho|Z\rangle v_i p_\beta(Z)\right]+\frac{\partial^2}{\partial Z^2}\left[\langle\rho\chi|Z\rangle p_\beta(Z)\right]=0. \tag{14.16}$$

Since the spatial-temporal development of $p_\beta(Z)$ is known [it is finally given by the temporal development of Z and the variance of Z, see (12.32) and (12.55)], the term $\langle \rho\chi|Z\rangle$ (the *conditioned scalar dissipation rate*) can be calculated from (14.16) and then inserted into (14.15).

This procedure is very complicated however, because to transport a single concentration c, several transport equations are required, since each of the "support points" of the c-distribution in the mixture fraction space $Q(Z_n) = \langle c|Z_n\rangle$ must be transported separately. Previously, at most two equations (for the average and variance) were used to transport a scalar. However, initial approaches to engine application already exist (see De Paola et al. (2008)).

14.3.2 Ignition

The simulation of ignition is an especially difficult problem, since here the time scales of flow and (especially in this case very complex) reaction kinetics interfere. Well functioning reaction-kinetic mechanisms do indeed already exist for many hydrocarbons like heptane, which is quite well suited to model the autoignition behavior of Diesel engines because of its comparable cetane number (about 50). But finally, this information does not help very much when the turbulence interaction is lacking.

One solution consists in ignoring turbulence interaction and introducing a source term based on laminar, detailed reaction kinetics. However, this approach is so imperfect that it is not worth the high calculation cost. Another often used alternative, which is at least less costly, consists in the use of phenomenological, reduced reaction kinetics; for example, the use of the adapted Shell model of Halstead et al. (1977) or even simpler phenomenological methods on the basis of a Wolfer equation (Wolfer (1938)). A typical modeling could look like this: one defines an indicator species c_I; if a predetermined threshold value $c_I^{(0)}$ is reached at a certain time, a local ignition occurs (i.e. the heat release model is activated). A transport equation is solved for c_I, for example with a source term according to Wolfer (1938):

$$\rho\left(\frac{\partial}{\partial t}+v_i\frac{\partial}{\partial x_i}\right)c_I-\frac{\partial}{\partial x_i}\left(D_{v_t}\rho\frac{\partial}{\partial x_i}c_I\right)=A_{id}\rho\frac{p}{p_0}f(\lambda)\exp\left(-\frac{E_{id}}{T}\right). \tag{14.17}$$

More elaborate methods with turbulence interaction work according to the flamelet concept. Unfortunately, the reaction kinetics is to slow to expect equilibrium. For this reason, we may proceed to describe only the source terms of a transport equation (e.g. for an ignition indicator) with a flamelet model (i.e. by means of mixture fraction averaging). As an indicator, CO has proven to be practical, among others, since the increase in CO-concentrations depicts the ignition process relatively monotonously. One difficulty consists however in the fact that now the CO source term is again dependent on the reaction progress (in our case, of CO concentration itself). Yet the distribution of progress variables over the mixture fraction is unknown, as previously discussed. Again, a CMC approach could be helpful here.

In conclusion, it can be established that no patent formula exists; we are directed rather to the utilization of more or less inadequate methods and model tunings. In the case of typical Diesel engine combustions with variations in ignition which are not all-too large, this is also not very critical; for HCCI combustions, this assertion is however no longer valid.

14.3.3 NO_x Formation

The simulation of NO_x formation is restricted usually to thermal NO_x and thus works with the Zeldovich mechanism, i.e. a transport equation is solved for the NO_x concentration

$$\rho\left(\frac{\partial}{\partial t}+v_i\frac{\partial}{\partial x_i}\right)c_{\mathrm{NOx}}-\frac{\partial}{\partial x_i}\left(D_t\rho\,\frac{\partial}{\partial x_i}c_{\mathrm{NOx}}\right)=Q_{Zeldovich}(c_{\mathrm{NOx}},c_{\mathrm{O}},c_{\mathrm{OH}},c_{\mathrm{H}},\lambda,p,T), \tag{14.18}$$

whereby the source term is calculated in accordance with the Zeldovich mechanism from the radical concentrations O, OH, and H (i.e. without ensemble averaging). The radical concentration N is seen as being in partial equilibrium. Since the time scale of NO_x formation is very large (larger than the turbulent time scales – this precisely is the other limiting case of chemical turbulence interaction), most of the NO_x is formed in the burned matter, and for this reason, turbulent temperature fluctuations are less relevant (the latter play the largest role in the burning zone). For this reason, NO_x formation is usually calculated in a laminar fashion (i.e. purely reaction-kinetically on the basis of ensemble-mean values).

This approach seems to be not quite justified however, since the errors made by non-execution of the ensemble averaging are not negligible. But then we wind up in the known situation that only the direct reactions (under certain assumptions of equilibrium or partial equilibrium) are clearly determined by the air-fuel ratio and can thus be integrated in the flamelet (i.e. by means of β-pdf over the mass fraction). A flamelet averaging of the reverse reaction again encounters the problem of unknown distribution of NO_x in the mixture fraction space. The approach of choice could thus again be a CMC approach.

Whether or not this approach is sufficient cannot as of yet be conclusively determined however, since the temperature-dependence of NO_x formation is extremely high and thus the precision limits of statements from 3D-simulation are being approached.

Figure 14.5 contains a calculated representation of a NO_x distribution in a passenger car Diesel engine along a ring curve on the spray level.

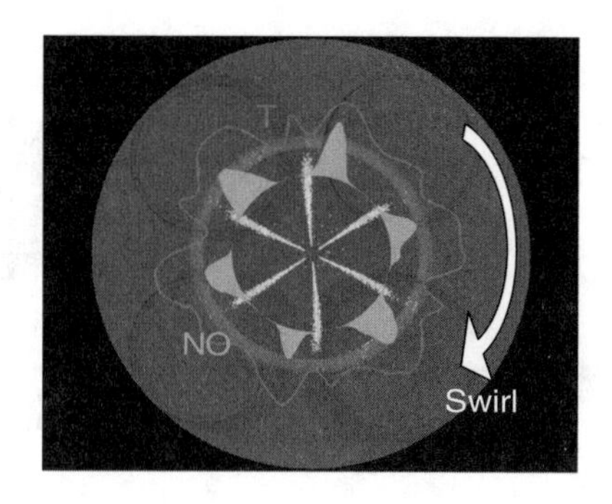

Fig. 14.5 Temperature and NO distribution in a passenger car Diesel engine. On the lee-side (averted from the swirl) of the jets, the highest temperatures and thus the highest NO concentrations are reached

14.3.4 Soot Formation

To simulate soot formation and oxidation, different approaches exist, see also Sect. 6.4. Firstly, there are the phenomenological models like Hiroyasu et al. (1983) or Nagle and Strickland-Constable (1962). A typical variant uses the formation model from the first source and from the second the oxidation model. Translated into fluid mechanics, a transport equation for the soot mass fraction is solved

$$\rho\left(\frac{\partial}{\partial t}+v_i\frac{\partial}{\partial x_i}\right)c_{Ruß}-\frac{\partial}{\partial x_i}\left(D_t\rho\frac{\partial}{\partial x_i}c_{Ruß}\right)=Q_{Hiroyasu}(\lambda,p,T)-Q_{Nagle-Strickland}(\lambda,p,T) \tag{14.19}$$

The prediction capability of this model is not very high however. Typically, the very high intermediary soot concentrations before the start of oxidation are only insufficiently reproduced.

A new flamelet-based method was suggested by Dederichs et al. (1999), which appears to be more predictive. In this case, the source terms of the soot transport equation are calculated in the flamelet method by means of an integral of type

$$Q_{Ruß}(\langle\chi\rangle;p,T)=\int dZ\,p_\beta(Z)\,Q_{Ruß}(Z,\langle\chi\rangle;p,T) \tag{14.20}$$

As functions of Z, the expressions $Q_{Ruß}(Z,\langle\chi\rangle;p,T)$ are approximated by β-functions. In this form, the Z-integration (14.20) takes on a simple form, since the product of two β-functions is again a β-function, and the integration can be performed analytically. Figure 14.6 shows a calculation result obtained with the flamelet model.

Absolute soot emission values are, as always, difficult to calculate. Relative statements (e.g. comparisons between different trough shapes) can sometimes be made from an assessment of the mixture distribution and its temporal development.

When evaluating calculative inaccuracies, it should not be forgotten that soot oxidation is an important phenomenon contributing to the real, measured soot

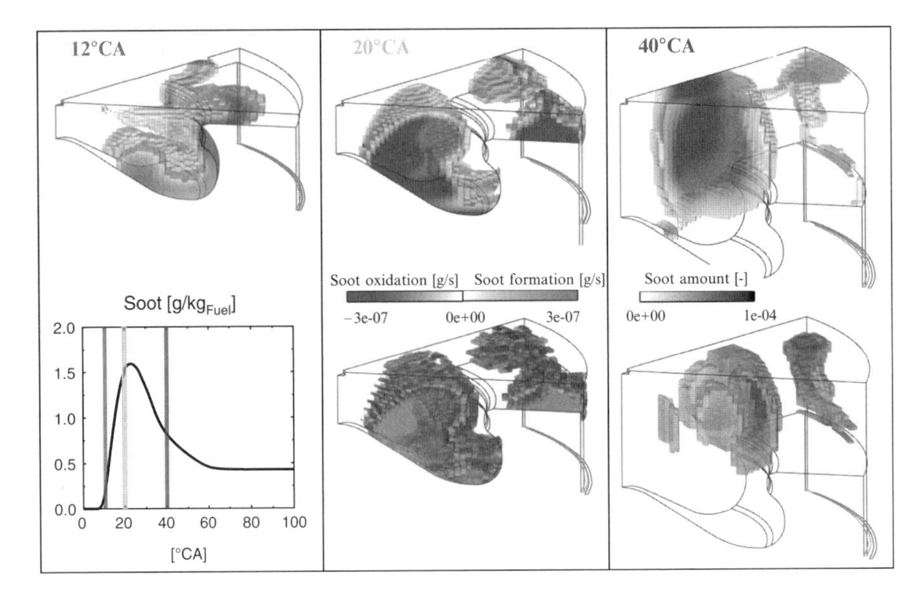

Fig. 14.6 3D simulation of soot distribution in a passenger car Diesel engine with the flamelet concept in different phases of combustion

emission. But it is again true that the soot distribution in the mixture fraction space is unknown. A CMC approach could help here as well.

14.3.5 HC and CO Emissions

Further emissions playing an increasingly important role in modern Diesel combustion processes, are unburned hydrocarbons (HC) and CO. Both are components of the seven-species modeling (HC only as fuel), i.e. the result of a combustion calculation with the seven-species model in principle provides HC and CO emissions. However, detailed reaction kinetics would have to come into the foreground for a correct calculation, since a "freezing" of the combustion occurs (i.e. we are confronted again with "large" chemical time scales), and this is of course ensemble-averaged. None of these phenomena can be described well in present combustion models however.

Furthermore, it should be remembered that the CO and HC remaining in the calculation at the end of combustion are typically "mixture formation artifacts", i.e. the consequence of insufficient injection and mixture formation models. This is especially true in when the "classical" Lagrangian spray model is used. Even when improved spray models are applied as discussed in Chap. 13, the predictive quality of the (turbulent) mixture states at the end of combustion will not be very high; deviations from the real process add up.

In summary, we can therefore say that HC and CO emissions will not be subject to capable calculation by a CFD code in the foreseeable future.

14.4 The Homogeneous SI Engine (Premixed Combustion)

Even the concept of the "homogeneous gasoline engine" is itself a fiction. In reality, such an engine is far from completely homogeneous; to what degree the homogeneity of the mixture and temperature distribution is fulfilled is primarily a question of the quality of the mixture formation process. However, fictions are the lifeblood of calculation – in the final analysis, we must carry out one approximation or another in order to concentrate on what is essential in the concrete case at hand. So-called TOEs ("theories of everything") or "global models" are nonsense. In the present section, we will therefore make the assumption of a perfectly homogeneous mixture.

This can of course be impractical in some cases, e.g. if it is exactly the quality of the mixture formation that is to be the object of investigation (for example, if there are high levels of HC and CO emission). Then it cannot be avoided to consider mixture formation. This is a big problem in the case of a port fuel injection because this involves a multicycle process – several cycles are required until equilibrium between injection and vaporization is reached – and complex physical mechanisms such as film dynamics, film vaporization and film detachment are also relevant. However, port fuel injection is becoming less and less relevant and the mixture formation is basically understood. Mixture formation of homogeneous direct injection (i.e. early injection during the intake stroke) is a much more important problem and, luckily, also more accessible to calculation. Yet then the discussions in the last section are of the highest relevance; due to the long life of the liquid phase and the complex flow structures, the problem is extremely (!) challenging from numerical/statistical/physical points of view (mesh structure and resolution, turbulent dispersion, wall film formation, oil film vaporization, multicomponent vaporization).

But now let us return to the SI engine with a perfectly homogeneous mixture. One might believe that simulation of flame front combustion of a gasoline engine with a homogeneous mixture would be relatively unproblematic, since the underlying physical processes are well-known and easy to describe. Unfortunately, precisely the opposite is true. No engine CFD code exists that offers a truly acceptable model for describing SI engine combustion. This is also related – as in the case of spray models – to the high numerical requirements involved in flame front combustion models. In the following, different current approaches will be discussed. First however, we will devote ourselves to the most important deficit of present-day combustion models – the non-consideration of two phases.

14.4.1 The Two-Phase Problem

As has already been discussed in Sect. 4.3, the wrinkled laminar flame front is very thin, usually only a few micrometers. We must now carry out an ensemble averaging. In it, the thin wrinkled laminar flame front (flame surface A_l) "blurs" and a thicker, "turbulent" flame front arises (the flame thickness is of the order of magnitude of the turbulent length scale), which is no longer correspondingly wrinkled (flame surface A_l). The propagation speeds s_l and s_t into the unburned mixture are different; their ratio corresponds to the reciprocal ratio of the flame surfaces

$$\frac{s_t}{s_l} = \frac{A_l}{A_t}, \tag{14.21}$$

so that in the averaged as well as the non-averaged image (turbulent and laminar), the same burning rate is calculated ($A_l s_l = A_t s_t$). Because of the finite thickness of the turbulent flame front, an exact definition of the flame front position is not obvious, we can for example utilize the position of the 50% conversion point. In Fig. 14.7, the relation between laminar and turbulent flame fronts is illustrated.

In principle, a step in density occurs in the laminar flame front (corresponding to the step in temperature), which corresponds to a step in the speeds. This situation is represented in Fig. 14.8a in the reference system of the flame front, i.e. the latter is at rest. With the laminar flame speed s_l, the unburned mixture enters into the stationary flame front, the burned mixture leaves it with another velocity v_v. Let ρ_v and ρ_u be the densities in the burned and unburned mixtures. Due to mass conservation

$$\rho_v \, v_v = \rho_u \, s_l. \tag{14.22}$$

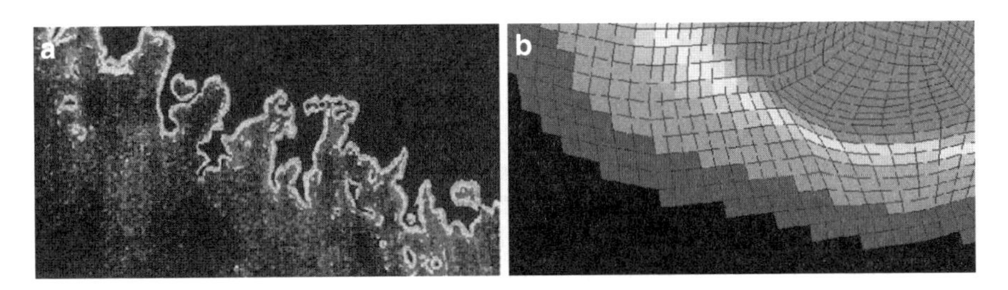

Fig. 14.7 Relation between laminar and turbulent flame fronts. (**a**) Experimental laminar flame front, position and state of the turbulent flame front are shown (**b**) The turbulent flame front in the averaged simulation

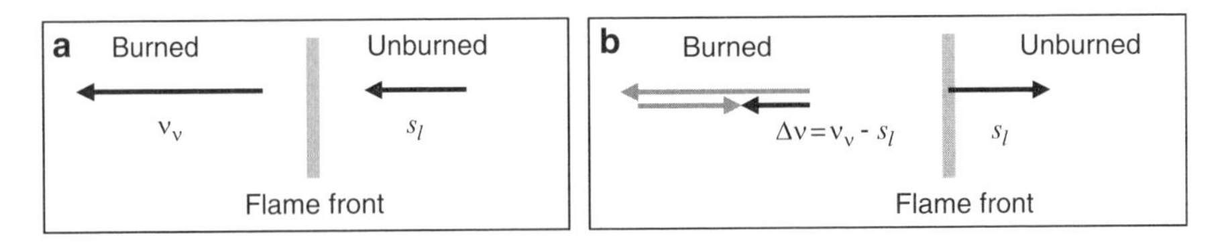

Fig. 14.8 Situation in the reference system (**a**) of the flame front and (**b**) of the unburned gases

In Fig. 14.8b, the situation is reproduced in the reference system of the unburned gas, i.e. a velocity shift transformation was simply carried out. We see that the speed makes a step between the unburned and burned mixture which is of quantity

$$\Delta v = v_v - s_l = \frac{\rho_u - \rho_v}{\rho_v} s_l! \tag{14.23}$$

In the turbulent case, an overlapping of burned and unburned states occurs through the ensemble-average, with different speeds, densities, temperatures, and turbulence levels. This is a two-phase flow! Unfortunately, premixed combustion is not handled this way in any of the engine CFD codes, which is the main inadequacy in their treatment of premixed combustion. Essential modeling progress is only to be expected by application of two-phase treatment (due to the obvious "jump relations" like (14.23), an equivalent single-phase treatment of an averaged state is conceivable, from which both phases can be "reconstructed").

Neglecting two-phase behavior leads to various problems, one of which is the production of artificial turbulence in the flame zone. This is easy to see. The most important term of turbulence production reads [see (12.36)]

$$P = \tau_{R,ij} \cdot S_{ij} = F\left(\frac{\partial v_k}{\partial x_l}\right) \quad \text{with} \quad F(0) = 0. \tag{14.24}$$

In the two-phase approach, this term would have to be interpreted correctly as the average of the term in the burned and unburned phases (v_v and v_u describe the velocity of the burned and unburned phases)

$$P_{2Phasen} = cF\left(\frac{\partial v_{v,k}}{\partial x_l}\right) + (1-c)F\left(\frac{\partial v_{u,k}}{\partial x_l}\right). \tag{14.25}$$

where c designates the *progress variable*, i.e. the statistical weight of the "burned" state. In the standard, single-phase approach, one works only with a mean velocity v, which represents an overlapping of the velocities of the burned and unburned phases

$$v_k = c v_{v,k} + (1-c) v_{u,k}. \tag{14.26}$$

With this, the term P is calculated. Upon the disappearance of the velocity gradients

$$\frac{\partial v_{v,k}}{\partial x_l} = \frac{\partial v_{u,k}}{\partial x_l} = 0 \tag{14.27}$$

no turbulence is produced in the (correct) two-phase model. In the single-phase standard model however, already a c-gradient in the case of a simultaneous phase step in velocity produces (artificial) turbulence (at vanishing velocity gradients)

$$P = F\left(\frac{\partial v_k}{\partial x_l}\right) = F\left((v_{v,k} - v_{u,k})\frac{\partial c}{\partial x_l}\right). \tag{14.28}$$

Since F is essentially a quadratic function in the velocity gradients, the artificial turbulence is all the stronger, the thinner the flame front, because

$$P \propto \left[\Delta v \frac{\partial c}{\partial x}\right]^2 \approx \left[\Delta v \frac{1-0}{l_F}\right]^2 = \left[\frac{\Delta v}{l_F}\right]^2. \tag{14.29}$$

is roughly valid (l_F denotes the flame thickness). The total turbulence production as an integral across the flame front then results in

$$P_{ges} \approx P\, l_F \propto \frac{(\Delta v)^2}{l_F}, \tag{14.30}$$

i.e. it diverges for $l_F \rightarrow 0$.

The error can assume dramatic proportions, when no model is used to stabilize the flame thickness. In this case, the flame produces turbulence, which thus assumes the highest values on the reverse side. This turbulence accelerates the flame [see for example the Damköhler relation (Sect. 4.3)], the reverse side more than the forefront. The flame thus becomes ever faster and thinner. A thinner flame front creates however even more turbulence, with which the circle closes.

For the solution of this problem, it is possible for example to suppress the turbulence production term in the flame front. Especially for thin flame fronts, the errors caused by this are tolerable. However, no path can avoid a two-phase formulation in the long run.

In the literature, approaches for a two-phase treatment of enthalpy or inner energy have been suggested, but not for velocity. But it is precisely the latter which is decisive for a correct calculation of turbulence production.

14.4.2 The Magnussen Model

The simplest combustion model for premixed flames is the Magnussen model, consisting of a transport equation for the progress variable c ($c = 0$: no conversion, $c = 1$: conversion complete). In analogy to the breakup model for diffusion flames, the reaction rate is proportional to the inverse turbulent time scale ε/k, whereby it becomes clear that we are concerned with a turbulent premixed combustion. Moreover, the reaction rate must be zero for $c = 0$ and $c = 1$, in the burned and unburned mixture. The Magnussen model thus reads

$$\rho\left(\frac{\partial}{\partial t} + v_i \frac{\partial}{\partial x_i}\right)c - \frac{\partial}{\partial x_i}\left(D_t \rho \frac{\partial}{\partial x_i} c\right) = \alpha \rho_u \frac{\varepsilon}{k} c(1-c), \tag{14.31}$$

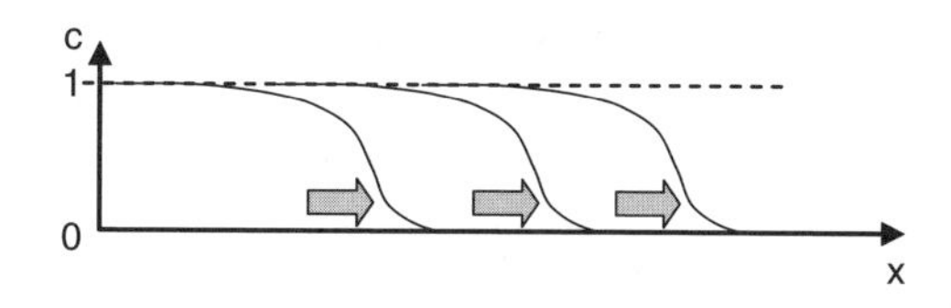

Fig. 14.9 Stable turbulent flame front profile

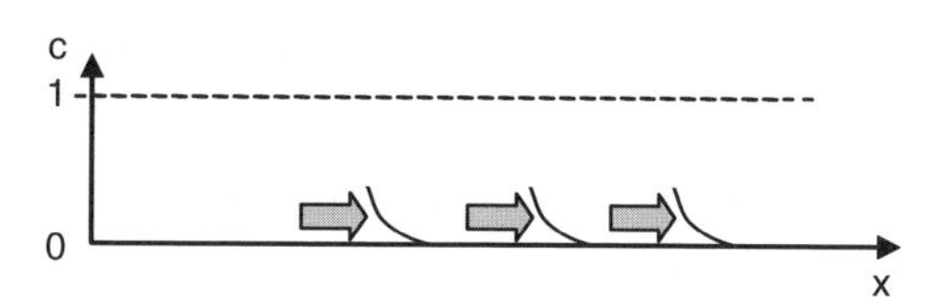

Fig. 14.10 Propagation of the face of the flame front

where α describes a model parameter. This model is scarcely used any longer, since is has serious insufficiencies; because of its simplicity, it is however well-suited to study the basic properties of an entire class of combustion models.

Equation (14.31) forms, after a starting time, a stable flame front profile that is independent of the exact initial conditions and runs through the combustion chamber like a dispersion-free wave with a defined propagation speed, see Fig. 14.9. This is a consequence of the non-linear source term. Such non-linear waves are known in various areas of physics and are called *solitary waves* or *solitons*. In contrast to this, the profile in the case of linear waves is not determined by the wave equation before, but is rather given by the initial conditions. Moreover, linear waves are usually subject to dispersion.

Yet problems result from this soliton property at the same time: profile determination takes place at all flame front locations numerically, i.e. by means of a solution of a non-linear differential equation sometimes, with a very poor mesh resolution. Turbulent flame thicknesses lie in the order of magnitude of the turbulent length scale, and these often amount to a mere 1–2 mm under engine conditions. With a mesh, all size of 0.5 mm, this means four mesh cells per flame front, and that this is not necessarily sufficient for the discretization of a nonlinear differential equation should be clear!

The question of propagation speed is answered in the so-called KPP theorem (Kolmogorov, Pichunov, Petrovski), see also Kolmogorov et al. (1937). The basic idea is that the propagation speed of the flame front can be analyzed with the help of the spreading speed of its "bow wave", i.e. its forefront (see Fig. 14.10). Valid in this area is namely $c \approx 0$, and from this, (14.31) can be linearly approximated in c. $\rho \approx \rho_u$ is additionally valid.

If we work in a single spatial dimension and assume constant turbulence values, we then obtain the equation

$$\rho_u \frac{\partial c}{\partial t} + \rho_u v \frac{\partial c}{\partial x} - \rho_u D_t \frac{\partial^2 c}{\partial x^2} = \alpha \rho_u \frac{\varepsilon}{k} c. \tag{14.32}$$

A stationary wave solution reads

$$c(x,t) = \gamma(x - (v + s_t)t) = \gamma(\xi). \tag{14.33}$$

Insertion in (14.32) provides

$$-s_t \frac{d\gamma}{d\xi} - D_t \frac{d^2\gamma}{d\xi^2} = \alpha \frac{\varepsilon}{k} \gamma. \tag{14.34}$$

This equation has exponential functions as a solution. In the sense of Fig. 14.10, a solution of type

$$\gamma = \exp(-\omega\xi) \qquad \omega \geq 0. \tag{14.35}$$

should be sought. With it, we obtain

$$\omega = \frac{s_t \pm \sqrt{s_t{}^2 - 4\alpha \frac{\varepsilon}{k} D_t}}{2D_t}. \tag{14.36}$$

This equation only has a real solution, if

$$s_t \geq 2\sqrt{\alpha \frac{\varepsilon}{k} D_t}.$$

The KPP theorem now asserts that the minimal speed value

$$s_{t,\min} = 2\sqrt{\alpha \frac{\varepsilon}{k} D_t} \tag{14.37}$$

is precisely that which arises when the flame front propagates from a limited location into a region with $c = 0$. In order to understand this, let us look again at the linearized equation (this time without the v-term for the sake of simplicity)

$$\frac{\partial c}{\partial t} - D_t \frac{\partial^2 c}{\partial x^2} = \alpha \frac{\varepsilon}{k} c, \tag{14.38}$$

the Greens function of which reads

$$c_G(x,t) = \frac{N}{\sqrt{t}} \exp\left(-\frac{x^2}{4D_t t} + \alpha \frac{\varepsilon}{k} t \right) \tag{14.39}$$

This result is not difficult to derive from the case $\alpha = 0$ of the pure diffusion equation. The Greens function describes asymptotic propagation behavior (only of the frontage of the flame front!), which proceeds from a point source – the constant pre-factor N is irrelevant. In order to find the propagation speed, we have to calculate the function $x(t)$, for which c_G is stationary, i.e.

$$-\frac{x^2}{4D_t t} + \alpha \frac{\varepsilon}{k} t - \frac{1}{2} \ln(D_t t) = \text{const.} \tag{14.40}$$

For large t, the logarithmic term is negligible, and we obtain approximately

$$\left(\frac{x}{t}\right)^2 = 4\alpha \frac{\varepsilon}{k} D_t = s_{t,\min}^2, \tag{14.41}$$

i.e. we have again found our minimal speed [see (14.37)] as the speed of a flame proceeding from a point source! If we set

$$D_t = c_\mu \frac{k^2}{\varepsilon},$$

we then obtain

$$s_t = 2\sqrt{c_\mu \alpha}\, u'. \tag{14.42}$$

This corresponds to the Damköhler relation (Sect. 4.3) in the limiting case $s_t >> s_l$.

With the help of the Greens function (14.39), we can also see that at suitable spatial pre-initialization (i.e. no point source) higher burning velocities can also be reached. For example, if we choose the initialization

$$c(x; t = 0) = \exp(-\beta|x|) \quad \text{with} \quad \beta < \frac{s_{t,\min}}{2D_t}. \tag{14.43}$$

This leads for $t \geq 0$ to

$$c(x; t) = \int_{-\infty}^{\infty} \mathrm{d}y \exp(-\beta|y|) \frac{N}{\sqrt{t}} \exp\left(-\frac{(x-y)^2}{4D_t t} + \alpha \frac{\varepsilon}{k} t\right). \tag{14.44}$$

For the calculation of flame propagation for $x > 0$, this expression can be substituted by

$$c(x; t) \approx \int_{-\infty}^{\infty} dy \frac{N}{\sqrt{t}} \exp\left(-\frac{(x-y)^2}{4D_t t} - \beta y + \alpha \frac{\varepsilon}{k} t\right)$$
$$= 2\sqrt{\pi D_t} N \exp\left(-\beta x + \left(D_t \beta^2 + \alpha \frac{\varepsilon}{k}\right) t\right), \tag{14.45}$$

because the integrand of (14.45) is a Gaussian function, the maximum of which lies at

$$y_{\max} = x - 2D_t t\beta$$

The width at half maximum scales with $\sqrt{t}$. Flame front propagation at large time periods should now be described, i.e. $x \cong s_t t$. From this follows for $y_{\max}$ [see the limit on β in (14.43)]

$$\frac{y_{\max}}{t} = s_t - 2D_t\beta > s_t - s_{t,\min} > 0$$

(the last inequality sign is valid as long as $s_t > s_{t,\min}$, which is guaranteed acc. to (14.46) and (14.47)). Together with the width at half maximum scaled with $\sqrt{t}$, we see that the range of the integrand of (14.45) that is considerably different than zero has, at least for large t, positive y-values. The equations (14.45) and (14.44) have identical propagation behavior.

The need for a stationary exponent leads for (14.45), lower line, to the relation

$$s_t(\beta) = \frac{x}{t} = D_t\beta + \frac{\alpha}{\beta}\frac{\varepsilon}{k}. \tag{14.46}$$

The minimum of this function (given by $\mathrm{d}s_t/\mathrm{d}\beta = 0$) reads in turn

$$s_{t,\min} = 2\sqrt{\alpha\frac{\varepsilon}{k}D_t}. \quad \text{for} \quad \beta = \sqrt{\frac{\alpha\varepsilon}{D_t k}}. \tag{14.47}$$

However, this also means that for smaller β values there are higher flame propagation speeds!

From this follows a further, very serious problem of the Magnussen model: it is very unstable at incorrect initializations, in general at c-values differing slightly from zero in front of the flame front (an initialization such as in (14.43) deviates indeed only slightly from zero). Metaphorically speaking, this is due to the fact that the propagation speed is determined by the forefront of the flame (i.e. small c-values), while the larger c-values "provide" for right profile within the flame front.

This numerical sensitivity comes to light especially near the wall. In principle, the KPP analysis should also be valid for the Magnussen model at the wall, i.e. the burning velocity (14.42) should drop, since the turbulence is reduced at the wall (dissipation increases considerably). Typically, 3D-simulations supply the exact opposite behavior: extreme, completely unphysical flame accelerations in the area of the wall occur. To understand this phenomenon, we need to analyze the flame speed according to (14.37). At the wall

$$\varepsilon \propto \frac{1}{y} \xrightarrow[y\to 0]{} \infty, \tag{14.48}$$

is valid, and as a result, the source term of (14.32) approaches infinity, while the diffusion term should tend towards zero, with a finite product. Because of a lack of numerical precision, diffusion does not approach zero however, and a certain numerical diffusion D_{num} remains, which is attached to the mesh resolution and numerical scheme. Consequently, the term (14.37) can exceed all limits in the area of the wall

$$s_{t,Wand} = 2\sqrt{\alpha \frac{\varepsilon}{k} D_{num}} \propto \frac{1}{\sqrt{y}} [\xrightarrow[y \to 0]{}] \infty. \tag{14.49}$$

To sum up, one can say that the Magnussen model produces a solitary wave, the profile and propagation speed of which are generated by means of a complex interaction of the source term and diffusion term. In the case of thin, turbulent flame fronts or in the area close to the wall, it suffers from serious numerical problems.

14.4.3 Flame Surface Density Model

The flame surface density model (also called the coherent flame model) provides an improved physical description. In it, turbulence does not directly accelerate the flame (as in (14.31)), but wrinkles it more strongly. However, a more strongly wrinkled flame burns faster.

For this purpose, an additional transport equation is solved for the flame surface density Σ (flame surface per volume unit), or instead, for the specific flame surface $\sigma = \Sigma/\rho$. This equation exists in the various versions, see Poinsot and Veynante (2001). One typical variant reads

$$\rho\left(\frac{\partial}{\partial t} + v_i \frac{\partial}{\partial x_i}\right)\sigma - \frac{\partial}{\partial x_i}\left[\rho D_t \frac{\partial \sigma}{\partial x}\right] = \alpha_F \frac{\varepsilon}{k} \rho\sigma - \beta_F \frac{s_l}{c(1-c)} (\rho\sigma)^2, \tag{14.50}$$

where α_F and β_F (may) contain, besides model constants, functional dependencies on the turbulent and chemical time and length scales. The first term on the right side describes the flame surface production from turbulence, the second term represents a sink caused by burn-out. In addition, another transport equation must be solved for the progress variable

$$\rho\left(\frac{\partial}{\partial t} + v_i \frac{\partial}{\partial x_i}\right)c - \frac{\partial}{\partial x_i}\left(\rho D_t \frac{\partial}{\partial x_i} c\right) = \rho_u s_l \rho\sigma. \tag{14.51}$$

The laminar burning velocity is calculated with (4.1). If an equilibrium between flame surface production and destruction exists in (14.50), then:

$$(\rho\sigma)_{eq} s_l = \frac{\alpha_F}{\beta_F} \frac{\varepsilon}{k} c(1-c). \tag{14.52}$$

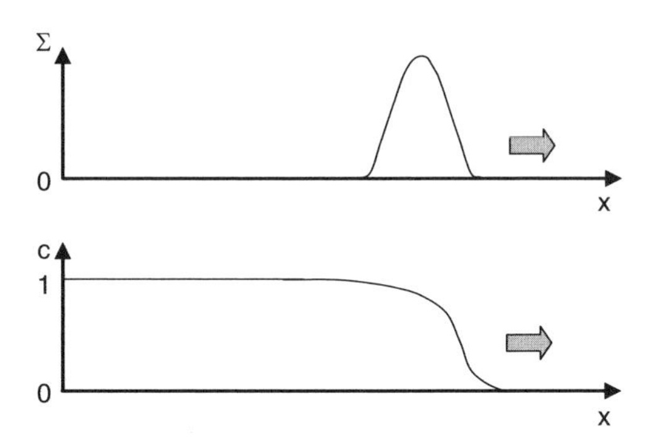

Fig. 14.11 Flame front profiles in c and σ

With that, we obtain the Magnussen model from (14.51)! This approximation is however only fulfilled for large values of α_F and β_F.

In the case of flame surface density models, the flame front is again well-described by a solitary wave (this time in the σ and c-field), with a well-defined propagation speed and well-defined c and σ profiles. The c field goes thereby from 0 (in front of the flame) to 1 (behind the flame), while σ starts at zero before the flame, increases to a maximum value in the flame, in order then to fall again to zero towards the rear flame front (see Fig. 14.11). To calculate the propagation speed of a planar flame front, we proceed similarly as in the Magnussen model, determining the propagation speed of the flame forefront; for this, (14.50) and (14.51) have to be linearized in σ and c, where $\rho = \rho_u$. The equations thus obtained read

$$\begin{aligned}\rho_u\left(\frac{\partial}{\partial t}+v\frac{\partial}{\partial x}\right)\sigma-\rho_u D_t\frac{\partial^2\sigma}{\partial x^2}&=\alpha_F\frac{\varepsilon}{k}\rho_u\sigma-\beta_F\, s_l(\rho_u)^2\left(\frac{\sigma}{c}\right)\sigma,\\ \rho_u\left(\frac{\partial}{\partial t}+v\frac{\partial}{\partial x}\right)c-\rho_u D_t\frac{\partial^2 c}{\partial x}&=(\rho_u)^2 s_l\left(\frac{\sigma}{c}\right)c.\end{aligned}\tag{14.53}$$

One recognizes that a stationary solution with $\sigma/c = \text{const.}$ exists

$$\begin{aligned}c(x,t)&=c_0\exp\left(-\omega(x-(v+s_t)t)\right),\\ \sigma(x,t)&=\sigma_0\exp\left(-\omega(x-(v+s_t)t)\right).\end{aligned}\tag{14.54}$$

Insertion into (14.53) provides

$$\begin{aligned}\omega\, s_t-D_t\omega^2&=\alpha_F\frac{\varepsilon}{k}-\beta_F\, s_l\,\rho_u\frac{\sigma_0}{c_0},\\ \omega\, s_t-D_t\omega^2&=s_l\,\rho_u\frac{\sigma_0}{c_0},\end{aligned}\tag{14.55}$$

which finally leads to

$$\omega=\frac{s_t\pm\sqrt{s_t^2-4D_t\frac{\alpha_F}{\beta_F+1}\frac{\varepsilon}{k}}}{2D_t}.\tag{14.56}$$

According to the KPP theorem, the propagation speed follows

$$s_t = 2\sqrt{D_t \frac{\alpha_F}{\beta_F + 1} \frac{\varepsilon}{k}}. \tag{14.57}$$

(Depending on the concrete formulation of the initial (14.50), this calculation may have a somewhat different outcome, one should only understand the result (14.57) as an example.)

It is thus clear that the flame front models have a mathematical behavior comparable to the Magnussen model, propagation speed and flame front profile are again the result of the interaction of diffusion and the source terms in the relevant transport equations. And a good numerical resolution of the flame front is indispensable. But it is therefore also obvious that the flame surface density front models do not repair what is precisely the worst inadequacy of the Magnussen model, its poor numerical behavior!

Also, behavior near the wall continues to be critical. However, modeling methods exist, in which the wall problem no longer exists or is at least markedly reduced. In this way, for Poinsot and Menevaux, the function α_F contains the so-called ITNFS function Γ (Γ = intermittent turbulent net flame stretch, see Poinsot and Veynante (2001)),

$$\alpha_F = \alpha_0 \frac{\varepsilon}{k} \Gamma\left(\frac{l_t}{\delta_0}, \frac{\sqrt{k}}{s_l}\right), \tag{14.58}$$

where δ_0 designates the laminar flame thickness.

The function Γ, however, tends with a decreasing turbulent length scale towards zero, i.e. at the wall ($l_t \to 0$), the product $\varepsilon/k\,\Gamma$ also becomes equal to zero, no numerically caused divergence appears any longer! This formulation would also clearly be of help in the case of the Magnussen model.[2]

Because of the resolution problems, flame surface density models are poorly adaptable to engine calculations; the available computational meshes are typically too coarse, especially in the case of late direct injection (high turbulence).

Another, occasionally even practiced option is flame front-adaptive mesh refinement. This is very costly however. But in every case it is sensible to have a separate mesh (in the case of premixed combustion structured as homogeneously as possible) available for combustion simulation which no longer contains the valve structures.

It is often asserted that turbulent flame fronts are not at all that thin, but rather show a considerable flame thickness, which is experimentally proven, if one superimposes the flames of several cycles. In this case, ensemble averaging and cycle averaging are being mixed up again. As previously shown (see Chap. 12), ensemble

[2]We should point out here that the ITNFS function has a physical meaning. But it has an extremely advantageous effect in the numerical sense as well.

averaging only contains the averaging over turbulent ("coherent") fluctuations, which are produced at identical boundary conditions in the flow by the chaotic behavior of the underlying fluid dynamics. Cyclic fluctuations contain additional fluctuations however, produced by fluctuations of the boundary or initial conditions (throttle, injection, residual gas, ignition, etc.). A turbulent combustion model includes only the turbulent, coherent fluctuations (thus the name "coherent flame model"); only these contribute to the formation of turbulent model flame front.

It is easy to see that one is not free in the choice of the ensemble under consideration. Typical incoherent fluctuations are, for example, fluctuations in the (effective) ignition time. Let

$$\langle\sigma\rangle_\varphi \quad \text{and} \quad \langle c\rangle_\varphi$$

be the ensemble-averaged quantities σ and c with reference to the coherent fluctuations, at a fixed ignition time (or angle) φ. We then obtain the total mean values with reference to the coherent and incoherent fluctuations

$$\begin{aligned} \langle\sigma\rangle_{ges} &= \int d\varphi\, f(\varphi)\langle\sigma\rangle_\varphi \\ \langle c\rangle_{ges} &= \int d\varphi\, f(\varphi)\langle c\rangle_\varphi \end{aligned} \tag{14.59}$$

Because of their non-linearity however, the transport equations (14.50) and (14.51) is not invariant under such a transformation! I.e. if $\langle\sigma\rangle_\varphi$ and $\langle c\rangle_\varphi$ fulfill the (14.50) and (14.51), then this is not valid for

$$\langle\sigma\rangle_{ges} \quad \text{and} \quad \langle c\rangle_{ges}.$$

Thus we are not at liberty to choose the ensemble for averaging "suitably", the flame surface density models (like the Magnussen model) already correspond to a fixed choice, namely, that of the "minimal" ensemble, which contains only the coherent, intrinsic, fluid mechanical fluctuations.

14.4.4 The G-Equation

In order to escape the numerical problems of the flame front and Magnussen models, a combustion model is needed with a formulation in which the turbulent flame speed appears explicitly. Moreover, the sensitivity with reference to the resolution of the turbulent flame front should be as small as possible. The G-equation is such a model:

$$\frac{\partial G}{\partial t} + v_i \frac{\partial G}{\partial x_i} = s_t |\nabla G| \tag{14.60}$$

or

$$\frac{\partial G}{\partial t} + (v_i + s_t \hat{n}_i) \frac{\partial G}{\partial x_i} = 0 \quad \text{with} \quad \hat{n} = -\frac{\nabla G}{|\nabla G|}. \tag{14.61}$$

This equation describes the propagation of a surface such that every surface element propagates with a propagation speed normal to it of amount s_t relative to the fluid. The surface is characterized by the amount of points, for which is valid $G(x) = 0$. Outside the flame surface, the variable G can be chosen arbitrarily, only that it be other than zero. The flame is thus first described in this image as an infinitely thin surface. Naturally, it cannot stay this way: aside from the physical realities, we should not permit a "step" in density and temperature into a CFD code. Thus, a finite flame thickness l_F has to be introduced. As a determination equation for l_F, the relation

$$l_F = b\, l_t \tag{14.62}$$

for example may be used for a stationary flame, whereby $b \approx 2$. This is not valid near the wall however, since the flow is laminarized there. Model variants also exist with their own transport equations for the flame thickness, see Peters (2000).

$G = 0$ thus describes the central position of the flame front. As a profile for ∇c, a Gaussian function can be used; c is then defined as[3]:

$$c(x) = erf\left(\frac{2\,\mathrm{d}(x)}{l_F}\right), \tag{14.63}$$

if $\mathrm{d}(x)$ describes the (positive or negative according to the position) distance of a given point to the flame front. However, we must still determine how operationally the distance of a spatial point from the flame front is to be calculated.

If s_t designates, as is usual, the turbulent flame speed relative to the unburned mixture, then (14.60) is valid only for the flame forefront, corresponding to the (14.53), i.e. for $\rho = \rho_u$. In the case of a finite reaction progress with assigned density $\rho < \rho_u$, an additional reverse flow [see (14.23)] must still be overcome, the difference velocity to the flame front (the speed of which relative to the unburned mixture amounts to s_t) is

$$v_\rho = s_t \frac{\rho_u - \rho}{\rho}. \tag{14.64}$$

In total therefore, one can write more generally

$$\rho \frac{\partial G}{\partial t} + (\rho v_i + \rho_u s_t \hat{n}_i) \frac{\partial G}{\partial x_i} = 0. \tag{14.65}$$

[3] $erf(x) = \frac{1}{\sqrt{\pi}} \int\limits_{-\infty}^{x} \exp(-x^2)\mathrm{d}x.$

The G-equation is a hyperbolic equation with a transport velocity different from $\vec{v}$

$$\vec{v} + \frac{\rho_u}{\rho}\, s_t \hat{n};$$

it thus needs in principle its own solution algorithm, which is not provided in the standard CFD codes. Even in the case that a solution method were available, (14.65) can behave problematically outside the flame front, since no special behavior is prescribed there. A recommendable approach is thus to make the demand

$$|\nabla G| = 1, \tag{14.66}$$

i.e. with this constraint ("gauging" of the G field) the G values outside the flame front correspond to the distances to the flame front (negative before and positive behind the flame front). With this choice, the local definition of the progress variables can be given, according to (14.63) is valid

$$c(x) = erf\left(\frac{2G(x)}{l_F}\right). \tag{14.67}$$

However, the property (14.66) is not preserved in time, the (14.65) has to be re-initialized! This means that after every time interval (or after a number of time intervals), the differential equation

$$\frac{\partial G(x,t,\tau)}{\partial \tau} = sign(G(x,t))(1 - |\nabla G(x,t,\tau)|), \quad G(x,t,0) = G(x,t), \tag{14.68}$$

has to be solved for $\tau \to \infty$, it converges then towards $|\nabla G| = 1$. This implies a considerable additional calculation effort indeed.

However, the great advantage of the G-equation is that we are concerned with a linear wave, at least for the planar case; the flame profile and propagation speed are uncorrelated. A low numerical resolution of the profile is thus not so critical.

Various formulae can be used for s_t, the flame speeds of the flame surface density model [e.g. (14.57)] as well. Frequently, phenomenological relations similar to the Damköhler relation (Sect. 4.3) are utilized, like

$$s_t = s_l\left(1 + A \cdot \left(\frac{u'}{s_l}\right)^n\right). \tag{14.69}$$

The laminar combustion velocity can be calculated, for example, acc. to (4.1), then the AGR-dependence of the combustion velocity is also given by this relation (Fig. 14.12).

A further problem is the G-equation's handling of the wall. The use of (14.62) at the wall entails problems, since the flame front would be very thin (laminarized).

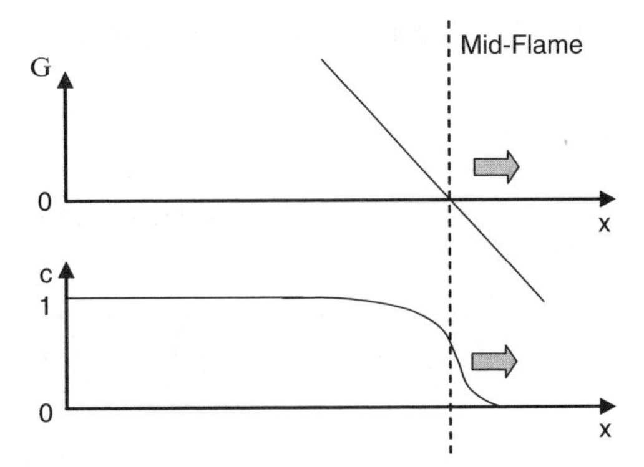

Fig. 14.12 Schematic representation of the simulation of flame front propagation with the G-equation with re-initialization

We can simply put a numerical lower limit on the flame front thickness or introduce an additional transport equation for the turbulent flame thickness l_F (as then l_F does not normally reach equilibrium).

Sometimes, besides the l_F equation, further transport equations are formulated for flame properties, for example for the flame surface density Σ (see also Peters (2000)). In the formulation of these transport equations, one should be careful that no violation of causality takes place; the propagation of flame properties must be compatible to the flame propagation itself. Only the flame variables at the flame location ($G = 0$) are physical and may have an influence on (physical) flame variables at a later point of time. One admissible transport equation for a flame property l_F reads[4]

$$\rho \frac{\partial l_F}{\partial t} + (\rho v_i + \rho_u s_t \hat{n}_i) \frac{\partial l_F}{\partial x_i} - \frac{\partial}{\partial x_{||}} \left(\rho D_t \frac{\partial}{\partial x_{||}} l_F \right) = 2\rho D_t - c_s \rho \frac{\varepsilon}{k} l_F^2, \qquad (14.70)$$

Where

$$\hat{n} = -\frac{\nabla G}{|\nabla G|}, \quad \frac{\partial}{\partial x_{||}} \left(\rho D_t \frac{\partial}{\partial x_{||}} l_F \right) := \frac{\partial}{\partial x_i} \left(\rho D_t \frac{\partial}{\partial x_i} l_F \right) - \hat{n}_i \frac{\partial}{\partial x_i} \left(\rho D_t \hat{n}_j \frac{\partial}{\partial x_i} l_F \right). \qquad (14.71)$$

It is also possible to select $G = 0$ not for the flame center but for the forward front of the flame. This has the advantage that the flame "sees" the conditions in the unburned, which somewhat defuses the two-phase problem discussed above.

[4]The physical information of a configuration should be invariable under a gauge transformation $l_F(x) \to l_F(x) + G(x)\Phi(x)$ for any $\Phi(x)$. This demand can serve to evaluate suitable forms of the transport equations (14.70). On this topic, also see Kraus (2006).

14.4.5 The Diffusive G-Equation

Unfortunately, no complete G-equation implementation with its own convection and re-initialization exists until now in engine CFD codes, STAR CD does however contain a diffusive G-equation for the progress variable (under the designation single-equation Weller model[5])

$$\rho\left(\frac{\partial c}{\partial t}+v_i\frac{\partial G}{\partial x_i}\right)-\frac{\partial}{\partial x_i}\left(\rho D_t\frac{\partial G}{\partial x_i}\right)=\rho_u s_t|\nabla c|. \tag{14.72}$$

In contrast to (14.65), this equation contains a turbulent diffusion term; moreover, it is used directly to calculate the progress variable c (i.e. a relation like (14.67) can now be dropped, it could however not even be applied any more, since a distance variable to the flame front is now lacking). The term proportional to s_t is treated as a source term.

The advantages of the formulation (14.72) are obvious: we are dealing here with a "conventional" scalar transport equation, which can be dealt with in the standard approach. Special effort for a particular convection or a re-initialization need not be made.

On the other hand, the solution quality reduces clearly with the source term treatment of the s_t term. And the re-initialization was necessary for the generation of a distance variable, which served itself for the calculation of flame thickness. It is here that the largest deficiency of the diffusive G-equation becomes apparent: it computes flames that are obviously too thick, which deliquesce under the influence of diffusion ($l_{F,Diff-G}\propto 2\sqrt{D_t t}$). This does not at all characterize the behavior of a flame front however, which of course forms a solitary wave with a stationary profile, as we learned above. The propagation speed remains untouched by the diffusion term however, and the same is true of the global conversion rate. Thus, excessively thick flame fronts are effectively less harmful as one might at first assume. At least in the planar, one-dimensional case, the G-equation is form-invariance with respect to the transformation (14.59), i.e. in using a G-equation formulation, one indeed has a certain right to permit incoherent, flame-thickening fluctuations as well.

In conclusion, a great practical advantage of thick flames should not remain unmentioned: the artificial turbulence production in accordance with (14.28) to (14.30) turns out very low. As long as the turbulence production problem is not solved, one can actually not at all avoid calculating with non-physically thick flames.

However, this means that there is a spatially erroneous flame distribution in the CFD calculation, especially towards the end of combustion. In particular when other physical phenomena running parallel to combustion are also investigated

[5]The actual Weller model is however a two-equation model that resolves the flame front in detail and resembles the coherent flame model (see Weller (1993)).

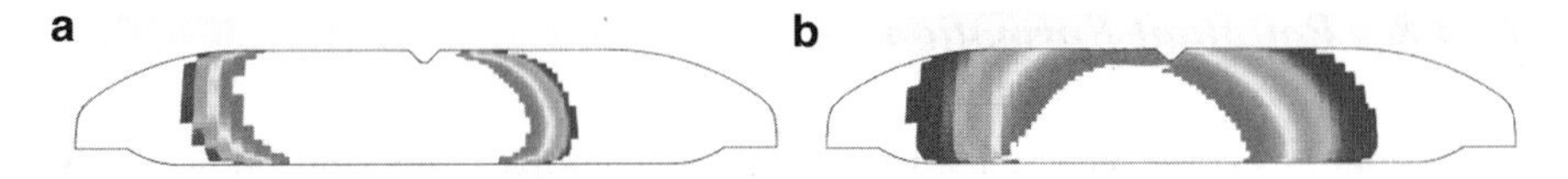

Fig. 14.13 Simulation of premixed combustion (**a**) with the G-equation and (**b**) with the diffusive G-equation

(e.g. knocking or mixture formation in stratified charge combustion), a false flame distribution causes problems (Fig. 14.13).

14.4.6 Ignition

In the description of spark ignition and early flame kernel formation, it is primarily a question of representing laminar flame propagation of the initial flame core and the transition to turbulent combustion. The actual ignition process with plasma formation etc. is generally not possible to simulate in a practical way. To achieve this, there is a series of quite phenomenological approaches, for example that of Herweg (1992). In general, ignition is a main source of cycle deviations because the current local conditions at the spark plug have a dominant effect on ignition delay. In an ensemble-averaged calculation, this is not really resolvable.

With respect to numerical conversion, it should be kept in mind that a numerically correct procedure is essential and that no discontinuities should be created in the marginal conditions. For example, an adaptive mesh refinement (e.g. 0.1 mm edge length) should be undertaken in the ignition area so that the ignition area can be distributed over several cells. Also, the progress variables in these ignition cells should not be set abruptly from 0 to 1 but should rather be continuously increased.

14.4.7 Knocking

First approaches to knock calculation with CFD exist already in the literature; because of the limitations of the available combustion and wall heat transfer models, these attempts must however be approached with caution.

To describe autoignition reaction, knock kinetics [for example the Shell model of Halstead et al. (1977), see Sect. (5.2.2)] must be solved in the CFD code. One must take care that this kinetics is only applied to the unburned phase. Every small increase in temperature because of an approaching main combustion (especially when we are dealing with a "numerical precursor") leads inevitably to knocking rates that are much too high. This is a serious problem especially in the case of thick, diffusive flames.

The consideration of the influence of turbulence is another, as yet unsolved problem. As a first approximation, fluctuations in temperature or in the mixture composition can be modeled by use of variance transport equations (see Mayer (2005)).

14.4.8 Pollutant Formation

Since a high-performance exhaust treatment concept exists for stoichiometrically operated SI engines in the form of the three-way catalytic convertor, the calculation of pollutant formation is of little importance in this case. Mayer (2005) has provided an exemplary investigation on NO_x formation. For HC or CO emissions, what was said in the case of the Diesel engine should also be valid here: details of mixture formation and wall interaction dominate emissions so much that these variables cannot be effectively calculated.

14.5 The SI Engine with Stratified Charge (Partially Premixed Flames)

In the case of a stratified charge combustion, at the moment of flame front propagation, rich ($\lambda < 1$) and lean ($\lambda > 1$) mixtures are found in the combustion chamber at the same time, behind the flame front furthermore rich zones exists with reducing agents (basically CO) and lean zones with oxygen. Thus a diffusion flame remains at the ($\lambda = 1$) isoline. This coupled structure of premixed and diffusion flame is also called a triple flame (see also Fig. 14.14).

It is not hard to develop a suitable method for stratified charge simulation from already existing modeling elements of turbulent premixed and diffusion flames. Astonishingly, stratified charge combustion models behave more "good-natured" in many cases than pure premixed combustion models, since no flame fronts with such sharp temperature and density gradients appear. Moreover, as mentioned in Sect. 14.1, stratified charge combustion is an advantageous case despite "mixed combustion", since both combustion modes (turbulent premixed and diffusion combustion) are dominated by the turbulent time scale.

As in the Diesel model, one should first set the base species to be used for the description of the local gas conditions, the seven species from Sect. 14.3.1 (fuel, N_2, O_2, H_2O, CO_2, CO, H_2) are practical here as well (as a first step, one can also work with three species, air, fuel, and product). If we use a G-equation to describe the

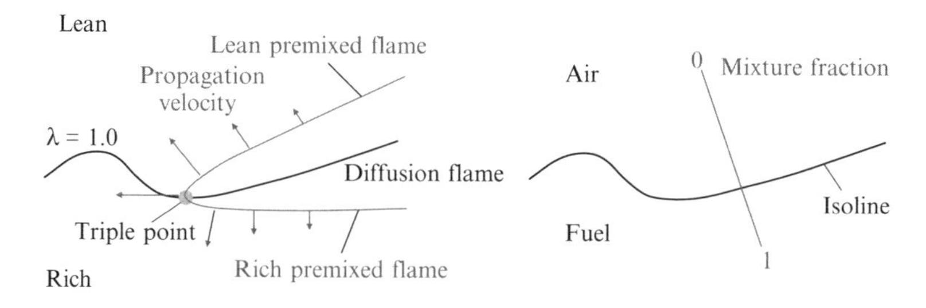

Fig. 14.14 Structure of a triple flame

premixed combustion, and a simple flamelet approach to describe the diffusion combustion, i.e. we transport the mixture fraction (also calculable from the species) and the mixture fraction variance, and then the instantaneous species mass fractions can be determined as follows

$$c_{(i)} = c_{(i),um} + c_{progr}\left(c_{(i),m} - c_{(i),um}\right), \tag{14.73}$$

where

$$c_{(i),m} = \int_0^1 c_{(i)}(Z) p_\beta\left(Z; \langle Z\rangle, Z''^2\right) \mathrm{d}Z \tag{14.74}$$

designates the equilibrium value for the current mixture state and

$$c_{(i),um} = \int_0^1 c_{(i)}(Z) p_\beta\left(Z; \langle Z\rangle, Z''^2 = \langle Z\rangle(1 - \langle Z\rangle)\right) \mathrm{d}Z \tag{14.75}$$

describes the corresponding unmixed state with maximum variance. A reaction is described by

$$\mathrm{d}c_{(i)} = \underbrace{\left(c_{(i),m} - c_{(i),um}\right) \mathrm{d}c_{progr}}_{\text{Premixed combustion}} + \underbrace{c_{progr}\, \mathrm{d}c_{i,m}}_{\text{Diffusion combustion}}, \tag{14.76}$$

i.e. it splits up naturally into premixed and diffusion combustion. Correspondingly, the source term in the transport equation of the internal energy reads

$$q\mathrm{d}t = \sum_k h_{(k)} \rho \, \mathrm{d}c_{(k)} \tag{14.77}$$

with the specific species formation enthalpies $h_{(i)}$. As alterative variants to (14.76), one can calculate the instantaneous species content from Z, Z''^2 and c and then compare it with the one that is only transported (with convection and diffusion). The difference of both quantities provides the reaction in the current time interval. Finally, the merely transported species are overwritten with the new total target values.

Calculation with the standard G-equation is easier that in the case of a homogeneously premixed combustion since even an implementation as a standard transport equation of a passive scalar without a diffusion term is typically quite favorable. A practical generalization of the diffusive G-equation reads:

$$\rho\left(\frac{\partial(Zc)}{\partial t} + v_i \frac{\partial(Zc)}{\partial x_i}\right) - \frac{\partial}{\partial x_i}\left(\rho D_t \frac{\partial(Zc)}{\partial x_i}\right) = \rho_u \langle s_t(Z) Z\rangle |\nabla c|. \tag{14.78}$$

One can work well with this too (the mixture fraction Z itself must also be transported). The turbulent flame speed of the premixed flame s_t can be obtained by flamelet-based averaging in the mixture fraction space

$$s_t = \frac{\langle s_t(Z)Z\rangle}{\langle Z\rangle} = \frac{\int_0^1 s_t(s_l(Z))p_\beta\left(Z;\langle Z\rangle, Z''^2\right) Z\,dZ}{\int_0^1 p_\beta\left(Z;\langle Z\rangle, Z''^2\right) Z\,dZ} . \tag{14.79}$$

We will now discuss a computationally efficient conversion (14.79). For it, we formulate the turbulent burning velocity in accordance with Herweg (1992) as

$$s_t = s_l\left[1 + A\sqrt{\frac{v'}{v'+s_l}}\left(\frac{v'}{s_l}\right)^n\right] \approx s_l + As_l^{1-n}(v')^n - \frac{A}{2}s_l^{2-n}(v')^{n-1}$$
$$\text{with} \quad n = \frac{5}{6} \quad \text{and} \quad A \approx 2,5 . \tag{14.80}$$

parameter A can be adapted to the problem at hand. According to Gülder (1984), the laminar burning velocity is given as

$$s_l = W\Phi^\eta \exp\left(-\xi(\Phi - \Phi^*)^2\right)\left[\frac{T_u}{T_{ref}}\right]^\alpha \left[\frac{p}{p_{ref}}\right]^\beta . \tag{14.81}$$

where T_u is the mixture temperature in the unburned part and $\Phi = 1/\lambda$ is the fuel-air ratio. The following parameters are also used:

$W[m/sec]$	η	ξ	α	β	$T_{ref}\,[K]$	$p_{ref}\,[bar]$	$\Phi*$
0.47	−0.33	4.48	1.56	−0.22	300	1	1.04

If the stoichiometric fuel-air ratio is given as

$$\varsigma = \frac{\rho_{Luft}}{\rho_{Kraftstoff}} \approx 15 \tag{14.82}$$

The mixture fraction Z results from Φ in the following way:

$$Z = \frac{\Phi}{\Phi + \varsigma} . \tag{14.83}$$

If we insert the β-pdf (Sect. 12.2.4) into (14.79), we obtain:

$$\begin{aligned}\langle s_t \rangle &= \frac{1}{\langle Z \rangle} \frac{\Gamma(a+b)}{\Gamma(a)\Gamma(b)} \int_0^1 s_t(s_l(\Phi))\, Z^{a-1}\,(1-Z)^{b-1}\, Z\, \mathrm{d}Z \\ &= \frac{1}{\langle Z \rangle} \frac{\Gamma(a+b)}{\Gamma(a)\Gamma(b)} \int_0^1 s_t(s_l(\Phi)) \left(\frac{\Phi}{\Phi+\varsigma}\right)^a \left(\frac{\varsigma}{\Phi+\varsigma}\right)^{b-1} \frac{\varsigma}{(\Phi+\varsigma)^2}\, \mathrm{d}\Phi \\ &= \frac{1}{\langle Z \rangle} \frac{\Gamma(a+b)}{\Gamma(a)\Gamma(b)} \frac{1}{\zeta} \int_0^1 s_t(s_l(\Phi)) \left(\frac{\Phi}{\Phi+\varsigma}\right)^a \left(\frac{\varsigma}{\Phi+\varsigma}\right)^{b+1} d\Phi. \end{aligned} \quad (14.84)$$

If one considers that s_t depends on s_l, then the integral must be solved for different values of q (for $q = 1$, $q = {}^1/_6$ and $q = {}^7/_6$):

$$\begin{aligned} I(q) &= \int_0^1 \Phi^{\eta q} \exp\left(-q\xi(\Phi - \Phi^*)^2\right) \left(\frac{\Phi}{\Phi+\varsigma}\right)^a \left(\frac{\varsigma}{\Phi+\varsigma}\right)^{b+1} \\ d\Phi &= \varsigma^{b+1} \int_0^1 \frac{\Phi^{\alpha+\eta q}}{(\Phi+\varsigma)^{a+b+1}} \exp\left(-q\xi(\Phi-\Phi^*)^2\right) d\Phi \end{aligned} \quad (14.85)$$

In the case of an integral of this kind, we can make use of an old physicist's trick, the so-called saddle point method, in order to find an approximate solution. This method makes use of the fact that the integral of a function F is limited downwards

$$J = \int_{-\infty}^{\infty} \exp(-F(x))dx. \quad (14.86)$$

is essentially determined by the profile of function F at its minimum. We approximate $F(x)$ by its Taylor row at the minimum x_0 to the second order

$$F(x) \approx F(x_0) + \frac{1}{2} F''(x_0) \cdot (x - x_0)^2 \,. \quad (14.87)$$

and, after inserting in (14.86), we obtain a Gaussian integral. After executing it, we find

$$J = \sqrt{\frac{2\pi}{F''(x_0)}} \cdot \exp(-F(x_0)). \quad (14.88)$$

Applied to (14.85), we obtain

$$I(q) = \varsigma^{b+1} \int_0^1 \exp\left(\underbrace{-q\xi(\Phi - \Phi^*)^2 + (a+\eta q)\ln\Phi - (a+b+1)\ln(\Phi+\varsigma)}_{-F(\Phi)} \right) d\Phi. \tag{14.89}$$

The minimum is obtained by means of the zero of $F_q(\Phi)$:

$$2q\xi\left(\Phi_{q,0} - \Phi^*\right) - \frac{a+\eta q}{\Phi_{q,0}} + \frac{a+b+1}{\Phi_{q,0}+\varsigma} = 0. \tag{14.90}$$

This equation is a cubic equation in Φ_0 and can be solved with the Cardano formulae. The following is valid for the second derivation:

$$F''(\Phi_{q,0}) = 2q\xi + \frac{a+\eta q}{\Phi_{q,0}^2} - \frac{a+b+1}{\left(\Phi_{q,0}+\varsigma\right)^2}. \tag{14.91}$$

After insertion of (14.91) into (14.88), we obtain:

$$I(q) = \varsigma^{b+1} \frac{{\Phi_{q,0}}^{\alpha+\eta q}}{\left(\Phi_{q,0}+\varsigma\right)^{a+b+1}} \exp\left(-q\xi(\Phi_{q,0}-\Phi^*)^2\right) \cdot \Delta_q. \tag{14.92}$$

With

$$\Delta_q = \sqrt{\frac{2\pi}{2q\xi + \frac{a+\eta q}{\Phi_{q,0}^2} - \frac{a+b+1}{\left(\Phi_{q,0}+\varsigma\right)^2}}}. \tag{14.93}$$

If we insert (14.92) and (14.93) into (14.84), we find

$$\begin{aligned} \langle s_t \rangle \approx{}& s_l(\Phi_{1,0}) \cdot \frac{Z_{1,0}}{\langle Z \rangle} \cdot p_\beta(Z_{1,0}) \cdot \Delta_1 + A \cdot s_l^{\frac{1}{6}}\left(\Phi_{\frac{1}{6},0}\right) \cdot \frac{Z_{\frac{1}{6},0}}{\langle Z \rangle} \cdot p_\beta\left(Z_{\frac{1}{6},0}\right) \cdot \Delta_{\frac{1}{6}} \cdot (2k)^{\frac{5}{12}} \\ &- \frac{A}{2} \cdot s_l^{\frac{7}{6}}\left(\Phi_{\frac{7}{6},0}\right) \cdot \frac{Z_{\frac{7}{6},0}}{\langle Z \rangle} \cdot p_\beta\left(Z_{\frac{7}{6},0}\right) \cdot \Delta_{\frac{7}{6}} \cdot (2k)^{-\frac{1}{12}}, \end{aligned} \tag{14.94}$$

Whereby

$$Z_{q,0} = \frac{\Phi_{q,0}}{\Phi_{q,0}+\varsigma}. \tag{14.95}$$

was set.

Figure 14.15 provides an example of such a 3D simulation of stratified charge combustion. More details can be found in Hermann (2008). Pollutant formation models (NO_x, soot) can be taken from Diesel combustion models.

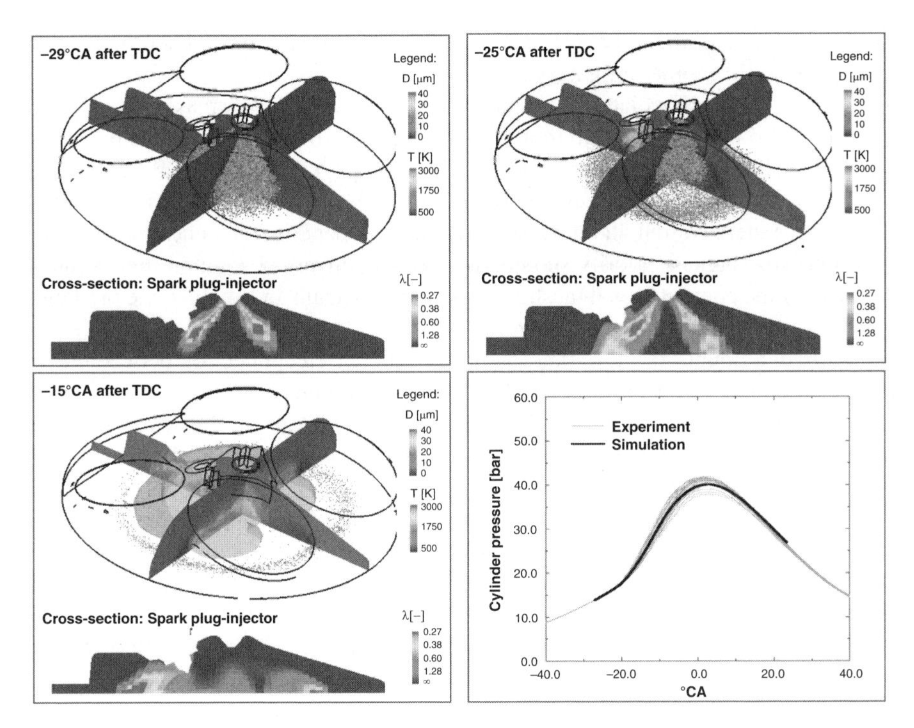

Fig. 14.15 *Left top and bottom*: simulation of stratified charge combustion in a DE SI engine with a spray-guided combustion process at 2,000 rpm and 2 bar p_{me}; *right bottom*: associated pressure curves, simulation vs. experiment (different cycles)

14.6 Fluid-Mechanical Simulation of Charge Exchange, Mixture Formation and Combustion: Future Prospects

In the last few years, flow-mechanical simulation has been developed into an indispensable tool for improving combustion processes. This was essentially made possible by the continuous increase in computer and computer cluster speed. The temporally exponential growth of processing power is described by the so-called *Moore's law*, which postulates that processing power doubles approximately every 2 years. Across several decades, this law has been verified, and nothing indicates any change in this trend. However, there are noticeable changes at least with respect to the realization of this increase of processing power. The tendency to produce increasingly fast individual processors with higher clocking has already fallen. The reason is the drastically increasing energy requirement – as well as the resultant cooling requirement – of the clusters. We can therefore expect heavily increasing numbers of processors with low energy requirements (similar to current laptop processors), which however places high demands on the network. This increases the urgency of improving the scalability of CFD codes significantly. A linear speed advantage at about 10,000 nodes per processor will become obligatory.

The question immediately becomes obvious of how the increased processing power should be invested: into more exact numerical discretization (i.e. finer grids or better solvers), into expanded and more precise physical models, or into shorter cycle times during variant calculations? This question cannot be answered in a sweeping generalization and must without doubt depend on the concrete problem. However, the highest priority should surely be a converged solution that is independent of the computational mesh. As was stressed again and again in the previous chapters, this is far from self-evident!

Without a doubt, in future there will be an increased connection between flow calculation and CAD programs. Simple flow programs can already be solved directly in CAD as a plug-in. For engine problems, this is however a much more long-term project. By parametrical constructions however, the preprocessing cost will be reduced considerably, which is also a prerequisite for automatic optimization of geometry. This trend is being supported by increasingly object-oriented codes which will then offer advantages in handling with a more flexible selection of certain geometry areas for mesh refinement or choosing marginal conditions.

In general, we will see an increased trend to calculate indirectly interpretable variables such as tumble or swirl numbers or "rich zones" as indicators of soot as well as parameters like spark advance requirement, knocking limit or emissions. Of course, this requires suitable, predictive physical models. CFD simulation should not rely exclusively on a kind of "virtual engine" which then simply reproduces the parameters of thermodynamic engines. Ultimately, the central advantage of the illustration of local flow processes that can then, suitably interpreted, give us directions for improvement.

Linking different calculation disciplines will be decisive for successful future operative use in the development of engines. Thermodynamic 1D-3D coupling is already state of the art today. But we can well imagine that CFD codes will make predictions on heat release rates and emissions, with which phenomenological models can then be calibrated in order in turn to be used for overall system improvement. Thus surely makes it necessary to have a sophisticated data management system.

After this general outlook, the focus of the following sections will by on physical aspects of this development.

14.6.1 Mesh Movement

Key point for fluid mechanical simulations are the computational meshes. Surface-based solvers in combination with arbitrary cell types (tetrahedra, hexahedra, prisms, polyhedra,...) will become established. For wall-dominated flows, as they are typical for engines, wall-adapted meshes with $y^+ < 100$ are mandatory.

The emerging trend towards low-Reynolds-number wall models, however, will require further cell layers for a better resolution of the wall area.

As for the treatment of the wall, problem-adapted meshing will also become importance for other processes, most mentionable the shear layers around the valves, the area of the injection spray and possibly also the premixed flame front. Altogether, still a further area of length scales will need to be resolved by the mesh (turbulent length scale of the internal nozzle flow relative to the cylinder bore). Today, these problem-adjusted meshes can only be realized at great cost. It seems certain that, in future, intelligent, automatic algorithms will be established for mesh generation that will take over at least some of this work from the user. It is however not yet clear which strategy will become successful in the case of mesh movement in engines. Till now, many codes (e.g. STAR-CD, kiva3v) have supplemented/erased defined cell layers. This minimizes interpolation errors, but it makes the movement of the mesh somewhat inflexible. Mapping techniques (e.g. FIRE), which creates meshes ahead of time for certain phases, between which the flow field is then projected, are an alternative. In the past, this was the less efficient and less accurate path. Meshes had to be created for different pistons and valve positions at great cost, and there were projection errors in the transition between the meshes. Automatic meshing strategies and interpolation processes of a higher order will make this method much more attractive however, permitting the use of specifically optimized meshes in the corresponding crank angle area. At the moment however, other innovative methods like that of overlapping meshes in different areas (pistons, valves, ...) are also being used.

14.6.2 Numerics

Future physical models (such as Euler spray models or turbulent combustion models) will require flow resolution with significantly reduced numerical diffusion. Turbulent combustion models based on variance equations of the mixture fraction or of temperature are even today incapable of producing practical results without the use of second order processes. These requirements are even more valid for alternative turbulence models like LES. Until now, the focus was clearly on stable and robust algorithms, unfortunately often at the expense of numerical diffusion. Yet even now, promising flow solvers are in development that provide results on complex, not completely optimal meshes on the one hand, but are also conservative and have TVD properties and thus do not produce local overshots [see (12.76)].

The interpolation capability in mesh changing will become more and more the focus, particularly with respect to alternative strategies of mesh movement. Naturally, there should be no integral or local discontinuities in the conserved quantities during mesh changing, but the gradients must also be conserved since these are indeed an important input for many source terms.

14.6.3 Turbulence

As opposed to the view often propagated at symposia that current RANS models could now be quickly replaced by LES, we assume that RANS models will remain the state of the art for at least the next 5 or even 10 years. The essential reason for this in our view is that no flow resolution with the necessarily low viscosity will be available for complex moving engine geometries. Yet this is the absolutely necessary prerequisite (see Fig. 12.5)! In the intermediate term, there will also be a transition to LES for engine calculations; without a doubt, this will however be focused on aerodynamic problems. The first useable multicycle calculations are currently being realized with research codes. But we are still far from high-quality spray or combustion calculations with LES. For this to be possible, the quality of the boundary conditions (e.g. deviations from the injection system) must be significantly increased among other things.

Cyclic deviations in the engine will remain, in our view, scarcely calculable until further notice (with LES as well, as discussed already in Chap. 12). In this connection however, a gainful linkage with diagnostic tools is conceivable. In the meantime, high-speed laser techniques in transparent engines such as high-speed PIV permit a more exact insight into flow processes in the cylinder and can be used to supplement CFD calculations.

14.6.4 Modeling Injection Processes

Combustion simulation for direct injection engines is usually highly influenced by the spray model. If the latter is incorrect, little can be expected of the result of the combustion simulation. For this reason, it is enormously important, particularly from the standpoint of combustion simulation, first to fix the spray model. This is especially true with regards to the fact that this step is generally feasible today; the building blocks required for this are at hand.

However, these building blocks have hardly been available if at all in most commercial engine codes till now. One could address the Lagrangian model for starters. If one could apply this approach on the modeling side in a practical way – that is, using spray-adapted meshes with high resolution near the nozzle as well as correspondingly high parcel numbers – quite reasonable results could be obtained. With respect to modeling, especially turbulent dispersion should be mentioned. As we have discussed, the turbulence of the liquid phase should definitely be transported as a parcel property, i.e. in principle a "droplet turbulence model" be introduced, just as gas flow requires a turbulence model. From the standpoint of computation, this is not or, depending on the model, hardly more costly than the standard model, but it is unfortunately not available in the current commercial codes. It is therefore urgent that this model innovation finds its way into the CFD codes. As mentioned in Chap. 13, we presently actually have only the chance of reformulating in the CFD code KIVA the O'Rourke model present there. But this is quite uncomplicated.

There is especially great potential in the application of an Eulerian formulation to spray modeling. In particular, the chance presents itself of letting the calculation of internal nozzle flow pass continuously into the spray calculation, providing the more accurate starting conditions for injection spray calculation. As we known today, flow processes in the internal nozzle flow have a major impact on the behavior of the spray.

As we attempted to show in Chap. 13, we consider it absolutely advisable not to introduce a completely independent, essentially empirically motivated approach but rather to proceed from the well-established Lagrangian model, transforming it into an equivalent Eulerian formulation. This was exactly the subject of Sect. 13.3 (under certain assumptions of approximation). There is a prospect that this model will be available in engine CFD codes in the near future (e.g. in STAR-CD 4). With respect to he approximations carried out (breakup only as average process, no collisions), further progress can surely be made. These approximations correspond to those of the Lagrangian model, but there are other boundary conditions there. For in the Lagrangian model, the approximations mentioned stem essentially from the fact that no statistical convergence is possible in the consideration of these models. Model implementation as such is self-evident and uncritical however. In the Eulerian model, the situation is opposite; model implementation is critical, while statistical convergence is automatically fulfilled. Of both cases however, the latter is clearly more favorable; difficult model implementation is much preferable to poor convergence behavior. There are certainly approaches to solving the first case, as we will discuss in the following.

14.6.4.1 Approximation I: Consideration of Breakup Only as an Average Value Process

As a consequence of this, we finally introduced the concept of droplet classes, whereby now the radius variance within a class should be relatively small so that we can work with an average radius.

A consideration of droplet size distributions forming due to breakup processes could be done in such a way that we also introduce generalized diffusion terms in the sense of a Fokker-Planck equation for the radius in the transport equation of the distribution function (13.71). In the generalized sense, a Fokker-Planck equation describes the distribution function of a variable x, the temporal development of which is influenced by the additive affect of a stochastic force F:

$$\frac{dx}{dt} = f(x) + F_{stochast}. \tag{14.96}$$

[(14.96) is then the associated Langevin equation.] One is now tempted to argue that such behavior is not applicable to breakups, since for the radius of a droplet after n breakup processes (with m_i daughter droplets in the ith breakup, $i = 1, \ldots, n$) is valid:

$$R_n = R_0 \cdot \frac{1}{\sqrt[3]{m_1}} \cdot \frac{1}{\sqrt[3]{m_2}} \cdot \ldots \cdot \frac{1}{\sqrt[3]{m_n}}. \tag{14.97}$$

The stochastic influence is thus multiplicative, not additive. Yet this changes after logarithmizing:

$$\ln R_n = \ln R_0 - \frac{\ln m_1}{3} - \frac{\ln m_2}{3} \ldots - \frac{\ln m_n}{3}. \tag{14.98}$$

That is, it is possible to formulate a Fokker-Planck equation for the logarithm of the radius to describe breakup processes:

$$\ln R_n = \ln R_0 - \frac{\ln m_1}{3} - \frac{\ln m_2}{3} \ldots - \frac{\ln m_n}{3}. \tag{14.99}$$

If we take into consideration that with the designations of (13.56) for the mean value and variance of *lnR*:

$$\begin{aligned} \langle \ln R \rangle &= \ln R_0 - \frac{1}{3} \int_0^t \frac{\ln m}{\tau} d\tilde{t} \\ \left\langle (\ln R)^2 \right\rangle - \langle \ln R \rangle^2 &= \frac{1}{9} \int_0^t \frac{(\ln m)^2}{\tau} d\tilde{t}, \end{aligned} \tag{14.100}$$

the following Fokker-Planck equation can be written for the mass-based distribution function ρ [i.e. the density – see the discussion concerning (13.58) and (13.59)]:

$$\frac{\partial}{\partial t}\rho(R,t) - \frac{\partial}{\partial \ln R}\left(\frac{1}{3}\frac{\ln m}{\tau}\rho(R,t)\right) - \frac{\partial}{\partial \ln R}\left[\frac{1}{18}\frac{\ln^2 m}{\tau} \cdot \frac{\partial}{\partial \ln R}\rho(R,t)\right] = 0. \tag{14.101}$$

Considering (13.58) as well as $d\ln R = dR/R$ we finally obtain the result that the diffusion term

$$-\frac{1}{R^2}\frac{\partial}{\partial R}\left[\frac{1}{18}\frac{\ln^2 m}{\tau} \cdot R\frac{\partial}{\partial R}\left(R^3 \cdot p(R,v,T,t)\right)\right] \tag{14.102}$$

should be added in (13.71). Mixed diffusion terms of type $\partial/\partial R \cdot \partial/\partial v$ and $\partial/\partial R \cdot \partial/\partial T$ must also still be included [similar to (13.68) to (13.70)].

These correspond to the phenomena that variously large and variously fast droplets vaporize, heat up and are slowed down at different rates.

The probability distribution of breakup is indeed a Poisson distribution, but for larger breakup numbers, this converges towards a normal distribution; and this is finally the prerequisite for the use of a Fokker-Planck equation. In addition, another result of this is that at a constant (i.e. especially independent of the radius) breakup rate, without influence by retarding or vaporizing processes, a *log-normal distribution* arises as the size distribution:

$$p(R;t) = \frac{1}{\sqrt{2\pi\alpha}} \exp\left(-\frac{\left(\ln\frac{R}{R_0} + \frac{\ln m}{3\tau} t\right)^2}{2\alpha} \right) \frac{dR}{R}$$

$$\alpha = \frac{\ln^2 m}{9\tau} t. \tag{14.103}$$

14.6.4.2 Approximation II: Negligence of Collision Processes

Let us first consider the (quasi)-elastic collisions. In this case it is sensible to take a closer look at the procedure when deriving the Navier-Stokes equations from the Boltzmann equation by means of the Chapman-Enskog approach (see e.g. Landau and Lifshitz 1981). The distribution function is written as the sum of a local Maxwell distribution p_0 and a small disruption p_1, whereby the local Maxwell distribution eliminates the (elastic) collision integral. Applied to (13.71) (which we write compactly as $\hat{L}p_0 = 0$), we can then write after re-introduction of the elastic collision term:

$$L(p_0) \approx L(p_0 + p_1) = I_{elast.\ Stoßt.}[p_0 + p_1] = I_{elast.\ Stoßt.}[p_1] \approx \frac{1}{\tau_{Stoß}} p_1, \tag{14.104}$$

where the approximation customary in the classic Chapman-Enskog approach was already introduced in the last step. From the relation, the correction term p_1 can then be calculated. This approach would still have to be combined with the procedure of Sect. 13.3.3 in order to close the moment equations.

The treatment of inelastic collisions is more difficult but could potentially be interpreted as a superimposition of elastic collisions and breakup (coalescence can also be interpreted as "inverse breakup" with increasing droplet radius).

14.6.4.3 Turbulent Gas Fluctuations as a Source of Turbulent Dispersion

In addition, we should make a note of the fact that the correlation of turbulent fluctuations of the gas velocity of (13.61) or (13.69) is very easy to select. It would of course be more correct and consistent to estimate in accordance with (12.34), i.e. to take non-diagonal terms into consideration, which are proportional to the shear strain tensor.

14.6.5 Modeling Combustion

With respect to the modeling of combustion itself, we can again say that the approaches described in this chapter are unfortunately not all realized in the usual engine CFD codes. This is true, for example, of the G-equation as well as the two-phase approach in flame front combustion (Sect. 14.4.1). Their inclusion would certainly be a major advance. Otherwise, it should be possible to describe all combustion regimes with very fast chemical processes (i.e. large Damköhler number), in which the turbulent time scale is dominant, with the approaches available today described in this chapter (these are precisely the combustion regimes "based" on the triangle of Fig. 14.1).

For processes more under the influence of reaction kinetics such as autoignition, Diesel-engine "premixed" combustion or pollutant formation (especially NO_x and soot), we still lack simple, high-performance methods of sufficient universality. The basic problem is the previously discussed distribution problem of the progress variables in the mixture fraction space specific to the respective problem.

If we approach these problems, we must differentiate whether we are dealing essentially with a diffusion flame-oriented combustion with strong mixture fraction gradients, or whether a premixed system with smaller large-scale and turbulent small-scale inhomogeneities of mixture and temperature distribution is to be described. In the former case, we need to derive an expansion of the classic diffusion flamelet approach. In case one of the progress variable approaches discussed in the literature should not become accepted for this or that specific problem (see e.g. Steiner et al. (2004) or Lehtiniemi et al. (2005)), one must presumably fall back on the briefly introduced conditional moment closure (CMC) method, which however can entail extremely high calculation costs. It is perhaps possible to find an intelligent combination with more simple approaches for some cases.

If the basic problem should correspond more to the second type, i.e. a relatively homogeneous mixture with minimal inhomogeneities (which can still have an influence on the process however), the concept of calculating a posteriori distribution functions by means of so-called *transported pdf models* is more appropriate (see Pope (1985)). In this case, a local mixture state (of the initial and end products of the reaction, i.e. incl. progress variables) is represented by means of an ensemble of homogeneous reactors. Every homogeneous reactor is represented by a parcel, which embodies a possible local species composition. Such a parcel moves with the gas velocity in the combustion space, mixes with the other homogeneous reactors in a computational cell (diffusion!), and its components react with themselves (laminar). There is a problem in the modeling of diffusion processes. This approach resembles the parcel concept of the Lagrangian spray model and thus also suffers from the same weaknesses: to reach statistical convergence, very extensive calculations are required and small structures like flame fronts cannot be resolved. Application is thus more suited to problems with spatial gradients that are not too sharp but have a high level of chemical complexity (e.g. SI engine knocking).

References

Bilger RW (1993) Conditional moment closure for turbulent reacting flow. Phys Fluids A 5(2):436–444

De Paola G, Mastorakos E, Wright YM, Boulouchos K (2008) Diesel engine simulations with multi-dimensional conditional moment closure. Combust Sci Technol 180(5):883–899

Dederichs AS, Balthasar M, Mauß F (1999) Modeling of NO_x and soot formation in diesel combustion. Oil Sci Technol 54:246–249

Gülder ÖL (1984) Correlations of laminar combustion data for alternative S.I. engine fuels, SAE 841000

Halstead MP, Kirsch LJ, Prothero A, Quinn CP (1975) A mathematical model for hydrocarbon autoignition at high pressures. Proc Roy Soc Lond A346, pp 515–538

Halstead MP, Kirsch LJ, Quinn CP (1977) The autoignition of hydrocarbon fuels at high temperatures and pressures – fitting of a mathematical model. Combust Flame 30:45–60

Hermann A (2008) Modellbildung für die 3D-Simulation der Gemischbildung und Verbrennung für Ottomotoren mit Benzin-Direkteinspritzung. PhD thesis, Universität Karlsruhe

Herweg R (1992) Die Entflammung brennbarer, turbulenter Gemische durch elektrische Zündanlagen – Bildung von Flammenkernen. Dissertation, Universität Stuttgart

Hiroyasu H, Kadota T, Arai M (1983) Development and use of a spray combustion modeling to predict diesel engine efficiency and pollutant emission. part 1: combustion modeling. Bull JSME 26:569–575

Klimenko AY, Bilger RW (1999) Conditional moment closure for turbulent combustion. Prog Energy Comb Sci 25:595–687

Kolmogorov AN, Petrovsky IG, Piskunov NS (1937) Study of the diffusion equation with growth of the quantity of matter and its application to a biology problem. Bull Univ Moscou Ser Int Sec A 1, 1–25 (translated in: Pelcé P (1988) Dynamics of curved fronts, perspectives in physics. Academic, New York)

Kraus E (2006) Simulation der vorgemischten Verbrennung in einem realen Motor mit dem Level-Set-Ansatz. PhD thesis, Universität Tübingen

Landau LD, Lifschitz EM (1981) Course of theoretical physics, vol 10, Physical kinetics. Pergamon, New York

Landau LD, Lifschitz EM (1980) Course of theoretical physics, vol 5, Statictical physics, part I. Reed Educational and Professional, Boston, MA

Lehtiniemi H, Amnéus P, Mauss F, Balthasar M, Karlsson A, Magnusson I (2005) Modeling diesel spray ignition using detailed chemistry with a flamelet progress variable approach, Fifth Symposium on towards clean diesel engines, Lund, Sweden, June 2–3

Mayer Th (2005) Dreidimensionale Simulation der Stickoxidbildung und der Klopfwahrscheinlichkeit in einem Ottomotor. PhD thesis, Universität Stuttgart

Nagle J, Strickland-Constable RF (1962) Oxidation of carbon between 1000-2000°C. Proceedings of the 5th conference on carbon, vol 1. Pergamon, London, pp 154–164

Patterson MA, Reitz RD (1998) Modelling the effects of fuel spray characteristics on diesel engine combustion and emissions. SAE-Paper 980131

Peters N (2000) Turbulent combustion. Cambridge University Press, Cambridge

Poinsot Th, Veynante D (2001) Theoretical and numerical combustion. R.T. Edwards, Flourtown, PA

Pope SB (1985) PDF methods for turbulent reactive flows. Prog Energy Comb Sci 19:119–197

Rao S, Rutland CJ (2002) A flamelet timescale combustion model for turbulent combustion in KIVA. 12th international multidimensional engine modeling user's group meeting at the SAE Congress

Steiner R, Bauer C, Krüger Ch, Otto F, Maas U (2004) 3D-simulation of DI-diesel combustion applying a progress variable approach accounting for complex chemistry, to be published at SAE 03/04

Weller HG (1993) The development of a new flame area combustion model using conditional averaging. thermo-fluids section Report TF/9307, Imperial College of Science, Technology and Medicine, London

Wolfer HH (1938) Der Zündverzug beim Dieselmotor, VDI Forschungsheft 392

Chapter 15
3D Supercharging Simulations

Bodo Durst

15.1 Introduction

The foundations for designing and dimensioning turbocharger components for supercharging combustion engines were already laid in previous chapters. With the aid of the methods discussed, we can determine not only such basic dimensions and geometrical parameters as diameters, cross-sectional profiles, radii, and forms, but also other values, such as nominal rotation speed. This ensures that the relevant turbo-machine fulfils the basic requirements placed on it with respect to these parameters. The methods discussed, however, are based on fluid-mechanical and thermodynamic simplifications, empirical approaches or, in part, on diagrams which offer visualizations of the basic relationships and allow graphic definitions. For example, many of the equations are used on the basis of the assumption that the stationary flow filament theory applies, i.e. that the flow can be regarded as purely one-dimensional. Also, it is usually assumed that the flow can be considered nonviscous. For initial approximations, these assumptions are certainly justified. However, flows in turbo-machines are generally much more complex than the above premises would suggest. In particular, in conjunction with the rotation of the impeller, the surfaces, which can be highly bent (see Fig. 15.1), cause three-dimensional and, in part, unsteady flow processes which, because of the typically high Reynolds numbers involved, are turbulent and highly vortex-dominated. For this reason, the geometry of a component should be verified and confirmed at an early stage preceding final manufacture with the aid of a three-dimensional flow simulation. Also, turbo-machines have since become so efficient that neither a further optimization nor the analysis of deteriorations in the development process is possible any longer without more penetrating insights into the detailed flow processes involved.

The present chapter will summarize the essential aspects of the three-dimensional flow simulation of radial compressors and radial-flow turbines and will demonstrate the aims and potential of the numerical analysis of turbo-machines.

G.P. Merker et al. (eds.), *Combustion Engines Development*,
DOI 10.1007/978-3-642-14094-5_15,

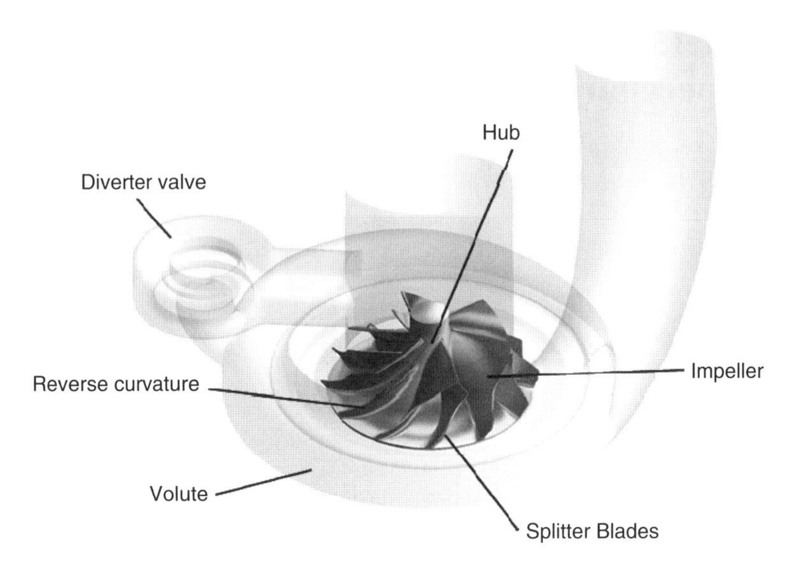

Fig. 15.1 Complex three-dimensional geometry of a compressor

15.2 Foundations of the 3D Simulation of Turbo-Machines

The general foundations of the numerical simulation of flow processes were already treated in Chap. 12. In principle, the relations discussed there with respect to the basic fluid-mechanical equations and their discretization, to turbulence modelling and wall functions, as well as to the numerical grid to be generated and the corresponding mesh resolution also apply to numerical calculations in turbo-machines. The following will thus focus only on the specific aspects involved in the case of turbo-machines.

15.2.1 The Treatment of Different and Various Moving Coordinate Systems

Basically, the two fluid-mechanically relevant parts of a turbo-machine that can be distinguished are the impeller and the housing. Whereas the housing is fixed statically to the engine and can be described very well in an absolute Cartesian coordinate system fixed at a point in space, the rotating impeller calls for a co-rotating, relative coordinate system. The advantage of this approach is that the inherently unsteady flow in the impeller can be viewed statically by representing it in a relative system. The flow speed in the relative system is defined by:

$$\vec{c} = \vec{w} + \vec{u}, \tag{15.1}$$

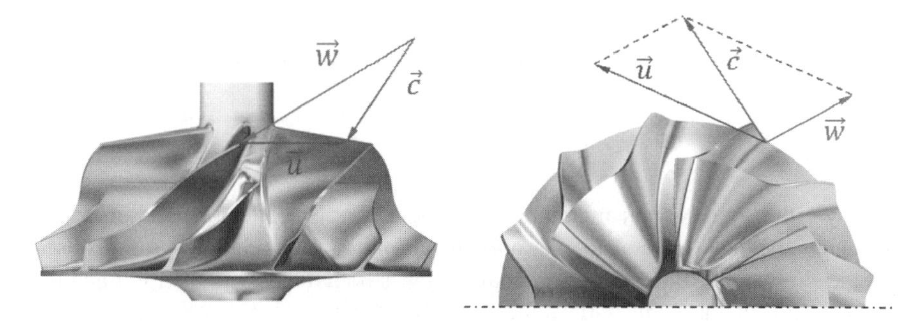

Fig. 15.2 Velocity triangles at the inlet and the outlet of a radial compressor

with $\vec{c}$ as the velocity vector in the absolute system, $\vec{w}$ the velocity vector in the relative system, and $\vec{u}$ the vector of the circumferential velocity of the impeller at the observed point (see Fig. 15.2). The latter is defined by the following vector product:

$$\vec{u} = \vec{\omega} \times \vec{r}, \tag{15.2}$$

with $\vec{\omega}$ as the rotation vector of the impeller and $\vec{r}$ as the position vector. The relative Velocity can thus be expressed as:

$$\vec{w} = \vec{c} - \vec{\omega} \times \vec{r}. \tag{15.3}$$

Expressed in the index notation selected earlier, we have the following equation:

$$w_i = c_i - \varepsilon_{ijk}\omega_j r_k, \tag{15.4}$$

with ε_{ijk} representing the identity tensor. Since the index notation used here is somewhat confusing, the symbolic notation will be used in the following.

Thus the momentum transfer equation given in Chap. 12 is as follows in the absolute system:

$$\frac{\partial(\rho\vec{c})}{\partial t} + \nabla \cdot (\rho\vec{c} \otimes \vec{c}) = -\nabla p + \nabla \cdot \tau + \vec{f}. \tag{15.5}$$

The operator $\otimes$ represents the so-called dyadic product of two vectors. After the transformation into relative velocity and rotational velocity, the equation is:

$$\frac{\partial(\rho\vec{w})}{\partial t} + \frac{\partial(\rho\vec{\omega})}{\partial t} \times \vec{r} + \nabla \cdot (\rho\vec{w} \otimes \vec{w}) = -\nabla p + \nabla \cdot \tau + \vec{f}' - 2\rho\vec{\omega} \times \vec{r} - \rho\vec{\omega} \times (\vec{\omega} \times \vec{r}). \tag{15.6}$$

The last two terms represent the influence of Coriolis force and centrifugal force, respectively. The remaining transport equations are represented and solved in the relative system in a similar way. The transformation into the relative system is also to be taken into account for total values. The total pressure, for example, is:

$$p_{\text{tot}} = p_{\text{stat}} + \frac{1}{2}\rho(\vec{w} \cdot \vec{w} - \vec{\omega} \times \vec{r} \cdot \vec{\omega} \times \vec{r}). \tag{15.7}$$

If only either stators (housing/guide vanes) or rotors (impeller or parts of it, e.g. a single vane duct) are considered, only the absolute or the relative system is used for the simulation. If the numerical analysis takes both the flow in the rotor and the stator into account at the same time, both coordinate systems ("multiple frames of reference") must be used together. The stator is calculated in the absolute system and the rotor in the relative system. In this case, there is an internal boundary surface for the simulation at the transition between the two partial areas. This boundary area can be treated in different ways depending on the type of simulation.

The definitions usually made available by the most commercial CFD programs in this context are:

- Frozen rotor interface
- Mixing plane interface
- Rotor/stator connection

In the case of the *frozen rotor interface*, the relative position of the rotor remains fixed relative to the stator and the boundary values resulting at the boundary surface are simply transferred from one coordinate system to other. This allows for an evaluation of asymmetric geometrical configurations which typically occur in automotive turbo-machines. Principally however, this definition of the interface only applies to static simulations, since the rotor is viewed as frozen. The rotation influences the flow via the co-rotating coordinate system. However, the advantage that the values transferred between the coordinate systems bordering the interface reflect the asymmetry of the stator area also means that the result of the simulation depends on the relative orientation of the rotor to the stator. Moreover, multiple static frozen rotor calculations with different rotor positions do not correspond to the result of the unsteady simulation.

For the *mixing plane interface,* too, the relative orientation of the rotor to the stator does not change during calculation. However, an averaging of values does take place in the circumferential direction of the boundary surface between the two coordinate systems. The flow values transferred between the neighboring nodes are thus distributed evenly on entering into the relative system, which leads to the result being independent of how the rotor happens to be located in relation to the stator. The assumption that flow values are equally distributed is usually inappropriate for automotive turbo-machines in particular, leading to an influence on flow in the stator caused by the averaging. Areas of application for mixture level connections are thus rather to be found among axial turbo-machines in which an azimuthally equal distribution can be approximated.

As opposed to the above two interfaces, the *rotor/stator connection* allows for the unsteady simulation of turbo-machine behavior. With this interface, the angular position of the impeller is altered over time. The true respective relative position between rotor and stator is thus observed in accordance with the angular speed. In this way, the rotor/stator interaction can also be analyzed, for example, between the impeller and asymmetric housing parts or between the impeller and guide vanes. As a result of the rotation of the rotor, the interface between the two bordering coordinate systems represents an internal boundary surface of the calculation model with non-conforming mesh topologies on both sides. This means that the grid points of one sub-mesh must not necessarily be identical to those of the other sub-mesh. The connection occurs in such cases with the aid of so-called control surfaces, which ensure a suitable value transfer from one network to another. Depending on the program used, such control surfaces are referred to as "generalized grid interfaces", "arbitrary interfaces", or simply "interfaces". The boundary surface in the case of rotor/stator connections represents a special instance of control surfaces, since the grid point configuration changes with each time step due to the rotation of the rotor. One speaks in such cases of a "sliding interface".

The first two interfaces are fast and robust possibilities for simulating flows in automotive turbo-machines and are suitable, for example, for improving turbo-machines with the additional aid of automatic optimization methods. However, both exhibit the disadvantages discussed above. The unsteady analysis using the rotor/stator connection is the only way to take transient effects into account correctly and also to assess the interaction between the rotor and the stator. The method is, however, much more demanding with respect to computing time and computer storage space and also places other demands on the computation process (e.g. with respect to the boundary conditions to be specified, which may themselves also have to be temporally variable). For this reason, such analyses have not yet become the standard in the industrial sector.

15.2.2 Grid Generation for Turbo-Machines

As mentioned already at the beginning of Sect. 15.2, similar requirements apply to 3D CFD simulations of turbo-machines as to other areas of 3D CFD simulations. Since the detailed analysis of losses usually goes hand in hand with predictions of flow separations with the highest possible accuracy, especial value is placed in the case of turbo-machine simulations on the quality of the calculation grid. This relates to parameters such as grid angle, expansion ratio or even the aspect ratio of the control volumes, but special attention must be paid to the sufficient resolution of the boundary layer and, depending on the turbulence and wall function model used, to the y^+ value of the first cells within the calculation area.

The numerical error of the solution is influenced by the grid angle. Good values lie between 20 and 160° in this case, acceptable values between 10 and 170°. Beyond this range, one must assume considerable influences on the quality of the solution.

In the case of one-dimensional meshes, the expansion ratio refers to the ratio between the distances of neighboring grid points. In the three-dimensional case, one can instead use the volume ratio between adjacent control volumes. This parameter also directly influences the numerical error of the solution. Ideally, the values for the expansion ratio range between 1 and 1.5. Values between 1.5 and 2.5 are still acceptable.

The aspect ratio of a control volume influences the convergence behavior of iterative solving methods, with whose aid the matrix equation resulting from the discretization of the transport equations is solved. In part, these solvers require that the aspect ratio remains limited. Aspect ratios of 10 or smaller are desirable here. Especially for the resolution of the boundary layer, however, this value cannot necessarily always be adhered to and that aspect ratios of 100 or more must be accepted in order to keep the number of grid points at an acceptable level. In this case, one must check the suitability of the respective matrix solvers for such meshes.

One goal that must always be borne in mind with numerical simulations is finding consistent solutions. This means that the calculation error should become smaller with an increasing refinement of the mesh. This can be checked by comparing the acquired solution using increasingly refined meshes. If the solution does not exhibit an asymptotic tendency towards a limit, this can be the general result of the grid still being much too coarse. One problem, however, may be that a mesh refinement causes the parameters listed above which describe the mesh quality to become worse instead of improving. One speaks in such cases of "non-scalable meshes".

In principle, hexahedral grids (cf. Fig. 15.3a) represent the best possibility for creating a good mesh quality with high calculation efficiency. Hexahedral elements are suited well to resolve the boundary layer near the wall and shear layers which form due to flow separation. The disadvantage of hexahedral grids lies in the very difficult process of grid generation, which can only be automated to a small extent. Manual mesh generation, however, offers the advantage that the control volumes can already be arranged in such a way that the high gradients to be expected in the flow can be resolved well.

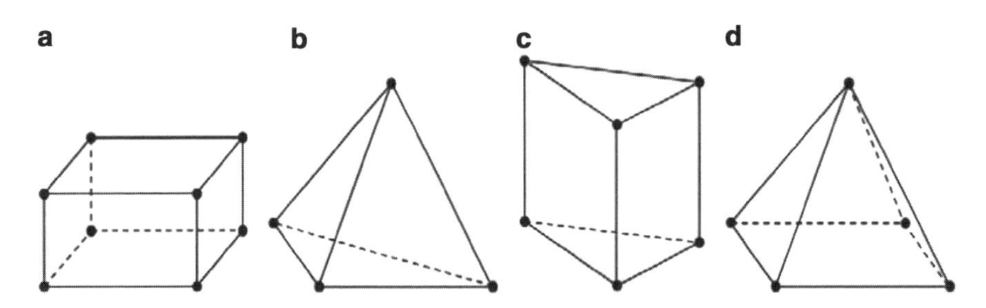

Fig. 15.3 Overview of typical element types for 3D CFD control volumes. (**a**) hexahedron; (**b**) tetrahedron; (**c**) prism; (**d**) pyramid

In contrast, tetrahedral networks (Fig. 15.3b) can be generated automatically very easily. On the basis of a surface mesh on the wetted geometry, the available volume is filled with tetrahedral control volumes. A disadvantage of tetrahedral networks, however, is that due to their relative arrangement to each other they are not entirely suited to resolve shear layers and boundary layers on walls. In addition, it is not possible to orient the arrangement of control volumes to the curvature of the streamlines. This means that considerably more grid points must be used to achieve a solution with a quality similar to that of hexahedral grids. The control-volume type alone raises the amount of calculation and computer storage space required for tetrahedral network to a value approximately 50% higher than for hexahedral networks.

While still less efficient than hexahedral networks, prismatic networks (Fig. 15.3c) are nevertheless better adapted to resolving the boundary layer than tetrahedrons. They can be created on the basis of a two-dimensional mesh, e.g. by extrusion, and can be easily automated. In conjunction with tetrahedrons in the internal area, prismatic meshes offer a very fast possibility for generating meshes in an unattended way and with little manual intervention.

In general, pyramid elements (Fig. 15.3d) only represent transition elements for connecting hexahedral elements with tetrahedral elements and should be used as control volumes as little as possible.

15.2.3 The Creation of Calculation Models and Boundary Conditions

In Sect. 15.2.1, in the context of the parts of a turbo-machine relevant to flow mechanics, we only distinguished between the impeller and the housing. As practical experience shows, it is a good idea to structure calculation models in a more refined way in order to define the boundary conditions and make evaluations more easily for analysis. Using the example of a compressor, the following will introduce a sensible approach to this which can be applied in a similar form to other turbo-machines, as well.

The compressor model shown in Fig. 15.4a includes a diverter valve. By separating the individual elements of the overall model, one can recognize that both a feeding and run-out section were attached to the housing. This is required for creating the necessary preconditions to ensure that the corresponding boundary conditions can be applied in such a way that a uniform flow profile develops at the entry to the truly relevant area of calculation. If the boundary conditions are specified too near the impeller, there is a danger that they will influence the solution too strongly.

The geometrical model of the compressor has been subdivided into the inlet zone of the housing, “casing in”, which includes the diverter valve, the so-called “impeller”, which includes the impeller and one part of the diffusor, and the

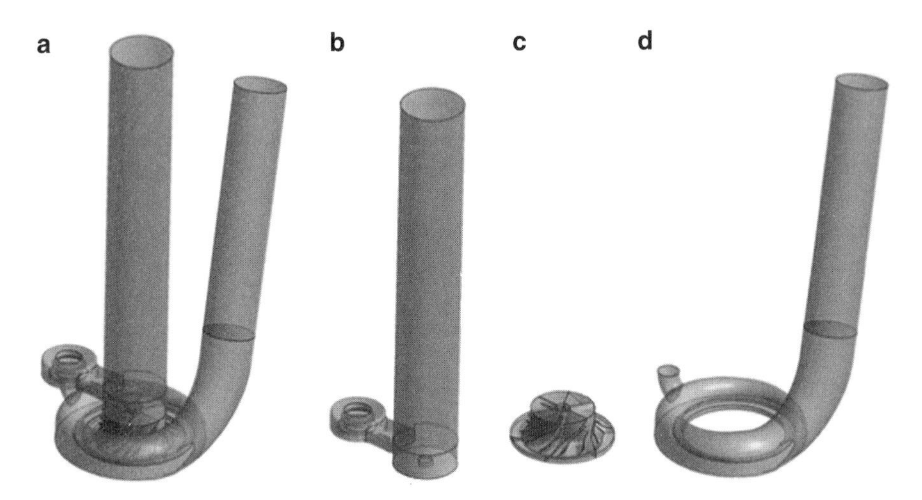

Fig. 15.4 Subdivision of compressor geometry into different subareas. (**a**) overall model; (**b**) casing in; (**c**) impeller; (**d**) casing out

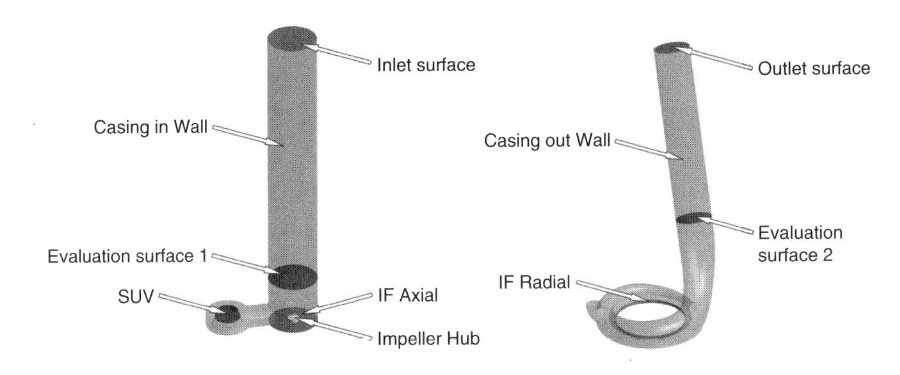

Fig. 15.5 Boundary and edge surfaces of compressor housing

outlet zone of the housing, "casing out", which includes the other part of the diffusor, the volutes and the run-out section.

The various boundary and edge surfaces resulting from the subdivision of the compressor housing are shown in Fig. 15.5. Also, one can see how additional evaluation surfaces are created in places in which comparisons should be made, e.g. with measured values. The surfaces which are also fixed walls in reality are defined in the CFD model as "non-slip walls", i.e. walls on which the flow must fulfill the no-slip condition. The boundary conditions specified at the inlet and outlet surfaces essentially define the flow of the turbo-machine. It must be borne in mind that the boundary conditions may not overdetermine the flow. This can happen, for example, if velocities are specified for both edges. Since the flow must fulfill the continuity equation, the velocity at the outlet results from the through-flow of the component. Thus, if the velocity is specified "incorrectly",

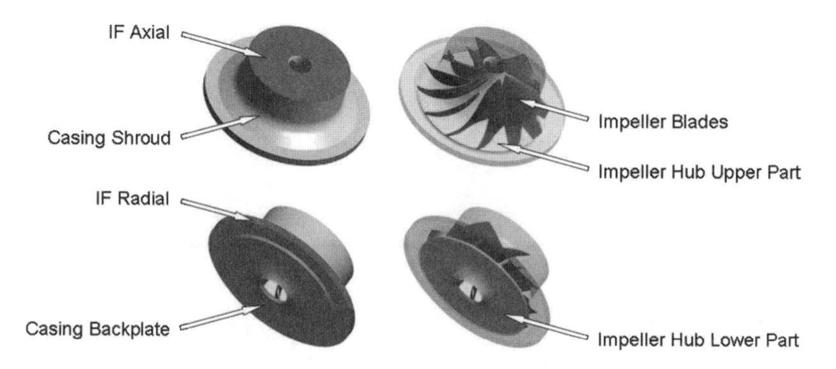

Fig. 15.6 Boundary and edge surfaces in the impeller area

i.e. unsuitable for the conservation of mass, the numerical solution cannot fulfill the continuity equation and does not converge.

In the interface area between the inlet housing and the impeller, the impeller hub protrudes into the inlet housing. This surface is also specified as a wall. In the absolute system, however, the rotational speed of the impeller must be applied to this wall.

For the subdivision of and the definition of the boundary conditions of the impeller area (see Fig. 15.6), a rotational speed contrary to the impeller rotation must be applied to the walls that are fixed in the absolute system (shroud, backplate) in order to compensate for the rotation of the coordinate system in the relative system.

The interfaces between the absolute and the relative system are designated in the figures as “IF axial” and “IF radial”. On these surfaces, the values are transferred between the two coordinate systems via the sliding interfaces.

15.3 Postprocessing: Analysis and Visualization of Results

How the results of a simulation of turbo-machine behavior are analyzed depends, of course, on the problem to be solved. The entire array of available possibilities for visualization offered in commercial CFD programs can be drawn upon for turbomachines, as well. If, for example, the streamlines are shown in the compressor geometry discussed in the previous section, the image shown in Fig. 15.7a is the result in the case of an operating point in the middle of the compressor map. If the operating point is near the pump boundary, one receives the visualization in Fig. 15.7b, in which the resulting reverse flow in the region of the inlet housing can clearly be recognized in the feeding section.

For a detailed flow analysis, one must introduce cutting planes or curved surfaces into the CFD model and perform the visualization on these surfaces. One distinguishes in this context between planes which either simply intersect the model

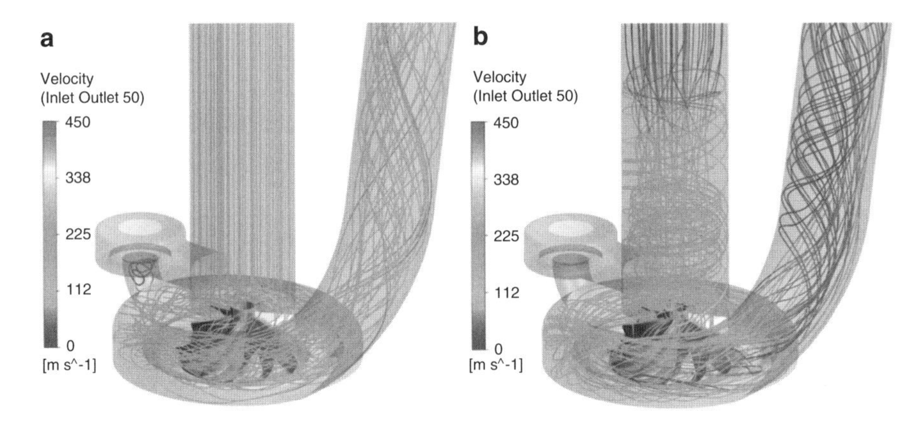

Fig. 15.7 Streamline plot for the visualization of compressor behavior. (**a**) operating point at center of compressor map; (**b**) operating point near pump boundary

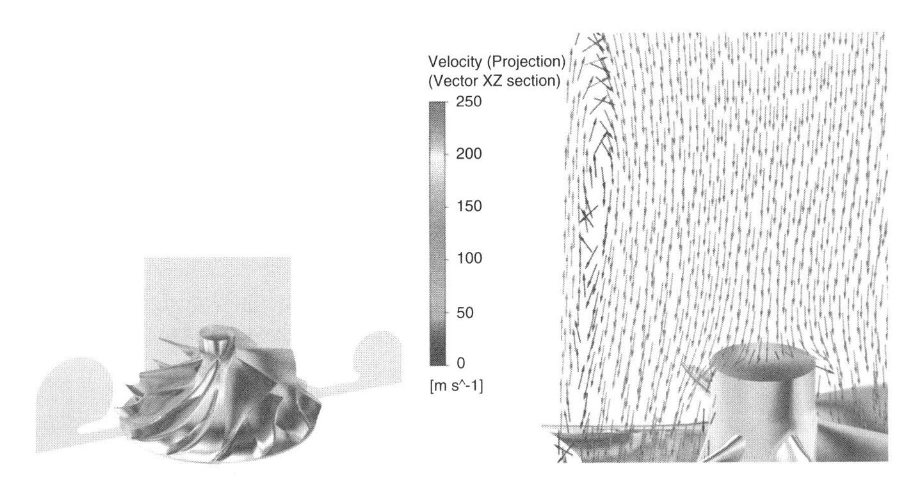

Fig. 15.8 Vector plot at cutting planes

at constant coordinate values or are positioned at an angle in space and surfaces which are curved and, for example, are oriented on the geometry being investigated. In Fig. 15.8, for example, the reverse flow region which causes the streamline pattern shown in Fig. 15.7b is visualized on a geometric x-z plane. As a result of the complex curved forms of the geometries observed, however, visualizations on curved surfaces have proven useful. Visualizations at fixed positions are subdivided into those in the span direction, i.e. between the hub and the shroud, those in the flow direction, and those in the circumferential direction.

Figure 15.9, for example, shows the separation vortex resulting in the region of an impeller duct on two different curved surfaces, one for a constant coordinate in the span direction (Fig. 15.9a), the other for a constant coordinate value in the flow direction (Fig. 15.9b).

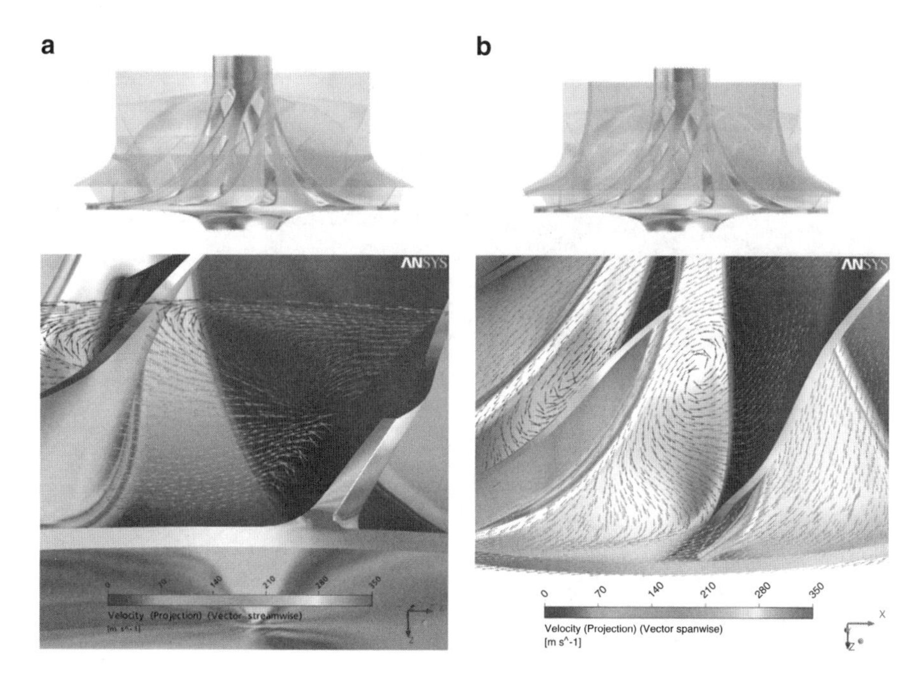

Fig. 15.9 Vector plot on curved surfaces. (**a**) constant in the span direction; (**b**) constant in the flow direction

For a simpler overview, the surfaces can also usually be unwound into a plane presentation. These are also referred to as blade-to-blade views (see Fig. 15.10).

15.4 Examples of Application

15.4.1 Analysis of Compressor Behavior

It is of course extremely important for turbo-machine manufacturers to ensure that a robust flow-machine design was created prior to its implementation in hardware. By means of 3D CFD simulation, one can pre-calculate entire compressor maps in the same way they are mapped out on turbocharger test benches. The areas of the compressor map which are of interest here are schematically illustrated in Fig. 15.11. Firstly, it is interesting to observe the dependence of the compressor pressure ratio for a given mass flow and constant compressor rotational speed – i.e. the distribution of the so-called speed lines. Also, one can derive useful initial information on the stable and unstable regions of the compressor map, i.e. so as to specify the approximate location of the surge line and predict the choke line. The reliable prediction of efficiency levels and of the position of the optimum point of efficiency also helps to best possibly ensure an optimized design.

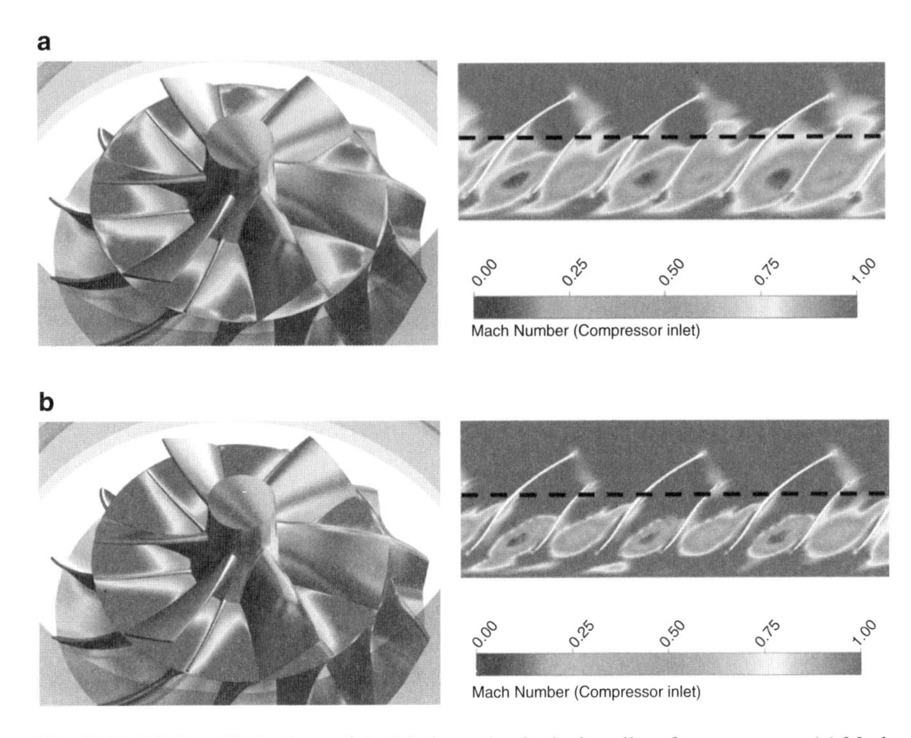

Fig. 15.10 Blade-to-blade views of the Mach number in the impeller of a compressor. (**a**) Mach number at center of compressor map. (**b**) Mach number near choke line

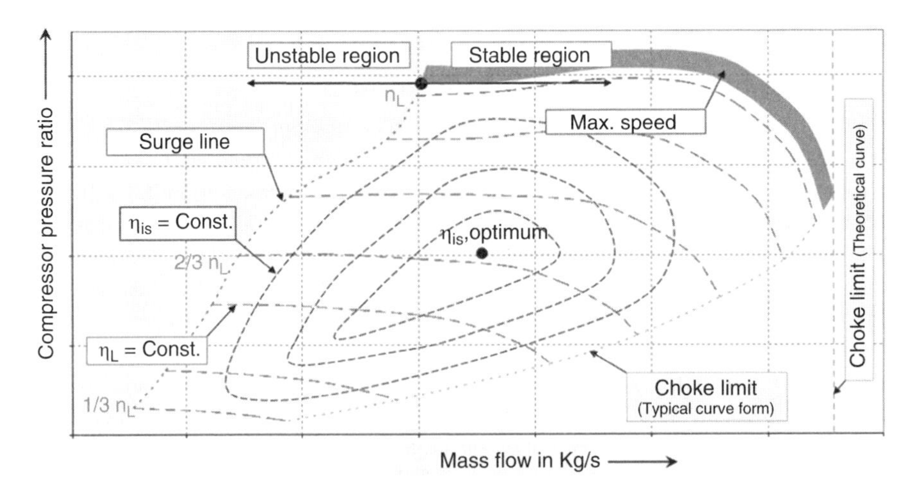

Fig. 15.11 Schematic diagram of a compressor map

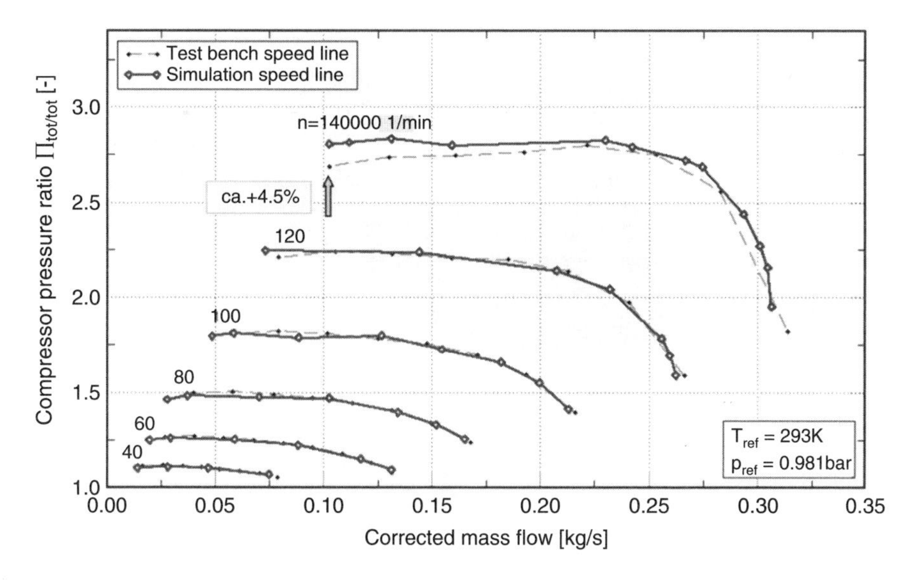

Fig. 15.12 Comparison of calculated and measured compressor map

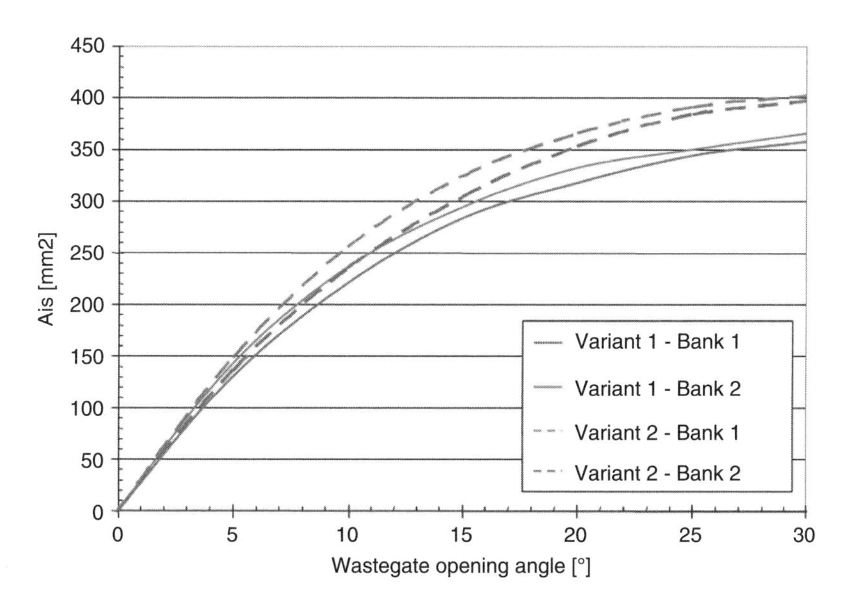

Fig. 15.13 Wastegate mass flows as a function of the throttle angle

For engine manufacturers, 3D simulations offer the possibility of creating maps or obtaining them from a charger manufacturer before the turbocharger is available for a real measurement and thus to execute 1D simulations of charge exchange so as to test the engine behavior in advance by way of calculation.

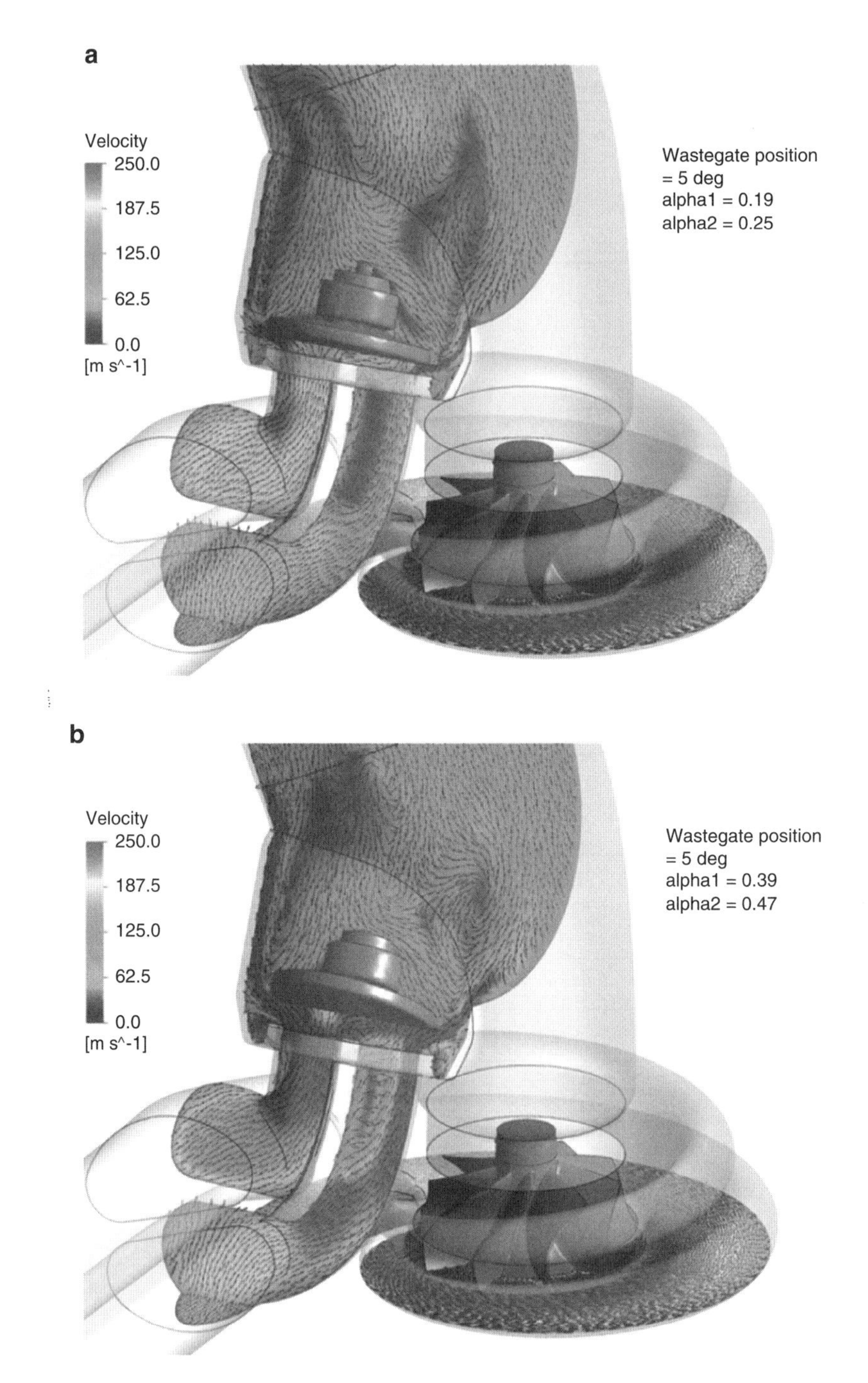

Fig. 15.14 Flow around the wastegate flap of a twin scroll turbine. (**a**) aperture angle 15°. (**b**) aperture angle 5°

The achievable quality which can be expected from the results of such a test is shown in Fig. 15.12. On can see that the relevant speed lines are excellently reproduced by the calculation. It is only the highest measured speed line that exhibits a maximum deviation of 4.5%. The location of the choke line also agrees very well with the measured values. The surge line can be estimated using the left endpoints of the speed lines. The calculations were discontinued for this map as soon as the static solutions began to display periodic convergence behavior. The pumping of a compressor is a strongly unsteady phenomenon which results from the interaction of the compressor and the air guide. This behavior cannot be simulated by means of steady state calculations of a compressor. However, since transient analyses are much too time-consuming for the fast assessments desired, this pragmatic approach is appropriate in this case.

15.4.2 Examination of Turbine Variants

The through-flow of the wastegate channels of a turbocharger turbine plays an important role with respect to an engine's charge cycle. The layout should be designed in such a way that the duct to creates the smallest possible loss of pressure and that, for twin scroll turbines, the two wastegate ports are subject to a through-flow which is as uniform as possible.

Using such a simulation, one can evaluate the wastegate mass flow as a function of the wastegate opening angle and the mass flow distribution to the two individual ports (shown in Fig. 15.13 as isentropic flow cross-section) or, as shown in Fig. 15.14, the flow field in the form of a velocity field in a vector plot which, for example, results from the flow around the wastegate flap itself.

Appendix
3D-CFD Simulation of IC-Engine Flow, Mixture Formation and Combustion with AVL FIRE®

Reinhard Tatschl

A.1 Introduction

Internal-combustion engine performance and emission characteristics are strongly determined by the complex interacting processes of in-cylinder flow, fuel spray injection and combustion. Due to the strong spatial and temporal variability of the in-cylinder temperature and mixture composition fields over a single engine-cycle, conventional cylinder-pressure analysis and tail-pipe emission measurements only provide limited information on the cause-and-effect relation between engine geometry details and injection/combustion system parameters on one side and the overall engine performance and emission characteristics on the other.

Specifically with respect to mixture formation quality and the related combustion efficiency and hence pollutant formation characteristics, a better understanding of the interacting processes of turbulent in-cylinder flow, fuel injection and combustion is of significant importance in view of the increasing legislative and environmental demands. Especially in the context of novel combustion concepts and the variety of alternative fuels to come in the near future, 3D computational fluid dynamics simulation of in-cylinder spray formation/propagation, combustion and pollutant formation processes is gaining increasing importance during engine and combustion system development.

In recent years CFD has been successfully established for the calculation of fluid flow, mixture formation and combustion in internal combustion engines as a complementary tool to in-cylinder pressure analysis and optical mixture formation and combustion diagnostics (see Sects 3.2 and 3.3, respectively). The accuracy of the calculation results and hence the potential contribution of the CFD simulation to major design decisions within the engine development process strongly depends on the achievable project turnaround times and the reliability of the models adopted for the treatment of the individual in-cylinder physical and chemical processes, such as cavitating injector flow, liquid fuel spray propagation, evaporation and mixing with the in-cylinder charge, auto-ignition, turbulent combustion and pollutants formation. As the result of intense world-wide research and development efforts over

G.P. Merker et al. (eds.), *Combustion Engines Development*,
DOI 10.1007/978-3-642-14094-5,

the last decades, a variety of models exhibiting different levels of complexity and sophistication is available today. For an overview of the theoretical background of the different models and modeling approaches for IC-engine CFD simulation the reader is referred to Chaps. 13 and 14.

The present chapter is aimed at providing an overview of the CFD workflow adopted for simulation of the in-cylinder physical and chemical processes governing engine performance and emission characteristics. The workflow description including the CFD solver and modeling details as well as all calculation results shown in the present chapter are based on the commercial CFD code AVL FIRE®.

First, a basic introduction of the overall CFD methodology, comprising of the pre-/post-processing tools and the flow solver is provided. Then the models adopted for simulation of injector flow, liquid fuel spray injection, ignition and combustion as well as pollutant formation are described. For illustration of the model applicability to different engine engineering tasks, calculation results are shown for compression-ignition as well as spark-ignition engines operated under different engine speed and load conditions. Where appropriate, the calculation results are compared with experimental data in order to demonstrate the level of accuracy that can be obtained with the models available today.

A.2 CFD Simulation Methodology

The CFD simulation of internal combustion engine processes can be divided into three major steps, i.e. the generation of the computational meshes required to cover the fluid domain over the crank-angle interval of interest, the specification of the initial and boundary conditions, of the flow solver settings and of the physical and chemical models adopted to simulate the governing in-cylinder processes and, finally, the post-processing and interpretation of the simulation results. For meshing of the moving IC-engine geometries AVL FIRE® offers appropriate tools and methods applicable to all types of spark-ignition engines (SI-engines) and compression-ignition engines (CI-engines), the related solver technology capable of handling turbulent multi-phase reactive flows on moving (contracting/expanding) computational meshes and the related post-processing capabilities required for solution monitoring and results analysis. All tools required to cover the above tasks to successfully perform an IC-engine CFD calculation are accessible via a fully interactive graphical user interface.

A.2.1 Pre-/Postprocessing

A.2.1.1 Mesh Generation

The pre-processor of AVL FIRE® offers different meshing tools for the generation of boundary fitted computational grids to represent the complex IC-engine geometries including moving parts, such as e.g. piston and intake/exhaust valves. The flexible automated meshing environment applies advanced hybrid grid generation techniques which enable fast, reliable and accurate meshing of the complex 3D domains. In order to generate the set of meshes required to cover the complete IC-engine cycle, a baseline computational mesh, usually generated for the piston and valve positions at gas-exchange top dead centre (TDC), is automatically adapted to reflect the geometrical conditions for the crank-angle interval of interest.

In order to meet the demands concerning accuracy and hence reliability of the numerical results, the quality of the computational grids has to fulfill certain requirements. Besides a reasonable overall spatial resolution for representation of the in-cylinder flow domain, including proper resolution of the near wall layer, an accurate modeling of valve and piston movement is required for modeling the engine configuration in the CFD simulation. Different approaches, varying with respect to their degree of automation, are available within the AVL FIRE® pre-processor.

The possibility to start from user-defined geometry sub-domains enables the efficient meshing of complex geometries by simultaneously maintaining a high grid quality. Topology, cell orientation and mesh resolution in critical areas such as small gaps can be easily controlled by the user. Adopting arbitrary and/or conform interfaces, the relevant geometry can easily be split into different domains, such as intake port, valve seat and combustion chamber, avoiding the mutual interaction of different parts with respect to topology and distortion due to mesh movement (Fig. A.1). Hence, the complexity of the meshing process is reduced to meshing of simple parts which can easily be exchanged for performing parametric studies related to geometrical and valve lift and timing variations. For meshing of the individual components specifically tailored methods are available. For the valve gap area, for example, a polar block-structured grid can be used which is characterized by best grid quality and mesh movement characteristics.

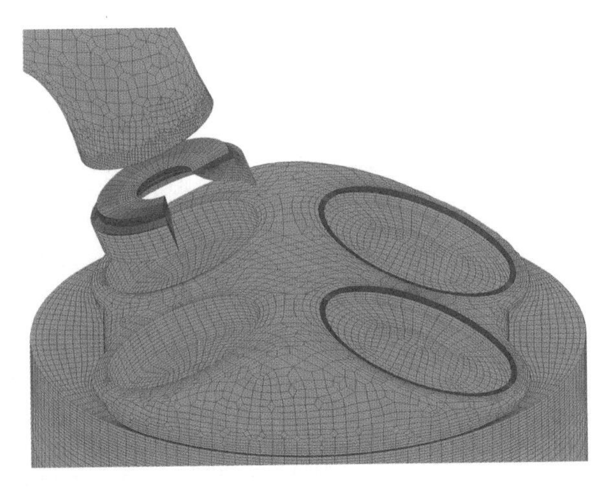

Fig. A.1 Computational mesh of an SI-engine configuration composed of a set of sub-domain meshes for the intake ports, valve seats and combustion chamber

In order to further reduce the complexity of the mesh generation process and hence to maximize ease-of-use and to minimize meshing times, AVL FIRE® also supports a fully automated grid generation approach. The automatically generated computational grids consist of mainly hexahedral elements with a few tetrahedrons, prisms and pyramids in the vicinity of local refinement areas. The boundary layers consist of hexahedral elements and prisms only. Sophisticated grid refinement techniques and algorithms for automatic resolution of geometrical details provide maximum flexibility for all relevant grid generation tasks (Fig. A.2).

Mesh movement in AVL FIRE® is based upon interpolation between two grids of identical topology and number of elements set up at two different positions of the moving elements, typically at different piston/valve positions. Thus it is possible to virtually generate grids for any position of the moving geometry between the extreme positions represented by the original two grids, enabling the expansion and compression of the computational grid according to different predefined or user-defined displacement functions. At some point during expansion/contraction of the mesh, the cell aspect ratio or other mesh quality criteria may exceed permitted values. For that reason so-called rezones are performed, a procedure that ensures an optimum cell density and quality over the entire crank-angle interval under consideration. The rezoning method can be applied to arbitrarily complex geometries with an arbitrary number of moving boundaries. Since the computational grid is generated for one representative surface model, artificially assembled valve and piston positions are allowed. Extreme positions such as top-dead centre, bottom dead-centre and maximum valve lift are detected automatically and respective grids are generated without requiring additional user input.

Due to the often symmetric arrangement of IC-engine injector/piston bowl configurations, as it is typically the case for modern diesel engines, the analysis and optimization of the spray injection and combustion processes is usually done by simulating an engine segment model only. The flow domain under investigation is

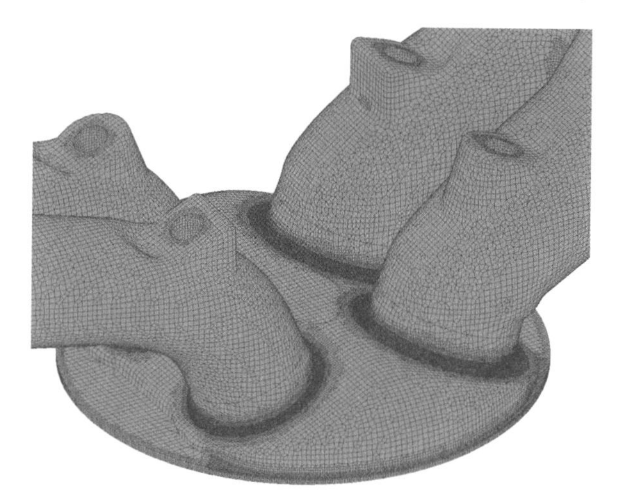

Fig. A.2 Automatically generated computational mesh of an SI-engine intake port/combustion chamber configuration

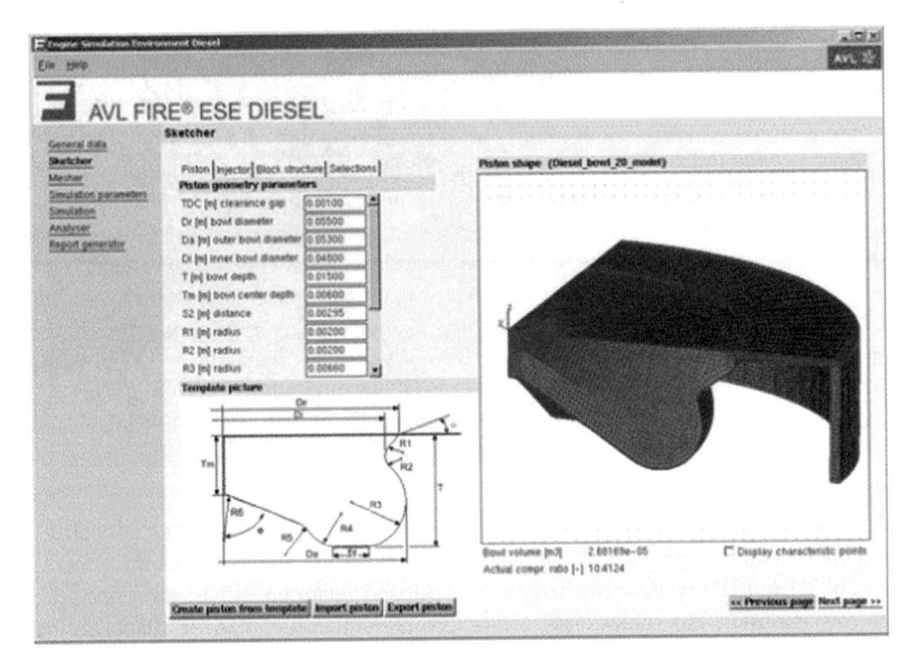

Fig. A.3 AVL FIRE® ESE Diesel – diesel engine meshing/simulation environment

then limited to the combustion chamber part around one single fuel spray applying cyclically symmetric boundary conditions for the crank-angle interval from inlet valve closure to exhaust valve opening. In this case the computational grid generation can be based on parameterized 2D curves, describing the combustion chamber and optionally the injector geometry. Based on this input the generation of the grids required to cover the simulation period between intake valve closing and exhaust valve opening is performed automatically within AVL FIRE® Engine Simulation Environment Diesel (ESE Diesel). The meshing process also takes into account the generation of a piston position independent mesh topology for the spray domain, a defined number and thickness of boundary layers and matching of the compression ratio without requiring user interaction (Fig. A.3).

The AVL FIRE® ESE Diesel user interface moreover offers the full functionality to set up and run the entire diesel engine spray injection and combustion calculation and to perform all relevant application specific post-processing activities.

A.2.1.2 Results Post-Processing

The AVL FIRE® post-processor offers numerous functions enabling detailed online job monitoring, results visualization and analysis. During the simulation the post-processor allows monitoring of the convergence behavior and the results

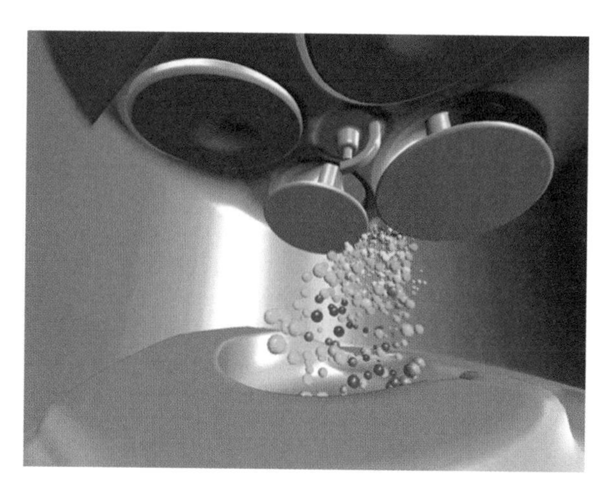

Fig. A.4 Visualization of DI gasoline engine spray simulation results

development by visualizing the residuals of the differential equations as well as selected results as graphs in 2D diagrams. The post-processor enables the generation of 2D and 3D cuts for visualization of scalar and vector quantities, spray droplets, etc. as iso-lines, iso-values, iso-surfaces; vector quantities can be displayed as vector plots (Fig. A.4).

In addition the display of streamlines and particle tracing is possible. In order to analyze results the post-processor offers the possibility of operations on 3D result data, macro and formula capabilities and import of selected test-bed data. For the purpose of result presentation the post-processor also offers shading, blending and texturing options, manipulation of object lighting, manipulation of color assignments, inserting of text and pre-defined text variables, generation of colored plots in various formats and generation of animations.

A.2.2 Flow Solver and Solution Algortihm

The CFD code AVL FIRE® solves the general conservation equations of mass, momentum and enthalpy plus additional transport equations for turbulence related quantities and for conservation of chemical species and – depending on the physical and chemical sub-models employed – additional scalar quantities, such as e.g. mixture fraction, reaction progress variable, flame surface density, etc.

The solution method is based on a fully conservative finite volume approach. All dependent variables for momentum, pressure, density, turbulent kinetic energy, dissipation rate, and scalar quantities, such as e.g. chemical species, are evaluated at the centers of the computational cells (Demirdzic et al. 1993). Major attention is given to the numerical accuracy of the algorithm due to the use of unstructured grids. A second-order midpoint rule is used for integral approximation and a second order linear approximation for any value at the cell-face. A diffusion term is incorporated into the surface integral source after employment of the spatial interpolation practice. For discretization of the convection term a variety of differencing schemes is available.

In order to offer full flexibility in terms of the structure and topology of the employed computational meshes, AVL FIRE® allows for each computational cell to consist of an arbitrary number of cell faces. Connectivity and interpolation practices for gradients and cell-face values are introduced to accommodate such 'polyhedral' calculation volumes.

The rate of change is discretized by using implicit schemes, namely an Euler implicit scheme and a three time level implicit scheme of second order accuracy. The overall solution procedure is iterative and is based on the Semi-Implicit Method for Pressure-Linked Equations algorithm (SIMPLE) applicable to turbulent flows at all speeds even including super-sonic flows.

For solving the large sets of linear equation systems evolving from the discretization of the governing equations, AVL FIRE® adopts efficient preconditioned conjugate gradient methods. The symmetric gradient method is employed for solving equations with symmetric matrix and the bi-conjugate method is adopted for equations with asymmetric matrix. Both methods (solvers) are used with either Incomplete Cholesky or Jacoby preconditioning technique. As a very efficient method for solving large sparse linear systems, AVL FIRE® also offers Algebraic Multigrid Methods (AMG).

Various initial and boundary conditions are available to match the setup of any simulation to the real flow problem. Focusing on applications related to internal combustion engine related applications the solver is capable of handling computational models including moving boundaries. Meeting the requirements of different fields of application, AVL FIRE® simulations may be set up in steady or transient (time stepping and crank angle) mode. For execution on multi-processor hardware, AVL FIRE® employs a domain decomposition parallelization approach, enabling efficient solution of flow problems comprising of a large number of computational cells (Fig. A.5).

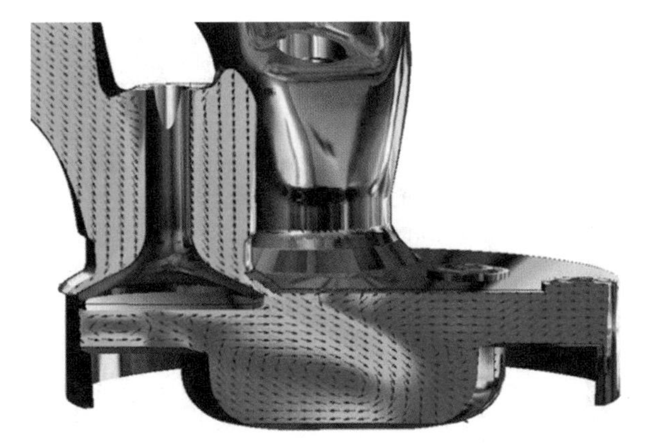

Fig. A.5 Flow velocities and turbulence intensity distribution during the intake stroke of a DI diesel engine

A.3 Turbulent Flow and Heat Transfer

Most IC-engine related fluid flow problems are turbulent flows. Hence, as a basis for precisely simulating the real flows, it is of utmost importance to be able to accurately model the phenomenon of turbulence. This is in particular necessary, since turbulence not only determines the details of the fluid flow itself, but also strongly influences the physical and chemical processes taking place during mixture formation and combustion. For example, in IC-engines the turbulent kinetic energy is a major influencing factor on the propagation and evaporation of liquid fuel spray droplets and the subsequent combustion of the air/fuel mixture. In addition to the well known, standard turbulence models, such as e.g. k-ε, Spalart-Allmaras, Reynolds Stress, etc., AVL FIRE® offers the k-ζ-f turbulence model, recently developed and validated for IC-engine related flow, heat transfer and combustion processes (Basara 2006).

For IC-engine flows the k-ζ-f model leads to more accurate results than the much simpler two-equation eddy viscosity models of the k-ε type by simultaneously exhibiting a high degree of numerical robustness. In combination with a hybrid wall treatment, as proposed by Popovac and Hanjalic (2005), combining the integration up to the wall with standard wall functions, the k-ζ-f turbulence model is universally applicable to computational meshes and flow situations of any reasonable y+ value near the wall.

Today, the k-ζ-f model is used as default model for turbulence and turbulent wall heat transfer modeling in most of the common IC-engine related applications performed with AVL FIRE®. Of particular advantage is its robustness to be used

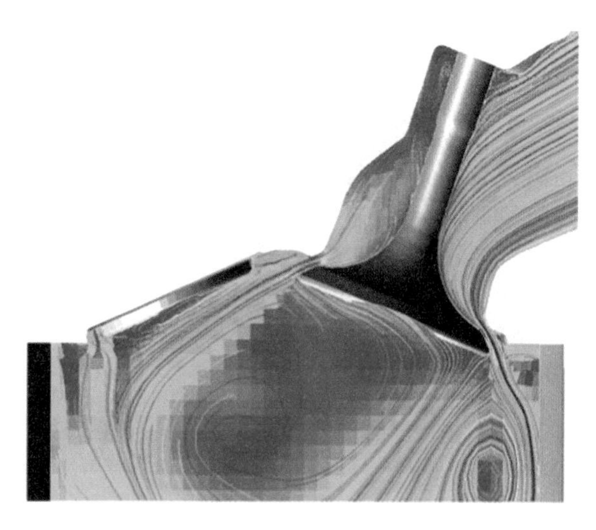

Fig. A.6 Turbulence intensity distribution during SI-engine intake flow adopting the k-ζ-f turbulence model (Tatschl et al. 2006)

for computations involving grids with moving boundaries and highly compressed flows as it is the case in IC-engines. In conjunction with the hybrid wall treatment the k-ζ-f model guarantees the optimum solution for any computational mesh regarding robustness, computing time and accuracy (Fig. A.6).

A.4 Multiphase Injector Flow

The major challenge in the simulation of injector flows is related to the presence of multiphase processes which in AVL FIRE® are modeled based upon the multi-fluid approach and adopting sub-models for the treatment of various phenomena, such as e.g. cavitation, flash boiling, etc. In these simulations the injection system characteristics is accounted for via offline or direct coupling of the CFD methodology to 1D hydraulic simulations (Chiavola and Palmieri 2006; Caika et al. 2009).

The multiphase treatment in AVL FIRE® is based on the Eulerian multi-fluid method which is the most general approach for simulating multiphase flows. The individual fluids are treated as continuous phases with conservation laws applied for each fluid. An ensemble averaging technique is adopted to remove the microscopic interfaces. This results in macroscopic conservation equations which are analogous to their single-phase counterparts but differ in that the new variable volume fraction and additional exchange terms between the phases are introduced (Drew and Passman 1998). In its present implementation in AVL FIRE® the adopted multiphase framework is capable of handling an arbitrary number of phases.

Due to the wide range of applicability of the multi-fluid method, the phase-exchange terms are flexibly modeled according to the different types of simulation problems, such as e.g. cavitation, flash boiling, etc. Mass exchange due to cavitation is approximated using the Rayleigh equation assuming a uniform pressure field for all phases and, depending on the selected cavitation sub-model, a mono-disperse or poly-disperse fuel vapor bubble size distribution (Wang et al. 2005). Optionally, the standard k-ε or k-ζ-f turbulence model extended for multi-phase flows is applied, with separate turbulence conservation equations solved for each phase. Adopting the multi-fluid approach it is possible to numerically simulate flows in diesel and gasoline injectors taking into account cavitation phenomena in both the nozzle hole and needle seat areas (Fig. A.7).

Based on the results provided by the simulation the locations of cavitation onset can be identified and the shape and extension of cavitation-induced fuel vapor containing regions can be investigated (see also Sect. 12.5.3). Moreover, the impact of different nozzle geometrical details and nozzle configurations, such as e.g. nozzle type (VCO, SAC, etc.) or variations of nozzle hole diameter, diameter/length ratio as well as injection strategies, can be easily analyzed (Chiatti et al. 2007).

Fig. A.7 Calculated cavitation characteristics in a diesel injector nozzle

The computational results also provide detailed information on the velocity, turbulence intensity and liquid to fuel vapor volume fraction at the injector nozzle exit. This information can be used for either a direct assessment of the break-up characteristics of the fuel spray exiting the nozzle (Tatschl et al. 2000a), or as input condition for a subsequent in-cylinder spray simulation for diesel (Chiavola and Palmieri 2006), and gasoline injector nozzles (Greif et al. 2003).

Moreover, the prediction of cavitation phenomena can be used in the context of the analysis and optimization of injector nozzles and injection system components with respect to the appearance of cavitation-induced erosion (Greif et al. 2005), and for the numerical study of thermal-fluid interaction in diesel injectors (Leuthel et al. 2008).

A.5 Fuel Spray and Wallfilm

In IC-engines adopting either direct or indirect fuel injection technology, the accuracy of the calculation results with respect to the fuel/air mixture distribution evolution and hence the accuracy of the subsequent combustion simulation strongly depends on the predictive capabilities of the models adopted for spray propagation, wallfilm formation and wallfilm transport (Bianchi et al. 2006, 2007; Musu et al. 2006). AVL FIRE® offers a comprehensive model suite for treatment of liquid fuel injection, spray atomization, droplet secondary breakup, evaporation, droplet/wall interaction, etc. The present section provides an overview of the models adopted in AVL FIRE® for simulation of diesel and gasoline spray propagation processes.

A.5.1 Discrete Droplet Spray Method

The spray model most commonly adopted in AVL FIRE® for IC-engine spray and mixture formation simulations is based on the Lagrangian *D*iscrete *D*roplet *M*ethod (DDM) (Dukowicz 1980). While the continuous gaseous phase is described by the standard Eulerian conservation equations, the transport of the dispersed phase is calculated by tracking the trajectories of representative droplet parcels (Fig. A.8). A parcel consists of a number of droplets and it is assumed that all the droplets within one parcel have the same physical properties and behave equally when they move, break up, hit a wall or evaporate (Fig. A.9). The calculation of the parcel movement is done with a sub-cycling procedure between the gas phase time steps taking into account the forces exerted on the parcels by the gas phase as well as the related heat and mass transfer. The coupling between the liquid and the gaseous phases is achieved by source term exchange for mass, momentum, energy and turbulence.

Droplet parcels are introduced in the flow domain with initial conditions of position, size, velocity and temperature. AVL FIRE® supports the introduction of droplets emerging from a nozzle as a spray at user defined conditions or alternatively at conditions provided as input from a preceding nozzle flow simulation. The second approach enables to fully take into account nozzle flow effects onto the primary atomization behavior of the liquid fuel exiting from the injector nozzle and hence onto the subsequent spray shape and droplet size distribution evolution characteristics (Tatschl et al. 2000b; Chiavola and Palmieri 2006).

For modeling the primary atomization process AVL FIRE® offers two basic options within the framework of the DDM. First, the introduction of a series of large blobs of the size of the nozzle hole diameter which represent the coherent liquid jet,

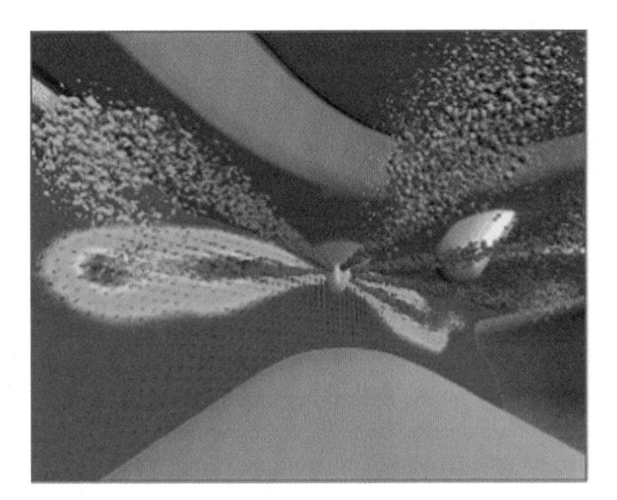

Fig. A.8 Spray droplet distribution and vapor concentration pattern during liquid fuel spray injection in a DI diesel combustion chamber

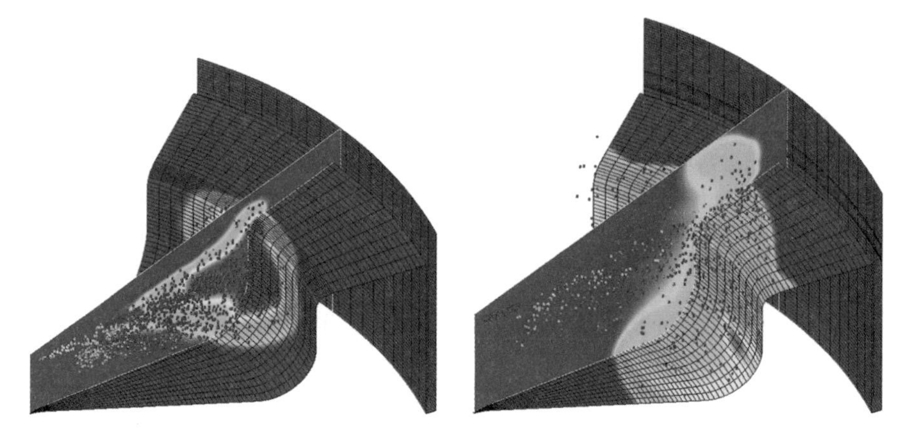

Fig. A.9 Fuel spray propagation and near wall fuel to air equivalence ratio distribution during DI diesel engine cold start at 20° crank-angle (*left*) and 30° crank-angle (*right*) ATDC

with their diameter being subsequently reduced according to the mass detachment rate calculated from the primary break-up model (Fink et al. 2009). The second approach is based on the calculation of the erosion of the liquid fuel core within a separate sub-model and release of the droplet parcels downstream of the nozzle on the core surface with the initial parcels already being considerably smaller than the nozzle diameter (von Berg et al. 2005).

Both models use detailed information from previous simulations of the cavitating nozzle flow to calculate the primary break-up rate and resulting ligament or droplet size. This allows a unique link to the injector flow conditions, which has proven to be decisive for accurate spray initialization and the consecutive mixture formation simulation in diesel engines (Masuda et al. 2005; Nagaoka et al. 2008). The interface to the injector flow is provided by a nozzle flow data file containing geometrical as well as detailed time resolved flow data in the nozzle exit cross section.

Secondary break-up is a consecutive process within the spray cone acting on the individual droplets until a stable droplet size is reached. Phenomena which are covered by the secondary break-up models in AVL FIRE® are break-up through deformation, bag break-up, boundary layer stripping, capillary wave stripping and catastrophic break-up by Raleigh Taylor instability (von Künsberg Sarre and Tatschl 1998).

The turbulent dispersion model in AVL FIRE® considers the interaction of the individual droplets with the local turbulent eddies of the flow field. Each interaction deflects the droplet according to the instantaneous velocity of the turbulent eddy and the particle inertia. Collision and coalescence models describe the probability and the effect of particles hitting each other based on a statistically sampled probability density function governing frequency and nature of collision events. A recent development also handles grazing collisions and splashing effects (Stralin 2006). Distortion and drag models take into account the deformation of the droplets due to aerodynamic conditions and the influence on the droplet drag coefficient.

For proper treatment of evaporation processes AVL FIRE® offers different approaches to model the droplet heat-up and liquid–gas mass transfer. In their original form they assume spherical droplets, quasistationary conditions at the droplet surface, homogeneous droplet temperature and internal circulation. They differ with regard to the Lewis number limitation and rule of reference conditions. Heat and mass transfer coefficients are set up according to physics for each model. Additionally, correction functions are available to overcome the above simplifications for transient heating, drop deformation or internal circulation. In addition, a multi-component evaporation model is available that enables handling of the evaporation process of droplets being composed by an arbitrary number of individual components (Brenn et al. 2007). In practice, real IC-engine relevant fuels are usually assumed to be represented by about four to six species in order to approximate their multi-component evaporation characteristics.

Wall interaction models describe the behavior of a droplet when hitting a wall, depending on parameters like droplet velocity, diameter, physical properties, wall surface roughness and temperature. At very low impact velocities the droplet sticks to the wall or is transferred to the wallfilm (Fig. A.10). With increasing impact velocity a vapor or gas boundary layer is trapped underneath the droplet and causes the liquid to rebound when approaching a wall. During the rebound parts of the kinetic energy are dissipated and the outgoing droplet normal velocity is usually lower than the incoming velocity. A further increase of the velocity leads either to the spreading or the splashing regime. Droplet splashing models have been recently extended by taking into account the wall temperature for determination of the applicable splashing regime (Birkhold et al. 2007).

Fig. A.10 Spray droplet distribution and wallfilm formation in a direct injected SI-engine

A.5.2 Eulerian Spray Model

The Discrete Droplet Model described above is especially suitable for dilute sprays, but has drawbacks with respect to the modeling of dense sprays close to the nozzle exit (where gas and droplet phases are intimately coupled) and with statistical convergence in the context of local grid refinement. An alternative approach to the DDM is based upon adopting an Eulerian–Eulerian method treating different size classes of the spray droplets as separate, interpenetrating phases. The respective model available in AVL FIRE® is based on the Eulerian multiphase approach that has been derived from ensemble averaging of the governing conservation equations (Alajbegovic et al. 1999). For each phase mass, momentum and energy conservation equations are solved as well as corresponding equations for turbulent kinetic energy and turbulent energy dissipation. Within each computational cell the droplet phases are characterized by a certain volume fraction and diameter as well as optionally by a droplet number density (von Berg et al. 2001).

All exchange processes related to droplet size or specific surface of the droplet phases are modeled accordingly, i.e. the special physics of the relevant processes are treated by interfacial exchange terms between the phases. Specific models are adopted for treatment of evaporation and break-up processes, as there are primary break-up considering the atomization of the bulk liquid phase into the droplet phases and secondary break-up considering the break-up of large droplets into smaller droplets. Drag forces and turbulent dispersion forces treat the interactions of the momentum between the gaseous and the liquid droplet phase, the evaporation model contributes to the interfacial mass and heat exchange (von Berg et al. 2003; Vujanović et al. 2008).

The spray inlet is fully resolved by the Eulerian spray grid and hence it is expedient to apply the inlet boundary conditions from a separate nozzle flow calculation, which stores the flow field data of the nozzle orifice at a separate data file. This nozzle data file contains the geometry of the nozzle outlet and the flow data at certain timesteps, which are used by the Eulerian spray model as boundary condition for the bulk liquid and, in case of cavitation, the vapor phase at the spray inlet. The nozzle flow turbulence is used in the primary break-up model for calculation of the break-up rate and the target diameter of the droplets. Coupling of the Eulerian dense spray model with the in-cylinder dispersed spray/combustion treatment is realized via an embedded grid approach adopting the AVL Code Coupling Interface (ACCI) (Edelbauer et al. 2006; Suzzi et al. 2007) (Fig. A.11).

A.5.3 Wallfilm Transport

The wallfilm model available in AVL FIRE® enables the simulation of the film formation, the film transport and its evaporation. It accounts for the wallfilm interaction with the gas flow and the liquid fuel spray and is considering the effects of

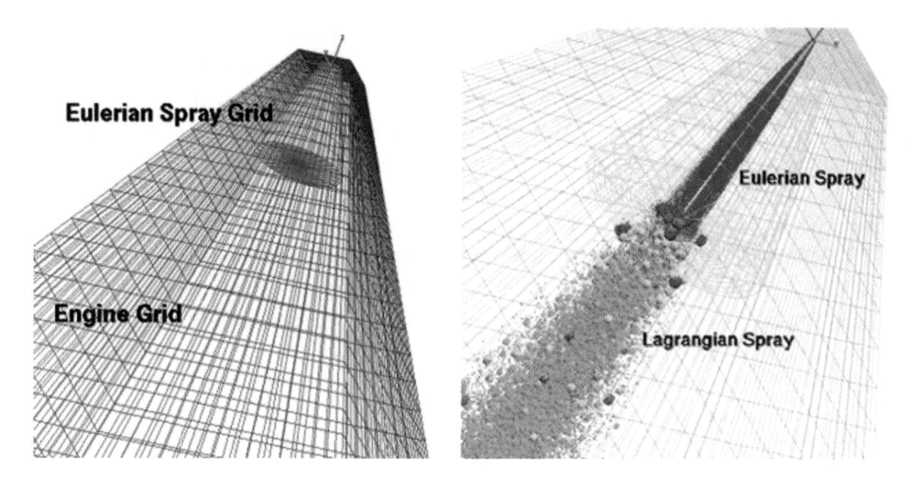

Fig. A.11 Schematics of the embedded grid approach (*left*) and representative result of the ACCI coupled Eulerian/Lagrangian spray treatment (*right*) (Edelbauer et al. 2006)

surface roughness on the film transport. The fundamental assumptions in the adopted modeling approach are that the gas and wall film flow are treated as separate single phases. The implementation of the wallfilm model is based on a 2D finite-volume method at the wall boundaries of the gas flow geometry (Stanton and Rutland 1998).

The coupling of the gas-phase and the liquid wallfilm is achieved via mass and momentum exchange between the film and the gas phase at the film surface adopting semi-empirical relations. The film surface is assumed to be parallel to the solid wall and the film thickness is assumed to be small in relation to the dimensions of the gas flow region. The wavy surface of the film is modeled by superimposing a film roughness over the mean film thickness. Due to the thin film assumption, wall friction and interfacial shear stress dominate the film behavior as compared to inertial forces and lateral shear. The relevant physical effects which influence the film formation and film flow, such as e.g. film entrainment (film rupture or shearing off at the surface due to high shear forces), interaction with impinging droplet spray, heat transfer between film and solid wall and gas phase, film evaporation (including evaporation of multi-component fuel films) and inter-phase shear force and gravitation are modeled via appropriate sub-models (Ishii and Mishima 1989; Birkhold et al. 2006, 2007).

A.6 Combustion

Modeling the combustion chemistry and the turbulence/chemistry interaction during hydrocarbon oxidation is the major challenge in simulating turbulent reacting flows. The chemical kinetic reactions of typical hydrocarbon fuels involve hundreds of intermediate species and their reaction paths usually comprise of several hundred up to several thousand of reaction steps. Neither are the reaction details and the related rate coefficients of relevant hydrocarbon fuels known in sufficient detail nor would it be possible to account for the entire set of reactions within an engineering environment due to excessively high computational demands.

The intimate coupling between the chemistry and the flow field quantities results from the high non-linearity of the chemical reaction rates under the presence of local stochastic fluctuations of the flow and composition fields. The common practice adopts an averaging procedure replacing the instantaneous flow field quantities by its mean and fluctuating components. Mathematically this results in the appearance of terms in the conservation equations which contain statistical correlations of fluctuating components that have to be expressed based upon known mean flow quantities.

A large variety of models of various levels of complexity and sophistication have been proposed in the last decades in order to appropriately model the complex hydrocarbon auto-ignition and combustion chemistry and the mean rate of reaction in the relevant conservation equations. Depending on the specific IC-engine combustion system and the underlying combustion regimes, tailored models are required in order to cover the relevant physical and chemical processes to be considered.

In the following, an overview of the models adopted in AVL FIRE® for calculation of ignition and combustion processes in compression-ignition and spark-ignition engine configurations is provided.

A.6.1 Diesel Combustion

For calculating the combustion process in diesel engines the combustion model adopted in AVL FIRE® distinguishes between the three major regimes relevant during compression-ignition combustion, namely auto-ignition, premixed flame and non-premixed diffusion combustion (Fig. A.12). The auto-ignition pre-reactions are calculated within the premixed charge of fuel and air, with the ignition delay governed by the local temperature, pressure, fuel/air equivalence ratio and the amount of residual gas. Local auto-ignition is followed by premixed combustion in the fuel/air/residual gas charge formed during the time period between start-of-injection and auto-ignition onset. The third regime is the one of diffusion combustion where the reaction takes place in a thin zone which separates fuel and oxidizer. In the adopted model it is assumed that the chemical time in the reaction zone is much smaller than the time needed for the diffusion process. Therefore, the rate of reaction during diffusion combustion is determined entirely by the intermixing of fuel and oxidizer (Fig. A.13). The amount of mixing is computed with a characteristic time-scale obtained from the solution of the adopted turbulence model (Colin and Benkenida 2004).

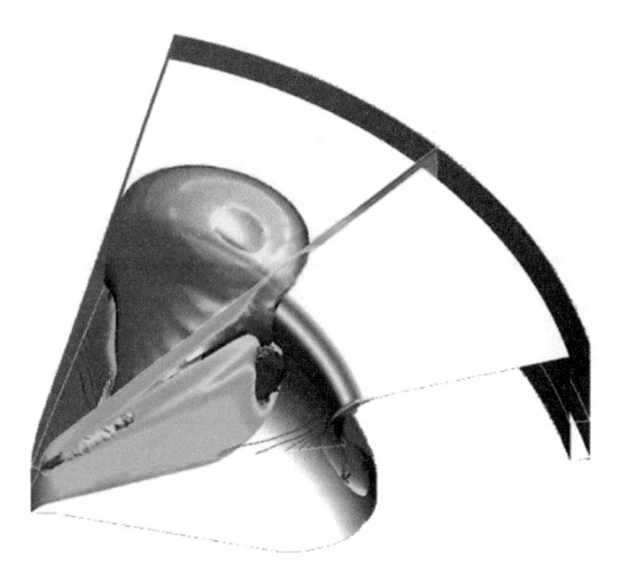

Fig. A.12 Spray droplets, temperature iso-surface and temperature distribution in a cut-plane across the spray axis during DI diesel combustion

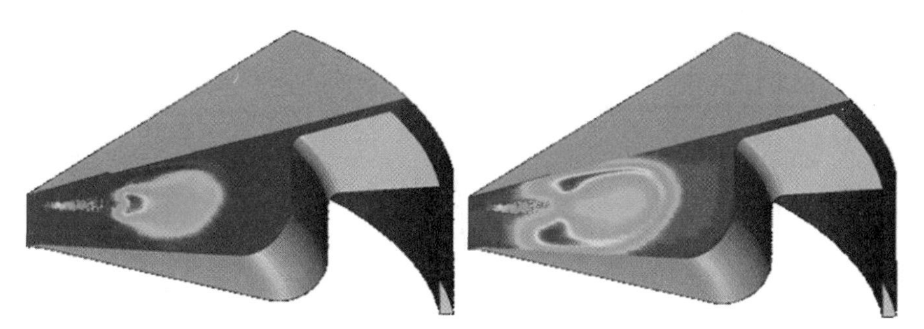

Fig. A.13 Fuel vapor concentration (*left*) and temperature distribution (*right*) in a DI diesel engine at 10°crank-angle ATDC

This distinct separation of the different ignition/combustion regimes makes the present model specifically applicable to conventional as well as alternative diesel combustion (Tatschl et al. 2007; Priesching et al. 2007), respectively. In the conventional case most part of the combustion can be assumed to be of the diffusion type (Fig. A.14), in the case of alternative combustion concepts a large amount of fuel is consumed within the premixed auto-ignition/combustion regime (Fig. A.15). Due to its generic implementation the model is applicable to all types of fuels relevant for today's and future IC-engine combustion concepts.

For prediction of the auto-ignition delay tabulated ignition data are used which are stored in look-up tables that are available in AVL FIRE® for different fuels, generated based upon chemical kinetic calculations adopting complex reaction schemes. The tabulated values are stored as functions of the parameters pressure,

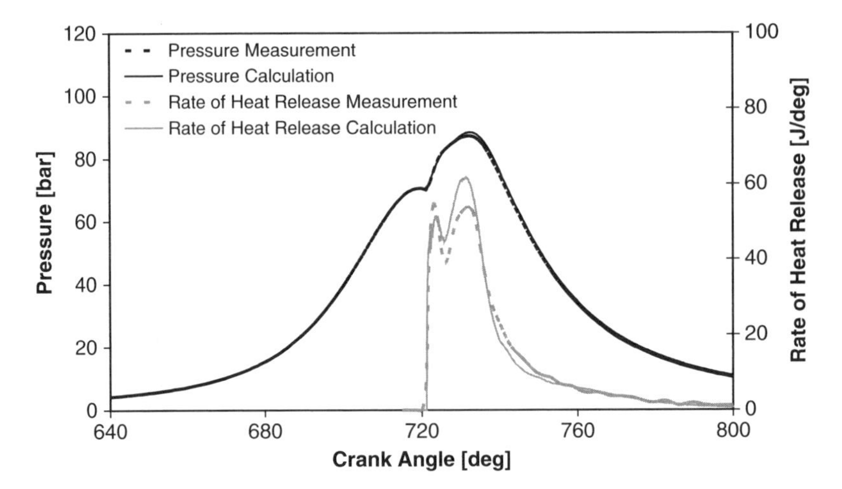

Fig. A.14 Comparison of calculated and measured cylinder pressure evolution and heat release rate for conventional diesel combustion (Tatschl et al. 2007)

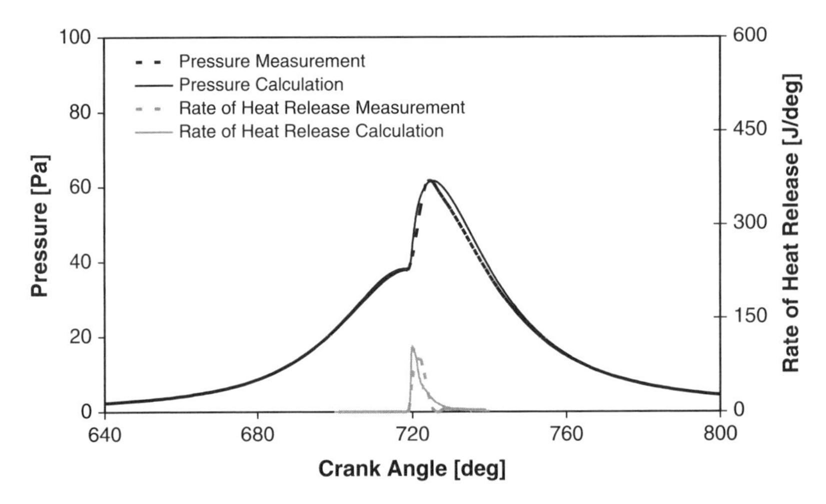

Fig. A.15 Comparison of calculated and measured cylinder pressure evolution and heat release rate for HCCI-type diesel combustion (Priesching et al. 2007)

temperature, fuel/air equivalence ratio and residual gas content. The range of these parameters has been chosen in a way to be able to cover the relevant in-cylinder pressure, temperature and charge-mixture composition conditions prior to combustion. For the actual determination of the auto-ignition delay time in the CFD simulation, a transport equation for an auto-ignition indicator species is solved with the formation rate derived from the tabulated values. Once the local value of the indicator species attains a certain threshold value, auto-ignition is initiated. Fuel consumption is then controlled by a characteristic chemical time-scale which ensures rapid combustion after auto-ignition (Colin et al. 2005).

The hydrocarbon oxidation process during high temperature combustion is separated into three major reaction steps. First the fuel is partly oxidized to CO and to CO_2, followed by CO oxidation, and finally a post-flame equilibrium chemistry approach is applied which results in the final species concentrations. The above described combustion reactions cover the relevant range of mixture composition from lean to rich and the different levels of residual gas content. In addition to the amount of heat that is released within the flame the procedure provides all relevant information about CO and radical species which are important for the subsequent calculation of the pollutants.

In addition to the above described combustion modeling approach, AVL FIRE® offers a variety of additional models for specific applications related to compression-ignition combustion analysis, such as e.g. an eddy-break-up type model which (in combination with a suitable auto-ignition model) is applicable to conventional diesel combustion (Tatschl et al. 1998; Dahlen and Larsson 2000; Cipolla et al. 2007), as well as to HCCI combustion concepts (Priesching et al. 2003). The characteristic-time-scale model offered in AVL FIRE® can be used in the context of the original approach following Kong et al. (1995) or in combination with detailed chemical kinetic mechanisms, as demonstrated for HCCI combustion in Priesching et al. (2003).

In order to handle detailed chemical kinetics within AVL FIRE® a stiff ODE chemistry solver is directly coupled with the basic flow solver algorithm of AVL FIRE®, ensuring stable and fast solution of the coupled equation systems. The effect of chemistry (level of elementary reactions) is considered such that at each time step a single zone reactor model is calculated for each computational cell with the turbulence effects taken into account via the characteristic time-scale approach. The detailed chemical kinetics mechanisms can be read in via the standard CHEMKIN data format, calculation of the source terms and the turbulence interaction model is fully accessible by the user via open software interfaces (Wang et al. 2006).

A.6.2 Gasoline Combustion

The premixed flame propagation characteristic during SI-engine combustion is mainly determined by the in-cylinder turbulent flow, the local mixture composition distribution and the in-cylinder pressure and temperature conditions in the unburned charge ahead of the flame. The local turbulent flame speed itself is governed by the fuel specific laminar flame velocity and by the turbulent fluctuations of the flow field, resulting in wrinkling of the flame. In direct injected SI-engines turbulent combustion takes place under partially premixed or even diffusion controlled conditions.

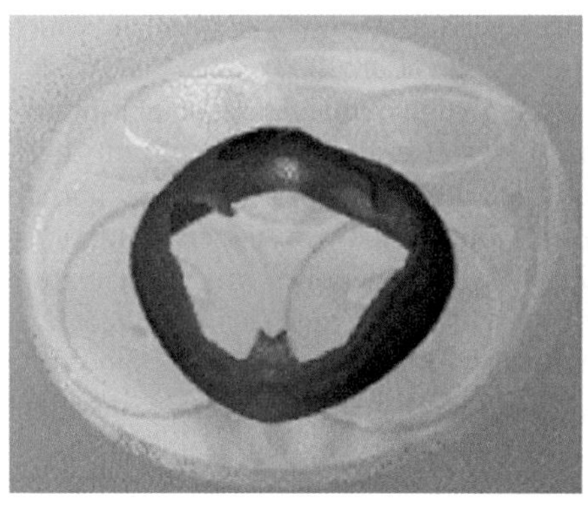
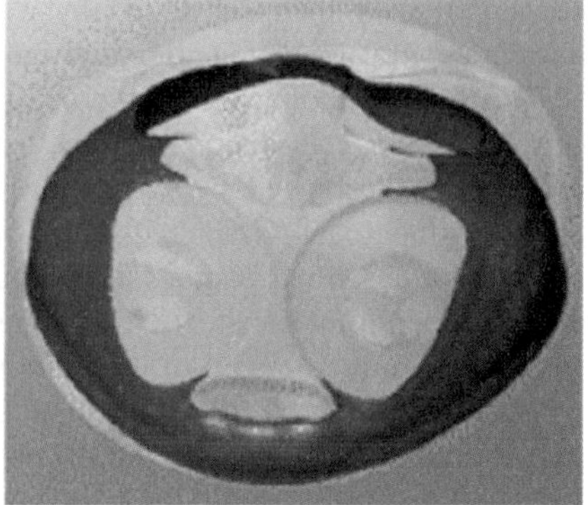

Fig. A.16 Flame front shape and location in a 4V SI-engine at 15° crank-angle (*left*) and 25° crank-angle (*right*) after TDC

For calculating SI-engine combustion the coherent flame modeling approach is adopted in AVL FIRE®. The model is based upon solving a transport equation for the flame surface density with source terms accounting for flame production due to wrinkling of the flame by turbulence effects and for flame surface annihilation due to chemical reaction (Fig. A.16). The actual version of the coherent flame model implemented in AVL FIRE® fully accounts for equivalence ratio and residual gas inhomogeneities on the flame propagation characteristics, thus being able to also handle stratified-charge combustion (Duclos et al. 1996; Georjon et al. 2000; Patel et al. 2003).

The laminar flame speed information required in the coherent flame modeling approach is obtained either from empirical correlations, such as e.g. Metghalchi and Keck (1982) or from detailed chemical kinetic calculations and tabulation of the laminar flame speed data as a function of temperature, pressure, equivalence ratio and residual gas content (Bogensperger et al. 2008). The detailed kinetic calculations and the tabulation of the laminar flame speed data are performed as a pre-processor step prior to the CFD calculation. The tabulated data once generated for a specific fuel type are then available for all further calculations within AVL FIRE®. A fast interpolation algorithm adopted for extraction of the tabulated flame speed information ensures CPU efficient use of the detailed chemical kinetic information within the combustion calculation.

In order to describe the fuel oxidation process in the flame, arbitrarily complex chemical kinetic schemes can be adopted in the combustion model. In practice, however, the number of species involved in a multidimensional simulation is limited by the available CPU resources. Hence, a reduced hydrocarbon oxidation scheme and an equilibrium chemistry approach are used in order to cover the high temperature oxidation process within the flame and in the post-flame region, respectively (Fig. A.17).

For modeling spark ignition and early flame kernel growth processes AVL FIRE® offers both semi-empirical and complex models that fully take into account

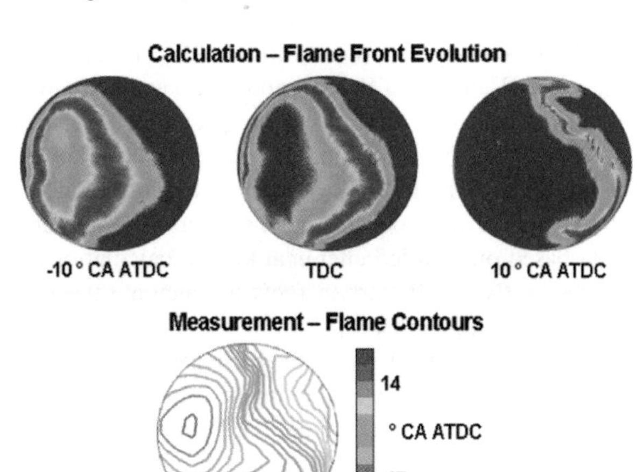

Fig. A.17 Comparison of calculated flame front position in the cylinder head gasket plane of a 2-valve SI-engine with measured flame contours

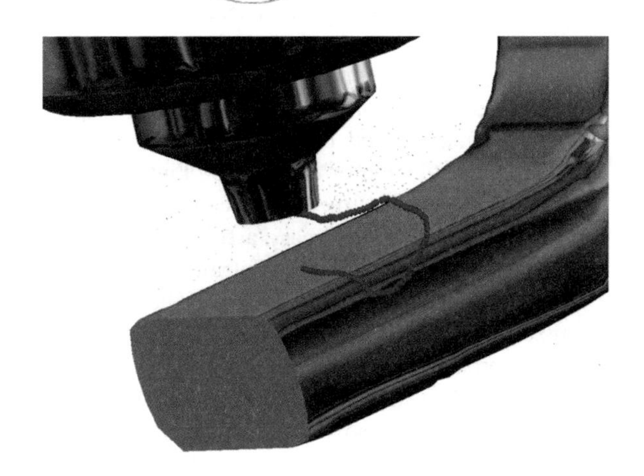

Fig. A.18 AKTIM calculation result of the interaction of the spark-induced plasma channel with the turbulent flow field

the electrical energy deposition, heat losses to the spark plug and the influence of charge-inhomogeneities and turbulence on the early flame kernel growth characteristics (Duclos and Colin 2001).

The Arc and Flame Kernel Tracking (AKTIM) ignition model is composed of four sub-models. One describes the secondary side of the electrical inductive system which provides the available initial electric energy and the duration of the glow phase. The spark itself is modeled by a set of particles placed along the spark path. The flame kernel is described by Lagrangian marker particles that are convected by the mean and turbulent flow and which receive energy from the electrical circuit and lose energy due to heat conduction to the spark electrodes. Hence, AKTIM allows a quite complete description of the spark ignition and early flame kernel growth processes and is well suited for detailed analysis of spark flame initiation and its impact on the subsequent flame propagation characteristics (Fig. A.18). In addition to the classical AKTIM approach AVL FIRE® also offers a streamlined AKTIM version, the Imposed Stretch Spark Ignition Model ISSIM.

For modeling the pre-reactions in the end gas that finally lead to the onset of knock in SI-engines different models are available in AVL FIRE®. The AnB knock model is based on describing the growth of a knock precursor representing the progress of the auto-ignition pre-reactions. The knock precursor production rate is based on an Arrhenius-like expression with the rate parameters being functions of the specific fuel type under consideration. Alternatively, tabulated auto-ignition data based on detailed chemical kinetic calculations are available for a number of representative hydrocarbon fuels to calculate the knock-precursor formation rate (Halstead et al. 1975).

In the kinetic knock model the pre-reactions in the end-gas are modeled according to a reduced kinetic scheme adopting so-called generic chemical species (Halstead et al. 1975). The kinetic knock model is capable of reflecting the complex processes during hydrocarbon auto-ignition, such as the appearance of cool flames and the negative temperature dependence of the ignition delay. Optimized sets of reaction rate parameters are available for primary and toluene reference fuels of different octane number (Fig. A.19).

In addition to the coherent flame combustion model AVL FIRE® also offers a range of alternative models for premixed-charge flame propagation simulation. The eddy-break-up model that is based on the assumption that the mean rate of reaction is independent of the reaction kinetics and hence the reaction progress is determined by the intermixing of cold reactants with hot combustion products, can be used for a fast and easy assessment of the flow/flame interaction characteristics in SI-engines (Ahmadi-Befrui and Kratochwill 1990).

For a more accurate simulation of both homogeneous and inhomogeneous premixed combustion processes in SI-engines, a generic turbulent flame speed closure approach is available in AVL FIRE®, enabling the user to easily implement his own turbulent flame speed correlations for the fuel and turbulence conditions of interest. The flame speed closure approach is based upon determination of the reaction rate as a function of local laminar flame speed data in combination with turbulent intensity and length scale information (Wallesten et al. 1998). Charge temperature, pressure and local fuel to air equivalence ratio influences are fully accounted for, as is the influence of local turbulence and residual gas effects.

For detailed, fundamental studies of flame propagation processes AVL FIRE® offers a transported joint-scalar probability density function (PDF) approach for

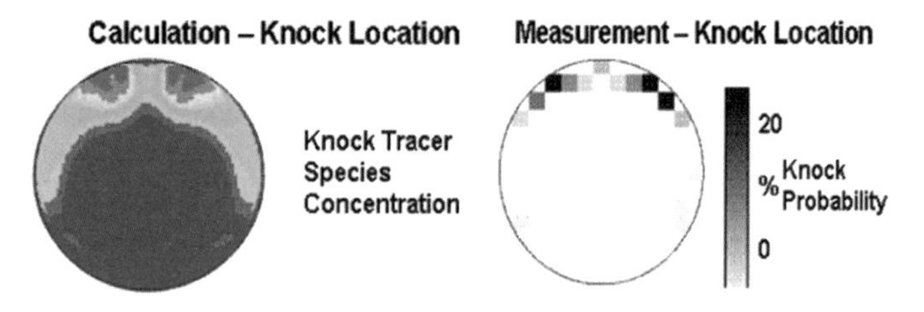

Fig. A.19 Calculated knock onset locations (*left*) vs. experimentally obtained knock probability distribution (*right*) in the cylinder head gasket plane of a 4V SI-engine (Tatschl et al. 2005)

modeling SI-engine combustion (Cartellieri et al. 1994; Tatschl and Riediger 1998; Amer and Reddy 2002). The transported PDF model does not rely on the usually made assumptions of infinitely fast chemical reaction, the presence of thin reaction sheets (flamelets) or on any presumed shape of the probability density function distribution. It fully accounts for the simultaneous effects of both finite rate chemistry and turbulence, thus obviating the need for any prior assumptions as to whether one or the other of the two processes determines the mean rate of reaction. The adopted model variant solves the joint scalar PDF transport equation for the enthalpy and the thermochemical quantities mixture fraction and reaction progress variable representing the reactive system. The PDF transport equation is solved adopting a Monte Carlo simulation approach with the mean velocity, pressure and scalar fields iteratively updated according to the finite-volume flow solver until full convergence of the set of equation systems is achieved.

A.7 Pollutant Formation

This chapter provides a brief overview of the models available in AVL FIRE® for calculation of NO and soot formation in IC-engines. The presented models are applicable to both compression-ignition and spark-ignition engines and can be combined with the available ignition and combustion models described in the previous chapter.

As a result of the application of the NO and soot formation models, details of the local interaction of mixture formation, combustion and pollutant formation are obtained. Based on the temperature and species composition distributions in different sections across the combustion chamber the governing processes can be easily interpreted. The impact of combustion system parameter variations on the details of the soot and NO formation mechanisms can be assessed on a local space and time resolved basis. The extraction of global pollutant formation data then serves as the basis for further assessment of the combustion system behavior under parameter variations. This enables the individual study of, for example, the influence of start of injection and injection pressure on NO and soot formation trends.

A.7.1 NO Formation

The nitric oxides which are formed under typical combustion conditions originate from three major sources. Thermal NO which is formed due to the dissociation of the molecular air-nitrogen, prompt NO formed by the attack of hydrocarbon fragments on the air-nitrogen and fuel NO formed from nitrogen containing components in the fuel. Usually in IC-engines the third one can be neglected because there is no significant amount of nitrogen in the gasoline/diesel fuel. The two other

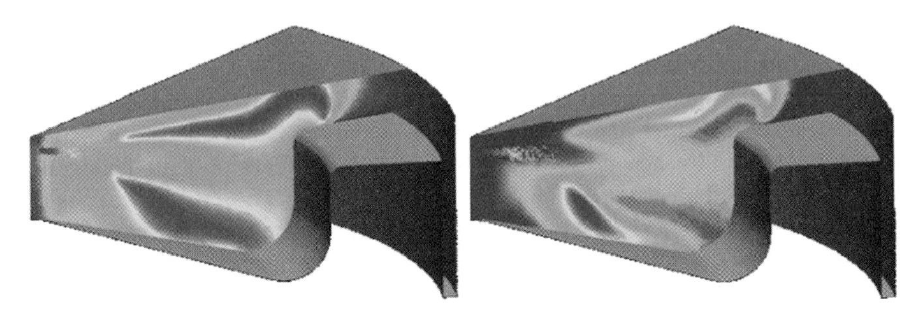

Fig. A.20 Temperature (*left*) and NO concentration (*right*) in a DI diesel engine at 20° crank-angle ATDC

mechanisms do contribute to the NO formation in engines whereby mainly thermal NO is formed but also some amount of prompt NO can appear.

According to the thermal and prompt NO formation mechanisms, NO can be formed in the reaction zone itself as well as in the post-flame region. In IC-engines the cylinder pressure rises during the combustion process and hence earlier burned gases are compressed to a higher temperature level as they have immediately after their combustion. As a consequence of the high temperatures in the post-flame region and the simultaneously long residence times, the thermal NO formation in the burned gases usually exceeds the NO formed in the flame front and hence represents the main source of the nitric oxides in IC-engines (Fig. A.20).

In AVL FIRE® the thermal NO formation is modeled according to the Zeldovich mechanism (Zeldovich et al. 1947), which is based upon the chemical equilibrium assumption, i.e. only the atomic nitrogen is accounted for as an additional intermediate species. Since its concentration does not depend on the hydrocarbon oxidation kinetics, the thermal NO formation can be described by the following reaction steps, usually known as the extended Zeldovich mechanism:

$$N_2 + O \underset{k_{1b}}{\overset{k_{1f}}{\leftrightarrow}} NO + N$$

$$N + O_2 \underset{k_{2b}}{\overset{k_{2f}}{\leftrightarrow}} NO + O$$

$$N + OH \underset{k_{3b}}{\overset{k_{3f}}{\leftrightarrow}} NO + H.$$

The first reaction represents the rate limiting step of the thermal NO formation mechanism. Very high activation energy (temperature) is necessary to decompose the stable triple-bond of the molecular air-nitrogen. Accordingly, this reaction is significantly fast at high temperatures. The nitrogen atoms in the second reaction are oxidized to nitric oxide by oxygen in near stoichiometric or slightly fuel lean conditions. The third reaction is usually negligible except in fuel rich flames. From the set of reactions it can be seen that the thermal nitric oxide formation from N_2 is mainly determined by the five chemical species O, H, OH, N and O_2 but not by the fuel used. Based upon the above reaction mechanism the overall NO formation rate can be written as (Bowman 1992):

$$\frac{dc_{NO}}{dt} = 2k_{1f}c_Oc_{N_2}\frac{\left(1 - \frac{k_{1b}k_{2b}c_{NO}^2}{k_{1f}c_{N_2}k_{2f}c_{O_2}}\right)}{1 + \frac{k_{1b}c_{NO}}{k_{2f}c_{O_2}+k_{3f}c_{OH}}}.$$

In order to solve the above equation, the concentrations of the O atoms and the free radical OH are required which can be obtained in AVL FIRE® either by adopting detailed or reduced chemical kinetics (depending on the combustion model used) or based upon empirical correlations.

Under specific operating conditions, such as alternative Diesel combustion, the rate of NO generated during combustion of hydrocarbon fuels can be higher than that predicted by the Zeldovich mechanism. This enhanced NO formation is attributed to the presence of hydrocarbon species which result from fuel fragmentation during the combustion process. Prompt NO is formed by the reaction of atmospheric nitrogen with these hydrocarbon fragments. The prompt NO mechanism forms, as its name suggests, NO from nitrogen much faster in the flame than the thermal NO mechanism. The model available in AVL FIRE® to predict prompt NO applies an overall reaction following De Soete (1975).

The effect of turbulent temperature fluctuations on the NO formation rates are accounted for in AVL FIRE® by integrating the kinetic rates with respect to the local fluctuating temperatures adopting the presumed probability density function (PDF) approach (Vujanovic et al. 2006) (Fig. A.21).

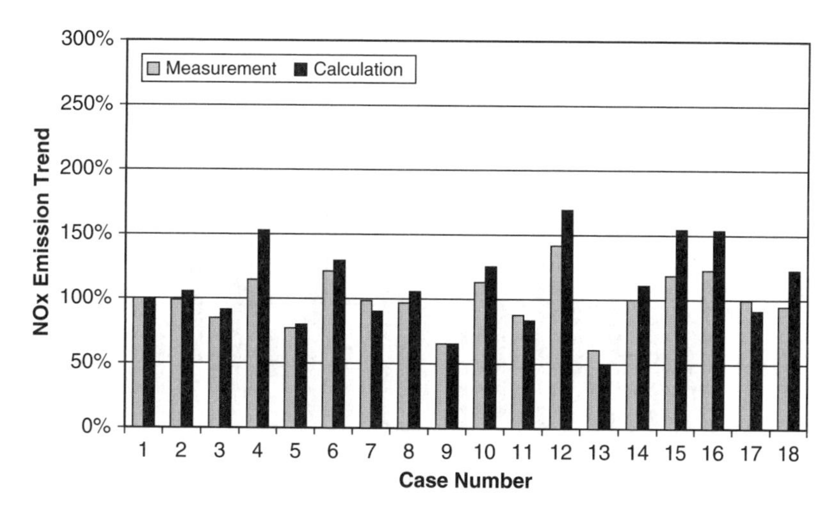

Fig. A.21 Comparison of calculated and measured NO emission trends in a DI diesel engine for combustion system parameter variations (Priesching et al. 2007)

A.7.2 Soot Formation

The formation of soot is the conversion of hydrocarbon rich, aliphatic compounds into an agglomerate of large aromatic hydrocarbons, their subsequent conversion to particles, the coagulation of primary particles, and the growth of solid soot particles due to the accumulation of gaseous components (Fig. A.22). Particle oxidation takes place as a result of the attack of atomic oxygen onto the carbonaceous particles under high temperature conditions (Frenklach and Wang 1990).

Soot modeling in AVL FIRE® is based upon combinations of suitably extended and adapted joint chemical/physical rate expressions for representation of the processes of particle nucleation, surface growth, particle coagulation and oxidation. Besides semi-empirical correlations, soot models based upon the flamelet approach or adopting reduced chemical reacting schemes for the description of the soot formation and oxidation processes are available.

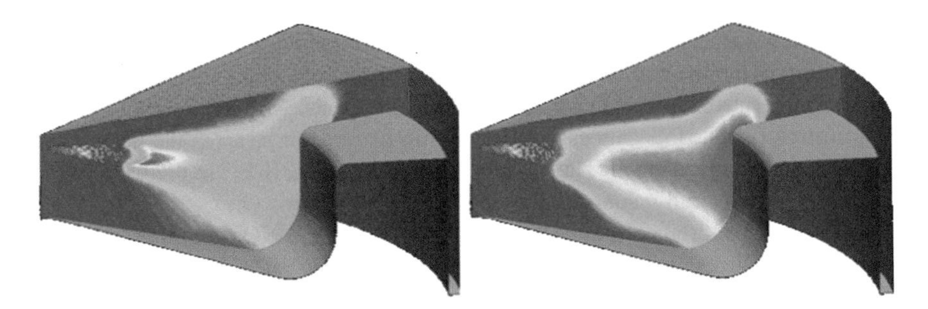

Fig. A.22 Fuel vapor (*left*) and soot concentration (*right*) in a DI diesel engine at 20° crank-angle ATDC

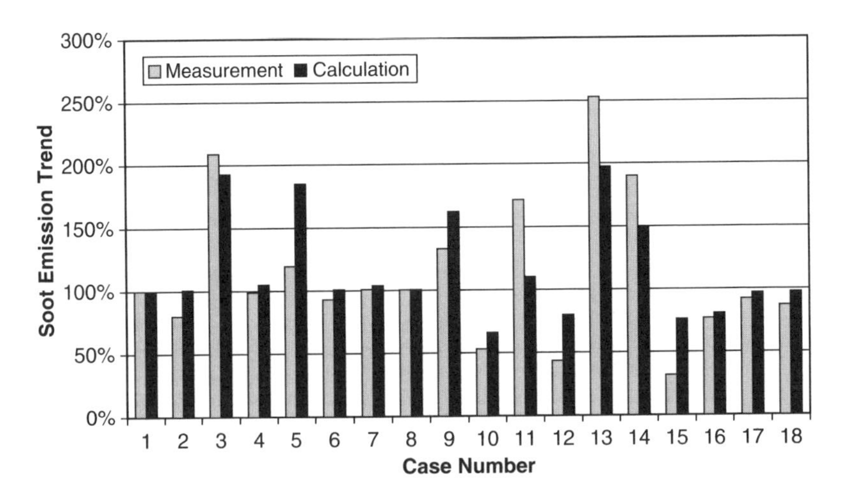

Fig. A.23 Comparison of calculated and measured soot emission trends in a DI diesel engine for combustion system parameter variations (Priesching et al. 2007)

The flamelet soot model formulation adopted in AVL FIRE® is based on a flamelet library approach for the soot source terms (Mauss 1998). Implementation of the soot model into the CFD framework is achieved via calculation of transport equations for the mixture fraction and its variance and integration of the instantaneous soot formation rates in mixture fraction space via a presumed β-PDF in order to obtain their mean values (Priesching et al. 2005). Since the source terms for the soot formation and oxidation are evaluated before the CFD simulation and stored in a library, this model works very time efficient despite the complexity of the underlying chemical kinetics (Fig. A.23).

The basis of the kinetic soot model in AVL FIRE® is a detailed chemical reaction scheme for the calculation of soot formation and oxidation. It combines the mechanisms of formation of polyaromatic hydrocarbons, soot precursor formation due to condensation of hydrocarbon molecules, soot particle growth by the HACA reactions and polyyne molecule addition, the mechanism of acetylene pyrolysis and pure carbon cluster formation, as well as the reactions of hydrocarbon oxidation [see e.g. Krestinin (2000), Agafonov et al. (2002)]. The model implemented in AVL FIRE® adopts a reduced number of species and reactions of the above described mechanism with suitably optimized rate coefficients for a number of IC-engine relevant fuels, and hence provides a computationally efficient approach for soot formation calculation within the context of a multi-dimensional CFD methodology.

References

Agafonov GL, Nullmeier M, Vlasov PA, Warnatz J, Zaslonko IS (2002) Kinetic modeling of solid carbon particle formation and thermal decomposition during carbon suboxide pyrolysis behind shock waves. Combust Sci Techn 174:1–29

Ahmadi-Befrui B, Kratochwill H (1990) Multidimensional calculation of combustion in a loop-scavenged two-stroke cycle engine. In: Proceedings of COMODIA 1990, Kyoto, Japan, pp 465–474

Alajbegovic A, Drew DA, Lahey RT (1999) An analysis of phase distribution and turbulence in dispersed particle/liquid flows. Chem Eng Comm 174:85–133

Amer AA, Reddy TN (2002) Multidimensional optimization of in-cylinder tumble motion for the New Chrysler Hemi. SAE 2002-01-1732

Basara B (2006) An eddy viscosity transport model based on elliptic relaxation approach. AIAA J 44(7):1686–1690

Bianchi GM, Brusiani F, Postrioti L, Grimaldi CN, Di Palma S, Matteucci L, Marcacci M, Carmignani L (2006) CFD analysis of injection timing influence on mixture preparation in a PFI motorcycle engine. SAE 2006-32-0022

Bianchi GM, Brusiani F, Postrioti L, Grimaldi CN, Marcacci M, Carmignani L (2007) CFD analysis of injection timing and injector geometry influences on mixture preparation at idle in a PFI motorcycle engine. SAE 2007-24-0041

Birkhold F, Meingast U, Wassermann P, Deutschmann O (2006) Analysis of the injection of urea-water-solution for automotive SCR DeNOx-systems: modelling of two-phase flow and spray/wall-interaction. SAE 2006-01-0643

Birkhold F, Meingast U, Wassermann P, Deutschmann O (2007) Modelling and simulation of the injection of urea-water-solution for automotive SCR DeNOx-systems. Appl Catal B 70: 119–127

Bogensperger M, Ban M, Priesching P, Tatschl R (2008) Modelling of premixed SI-engine combustion using AVL FIRE® – a validation study. In: Proceedings of international multidimensional engine modelling user's group meeting, Detroit, MI

Bowman CT (1992) Control of combustion-generated nitrogen oxide emissions: technology driven by regulation. In: Proceedings of Combustion Institute 24, pp. 859–878

Brenn G, Deviprasath LJ, Durst F, Fink C (2007) Evaporation of acoustically levitated multicomponent liquid droplets. Int J Heat Mass Transfer 50:5073–5086

Caika V, Sampl P, Tatschl R, Krammer J, Greif D (2009) Coupled 1D-3D simulation of common rail injector flow using AVL HYDSIM and AVL FIRE® SAE 2009-24-0029

Cartellieri W, Chmela F, Kapus P, Tatschl R (1994) Mechanisms leading to stable and efficient combustion in lean burn gas engines. In: Proceedings of COMODIA 1994. Yokohama, Japan, pp 17–24

Chiatti G, Chiavola O, Palmieri F (2007) Injector dynamic and nozzle flow features in multiple injection modeling. SAE 2007-24-0038

Chiavola O, Palmieri F (2006) Coupling codes for nozzle flow modelling in diesel injection system. In: Proceedings of ICES2006, ASME International Engine Division 2006 Spring Technical Conference. Aachen, Germany

Cipolla G, Vassallo A, Catania AE, Spessa E, Stan C, Drischmann L (2007) Combined application of CFD modelling and pressure-based combustion diagnostics for the development of a low compression ratio high-performance diesel engine. SAE 2007-24-0034

Colin O, Benkenida A (2004) The 3-zones extended coherent flame model (ECFM3Z) for computing premixed/diffusion combustion. Oil Gas Sci Technol Rev IFP 59(6):593–609

Colin O, Pires da Cruz A, Jay S (2005) Detailed chemistry-based auto-ignition model including low temperature phenomena applied to 3D engine calculations. In: Proceedings of Combustion Institute 30, pp 2649–2656

Dahlen L, Larsson A (2000) CFD studies of combustion and in-cylinder soot trends in a DI diesel engine – comparison to direct photography studies. SAE 2000-01-1889

De Soete GG (1975) Overall reaction rates of NO and N2 formation from fuel nitrogen. In: Proceedings of Combustion Institute 15, pp 1093

Demirdzic I, Lilek Z, Peric M (1993) A collocated finite volume method for predicting flows at all speeds. Int J Numer Meth Fluid 16:1029–1050

Drew DA, Passman SL (1998) Theory of multi-component fluids. Springer, New York

Duclos JM, Colin O (2001) Arc and kernel tracking ignition model for 3D spark-ignition engine calculations. In: Proceedings of COMODIA 2001. Nagoya, pp 343–350

Duclos JM, Bruneaux G, Baritaud TA (1996) 3D modelling of combustion and pollutants in a 4-valve SI engine; effect of fuel and residual distribution and spark location. SAE 961964

Dukowicz JK (1980) A particle-fluid numerical model for liquid sprays. J Comput Phys 35: 229–253

Edelbauer W, Suzzi D, Sampl P, Tatschl, R, Krüger C, Weigand B (2006) New concept for on-line coupling of 3D-eulerian and lagrangian spray approaches in engine simulations. In: Proceedings of ICLASS 2006, Kyoto, Japan

Fink C, Frobenius M, Meindl E, Harndorf H (2009) Experimental and numerical analysis of marine diesel engine injection sprays under cold and evaporative conditions. In: Proceedings of ICLASS 2009. Vail, CO, USA

Frenklach M, Wang H (1990) Detailed modeling of soot particle nucleation and growth. In: Proceedings of Combustion Institute. 23, pp 1559–1566

Georjon T, Bourguignon E, Duverger T, Delhaye B, Voisard P (2000) Characteristics of mixture formation and combustion in a spray-guided concept gasoline direct injection engine: an experimental and numerical approach. SAE 2000-01-0534

Greif D, Alajbegovic A, Monteverde B (2003) Simulation of cavitating flow in high pressure gasoline injectors. In: Proceedings of ICLASS 2003. Sorrento, Italy

Greif D, Morozov A, Winklhofer E, Tatschl R (2005) Experimental and numerical investigation of erosive effects due to cavitation within injection equipment. In: Proceedings of ICCHMT 2005, Paris-Cachan, France
Halstead MP, Kirsch LJ, Prothero A, Quinn CP (1975) A mathematical model for hydrocarbon auto-ignition at high pressures. Proc R Soc Lond A 364:515–538
Ishii M, Mishima K (1989) Droplet entrainment correlation in annular two-phase flow. Int J Heat Mass Transfer 32(10):1835
Kong S-C, Han ZY, Reitz RD (1995) The development and application of a diesel ignition and combustion model for multidimensional engine simulations. SAE 950278
Krestinin AV (2000) Detailed modeling of soot formation in hydrocarbon pyrolysis. Combust Flame 121:513–524
Leuthel R, Pfitzner M, Frobenius M (2008) Numerical study of thermal-fluid interaction in a diesel fuel injector. SAE 2008-01-2760
Masuda R, Fuyuto T, Nagaoka M, von Berg E, Tatschl R (2005) Validation of diesel fuel spray and mixture formation from nozzle internal flow calculation. SAE 2005-01-2098
Mauss F (1998) Entwicklung eines kinetischen Modells der Rußbildung mit schneller Polymerisation. D 82 (RWTH Aachen), ISBN 3-89712-152-2
Metghalchi M, Keck JC (1982) Burning velocities of mixtures of air with methanol, isooctane and indolene at high pressure and temperature. Combust Flame 48:191–210
Musu E, Frigo S, De Angelis F, Gentili R (2006) Evolution of a small two-stroke engine with direct liquid injection and stratified charge. SAE 2006-32-0066
Nagaoka M, Ueda R, Masuda R, von Berg E, Tatschl R (2008) Modeling of diesel spray atomization linked with internal nozzle flow. In: Proceedings of THIESEL 2008. Valencia, Spain
Patel SNDH, Bogensperger M, Tatschl R, Ibrahim SS, Hargrave GK (2003) Coherent flame modeling of turbulent combustion – a validation study. In: Proceedings of second MIT conference on computational fluid and solid mechanics. Boston, MA
Popovac M, Hanjalic K (2005) Compound wall treatment for RANS computation of complex turbulent flows. In: Bathe K (ed) Proceedings of third MIT conference on computational fluid and solid mechanics, vol 1. Elsevier, Amsterdam, pp 802–806
Priesching P, Wanker R, Cartellieri P, Tatschl R (2003) Detailed and reduced chemistry CFD modeling of premixed charge compression ignition engine combustion. In: Proceedings of international multidimensional engine modeling user's group meeting 2003, Detroit, MI
Priesching P, Tatschl R, Mauss F, Saric F, Netzell K, Bauer W, Schmid M, Leipertz A, Merola SS, Vaglieco BM (2005) Soot particle size distribution – a joint work for kinetic modeling and experimental investigations. SAE 2005-24-053
Priesching P, Ramusch G, Ruetz J, Tatschl R (2007) 3D-CFD modelling of conventional and alternative diesel combustion and pollutant formation – a validation study. JSAE 2007-72-85
Stanton DW, Rutland CJ (1998) Multi-dimensional modelling of heat and mass transfer of fuel films resulting from impinging sprays. SAE 980132
Stralin P (2006) A lagrangian collision model applied to an impinging spray nozzle. SAE 2006-01-3331
Suzzi D, Krüger C, Blessing M, Wenzel P, Weigand B (2007) Validation of Eulerian spray concept coupled with CFD combustion analysis. SAE 2007-24-0044
Tatschl R, Riediger H (1998) PDF modelling of stratified charge SI engine combustion. SAE 981464
Tatschl R, Pachler K, Winklhofer E (1998) A comprehensive DI diesel combustion model for multidimensional engine simulation. In Proceedings of COMODIA 1998. Kyoto, pp 141–148
Tatschl R, von Künsberg Sarre Ch, Alajbegovic A, Winklhofer E (2000) Diesel spray break-up modeling including multidimensional cavitating nozzle flow effects. In: Proceedings of ILASS Europe 2000. Darmstadt, Deutschland
Tatschl R, Wiesler B, Alajbegovic A, von Künsberg Sarre Ch (2000) Advanced 3D fluid dynamic simulation for diesel engines. In: Proceedings of THIESEL 2000. Valencia, Spain, pp 113–121

Tatschl R, Winklhofer E, Fuchs H, Kotnik G, Priesching P (2005) Analysis of flame propagation and knock onset for full load SI-engine combustion optimization – a joint numerical and experimental approach. Proceedings NAFEMS World Congress, Malta

Tatschl R, Basara B, Schneider J, Hanjalic K, Popovac M, Brohmer A, Mehring J (2006) Advanced turbulent heat transfer modelling for IC-engine applications using AVL FIRE®. In: Proceedings of international multidimensional engine modeling user's group meeting. Detroit, MI

Tatschl R, Priesching P, Ruetz J, Kammerdiener Th (2007) DoE based CFD analysis of diesel combustion and pollutant formation. SAE 2007-24-0049

von Berg E, Alajbegovic A, Tatschl R, Krüger Ch, Michels U (2001) Multiphase modeling of diesel sprays with the Eulerian/Eulerian Approach. In: Proceedings of ILASS Europe 2001. Zürich, Switzerland

von Berg E, Edelbauer W, Alajbegovic A, Tatschl R (2003) Coupled calculation of cavitating nozzle flow, primary diesel fuel break-up and spray formation with an Eulerian Multi-Fluid-Model. In: Proceedings of ICLASS 2003, Sorrento, Italy

von Berg E, Edelbauer W, Tatschl R, Alajbegovic A, Volmajer M, Kegl B, Ganippa LC (2005) Coupled simulation of nozzle flow, primary fuel jet break-up, and spray formation. J Eng Gas Turbines Power 127:897–908

von Künsberg Sarre Ch, Tatschl R (1998) Spray modelling/atomisation – current status of break-up models. In: Proceedings of IMechE seminar on turbulent combustion of gaseous and liquids. Lincoln, UK

Vujanovic M, Baburic M, Duic N, Priesching P, Tatschl R (2006) Application of reduced mechanisms for nitrogen chemistry in numerical simulation of a turbulent non-premixed flame. In: Proceedings of CMFF 2006, 13th international conference on fluid flow technologies, Budapest, Hungary

Vujanović M, Edelbauer W, von Berg E, Tatschl R, Duić N (2008) Enhancement and validation of an Eulerian-Eulerian approach for diesel sprays. In: Proceedings of ILASS-Europe 2008. Como, Italy

Wallesten J, Lipatnikov AN, Chomiak J, Nisbet J (1998) Turbulent flame speed closure model: further development and implementation for 3-D simulation of combustion in SI engine. SAE 982613

Wang DM, Han J, Greif D, Zun I, Perpar M (2005) Interfacial area and number density transport equations for modeling multiphase flows with cavitation. In: Proceedings of ASME FEDSM 2005, 9th international symposium on gas-liquid two-phase flow, Houston, TX

Wang Z, Shuai S-J, Wang J-Y, Tian G-H, An X L (2006) Modeling of HCCI combustion: from 0D to 3D. SAE 2006-01-1364

Zeldovich YB, Sadovnikov PY, Frank-Kamenetskii DA (1947) Oxidation of nitrogen in combustion (Trans: Shelef M). Academy of Sciences of USSR, Institute of Chemical Physics, Moscow-Leningrad

Index

G.P. Merker et al. (eds.), *Combustion Engines Development*,
DOI 10.1007/978-3-642-14094-5, © Springer-Verlag Berlin Heidelberg 2012

G

N

O

P

Printing: Ten Brink, Meppel, The Netherlands
Binding: Stürtz, Würzburg, Germany